Shamrock

Daniel Maclise's conceit of shamrocks and rustic trellis decorating the engraved title-page of the Longman edition of Thomas Moore's collected songs, 1845. (Reproduced by courtesy of Aidan Heavey)

SHAMROCK
Botany and History of an Irish Myth

E. Charles Nelson

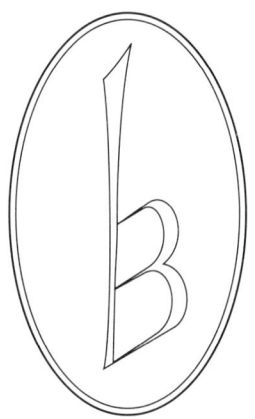

BOETHIUS PRESS
Aberystwyth · Wales
Kilkenny · Ireland
1991

© 1991 Text E.C. Nelson
Original illustrations
Bridget Flynn and Wendy Walsh
Designs and settings
Boethius Press

British Library Cataloguing in Publication Data
Nelson, E. C. (E Charles)
Shamrock.
1. Ireland. Plants.
I. Title
583.32209415

ISBN 0 86314 200 1
hard covers with bibliography

ISBN 0 86314 199 4
paper covers

Designed, typeset and printed by
Boethius Press (UK) Ltd
3 The Science Park, Aberystwyth, Wales
for Boethius Press
Kilkenny, Ireland.

for
Annaghmakerrig
where I picked shamrock
on St Patrick's Day
1987

...post obitum meum exagaellias
relinquere fratribus

Patrick, *Confessio*, 14

Contents

Foreword by Bernard Loughlin
Director, Tyrone Guthrie Centre, Annaghmakerrig

Preface and Acknowledgements

Prologue		3
Act I:	Enter an Ancient Briton	8
Act II:	Enter the Biographers	12
Act III:	Enter the Scribes	14
Act IV:	Enter the English Chroniclers and Dramatists	18
Act V:	Enter the Herbalists	23
Act VI:	Enter a Map-Maker and several Poets	27
Act VII:	Enter the Shamrock-Wearers	33
Act VIII:	Enter the Reverend Dr Threlkeld	40
Act IX:	Enter the Story-Tellers	45
Act X:	Enter the Decorated Courtiers	49
Act XI:	Enter the Patriots	55
Act XII:	Interlude for a Monopolylogue	65
Act XIII:	Enter King George IV	71
Act XIV:	Enter the Shamrock Hunters	77
Act XV:	Enter the Shamrock Makers	82
Act XVI:	Enter Artisans and Artists	90
Act XVII:	Soldiers, Exeunt	115
Act XVIII:	Take a bow, Mr Hartland	121
Act XIX:	Enter and Exit the Antiquarians	130
Act XX:	Enter William Butler Yeats	134
Act XXI:	Vox Hiberniae	139
Epilogue		145
Notes and References		147
Appendix I: Survey of shamrock specimens		153
Appendix II: Chronology		158
Appendix III: (library issue only) Bibliography		160

Colour Illustrations

Plate 1: *Trifolium dubium*, lesser clover, seamair bhuí   88

Plate 2: *Trifolium pratense*: red clover: seamair dhearg   92

Plate 3: *Trifolium repens*: white clover: seamair bhan   93

Plate 4: *Oxalis acetosells*: wood sorrel: seamsóg   96

Plate 5: *Medicago lupulina*: black medick, dumheidic   97

Foreword

It is appropriate that the royalties from this marvellous book on the myth and metaphor of shamrock should go to help support The Tyrone Guthrie Centre, for are not the artists who live and work at Annaghmakerrig in some sense the myth-makers of the present day, as all the busy and inventive chroniclers and scholars described in these pages were of theirs?

Whatever you make of the symbology and botany of the humble clover and its various vegetable rivals that have for centuries disputed for supremacy as our authentic national emblem, you will be richly entertained by the account Charles Nelson gives of the shamrock's precedence and progeny.

Certain it is that after reading this book you will never watch wither whatever weed adorns your lapel on the 17th of March with the same happy, drunken insouciance of heretofore. Whether it is spotted medick or white clover you're poisoning with ethyl alcohol, as well as yourself of course, there are saints and scribes, panegyrics and jeremiads, facts and fictions go leor here with which to beguile the intervals between slugs of the vile stuff in Ballybrack or Brooklyn, Ahoghill or Adelaide, Belfast or Birmingham, wherever the loyal sons and daughters of Saint Patrick gather to celebrate their own banishment in the wake of the snakes.

Since The Tyrone Guthrie Centre is something of a pan-Ireland, cross-border venture itself, I was particularly struck in reading this book by how universal a symbol the shamrock has become: as likely to turn up in Protestant as in Catholic iconography; entwined about the feet of King Billy's charger as wreathed around the brow of Daniel O'Connell; that it is as much Orange as Green, Northern as Southern, Ascendancy as Gaelic.

We should thus perhaps be encouraged to rejoice more in the colourful ambiguity which bestrews so much of Irish history as it is encapsulated in the sub-history of our great national chimera and shibboleth, the shamrock, whose roguery entertains us all, irrespective of class or creed, if only for a day.

Indeed, while the down to earth English may say, after Gertrude Stein, 'a rose is a rose is a rose' and the Welsh might instance the evident practicality of the leek as their national greenstuff, the shamrock will forever be just whatever three-lobed, low, leafy thing comes to hand on the day of our patron saint. And is not this a little what Irishness itself is, especially in the diaspora?

Finally, since it seems the ancient Irish ate shamrock as a foodstuff, at least according to English travellers of the time, who have always been a bit funny about other people's grub anyway, as is witnessed by the Fish 'n' Chipperies of the Costas, it is doubly appropriate, metaphorically speaking, that the royalties of this treatise on shamrock will be turned back into fodder for the artists of Ireland.

Therefore thank you for buying it, Charles Nelson for writing it and bestowing the proceeds upon Annaghmakerrig, and thank Christ for St Patrick who invented the whole shamrock industry in the first place. Or did he?

We will never know, but then who needs to know anyway? The excuse for drink and revelry is enough.

Bernard Loughlin
Annaghmakerrig

Preface and acknowledgements

It was doubtless foolhardy even to contemplate writing a brief article on an object that is neither real nor imaginary for the pit-falls are many and the consequences unimaginable. I did unwisely pen such a piece in 1985 for the now extinct magazine *Field and Countryside*, and I thank Pamela Brogan for publishing it. Kevin Myers with his usual barbed panache accused me of producing a work of such indelicacy that it should be locked away. Thus to Kevin and Pamela must go my foremost thanks for the stimulus necessary to take up the task of writing even more about the sacred vegetable of Ireland. Other journalists, who will not be named, illuminated my work on several occasions with singular but presumably unintentional fatuities, especially during 1988; I cannot exclude them from my expressions of gratitude.

My sometimes faltering footsteps have been returned to straight and narrow paths by many friends and colleagues: Donal Synnott, as ever, carefully explained and criticised; Dr Éamonn Ó hÓgáin educated me in the finer points of the Irish language of which, alas, I am profoundly ignorant; Lothian Lynas supplied me with curiosities from the Americas; Judy Cassells, Valerie Ingram, Jennifer Lammond and Susyn Andrews burrowed into their local botanical libraries and emerged with countless treasures; my other accomplices, all of whom have helped to unearth shamrocks from obscure pages, from among personal bric-a-brac, or dusty museums included Helen Dillon, Paul and Catherine Hackney, Mrs Joan Gant, Aidan Heavey, Jeanne Sheehy, David Smythe, David Davison, Dr Maurice Craig and Ruth Brennan; and for their assistance (including occasionally the 'burial' of phantom shamrocks), I extend thanks to Donal F. Bagley (Chief Herald of Ireland), Dr Mary Boydell, William Crampton, Mary Davies, Mrs Brigid Dolan (Royal Irish Academy, Dublin), Gina Douglas (Linnean Society, London), Revd Paul Duffner o.s.a., Mairead Dunlevy (National Museum of Ireland, Dublin), D. W. Dunstan (State Library of Tasmania, Hobart), Patrick Forde, Colm Gallagher, William Garner, Revd J. A. Hunt o.s.a., Dr Peter Wyse Jackson, Kenneth James (Ulster Museum, Belfast), Douglas Kent, Mrs Lynn Miller J. S. B. McQuitty, Lieut Col. N. O'Byrne, Séamus Ó Brógáin, Dr Séamus Ó Catháin (University College, Dublin), Dr Anne O'Dowd (National Museum of Ireland, Dublin), P. Ó Snodaigh (National Museum of Ireland, Dublin), Aodh O'Tuama, Mrs Siobhan de hÓir, Jennifer Ramkalawon (National

Portrait Gallery, London), John Rodgers, Peter Ryan and Mary Preece (Department of the Taoiseach, Dublin), Prof. Alistair Rowan (University College, Dublin), Maura Scannel (National Botanic Gardens, Dublin), Miss B. Tomlinson, Commander D. Turkington, Mrs Elaine Urquhart (University of Ulster, Coleraine), Roy Vickery (Natural History Museum, London), Pat Walsh (Department of Agriculture and Food, Dublin), Timothy Wilson, and the staff of the National Gallery of Ireland and the National Library of Ireland, Dublin. None should be held responsible for the way I have used or interpreted their labours for which I am most grateful. All truths and heresies and any errors contained herein are my sole responsibility.

I must also pay tribute to Bridget Flynn for her fine watercolours, Wendy Walsh for the illustration of shamrock pea, and the many good people who supplied me with true shamrocks, of their own gathering, from the highways and gardens of Ireland during 1988; and to the staff at the National Botanic Gardens at Glasnevin, who patiently tended these national treasures.

I am more than indebted to Leslie, Jen and John Hewitt for putting this book into print after another (American!) publisher turned it down fearing an onslaught of irate Hibernians. And last, but not least, to Bernard Loughlin my thanks, not only for giving me board and lodging at Annaghmakerrig on a number of occasions, but also for the courage to write a foreword; and to the other residents in that drumlin-encircled retreat who broadened my horizons, my salutations.

I should conclude with my excuses. Writing about shamrock has been both easy and difficult because of the superfluity of material available; most perplexing has been the need to select examples of both text and illustrations, for in the nineteenth and the present centuries shamrock has been all-smothering. I beg any bewildered readers who are disturbed by the omission of their favourite shamrock-encrusted artefacts or their preferred quotations, to bear this cornucopia in mind; all the same, if anyone feels that I remain blissfully ignorant about a crucial aspect of this enchanted weed, or that I have overlooked some critical reference, I invite them to contact me. Should this book be worthy of a revised edition, my laches and any lacunae but none of the heresies will be rectified—*Deo volente*.

E. Charles Nelson
Dublin, July 1989

Shamrock

Fig. 1: Transfer-printed pottery souvenir plate (late 19th century) decorated with shamrocks. (Reproduced by courtesy of Helen Dillon; photograph by David Davison)

PROLOGUE

Shamrock, **seamróg**, **seamair óg**, the young clover: nothing else is so evocative of Ireland, nor is any other Irish thing so misunderstood.

> Through Erin's Isle,
> To sport awhile,
> As Love and Valour wander'd,
> With Wit the sprite,
> Whose quiver bright
> A thousand arrows squander'd,
> Where'er they pass
> A triple grass
> Shoots up, with dewdrops streaming,
> As softly green
> As emeralds seen
> Through purest crystal gleaming.
> Oh! the Shamrock, the green, immortal Shamrock!
> Chosen leaf
> Of Bard and chief,
> Old Erin's native Shamrock![1]

Thomas Moore, who long ago was given the title 'national lyricist of Ireland', composed those lyric lines towards the beginning of the last century and thus greatly helped to enshrine the shamrock, and to gild it with a sentimental sheen that it has never lost. And he wrote this foot-note[2] to the ballad in which he recalled that Saint Patrick

> …is said to have made use of that species of the trefoil, to which in Ireland we give the name Shamrock, in explaining the doctrine of the Trinity to the pagan Irish. I do not know if there be any other reason for our adoption of this plant as a national emblem. Hope, among the ancients, was sometimes presented as a beautiful child, "standing upon tip-toes, and a trefoil or three-coloured grass in her hand."

The romance of the shamrock, epitomized by Thomas Moore's honied words, has been vigorously championed and polished by succeeding generations, so that even in this sceptical, scientific age there are many who cling tenaciously to myths that are, to put it bluntly, little more than modern inventions.

In Ireland the chief manifestation of the shamrock is on St Patrick's Day, 17 March, when many people wear sprigs in tacit acceptance of its link with the saint. Nowadays these tufts of greenery are purchased, as often as not, with the kiwi fruits, lettuces and other groceries in the local supermarket at about fifty pence a bunch. Few of us gather wild shamrock from the hills and fields as our forebears did because as town and city dwellers we do not have the old knowledge about the shamrock; no longer are we custodians of the traditional lore which used to be handed down from immemorial generation to generation.

For the exiles of Ireland's scattering, the three heart-shaped leaflets of the shamrock have special significance:

> Sing a song of Ireland,
> Blue lakes and sparkling rills,
> Gray rocks and misty moorlands,
> Of shamrocks and green hills.[3]

They immediately provoke romantic visions, conjuring up scenes of emerald fields patterning egg-shaped drumlins, blue skies speckled with fluffy white clouds from which very occasionally falls a soft rain, wild and majestic mountains cloaked with purple heather, white-washed cottages with thatch and wispy chimney smoke, glowing turf in a warm hearth, and all the other perfect picture-postcard images of a faraway island, the land of their fathers.

Thus the shamrock is a distinctive symbol, identifying the wearer as Irish, or as a descendent of an exile who, in all probability, left Ireland bearing a sod of snake-repelling turf cut from the family's turf-bank and a bunch of sacred shamrock.

> Farewell my old acquaintance, my friends both one and all,
> My lot is in America, to either rise or fall.
> From my cabin I'm evicted and likewise compelled to go
> From that lovely land called Erin
> Where the green shamrocks grow.[4]

The shamrock's message is still unequivocal and universal. It is emblazoned on the tail-fins of Irish aircraft and on the funnels of Irish ships; great swathes of shamrocks decorate fine Irish porcelain; the word and the motif advertise Irish moss-peat—conveniently milled and packaged—to the gardeners of other lands; myriad souvenirs need nothing more than a green trefoil to indicate where they were purchased and it is possible to buy a packet of shamrock seed—'a little bit of Ireland'—as a souvenir.

So what is shamrock, and does it really only grow in Ireland? What is the correct botanical name for the plant that Saint Patrick plucked and used to explain the Holy Trinity to our pagan forebears?

Questions like those are asked, over and over again, by many people curious to know more about the plant that is so closely associated with Ireland

Fig. 2: Shamrocks and turf cutters, a monochrome postcard (Valentine X. L. series Real Photo Card) in circulation in 1913.
(Reproduced by courtesy of Mr and Mrs R. Nelson)

and the Irish. Around St Patrick's Day, every year, newspapers and magazines in Ireland and overseas are filled with fanciful columns written by imaginative authors (the vast majority of whom should know better) repeating colourful legends and sentimental nonsense dressed up as fact. But those modern scribes are not blazing a new trail; they have innumerable, equally imaginative predecessors!

> There's a dear little plant that grows in our isle
> 'Twas St Patrick himself sure, that set it,
> And the sun on his labour with pleasure did smile,
> And the dews from his eye oft did wet it.
> It thrives through the bog, through the brake, through the mireland
> And he called it the dear little Shamrock of Ireland,
> The sweet little Shamrock, the dear little Shamrock,
> The sweet little, dear little, Shamrock of Ireland.[1]

In fact, the 'sweet little, dear little shamrock' has been more thoroughly subjected to scrutiny by botanists and historians than most other Irish plants and there are several scrupulous articles, hidden away in the pages of learned journals, documenting in meticulous detail its origin and evolution. But, for unfathomable reasons, sentiment and romance hold sway over pedantic explanations of Irish myths, folk-lore and botany, and the true story of the shamrock is known to relatively few people.

Almost one hundred years ago, Nathaniel Colgan, a retiring, modest man and a keen amateur naturalist who earned his living as a clerk in the Dublin Metropolitan Police Court, carefully set down a chronology of the shamrock in literature. He also investigated its identity, and to him must still go thanks

Fig. 3: A particularly grotesque but colourful card in circulation in 1903 (The "Clarence" series) incorporating all the symbols of 'romantic' Ireland—setting sun, round tower, greyhound, "Dear harp of my country" and some pathetic shamrocks. (Reproduced by courtesy of Mrs Joan Gant)

for helping to remove the voluminous shrouds of sentimental cobwebs that have obscured this singular plant. Mr Colgan's painstaking research was reported in the *Journal and Proceedings of Royal Society of Antiquaries of Ireland* and in the *Irish Naturalist* between 1892 and 1896. Little new has been unearthed in the intervening nine decades to cause us to revise his well-founded conclusions, and it is most unlikely that the passing of nine more decades will have any effect either. Nathaniel Colgan's historical treatise is as immortal as the green shamrock.

My purpose is to attempt to explain the shamrock, as best I can, and in undertaking that task I must acknowledge my immense debt to Nathaniel Colgan. Some will say that it is reprehensible to tear away the myths and sentimental veneer, but I believe the true story is just as fascinating, and well worth the telling. After all, it is an integral part of our cultural and scientific history.

> Ireland's the land of the harp and the shamrock.
> Ireland's the land of the true and the good.
> Ireland's the land of the true Irish patriots
>   who shed for their country, the last drop of blood.
> Here's to the lake, to the vale and the green moss,
> The harp and the shamrock, the green flag and cross,
> And here's to the heroes that old Ireland can boast,
> May their names never die; it's an Irishman's toast.[5]

ACT I

Enter an Ancient Briton

Let's begin at the beginning with the principal character in this drama, Patrick, the patron saint of Ireland.

> I, Patrick, …had as my father the deacon Calpornius, son of the late Potitus, a priest, who belonged to the town of Bannavem Taberniae; he had a small estate nearby…[1]

Thus Patrick himself set down clearly his ancestry in the opening passage of his *Confession*, a letter which he composed during his elder years. With the *Epistle to the soldiers of Coroticus* which he had written some years earlier, the *Confession* provides a remarkable picture of a man who lived about fifteen centuries ago. The original manuscripts do not survive today, but we are most fortunate that the texts of both were copied and thus preserved, and from these transcripts we can sketch a fragmentary outline of the life and career of the ancient Briton who became Ireland's patron.

Patrick was born in Britain during the last years of the crumbling Roman Empire about the turn of the fourth century; the year of his birth is not known, nor are any other dates during his long life recorded with anything approaching accuracy. His father was a deacon in the early British Church and also a village councillor, a member of the local Romano-British administration, and it is more than likely that Potitus had held the same position as it was an hereditary office. But although Patrick gave the name of the village, Bannavem Taberniae, we do not know where it was situated within Roman-occupied Britain. The precise location need not concern us, but it does continue to exercise Patrick's biographers. Some scholars have suggested that Bannavem Taberniae lay at the western extremity of Hadrian's Wall in the northwest of England not far, perhaps, from the modern city of Carlisle, but it is as probable that Bannavem Taberniae was in the southwest near the shores of the Bristol Channel.[2]

When he was sixteen years old, Patrick was captured at Bannavem Taberniae by a party of Irish pirates who carried him off with many other people. The captives were sold as slaves, and Patrick was purchased by an Irish farmer. For six years Patrick was enslaved; his life was comfortless and perilous, but the young man survived. He does not complain in his letters about his treatment, so it has been surmised that his anonymous owner was not brutally cruel. In later years Patrick did remember the bitter cold,

Fig. 4: Sliabh Mis, Slemish Mountain near Ballymena, Co. Antrim, which is the remnant of an ancient volcano. (Photograph by Robert Welch, 1907 (no. 562); reproduced by courtesy of Ulster Museum, Belfast)

the rain and the snow, all of which he endured while tending sheep on the exposed hillslopes near the ocean.

The traditional place of Patrick's enslavement is Sliabh Mis (Slemish), a striking, isolated hill situated in County Antrim in the far northeastern corner of Ireland. But some believe that he may have been held captive at the Wood of Voclut because it is the only Irish locality that is named by Patrick in his *Confession* and *Epistle*. Patrick wrote that the Wood of Voclut was near the western sea and it is as difficult to find Voclut on modern maps of Ireland as it is to locate Bannavem Taberniae on a map of Britain. Again scholars have their own cherished opinions, and some equate Voclut with the village called Fochoill (in English, Foghill) which does lie not far from the Atlantic Ocean to the east of Killala in the north of County Mayo near the border with Sligo. Like the many other mysteries in the Patrician story the riddle of Voclut's precise location will never be solved.

Although his grandfather was a priest and his father a deacon, the youthful Patrick was not a faithful Christian when he was captured and brought to Ireland. However during his years as a slave he was converted, not by local missionaries but by his own meditations on the religious beliefs which his family had already espoused. After his conversion he would wake before dawn to pray and his faith grew deeper as the time passed.

One night the young slave had a dream. A voice told him that he should fast for soon he would be returning to his home country. Later the voice said "Your ship is ready", so Patrick fled from the farm where he was imprisoned and travelled about two hundred miles to a place that he had never visited before. A ship was indeed just about to set sail, and Patrick enquired of the master if he could have a passage, but he was angrily rebuffed. Yet as he turned to go back to a miserable hut where he had been hiding, one of the crew shouted to him and he was allowed on board.

After three days the ship docked in a mysterious place, perhaps in northern Gaul (France)—Patrick omitted to give any names. Then followed, according to his own strange narrative, a journey lasting twenty-eight days 'through a desert'; no-one has succeeded in explaining this, and we must pass it by. It seems likely that he stayed in Gaul for several years before eventually returning to Britain and his parents' house. Patrick had been away for more than six years, and his parents greeted him warmly.

Patrick settled down again to a more comfortable life on his family's estate at Bannavem Taberniae, but one night he had another visionary dream. He dreamt that a man called Victoricus came bearing letters, one of which he gave to Patrick. In this, *Vox Hiberniae* (The Voice of Ireland) beseeched the 'holy youth to come and walk once more among us'. He awoke and obeyed the vision, going to Ireland for the second time to commence his mission there. The year of Patrick's return to Ireland is not known—the traditional date is 431 A.D.

Patrick was probably not yet thirty years old when he took up his ministry, and for perhaps another thirty years he laboured to convert the pagan Irish. He was ordained as a deacon in the Church and at one stage in his later life was rejected as a suitable candidate for a bishopric because of a sin he confessed to having committed in his boyhood. But eventually this British missionary was consecrated and served as the only bishop to the Irish. Patrick carried out his episcopal duties diligently, baptising new converts and preaching the gospel to those who yet remained pagans.

We have no indubitably authentic records of the deeds performed by Patrick during his life, apart from what may be inferred by careful reading of his *Confession* and the letter to the barbaric soldiers of Coroticus. Even that letter says only that Patrick had recently—we have no date—converted a group of Irish men and women who were suddenly attacked by Coroticus' men; some of the converts were murdered and others were taken captive. Patrick wrote two letters to the soldiers, admonishing them and demanding the safe return of the prisoners—it was his second letter that was copied.

Patrick died, perhaps in the latter half of the fifth century, but there is no account known today telling when, where or how his life ended, nor where his body was buried. Soon he was almost entirely forgotten.

That is as much as is known about Patrick. It amounts to little, but in reality we know much more about Patrick than we do about most other Ancient Britons or, indeed, about any other escaped slave during the Roman period. And it must be stressed that Patrick's own *Epistle* and *Confession* are the only authentic sources about his life. May I underline the silence about the shamrock; nowhere in Patrick's letters is there any mention of trefoils of any botanical variety.

ACT II

Enter the Biographers

After Patrick's death, for perhaps a century and a half, he was remembered only by a handful of people and he was not then venerated as a saint. Somehow transcripts of his two letters did survive, and it is possible that a few other sketchy written accounts of his life and deeds were also kept.

A measure of the obscurity into which this now-familiar saint fell can be gauged by noting the silence about Patrick in the earliest histories of the Christian church in Ireland and Britain. The British chronicler Gildas makes no mention of the saintly Irish bishop, nor does the Venerable Bede whose history of the church in England, completed in 731, includes a substantial amount of information about the progress of Christianity in Ireland.

However, during the seventh century memories had stirred in Ireland. About 632, in a letter to the Abbot of Iona, a monk named Cummain wrote of 'our father Patrick' who had introduced an Easter Cycle to Ireland. There is a prayer asking for Patrick's blessing in the *Book of Durrow*, one of the earliest Irish manuscripts compiled by an anonymous scribe shortly after the year 600.[1] Late in the same century, Adomnan wrote his *Life of St Columba* and in a second preface to it he mentioned a disciple of Bishop Patrick. These incidental references to Patrick seem to indicate that after a lapse of about five generations his name was again familiar at least in monastic scriptoria.

The first biographer of Patrick was Muirchú who wrote a life of the saint about 700 (perhaps some years earlier) at the request of Aed, Bishop of Sletty. Muirchú knew about Patrick's *Confession* and his *Epistle to the soldiers of Coroticus* and had a variety of other sources for his work, but he admitted that these were of uncertain value and furthermore that he himself was inexperienced and had a faulty memory. So Muirchú's biography of Patrick was, like so many that followed, an unreliable book in which fiction and fact were inextricably interlaced.

The surviving version of Muirchú's *Life of St. Patrick* is contained in *The Book of Armagh* (*Canóin Phádraig, Liber Ardmachanus*) dating from about 800; it is a rather plain little hand-written book, without elaborate and colourful decorations, but it is one of the most important of the early Irish manuscripts. It is the work of Ferdomnach, who was described as 'the wise and very best scribe', and probably also of a few of his pupils. *Canóin Phádraig* also contains

Fig. 5: Saint Patrick as a bishop vanquishing the snakes; this Irish stamp incorporates a reinterpretation of T. Messingham's famous engraving from *Florilegium Insulae Sanctorum*, 1624. The stamp was issued in 1961 to mark the reputed 1,500th anniversary of the saint's death.

a version of Patrick's *Confession*, and a most unreliable chronology of episodes from Patrick's life concocted by another Irish writer, Tírechán. While Muirchú's work was intended as a biography, Tírechán's compilation is merely a cocktail of 'facts' about the saint which was written for Ultan of Ardbraccan using materials gathered while Tírechán travelled through the countryside. Tírechán's *Episodes from Patrick's life* is hardly credible, but it does clearly show that by the beginning of the eighth century, fable and legend already enveloped the real Patrick.

Muirchú's biography contains some imaginative incidents; for example, he related that Coroticus ultimately turned into a fox! Tírechán excelled that, stretching history beyond all reasonable bounds, and listed over four hundred bishops that Patrick consecrated! The weaving of the tapestry of myth began in style. What is more, the embroidery of history that Muirchú and Tírechán commenced, others enriched with great enthusiasm, and the stories became progressively more extraordinary. Patrick was credited with building hundreds of churches, ordaining thousands of priests, performing miraculous feats and proclaiming the nastiest of curses. His footprints became embedded in rocks, and snakes and other venomous creatures were expelled from Ireland for ever.

But to give these earliest story-tellers some credit, they did not invent any myths about shamrocks. They remain silent; there is nothing about trefoils in Muirchú's and Tírechán's flawed chronicles. Almost an entire millenium must pass before we encounter the sacred plant.

ACT III
Enter the Scribes

Thus far we have found no mention of shamrocks, neither in Irish, in Latin nor in English. The word does not appear in any of the numerous *'lives'* of Saint Patrick written before the close of the first millenium. Indeed it is not known in any book before the fourteenth century and even then it does not appear as we would expect it. Yet it surely was in the vocabulary of the Irish people, for the earliest written references take the form of poems in the Irish language, verses that must have been held long since in men's memories.

But before noting the earliest appearances of the word, it is essential to explain how **shamrock**, a modern, English, word evolved.

In Irish, both **seamair** [pronounced *shamir*] and its variant **seamhair** [pronounced *shower*] signify clover—this is common usage and the interpretation is generally accepted by scholars, both lexicographers and botanists. From that word can be formed the collective noun **seamrach** [pronounced *shamarach*] which also means clover; the same word may be used as an adjective when it has the meaning 'covered with clover' or 'clover-grown'. Of more interest is another derivitive, **seamróg** [pronounced *shamarogue*], which is formed from **seamair** and the suffix **-óg**. **Seamróg** is the diminutive of **seamair** and may be translated as 'young clover' or 'little clover', but it may also signify a relationship, the implication being that the plant named **seamróg** is believed to be connected with that called **seamair**. The suffix **-óg** is not uncommon in Irish—it also occurs, for example, in the place-name **Cnapóg** formed from **cnap** and meaning 'a little heap of rock' (the modern English name for the place is Knappogue, which is in County Clare, and has a restored mediaeval castle).[1]

Shamrock is merely the anglicized version of **seamróg**, and literally means a little, a young clover.

In early Irish manuscripts, **seamair** and **seamrach** are sometimes combined with **scoth**, meaning flower or tuft, to form a poetic compound word **scothsheamrach** denoting clover-flowered. Mediaeval scribes did not spell words consistently, nor did they use the same spelling as modern Irish dictionaries, so many variations on **scothsheamrach** occur, yet this is the word that first appears in the written record.

The *Metrical Dindsenchas* is a miscellaneous collection of poems, which relate ancient legends, myths and other lore about the names of places in

Fig. 6: 'Ni muir itir acht mag scothach scothemrach': detail from Daniel Maclise's startling canvas 'Marriage of Strongbow and Eva' (1854)—among the herbage is clover, meticulously depicted—or is it shamrock? (Reproduced by courtesy of National Gallery of Ireland, Dublin)

Ireland. The surviving manuscripts are mediaeval, although the individual poems are mostly of earlier but indeterminate age. Several of these topographic verses contain words of interest to us. One poem about Temair Lúachra, which lies near Castleisland in County Kerry, has this line—**nos fuilngtis a scothshemair**—which can be translated as '…its flowering clover was beneath their feet.'

A second poem, perhaps composed to celebrate a fair in 1006 at Tailtin, the modern Teltown which is situated half way between Navan and Kells in County Meath, also contains a phrase of note—**ba mag scothach scothsemrach**—and that may be rendered as '…it became a plain blossoming with flowering clover'. There is no reason why the word shamrock should not be substituted for clover as the meaning is not altered in any way; indeed the translation would be more exact if shamrock was employed.

About the year 1339, the *Book of Leinster* (*An Leabhar Laighneach*) was written. In it there is also a phrase signifying 'clover-covered plain'—**mmaig siamrach** could with equal validity be given in English as shamrocked plain!

Among mediaeval Irish manuscripts, *The Speckled Book* (*An Leabhar Breac*) stands apart. It is the work of a single scribe, almost certainly Murchadh Ó Cuindlis, who had as patrons the Mac Aodhagáin (MacEgan) family of County Galway. The book is a marvellous miscellany of good stories, many with a religious theme, which the scribe and his patrons clearly must have enjoyed. Ó Cuindlis frequently added marginal notes that further enliven the book and provide little glimpses into a writer's life in early fifteenth century Ireland. *An Leabhar Breac* was completed sometime before 1411; in other words the stories were not written down until one thousand years after the young Patrick endured the cold and rain of Ireland.

For the shamrock story, this manuscript volume holds interest because of a short tale about saints Scuithín and Barra:

> *Fecht dorala he do Barra Chorcaige ocus eisium oc imthecht in mara ocus Barra hillúing.*
> *"Cid fodera in muir do imtechtt duit?" ol Barra.*
> *"Ni muir itir acht mag scothach **scothemrach**," ol Scuthín, ocus tocbaid a laim scoth chorcra ocus cuirid uad do Barra isin luing.*
> *Ocus atbert Scothín, "Cid fodera long do shnám for in mag?"*
> *Lasin nguth sin sínid Barra a láim isin muir ocus atnaig bradan esti ocus telcid cu Scothín.*

> On one occasion he [Scuithín] met Barra of Cork, when he [Scuithín] was walking on the sea and Barra was in a ship.
> "Why do you walk on the sea?", asked Barra.
> "It is not the sea," answered Scuithín, "but a plain, flowery and shamrock-bearing". He picked a purple flower and threw it to Barra in the ship asking "Why does a ship swim on the plain?"
> Then Barra put down his hand into the sea and hooked out a salmon, which he threw to Scuithín.[2]

The original word employed by Murchadh Ó Cuindlis, which has been translated as shamrock-bearing was **scothemrach**, which is an acceptable variant of **scoithshemrach**.

Jumping ahead several centuries, it is worth noting a few other Irish sources in which the words **seamrach** and **seamróg** are used. In 1607 Tagdh Ó Cianáin fled Ulster with other Catholic Irish noblemen and travelled through France and Italy to Rome. His diary of the events now known as The Flight of the Earls was begun on 7 September 1607 and the last entry is dated 24 September 1609; it is a fascinating document telling of their journey through Europe, the cities and towns that they visited and the kind of countryside through which they passed. After travelling along the Adriatic coast passing Rimini, Venice, and Pesaro, Tagdh Ó Cianáin and his companions reached Senigallia, a fortified city belonging to the Duke of Urbino, picturesquely situated on a lovely river. Tagdh wrote:

> *Tegh osta ba lor feabhus [agus] deissi sechtair na cathrach. Ruiber roi-dhess go ffaith[ch]i noininigh sgoth-shemraigh chomthroim chomh-fhairsing a n-imfhoixe in ósta-thighe sin tra.*[3]

The Revd Paul Walsh, who edited Tagdh Ó Cianáin's journal for publication, translated this passage:

> There is an excellent and pretty hostel outside the city. There is a very fine river and a daisy-covered, clover-flowered, level, wide green near that hostel.[3]

Not long afterwards the word **seamróg** begins to crop up in dictionaries and glossaries. Pilib Ó Súilleabháin Béarra, a native of the County Cork, went as an exile to Spain when he was a boy and lived there for the rest of his life. He compiled his *Zoilomastix* in the 1620s; it was not published in Ó Súilleabháin's lifetime, but the manuscript survived and the work contains chapters on plants that include (among other fascinating matters) references to gardens in Connaught. For our purpose the section on *Trifolium* is important, because Ó Súilleabháin gives the Greek, Spanish and Irish names as well as the Latin one, and the Irish names which he knew were **seamróg** and **seámur**.[4] Other lexicographers followed suit: Risteard Pluincéad's Latin-Irish dictionary of 1662 noted that *trifolium* was equivalent to **seamar** and distinguished *trifolium acetosum* as **seamsóg**; in 1768 Ó Briain listed **seamróg** and stated it was clover, and so did, for example, Connellan in 1814 and Ó Raghallaigh in 1821.

While **seamróg** continues to feature in Irish glossaries and dictionaries until the present day, it plays no more significant part in the literary horizon of its homeland. All the critical texts, all the arguments about the shamrock and its origins, all the errors and the myths, are in English or, less frequently, Latin. To the Irish, **seamróg** was just one thing, a little clover.

But we have now got ahead in the story, and must return to the sixteenth century and to English historians and playwrights.

ACT IV

Enter the English Chroniclers and Dramatists

One and a half centuries after Murchadh Ó Cuindlis completed *The Speckled Book*, the word **shamrock** first makes its appearance in an English text. It is concealed in the handwritten pages of *The First Boke of the Histories of Irelande* compiled by the eminent Elizabethan orator, scholar and Jesuit, Edmond Campion, who was hanged at Tyburn in 1581. He journeyed to Ireland in August 1570 and lived for several months more or less as a recluse in the house of James Stanihurst, the Recorder of Dublin and Speaker of the Irish House of Commons. But Campion was hounded by the authorities, because they distrusted him as a papist, and after a while he had to go into hiding and move from house to house; it was during this time that he drafted his *Boke of the Histories of Irelande* which has been described as a pamphlet composed to prove '…that education is the only means of taming the Irish.'

Edmond Campion completed the manuscript on 9 June 1571 at Drogheda, and in it there is this descripton of the wild Irish:

> Prowd they are of longe crisped gleebes, and the same doe nourishe with all theire cunnyng. To crop the front thereof they take yt for a notable peece of vilany. Shamrotes, watercresses, rootes, and other hearbes they feed upon. Otemeale and butter they cramme together. They drincke whea, milke, and bieffe brothe.[1]

As Nathaniel Colgan has pointed out, the use of **t** instead of **ck** is not a problem; sixteenth century English authors were just as inconsistent when they spelled words as their Irish counterparts, and the spelling they employed is often quite different from that which we use today. Be that as it may, the sound of shamrote does closely reproduce that of the Irish **seamróg**.

Edmond Campion did not publish his own book, but he allowed one of his pupils, Richard Stanihurst, to have unrestricted access to the manuscript and permitted him to edit it for publication. Stanihurst was the son of James Stanihurst and was thus a Dubliner, yet he had studied under Campion at the University of Oxford between 1563 and 1568. What Richard Stanihurst did was this: he extracted great chunks of Campion's text and used these almost unaltered, but he also greatly enlarged and embellished other portions of his tutor's work. Stanihurst did most carefully acknowledge the work of his 'fast friende, & inwarde compagnion, M. Edmond Campion', so he

Fig. 7: Edmond Campion (artist unknown; reproduced by courtesy of National Portrait Gallery, London)

was not a plagiarist and may well have discussed the additions and embellishments with Campion. Thus Edmond Campion's statement that shamrocks were eaten by the Irish was published under Richard Stanihurst's name in the famous *Chronicles* of Raphael Holinshed which were printed in London in 1577.

The passage from Campion's *First Boke* which includes the reference to shamrock appeared in the eighth chapter (headed 'The Disposition and Maners of the Meere Irish, commonly called the wyld Irishe') of *A Treatise contayning a playne and perfect Description of Irelande* by Richard Stanihurst, and this is what was published:

> Proud they are of long crisped bushes of heare which they terme glibs, and the same they nourish with all their cuning; to crop thee front thereof they take it for a notable piece of villany. Water cresses, which they terme shamrocks, rootes, and other herbes they feede vpon, otemeale and butter they cramme together, they drinke whey, mylke, and biefe brothe.[2]

Thus the precocious pupil altered his master's text! The definite equation—watercresses are shamrocks—is not in Campion's original manuscript.

But was Stanihurst correct to state that shamrock was the name used by the 'wyld Irishe' for watercress? It is now impossible to tell if he knew the Irish names for plants and thus decided that the meaning of his tutor's text would be made clearer by inserting the three words 'which they terme', just as he had done with the explanatory phrase 'bushes of heare' before the word glib. Stanihurst could simply have misunderstood the original—in that case his editing has created great confusion.

This problem has exercised many scholars for far too many years, although the evidence of Irish texts and glossaries strongly suggests that Stanihurst made an error. Nathaniel Colgan explained that the Irish name for watercress is **biolar**, and that name appeared in an Irish glossary (in its earlier form **biror**) as far back as the tenth century, so there is no linguistic basis for Stanihurst's strange equation. Many writers who copied Richard Stanihurst's embellished text did not realize that and so watercresses have become the red-herrings in the saga of the shamrock.

In strict chronological sequence, Stanihurst's edition of Campion's manuscript was followed by the strange poem of John Derricke, which bears the title *Image of Ireland with a Discoverie of Wood Kearne*. The 'Epistle Dedicatorie' is dated 16 June 1578, but the work was not printed for another three years. Only in the 'Epistle' is there a reference to shamrock, where Derricke made this allusion

> For (in verie troth) my harte abhorreth their dealynges and my soule dooeth deteste their wilde shamrocke manners.[3]

What does Derricke mean by 'wilde shamrocke manners'? Again we can turn to Nathaniel Colgan who discussed this phrase and suggested that it was intended as a scornful term, an expression of contempt, presumably derived from the reported habit of the wild Irish—but not the civilized English!—of eating something called shamrock. Derricke did not need to know what shamrock was, and he probably had never seen one. But where did the report of the shamrock-eating Irish come from? Undoubtedly he had read Raphael Holinshed's recently published *Chronicles* containing Edmond Campion's—*alias* Richard Stanihurst's—account of the manners and customs of the Irish.

Raphael Holinshed's *Chronicles* attracted considerable attention and a second edition was published one decade after the first. In particular the *Chronicles* provided a marvellous source of material which could be used by playwrights, the contemporaries and successors of William Shakespeare. The Bard of Avon never used the word shamrock, but a clutch of other authors did when treating Ireland and Irish characters.

In 1596 a play, entitled *The famous Historye of the life and death of Captaine Thomas Stukeley*, was performed in London; its anonymous author certainly made use of Richard Stanihurst's work although it is evident that he was not unfamiliar with Irish oaths!

It is night. Near Dundalk, Irish troops are about to attack the occupied town; their leaders, Neale Mackener [McKenna], O'Neale and O'Hanlon are discussing strategy. "Where shall we enter to suprize Dundalke", asks Mackener. In the still darkness they hear an English soldier cough.

Mackener: Be whist I heare one stir
O'Neale: Some English soldior that hath got the cough, Ile ease that griefe by cutting off his head.
Mackener: These English churles die if they lacke there bed, and bread and beere porrage and powdred beefe.
O'Hanlon: O Marafasrot shamrocks, are no meat, Nor Bonny clabbo, nor greene Water-cresses, Nor our strong butter, nor our swelld otmeale, and drinking water brings them to the Flixe.
O'Neale: It is these nicenes silly puting fooles
Mackener: There be of them can fare as hard as we, and harder too, but drunkerds and such like as spend there time in ale house surfetting, And brothell houses quickly catch their Bane
O'Neale: One coughes again, lets slip aside unseene, to morrow we will ease them of their spleen.[4]

Captain Thomas Stukeley was printed and published in 1605, nine years after its premiere. Apart from the use of shamrock, it is fascinating to note the Irish curse **marbhfháisc ort!**; the author has rendered it as 'O Marafasrot', which may be politely rendered as 'bad cess to you!' 'Bonny clabbo' is an attempt at anglicizing **bainne clábair**, literally 'milk mud', that is thick milk or in modern parlance buttermilk.[5]

But I have strayed from shamrocks! In 1605 or 1606, Edward Sharpham's play *The Fleire* was first performed at Blackfriars by the Children of the Revels; four editions were printed, the first being issued in 1607. In Act III, satirical reference is made to '...Maister Oscabath the Irishman...and Maister Shamrough his Lackey...'—the first name is a corruption of **uisce beatha**, the water of life, whiskey!

Three other Jacobean plays contain mention of shamrock, the first being the anonymous *The Welsh Embassadors* which was performed in London during 1623 and published the same year. The best known is Thomas Dekker's *The Second part of the Honest Whore* from which comes this oft-quoted phrase: 'Longed you for shamrocke?' The earliest extant edition of that play was published in 1630.

James Shirley, an English dramatist who came to Ireland about 1636 under the patronage of the sixteenth Earl of Kildare, wrote a tragedy entitled *Saint Patrick for Ireland* which was performed at the theatre in Werburgh Street, Dublin, during 1639 and is generally considered to be the first truly Anglo-Irish drama. It contains this scene: two soldier are in a wood when Rodamant enters wearing on his wrist a bracelet that renders him invisible.

> First soldier: I see nothing but a voice; shall I strike it.
> Second Soldier: No, 'tis some spirit; take heed, and offend it not. I never knew any man strike the devil, but he put out his neck-bone or his shoulder-blade. Let him alone it may be the ghost of some usurer that kick'd up his heels in a dear year, and died upon a surfeit of shamrocks and cheese-parings.[6]

There can be little doubt that Edmund Campion was the sole original source for the information on the shamrock employed by the authors of these poetic and dramatic works; none of the writers displays any real inkling of the true nature of shamrock, nor were they aware, it seems, of long-available botanical texts containing other, more accurate accounts about this Irish staff-of-life.

ACT V
Enter the Herbalists

With the word shamrock ensconced in English literature, but with a thoroughly unreliable set of meanings for it, the problem of equating the name with a real plant remains. Edmond Campion had simply reported that the Irish ate shamrocks, and Richard Stanihurst had in error stated that watercress was the same thing by another name, so we have ample reason to be sceptical.

It is clear from a herbal written by two French botanists and published in London during 1570, one year before Edmond Campion completed the manuscript of his book, that information—but not always accurate information—about the native plants of Ireland was slowly filtering across the Irish Sea to England.

Matthias de l'Obel was a native of Lille in Flanders and a trained physician. He travelled to England in 1569 and eventually settled there, working for a time as superintendent of the physic garden on Lord Zouche's estate in Hackney and eventually obtaining from King James I the title of *Botanicus Regius* (King's botanist). Matthias de l'Obel is commemorated in the popular blue-flowered bedding-plants called *Lobelia*.

Before settling in London, l'Obel had studied at the famous university at Montpellier where he met Pierre Pena. This young man, who was born near Aix in Provence, had forsaken a military career, entered Montpellier University and turned to medicine. Discovering that they shared a passion for botany, Pena and l'Obel became great friends; they went on plant-hunting trips together, and collaborated in writing the book *Stirpium Adversaria Nova*. This herbal included notes about plants they had collected themselves since their arrival in England, and it was ready for publishing late in 1570 or early in 1571 and the authors dedicated it to Queen Elizabeth I.

Stirpium Adversaria Nova was written in Latin. It contains a strange passage about the Irish passion for a meal of clover—meadow trefoil—which has been translated as follows:

> The Meadow Trefoil…there is nothing better known, or more frequent than either, or more useful for the fattening, whether of kine, or of beasts of burden. Nor is it from any other than this that the mere Irish, scorning all the delights and spurs of the palate grind their cakes and loaves which they knead with butter, and thrust into their groaning bellies, when, as sometimes happens, they are vexed and high maddened with a three days' hunger…[1]

Fig. 8: John Gerard, c. 1598, holding in his hand a sprig of potato in blossom, then cultivated merely as a botanical curiosity; this portrait appeared in the revised (1633) edition of *The Herball*. (Engraving by John Paye; reproduced by courtesy of Hunt Institute for Botanical Documentation, Pittsburgh)

The two authors acknowledged that they had obtained their information from 'certain gentlemen of our acquaintance' who had served for several years with the English army in Ireland. They do not employ the word shamrock anywhere, and it would be stretching reason, for the moment, to add this account to that of Edmond Campion and obtain the equation **meadow trefoil = shamrock**.

Perhaps it was from similar contacts that another English herbalist obtained his less colourful account of the same meadow trefoils.

John Gerard lived at Holburn near the city of London. He was an apothecary and herbalist, an enterprising gardener and an inveterate collector of facts about plants. His reputation has taken something of a battering this century from botanical historians who accuse him of being a thief and a rogue, and especially of using earlier authors without acknowledging his sources. Be that as it may, Gerard's account of the shamrock owes nothing to earlier sources, unless he had access to books and manuscripts that have totally vanished.

The Herball or Generall Historie of Plantes by John Gerard was published in London probably during the last weeks of 1597; its text is in English and its pages are illuminated with superb wood-cuts. Chapter 477, on page 1017, contains the account 'Of Three leafed grasse, or Medow Trefoile', and opens with this paragraph

> There be diuers sortes of Three leafed grasses, some greater, others lesser; some bring foorth flowers of one colour, some of another; some of the water, and others of the land; some of a pleasant smell, others stinking: and first of common Medow Trefoiles, which are called in Irish *Shamrockes*.[2]

Two marvellous illustrations grace the lower half of the page; on the left the 'Medow Trefoile' named in Latin, then as today, *Trifolium pratense*, and on the right 'Medow Trefoil with white flowers' or *Trifolium pratense flore albo*. The plant on the left is the familiar claret-blossomed red clover, and the one on the right is the abundant, weedy white clover which is now simply called *Trifolium repens*. Gerard leaves us in no doubt that he understood shamrock to be clover; indeed his account suggests that he understood the Irish used **seamróg** as a generic name covering both white and red clover.

There is nothing else in John Gerard's herbal about shamrock; no tales about saints, nor lurid accounts of hordes of ravenous warriors stuffing themselves with shamrocks. Nathaniel Colgan commented about this silence because Gerard was not at all reluctant to relate exotic stories—his famous

Of Three leafed graſſe, or Medow Trefoile. *Chap. 477.*

❊ *The kindes.*

There be diuers ſortes of Three leafed graſſes, ſome greater, others leſſer; ſome bring foorth flowers of one colour, ſome of another; ſome of the water, and others of the land; ſome of a pleaſant ſmell, others ſtinking: and firſt of the common Medow Trefoiles, which are called in Iriſh *Shamrockes*.

1 *Trifolium pratenſe.*
Medow Trefoile.

2 *Trifolium pratenſe flore albo.*
Medow Trefoile with white flowers.

Fig. 9: '…called in Irish *Shamrockes*': lower portion of page 1017 from John Gerard's *The Herball*, 1597. (Reproduced by courtesy of National Botanic Gardens, Glasnevin)

one is about goose barnacles turning into barnacle geese—but he cannot have known any about shamrocks.

If only the precise meaning of the Irish word **seamróg**—the young clover—and John Gerard's simple statement about clovers had been accepted at face value we might have been spared the confusion of the next century. But we would also have been deprived of the marvellous tapestry of fable and folk-lore that was to avalanche from the pens of writers throughout the next four hundred years!

ACT VI

Enter a Map-Maker and several Poets

The next person to use the word shamrock in a published work was the translator Philomen Holland, who lived in Coventry and was 'fond of parading' a degree in medicine awarded either by a Scottish or foreign university. His medical practice was small and he really spent his time translating the classics. Dr Holland also produced an English edition of William Camden's floridly titled *Britannia: sive florentissimorum regnorum Angliae, Scotiae, Hiberniae…descriptio.* This pocket-sized Latin text purported to be a geographical and historical account of the British Isles and included a separate section about Ireland (*Hibernia*); it first appeared in 1587, about the same time as the second edition of Raphael Holinshed's *Chronicles*. Camden's *Britannia* was a best-seller and went through four editions in seven years, each one revised under the author's supervision. There is no mention of shamrocks in the Latin text of the original edition, but they do appear in the English translation which was published by Holland in London during 1610. Camden collaborated with his translator; Philomen Holland asked William Camden to clarify the meaning of difficult phrases, and it is reported that Camden corrected the proofs. Thus between them, the two writers produced an expanded edition of *Britannia*.

In the part dealing with *Hibernia*, there are a few sentences towards the end about the feeding habits of the Irish.

> As for their meates, they feed willingly upon herbs, & watercresses especially; upon Mushroomes Shamroots & rootes: so that Strabo not without good cause said they were…*eaters of herbs*, for which in some copies is falsly read…*Great Eaters*. They delight also in butter tempered with ote meale, in milke, whey. beefe-broth, and flesh sometimes without any bread at all.[1]

Apart from the invocation of Strabo, and the interpolation of mushrooms into the list of herbs devoured, this is clearly nothing more than a copy of the Blessed Edmond Campion's catalogue. Camden did not visit Ireland, but while a master at Westminster School he did teach, and convert to the protestant faith, '…divers gentlemen of Ireland, as Walshes, Nugents, O'Raily, Shees…and others bred popishly and so affected.' Could he have learnt about the Irish diet from those boys? I doubt it; the similarities between Stanihurst's edition of Campion, and Holland's translation of Camden are too transparent for any attribution of originality to be granted to the latter.

Copying Camden's copy of Campion became a minor industry. Within a few months, John Speed, an English antiquarian and map-maker, had lifted the information and it appeared, little disguised, in *Theatre of the Empire of Great Britain* published in 1611. This fine book was the first printed atlas of the British Isles and it contained five famous maps of Ireland (one of the whole island showing the individual counties, and separate maps of the four provinces). Such was the success of Speed's atlas that it went through many editions both folio and pocket-sized.

John Speed's map of 'The Kingdome of Irland', embellished with two cherubs holding a harp, also has six vignettes portraying the natives—The Wilde Irish man, replete with glib, and The Wilde Irish woman are tellingly represented. The text accompanying the map includes a sentence on the food of this dishevelled, uncherubic couple:

> Their dyet in necessitie was slender, feeding vpon water-cresses rootes, mushromes, shamrogh, butter tempered with oat-meale, milke, whey, yea, and raw flesh...[2]

What was Speed's source? The mushrooms indicate Philomen Holland's translation of Camden's *Britannia*. Speed, like so many other writers, never set foot in Ireland, copying even his maps from the works of other men, so his information about the Irish diet, including shamrock, had also to be second-hand.

It is obvious that there was a widespread belief in England during the reign of King James I that the native Irish were uncouth and ate wild plants—such ideas were propagated by these popular books and the Jacobean dramatists already mentioned were among the authors who accepted this popular notion. The same habits were noticed in a series of satirical verses by the poet and pamphleteer George Wither. His *Abuses, stript and whipt, or Satirical Essayes*, published in 1613, was a run-away success going into at least five editions that year, but it also earned for its author a prison sentence—Wither was released from the Marshalsea Prison after a few months thanks to a poetic appeal which he addressed to the king from his prison cell and, it is believed, to the intervention of the Princess Elizabeth.

In the first volume, in the eighth satire on covetousness, George Wither wrote this:

> Or e're I'd coin a lye, be it ne'er so small
> For e're a bragging Thraso of them all
> In hope of profit, I'de give up my play,
> Begin to labour for a groat a day,

> In no more clothing than a mantle goe
> And feed on Sham-rootes as the Irish doe.³

In another verse on the subject of vanity, the Irish vegetarian diet is again invoked:

> But see whereto this dainty time hath brought us
> The time hath been that if a Famine caught us
> And left us neither Sheep, nor Oxe, nor Corne,
> Yet unto such a diet were we borne,
> Were we not in our Townes kept in by th' foe?
> The woods and field had yielded us enough
> To content Nature: And then in our needs
> Had we found either leaves or grasse, or weeds,
> We could have liv'd as now at this day can
> Many a fellow subject Irish-man.³

But the question about the identity of shamrock remains unanswered: are shamrocks definitely the same as watercresses? Perhaps no-one has contributed more to this confusing conundrum than a contemporary of John Gerard by the name of Fynes Moryson, one-time secretary to the Lord Deputy of Ireland. He first came to Ireland in 1599 in 'the hopes of preferment', and while living in this country made some observations which he later set down in his *An Itinerary…thorow Twelve Dominions* published in London in 1617, four years after Wither's satirical poems. Reporting that the 'wild Irish do not thresh their Oates, but burn them from the straw, and so make cakes thereof, yet they seldom eat the bread', Fynes Moryson claimed that they

> impute covetousness and base birth to him that hath any corne after Christmas, as if it were a point of nobility to consume all within these festival days. They willingly eat the herbe Schamrock being of a sharpe taste which as they run and are chased to and fro they snatch like beasts out of the ditches.⁴

Various scholars have discussed at considerable length the significance of Moryson's description of the herb with 'a sharpe taste', and have attempted to reconcile shamrock with watercress because of its peppery tang, or with the sharp, lemon-flavoured wood sorrel.

Moryson's paragraph is strange, to say the least—can we rely upon it at all? He credited his information to 'a Bohemian Baron' who had lived in the English and Scottish royal courts before having 'out of his curiositie to return [to Bohemia] through Irelande in the heate of the rebellion'. Thus Moryson was reporting something he himself had not heard when in Ireland—his information was second-hand. Moreover his informant took an extraordinary round-about way home, deliberately venturing to travel through a rebellious, unsettled country. Even if the account contains truths about the state of Ireland at the turn of the sixteenth century, about a desperate people, brutally treated by English troops, starved and forced to eat wild plants, it is most unlikely that the Bohemian baron had any opportunity or desire to ask the famished wretches what they were eating, and even less to discover what they called the plant. So implausible is the scenario that any arguments about its botanical merits are devoid of sense.

The last of the minor poets deserving mention, for the present, is John Taylor, the so-called Water Poet, whose *Sir Gregory Nonsence, his Newes from No Place* was likened by Nathaniel Colgan to Lewis Carroll's verses about the walrus and the carpenter. In 1630 Taylor published a volume of his collected works, and from it, from *Sir Gregory Nonsence*, Colgan extracted these lines:

> The mold-warp all this while in white broth bathed,
> Did Caroll Didoes happinesse in love,
> Upon a gridiron made of whiting-mops,
> Unto the tune of John Come kisse me now,
> At which Avernus' Musicke 'gan to rore,
> Inthroned upon a seat of three-leav'd grasse,
> Whilste all the Hibernian Kernes in multitudes,
> Did feast with Shamerags steev'd in Usquebagh.[5]

The final line perhaps refers to the drowning of the shamrock, although Colgan dismissed this, arguing that because '…the national badge was still below the literary horizon there is really no shamrock to drown'. We will return to that matter anon.

The next player on the shamrock stage is Edmund Spenser, from whose fertile imagination came the great poem *The Fairie Queene* which was composed in County Cork. He lived in Ireland for almost twenty years, much longer than either Edmond Campion or Fynes Moryson. Spenser landed at Dublin on 12 August 1580 and remained here, apart from two brief visits to England, until the end of 1598, in fact to within one month of his death on 16 January 1598/9. In the early days of his Irish living, Edmund Spenser was secretary to Lord Grey, Lord Deputy of Ireland, and he accompanied Lord Grey to Kerry in November 1580. It was this journey, and the observations which he was able to make of the state of war-torn Munster, that were described by the poet in his *A view of the present state of Ireland* which was not completed until 1596. By that time Spenser had acquired

Fig. 10: Edmund Spenser (Engraved by J. Thomson, reproduced by courtesy of National Portrait Gallery, London)

land in Munster at Kilcolman near Doneraile in County Cork, and he lived there during the latter part of his life.

A view of the present state of Ireland is a strange work taking the form of a conversation between Irenæus and Eudoxus. It is not a factual history, but a long piece of imaginative blank verse and there are numerous references in it to Richard Stanihurst. Thus, despite the incorporation of details about Munster, much of Spenser's text must be regarded as 'second-hand'. It was not published in Edmund Spenser's life-time, but was inaccurately edited by Sir James Ware and included as an appendix to Ware's *Historie of Ireland* issued in 1633.

Spenser penned this harrowing vignette of starving people, the wreckage of the Munster war.

> Out of everye corner of the woode and glenns they came creepeinge forth upon theire handes, for theire legges could not beare them; they looked Anatomies [of] death, they spake like ghostes, crying out of theire graves; they did eate of the carrions…and if they found a plott of water-cresses or shamrockes theyr they flocked as to a feast for the time, yett not able long to contynewe therewithall.[6]

I can detect nothing new in this passage except the poetic imagery of the anatomies of death. Whether Spenser wished to suggest that shamrock was an alternative name for watercress, or whether he is denoting two different plants is largely irrelevant. Spenser had read Holinshed's *Chronicles* and knew what Campion and Stanihurst had written.

Sir James Ware, Spenser's first editor, is an interesting figure in the drama. He was an antiquary and historian, a native of Dublin and graduate of Trinity College, Dublin, and he published an edition of Edmond Campion's history as well as that of Edmund Spenser, so he was not unfamiliar with shamrocks. It is therefore note-worthy that in one of his own books, *De Hibernia et Antiquitatibus ejus Disquisitiones* of 1654, Ware provided a summary of the opinions of his predecessors about the eating habits of the ancient Irish. He accepted that the principal foods in their meagre diet were butter, milk and herbs especially '…the meadow trefoil, the water-cress, the common sorrel, and the *Cochlearia* which is now named scurvy-grasse…'.[7] Ware's original text was in Latin and he employed the accepted, contemporary Latin names for those plants—he did not use the word shamrock, although he obviously knew it having published the writings of Campion and Spenser.

ACT VII

Enter the Shamrock-Wearers

The seventeenth century was a period of transition for Ireland. The Cromwellian conquest in the mid-1600s and the subsequent confiscation of land hastened the influx of new, English landowners. By the close of the century, the old Gaelic order had largely passed away; the Irish language was no longer the every-day language of the landed families and the aristocracy most of whom were now protestant by faith and anglophone, but the ancient arts of poetry and music, and the native tongue remained the property of the common people who adhered, despite the Penal Laws, to the Catholic Church. At the end of the 1600s Dublin was a bustling capital city with magnificent new civic buildings, a thriving university and an intellectual elite who espoused the tenets of Protestant Anglo-Ireland. One other trait may be noted: Dr Edward MacLysaght in his masterly *Irish life in the seventeenth century* remarked that during this period '…the modern idea of nationality, though not of nationalism, had so far developed that for the first time in history we find a war in which the Irish are on one side and the British on the other…'[1] Thus, Ireland entered the modern era.

At the beginning of the 1600s shamrock was still supposed by English authors to be a plant which the Irish ate, either every day, or in times of dire necessity—the difference is not significant because it arises out of the prejudices of the writers. English ignorance of Ireland and the customs of the Irish is nicely illustrated by a few sentences from a letter dated 20 March 1638 and addressed to the Secretary of State, Sir Francis Windebank, by the Lord Deputy of Ireland, the Earl of Stafford (Thomas Wentworth). Stafford had just interviewed the young Earl of Antrim who was proposing to raise an army in Ulster and invade the Scottish islands so that he could assert his claim to the estates of the Marquis of Argyll. Lord Antrim was asked about provisions for his farcical campaign and responded that he had made no arrangements because

> …he conceived they should find sufficient in the Enemy's Country to sustain them, only his Lordship proposed to transport over with him ten thousand live Cows to furnish them with Milk… I [Stafford] told his Lordship, that seemed to me a great Adventure he put himself and Friends upon: For in case (as was most likely) the Earl of Argyle should draw all the Cattle and Corn into Places of Strength, and lay the Remainder waste how would he,

> in so bare a Country, feed either his Men, his Horses or his Cows? ... To that his Lordship replied, they would do well enough, feed their Horses with Leaves of Trees and themselves with Shamrocks.[2]

The ingenuous young earl was Irish but he had been raised at the Court of King Charles I, and therefore was entirely ignorant about the painful realities of Irish life. He was merely reflecting in his outrageous idea of feeding an army on shamrocks the prevalent state of knowledge about Ireland and her people.

Equally credulous was an Oxford doctor, Henry Mundy, who wrote a pamphlet about diet which was first published in 1680. Nathaniel Colgan drew attention to the doctor's ideas,[3] and noted that Mundy was an ardent vegetarian. In his booklet, written in Latin and bearing the title *Commentarii de Aere vitali, Esculentis ac Potulentis*, this English physician vigorously asserted that a vegetarian diet was far superior to one containing flesh. As an example of the benefits of the former he cited the Irish who '...nourish themselves with their shamrock (which is the purple clover) [and thus] are swift of foot and of nimble strength': the original text from *Commentarii...* is

> *Sunt item Hiberni qui suo Chambroch (quod est trifolium pratense purpureum) aluntur celeres et promptissimi roboris.*[4]

Dr Mundy's obscure text was known to at least one eminent eighteenth-century botanist, Carl Linné of Uppsala in Sweden who was to revolutionize the system of naming plants and animals. Linné even used Mundy's unique spelling of shamrock in his *Flora Lapponica*, published in Amsterdam in 1737; more than that he quoted from *Commentarii de Aere...* that entire sentence almost unaltered:

> *Hiberni suo Chambroch, quod est Trifolium pratense purpureum, aluntur celeres et promptissimi roboris.*[5]

Henry Mundy had little or no knowledge of Ireland, nor of the Irish diet; he must have taken his information from older sources such as William Camden's *Britannia...* in the same way that he quoted Ovid. Thus we cannot grant any significant place in our shamrock history to Mundy's definite equation of the common red clover and the shamrock.

In contrast to the Oxford physician, Sir Henry Piers lived in this country, at Tristernaght Abbey, County Westmeath. He was possibly the last person to make any new statement about the eating of shamrock, but strangely his observation was not about the habits of the people of Westmeath rather of those who lived in the rock-strewn barony of Thomond in County Clare, the area now more familiarly known as The Burren. There, according to

Fig. 11: A modern version of St Patrick's cross, made in 1907 in Co. Kildare from white paper (11cm diameter) and ribbons (the broadest pair are deep red, with solitary light green and dark green ribbons) and rosettes (green at ends of red ribbons; red at ends of light green ribbon, yellow at ends of dark green ribbon). (Reproduced by courtesy of National Museum of Ireland, Dublin)

Sir Henry, '…butter, new cheese, and curds and shamrocks are the food of the meaner sort' for the entire summer. Although compiled in 1682, two years after Mundy's booklet, Sir Henry Piers' *Chorographical description of the County of Westmeath* was not published until 1774, so it could not have been pillaged by authors in the seventeenth and eighteenth centuries.

There is evidence that neither Munday nor Piers were reporting a living custom, but antiquated hearsay and myths. Two other writers recorded a completely new use for shamrock during the same decade, the sixteen eighties.

THE

Irish Rendezvouz,

OR A

DESCRIPTION

OF

T——ll's Army

OF

TORIES

AND

BOG-TROTTERS.

In Dogrel VERSE.

LONDON,
Prined for *Randal Taylor* near *Stationers-hall.* 1689.

Fig. 12: Title page of the first publication to contain a reference to the wearing of the shamrock (Reproduced by courtesy of Beinecke Library, Yale University, New Haven)

Thomas Dineley (otherwise Dingley), an English gentleman, had studied under James Shirley, the dramatist and author of *St Patrick for Ireland*. Dineley was an insatiable traveller, visiting the Low Countries, France and, in 1680, Ireland where he remained for about twelve months, perhaps in a military capacity. He wrote an account of his adventures and observations in 1681, and while this too remained as an unpublished manuscript until 1856, there are good reasons to believe that his diary contains accurate information about the Irish. His comments about customs and manners occasionally are a bit perfunctory, and some of the reports evidently did not arise from his own personal experiences, but he penned these famous lines about the celebrations of Ireland's patron saint.

> The 17nth day of March yeerly is St Patricks, an immoveable feast, when ye Irish of all stations and condicions were crosses in their hatts, some of pinns, some of green ribbon, and the vulgar superstitiously wear shamroges, 3 leav'd grass, which they likewise eat (they say) to cause a sweet breath.[6]

Taken by itself that sentence may seem insubstantial proof that in the 1680s Irish men and women wore shamrocks on St Patrick's Day, but it has to be considered with one other, much stranger piece of writing.

In 1689, a 'Dogrel VERSE' entitled *The Irish Rendezvouz, or a description of T---ll's army of Tories and Bog-trotters*, was anonymously published for Randal Taylor of London. It is a satirical polemic against Tyrconnell, and begins with a direct allusion to the wearing of shamrock:

> Of Irish Tories and Scallogues,
> Arch as e'er trotted Bog in Brogues,
> And of their Mantles and their Trouses,
> Wherein full many a hungry Louse is,
> I mean to sing; Help me, St. *Patrick*,
> Or else I swear I'll serve thee a Trick.
> For if I wear thy Cross and Shamarogue
> Next Seventeenth of *March*, I am a Rogue:
> Nor will I more believe the Story
> Of thy Fam'd Northern Purgatory;
> But think the Fable, by your leave, is
> As false as that of *Guy* or *Bevis*.[7]

That is the *first* printed record of the tradition of wearing shamrocks on St Patrick's Day but, of course, there is no explanation of the custom nor of its association with that particular saint's day. Yet in two footnotes that anonymous rhymster stated that the cross was '…worn by all the Irish on St. Patrick's day', and the shamarogue was '…a little three-leav'd Grass, worn at the same time at the Foot of the St. *Patrick*'s Cross.'[7]

It is amusing to note that in the same year, 1689, James Farewell composed and printed another satirical poem in which the words shamrock and Shamrockshire are frequently employed. *Irish Hudibras, or Fingallian Prince...* takes the form of a travesty of the sixth book of Virgil's *Aeneid*—Nathaniel Colgan did not enjoy reading it and ventured to suggest that the poem was the work of 'a foul imagination' and was now 'deservedly scarce'. Whatever James Farewell's state of mind, he used shamrock—almost invariably spelled shamrogue—in the role of the golden bough which magically assisted Nees (the Hibernian Aeneas) in his journey through the Underworld. Shamrockshire is also employed on numerous occasions, as for example in this couplet:

> Not one, like Him, will e'er appear
> Again, to grow in Shamrogeshire...[8]

Nathaniel Colgan quoted several extracts from this metrical lampoon merely to give a flavour of the doggerel, and he carefully omitted some quite unpleasant allusions; for example he replaced by dots the second line of this section which alludes to the ghosts of Irish heroes:

> Stalking about the Bogs and Moors
> Together with their Dogs and Whores;
> Without a Rag, Trouses, or Brogues
> Picking of Sorrel and Sham-rogues...[8]

For Farewell's purpose, the ridicule of the Irish, it was appropriate that the shamrock should be eaten. Thus, taking his cue from Edmund Spenser, the poet recounts that after Nees and his companions landed on the shore of Lough Erne they searched for food and discovered

> Springs, happy Springs, adorn'd with Sallets,
> Which Nature purpos'd for their Palats;
> Shamrogs and Watergrass he shows,
> Which was both Meat, and Drink, and Close...[8]

And later—

> Thus hotly they pursued the scent,
> Cramming their gorges as they went;
> Until they cropt the very weed
> Where every day they used to feed.
> Nees, when the shamrog he did spye
> Cries out, I have it in my eye
> Is vid me fait. And so he run,
> To bring the presents to the nun.[8]

The author's name does not appear on the published pamphlet, but he generously provided marginal glosses to explain the numerous Irish words that sprinkle his dire rhymes; shamrogue is interpreted as 'three-leav'd grass', shoges are 'spirits' and boglanders 'clowns'!

Nees, the hero, prepares to enter the Underworld and seeks advice from the nun, who gives it:

> If thou'dst be damn'd before thy day,
> Take a fool's counsel first, I say.
> Within a Wood, near to this place,
> There grows a Bunch of Three-leav'd-grass,
> Call'd by the Boglanders, Shamrogues,
> A present for the Queen of Shoges,
> Which thou must first be after-fetching,
> But all the cunning's in the catching…

James Farewell's caustic poem adds nothing to our knowledge of shamrock. One could argue that the crude equation of the shamrock with the fabulous golden bough was intended to signify that the two had the similar mythical yet mystical properties, and further that Farewell's frequent use of shamrogue and Shamrogueshire means these words were well-known to English readers. Such discussions are unnecessary, for the use of shamrock as an adornment on St Patrick's Day is now established, and its former role as the staple food of Ireland has been clearly assumed by the potato:

> Bring me a Bunch of Suggane Ropes,
> Of Shamroges, and Pottado-Tops…[8]

ACT VIII

Enter the Reverend Dr Threlkeld

Thirty-seven more years elapsed before there was any significant new information published about shamrock.

Meanwhile the Welsh antiquarian and naturalist Edward Lhuyd visited Ireland, touring the north and west as far as the Aran Islands and County Clare, and in a letter to Dr Tancred Robinson dated 15 December 1699, he remarked that '…Their Shamrug is the Common Clover'.[1] Yet Lhuyd contradicted himself a few years later when, in an Irish dictionary incorporated within his *Archæologia Britannica*, he defined **seamróg** as wood sorrel.

The shamrock and the wearing of crosses on St Patrick's Day were mentioned by several others during this same period, Sir Richard Steele and Dr Jonathan Swift being the most noteworthy. Steele used the columns of *The Spectator* on 12 August 1712 to set down some jocular analogies between plants used as national emblems and peoples of different nationalities:

> I reflected further on the intellectual Leaves…and found almost as great a Variety among them as in the vegetable World. I could easily observe the smooth shining *Italian* Leaves; the nimble *French* Aspen always in Motion…the *English* Oak, the *Scotch* Thistle, the *Irish* Shambrogue, the prickly *German* and *Dutch* Holly, the *Polish* and *Russian* Nettle, besides a vast Number of Exoticks imported from *Asia*, *Afric*, and *America*.[2]

On the subject of St Patrick's crosses, we must turn to Dr Swift. Writing from London to his beloved Stella in the spring of 1713, the future Dean of St Patrick's Cathedral, Dublin, complained that

> The Irish folks were disappointed that the Parliament did not meet to-day because it was St Patrick's Day, and the Mall was so full of crosses that I thought all the world was Irish.[3]

As I will return to the subject of crosses later, and to the wood sorrel mentioned by Edward Lhuyd, let's proceed with the history of shamrock.

On Thursday 26 October 1726 a small pocket-sized book about the wild plants of Ireland was published in Dublin.[4] It caused a ripple of sensation, even provoking some strident comments in a local newspaper. The provocation arose not from the author's castigation of those who said the Irish were uncultured, nor from his little sermons about excessive drinking,

Fig. 13: Edward Lhuyd (reproduced by courtesy of Ashmolean Museum, Oxford)

but from his remarks about potatoes and patriotism. No comments were passed about his paragraph on shamrock.

The book was the work of the Reverend Dr Caleb Threlkeld who lived at Francis Street, The Coombe, in the heart of old Dublin within the shadow of St Patrick's Cathedral. He was a native of Cumberland, not an Irishman, and had come from the English Lake District in 1713 to practise as a physician and to preach. Dr Threlkeld was a Protestant dissenter, a non-conformist, with fundamentalist views about the Christian faith and morality which he forcefully expounded in his botanical book and from the pulpit of a Dublin conventicle on Sundays.[4]

Caleb had been a student of philosophy at the University of Glasgow where also he developed his love of botanizing, perhaps to such an extent that he left without obtaining a degree. After serving as a cleric in his home village in Cumberland, the Revd Threlkeld went to Edinburgh University whence he received his doctorate in medicine. As a physician, Dr Threlkeld was kind and generous, treating the poor of Dublin without seeking payment, and when he died in 1728 he was mourned by many who had benefited from his charitable ways.

During the years he lived in The Coombe, Caleb Threlkeld often walked through the city streets noting what plants were being sold by the street-vendors and herb-women. They peddled royal ferns as 'good for obstructions of the liver', berries of the wild fraochán (bilberry) in season, and excellent artichokes better than any sold in England. He sometimes brought his young son with him, and would occasionally venture into the countryside around Dublin where white-flowered water-buttercups floated on the '…rolling streams of the Aon na Liffy'. The botanical observations made during these gentle perambulations were recounted in his little book, whose pretentious title *Synopsis Stirpium Hibernicarum…* belies the fascinating asides that it contains. Here is Dr Threlkeld's account of clover

> TRIFOLIUM PRATENSE ALBUM, *White Flowered Meadow Trefoyl*. The *Meadow Trefoyls* are called in *Irish* **Shamrocks**, as *Gerard* writes in his *Herbal…* The Word *Seamar Leaune* and *Seamar-oge*, being in signification the same, the first signifying the *Childs Trefoyl*, the other the *Young Trefoyl*, to distinguish them from the *Seamar Capuil*, or *Horse Trefoyl* as I suppose.[5]

So Caleb Threlkeld agreed with John Gerard and assigned the Irish name **seamróg** to the plants which today we call clovers. But he does not end there.

> This Plant is worn by the People in their Hats upon the 17. Day of *March* yearly, (which is called St. *Patrick*'s Day.) It being a Current *Tradition*, that by this *Three Leafed Grass*, he emblematically set forth to them the *Mystery of the Holy Trinity*.[5]

Thus an Englishman, a dissenter too, set down for the first time the tradition—his own word—about St Patrick and the shamrock as an explanation of the Holy Trinity. I think it is not unreasonable to suggest that he used the same parable in explaining the Holy Trinity to Dublin congregations and it is interesting that he did not say it was an historical fact.

Never one to pass an opportunity by, and not entirely approving of the bacchanalia associated with St Patrick's Day, the Reverend Dr Threlkeld added his own short sermon—it is easy to imagine him thundering the same words from a pulpit!

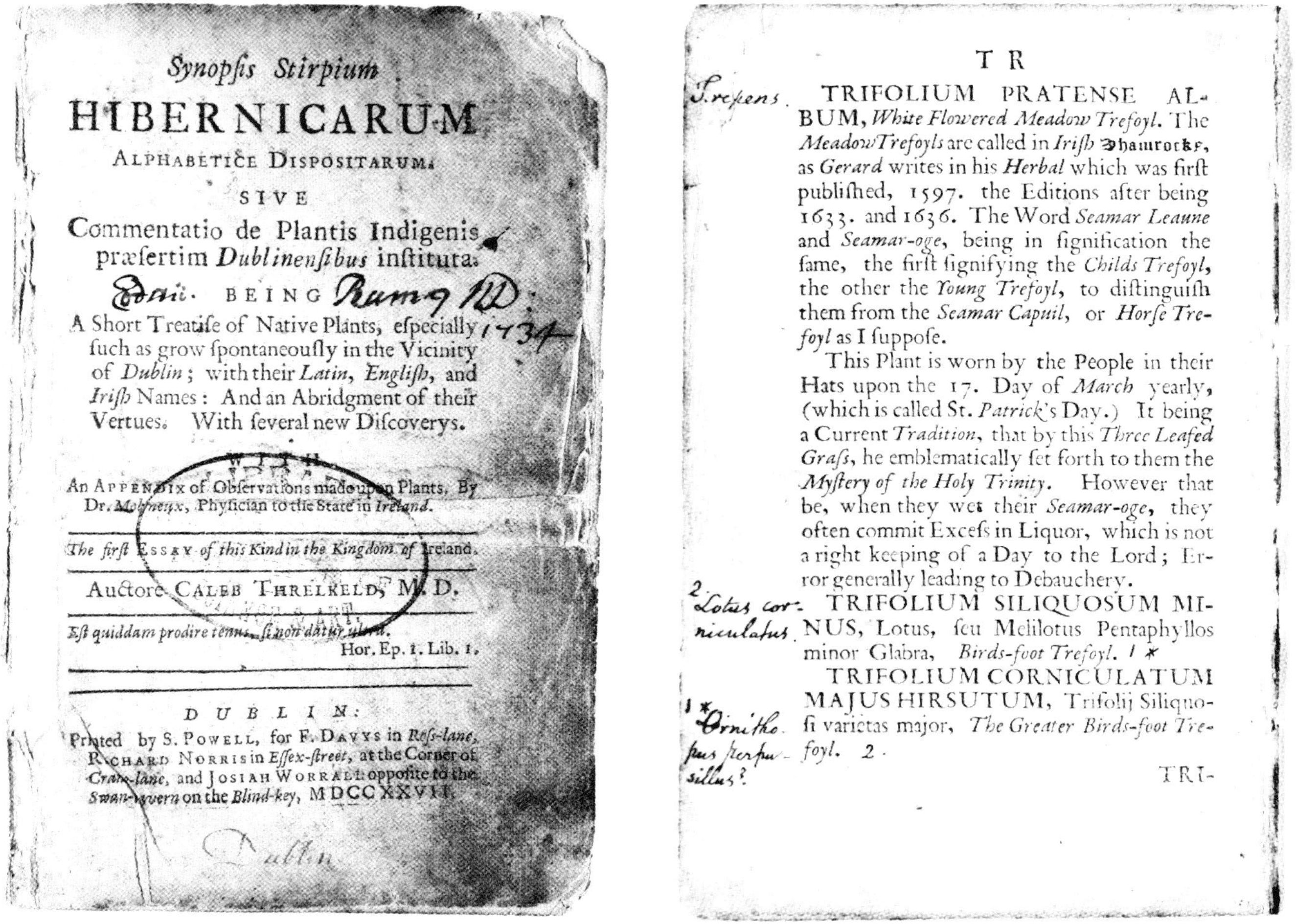

Fig. 14: In this work appeared for the first time on page 160 (above), an exposition of the traditional link between shamrock and Saint Patrick—Caleb Threlkeld's book was first issued in Dublin on Thursday 27 October 1726. (Reproduced by courtesy of National Botanic Gardens, Dublin)

> However that be, when they wet their *Seamar-oge*, they often commit Excess in Liquor, which is not a right keeping of a Day to the Lord; Error generally leading to Debauchery.[5]

So Caleb Threlkeld was also the first clearly to record that other tradition, the wetting or drowning of the shamrock, although in 1630 John Taylor, the Water Poet, had implied it with his lines

> Whilst all the Hibernian Kernes in multitudes
> Did feast with Shamroges steev'd in Usquebagh.[6]

As I have already mentioned, Nathaniel Colgan was cautious, perhaps overly so, in treating Taylor's doggerel as 'an early hint at the practice of drowning the shamrock'. In fact there are even earlier records of the celebrations to honour the patron saint of Ireland which indicate that the 'drowning of the shamrock' (perhaps without the shamrock!) was the common practice. The Lord Deputy of Ireland, Sir Henry Sidney observed the celebrations on 17 March 1576/7, and later wrote that the Irish

> ...surfeited upon their patron day, which day is celebrated for the most part of the people of this country birth with gluttony and Idolatry as far as they dare.[7]

W. H. Grattan Flood noted that the papers of the Confederate period (1642-3) contain an account of inordinate feasting and drinking on the patronal festival.[7]

Returning to Dr Threlkeld, he felt compelled to write that the tradition about St Patrick and the shamrock was a current tradition. Why? Was it a recent invention, an explanation of the quite novel fashion for wearing shamrock on St Patrick's Day? That is another impossible question to answer as we must rely principally on the written evidence which gives us no help. Perhaps there are other, more transparent clues in different forms.

ACT IX

Enter the Story-Tellers

Caleb Threlkeld reported only that the traditional story about Saint Patrick preaching the doctrine of the Holy Trinity, illustrated by the shamrock, was current in Dublin by 1726. No doubt the legend was well-known to both botanists and the common people so he felt it was not necessary to repeat it. I have not been able to determine beyond doubt when a 'complete' version of the fable was first printed, but it seems probable that more than half a century elapsed between Dr Threlkeld's statement and the publication of any expanded version of the Patrick tradition.

A report of the Saint Patrick's Day celebrations conducted by the Friendly Sons of St Patrick which appeared in the *New York Gazette* on the following day, 18 March 1789, contained one of the earliest, expansive accounts of the legend. I have been unable to trace this; my source for it is W. H. Grattan Flood's article, 'Vindication of the shamrock legend' published in *The Month* under the banner 'Critical and historical notes'—anything less critical would be hard to find for Flood tried to vindicate an indefensible legend and his sources are often suspect!

In 1794, Edward Jones gave one version of the legend in a book about Welsh bards.

> When Saint Patrick landed near Wicklow the inhabitants were ready to stone him for attempting an innovation in the religion of their ancestors. He requested to be heard, and explained unto them that God is an omnipotent, sacred spirit, who created Heaven and Earth, and that the Trinity is contained in the Unity; but they were reluctant to give credit to his words. St Patrick, therefore, plucked a trefoil from the ground and expostulated with the Hibernians:- "Is it not as feasible for the Father, Son, and Holy Ghost, as for these three leaves thus to grow upon a single stalk?" Then the Irish were immediately convinced of their error, and were solemnly baptized by St. Patrick.[1]

Mr Jones' version of the event would not be generally recognized in Ireland today, for it is more usual to place the Trinity sermon at Tara on Easter Sunday after the dramatic contest between the saint and the druids, and Patrick's defiant lighting of a Paschal Fire. But myths are myths, boundless by nature, and not everyone in Ireland would agree that Patrick preached about the Trinity at Tara. Local claims assert that it took place in Ulster, Leinster and

Fig. 15: Saint Patrick holding a large shamrock! This fanciful portrait was published in the Revd James O'Leary's *The most ancient lives of St Patrick* (1877)

Munster; as recently as 1905 the Most Revd Dr John Healy, Archbishop of Tuam, recorded that

> ...a very widespread, living tradition tells another well-known story of Patrick's preaching, either on the Rock of Cashel or on Tara Hill. When trying to explain the mystery of the Holy Trinity to his hearers, he saw the trefoil growing on the green sod beneath his feet, and taking it up in his hand, he pointed out how the triple leaf sprang from the single stem, even as the Three Divine Persons, really distinct from each other, were yet One in the unity of the Godhead. It was, of course, an imperfect, but yet, for a simple people, a very apt illustration of the great Mystery he was trying to explain.[2]

Archbishop Healy's *The life and writings of St Patrick* has been described as 'pre-critical' by the great Patrician scholar Ludwig Bieler, but yet Dr Bieler granted that this pious book had merit as an inexhaustable source for local traditions. Certainly, Dr Healy should be given credit for honesty because he appended this sentence to the paragraph about the Trinity sermon:

> We can find no trace of this story in the ancient Lives of the Saint; still it has caught the popular imagination, and made St Patrick's Shamrock the immortal symbol of Ireland's faith and nationality.[2]

Given that there are so many versions of the traditional account, what historical conclusions can be drawn from the shamrock stories?

Even writers who claim that Patrick *did* preach a sermon about the Holy Trinity, using the shamrock as a parable for the Three Persons in One God, have to admit that none of the mediaeval biographers and chroniclers mentioned the dramatic yet simple event. Muirchú reported Patrick's visit to Tara, his contest with the druids and the miracle of the burning house, and finally the conversion of King Laoghaire, but not the sermon about the Holy Trinity. Prudent scholars, from John Bury to Ludwig Bieler, remain sceptical about the truthfulness of most of Muirchú's biography while admitting that ancient legends and traditional stories can sometimes conceal grains of historical fact. Nobody has devised a method of separating those hidden facts from the smothering fiction. The task is hopeless.

From the tapestry of myth all we can distil is this suggestion. If Patrick's mission was to succeed he needed safe-passage through Ireland and the conversion of the chieftains and kings would have ensured not only his personal safety but also the subsequent conversion of their peoples. So Patrick surely would have met the local rulers, even the High King, perhaps not on Easter Day nor at Tara. The story about the encounter with King Laoghaire probably records an actual event, but the story about the Trinity sermon is not among the fables that embroidered the account of that summit meeting in Muirchú's time. The shamrock had not been invented in 700 AD.

The earliest Irish manuscripts containing mentions of shamrock are essentially poetic and do not link **seamróg** with Saint Patrick, and the first English texts indicate only that the plant was eaten. The contrast between the separate writings could not be more stark. Irish-speaking authors used the word to describe landscapes: 'clover-flowered plains', 'daisy-covered, clover-blossomed greens'. On the other hand, Edmond Campion and Richard Stanihurst introduced shamrock to English readers as a food for starving savages; later scribblers, William Moffet, for example, softened this until it became a kind of exotic salad:

> Besides all this, vast bundles came
> Of sorrel, more than I can name.
> And many sheaves I hear there was
> Of shamrocks, and of water-grass,
> Which there for curious sallads pass.[3]

Can we fix a date for the transmogrification of the wild Irishman's salad into a symbol of the Trinity, for the evolution of the legend of Saint Patrick and the shamrock?

ACT X
Enter the Decorated Courtiers

The veneration of Patrick as a saint which had its beginning in the time of Muirchú, blossomed into the Middle Ages. Penitent pilgrims from all over Christendom made the arduous and dangerous journey across Ireland to the remote, bog-encircled Lough Derg and the shrine of Saint Patrick's Purgatory. Thus Patrick was well-known throughout Europe in mediaeval times. However, there is no evidence that the festival held on the traditional day of his death, 17 March, was anything other than a strictly religious one before the commencement of the second Christian millenium for, because of the way the date of Easter is calculated, St Patrick's Day invariably falls within the forty days of Lent, a period for abstinence and fasting.

By the late twelfth century the Lenten restrictions had been relaxed for this patronal festival. The English Cistercian historian, Jocelin of Furness, gave some account of the celebrations in a biography of Saint Patrick which he wrote shortly after 1186, the year when relics reputed to be of Patrick, Brigid and Columkille were discovered by Malachy, Bishop of Down. Jocelin reported that on St Patrick's Day many Irish people ate meat which was first plunged into water, then dressed to eat, and called 'Fishes of Saint Patrick.' From this practice, whether sanctioned by the Church or not, perhaps developed the exuberant feasting and unstinted imbibing of later centuries.

Secular celebrations in honour of the patron saint of Ireland are known to taken place as long ago as the reign of Queen Elizabeth I and to have been merry occasions. The day was kept as a holiday in Dublin, and perhaps elsewhere.

During the latter decades of the seventeenth century the first reports appear of the wearing of St Patrick's crosses, a custom that continued into the present century but which has now died out. Thomas Dineley's observations about the crosses have already been quoted and his account continues

> The common people and servants also demand their Patrick's groat of their masters, which they goe expressly to town, though half a dozen miles off, to spend, where sometimes it amounts to a piece of 8 or cobb a piece, and very few of the zealous are found sober at night.[1]

Some crosses were made of green ribbon, noted Dinely, but in the next year soldiers of an Irish regiment stationed in England were seen wearing crosses of red ribbon on St Patrick's Day: to this day the red saltire, incorporated within the Union Flag, is still called St Patrick's Cross.

These red or green crosses of ribbon worn in honour of Ireland and her patron saint were not confined to Irish breasts. By this time too the English Court observed the custom of the day:

> Their Majesties, with the Prince of Wales and the Princesses, being the Feast of St. Patrick, the tutelary saint of Ireland, wore Crosses in honour of that Nation.[2]

That was the spring of 1726: six months later Dr Caleb Threlkeld's botanical book was published. At the Court of St James between 1725 and 1760, St Patrick's Day was a Collar Day, one when knights attending ceremonies wore the collars of their order; thus in 1752 this report appeared in *Pue's Occurrences*:

> St Patrick's Day was observed as a Collar Day at Court. The natives of Ireland adorn their hats with Shamrogue, which is composed of a sort of grass called trefoil, which allusion is taken from St. Patrick's first propagating Christ there, and establishing the doctrine of the Trinity.[2]

None of these accounts indicates that anyone other than the Irish-born deigned to wear shamrock—the fashionable adornment was, it seems, the more elegant cross. As for the colour, green was not universally accepted as the distinctive colour of Ireland and the Irish until the end of the eighteenth century, although it has an ancient and immemorial association with this island.[3]

It is also worth noting that the Welsh custom of wearing leeks on St David's Day, 1 March, has a more ancient history that that of wearing shamrocks: there is a reference in William Shakespeare's *Henry V* to wearing a leek, and a book published in Douay (France) during 1632 contains the lament that in

> ...these our unhappy days, the greatest part of his [St David's] solemnitie consisteth in wearing of a green Leeke, and is a sufficient theame for a zealous Welshman to ground a quarrell against him that doth not honour his capp with the leke ornement that day.[4]

And just as the English Court kept St Patrick's Day, in 1755 the *Gentleman's Magazine* reported that the King and other members of the Royal Family wore leeks in honour of St David on 1 March. Did the shamrock tradition arise, I wonder, as an Hibernian response to the leek when Irishmen were admitted to the Court of St James?

Remarkable and amusing evidence for the wearing of leeks and shamrock in one's hat comes from a painting executed by Sir Joshua Reynolds in 1751. A cartoon for the painting is also extant and it takes the form of an unflattering caricature showing four gentlemen-musicians—the figures are lampooned,

Fig. 16: Lord Charlemont (centre) bedecked with shamrock—a detail from
Joshua Reynold's cartoon (1751) (Reproduced by courtesy of
National Gallery of Ireland, Dublin)

not least by the head-gear. On the left stands Sir Thomas Kennedy of Culzean, a Scot although not attired in a kilt, but clearly identified by the thistles in his hat. On the right sits Mr Phelps, undoubtedly Welsh, and sporting an outrageous leek. At the rear stands the Englishman from whose hat flops a wilted rose—he is the Hon. John Ward of Staffordshire. And seated in the middle, playing a recorder, is Lord Charlemont representing Ireland with a veritable meadow of shamrock. The title of Sir Joshua Reynold's complete picture is 'The Parody on the School of Athens'.[5]

The safe conclusion is that during the last half of the seventeenth century the custom of wearing shamrock on St Patrick's Day was being propagated in Ireland. By 1726 it was a common practice. Crosses made of red or green ribbon were also worn, even at Court, during this same period. Slowly the latter habit declined until only children wore St Patrick's crosses, but the wearing of shamrock increased steadily in popularity.

Incidentally the Most Illustrious Order of Saint Patrick, the Irish order of knighthood, was not instituted until 1783 so none of the collars evident at the Court before that year would have had any connection with Ireland's patron.[6] King George III established this new order supposedly 'for the

Fig. 17: Prelate's badge, made in 1819 for Archbishop Stewart,
of the Most Illustrious Order of St Patrick.
(Reproduced by courtesy of Ulster Museum, Belfast)

dignity and honour of Our Realm of Ireland' but it was a blatant scheme for enhancing vice-regal patronage, for bribing members of the Irish Parliament following the granting of 'independence' to it in 1782. The Order of St Patrick was conceived and founded in about six months; the first knights being instituted on 11 March 1783 at a ceremony in Dublin Castle, and then on 17 March, St Patrick's Day, they were installed during a lavish ceremony in St Patrick's Cathedral, Dublin.

The Order's insignia is of significance to our story, for while the robe and ribbon of the Knights of Saint Patrick were sky-blue, not green, its badge included familiar elements. The cross of St Patrick, a red saltire on a white ground, was charged with a green trefoil bearing a crown on each leaflet and encircled by the motto *Quis separabit* and the date MDCCLXXXIII. The trefoil superimposed on the saltire had oval and pointed leaflets, not heart-shaped ones, but undoubtedly it is a shamrock. Moreover the shamrock was not confined to the centre of the cross; several surviving pieces of regalia—for example a special badge for the Prelate of the Order—are quite remarkable having an outer border of linked trefoils in green enamel—the heart-like leaflets mean that the perimeter foliage is indubitably composed of shamrocks! The Prelate's badge, made in 1819 for Archbishop Stewart, shows that shamrocks, in recognizable modern form, were certainly in vogue by that period. The collar of the Most Illustrious Order of St Patrick was composed of a series of knots (in gold), alternating with a red and white rose, and a harp, symbolizing the amicable links between Ireland and England. One collar, hallmarked Dublin 1821, is awash with shamrocks for the roses are also rimmed with them!

The inclusion of a shamrock in the badge of this hastily ordained, and now defunct, Order apparently indicates royal approval for this uniquely Irish device. Yet the Knights of St Patrick were not the first fraternity to employ the shamrock as an every-day badge.

Fig. 18: Guidon (two sides) of Royal Glin Hussars
(Reproduced by courtesy of the Knight of Glin.)

ACT XI
Enter the Patriots

Through the first half of the eighteenth century, the shamrock was manifest only on one day each year, St Patrick's Day. It was not yet a badge evident throughout the year, a badge to be worn with pride. As late as the 1750s the shamrock was still often a matter of fun, mentioned in doggerel verse, and useful in satirical portraits. All this was to change during the turbulent decades that brought the century to its conclusion, when two groups at opposite political poles within Irish society adopted the shamrock as elements in their badges. The first to employ it were the Volunteers; the United Irishmen followed suit.

The Volunteers were organized bands of vigilantes formed to protect the property of the ruling class, the protestant Anglo-Irish, after the English government was forced to withdraw regular troops from Ireland in the late 1770s due to the dire need for soldiers to protect Britain and to fight in the American War of Independence. The Irish aristocrats and land-owners were alarmed and established their own private militias in the years following 1779.[1]

The various 'self-created troops and companies' of Volunteers had uniforms with military accoutrements, and carried colours some of which survive to testify to the movement's adoption of shamrock as an emblem. The guidon of the Royal Glin Hussars, formed in July 1779 by the Knight of Glin, has on one side a crowned harp encircled with twining strands of shamrock, and on the other side a mounted hussar similarly wreathed. Shamrocks also emblazon the guidon of the Tullamore True Blue Rangers which in its design closely resembles that of the Glin Hussars. The flags of the Castle Ray Fencibles, the Limerick Volunteers and the Braid Volunteers, among others, display the shamrock motif. The cross belt plate of one of the companies formed in County Fermanagh, the Maguires-Bridge Volunteers, has a wreath of shamrocks, and that of the Goldsmiths' Corp established in Dublin and dated 17 March 1780, features a sword and shamrock.

On St Patrick's Day 1780 the Loyal Volunteers of Cork paraded on the Mall wearing shamrock cockades:

> A noble train, most gloriously array'd
> To hail Saint Patrick, and a new free trade…[2]

A dinner followed at which a liberal quantity of alcohol was consumed and the after-dinner songs included this one:

Fig. 19: Crude shamrocks and a harp surmounted by a crown engraved on a cross-belt plate of the Maguires-Bridge Volunteers, a County Fermanagh company (Reproduced by courtesy of Ulster Museum, Belfast)

> Saint Patrick he is Ireland's Saint,
> And we're his Volunteers, sir;
> With hearts that treason cannot taint
> Their fire with joy he hears, sir.
> None need be told
> Our Saint so bold
> Will think that dog a damn'd rogue,
> Who on his day
> Would keep away,
> And does not mount his shamrock.
> O rally, O rally, O rally round, then:
> Who on this day
> Has kept away,
> Be sure they are not sound men.[2]

There are two more verses!

Shamrocks were not confined to regalia, drinking songs and Volunteer banners; a glass made in 1783 at Waterford was engraved with a fine wreath of shamrocks, and a beautiful goblet bearing a marvellous representation of a horseman and the name of the Castleknock Light Dragoons also has decorative shamrocks.[3]

The Volunteers soon became a political force representing the aspirations of the Protestant party and the English government had to bow to their

Fig. 20: 'Success to the Castleknock Light Dragoons': a goblet engraved with shamrocks and a splendid dragoon, made c. 1779-1793 in Dublin. (Photograph by courtesy of Mary Boydell; reproduced by courtesy of National Museum of Ireland, Dublin)

pressure. In 1780, trade regulations, which hitherto had crippled Irish exports, were relaxed. Later, coupled with the work and persuasive oratory of Henry Grattan, the Volunteers movement ultimately brought about the repeal of the act which had prevented the Irish Parliament from making its own laws. Even so, these private bands of armed men continued to represent a formidable force and the authorities had to suppress the Volunteers gradually; by the 1790s most of the companies were disbanded.

Although the Irish Parliament was now independent and capable of legislating for Ireland, the country was still ruled by the Protestant Ascendancy, the small minority of wealthy land-owners. While the Penal Laws had fallen into disuse—for example, Roman Catholics could be admitted into the army—the vast majority of the people were impoverished and deprived of opportunities of economic advancement. That majority now sought political power and a new movement succeeded the Volunteers.

The Society of United Irishmen, inspired by the French Revolution of 1789, sought to overthrow English power and to establish Ireland as an independent, egalitarian republic. It was formed in Belfast during 1791, but its headquarters were later transferred to Dublin and committees constituted throughout Ireland. The movement's leaders were intelligent, idealistic young men; paradoxically many were Presbyterians or members of other dissenting protestant churches. It was the Society of United Irishmen which established green as the revolutionary colour—they wore green uniforms and during the uprising of 1798 carried 'the National Flag of Sacred Green'. One contemporary writer explained that green had been '…adopted by the Irish malcontents as the revolutionary colour in imitation of the shamrock.'[4] Green thus became seditious, and the wearing of green emblems was seen as an act of defiance.

While the Volunteers appear not to have had a single distinctive badge, except possibly the crowned harp, the Society of United Irishmen had a seal composed of a harp (uncrowned) with a pike atop which was a Liberty cap, and the mottos *Equality* and *It is new strung and shall be heard* (a reference to the harp). Shamrocks and the motto *Erin go brath*—the correct form today is *Érinn (Éire) go brách*, which may be rendered as 'Ireland for ever'—were also closely associated with the society.

A triple-faced seal made from rock-crystal that belonged to one of the Coulters of Dundalk probably dates from this period; two of its faces have shamrocks, once associated with the motto *Eire go brát* and a harp, and on

Fig. 21: The Coulter family of Dundalk possessed this seal—two of the three faces are shown, both with shamrocks. (Photograph by David Davison; reproduced by courtesy of Brian Coulter)

the second face an arm wielding a cudgel rises from a shamrock wreath and there is an Irish inscription *Ttuigean Tu Gaoidilge* (in Irish today that would read *an dtuigeann tú Gaeilge*, meaning 'do you understand Irish?').[5]

At Ballinahinch in County Down during the 1798 rebellion, the insurgents wore accoutrements with the harp entwined with shamrock or bay. A green flag, bearing in the centre a gold harp with silver strings surmounted by a scarlet Cap of Liberty and with two sprigs of shamrocks beneath, was suggested to General Hoche in June 1796 as he prepared the French expeditionary force that was destined to land at Bantry Bay. Wolfe Tone, one of the leaders of the United Irishmen, proposed that the Irish soldiers accompanying this French invasion fleet should wear '…green jackets, with green feathers, and [display] a green standard with the harp surmounted by the cap of Liberty.'[6]

This revolutionary society also spawned its fair share of verses, good and bad. In *The Press* on 21 October 1797 appeared 'a fable' entitled 'The London Pride and Shamrock' by Trebor—as that name is Robert written backwards it is sometimes assumed that Robert Emmet wrote it.

> Full many a year, close side by side,
> A Shamrock green, and London Pride,
> Together how they came to grow,
> I do not care, nor do I know;
> But this I know, that overhead
> A laurel cast a wholesome shade.
> The Shamrock was a lovely green
> In early days as e'er was seen...[7]

And so it goes on to tell how the shamrock was overshadowed and choked by the London pride which was eventually scattered by a flash of lightening and black clouds.

> But soon succeeds a heavenly calm,
> Soft dews descend and showers of balm;
> The sun shoots forth his kindest ray,
> And Shamrock strengthens every day,
> And rais'd by heaven's assistance bland
> Bids fair to spread o'er all the land;
> The guards, the blasted laurel's roots,
> The nurtured laurel upward shoots,
> And grateful wreathes its dark green boughs
> To grace great Shamrock's aged brows.[7]

But the shamrock, again, was not confined to poetry and flags. On Monday 4 February 1799 a weekly journal was launched with the title *The Shamroc*. Its banner was a star—probably mocking the insignia of the Order of St Patrick—inside which was a shamrock and underneath the motto *In hoc signo vinces*! *The Shamroc* eagerly displayed its revolutionary, anti-Union sympathies.

The United Irishmen were crushed, their leaders imprisoned or executed and many of the ordinary members and sympathizers deported. The 'wearing of the green', now a mark of Irish nationalists, was exported world-wide. The scattering of exiles and the revolutionary significance of green are incorporated in the well-known ballad 'The Wearing of the Green' which Frank O'Connor[8] has called the Irish Marseillaise; it was probably written by a Presbyterian Ulsterman as the end of the 1790s after the United Irishmen were routed, and there are now innumerable versions of it. An early one begins:

> I met with Napper Tandy, and he took me by the hand,
> Saying, how is old Ireland? and how does she stand?
> She's the most distressful country that ever yet was seen,
> They are hanging men and women for the wearing of the green.
>   O wearing of the green, O wearing of the green
>   My native land, I cannot stand, for wearing of the green

THE SHAMROC,

IN HOC SIGNO VINCES

Fig. 22: The masthead of the first issue of *The Shamroc*, 4 February 1799.
(Reproduced by courtesy of National Library of Ireland, Dublin)

> My father loved you tenderly, he lies within your breast,
> While I, that would have died for you, must never be so blest;
> For laws, their cruel laws, have said that seas should roll between
> Old Ireland and her fathful sons who love to wear the green.
>   O wearing of the green, O wearing of the green
>   My native land, I cannot stand, for wearing of the green[9]

The now-familiar ballad did not take form until 1864 when it was rewritten by Dion Boucicault for his play *Arrah-na-Pogue*. Boucicault's version begins:

> O Paddy dear and did ye hear the news that's going round?
> The shamrock is by law forbid to grow on Irish ground;
> Saint Patrick's Day no more we'll keep, his colours can't be seen,
> For there's a bloody law against the wearing of the green.

[and it continues]

> So if the colour we must wear be England's cruel red
> Let it remind us of the blood that Irish men have shed;
> And pull the shamrock from your hat, and throw it on the sod,
> But never fear, 'twill take root there, though underfoot 'tis trod.

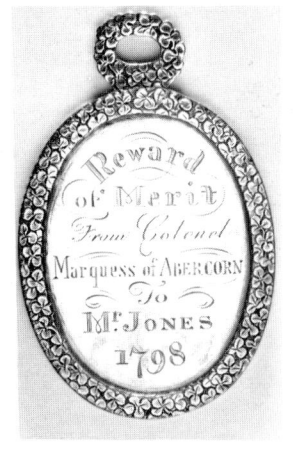

Fig. 23: Elegant shamrock encrusting a Tyrone Volunteer's medal, 1798. (Reproduced by courtesy of Ulster Museum, Belfast)

Fig. 24: Cross-belt plate of the Antrim Militia. (Reproduced by courtesy of Ulster Museum, Belfast)

> When laws can stop the blades of grass from growin' as they grow,
> And when the leaves in summer-time their colour dare not show,
> Then I will change the colour too I wear in my caubeen;
> But till that day, please God, I'll stick to Wearing of the Green.[10]

Yet there was a distinction between the flaunting of revolutionary, seditious green and the wearing of a shamrock of St Patrick—this a dichotomy apparent throughout the late eighteenth and the early nineteenth century and is perhaps best illustrated by the insignia of the Militia which, under the control of the government, assisted in quelling the 1798 insurrection. A medal, given as a 'Reward of Merit' for service in 1798, inscribed with the name of the Tyrone Volunteers, a Militia company, has a surround of embossed shamrocks. Shamrock also appears on a medal of the Armagh Volunteers. The cross-belt plate of the Antrim Militia has a trefoil on it, and so on.

Why did opposing factions in the late 1700s employ the same emblem, the shamrock? In their separate ways both were composed of Irish patriots, although their views of what constituted the Irish nation were dramatically different. Both wished to display their allegiance to Ireland, and the best way to do this was to parade a symbol which linked them with Ireland's past. At that time no-one would have questioned the story that Saint Patrick had preached a sermon while holding in his hand a shamrock. The shamrock was distinctly Irish, a perfect, sacred badge.

Fig. 25: Royal arms of the United Kingdom of Great Britain and Northern Ireland

(left) This form in current (1989) use by such establishments as The Royal Botanic Gardens, Kew, is a strange concoction of four exaggerated daffodils (Wales) smothering three roses (England) while the two thistles (Scotland) are relegated almost to invisibility either side of 'ET MON', with two pathetic shamrocks (Ireland) between; it had its origins about 1965.

(below) Edward Bowden (1903-1989) designed this version, with its wildly puzzled lion and startled unicorn for *The Observer*'s masthead in 1962. It was abandoned in 1989; the new masthead has no shamrocks!

Thus the shamrock's role as an emblem for Ireland was now solidly established and this is confirmed by its incorporation into that impossible botanical hybrid of rose, thistle and trefoil, sometimes with added daffodil or leek. Most examples of this triple conceit date from the beginning of the nineteenth century, just after the passing of the Act of Union and the abolition of the Irish Parliament in 1800. Wreathes of roses, thistles and shamrocks emblazon the flags of the Irish Yeomanry; there are several examples known, including a guidon of the Belfast regiment showing the crowned harp, and a shield encircled by a rose, thistle and shamrock garland. The triple motif is found engraved, sometimes crudely, on glassware, goblets and decanters, produced in Ireland about 1800. A presentation regimental sword dated 1803 and inscribed to Andrew Savage Esq. of Portaferry, Royal North Downshire Militia, has shamrock, rose and thistle cast into the design of the guard.

This fabulous arrangement of the three plants was symbolic of the new United Kingdom of Great Britain and Ireland, the rose representing England (and Wales!), the thistle for Scotland, and Ireland's shamrock. Yet the incorporation of the shamrock into the Royal arms does not appear to have happened following any specific event such as the Act of Union in 1800. As portrayed in Charles Hasler's excellent book [11] about the graphic and decorative development of the Royal arms, the shamrock insinuated its way into the compartment of the armorial bearings—the portion below the shield in the full achievement on which the supporters stand—only in the last few years of the eighteenth century. There is no evidence of shamrocks during the 1780s and early 1790s in any Royal coats-of-arms, but in 1799 the masthead of *The Times* newspaper was redesigned and a shamrock leaf was added to accompany the rose and thistle. Thereafter the shamrock was almost invariably included in that particular masthead and it soon appeared in other compartments. There the shamrock survives to this day with the rose, the thistle and the daffodil.

> I care not for the thistle, and I care not for the rose;
> When bleak winds round us whistle, neither down nor crimson shows.
> But like hope to him that's friendless, when no joy around is seen,
> O'er our graves with love that's endless blooms our own immortal green.
>   O wearing of the green, O wearing of the green
>   My native land, I cannot stand, for wearing of the green.

ACT XII

Interlude for a Monopolylogue

The next act in the shamrock's rise was its acquisition of the still-bright romantic sheen. The ordinary Irish man and Irish woman wore it on St Patrick's Day, and as far as anyone could recall that had been the practice for generations, since time immemorial. The historiographers had affirmed that Patrick had plucked the shamrock on the Hill of Tara—or the Rock of Cashel, or near Wicklow Town—and had then preached about the Holy Trinity. Doubtless the clergy, who were his successors, preached sermon after sermon about this deed and its significance. The courtiers and patriots had taken the shamrock and used it on their banners and seals and praised it in their songs; they made it into a national badge, and linked it with green, the revolutionary colour.

Fig. 26: Mrs Rosoman Mountain: a miniature signed S.G. (Reproduced by courtesy of National Portrait Gallery, London)

1801, the first year of a new century, was a watershed in Irish political life. The Act of Union came into force depriving Ireland of an independent parliament and the red saltire of Saint Patrick was incorporated into the Union Flag. No-one had won in that contest except the British administration—the Ascendancy had been stripped of much of the power they had held, and the Republicans had been crushed.

The shamrock was not forgotten, and two of the best remembered ballads about it were written in the next few years.

Mr Andrew Cherry collaborated with Mr William Shield to create for Mrs Rosoman Mountain an entertainment entitled *Travellers at the Spa* which was performed by that redoubtable lady at the Opera House in Capel Street, Dublin, during 1806. This monopolylogue—a one-man show!—was '…entirely recited and sung by me,' wrote Mrs Mountain to Thomas Crofton Croker, 'and attracted crowded houses in defiance of the denouncement of Mr Jones, the manager of the Crow Street theatre who threatened and did in part proceed against me.' She was extremely proud of her show, because '…talent (however humble) triumphed over oppression.'[1] What made *Travellers at the Spa* so notable was the inclusion of this ambrosial ballad.

> There's a dear little plant that grows in our isle,
> 'Twas St Patrick himself, sure that set it;
> And the sun on his labour with pleasure did smile,
> And the dew from his eye often wet it.
> It thrives through the bog, through the brake, through the mireland;
> And he called it the dear little shamrock of Ireland,
> The sweet little shamrock, the dear little shamrock,
> The sweet little, green little, shamrock of Ireland.
>
> This dear little plant still grows in our land,
> Fresh and fair as the daughters of Erin,
> Whose smiles can bewitch, whose eyes can command,
> In each climate that they may appear in;
> And shine through the bog, through the brake, through the mireland;
> Just like their own dear little shamrock of Ireland.
> The sweet little shamrock, the dear little shamrock,
> The sweet little, green little, shamrock of Ireland.
>
> This dear little plant that springs from our soil,
> When its three little leaves are extended,
> Denotes from one stalk we together should toil,
> And ourselves by ourselves be befriended:
> And still through the bog, through the brake, through the mireland;
> From one root should branch, like the shamrock of Ireland.
> The sweet little shamrock, the dear little shamrock,
> The sweet little, green little, shamrock of Ireland.[2]

Fig. 27: Thomas Moore (1779-1852). To mark the centenary of his death the Irish Post Office issued two stamps on 10 Nevember 1952, the design based on a portrait by Sir M. Archer Shee.

The other classic and unforgettable lyric about the shamrock was not written for a one-man show, but was published along with other ballads extolling romantic Ireland in Thomas Moore's collection of *Irish Melodies*; this particular piece was included in the fifth fascicle of the *Melodies* and the original manuscript of it, dated 2 October 1812, was sold in 1914 by Maggs Bros.[3] for seven pounds and ten shillings. The original version included just two verses, the third one being added sometime before 1845.

Merely for the sake of completeness Moore's sentimental trivia is reprinted here in its final form and in that most sumptuous version decorated by Daniel Maclise for the Longman edition of Moore Songs, 1845.

Oh the Shamrock.

Through Erin's Isle,
To sport awhile,
As Love and Valour wander'd,
With Wit, the sprite,
Whose quiver bright
A thousand arrows squander'd.
Where'er they pass,
A triple grass "
Shoots up, with dew-drops streaming,
As softly green
As emeralds seen
Thro' purest crystal gleaming.
Oh the Shamrock, the green, immortal Shamrock!
Chosen leaf
Of Bard and Chief,
Old Erin's native Shamrock!

Says Valour, "See,
"They spring for me,
"Those leafy gems of morning!"—
Says Love, "No, no,
"For *me* they grow,
"My fragrant path adorning."

But Wit perceives
The triple leaves,
And cries, "Oh! do not sever
"A type, that blends
"Three godlike friends,
"Love, Valour, Wit, for ever!"

Oh the Shamrock, the green, immortal Shamrock!
Chosen leaf
Of Bard and Chief,
Old Erin's native Shamrock!

So firmly fond
May last the bond,
They wove that morn together,
And ne'er may fall
One drop of gall
On Wit's celestial feather.
May Love, as twine
His flowers divine,
Of thorny falsehood weed 'em;
May Valour ne'er
His standard rear
Against the cause of Freedom!
Oh the Shamrock, the green, immortal Shamrock!
Chosen leaf
Of Bard and Chief,
Old Erin's native Shamrock!

D. Maclise R.A.   T. P. Becker

ACT XIII

Enter King George IV

The elderly King George IV, corpulent, infirm and speechlessly drunk, arrived at Howth on 12 August 1821 '…at half-past four o'clock in the afternoon, amidst the acclamations of the assembled people'.[1]

> Hail happy day that brings our royal host
> To land with rapture on old Erin's coast,
> That to the Shamrock of our native isle,
> The Rose of Britain doth once more entwine.[2]

Five days later, on 17 August His Majesty made 'a splendid public entry into Dublin…Great rejoicing followed and the city was illuminated for two nights'. Throughout his regal progress he was suffering from a 'distressing looseness' so a sanitary engine was prepared for his comfort.[3]

The visit was marked not just by an illuminated city, but also with a veritable torrent of verses, most of it so awful as to be utterly forgettable and almost all of it is now, thankfully, totally forgotten. Not unremembered, at least in some circles, was the king's ceremonial entry into Dublin; the portly monarch dressed in scarlet stood up in his carriage and '…repeatedly pointed to the shamrock which decorated the front of his hat, doubtless denoting it as the symbol of the sentiment which then beat in his bosom.'[4] That was a signal act which gave the badge of the patriots a publicly renewed royal seal of approval.

During the latter, mad years of his father, King George III, Prince George acted as regent and, despite being perpetually in debt, he was an instigator of fashion and an arbiter of taste—the Regency was marked by a new frothy style of architecture, supremely exemplified by the Prince Regent's Pavilion at Brighton, which precipitated the eclipse of the stately 'Classical' style that was so evident in Dublin's finest buildings and street-scapes. After ascending the throne, the dissolute and unpopular King George IV continued to indulge his whims and fancies all the more profligately; his coronation was lavish, the robes costing £24,000, and the crown £54,000, so the penurious new king had to rent the diamonds and other gemstones. A diadem, encrusted with hired diamonds, was fashioned of four crosses alternating with four floral motifs comprising a thistle, a rose and two shamrock—three kingdoms in one![4] This elegant coronet survived the decadent sovereign

Fig. 28: King George IV's entry into Dublin through a specially erected triumphal arch at the northern end of Sackville Street (now O'Connell Street)—detail from William Turner's painting of the scene (Reproduced by courtesy of National Gallery of Ireland)

for which it was made, and was worn by his neice, Queen Victoria, who sensibly had it reset with gems which were owned, not rented, and she left the diadem to the crown. Queen Elizabeth II wore this same diamond diadem to Westminster Abbey for her coronation in June 1953, and she is shown wearing it in such official portraits as that on postage stamps of the United Kingdom. Thus before his stupendous entry into Dublin, behatted and beshamrocked, King George IV had diamondized himself with very 'dear little shamrocks'.

It is hard to know if George IV's shamrock was a carefully planned piece of public relations—it most probably was. Be that as it may, the sprig adorning the king's hat served to enhance further the image of the sacred trefoil. The result was an explosion of shamrocks. They were engraved on glass goblets commemorating the Royal visit. Green, enamelled shamrock superflously encircled the pink and white roses in the beautiful collar of the Most Illustrious Order of St Patrick which was worn by the king at the installation ceremony in St Patrick's Cathedral. Plaster shamrocks were twined into wreaths decorating the second circle in the newly-completed Theatre Royal which the king visited on 22 August.[5]

Among the verses written for the royal visit were some by Dr Whitley Stokes, Fellow of Trinity College, Dublin, and Professor of Natural History. Stokes had been a supporter of the United Irishmen; indeed he was expelled from the College for several years because of this and no doubt the 'bitter portion of my youth' refers to that period.

> Sweet plant, beloved with steadfast truth
> And water'd by my tears;
> The bitter portion of my youth,
> The solace of my years!
>
> Sweet useful plant, too long oppress'd,
> Beneath the feet of pride
> Expand at last thy lovely breath,
> And shake the dust aside.
>
> Blest be thy pliant stem, that bore
> Misfortune's iron hand;
> When pride and passion fly the shore
> The meek possess the land!
>
> Belov'd, revive! the King appears,
> To wipe our tears away;
> The sorrows of a thousand years
> Have vanish'd in a day![6]

Like so many other pieces of poetry this was revised, rewritten and expanded; a later 'edition' concluded thus:

> His aged head thy grateful breast
> Shall soothe to safe repose
> Free from the thorns that still infest
> The Thistle and the Rose.[7]

That last line prompts me to recall that what King George IV did for the shamrock when he made his visit to Ireland in 1821, was to be repeated, but with a different species, when he visited Scotland in 1822. Sir Walter Scott, acting as master-of-ceremonies, arranged for thistles to be carried aloft in the procession that greeted the king on his arrival in Edinburgh. King George also made the kilt fashionable! However there is one notable difference between the Scottish thistle and the shamrock; like the Welsh leek, the thistle has a longer, more easily documented history as an emblem, although the exact species employed is as obscure as that intended by the shamrock, and the legend justifying its use is just as misty. The first evidence of thistles as a royal symbol in Scotland is a coin struck about 1470 during the reign of King James III, at least two centuries before there is any sign of shamrock as an emblem in Ireland.[8]

But to return to Erin, a young gentleman of Clongowes Wood College in County Kildare was moved to write two stanzas about the visit of King George IV to Dublin

> What radiance burst across our Isle,
> One night of woes redeeming,
> And lights on Erin's cheek a smile,
> So late with sorrow streaming.
> 'Tis he, that star, when hope was nigh,
> To wreck in sorrow's ocean,
> Arose with cresset light on high,
> And woke our heart's devotion.
>
> Wreath Erin wreath a garland bright,
> Of song and glory blended,
> Pure as the gems of starry light,
> That decks his brow so splendid,
> There let green Erin's Shamrock shine,
> In rays of triple gleaming,
> And Scotand's Thistle round entwine,
> The Rose between them bearing.[9]

The King left from Dún Laoghaire, renamed Kingstown to mark that event.

Kingstown was the landing place for Queen Victoria on several of her visits to Ireland. In the summer of 1849 triumphal arches, replete with the now obligatory shamrocks proclaimed *Céad míle fáilte* to Dublin. At twenty minutes to 11 o'clock on Monday 6 August 1849, the royal cortege reached one such edifice spanning Upper Baggot Street; this monstrous, iron arch was designed by Thomas Turner, whose father Richard was the famous Ballsbridge iron-master, builder and designer of glasshouses. *The Illustrated London News* reported:

> ...the iron-work was executed at the works of [Richard Turner]. The arch was 127 feet in width, and 92 feet high...constructed of wrought iron, and was of most tasteful style, bearing in one compartment the letters "V.R.", and on the corresponding one "A.C.", the whole exquisitely decorated with roses and floral wreathes, and surmounted by an immense shamrock branch. Over this stood an architrave ornamented with artificial flowers and laurel supporting the Royal arms... The whole was capped by an imperial crown of beautiful workmanship, 10 feet in diameter, with the usual national emblems, the shamrock, rose and thistle in their natural colours...[10]

That night the city was loyally illuminated; the Hibernian Gas Company 'after supplying so much light to others' had the front of its premises in Foster Place decorated with 'the mystic letters V A', a star and a shamrock. 'Variegated' oil lamps displayed shamrocks for several other company buildings, while Nelson's Pillar was lit by electric light, 'the most perfect and powerful display of this beautiful light'.

After all of that, however, Victoria was not as quick to espouse the trappings of Ireland's past as her uncle had been.

Fig. 29: James Ebenezer Bicheno F.R.S.: shamrock scholar and later
Colonial Secretary in Van Dieman's Land
(Portrait in oils by E. U. Eddis; reproduced by courtesy of
Linnean Society, London)

ACT XIV

Enter the Shamrock Hunters

The dawning of the 'Age of Reason' had its effect on the shamrock. Botany had shaken off its ancient link with medicine and had become a respectable academic pursuit in its own right. Within natural history as a whole, the principal task of the early nineteenth century was the cataloguing and naming of the innumerable plants and animals that were being discovered by ardent naturalists in newly explored territory overseas. It was not long before the shamrock too came under scrutiny; botanists and historians started to enquire which was the actual species that Saint Patrick had picked as he preached his sermon, and they attempted to discover if there really was a uniquely Irish plant that could rightfully bear the name shamrock.

Most people were prepared to accept the shamrock's credentials, but the academic fraternity began to worry about it; sentiment and romance were no longer regarded as sufficient justification for the emblem of Ireland, and neither the royal seal of approval nor pious history could make the shamrock irreproachable. Scholars looked for evidence to confirm its ancient Patrician link. Some of the most distinguished academic institutions gave time to discussions about the shamrock, and eminent botanists and historians delved into its past and its identity.

The first of the shamrock sleuths was Miss Louisa Catherine Beaufort, one of the daughters of the erudite Vicar of Collon, the Revd Daniel Augustus Beaufort LL.D. Miss Beaufort was widely-travelled and had written, anonymously, a little book entitled *Dialogues on entomology*. The shamrock does not appear in it but in a prize essay on the ancient architecture of Ireland that was read before the Royal Irish Academy on 22 October 1827, the first contribution to the Academy's Polite Literature and Antiquities section by a woman. She proposed that the antique buildings of Ireland bore the hallmarks of the island's early religious character which was, in the author's opinion, essentially oriental. To support her ideas Miss Beaufort noted that the word shamrock, signifying a trefoil, bore a resemblance to words in ancient Persian and Sanskrit:

> In the honoured Irish shamrock, we have a curious coincidence, the trefoil plant (shamroc or shamrakh in Arabic) having been held sacred in Iran, and considered emblematical of the Persian Triad. This recalls the well known legend of St Patrick's having illustrated the doctrine of the Holy Trinity, by the three-lobed shamrock leaf.[1]

These fanciful theories were endlessly repeated by later writers when the shamrock's historical base was being attacked in other ways.

In May 1831 a paper 'On the plant intended by the shamrock of Ireland' was included among other academic articles in the *Journal of the Royal Institution of Great Britain*. The author was a London botanist, James Ebenezer Bicheno, Fellow of the Royal Society and Secretary of the Linnean Society; in fact he had read it to a meeting of the Linnean Society over a year previously on the eve of St Patrick's Day, 16 March, 1830.

Following a brief visit to Ireland in the autumn of 1829, Bicheno had published a book, titled *Ireland and its economy*, but it contained nothing about the shamrock. In the dissertation he presented to the Linnean Society, Bicheno set out to prove 'by abundant testimony' that the shamrock worn on St Patrick's Day by the London Irish was not the plant that the saint had used. He had observed that Irishmen wore clover in their hats, mainly the common white clover (*Trifolium repens*), and 'very starved specimens' at that; he had also identified black medick (*Medicago lupulina*) 'and even chickweed' in Hibernian bonnets. But, taking as authorities such authors as Edmond Spenser, Fynes Moryson and George Wither, and claiming that clovers were only introduced into Ireland in the middle of the seventeenth century, Bicheno argued that the true shamrock must have been a woodland plant that had a tripartite leaf and blossomed in the middle of March—the only possible species was the wood sorrel (*Oxalis acetosella*). His paper concluded thus:

> ...I apprehend it can hardly be doubted, that the Oxalis acetosella is the original shamrock of Ireland. It possesses, in the first place, all the qualities to recommend it as appropriate for the national feast, and is even more beautifully three-leaved that the clover. It is abundant, and comes at the proper season, being one of the earliest plants, and pushing forth its delicate leaves and blossoms with the first spring... I think it cannot be questioned that St Patrick, who is said by Gibbons to have descended and to have derived his name from the patricians of Rome, exercised a good taste, worthy of his noble birth, when he selected so beautiful an emblem for his favourite island.[2]

Bicheno's paper was delivered to a reputable academic society and it was meant to be a serious discussion of an important botanical issue. But he employed arguments that were spurious and his historical facts were simply untrue. The whole thesis must now be regarded as an entertaining farago of absurd nonsense; unfortunately James Bicheno's article has been quoted again and again by subsequent writers seeking justification for their own works of shamrock-fiction and sadly incapable of discriminating between sense and nonsense.

Fig. 30: 'May no League ever arise to besmirch the tender grace of this flower, or extirpate it from our woods.'
Wood sorrel, *Oxalis acetosella*, seamsóg.
(Illustration by W. G. Smith from *The Gardener's Chronicle*, 29 June 1886, reproduced by courtesy of National Botanic Gardens, Glasnevin)

Numerous brief summaries of James Bicheno's paper appeared in natural history journals after it was read to the Linnean Society, and it was flayed within a few months of its publication by 'a facetious essayist' (a description used by Thomas Crofton Croker) in the *Dublin Penny Journal*.

> Other countries may boast of their trefoil as well as we; but nowhere on the broad earth, on continent or in isle, is there such an abundance of this succulent

material for making fat mutton. In winter as well as in summer, it is found to spread its green carpet over our limestone hills, drawing its verdure from the mists that sweep from the Atlantic. The seed of it is everywhere. Cast lime or limestone gravel on the top of a mountain, or on the centre of a bog, and up starts shamrock. St Patrick, when he drove all living things that had venom (save man) from the top of Croagh Patrick, had his foot planted on a shamrock; and if the readers of your journal will go on a pilgrimage to that most beautiful of Irish hills, they will see the shamrock still flourishing there, and expanding its fragrant honey-suckles to the western wind. I confess I have no patience with the impudent Englishman, who wants to make us believe that our darling plant, associated as it is with our religious and convivial partialities, was not the favourite of St Patrick, and who would subtitute in the place of that badge of our faith and our nationality, a little sour, puny plant of wood-sorrel!

This is actually attempted to be done by that stiff, sturdy Saxon, Mister Bicheno. Though Keogh, Threlkeld, and other Irish botanists assert that the Seamar oge, or Shamrog, is indeed the trifolium repens; and Threlkeld expressly says that 'the trefoil is worn by the people in their hats upon the 17th of March, which is called St Patrick's day, it being the current tradition, that by this three-leaved GRASS he emblematically set forth the Holy Trinity. However that be, when they wet their Seamar Oge, they often commit excess in Liquor, which is not a right keeping of a day of the Lord!' The proof the Englishman adduces, is the testimony of one Spenser, another Saxon, who, in his 'View of Ireland', described the people, in a great famine as creeping forth and flocking to a plot of shamrocks, or watercresses, to feed on them for the time; and he also quotes an English satirist, one Wytthe, who scoffingly says of those

'Who, for their clothing, in mantle goe,
And feed on shamroots as the Irish doe.'

But we are not so easily led, Mr Saxon; we, Irishmen, are not quite disposed to give up our favourite plant at your bidding. In time of famine, the Irish might have attempted to satisfy hunger with trefoil, as well as they did two years ago, when such a thing as seaweed was eaten,—for hunger will break through a stone wall. But do not the Welsh put leeks into their bonnets on St David's day? and now and then they may eat their leek as Shakespeare has it, as a relish either for an affront or for other sort of food; and small blame to an Irishman if, when he feels that queer sensation called hunger, he chews a plant of clover! I, for one, when going into good company, would rather have my breath redolent of the honeysuckle plants, than spiced with the haut goût of garlic! Yet no Welshman would like to live upon leeks, no more than a poor Irishman would upon grass or trefoil; for there is, doubtless, as little nourishment for man in the one as the other. But, to do Mr. Bicheno justice, he had another argument in favour of the wood-sorrel being the favourite plant of our country, which is far more to an Irishman's mind. He says that wood-sorrel, when steeped in punch, makes a better subtitute for lemon that trefoil. This has something very specious in it; if any thing would do, this would; but let the Saxon do his best. Even on his own ground—even in London—he would find it very hard to convince our countrymen, settled in St. Giles's, that

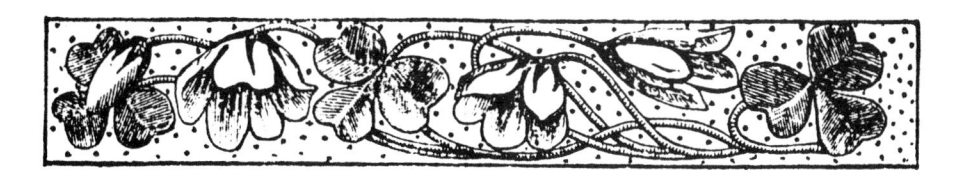

Fig. 31: Head-piece from Madden's *Literary remains of the United Irishman*. printed in Dublin, 1887: the plant is wood sorrel (*Oxalis acetosella*), and by association it is also shamrock. This head-piece exemplifies the nineteenth century confusion about the identity of true shamrock. (Reproduced by courtesy of Aidan Heavey)

> the oxalis acetosella, the sour, puny, crabbed wood-sorrel, is the proper emblem of Ireland. No; 'the shamrock—the green shamrock,' for me![3]

So James Ebenezer Bicheno was chastised. He had raised the spectre of wood sorrel and others, like sheep, gleefully followed his path. To this ramble, the most breath-taking contribution was a recent exchange, between the Department of Agriculture and Food in Dublin and the Director of Plant Quarantine in Australia about exporting shamrock from Ireland, which ended abruptly when this dispatch was received from the First Secretary of the Australian Embassy in Dublin:

> I have been asked to pass on to you the following message from the Australian Department of Primary Industry and Energy:
>
> "Shamrock is listed as *Oxalis Acetosella* not *Trifolium SPP* as you indicated in your telex…"[4]

The Australians evidently follow James Bicheno's argument without question. And why not—did not Mr Bicheno serve as Colonial Secretary of Van Dieman's Land (Tasmania) from 1842 until his death in Hobart on 25 February 1851?

And yet I think it is necessary to counter-pose this question—why in the late 1980s was anyone trying to export to Australia plants or seeds of something readily available there?

ACT XV
Enter the Shamrock Makers

The second half of the nineteenth century was to see the apotheosis of the shamrock, its acceptance in all quarters as the badge of Ireland and the Irish, its adoption as an artistic motif worthy of use by Church and State.

After James Bicheno's paper was published, and summarily dismissed by the Irish, there were several quiet years as far as the literature of shamrock is concerned. The botanists who were based in Ireland had no desire to engage in fruitless discussion about the shamrock; without exception they followed Caleb Threlkeld. John Keogh, John Rutty, Walter Wade, Katherine Baily, James Townsend Mackay, David Moore and Alexander More all applied the Irish name **seamróg** to one of the clovers.

But others knew better, especially in Britain. They were not content with Irish certainty that what was traditionally worn by the people of Ireland on St Patrick's Day was genuine shamrock. From 1851 until 1896 there was an endless procession of articles about the plant identified as the shamrock, queries and answering notes. These appeared in such journals as *The Gardener's Chronicle*, which from 1841 was the principal weekly periodical for British and Irish horticulturists, and later in its companion *The Garden*. The notes were frequently anonymous, and variously supported the wood sorrel as the shamrock or the more traditional clover. Some of the contributions were sensible; here is one item from a well-known gardener, Miss C. M. Owens of Gorey

> It is to be hoped that historical accuracy may not require the watercress to be adopted as the "Irish Shamrock", for it would at once destroy the point of the pretty legend which connects it with the teaching of St. Patrick on the doctrine of the Trinity, as he is said to have used the "three leaves in one" of the Shamrock as an illustration of the doctrine. The Oxalis would of course suit this legend as well as the Trefoil; the latter, however, has the advantage of being sanctioned by custom.[1]

When a new editor took over the reins of the monthly botanical journal *The Phytologist* in 1855, he allocated lots of space to the argument about the identity of Ireland's badge. Indeed this journal contains some of the dottiest contributions on the shamrock ever published—and that includes the work of the present century!

The opening salvo by H. B. (Mrs Harriet Beisly of Sydenham in Buckinghamshire) sought to derive the word shamrock from Hebrew, just as Miss Beaufort had attempted to connect it with ancient Persian.[2] This provoked an intelligent retort from the Revd William M. Hind, a native of Belfast who eventually became rector of a Suffolk parish. Botany was one of the vicar's pastimes—he was to write a flora of Suffolk—and he suggested to the readers of *The Phytologist* that an 'equally probable' derivation for the name could be found in the shamrock's native land; 'The word is Shamrog in Irish—a diminutive of Shamar, trefoil—and literally signifies young trefoil.'[3] But common sense and simple facts have never characterized the literature of the shamrock.

Thomas H. Porter, D.D., M.R.I.A., of Ballymully Glebe, Tullaghogue, in the county of Tyrone, entered the fray in the *Ulster Journal of Archaeology*, wherein he published a lengthy paper seeking to establish the pagan origins of the shamrock's sanctity. He drew heavily on Homer, Herodotus and Pliny, and chastised Patrick for theological ineptness, for resorting '…to such a poor attempt at argument or illustration' because '…no one rightly instructed respecting the Holy Trinity could admit any material resemblance whatsoever as an adequate or suitable representation of the Trinity in Unity.'[4] Not content to publish Dr Porter's extravagant paper, the editor added four pages of his own dense argument about the derivation of the name.

The shamrock continued to excite comment through the 1860s with articles on such abstruse subjects as 'Druidical or Ancient British Botany'. Mrs Beisly was still perplexed in September 1860; her son, Sidney, also joined in the literary exchanges. In the following month a new facet of the shamrock emerged when Charles Eyre Parker of Torquay asked why the wood sorrel was now considered to be 'the real Shamrock of Ireland' instead of the trefoil, pointing out that fine carvings of the latter were to be found on ancient stalls in the cathedrals at Exeter, Canterbury, Salisbury and Bristol.[5] To this R. replied with novel simplicity:

> The humble white clover, trifolium repens, the Saint could easily procure, as it grows everywhere; he had merely to stop and pick up an illustrative example of "three in one" and one developed into three. He might have hunted for days ere he had the hap to light upon Wood Sorrel. It is unphilosophical to reason on what is not fact. It is true that botanists deny the identity of Shamrock and Trefoil. A few of the fraternity, who wish to appear wiser than their brethern, may advocate the claims of Wood sorrel, but some affect paradoxes.[6]

William Jackson Hooker Ferguson, curator of the Belfast Botanic Garden, was not content with that; he quoted Fynes Moryson in reply and 'safely

Fig. 32: James Edward Britten F.L.S.: herbarium keeper, leader of a boys' club and shamrock scholar. (Reproduced by courtesy of Hunt Institute for Botanical Documentation, Pittsburgh)

concluded' that wood sorrel was the true shamrock.[7] On and on the argument went—Epsilon's rejoinder was to include any trifoliate plant as a possible vegetable for saintly sermons, '…buckbean or Medicago lupulina or the barren Strawberry, or the fertile one either'.[8]

The contributions to the pages of *The Phytologist* were mirrored by items in *Notes and Queries*—the same writers crop up in both periodicals during 1861, 1862, 1863 and 1864, and the same questions and replies were printed. Then there was a lull.

The first critical and scientific attempt to study the phenomenon of shamrock, to set down the precise identity of the plant worn by native-born Irish men and women on St Patrick's Day, was undertaken by James Britten, an English botanist, in the late 1870s. He incorporated the results into an admirable *Dictionary of English* plant-names, published in 1878, which he compiled in collaboration with Robert Holland.[9]

To start with, Britten and Holland noted that the story about Saint Patrick using the shamrock to illustrate the Holy Trinity was purely a traditional tale. Thus they were not led into the error of mistaking myth for fact, and did not make fatuous statements about a plant the saint actually picked during some legendary event at an unknown locality. Instead, these exemplary authors recorded simply this: in the late 1870s *Trifolium minus* (now named *Trifolium dubium*) was the plant '…most in repute as the true shamrock'. That was the species sold most frequently in Covent Garden as shamrock for St Patrick's Day, and moreover it was worn in counties Antrim, Carlow, Cork, Down, Dublin, Fermanagh, Kerry, Limerick, Meath, Waterford, Westmeath, Wexford and Wicklow. They noted an alternative species—occasionally *Medicago lupulina* was marketed as shamrock both in London and Dublin. James Britten and Robert Holland only received one specimen of wood sorrel, from Waterford![9]

More than forty years after their dictionary was published, Britten recalled the reason for his research. In the mid 1870s he ran a lads' club in Isleworth, London, and all the boys attending it were Irish. About 1875, as St Patrick's Day approached, he decided that the boys should have shamrocks for the occasion, and so he went to Kew Gardens to gather some. James Britten was not a member of the staff there, so keeping one eye open for irate gardeners, he carefully collected as much of the common white clover as he could. Next morning, St Patrick's Day, he proudly brought his shamrocks to the club to give to the boys only to discover that most of the lads were already wearing

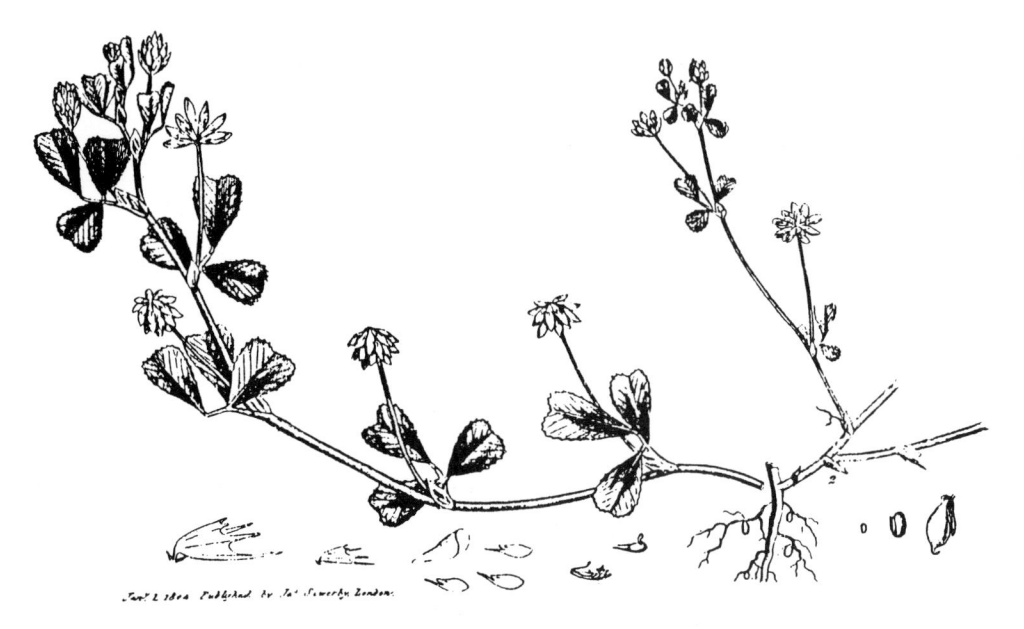

Fig. 33: Lesser trefoil, *Trifolium dubium* (formerly *Trifolium minor*), otherwise one of the shamrocks: hand-coloured engraving from J. Sowerby's *English botany* (1804) (By courtesy of National Botanic Gardens, Glasnevin)

shamrock which had been sent over by post from Ireland. Those who did not have any 'repudiated his offering with contumely'. To the Irish boys of Isleworth, the plant their club-leader had pilfered from Kew could never be the true shamrock for that only grew in Ireland. So James Britten quickly discovered the facts about real shamrock, and from that year onwards he received a box of the genuine article from an Irish convent—'...and every year', Britten wrote, 'it is *Trifolium minus* that arrives'![10]

The only trouble with James Britten's fastidious work was that few people, botanists included, noticed it. One decade later a group of Irish naturalists spent the eve of St Patrick's Day arguing about the identify of the true shamrock.[11] They did not come to any definitive conclusion. One of the company was Nathaniel Colgan and he went home none the wiser but quietly determined that the only way to settle the argument was a careful study of the whole, vexed problem. His first step was to become acquainted with the literature, but this only added to his confusion by establishing the rival claims of four different plants—red clover (*Trifolium pratense*), white clover (*Trifolium repens*), black medick (*Medicago lupulina*) and wood sorrel (*Oxalis acetosella*).

From his survey of published accounts, Colgan reached the irresistible conclusion that the question had '...never been seriously studied by a competent

Fig. 34: Nathaniel Colgan, c. 1883: clerk, scholar and author, shamrock sleuth. (*Irish Naturalist 28* (1919); reproduced by courtesy of National Botanic Gardens, Glasnevin)

botanist, perhaps because the subject was considered too trivial for serious treatment'. Moreover he recognized that 'every Irishman and every Englishman long domiciled in Ireland' knew well that the Irish peasant was extremely particular and careful in choosing his shamrock—to wear the wrong plant invited derision.[11] So he decided to acquire specimens of the true shamrock and if necessary to grow these in his garden until they flowered and could be identified precisely—as every one knows, shamrock has no flowers when adorning the hats or breasts of Irish men and women on St Patrick's Day.

From eleven counties, Antrim, Armagh, Carlow, Clare, Cork, Derry, Laois (in those days called Queen's County), Mayo, Roscommon, Wexford and Wicklow, Nathaniel Colgan obtained thirteen sprigs of shamrock, but he could not identify them as the leafy shoots gave no definitive clues to the species involved. So they were carefully labelled and planted—the shamrocks blossomed about two months later and Colgan was somewhat surprised to find that a fifth species, not one of his four original candidates, was the most frequent—remember he did not at this time know about the survey published by James Britten. Eight of the samples turned out to be lesser clover (then called *Trifolium minus*) and five were white clover. Later a fourteenth plant reached him from Iar Connacht—it was lesser clover too!

Plate 1: 'No warrior-cresset thou', the 'dear little shamrock'—*Trifolium dubium*, lesser clover, seamair bhuí, in full bloom (E. C. Nelson)

Nathaniel Colgan concluded that lesser clover had '…a decidedly stronger claim to be regarded as the shamrock of modern Ireland' than any other trefoil. If the remaining twenty-one counties had been represented in his study, he believed the nine-to-five proportion would have been weighted even more strongly in its favour.

Like any good scientist, Mr Colgan wrote a paper outlining the results of his research and it was published in the August 1892 issue of the *Irish Naturalist*.[11] In a postscript he reported a visit to the Aran Islands in Galway Bay during late May that year:

> I made enquiries for the true Shamrock from the Irish-speaking islanders. Several of them, searching for the plant in my presence, passed over *Trifolium repens* as too coarse, and though apparently inclined to fix on *Trifolium minus*, seemed so staggered by the appearance of its flowers that they gave up the search in the belief that it was too late for the Shamrock.[11]

What did the Irish shamrock-hunter seek? What qualities marked the true shamrock. Few writers have set down the criteria by which the 'rale shamrogue' can be judged—there is no class for it in horticultural shows!—but one author recording the folk-tradition of County Kildare did provide a set of guidelines. Opinions differ, he stated, but the ones sold in Dublin '…are two one-rooted varieties: one having a small *pink* clover blossom, and the other…a yellow flower.'[12] According to this writer, the old people affirmed that the 'rale errib' produced branches from the main root, and these branches took root themselves as they crept along the ground. The flower, they said, '…resembles a small *white* clover blossom.' This errib was not thought to be on sale in Dublin because of the problems in grubbing it up, and anyway the best place to go looking for it was along roads because its creeping branches extended out from the grassy sod. No-one could describe the common white clover more accurately.

The other vexacious problem is that of marks on the leaves. Most modern shamrock-wearers will reject plants with white or black marks on the leaves—these occur almost invariably on *Trifolium repens* and *T. pratense*.

Nathaniel Colgan's article provoked considerable interest among other Irish naturalists. Robert Lloyd Praeger sent a note to the *Irish Naturalist* and reported that flowering plants or luxuriant specimens of lesser clover were discarded as impostors in County Down.[13] He suggested that the shamrock study should be expanded to include every barony in Ireland.

This was indeed tried. Nathaniel Colgan had a notice printed requesting shamrock sprigs and he sent a copy to Roman Catholic parish priests throughout Ireland—no explanation for omitting the protestant clergy is given! This produced a good crop of plants. Colgan reported the results of the larger survey in the *Irish Naturalist* in 1893.[14] Sixteen counties produced *Trifolium repens*, but only thirteen contributed *T. minus*. However when James Britten's earlier results were added in, the number of counties for *T. minus* increased to twenty-one, giving it supremacy again.

Nathaniel Colgan was careful in drawing conclusions from his work and that of James Britten. He did not claim any plant as the one, true and only shamrock, nor did he invoke legends about St Patrick.

> *Trifolium repens* [white clover] can no longer claim pre-eminence as the true Irish shamrock. It must hereafter be content to share that honour, at least evenly, with its rival *T. minus*… While conceding that in the present day the neater *Trifolium minus* [lesser clover] is equally in favour with *T. repens* as our national badge, some may be disposed to argue that the true Shamrock of earlier times, before modern culture has spread abroad a taste for the elegant and the delicate, was, nevertheless, the coarser *T. repens*.[14]

ACT XVI
Enter Artisans and Artists

Nathaniel Colgan wrote his treatise about the shamrock in the final years of the romantic nineteenth century; to him the contemporary fashions were graceful and delicate. That is one way of characterizing the eruption of shamrock over Ireland during the late 1800s, but coarse and vulgar swathes of shamrocks—either by themselves or in those fabulous entaglements with roses and oak, with thistles and leeks, with round towers and greyhounds—were contrived to clamber over and cling to churches, books, jewellery and all sorts of bric-a-brac! Shamrock proliferated just like one of its weedy prototypes, the common white clover, which spreads inexorably stolon by stolon across a neat and weedless garden!

Apart from its appearance as a badge on regimental flags and other items of political and military origin, the shamrock as a purely decorative device is relatively rare before 1800; the earlier exceptions almost invariably have their roots in the Volunteer movement, the Order of Saint Patrick and the United Irishmen. After 1800, and especially following the visit of King George IV, shamrocks overflowed on almost every conceivable object.

In Glin Castle, County Limerick, the ancestral seat of the Knights of Glin, plasterwork with peltae bearing trefoils decorates cornices in the hallway. This is an unusually early example of the use of trefoils—otherwise shamrocks—as decoration in Irish buildings, and it may be linked with the Fitzgerald family's espousal of the cause of the Volunteers. The moulding dates from the early 1780s, and it is surely significant that in Glin Castle today is preserved the shamrock-emblazoned guidon of the Royal Glin Hussars, one of the Volunteers regiments.

As already noted, on the other side of the political fence, shamrocks were used by the United Irishmen, and like the Volunteers its principal manifestation is on flags. This movement did not spawn any artefacts like the mouldings at Glin Castle.

Shamrocks are rare in architectural embellishments before the 1820s, but near Glin Castle in County Limerick, at Ash Hill Towers, shamrocks were incorporated in stucco and this probably by the same craftsman who worked at Glin. In Dublin, one of the houses in Mountjoy Square had a frieze with shamrocks.[1] The most perplexing shamrocks are the crude ones that decorate

Fig. 35 (above): Plasterwork incorporating shamrocks, Glin Castle, Co. Limerick, c. 1780 (Reproduced by courtesy of the Knight of Glin)

Fig. 36 (left): Carved stone urn on the gate pier of the now-disused Roman Catholic church in Oranmore, Co. Galway, with carved shamrock, c. 1803. (E. C. Nelson)

one of the stone urns atop the gate-pillars in front of the disused Roman Catholic church in Oranmore, County Galway; and not just shamrocks are evident—a pathetic but recognisable thistle, and an even cruder, hardly discernible rose are there too. The austere little chapel was built 'Anno Domini 1803', and presumably the urn is contemporary, but why this strange confection should have been carved to decorate these particular entrance gates is quite obscure.

Plate 2: *Trifolium pratense*: red clover: seamair dhearg

Plate 3: *Trifolium repens*: white clover: seamair bhan

Fig. 37: Shamrock-spattered portable Irish harp made by John Egan, 1819 (Reproduced by courtesy of Metropolitan Museum of Art, The Crosby Brown Collection of Musical Instruments 1889 (89.4.1083))

I am not aware of any political significance in the cartouche with some crude trefoils—most probably meant to be shamrocks—on James Malton's representation of the ancient arms of Dublin published in 1792 among his famous set of prints, *Picturesque and Descriptive views of the City of Dublin*. But the goblets and decanters engraved with roses, thistles and shamrocks and dating from the beginning of the nineteenth century were celebrations of the Act of Union.

Shamrocks really begin to proliferate in the 1810s. Why? In 1813 Thomas Moore's ballad about the shamrock, 'chosen leaf of bard and chief', was published. Can it be coincidence? In the next few years '…shamrocks are found spattered over many of the harps produced in Dublin'[3] such as those crafted by John Egan; several examples are known with delicate shamrock marquetry on the sounding box.

Coins and tokens issued by Irish banks and companies, and by the Royal Mint have been decorated, sometimes exuberantly, sometimes discretely, with shamrock. Sir Joseph Banks, that great promoter of things botanical and for many years "director" of the Royal Gardens at Kew, suggested in April 1804 that Irish coinage should have on the reverse a device composed of an Irish harp 'surmounted with a wreath of shamrock'; the Privy Council Committee on Coin rejected this scheme in favour of Hibernia, but the particular coinage was never produced.[2] However, tokens issued in 1813 by the Bank of Ireland were festooned with shamrocks, their design being specified in the parliamentary act (53 George III 106) which permitted the minting—

> …the said last-mentioned Tokens containing on the one Side thereof His Majesty's Head…and on the reverse Side thereof respectively, with a Wreath of Shamrock Leaves, the Words and Figures "Bank Token 10 Pence Irish 1813", or "Bank Token 5 Pence Irish 1813"…

Other tokens issued by private companies were adorned with shamrocks—one example is the railway token (second class!) for the Dublin and Kingstown Railway.[4] The last occasions on which this particular emblem appeared on any coin issued in Ireland were in 1822 and 1823 when three shamrocks embellished the harp on the halfpenny and penny of King George IV; these coins were withdrawn in 1826. In 1821 the currencies of Ireland and Great Britain were amalgamated,

Fig. 38: Bank of Ireland ten pence token, 1813. (Reproduced by courtesy of National Museum of Ireland, Dublin)

Plate 4: *Oxalis acetosells*: wood sorrel: seamsog

Plate 5: *Medicago lupulina*: black medick, dumheidic

Fig. 39: Penny issued in Ireland during reign of King George IV, 1822, with three shamrocks on the harp. (Reproduced by courtesy of National Museum of Ireland, Dublin)

and thereafter the Imperial coins, issued by the Royal Mint in London, circulated in Ireland.

Worried by imperfections in the designs for the new Imperial coinage of George III, Richard Sainthill, partner in a firm of Cork-based wine importers and one of the great numismatists of the nineteenth century, wrote from London to Dublin to a member of the Pym family in December 1816 reporting that he had visited the Royal Mint and was dismayed that the officials had no understanding of form of shamrock which was destined for inclusion within the dessign of the new coins. Sainthill requested specimens of the national badge so that the Mint could ensure its more accurate representation on the coinage. On 30 December Susan Pym despatched real shamrock leaves—we cannot now discover the species sent—to Sainthill and he passed these on to Mint officials. It seems that the samples came too late for any alteration to be made in the designs of the half-sovereign (10 shilling piece), and the already-struck halfcrown (2s. 6d.), shilling and sixpence, but Sainthill was hopeful that

Fig. 40: Coins of the United Kingdom, 1820-1953, all with shamrocks in their designs.
Top row (left to right): crown (1827), half-crown (1820), florin (1953), florin (1937)
Bottom row (left to right): shilling (1821), sixpence (1953), shilling (1893)
(Courtesy of Ulster Museum, Belfast)

Fig. 41: Pin-cushion made in 1817, embroidered with shamrocks
(Courtesy of Helen Dillon)

in the future whenever a shamrock was represented on the United Kingdom's coins it would be copied from Susan Pym's specimen; 'It will consequently be King George's money and Queen Susan's shamrock', he quipped.[5] Whether this happened I cannot tell, but the shamrocks on the coins of George III issued during 1817 and afterwards do bear splendid trefoils. In the present century, shamrocks have appeared on coins of the United Kingdom—for example on the pre-decimal sixpence and florin of Queen Elizabeth II.

Other shamrock-laden items from the first two decades of the nineteenth century are known, but in quantity and quality they do not match the myriads of the latter years. A tiny cushion minutely stitched with shamrocks in 1817 has a verse that clearly indicates the wishes of the maker inscribed on the central panel.

Fig. 42: Sprigs of oak and shamrock on the main entrance porch of Seaforde House, Co. Down; c. 1855 (the shield bears the arms of the Forde and Meade families). (E. C. Nelson, by courtesy of Patrick Forde)

> *To Erin's land,*
> *With liberal hand,*
> *Extend your timely care,*
> *The poor to feed,*
> *In this their need,*
> *From your abundance spare.*

An early example of book decoration is a manuscript about the coats-of-arms of Irish families by Patrick Kennedy, Pursuivant of Athlone (one of the heraldic officers in Dublin Castle); he completed his *Book of Arms* in 1816 but it was not published until 1967. The manuscript contains sketches of Irish arms and the hand-drawn title-page has a border of shamrocks.

As an aside it should be stated that shamrocks do not appear in ancient Irish heraldry—even trefoils are extremely rare and these devices with pointed leaflets cannot really be equated with the Patrician shamrock's heart-shaped

Fig. 43 (left): The 'seal' of the Royal Irish Academy as it appears on current publications—the shamrock lawn is not part of the Academy's armorial bearings.

ones. Occasionally shamrocks were employed as decorative adjuncts to the arms; for example shamrocks and palms appear on the bookplate of Leonard MacNally, the barrister who defended the leaders of the United Irishmen and at the same time betrayed them by passing on information about their plans to the authorities. Oak and shamrock support a spurious coat-of-arms carved about 1850 on the facade of the ancestral home of the Forde family in County Down. The extraordinary background of shamrocks in the seal of the Royal Irish Academy dates from about 1860—this trifoliate pasture has nothing to do with the official coat-of-arms of the Academy and is perhaps just an example of excessive zeal for shamrocks in the latter years of last century. The coat of arms of University College Dublin, dating from 1906, has two shamrocks but armorial bearings incorporating shamrocks would not be granted today by the Chief Herald of Ireland.

Shamrock-passion began with King George IV's visit of 1821. Growing national pride, coupled with increased political awareness and especially opposition to rule from London, stimulated interest in Ireland's past, its antiquities and its artistic heritage. The Celtic Revival slowly gathered momentum through the middle years of the nineteenth century, and the whole panoply of Celtic imagery was quarried for distinctive Irish elements. The shamrock was just one small part of it.

In the last half of the nineteenth century, shamrock was used as a decorative motif on buildings. Many examples could be cited, but suffice it to note three here.

Shamrocks, complete with immature flower buds, were carved by the O'Shea brothers on the architraves of windows and on one of the entrances to the building that was originally the Kildare Street Club in Dublin; this same building is famously adorned with monkeys playing snooker and the shamrocked doorway now leads into the offices of the Chief Herald of Ireland.

Fig. 44: The official document granting armorial bearings to the College of the Holy and Undivided Trinity…near Dublin, with a fanciful 'tree' of roses and shamrocks embellishing it—this conceit is not part of the arms of the College, merely a decoration. (Reproduced by courtesy of The Provost and Fellows, Trinity College, Dublin)

Fig. 45: The west door of the Roman Catholic Church, Clara, Co. Offaly; the angel on the tympanum bears a shield carved with a harp and two shamrocks (E. C. Nelson)

Shamrocks also appeared in and on churches, especially those built when the rage for trefoils was most contagious. There are indubitable shamrocks on the tower of St Brendan's Parish Church, Birr, County Offaly, a 'Gothick essay' by the second Earl of Rosse (1758-1841) who was an amateur architect and an advocate of Catholic Emancipation. A decorative boss incorporating the harp and the shamrock can be seen in St Patrick's Roman Catholic Cathedral, Armagh, built between 1853 and 1873. The tympanum on the west door of the Roman Catholic parish church in Clara, County Offaly, is carved with an angel who carries a shield decorated with a harp and shamrocks, and there are more shamrocks inside. But the most remarkable ecclesiastical shamrocks, arranged in quartets and interspersed with harps, round towers and greyhounds, are those on the reredos in the Roman Catholic parish church at Kilcock, County Kildare.

Overseas, shamrocks appeared wherever Irish emigrants set up their homes. For many centuries there has been a small Irish community in Rome, but it had no church dedicated to Ireland's patron saint. In 1883 a successful appeal was launched by the Revd. P. J. Glynn, Prior of the Irish Augustinians,

Fig. 46: The spendid mosaic above the high altar in St Patrick's Church, Rome: here is depicted St Patrick, clutching a shamrock, preaching his sermon on the Holy Trinity—Three in One and One in Three. (Reproduced by courtesy of Revd J. Hunt o.s.a.)

to build an Irish national church in the Eternal City. St. Patrick's Church is situated within walking distance of the city centre. It was designed by Aristide Leonori in Lombardo-Gothic style. In the apse above the high altar is a colourful and highly romanticized mosaic depicting a bearded, mitred Patrick standing on the Hill of Tara explaining the doctrine of the Holy Trinity to the court of King Laoghaire. Patrick has a crozier in his left hand and with his right hand holds aloft the solitary leaf of a shamrock.

Shamrocks also haunt gravestones and monuments throughout Ireland as at Clogheen, County Tipperary, in memory of Father Nicholas Sheehy and in Prospect Cemetery, Glasnevin, on the monument to John Keegan Casey who was imprisoned as a Fenian and also wrote ballads and published a collection of poetry under the title *A wreath of shamrocks* in 1866. A

Fig. 47: A common sight in Ireland, a modern grave-marker mimicking a Celtic high-cross, with poorly-carved shamrocks; this example is one among many at Kilmacduagh, Co. Galway. (E. C. Nelson)

marvellous high cross stands in the centre of Monasterevin as a memorial to Francis Prendergast—it is covered in shamrocks. The Parnell monument at the north end of O'Connell Street in Dublin has shamrock ornament about the base, while the figure of Hibernia on the O'Connell monument at the opposite end of that street has a chaplet of shamrocks on her head. Headstones in the form of high crosses usually have scroll-like ornamentation but sometimes shamrocks fill the panels; several like this stand at Maynooth. Exquisitely carved shamrocks entwine the columns on the mock Hiberno-Romaesque arch marking the grave of philanthropist and anti-slavery campaigner, James Haughton, in Dublin's Mount Jerome Cemetery.

Returning to the mundane, rising suns, round towers, harps and shamrocks are not infrequent on secular buildings such as pubs. The famous but now demolished Irish House on Wood Quay in Dublin was a stunning example of contagious shamrockery. Still standing and exuberant are the Harp and Lion Bar and the Central Hotel in Listowel, County Kerry; from the facade of the hotel a reclining, portly Erin broods, her arm resting on a harp, shamrocks and a hound at her feet and a round tower by her side.

Fig. 48: Lamp standard in Dublin (David Davison)

Lamp standards adorned with shamrocks were erected in the 1890s thoughout the city of Dublin. There were shamrocks all over the livery worn during the late nineteenth century by the coachman of the Lord Mayor of Dublin.[6]

Elegant Victorian ladies wore poplin gowns made in Limerick which were hand-embroidered with lawns of shamrock![7] If that was not sufficient to proclaim one's Irish connexion, the ladies could also wear necklaces and bracelets fashioned from silver and 'Irish diamonds' (iron pyrites) arranged in shamrock patterns, or brooches made from bog-oak carved as harps, or showing Hibernia seated on shamrock-speckled mounds, or just with shamrock alone.[8] Solid gold shamrock bangles cost a mere twenty-five shillings in the late 1800s!

Not content to wear shamrock from head to foot, you could sit at oak tables, the bases of which were guarded by shaggy-coated wolfhounds reclining in a forest of harps, oak branches and shamrocks. On it, shamrock-laden napkins and table-clothes of the finest damask linen could add to the

Fig. 49: Day dress (c. 1867) in green poplin with separate bodice and skirt, embroidered with cream (on skirt) and green (on bodice) shamrocks. (Reproduced by courtesy of National Museum of Ireland, Dublin)

vegetable array, and, of course, the porcelain tea-service—the finest Parian ware from the Belleek Pottery in County Fermanagh—was hand-painted with green shamrocks, pink roses and golden heads of corn. About the only thing that was not shamrocked was the tea itself! At dinner, wine was poured from swan-necked decanters resplendent with engraved shamrocks and, depending on the political persuasion of the host, either the motto *Erin go bragh* (accompanied by harps, wolfhounds and other Celtic symbols) or *The immortal memory of King William III* (with the obligatory horse and royal rider). The shamrock crossed the political divide. The wine was sipped from goblets similarly emblazoned.

Shamrocks hybridized with thistles and six-petalled roses were etched into glass and inlaid on desks, tables, chairs and wooden boxes by craftsmen, especially in Killarney—they used local woods, including yew and arbutus as well as real and faked bog oak, for the inlay. Regardless of all the

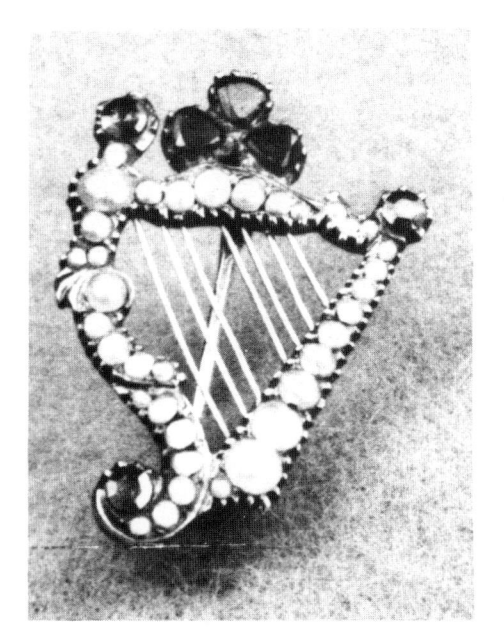

Fig. 50: Two brooches with shamrock and harp:
(left) a brooch of pearls and emeralds made for Sarah Curran (c. 1800)
(Reproduced by courtesy of Cork Public Museum):
(right) a silver brooch (date unknown) (Reproduced by courtesy of Mrs J. Gant)

Fig. 51: An unusual example of Belleek Parian ware porcelain, a flask with various emblems embossed on it, including rose, thistle and shamrock, with the distinctive Belleek trade-mark (round-tower, hound, harp and shamrock) too. (Reproduced by courtesy of National Museum of Ireland, Dublin)

Fig. 52 and 53: Decanters made in Ireland and engraved with shamrocks, and intended for use in houses of different political affiliations! (Reproduced by courtesy of Mary Boydell)

shamrockery, these pieces of so-called Killarney ware represent some of the best furniture ever produced in this country. They were fashionable in the late 1800s and were exported as souvenirs.[9]

Among other souvenirs that tourists purchased were shamrock-encrusted postcards. The postcard was invented in the 1880s and the hey-day of picture postcards was the early years of the present century. Being cheap to buy and cheap to post, everyone used them. Little has changed in the world of Irish postcards in the century of their existence—shamrocks still wreath many.

Fig. 54: Central motif from a table made in Killarney, showing Muckross Abbey, wreathed with shamrocks (photograph by E. C. Nelson, reproduced by courtesy of Val and Helen Dillon)

Some may even have 'genuine' shamrocks, dried and pressed, glued to them and the manufacturers' favourite plant is yellow clover. There were and are special cards issued for St Patrick's Day, and now as then the designs range from the ludicrous to the appalling; in recent years shamrocks, water-buttercups and crocuses were combined in a lough-side setting to illustrate Thomas Moore's ballad; at the beginning of the century shamrocks swathed the feet of a young girl playing a harp in front of a rising sun.

Bouquets of shamrock are not confined to postcards, household items and personal adornments. Books sprouted shamrock too. There are no shamrocks evident on tasteful leather binding of the eighteenth century, not even on the sumptuous bindings done for the Irish Parliament by the 'Parliamentary Binders' in the 1700s; those superb books were destroyed when the Four Courts were burnt in 1922, but rubbings of the patterns survive.[10] Such good taste did not survive the Act of Union! By the second decade of the nineteenth

Fig. 55: A finely crafted leather binding with shamrock tooling.
(Reproduced by courtesy of Aidan Heavey)

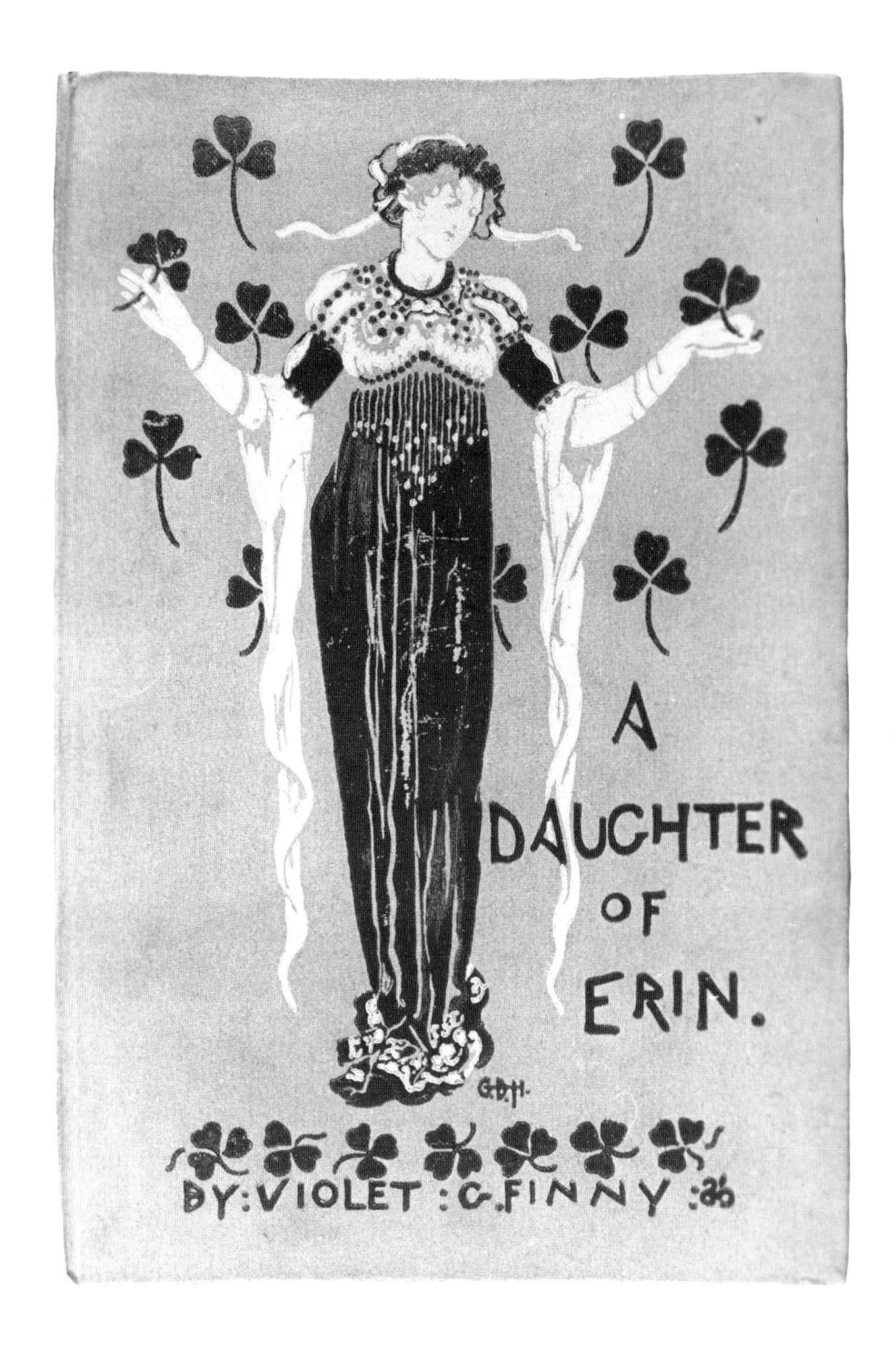

Fig. 56: A Victorian dust-jacket (Reproduced by courtesy of Mrs J. Gant)

century, with Mrs Mountain's 'sweet little shamrock' and Thomas Moore's 'chosen leaf' ringing in their ears, book-binders began to incorporate shamrocks into their designs. In 1815 George Mullen of Dublin used shamrocks large and small, individually and in a sinuous combination with oak leaves and acorns, to decorate the leather binding of Lord Castlehaven's *Memoirs of the Irish Wars*. But the great proliferation of shamrocks came in the last half of the century when almost any book with an Irish association had at least one shamrock on its dust-jacket, binding, end-papers or pages. Even official publications of the United Kingdom government destined for Ireland or about Ireland had shamrocks somewhere in them—these largely vanished when the Irish Free State was established in 1922, but they returned again in the 1970s!

The Irish Nation: its history and its biography published by A. Fullarton and Co. in 1875 has shamrocks bordering Hibernia who is seated inside a shamrock-shaped bubble. A magnificent copy of Edward Newman's *Nature-printed ferns* bound by Marcus Ward and Co. of Belfast, now in the National Botanic Gardens, Glasnevin, has luxurious shamrock-patterned end-papers in gold and emerald-green. About 1895, Violet G. Finny wrote a forgettable novel about *A daughter of Erin*, and her London publisher decorated the cover with a vision of Hibernia, in violet, surrounded by a shower of shamrocks. There are hundreds of other examples, running the full gamut from the repulsive and ridiculous, to the beautiful and even sublime; the green leather bindings by Zwehnsdorf on a two-volume set of Henry Jephson's research papers with its elegant and quite restrained gold-blocked shamrocks represent one of the best; one is used as comparison, in photographic facsimile, on page 1 of this modest book.

Inside the covers, on the printed pages, shamrocks also swarmed. A tangled mass of tendril-sprouting trefoils, drawn by S. Holden, forms an overpowering heading for the introduction to the illustrated edition of William Carleton's *Traits and stories of the Irish peasantry*. The finest shamrock-adorned book is undoubtedly Daniel Maclise's magnificent edition of Thomas Moore's *Irish Melodies* which was published in 1845.[11] The pages decorated by steel-engravings based on Maclise's original drawings are a sheer joy, inventive, sensuous and romantic. Not every page has the Irish symbols of shamrock and harp, so that it is not overladen with these now hackneyed things. A bearded king, surrounded by comely maidens, listens to the harper as he sings about 'The harp that once through Tara's halls…',

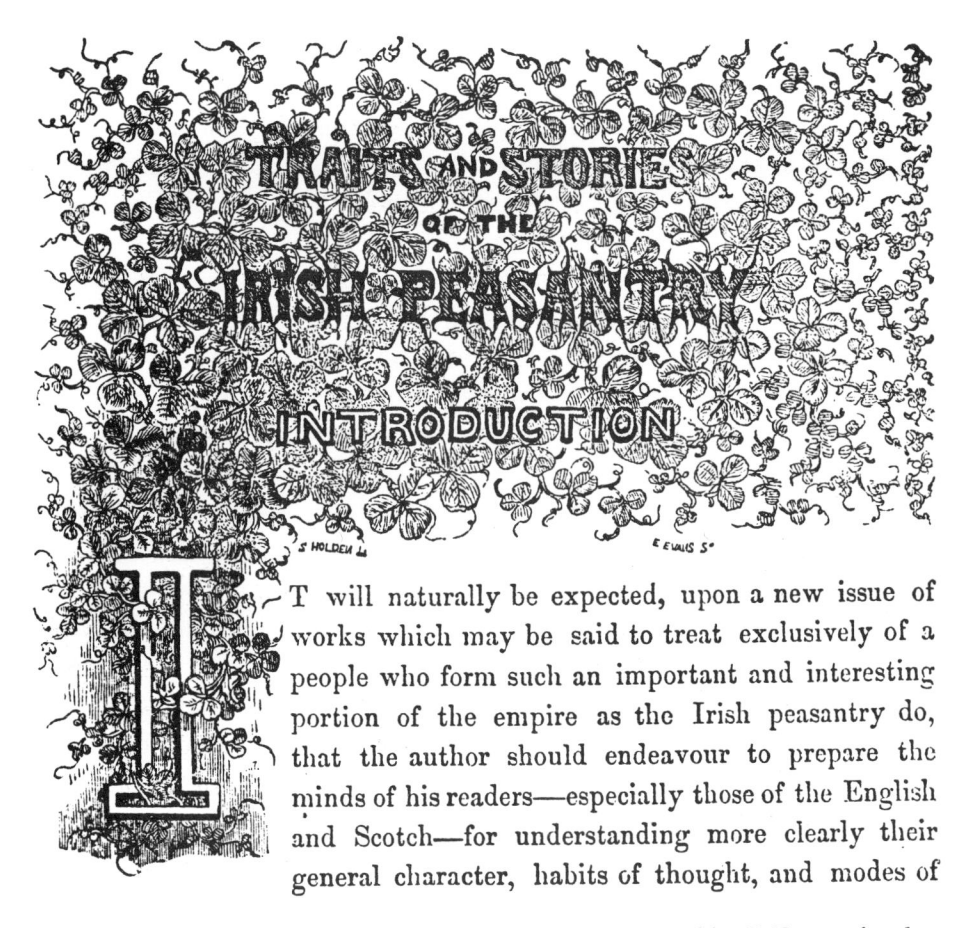

> **TRAITS AND STORIES OF THE IRISH PEASANTRY**
>
> **INTRODUCTION**
>
> IT will naturally be expected, upon a new issue of works which may be said to treat exclusively of a people who form such an important and interesting portion of the empire as the Irish peasantry do, that the author should endeavour to prepare the minds of his readers—especially those of the English and Scotch—for understanding more clearly their general character, habits of thought, and modes of

Fig. 57: The head-piece, drawn by S. Holden, and engraved by E. Evans, for the first page of William Carleton's *Traits and stories of the Irish peasantry*. (Reproduced by courtesy of Aidan Heavey)

and the border below has bluebells and lilies with cherubs. A chisel-faced, balding sage peering through a telescope at some distant star adorns the opening page of 'Now all the world is sleeping, love..', and a long-haired, flower-bedecked maiden, standing in a rock-pool surrounded by a waterfall of seaweeds, shows how Maclise responded to these lines by Thomas Moore:

> 'Tis believ'd that this Harp, which I wake now for thee,
> Was a Siren of old, who sung under the sea:
> And who often, at eve, thro' bright waters rov'd,
> To meet, on the green shore, a youth whom she lov'd.[12]

Shamrocks adorn the pages printed with the words of the famous ballad, and the title-page is well endowed with them (see Frontispiece and pp. 68-70).

ACT XVII

Soldiers, Exeunt

Oh! Dublin boys, and did ye hear
The news that's goin' round,
How the Queen of Ireland's settin'
Of her foot on Irish ground?
  [Chorus]
  'Tis the most enchanted country
  That iver ye have seen,
  Where St. Patrick will be honoured
  By the wearin' o' the green!

So, pluck the Shamrock from the soil,
And bear it on your crest,
'Twas won in fiercest battle
By the bravest an' the best!

We're the most uplifted regiments,
Bedad, we're mortal keen!
The Shamrock's in our forage caps
By order of the Queen!

They may speak of bouldest England,
Or of gallant little Wales,
Of lion-hearted Scotland,
But ould Ireland never fails!

Our boys will take the biscuit—
There's no need of Blarney stone
To magnify their valour,
For they've won by that alone!

With the trefoil in our helmets,
On this glad St Patrick's Day
Tune Erin's harps in concord,
And sing a mirthful lay!
  [Chorus]
  'Tis the most enchanted country
  That iver ye have seen,
  Where St. Patrick will be honoured
  By the wearin' o' the green![1]

Those verses were cobbled together by none other than the Duchess of Buckingham and Chandos. The awful pastiche was just one of many published after the announcement on 7 March 1900 that Queen Victoria had commanded that henceforth on Saint Patrick's Day 'it was Her Majesty's

Fig. 58: Lieutenant Collins, Royal Irish Rangers, collecting shamrock on Slemish Mountain, for St Patrick's day, 1985. (By courtesy of Lieutenant-Colonel Niall O'Byrne; reproduced by permission of The Royal Irish Rangers)

pleasure' each soldier in her Irish regiments should wear a sprig of shamrock in honour of the bravery displayed by Irish troops during the Boer War. The Imperial decree followed immediately after news reached Europe about successes in the campaign in southern Africa, especially the relief of Ladysmith on 28 February 1900. Much of the credit for the advances in the war was given to Lord Roberts, Knight of St Patrick and one of several commanders who were of Irish descent—the others included Kitchener, French and Mahon.

Queen Victoria also created the Regiment of Irish Guards as a tribute to Irish gallantry in the Boer War; the regiment was established on 1 April 1900 and its cap-badge is the star of the Order of St Patrick.

The war continued in Africa, and other pieces of verse, some even worse than those of the Duchess, were penned. Rudyard Kipling was invited to contribute to the inaugural issue of a newspaper published in South Africa and distributed to the troops fighting there.

> Oh, Terence, dear, and did you hear
> The news that's going round?
> The Shamrock's Erin's badge by law
> Where e'er her sons are found.
>
> From Bloemfontein to Ballybrack
> 'Tis ordered by the Queen,
> We've won our right in open fight,
> The wearing of the green.[2]

Shamrock obviously had to be supplied for the use of Irish soldiers on St Patrick's Day. William Baylor Hartland, the Cork nurseryman, ventured to suggest in a letter to the editor of the *Cork Daily Herald* that shamrocks raised from his seeds would '…grow freely on the Veldt among the rocks and stones in South Africa'. He did indeed supply the Imperial War Office with shamrock seed.

The decree making the 'wearing of the green' a matter of honouring brave men, was linked with an announcement that Queen Victoria was to pay a visit to Ireland; she had not been to the country since 1861. The occasion was marked by the usual festivities and the usual indifference. The Queen, in accord with her recent elevation of the shamrock to a memorial for brave men, was seen in Dublin wearing a sprig of shamrock, and one of her dresses and the matching parasol were embroidered with silver shamrocks. The military 'tradition' of wearing shamrock on St Patrick's Day, which she inaugurated, continues to this day both in the armed forces of the United Kingdom, and in the defence forces of the Republic of Ireland.

The shamrock has never divided the Irish people; in many instances it might be said to unite them for irrespective of their political creeds they have been willing to use it as a badge. Thus while shamrock emblazoned the Knights of St Patrick, it also adorned the collars and cuffs of the uniforms of the members of the '82 Club which was founded in Dublin to commemorate the 'independence' gained by Ireland in 1782. An embroidered border of shamrocks encircled the cap presented to Daniel O'Connell at the Monster Meeting in Mullaghmast in 1843.[3] The shield-shaped badges of the Ulster Unionist Convention included, in the lower half, a field of shamrocks set on a yellow ground around the Red Hand of Ulster, with the crowned harp and the Union Flag in the upper quadrants. The Fenians and the Land League used abundant shamrocks on their flags and banners. Anti-Home

Fig. 59: Badge of the Ulster Unionist Convention, 1892.
(Reproduced by courtesy of Ulster Museum, Belfast)

Fig. 60: There are discreet shamrocks on Ulster's robe in this propaganda post-card which was in circulation in 1915.
(Reproduced by courtesy of Mr and Mrs R. Nelson)

Fig. 61: Badge of the Young Citizen Volunteers, formed in Belfast during 1912.
(Reproduced by courtesy of Ulster Museum, Belfast)

Fig. 62: The shamrock garden at Mount Stewart, designed by Lady Londonderry. (E. C. Nelson)

Rule propaganda—for example postcards—had shamrocks discretely incorporated into the design.

Posters urging young Irish men to enlist for service during the First World War did not have Lord Kitchener's finger pointing at them. Instead, the 'Real Irish spirit' saying that "I'll go too" was conjured up by a silhouette of a ruined abbey with a round tower and a sprig of green shamrocks. The postcards sent home from the blood-soaked battlefields of Flanders, such a long way from Tipperary, had evocative trefoils too.

It is amazing to see how the shamrock crosses the political divide, and perhaps nowhere is that more tellingly shown than by the badge of the Young Citizen Volunteers established in Belfast during 1912 to fight for the Unionist cause. A shamrock, crowned, is charged with the Red Hand of Ulster. That same combination reappears in the Shamrock Garden created at Mount Stewart[4] on the Ards Peninsula by Edith, Lady Londonderry who was undoubtedly a staunch supporter of the Unionist cause; within a trefoil hedge she laid out paved shamrock within shamrock enfolding a begonia-flowered Red Hand and over it set a silent, living harp of Irish yew. That surely is the ultimate horticultural tribute to the shamrock—and not a living one in sight!.

ACT XVIII

Take a Bow, Mr Hartland

When the adoration of the shamrock reached its zenith in the late nineteenth century, the clamour for it was irresistible so that it was only a matter of time before the oldest profession realized the potential market! Mind you, gardeners and nurserymen were a bit slow off the mark, but when they seized the bit between their teeth they did not let go.

For many years the two botanical gardens in Dublin, the Royal Botanic Gardens at Glasnevin and the Trinity College Garden at Ballsbridge, grew shamrocks so that they could be shown to the inquisitive visitors, or sent overseas to acquisitive gardeners. We have no definite records of what plant was grown at Glasnevin but it seems that black medick (*Medicago lupulina*) was the principal shamrock. At Trinity College, in the early 1890s, Frederick Burbidge kept a small stock of shamrock from which specimens could be sent to English inquirers; the College shamrock was the 'real shamrock' as certified by one of the gardeners, a Tipperary man! This plant was lesser clover (*Trifolium dubium*).[1]

Glasnevin and Trinity College botanical gardens could not supply a large market, and by the late 1800s there was certainly substantial foreign demand for shamrock. The efficient postal service, which could ensure that a letter leaving Dublin at six o'clock in the evening was delivered in Kew by eight o'clock the next morning, meant that shamrock could be mailed on 16 March from Ireland and arrive in Britain in good condition on Saint Patrick's Day. But that market was a one-day wonder; the lucrative one was in Irish America.

When did a gardener first market shamrock? Out in the west of Ireland, shamrocks were being sold in the mid-1860s—an English visitor attending a '…bazaar, or fancy fair, held for some charitable purpose by the upper class in their county town' was offered a sprig of true shamrock by one of the 'presiding ladies'.[2] Tourists visiting Killarney, the prime tourism centre in Ireland during the latter years of the nineteenth century were frequently offered shamrock—in flower!—during the summer by the local guides and jarveys.[3] Those notorious characters were also known to sell fronds of the Killarney fern, and with the help of avaricious botanists and gardeners they succeeded in all but exterminating that delicate plant. A single piece of the local fern cost five shillings in 1844; the shamrock, perhaps, was not so expensive!

Fig. 63: William Baylor Hartland, c. 1890

Undoubtedly market-gardens about Dublin grew shamrock for the city hawkers by the late 1800s, for although most Irish men and women would have gathered their own shamrocks in the wild, the town-dwellers did not have access to pastureland where the clovers grew. By the 1890s, selling shamrock was a widespread business for nurserymen. Nathaniel Colgan mentioned that he had obtained a sample from a nursery in County Louth which advertised the plant at 'a not unprofitable price'.[4]

In 1889 when the ebullient William Baylor Hartland moved to Ard Cairn in Cork, he was faced with old sheep-pasture which had to be levelled and dug before he could recommence his nursery business. One particular hollow he had filled in, and during the autumn this was dug again, dressed with bone-meal and soot, and left fallow. In the spring on this spot appeared a crop of '…tufts of lovely Shamrock of the most verdant green, and having

W. B. HARTLAND & SONS.

Hartland's Ard-Cairn Russet, as named by Mr. J. Wright, of Pomological Fame.

We sent fruit to the late Doctor Masters, and are indebted to him and the *Gardeners' Chronicle* for the notices and illustration, March, 1907.

"To W. Baylor Hartland & Sons, of Ard-Cairn, Cork, we are indebted for samples of this Apple, received at the end of February. The fruit is about 2¾ inches long by 2½ inches in width, ovoid-conic, truncated and slightly five-angled at the top, with a shallow eye, short funnel-shaped tube, and unfolded calyx segments. Stalk about half-an-inch long, set in a shallow basin. Skin golden brown, more or less covered with russet. Flesh firm, yellowish white, with a sweet, rich flavour. In view of St. Patrick's Day, our Artist has added a few Shamrock leaves."

Fine Trees, 2/-, 2/6, 3/6 each.

THIS QUOTATION CANCELS ALL PREVIOUS OFFERS.

Where large quantities are required ask for special quotation.

Fig, 64: Even advertisements for Mr Hartland's apples had shamrocks to assist; from *Descriptive catalogue of fruit trees, shrubs, etc.* (c. 1909)

a 17th-of-March-like appearance—so full of virgin purity...' Mr Hartland dug up some of these tufts and replanted them about one foot apart in neat rows; they spread forming clumps over one and a half feet across, and from them he harvested seed. Ever mindful of business opportunities, William Baylor Hartland then decided to offer the seed to the Americans. In 1893, Lady Aberdeen asked him to supply one thousand pots of shamrock for the Irish village in Chicago. The first consignment was due there on 16 April and one writer was prompted to comment that

> so far as we know it is a rare thing to send living market-plants of Shamrock from Ireland to America, and if they will endure the voyage well this may be the beginning of a thriving industry.

Furthermore he suggested that if growing shamrock 'straight from the old sod' became fashionable among Irish Americans the market would be limitless.[5]

Hartland kept back a little of his valuable shamrock seed, and in 1895 planted one acre of young tufts on top of his daffodils so that he could harvest even more seed that autumn. In the Autumn 1898: Spring 1899 catalogue listing *Well-ripened Irish-grown Daffodils and Rare Tulips*, Hartland offered 'A St. Patrick's Day Present for the Irish Abroad'; packets containing 3,000 shamrock seeds cost only sixpence, and the product was 'the true guaranteed National Type from "Ould Ireland".' A special advertising leaflet was printed and despatched from Cork with the bulb catalogues issued in the early 1900s—seed of *Trifolium minus*, 'the dear little Shamrock of Ireland', was offered at 1/6d and 2/6d per packet (but the number of seeds was no longer given!). The leaflet concluded with these phrases

> "THERE'S A GOOD TIME COMING!!!
> A GOOD TIME COMING!!!"

Whether William Baylor Hartland was the first to offer packets of shamrock seed for sale cannot easily be determined; undoubtedly other nurserymen sold shamrock plants about the same time. But the market was never lucrative because it was too seasonal and could easily be over-supplied. In 1919 'amateur cultivators' were producing shamrock for the military market—the Countess of Limerick was reputed to have been a major supplier (although undoubtedly she herself never tilled the soil nor sowed the seed).[6]

A further disadvantage, as far as the trade was concerned, was the weedy aspect of the traditional plants, the clovers and black medick. Even the yellow clover, *Trifolium dubium* (formerly *T. minus*), which was emerald green and neat for Saint Patrick's Day, became less acceptable, even unrecognizable

"The dear little, sweet little, dear little, Shamrock of Ireland."

TRIFOLIUM MINUS.
TRUE.

Seed 1/6 and 2/6 per packet.

THE EDITOR OF THE "CORK DAILY HERALD."

Sir,—Simply a line to say how I got the above, and why for ten years I have been posting seed to Irishmen in all parts of the world. In the year 1889, when making alterations at Ard Cairn, there was a hollow in the centre of one of the fields, where water lodged, and we had to carry soil from a height to fill up same, doing so to get a site for a range of glass. The men got down six feet upon a bed of the most sterile soils—in fact, where we found "iceberg" drifted limestone "boulders," that when touched with the sledge fell to pieces, and disclosed a great variety of fossils. Upon this soil, after it had been well picked, I had soot, lime, salt, bone meal, etc. sprinkled, thus leaving it roughly exposed to the frost and snow of winter. In April following the first bit of green to appear was "Trifolium Minus" and common sorrel. I had all the shamrock plants put out on the tops of the daffodil beds for seed purposes, which plants have been recognised at different botanical gardens as the "true trifolium minus." It would grow freely on the Veldt among the rocks and stones in South Africa.—Yours,

W. BAYLOR HARTLAND.

Ard Cairn, Cork,
March 17th, 1900.

"It will be observed by the above fact that, like the "poor Irish," it thrives only on starvation. "Good food, justice, and kindness kills us all."

"May the Rose, Leek, and Thistle long flourish and twine round the Sprig of Shillelagh and Shamrock so Green."

"THERE'S A GOOD TIME COMING !!!
A GOOD TIME COMING !!!"

WM. BAYLOR HARTLAND,
Seedsman, Bulb and Begonia Grower,
CORK, IRELAND.

Fig. 65: A leaflet advertising shamrock issued by William Baylor Hartland, c. 1901, printed by Guy & Co. Ltd, Cork. (Reproduced by courtesy of University of Ulster)

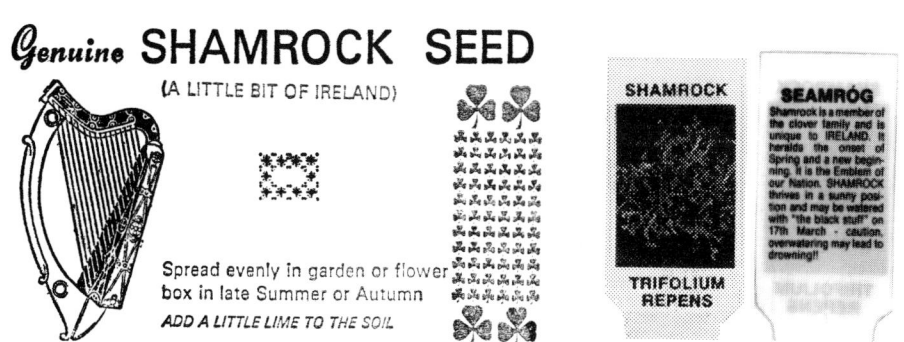

Fig. 66: Some modern 'packaging' for clover seeds, and a plant label 'colourfully' explaining the reason for charging a higher price than usual for a lawn weed!

once flowers started to develop. There were, and are, people who will substitute other trifoliate plants for the clovers, the principal imposters being species of *Oxalis* although not, in general, the native Irish wood sorrel, *O. acetosella*, which James Bicheno had attempted to foist on Ireland.

One experiment to 'improve' the shamrock was suggested in 1925 by the Celtic Fellowship of New York, whose members, bless them, thought that

Fig. 67: One beautiful plant not yet exploited as a shamrock is *Parochetus communis*, a blue-blossomed pea from the mountains of East Africa and the Himalaya; it is sometimes called shamrock pea. (Wendy Walsh 1988)

'the uncared for and unadorned trefoil' was a most unsuitable emblem for a 'free and pacifically united people'. The Fellowship approached the famous Californian plant-breeder Luther Burbank and asked him to undertake this task. He responded saying that he had already tried and failed, and suggested that Iceland poppies or his own famous breed of daisy, the Shasta daisy, might be substituted for the traditional clovers thereby providing prettier flowers. Rebuffed by Burbank, the Fellowship wrote to *The Irish Times* asking for comments—with commendable restraint the newspaper merely printed the story. The Irish would never have stood for such a sacriligious suggestion even from their benighted American cousins![7]

In 1949 for the first time shamrock was sent by air to the United States from the new airport at Shannon. Now-a-days airmail is a regular thing and, of course, tourists can still buy packets of clover seed labelled 'true shamrock'—these contain (if one is lucky) about thirty seeds (not Hartland's 3,000) and cost £1 (that is the price on packets being sold in Dublin, July 1987). And there are other novelties which I need not list apart from the 'real' shamrock leaves entombed in 24 carat gold—'Take a piece of Ireland home with you' suggests the advertisement. In the future, might we see shamrock packed in sealed plastic containers, and irradiated for longer shelf life, alongside the strawberries and bean shoots, all grown in some foreign country?

Shamrock can be cultivated with ease in Irish gardens, indeed in many it need not be planted for it is already present as a lawn-weed! But if one is particular about the shamrock worn on St Patrick's Day, a pinch of seed sown in a pot in the early autumn will yield a nice sprig in mid-March of the succeeding year. Left to grow, it will blossom—white, red or yellow depending on the species of clover that produced the original seeds. Of course, it is also possible to cultivate the various sorrels, species of *Oxalis*, and to pass them off as shamrocks. The American nursery trade does this, but few Irish nurseries would survive the derisive remarks of patrons.

One minor aspect of the shamrock which deserves at least a mention is the four-leaved one called in Irish **seamróg na gceithre gcluas**. It was regarded as a sign of great good luck throughout Ireland and in the Western Isles of Scotland, yet to some people the lucky one was the five-leaved shamrock. Moreover, I learnt recently that there is a *pisrog* associated with four-leaved clover and shamrock—they must never be brought into a church, for then they become harbingers of ill-fortune.[8]

In Scotland the traditonal lore of the **seamrag nam buadh** includes the provision that it must be found without looking or searching, a happenstance. If it is seen, the lucky leaf can be cherished as an invincible talisman. Among the recorded folk-lore about the four-leaved shamrock is this song in Scots Gaelic:

Sheamarag nam buadh	Thou shamrock of good omens
A fas fo bhruaich	Beneath the bank growing
Air na sheas Moire shuairce,	Whereon stood the gracious Mary
Mathair De.	The Mother of God.
Tha na seachd sonais,	The seven joys are,
Gun sgath donais	Without evil traces,
Ort, a mhoth-ghil	On thee, peerless one
Nan gath grein—	Of the sunbeams—

Sonas slainte,	Joy of health
Sonas chairde,	Joy of friends
Sonas taine,	Joy of kine
Sonas treuid,	Joy of sheep
Sonas mhac, us	Joy of sons, and
Mhurn mhin-gheal	Daughters fair,
Sonas siocha	Joy of peace,
Sonas De!	Joy of God!
Ceithir dhuilleagan na luirge dirich,	The four leaves on the straight stem,
Na luirge dirich a fraimh nam meanglan ceud,	Of the straight stem from the root of the hundred rootlets,
A sheamarag gheallaidh La Fheill Moire,	Thou shamrock of promise on Mary's Day,
Buaidh us beannachd thu gach re.	Bounty and blessing thou art at all times.

A four- and five-leaved variant of the white clover, *Trifolium repens* 'Purpureum', is cultivated in gardens; its leaflets are deeply stained with purple and appear almost black, and each leaf can have three, four, five or perhaps six leaflets. I do not know the origin of the common name St Patrick's tears, but Patrick O'Kelly, a plant-collector from Ballyvaghan in County Clare, listed it[9] under that sobriquet in the 1890s—he did not sell it as 'lucky clover'. Most of the modern products incorporating 'genuine' four-leaved shamrock—key-fobs, etcetera—are made from a small fern called *Marsilea* which has even less to do with Ireland and Saint Patrick than the white clover.

One little-known Patrician plant may also be mentioned. In the mid-1800s at the little town of Saint Patrice, a short distance from the French city of Tours on the banks of the River Loire, grew a thorn bush identical with the native Irish blackthorn or sloe—*draighean* in Irish, *Prunus spinosa*. Every year it came into blossom at the end of December, flowering from Christmas until the first of January. No other bush in the region bloomed at that time, four months before it should. This blackthorn was the subject of local curiosity and indeed devotions. The old people said that St Patrick was on his way to Ireland when he came to this spot and rested under a bush. It was Christmas and the weather was bitterly cold. But to honour the saint the shrub extended its branches, shook off the snow that lay on them and burst into flower as white as that snow. Since that time the blackthorn blossomed at Christmas. I do not know if the bush bearing *Les Fleurs de Saint Patrice* still grows by the Loire, but it was certainly alive in 1850, and James Britten was sent a sample of its flowers.[10]

ACT XIX
Enter and Exit the Antiquarians

James Britten and Nathaniel Colgan established the identity of the plant widely used and generally accepted as the shamrock of Ireland in the late 1800s. In the century that has elapsed nothing has altered; the shamrock of the 1980s may be packed in plastic bags but it is still usually the same lesser clover (*Trifolium dubium*). And today an Irishman seen wearing any other plant is just as likely to be ridiculed.

Britten's and Colgan's works were timely for they coincided with the beginnings of a new critical examination of Patrick's story, as scholars began to dust the cobwebs of myth from the historical figure. But for some the time-honoured legends were sacred, and no matter how carefully the historians explained their work and stated their conclusions, the myth of the shamrock sermon on the Hill of Tara remained entrenched.

For a period of about forty years, between 1890 and 1930, several eminent antiquaries and clerics fought rearguard actions on behalf of the shamrock and the myth. The first to enter the fray was Dr W. Frazer, Fellow of the Royal Society of Antiquaries of Ireland and a duly elected Member of the Royal Irish Academy. He published a paper on the chronology of shamrock in the *Journal of the Royal Society of Antiquaries* one year before Nathaniel Colgan's more expansive and better documented account. Frazer set down evidence for the antiquity of the shamrock, including such well-known authorities as Spenser, Moryson, Wither, Threlkeld and Keogh. He mentioned Britten and Colgan, and even admitted that he had failed to unearth any early accounts of the Patrick legend. 'It is therefore, with regret and disappointment,' wrote Frazer, that '...I am forced to conclude St Patrick never handled a shamrock during his protracted life to explain any of his teaching.'

Dr Frazer was determined and did not rest there. He proceeded to demonstrate that the shamrock had as ancient a usage as a national symbol as the rose and the thistle. His arguments must now be carefully assessed.

He noted that trefoils were used as mint-marks on the coins of King Henry IV between 1399 and 1413, that trefoils and roses were embossed on the coins of King Henry VI, and that there were shamrocks on the Irish crown groats of King Edward IV. It is true that mint-marks comprising three discs arranged in a triangle, and thus resembling a trefoil, occurred on coins circulating in Ireland during the reigns of those kings. Sadly Dr Frazer was blinkered, and fell into the trap that he accused others of entering:

> nothing is easier than to construct such theories, and the less evidence to support them, the more confidently they are advanced.[1]

To affirm that these mint-marks are shamrocks is stretching things too far; in reality they are not shamrocks, just three discs arranged in a triad, and they have no connexion with the Patrick legend.

The other evidence he advanced, in a separate paper, was the occurrence in mediaeval Irish churches of shamrock-patterned tiles. Dr Frazer published illustrations of a series of these tiles and he stated that

> …their popularity for decorative purposes must be attributed to the universal belief of the association of the trifoliate leaf with the alleged teachings of St. Patrick, from which originated the general acceptance of it as a special symbol of Ireland and Irishmen—a position may it long maintain, uninfluenced by the carping disparagements of critical archaeology, for at the lowest estimate its claims are as good as either rose or thistle can produce.[2]

The tiles do have trefoil-like motifs resembling the club in a pack of playing cards, but this was not a uniquely Irish pattern for the motif is commonly encountered in churches throughout Britain and western Europe. Such tiles, no matter how ancient nor how piously employed, do not prove the antiquity of the shamrock myth. Sadly Frazer's error has not served as a lesson to modern writers who still have the audacity to state that trefoil tiles in Irish churches have a special significance.

Dr Frazer also drew attention to coins bearing the image of a bishop preaching to a group of people and holding in his hand an object which is a grossly enlarged trefoil. He dated these about 1640, the Confederate period;

Fig. 68: The so-called 'Patrick' coin minted c. 1669-1670; the scene depicted on the coin surely is the occasion of St Patrick's sermon on the Holy Trinity; the saint, as bishop, holds a hugely enlarged trefoil—shamrock—in his hand. (Reproduced by courtesy of National Museum of Ireland, Dublin)

the current opinion among numismatists is that this coin was minted most probably not later than 1674. If the bishop is intended as a representation of St Patrick, which is most probable, the coin is the earliest evidence for the link between the shamrock and Ireland's patron saint. Dr Romauld Bauerreiss has pointed out that the shamrock does not appear in icons of Patrick until after Caleb Threlkeld's book was published in 1726, and has cogently argued that the shamrock is an iconographical misunderstanding of mediaeval representations which show Saint Patrick holding a Tau cross, a symbol of the Holy Trinity during the Middle Ages.

The great difficulties of interpreting mediaeval icons can be vividly demonstrated by Fiacail Phádraig (Shrine of St Patrick's Tooth) and the Domhnach Airgid Shrine. Fiacail Phádraig was made before 1376, and bears an image of St Patrick holding a cross—but the ornamentation includes trefoils. The Domhnach Airgid Shrine also has an image of Patrick on it—he is shown presenting the shrine to St Macartan, and beneath is a frieze with what seem to be four-petalled flowers and trefoils.[4]

It is not at all difficult to find trefoils, or tripartite motifs, in objects of the Early Christian Period and the Middle Ages. That most glorious of all Irish treasures, the *Book of Kells*, is full of such patterns; almost every decorated page has trefoils! They appear on the feathers of the winged lion of St Mark, and in the halo of Man (St Matthew) on folio 27v, one of the carpet-pages showing symbols of the Evangelists; folio 129v has another set of these symbols also with trefoils abounding. One of the angels (lower right) on the Madonna and Child page (folio 7v) holds an object formed from two linked trefoils. The decorated page of the Genealogies of Christ (folio 200r) has trefoils on each of the extended letters **t** of the repeated Latin word *fuit*. But are any of these shamrocks? Most emphatically, they are not shamrocks! Do any of these geometric patterns prove the Patrician legend? Again, most emphatically, they prove nothing of the kind.

The simple explanation is that a trefoil is an easily drawn decorative element and it is common throughout western European art. The exquisite La Tene ornamentation that characterizes pre-historic and early Christian Irish art abounds with three-fold designs. No-one would suggest that the famous triple spiral inside the tumulus at Newgrange, dating from about 3500 years before Patrick arrived in Ireland, is a shamrock—in recent years it has, of course, been bastardized into one by the National Botanic Gardens, among other organizations! The presence of trefoils on Irish artefacts from

Neolithic times or from Early Christian and Mediaeval periods contributes nothing whatsoever to discussion of the Patrick legend. That so many scholars should have been misled by them is a more significant fact.

The Patrick coin is the earliest representation of the saint holding a shamrock, and if the date for minting, about 1670, is correct, the coin does corroborate other evidence for the evolution of the shamrock myth sometime during the late seventeen century. All the earlier icons of Saint Patrick show him without shamrocks, sometimes with a crozier, sometimes with a Tau cross, and occasionally treading on snakes for the legend about the banishing of venomous reptiles from Ireland is much more ancient than that of the shamrock. What is perhaps most remarkable is that the earliest ecclesiastical representations of Patrick complete with shamrocks are to be found in Germany, at Waeschenbeuren and at Eggenrot, both of which towns possess statues dating from the first half of the eighteenth century.[5]

In March 1921 James Britten published a short, scholarly article in the Jesuits' journal *The Month*, reiterating the known history of the shamrock. W. H. Grattan Flood responded with the old chestnuts; his 'Vindication' of the shamrock was a pathetic attempt to maintain the myth. Flood accepted Britten's account of the cult of the shamrock since the time of Edmond Campion but, like Dr Frazer before him, he sought to prove the improbable. Noting the trefoil-like elements in the decoration of the *Book of Kells*, Flood produced this extraordinary *non sequitur*:

> In the *Antiphonary of St Gall*, written by Irish monks, about the year 870, the prominent feature of the initial letter of the Easter Sequence, commencing *Laudes Salvatori voce modulemur supplici*, is the Shamrock with interlaced ornament. And, be it noted, that this prominence of the Trefoil is given to the Sequence for the feast of Easter, on the eve of which St Patrick preached his "Shamrock" sermon at Tara.[6]

Thus, Grattan Flood concluded, two great Irish manuscripts contain proof 'strong as Holy writ' for the Patrician legend of the shamrock; piously he hoped that the myth will '…henceforth be admitted as fully sustained, based as it is on unimpeachable evidence going back over a thousand years'. The evidence, alas, was far from unimpeachable.

ACT XX
Enter William Butler Yeats

The war of the shamrock did not end in 1921. The formation of the Irish Free State in 1922 was to prolong the skirmishes for two more decades.

In 1926 the Government of Saorstát Éireann set up a commission to design a set of coins for the new state to replace those of the United Kingdom. The commission was chaired by William Butler Yeats, and it was provided with a set of clear guidelines about its task. Member of the public were asked to recommend symbols that might appear on the new coins—round towers, wolfhounds and 'a kneeling angel pouring money from a sack' were suggested, and Joseph Brennan, Secretary to the Ministry of Finance and later Chairman of the Currency Commission, put forward the idea that at least one of the new coins should have a shamrock scroll on it, similar to the Bank of Ireland token of 1813.[1] The first meeting of the committee on 17 June 1926 made three decisions: first, that the harp would appear on all of the coins; second, that all inscriptions would be in Irish; third, that effigies of modern persons would not be used. Various learned institutions were consulted including the Royal Society of Antiquaries of Ireland which cautioned against shamrocks and cited Nathaniel Colgan's historical survey of the shamrock as the definitive work. After further meetings the committee issued a report in which the following passage appeared

> The Committee is strongly of the opinion that hackneyed symbolism (round towers, shamrocks, wolf-dogs, sunbursts) should be entirely avoided; even the shamrock as a symbol, has no dignity of age behind it, being not much more than a hundred [sic.] years old.[2]

There was little or no opposition at this stage. The final designs were leaked to the press and published eighteen months before the coins were issued, and still there was hardly more than one murmur of protest. But on issue, Ireland was outraged! Intemperate, offensive and plainly ignorant criticisms were launched against the beautiful coins by priests and people alike. The committee was accused of adopting pagan symbols, of 'a turning down of God'.

The squall, as it was described by Dr Thomas Bodkin, Director of the National Gallery of Ireland, blew fiercely for a while and then vanished, but the final puff came from the Revd James Forrestall who repeated all Grattan Flood's material in a lengthy, superfluous article.

Fig. 69: Stamps issued by the Irish Post Office: the first Irish definitives, and one commemorative issue (Holy Year (*Annus sanctus*) 1933-1934), each with shamrocks in the design.

It seems almost commonplace to say that the shamrock is and always has been Ireland's national and religious emblem, yet there have been men, eminent authorities amongst them, who have denied this; who tell us that the 'shamrock legend', as they call it, has only the tradition of a century and a quarter to support it. How such a theory could have been held appears incredible, but it is, nevertheless, a fact, for in 1928 the Finance Committee refused to sanction the shamrock as a design for the new coinage on precisely these grounds… It is true, I know, that certain of our contrymen have questioned the tradition, but I say, with all the respect due to their genius, that they are wrong and evidently so… It will be shown that the tradition can be traced back, on at least probable ground, to the seventh or eighth century, and in incontestable evidence to the seventeenth.[3]

There is a millennium of a difference between the seventh and the seventeenth century, between the historical, the real Patrick and the romanticized and mythical saint, but Father Forrestall seemed not to worry. He proved to his own satisfaction that the tradition was grand and noble. *Vivat trefolium* were his final words. Amen.[3]

The Finance Committee showed commendable judgement and good taste, two qualities that were not, it seems, evident in the office of the Postmaster-General of Saorstát Éireann. On 1 February 1922 an invitation was issued for the public to submit designs for postage stamps to replace those bearing the portrait of King George V. A miscellaneous series of prototypes were prepared and many included all the symbols that the Finance Committee abhorred. The firm of O'Loughlin, Murphy & Boland designed a set at the request of the Minister of Posts and Telegraphs, and the halfpenny essay sported a round-tower, harp, rising sun and shamrocks—only the wolfhound was missing! The two-penny stamp had a central shamrock, and the seven-penny a harp on a field of shamrocks. The final designs were not so

Fig. 70: Stamps of the United Kingdon including shamrocks: George VI (9d.), Elizabeth II (coronation issue: 4d., and definitive: 1d.). The pre-decimal Northern Ireland regional issue does not include shamrock (the Ulster badge and flax are in the design).

over-powering but shamrocks were used. The 1d., 1½d., and 2d stamps, designed by J. Ingram, had a map of Ireland with shamrocks in the upper corners; the 2½d., 4d., and 9d. issues by Miss M. Girling had a shield with the arms of the four provinces set against a field of shamrocks, and her celtic cross design, used for the 3d. and 10d. stamps, also included four shamrocks. These stamps remained in circulation from 1922 until 1967.[4]

Shamrocks also appeared on a few other Irish stamps. On 18 September 1933 a commemorative issue, designed by R. J. King, marked the Holy Year of 1933-1934; it showed a cross with two attendant angels, and hidden in the surround were four shamrocks. In 1967 a pair of stamps marked the centenary of the Fenian Rising, and these reproduced two essays prepared in New York at the time of the Rising; the one cent stamp had a harp with shamrocks, as had the three cent one. The only other evidence of shamrocks is on postal stationery, on the typographed stamps on envelopes and post cards issued between 1922 and the present day. The recently formed semi-state company An Post which runs the postal service in the Republic of Ireland has resurrected the shamrock to decorate such pre-paid items as the popular St Patrick's Day cards.[4]

Stamps of the United Kingdom have also sprouted shamrocks, roses, thistles and daffodils (as well as a few leeks). The national vegetables first appeared in the design of an embossed 6d. stamp issued during Queen Victoria's reign. A shamrock-like element, with a stylized thistle and rose was included in the border of the 6d. Jubilee issue of 1892. They disappeared during the reign of King Edward VII, to reappear in Bertram Mackennal's 9d. stamp for King George V. There were no national emblems on the austere and short-lived issues of King Edward VIII, but they blossom again on those of King George VI. The frames of the definitives were designed by E. Gill and included a single rose, thistle, daffodil and shamrock. Wreathes and solitary emblems also appeared on the first stamps of Queen Elizabeth II, including the 4d. coronation commemorative designed by M. Goaman, a stamp designer of some repute. Preliminary essays for the half-crown stamp displaying Carrickfergus Castle did have shamrocks, but the final design had no recognizable vegetable. Strangely, shamrock was not employed in the regional stamps available in Northern Ireland—flax and the Red Hand of Ulster ornamented the pre-decimal issues, and the current decimal stamps just have the Red Hand.

Similarly the current coinage of the United Kingdom does not include ostentatious shamrocks. The 1986 £1 coin representing Northern Ireland bears the flax plant, although the earlier Scottish and Welsh (1984 and 1985) ones had thistle and leek respectively. Two minute shamrocks can be seen on the 1983 and 1988 coins within the Royal coat of arms.

Many people in Ireland and abroad believe that the shamrock is the official national emblem, that it has constitutional status as a symbol within the Republic of Ireland. But the emblem of the Irish state is the harp, and it is this twelve-stringed harp that adorns government documents; Dr Hayes McCoy suggested that its retention alone of all the traditional badges of Ireland—wolfhound, round tower, harp and shamrock—was that 'it gave satisfaction to display it without the crown'.[6] Yet the shamrock does remain as one of the emblems for Northern Ireland; within the compartment of the Royal coat-of-arms of the United Kingdom of Great Britain and Northern Ireland, shamrock still entwines with rose, thistle and daffodil—might that not have been the reason for its disfavour in the early days of Saorstát Éireann?

In the Republic, shamrocks are by no means scarce in spurious badges and even more spurious coats-of-arms, but they do not feature in official devices displayed by the state. A striking contrast again can be seen in Northern Ireland where the badge of the police, the Royal Ulster Constabulary, incorporates the harp, the crown and a wreath of shamrocks. This badge has a long pedigree, going back to the 1830s when the police barracks in every town had an iron badge above the door with the word 'Constabulary' and a shamrock leaf surmounted with a crown. On

Fig. 71: The crest of the Royal Ulster Constabulary. (Reproduced by courtesy of Divisional Commander D. Turkington, with permission of the Chief Constable, Royal Ulster Constabulary)

6 September 1867, Queen Victoria decreed that the constabulary should have the title Royal Irish Constabulary and wear the harp and crown badge. The Royal Irish Constabulary was disbanded in 1922; in the Irish Free State, the Garda Siochana was established with a new badge, while in Northern Ireland the newly formed Royal Ulster Constabulary retained the badge with the harp, the crown and shamrock wreath.[7]

However, the shamrock is so firmly entrenched as an emblem of the Irish, that in recent years the Irish government was obliged to seek to protect the shamrock—or perhaps more accurately a green trefoil—as a specifically Irish badge. It was registered under the international convention, which regulates the use of trademarks, as a national symbol and thus nowadays a green trefoil cannot be registered as a commercial trademark for use on goods that originate outside the Republic of Ireland; the Irish harp is similarly protected.[8]

ACT XXI

Vox Hiberniae

My investigation of 'the dear, little shamrock' would not have been complete without some attempt to assess its status in this present age; indeed this history could not be complete without some commentary about the myth's current wellbeing. Emboldened, late in February 1988 I sought to establish if the tradition of gathering this emblematic weed in wild places had altered in any way since the time of Nathaniel Colgan, and appealed through newspapers for living wild shamrocks to allow me to repeat his pioneering experiments. Some of the results of the simple request were not unexpected: some were beyond belief, with newspapers proclaiming the imminent extinction of shamrock, a state of affairs that is hardly possible.

As already explained, Nathaniel Colgan's prototype survey in the 1890s established that the native Irish used at least four different plants to represent shamrock; a minority opinion supported the claim of wood sorrel to be the true '…dear little, sweet little shamrock of Ireland.' Of course all Colgan's plants had a characteristic trifoliate leaf, thereby fitting the folk perception of a shamrock.

My appeal was carefully worded, for plants of shamrock gathered in 'traditional localities'. Letters seeking assistance were sent to local and national newspapers throughout Ireland and also to radio and television stations; many of the provincial and national newspapers published the request and a live television appearance—co-starring a genuine shamrock from the National Botanic Gardens which turned out to be white clover!—prompted numerous donations. Between the beginning of March and mid-April 1988, two hundred and eleven living shamrocks, two packets of shamrock seed and ten dried carcasses were received from thirty-one Irish counties—only the citizens of Monaghan failed to respond, while the most prolific counties were Dublin (25) and Tipperary (23).

The shamrocks were potted and grown in the National Botanic Gardens until they blossomed; by early June 1988 almost all of the samples had bloomed and on 8 June each one was examined and its correct botanical name determined.

What was revealed? That *Trifolium dubium* (= *T. minus*) (lesser clover, seamair bhuí) is still the plant most widely considered truly to represent shamrock is beyond dispute—almost half of the shamrocks sent were yellow clover. *T. repens* (white clover, seamair bhán) accounted for a little more than one third of the shamrocks. *Medicago lupulina* (black medick, dúmheidic) was collected infrequently, while *T. pratense* (red clover, seamair dhearg) and wood sorrel (*Oxalis acetosella*, seamsóg) were only sent by nine and six correspondents respectively.* When this state of affairs is compared with that reported by Nathaniel Colgan it is evident that no significant change has taken place during almost one century in the folk-botany of shamrock. Analysis of the 1988 sample suggests that there are some regional trends deserving notice; lesser clover came mainly from southern and eastern counties, while white clover was predominant among the shamrocks received from the western and northern counties; of the nine red clovers, four originated in County Sligo, and most of the black medick plants were sent from the southern counties.

Whence came the dear little shamrock plants? Their sources, as told by their collectors, were diverse: anywhere and everywhere, from 'a mossy bank about two fields from the foot of the Sugar Loaf Mountain and just up over Glen o' the Downs' where my elderly father had 'always instructed us that the real shamrock would be picked in moss', and from amongst the hazels on the southern facing banks of the Annalong River Valley in the Mourne Mountains. Others were plucked from flowerbeds in back and front gardens 'despite being "weeded" out annually', from 'the lawns of our house', from 'a very neglected flower bed bordering a very neglected, very small front lawn'. They were dug up in vegetable gardens and 'ground which until last summer was covered by a Dutch greenhouse' although the plants (so the donor stated) were 'not the shamrock of the "old days" which had a tufted habit with tiny, dark green slightly bronzed leaves'! Quite a few people picked shamrock on walls: from 'the top of an old stone wall' in Gorey, for example, and near Callan in County Kilkenny

> real old shamrock [was] gathered by locals year after year...from the top of a limestone wall about one hundred years old [until] a couple of years ago local work groups cleaned the wall and nearly scraped away our heritage. Luckily a part of the wall escaped & the shamrock is growing there away.

Other folk gathered their plants from 'an old railway line that runs at the back of my house', and 'about 700 yds from the sea in the gravelled drive leading to our house'. 'I have always found the best shamrock on the grass verges of

* For a summary of the results see Appendix I.

country roads that have not been tarred', stated Mrs Brigid Dwyer of Dundalk, and Mrs Coman's father collected it 'for at least 40 years…on the verge of a byroad on a bridge over the Nenagh river'. For the historically inclined shamrock-hunters, you can find genuine shamrock at Sarsfield's Rock in County Limerick—that is 'where Sarsfield blew up the Williamite seige train around 1690', affirmed young Master William Beary of Pallasgreen. It can be obtained, of course, on the Hill of Tara, County Meath, and on the top of a stone wall near St. Declan's Well in Ardmore, near St. Brigid's well at Croagh, County Limerick, and 'on the Banks of the Boyne where the battle was fought'.

Most of the plain people of Ireland envisage the true shamrock with clean, unblemished leaves, heart-shaped and emerald. This is affirmed by Mrs Dwyer of Dundalk who sent 'wild shamrock from the lawns of our house' where there was also 'a lot of clover with the white marks, and a larger leaf.' With her agreed Mrs Brien of Ballinamallard in County Fermanagh: 'It is not shamrock [if it has] a white spot; clover has a hairy leaf, shamrock has a shiny leaf'. Mrs Una Phillipson's lawn also had clover, 'lots of clover but it has a spot on the leaves.' But that is not what others would aver, and the contrary opinion came from a rather puzzled Mrs Piggott who sent me some shamrock from the Hill of Tara given to her in Ireland some fifteen years previously:

> I was told the red marking on the leaves represented the blood of Christ and was the true shamrock. But I must confess, seeking the true shamrock proved difficult as I was shown at least three different plants in different [places] in Ireland, one of which…bears a tiny yellow flower I found growing in my own lawn on my return! I like the idea of the Tara shamrock which blooms with the ordinary white clover flower.

'My mother always refused a three-leaved clover with its definitive white streak on the leaves and we had to find this little trefoil before we could say we had a shamrock' according to Mrs Mary Hanna of Newcastle in the County Down. Also in that county, Michael Drake's father 'never allowed us to pick this one for St Patrick's Day because it had a 'white line on it.'

There are other ways of telling shamrock from clover according to Ireland's folk-taxonomists: Marian Delany affirmed that 'it is very easy to distinguish' them:

> I tried everywhere to find a shamrock about here. I could find nothing but clover. The shamrock is very distinct from clover. It grows in a bunch forming a wreath, puts down only one root for every bunch. The clover is different it creeps along the ground putting down roots as it goes along. Sometimes with hairy or velvet type leaf. The shamrock has very clear and well-shaped leaf.

For the majority of people to whom shamrock appears only once a year, on St Patrick's Day, as a leafy trefoil, the true plant never has a blossom. Yet there are some others who recognize that later each year the shamrock will bloom. Evelyn King of Drimoleague, County Cork, described that blossom as 'a ball-like fluffy thing, yellow in colour', while a 'yellow tassel like flower' was noticed by Mrs Dwyer in County Louth. Most correspondents who mentioned flowers noted that these were yellow, and according to John J. Daly of Fermoy, County Cork, while

> the leaf may look like the clover…the flowers are totally different. The shamrock has a spray of tiny golden yellow blossom, small leaved, and will creep if allowed. If controlled it will become a big spray. There can be no comparison between the little yellow flowers and that of the red and white clover.

It seems to some folk that the shamrock is not as abundant as it used to be—indeed there are quite a few who believe (prompted, I am sure, by journalists) that at this latter end of the twentieth century the 'dear little shamrock' is on the verge of extinction because of the use of commercial weed killers. Nothing could be further from the truth as every gardener will aver, for when they are not transmuted into shamrocks the clovers used as shamrock will frequently be cursed as unwelcome, almost ineradicable weeds. Let me reiterate that my query was about the custom of gathering shamrock in the wild, not about it abundance, but I did receive comments on its scarcity. In 1988, J. Nolan noted 'a scarcity of real shamrock this year in spite of the mild winter and locals say the use of fertilisers and pesticides are slowly eroding the natural plants' about Callan, County Kilkenny. 'I've always picked [shamrock], altho' it doesn't seem to be in as much supply as it did years ago, except in places where there are no sheep' wrote Mrs Rose Normanly, from County Sligo. For three years Colonel Chavasse could not find shamrock at Cappagh in County Waterford, but 1988 was a plentiful year. Mrs Una Phillipson 'was quite excited to find … the real thing' in 1988 for the first time and wondered 'if the mild winter which kept the soil warm encouraged it to grow'. About Bangor Erris, County Mayo, 'as you know the shamrock will grow at any time of the year', reported Tony Corrigan.

There are keen shamrock gardeners everywhere, and many have their own recipes for success: for example Ruth Armstrong of Castlepollard, County Westmeath, 'sets it in gravel and it reseeds itself every year.' For E. Barry in Lismore, it grows well 'on my window sill in a flower pot.' Mrs C. McCabe of Newtownabbey obtained shamrock 'some twenty years ago in a pot with a little plant in Belfast…[and] it has been rampant in my greenhouse & outside ever since.' 'Shamrock has been growing at our place since my grandfather came to Monalow [Blackrock, County Dublin] in 1901 and of course it was here before that. My father always told me', recalled Mrs Alice Farrell, 'that it was never without shamrock as long as he could remember. He also said never to mow the grass at the top of the garden until the seeds have blown and I have always done this. Most people are surprised when I tell them that shamrock is growing naturally in the garden.'

As for the vexed question of growing shamrock beyond the shores of Ireland, quite a few correspondents sent specimens from 'across the water'. Mrs Burns of Cheshire had been given some shamrock seeds by 'my Irish assistant cook about twenty years ago; she had just returned from holiday at her home which, I seem to remember, was near Beleek', and the shamrock was still thriving. 'The sample I send to you was found growing among rose bushes in my garden yesterday' according to Mrs McKeigul who lived in London; 'Is it shamrock? My explanation is the area was covered with "Irish peat" last year.' But while Irish peat may have such a beneficial effect, most overseas growers, including M. E. Foster of Knaresborough, had been given plants; that person informed me that

> I have four pots of it inside in flower. My mother had the original given to her about fifty years ago by a lady who was Irish and she visited Ireland and came back with it. It is like wood sorrel but we enjoy it and I like to think it is the real shamrock. It used to grow through the bricks in the back yard of my Mother's house and nothing killed it. It was always a lovely show until the year she died and it did no good afterwards.

For the really intrepid there can be surprises. Mary McCartin found shamrock growing on hard dry ground of Malta, and she could not believe her eyes. But, that 'initial disbelief was finally converted to belief, because I found it near a small fort used in both "World wars", where Irish troops were sometimes stationed' and she had a 'lovely, sentimental dream of the Irish mothers sending shamrock to their soldier sons. After the "17th" passed, the men may just have thrown out the shamrock on the path outside and they *did* "take root and flourish".' But Mrs McCarthy's daughter 'declared it to be a "weed", but I still maintain it to be a "Med. cousin", at least, of our own Dear little shamrock.' The true eccentrics, the men and women who go to faraway lands, will also pine for shamrock about 17 March each year; Peter Beatty had since leaving Ireland as a boy…worn many different alleged shamrock on the 17th, however the very best I have ever encountered grows in Brunei, Borneo, the vivid green colour & heart shaped leaf, growing close to the soil was a treat for me, so very far from my native land.

Necessity is, of course, the mother of invention!

Anyone horrified at the thought of wearing a species of shamrock found in Borneo would be truly scandalized by the thought of shamrock being other than green. Yet Brother Nagle found 'a small reddish leafed trefoil outside Clonmel for the American cousins' in 1932, and Katherine McDonald sent me a 'brown shamrock which my family had down through the years.' 'Here is a black shamrock that is growing on a hedge at our house' at Corrowallen in County Leitrim, according to Ann Connolly. From Kathleen Feely, a resident of Manorhamilton, I received 'The Derry [shamrock which] is a three, four and five leafed with dark and green leaves'; she had it growing outside in an old bucket and it had 'to be protected from frost by covering with old dead leaves' These brown and black and many-leaved shamrocks were all *Trifolium repens* 'Purpureum', a cultivated form of the white clover which at the turn of the last century was marketed, as I have noted elsewhere, from County Clare by Patrick B. O'Kelly under the name 'St Patrick's tears'—well may the saint have wept at this mutant of his dear little shamrock.

As we all know shamrock is worn, and then drowned. Mr Barry of Lismore had for the 'past forty years…sent this type to my friends all over the world, also some friends call to my house for some. On Patrick's day I always wash and clean a nice spray and place it in my drink at the local. Strangers get a kick out of it when they see my drink.' Charles Cunningham of Annalong related that the shamrock he sent me was 'what my grandfather, Charles Cunningham, and his neighbours Art & Hugh McCartan, Jimmy Nevin, Hugh Rogers and James McVeigh, wore… They are all dead now but my family still stick to the custom… It was eaten on St Patrick's night, washed down with whiskey, literally drowned.'

EPILOGUE

I have laid down the history of the shamrock, noting the former authors who have discussed it and argued its case. I have pointedly dismissed some of the least critical, less scholarly essays. What are my own conclusions and what is the status of the 'sweet little shamrock' in the late twentieth century?

For the past eighty years historians have shone a stringently critical beam on the biographies of Patrick and concluded that he was a real person, but nowhere in the earliest written records of his life can anyone trace an account of a sermon illustrated by use of a trefoil. There is universal agreement that the story is a folk tale, part of a colourful tapestry that the Irish have so skilfully woven for their own enjoyment during many hundreds of years.

Moreover the earliest examples of the use of the word shamrock—whether in Irish or in English—display no links with Saint Patrick. In English texts shamrock is first noted as a food-plant eaten by starving, persecuted Irish peasants. In Irish manuscripts the word is employed poetically to describe flower-speckled landscapes. At the most basic level, the word **seamróg**, which is directly derived from the Irish word for clover, **seamair**, means little or young clover, and that is a precise description of the plant worn with pride by Irish men and women at home and abroad on Saint Patrick's Day.

About a century and a half after shamrock appeared in an English text, it is possible to detect a change in meaning. No longer is it a famine food eaten by wretched wild men; it has become an emblem with the same status as the Scottish thistle, the Welsh leek and the English rose. A short time later the legend of the shamrock sermon is printed for the first time.

From 1726 to the present day the use of the shamrock as a badge by the Irish has been justified by reference to a myth. That does not invalidate the badge, nor indeed does it detract from the parable. But to impute other significance to this Hibernian trefoil, to attempt to prove what cannot be proved, to find shamrocks on every Early Christian artefact, is spurious scholarship and a shocking waste of time.

And what is the shamrock? The most eloquent solution to that conundrum is the simplest. Shamrock is **seamróg**, young clover—the clover James Britten surreptitiously removed from the Royal Botanic Gardens at Kew was as valid as the clover posted all over the world by the resident Irish to the scattered exiles abroad. Towards the beginning of this century the Sisters of Charity at Ballaghadereen in County Mayo produced a leaflet about shamrock which, as James Britten remarked, showed laudable impartiality, except perhaps concerning the superiority of natives of Connaught to whom were confided the most sacred traditions of the Irish people. Thus the Connaught folk held white clover in 'great veneration and [wore] it exclusively', but they allowed the rest of the Irish people to wear that other great and genuine shamrock which was the 'spray of yellow trefoil', *Trifolium dubium*. Surely that sets a seal on the enigmatic shamrock. Both varieties are real genuine shamrock, stated the Sisters of Charity.[1]

A shamrock is a young clover. It does not matter where it is grown nor whence it originally came; it matters not whether the seed came from New Zealand or from the County Cork. It does not even matter which species is worn for the tradition declares no botanical preference.

Shamrock is clover, nothing more, nothing less. That is what the word originally meant, what it still means, and what it will mean until the end of time.

> And yet no warrior-cresset thou,
> A higher, holier spell is thine;
> Sign of her early faith, the Church
> Still claims thee for her hallow'd shrine;
> Symbol to her of mystic truth,
> Link of the golden chain first given;
> Dew-drop embosoming a star,
> Silent, but eloquent of Heaven.
>
> Well may the child of Erin deem
> His shamrock precious in his eyes;
> Its spell can wake the hidden spring,
> Bid Hope from Memory arise:
> And whisper that the 'Isle of Saints'
> Shall know a purer sanctity,
> When Glory shall illume the land,
> And Truth shall make her children free.

Notes and References

In the following, bibliographic citations have generally been abbreviated to the author's surname, brief title and date; full details of each shortened citation are given in the expanded bibliography of the shamrock, published as Appendix III in the library issue of this book (ISBN 0 86314 200 1). When a full reference is cited here, it is not repeated in the bibliography of the shamrock.

Prologue
1. Moore, Oh! The Shamrock, [1812]
2. Moore, *Irish Melodies*, 1845
3. quoted by Moldenke & Moldenke, *Journ. New York Bot. Gdn* 47 (1946): 46-60.
4. quoted by Glassie, *Passing the time*, 1982.
5. *ibid*.

Act I: Enter an Ancient Briton
1. A. B. E. Hood (ed.) *St Patrick, his writings and Muirchu's Life*. 1978. London & Chichester.
2. e.g. Bieler, *Life and legend of St Patrick*, 1949; Thompson, *Who was St Patrick*, 1985.

Act II: Enter the Biographers
1. V. Roth, Early insular manuscripts: ornament and archaeology, with special reference to the dating of the Book of Durrow, in M. Ryan (ed.) *Ireland and insular art A.D. 500-1200*. 1987. Dublin. pp. 23-29.

Act III: Enter the Scribes
1. For guidance in this chapter, I am most grateful to Dr Éamonn Ó hÓgáin.
2. Colgan, The shamrock in literature, *Journ. Proc. Royal Soc. Antiq. Ireland* 26 (1896): 211-226, 349-361.
3. Ó Cianáin, *The Flight of the earls*, 1916.
4. O'Donnell, *Selection from the Zoilomastix*, 1960.

Act IV: Enter the English Chroniclers and Dramatists
1. Campion, *First Boke*, 1571.
2. Stanihurst, *A treatise*, 1577.
3. Derrick, *Image of Ireland*, 1581.
4. Anon., *Captain Thomas Stukeley*, 1605.

5. Éamonn Ó hÓgáin, pers. comm.

6. Shirley, *Saint Patrick*, 1636.

Act V: Enter the Herbalists

1. translation from Colgan, The shamrock in literature, 1896. (fn. III 2)

2. Gerard, *The Herball*, 1597.

Act VI: Enter a Map-Maker and several Poets

1. Camden, *Britannia*, 1610.

2. Speed, *Theatre*, 1611.

3. Wither, *Abuses*, 1613.

4. Moryson, *An Itinerary*, 1617.

5. Taylor, *All the workes*, 1630.

6. Spencer, *View*, 1633.

7. Ware, *De Hibernia*, 1654.

Act VII: Enter the Shamrock-Wearers

1. E. MacLysaght. *Irish life in the seventeenth century*. 1979. Dublin, Irish Academic Press. pp. 3-4.

2. quoted by Colgan, The shamrock in literature, 1896 (fn III 2)

3. Colgan, *Journ. Bot.* 32 (1894): 109-110.

4. Mundy, *Commentarii*, 1680.

5. Linnaeus, *Flora Lapponica*, 1737.

6. *Journ. Kilkenny Arch. Soc.* 1 (1856): 183.

7. *The Irish Rendezvouz*, 1689: I am grateful to the Librarian, Beinecke Library, Yale University, New Haven, Connecticut, for a xerox of this rare work.

8. Farewell, *Irish Hudibras*, 1689.

Act VIII: Enter the Reverend Dr Threlkeld

1. Lhuyd, *Philosophical Transactions* 27 (1712): 524.

2. Steele, *The Spectator*, 1712 (no. 455), quoted by Macalister, *Journ. Royal Soc. Antiq. Ireland* 56 (1926): 126

3. Swift, Journal to Stella for 17 March 1712-13: quoted by Colgan, The shamrock in literature, 1896 (fn III 2).

4. Nelson, *Journ. Soc. Bibliography Nat. Hist.* 9 (1979): 257-273.

5. Threlkeld: *Synopsis Stirpium Hibernicarum*, 1726 p. [60].

6. Taylor, *All his workes*, 1630.

7. quoted by Flood, *The Month* 137 (1921): 541-545.

Act IX: Enter the Story-Tellers

1. quoted by Colgan, The shamrock in literature, 1896 (fn III 2).
2. Healy, *Life and writings of St Patrick*, 1905.
3. Moffet, *Irish Hudibras*, 1755.

Act X: Enter the Decorated Courtiers

1. *Journ. Kilkenny Arch. Soc. 1* (1856): 183.
2. quoted by Flood, *The Month 137* (1921): 541-545.
3. Alter, *Studia Hibernica 14* (1974): 104-123; Ó Cúiv, *Studia Hibernica*: 107-119; Hayes-McCoy, Saint Patrick, a national souvenir, 1961.
4. Jones, *Coat of Arms* (1960): 141-144 and (1961): 268-273.
5. Sheehy, *The rediscovery of Ireland's part*, 1980; O'Connor, *Bull. Irish Georg. Soc. 26* (1983): 14-15.
6. Galloway, *The most illustrious order of St Patrick*, 1983.

Act XI: Enter the Patriots

1. Hayes-McCoy, *Irish flags*, 1979.
2. quoted in Croker, *Popular songs*, 1839.
3. MacLeod, *Irish Volunteer glass*, n.d.; Boydell, *Country Life 155* (1974): 1280-1281.
4. Gordon, *History of the rebellion*, 1803.
5. E. C. Nelson and A. Probert, *Thomas Coulter of Dundalk*. (Boethius Press, 1991).
6. see Hayes-McCoy, *Irish flags*, 1979.
7. Madden, *Literary remains of the United Irishmen*, 1887.
8. O'Connor, *A book of Ireland*, 1959.
9. Zimmermann, *Songs of Irish rebellion*, 1967.
10. Boucicault, *Arrah-na-pogue*, 1864.
11. Hasler, *The royal arms*, 1980.
12. Zimmermann, *Songs of Irish rebellion*, 1967.

Act XII: Interlude for a Monopolylogue

1. Crocker, *Popular songs*, 1839.
2. *ibid.*
3. Ardagh, *Irish Book Lover 21* (1933): 37-40.

Act XIII: Enter King George IV

1. Anon., *The Royal Visit*, 1821.
2. *Limerick Chronicle*, 29 August 1821: Mary Davies pers. comm.
3. M. Craig, *Dublin 1660-1860*. 1969. Dublin. pp. 307-308.

4. Menkes, *The royal jewels*, 1988.

5. Sheehy, *The rediscovery of Ireland's part*, 1980.

6. in Anon., *The Royal Visit*, 1821.

7. for the later version, see Fallon, *Sketches of Erinensis*, 1979.

8. Dickson & Walker, *Glasgow Nat.* 20 (1981); Walker, *The thistles of Scotland*, 1983.

9. Anon., *The Royal Visit*, 1821.

10. *Illustrated London News* (11 August 1849)—see engraving p. 104.

Act XIV: Enter the Shamrock Hunters

1. Beaufort, *Trans. Royal Irish Acad.* 15 (1828): 104-105.

2. Bicheno, *Journ. Royal Inst. Gt Britain* 1 (1830): 453-458.

3. quoted in Crocker, *Popular songs*, 1839.

4. G. L. Robins, *in litt.* 1 March 1988.

Act XV: Enter the Shamrock Makers

1. Owens, *Gard. Chron.* 25 (ser. 2) (1886): 767.

2. H.B., *Phytologist* 1 (1856): 366.

3. Hind, *Phytologist* 1 (1857): 519

4. Porter, *Ulster Journ. Arch.* 5 (1857): 11-16.

5. Parker, *Phytologist* 4 (1860): 319. Carvings of this nature and other examples of ecclesiastical trefoils, having no connection with Ireland, should not be interpreted as shamrocks: such facile equations, alas, are still being made (see e.g. Carville, *Norman splendour*, 1979.)

   Despite an otherwise careful analysis, Hopkin (*The living legend of St Patrick* (1988): 110) falls into this epithetic trap, misnaming trefoils decorating sixteenth century lintels as shamrocks, yet she suggests, correctly I believe, they are not shamrocks. I have not seen the lintels Hopkin noted; they are in Kinsale Museum and one is dated 1574.

6. T., *Phytologist* 5 (1861): 94-95.

7. Ferguson, *Phytologist* 6 (1862): 30.

8. Epsilon, *Phytologist* 6 (1862): 254-255.

9. Britten & Holland, *Dictionary*, 1878.

10. Britten, *The Month* 137 (1921): 193-205.

11. Colgan, *Irish Nat.* 1 (1892): 95-97.

12. *Journ. Co. Kildare Arch. Soc.* 5(6) 441-443.

13. Praeger, *Irish Nat.* 1 (1892): 125.

14. Colgan, *Irish Nat.* 2 (1893): 207-211.

Act XVI: Enter Artisans and Artists

1. William Garner, pers. comm.

2. Colm Gallagher, pers. comm.

3. Dolley, *Numismatic Soc. Ireland Occ. Paper 10* (1970).

4. Seaby, *Coins and tokens of Ireland*, 1970.

5. Colm Gallagher, pers. comm. (Original mss. in archive of the Religious Society of Friends, Dublin (Pym Papers, LL 134, LL 135).)

6. Sheehy, *The rediscovery of Ireland's past*, 1980.

7. Reynolds, *Some Irish fashions and fabrics*, n.d.

8. McCrum, *Irish Arts Review 2* (1985): 18-21. Irons, *Irish Arts Review 4(2)* (1987): 54-63.

9. see e.g. Sheehy, *The rediscovery of Ireland's past*, 1980; *Irish Arts Review 1* (1984): 50.

10. Craig, *Irish bookbindings*, 1979.

11. Turpin, *Irish Arts Review 2* (1985): 23-27.

12. Moore, *Irish melodies*, 1845.

Act XVII: Soldiers, Exeunt

1. Buckingham & Chandos, 'St Patrick's Day', 1900.

2. Kipling, 'The wearing of the green', 1900.

3. Hayes-McCoy, *Irish flags*, 1979.

4. Londonderry, *Mount Stewart*, n.d.; Bowe & George, *The gardens of Ireland*, 1986; Thomas, *Gardens of the National Trust*, 1979.

Act XVIII: Take a bow, Mr Hartland

1. Britten, *The Garden 85* (1921): 139-140; Colgan, *Irish Nat. 2* (1893): 207-211.

2. P.N.R., *Hardwicke's Science Gossip* (1869): 167.

3. Sawvel, *Education 15* (1894): 140-148; Phillips, *Gard. Chron.* (1895): 272.

4. Colgan, *Irish Nat. 2* (1893): 207-211.

5. Hartland, *Gard. Chron. 16* (series 3) (1894): 759-760; Hartland, *Daffodils and rare tulips, Autumn 1898. Spring 1898* (1898); Morris, *Moorea 4* (1985): 27-41.

6. Britten, *The Month 137* (1921): 193-205.

7. Painting the lily, *Irish Times* 14 July 1925.

8. Carmichael, *Carmina Gadelica*, 1928.

   The *pisrog* was brought to my notice by Roy Vickery (*in litt.* 29 October 1990): he heard it from Mrs Mary McGovern of Ballaghadereen, Co. Roscommon (*in litt.* 10 October 1984). She brought a four-leaved clover in her prayer-book to church, and 'a lady sitting near me whispered "Get rid of those quickly—they bring bad luck".'

9. P. B. O'Kelly, *A complete list of the rare perennial plants...of the Burren... Ballyvaughan*; E. C. Nelson, 'A gem of the first water': P. B. O'Kelly of The Burren; *Kew Magazine 7* (1990): 31-47.

10. Morris, *The life of St Patrick*, 1878 (2nd edition): 168-172; Britten, *The Month 137* (1921): 193-205; O'Byrne, *Irish Month 66* (1938): 154-159.

Act XIX: Enter and Exit the Antiquarians

1. Frazer, *Journ. Proc. Royal Soc. Antiq. Ireland* 4 (1894): 132-135.
2. Frazer, *Journ. Proc. Royal Soc. Antiq. Ireland* 4 (1894): 133-138 (and see fn. XV 5).
3. Bauerreiss, *Seanchas Ardmhacha* 4 (1962): 92-94.
4. see e.g. M. de Paor, The relics of St Patrick, *Seanchas Ardmhcha* 4 (1962): 87-91.
5. G. Mesmer, The cult of St Patrick in the vicinity of Drackenstein, *Seanchas Ardmhcha* 4 (1962): 68-74.
6. Flood, *The Month* 137 (1921): 541-545.

Act XX: Enter William Butler Yeats

1. Colm Gallagher, pers. comm.
2. McCauley, *Coinage of Saorstát Éireann 1928*, 1928.
3. Forrestall, *Irish Ecclesiastical Record 36 (series 5)* (1930): 63-74.
4. Feldman, *Stamps of Ireland*, 1976.
5. Hayes-McCoy, *Irish flags*, 1976.
6. Fr P. Gaynor, *The faith and morals of Sin Fein*, [c.1917]; Fr Gaynor did argue this in the rather exaggerated language of his pamphlet, exclaiming that Queen Victoria had destroyed the national symbolism of shamrock by making her soldiers wear it.
7. For assistance with the history of the constabulary crests, I am grateful to R. Sinclair, Curator, Royal Ulster Constabulary Museum; see *Arresting moments*, Belfast (1987).
8. P. Ryan (Roinn an Taoisigh), pers. comm.

Act XXI: Vox Hiberniae

1. Nelson, *Linen Hall Review 6(1)* (1989): 12-13.
2. A fuller report of this study was published in *Ulster Folklife* 37 (1991): 32-42: this was originally submitted to The Folklore Society of Ireland for possible publication in its journal *Béaloideas*, but the society's editorial board responded stating that the article '…is not deemed suitable for publication in this journal as its main focus is not folkloristic' (P. Ó Héalai, *in litt.* 13 October 1988). This pithy contribution to modern shamrockery is risible, especially for the board's uniquely eccentric interpretation of 'folklore' which one (albeit English) dictionary defines as 'the beliefs, legends and customs current among the common people…'.

    'What on earth is not 'folkloristic' about our benighted shamrock?' (E. C. Nelson, *in litt.* 21 October 1988)!

Epilogue

1. Britten, *The Month* 137 (1921): 199—I have been unable to trace the original document.
2. An Irish Lady, a poem specially written for Anne Pratt, *Flowering plants of Great Britain*, [1855] (see vol. 2, p. 103).

APPENDIX I

Survey of shamrock specimens

Table 1: The proportion of the various species gathered as shamrock during March and April 1988.

	Botanical name	Percentage
1.	*Trifolium dubium* Sibth.	46·4
2.	*Trifolium repens* L.	35·0
3.	*Trifolium pratense* L.	4·0
4.	*Medicago lupulina* L.	7.3
5.	*Oxalis acetosella* L.	4·5
	Others*	4·5

Total number of plants surveyed (dead plants excluded) 221.

* These included:
   1 plant of *Trifolium fragiferum* L. (strawberry clover)
   3 plants of the weedy *Oxalis corniculata* L.
   4 plants of *Trifolium repens* 'Purpureus' (the purple-leaved cultivar of white clover
unidentified garden forms of exotic *Oxalis* spp.

Table 2: Comparison of 1893 and 1988 surveys.

	Botanical name	Percentage (1893)	(1988)
1.	*Trifolium dubium*	51	46
2.	*Trifolium repens*	34	35
3.	*Trifolium pratense*	6	4
4.	*Medicago lupulina*	6	7
5.	*Oxalis acetosella*	–	3

(N.B. in 1893 only 35 plants were received.)

Table 3

Comitial distribution of species gathered as shamrock in 1988. Species names are abbreviated to numbers given in Table 1 and miscellaneous plants are omitted. Column figures represent the total number of plants received (including those which died before flowering).

County [total number of plants received]	Species 1	2	3	4	5
Antrim [6]	1	4			
Armagh [4]	2	1			
Carlow [6]	3	1		1	
Cavan [3]	1	2			
Clare [9]	3	2	1	1	
Cork [16]	12	2			
Donegal [10]	2	5		1	1
Down [9]	3	2			2
Dublin [25]	13	7		1	1
Fermanagh [2]		1			
Galway [17]	3	11	1	1	1
Kerry [10]	5	1		2	
Kildare [2]	2				
Kilkenny [4]	2	1		1	
Laois [3]	1	2			
Leitrim [4]	1	2			
Limerick [8]	5	1			
Londonderry [3]	1	1			
Longford [3]		2			1
Louth [5]	3	1			
Mayo [10]	4	4	1		
Meath [4]		2		1	
Monaghan [0]					
Offaly [9]	2	2	1	2	
Roscommon [4]		3		1	
Sligo [10]		5	4	1	
Tipperary [23]	14	3		2	
Tyrone [1]		1			
Waterford [7]	6	1			
Westmeath [6]	2	4			
Wexford [13]	8	2	1		
Wicklow [7]	4	2			
Totals [243]	103	78	9	16	6

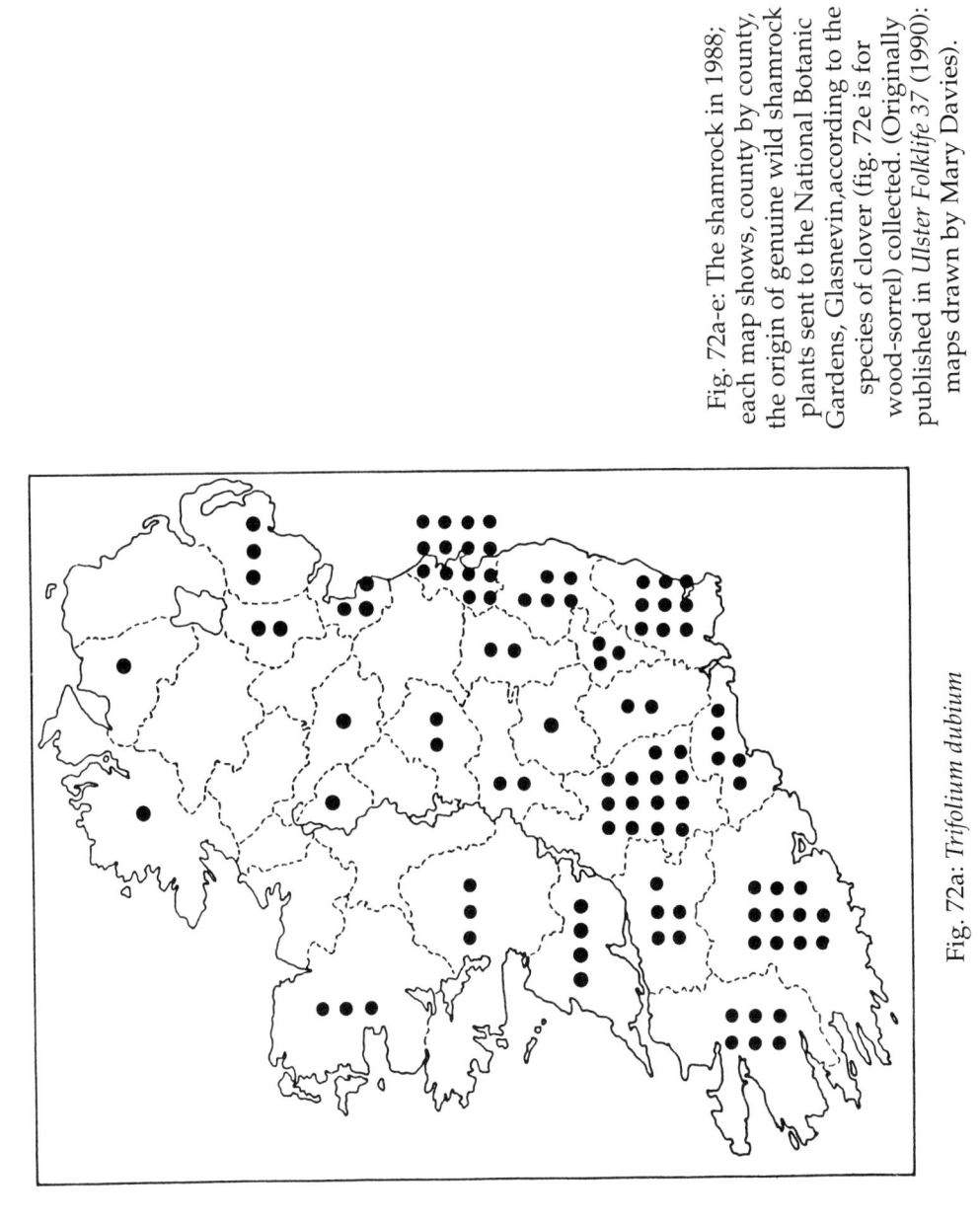

Fig. 72a: *Trifolium dubium*

Fig. 72a-e: The shamrock in 1988; each map shows, county by county, the origin of genuine wild shamrock plants sent to the National Botanic Gardens, Glasnevin, according to the species of clover (fig. 72e is for wood-sorrel) collected. (Originally published in *Ulster Folklife* 37 (1990): maps drawn by Mary Davies).

Fig. 72b: *Trifolium repens*

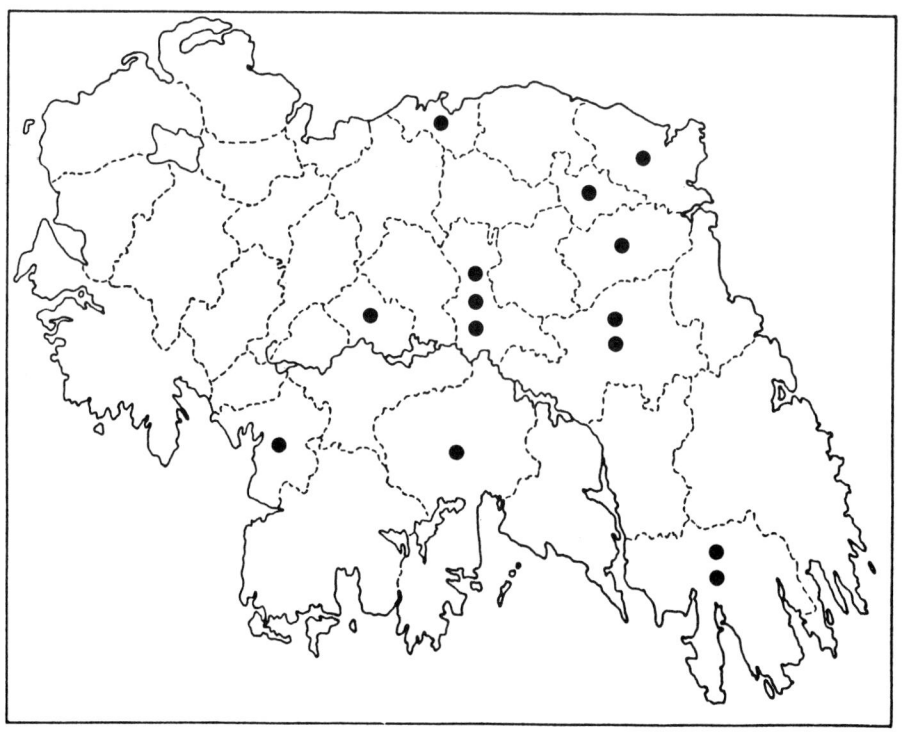

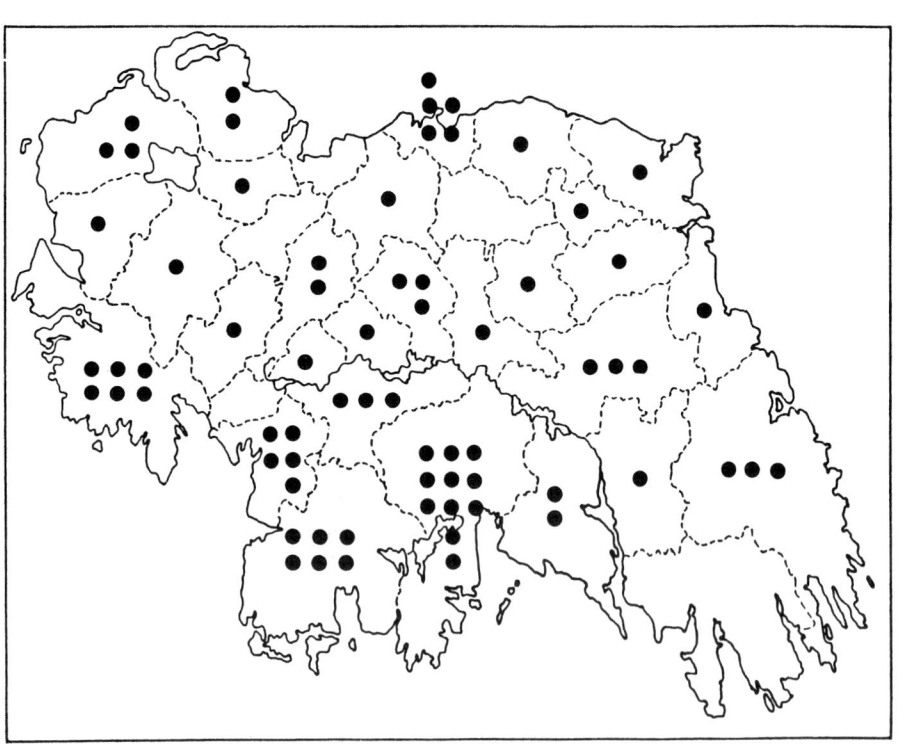

Fig. 72c: *Trifolium pratense*

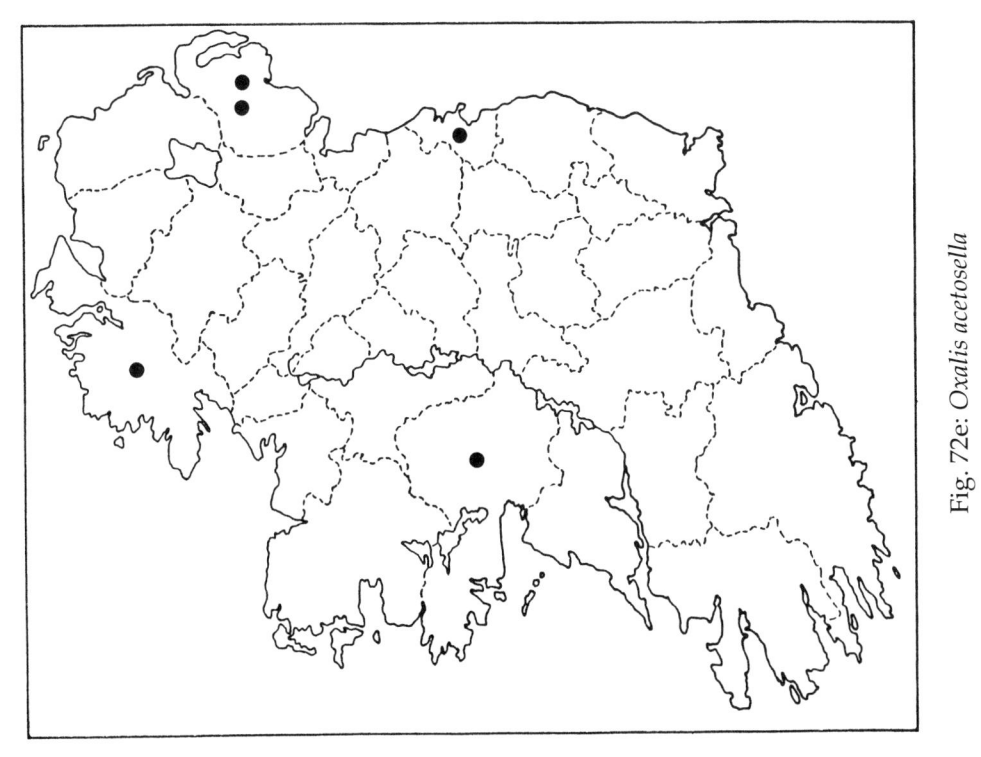

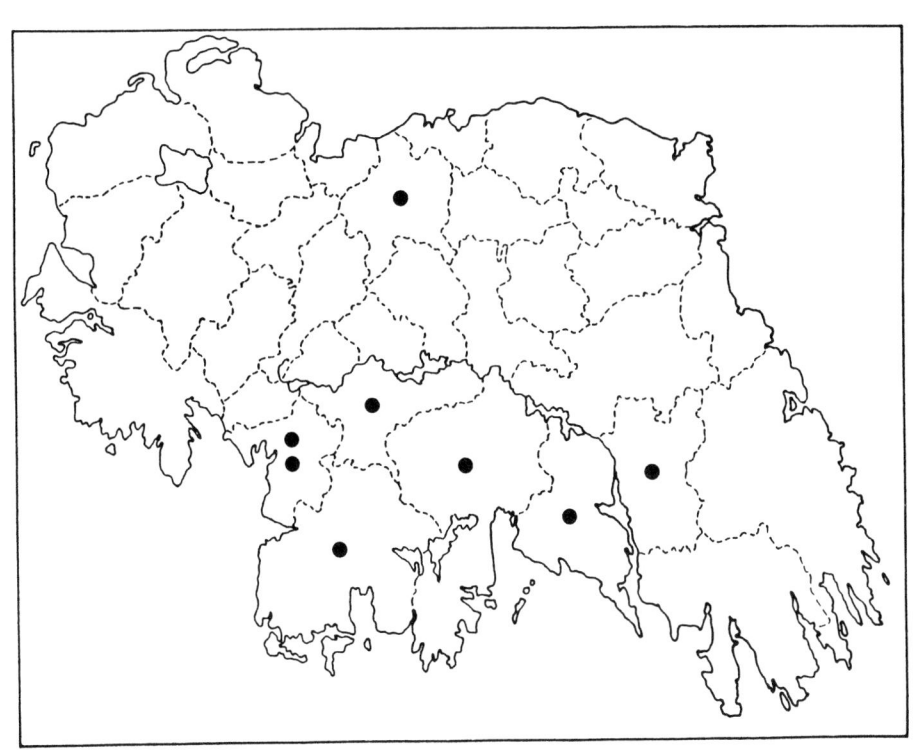

Fig. 72d: *Medicago lupulina*

Fig. 72e: *Oxalis acetosella*

APPENDIX II

Chronology

Chronology of use of the word *seamrog/shamrock*, and of published references in Irish and English, 1300-1812.

c. 1300		*Metrical Dindseanchas*
c. 1339		*An Leabhar Liagen*
c. 1410		*An Leabhar Breac*
[1570		M. L'Obel & P. Pena]
1571	[publ. 1577]	E. Campion
1576/7		Sir H. Sydney
1577	[publ. 1587]	R. Stanihurst
1578	[publ. 1581]	J. Derricke
1587		R. Holinshed
c. 1595	[publ. 1633]	E. Spenser
1596	[publ. 1605]	*Captain Thomas Stukeley*
1597		John Gerard
1605	[publ. 1630]	T. Decker
1606	[publ. 1607]	E. Sharpham
1607-09		T. Ó Cianáin
1610		W. Camden
1611		J. Speed
1613		G. Wither
1617		F. Moryson
1623		*The Welsh Emnbassador*
c. 1625	[publ. 1960]	P. Ó Súilleabháin Béarra
1630		J. Taylor
1632		P. Holland
1633		E. Spencer *in* J. Ware
1636		J. Shirley
1638	[publ. 1739]	Earl of Stafford
1643		J. Taylor
1654		J. Ware
1662		R. Pluincéad

1680		H. Mundy
1681	[publ. 1856]	T. Dinely
1682	[publ. 1786]	H. Piers
1686		J. Ray
1689		*The Irish Rendezvouz*
1689		J. Farewell
1707		E. Lhuyd
1712		E. Lhuyd
		R. Steele
1724		J. Farewell
1726		Caleb Threlkeld
1735		J. Keogh
1737		C. Linnaeus
1739		T Ó Neachtain
1741		*Gentleman's Magazine*
1745		C. Linnaeus
1755		W. Moffet
1768		J. O'Brien
1770		R. Griffith
1772		J. Rutty
		S. Whyte
1775	[publ. 1788]	R. B. Sheridan
1777		J. Lightfoot
1780		H. Pilon
		W. Shaw
1781		C. Johnston
1794		W. Wade
		E. Jones
1798		C. Jackson
1799		*The Shamroc*
1800		J. Buckeridge
1803		J. Gordon
1806		A. Cherry
1808		W. Wade
		T. Crosby
1812		T. Moore

INDEX

Adomnan, 12
Antrim (county), 9,8 ,87, 154
Atrim Militia, 62
Antrim, Earl of, 33
Aran Islands (County Galway), 40, 88
Argyll, Marquis of, 33
Armagh (county), 87, 103, 154
armorial bearings, 63-64, 100-102
Australia, 81

Baily, Katherine, 82
Banks, Joseph (Sir), 95
Beaufort, Louisa Catherine, 77, 83
Bede, The Venerable, 12
Beisley, Harriet (Mrs), 82
Belfast Botanic Garden, 83
Belleek Parian ware, 107-108
Bicheno, James Ebenezer, 76, 78-79, 81, 126
Bieler, Ludwig (Dr), 47
biolar—see also *Nasturtium officinale*—20
biror—see also *Nasturtium officinale*—20
Birr (County Offaly), 103
black medick—see *Medicago lupulina*
black shamrock, 145
Bodkin, Thomas (Dr), 134
Boer War, 117-118
Book of Kells, 132-133
Boucicault, Dion, 61
Boyne, River, 141
Britten, James Edward, 84-87, 89, 130, 133, 145-146
Buckingham and Chandos, Duchess of, 115
Burbank, Luther, 127
Burbidge, Frederick W., 121
Burren (County Clare), 35
Bury, John, 47

Camden, William, 27-28, 34
Campion, Edmond, 18-25, 27-28, 30, 48, 133
Carleton, William, 113
Carlow (County), 85, 87, 154
Carroll, Lewis, 30
Cashel, Rock of (County Tipperary), 47, 65
Castleknock Light Dragoons, 56-57
Cavan (county), 154
Celtic Fellowship of New York, 126-127
Charlemont, Earl of, 51
Cherry, Andrew, 66
chickweed worn as shamrock, 79
Clara (County Offaly), 103
Clare (county), 34, 87, 129, 144, 154
clover—see also *Trifolium*—42, 145-146
    as food of Irish, 23
    four-leaved, 128-129
    red—see *Trifolium Pratense*
    white—see *Trifolium repens*
    yellow—see *Trifolium dubium*

Cochlearia (scurvy grass), 32
coins and coinage, 95, 98-99, 130-131, 134-135
coins: Patrick coin, 131-133
Colgan, Nathaniel, 6-7, 18, 20-21, 26, 30, 34, 38, 44, 86-90, 122, 130, 139, 140
Connaught, 87, 146
Connellan, 17
Cork (county), 17, 30, 32, 55, 85, 87, 118, 122, 142, 154
Coulter family, 58
Croker, Thomas Crofton, 66, 79
crosses: St Patrick's, 35-36, 40, 49-51
Curran, Sarah, 108

daffodil, 62-64, 136
Dekker, Thomas, 22
Derg, Lough (County Donegal), 49
Derricke, John, 20-21
Derry (county), 87, 144, 154
Dineley, Thomas, 36, 49
Donegal (county), 154
Down (county), 86, 88, 100-101, [140], 141, [142], 144, 154
Drogheda (County Meath), 18
Dublin (county), 85, 89, 90, 105-106, 139, 143, 154
dúmheidic, 140
Dundalk (County Louth), 21

Egan, John (94-95
Elizabeth II, 73, 99, 136
emblems—see under daffodil, flax, harp, leek, rose, thistle
Emmet, Robert, 59
Erne, Lough (County Fermanagh), 38

Farewell, James, 38-39
Fenian Movement, 117, 136
Ferdomnach, 12
Ferguson, William Jackson Hooker, 83
Fermanagh (county), 55-56, 85, 141, 154
flax, as emblem for Northern Ireland, 136-137
Fleurs de Saint Patrice, Les, 129—see also *Prunus spinosa*
Flood, W. H. Grattan, 44-45, 133
Forde family, 100-101
Forrestall, James (Revd), 134-135
Frazer, W. (Dr), 130, 131

Galway (county), 15, 40, 91, 105, 154
George III, 51, 71, 95-96, 99
George IV, 71-74, 95, 101
Gerard, John, 24-26, 29, 42
Gildas, 12
Glasnevin, National Botanic Gardens, 121, 132, 139, 155
glass decorated with shamrocks, 57, 109

INDEX

Glin Castle (County Limerick), 90-91
Glin, Knight of, 55, 90-91
green, as national colour, 58-62, 65, 116

harp, 55, 58-59, 94-95, 98, 106, 109-111, 135, 138
Hartland, William Baylor, 117, 124-125, 128
Healy, John (Most Revd Dr), 47
Hind, William A. (Revd), 83
Holden, S., 20-21, 27
Holinshed, Raphael, 20-21, 27
Holland, Philomen, 27
Holland, Robert, 85

Irish Guards, 116
Irish Rendezvouz, The, 37-38

Jocelin of Furness, 49

Kennedy, Patrick (Pursuivant of Athlone), 100
Keogh, John (Revd), 80, 82, 130
Kerry (county), 15, 30, 85, 105, 154
Kew, Royal Botanic Gardens, 64, 86, 95, 121, 145
Kildare (county), 35, 89, 104, 154
Kildare Street Club (Dublin), 103
Kilkenny (county), 140, 142, 154
Killarney (County Kerry), 107, 121
Kilmacduagh (County Galway), 105
Kipling, Rudyard, 116

Land League, 117
Laois (county), 87, 154
leek, 50-51, 62-64, 136-137, 145
Leitrim (county), 144, 154
Lhuyd, Edward, 40-41
Limerick (county), 85, 90-91, 141, 154
Linné, Carl, 34
Linnean Society, London, 79
Londonderry (county), 87, 144, 154
Londonderry, Edith, Lady, 120
Longford (county), 154
Louth (county), 122, [141], 142, 154

Mackay, James Townshend, 82
Maclise, Daniel, ii, 15, 67-70, 113-114
MacLysaght, Edward (Dr), 33
MacNally, Leonard, 101
Maguires-Bridge Volunteers, 55-56
Malachy (Bishop of Down), 49
Malton, James, 95
Marcus Ward & Co. (Belfast), 113
Marsilea, 129
Mayo (county), 9, 87, 142, 146, 154
Meath (county), 15, 85, 141, 154
Medicago lupulina, 79, 86, 121, 140, 153, 157
Moffet, William, 48
Monaghan (county), 139, 154
Moore (David) (Dr), 82
Moore, Thomas, ii, 3, 67-70, 95, 110, 113-114
More, Alexander Goodman, 82
Moryson, Fynes, 29-30, 78, 83, 130

Mount Stewart (County Down), 120
Mountain, Rosoman (Mrs), 65-66, 113
Muirchu, 12-13, 47-49
Mundy, Henry (Dr), 34-35

Nasturtium officinale, 20, 29, 32
New York, 45, 127, 136
Newgrange (County Meath), 132

Northern Ireland, 137-138

Ó Briain, 17
Ó Cianain, Tagdh, 16-17
O'Connell, Daniel, 105, 117
O'Connor, Frank, 60
Ó Cuindlis, Murchadh, 15-16, 18
O'Kelly, Patrick B., 129, 144
O'Leary, James (Revd), 46
Ó Raghallaigh, 17
Ó Súilleabháin Béarra, Pilib, 17
Obel, Matthias de l', 23
Offaly (county), 103, 154
Oranmore (County Galway), 91
Order of Saint Patrick, Most Noble, 52-53, 60, 73, 90, 118
Owens, C. M. (Miss), 82
Oxalis acetosella, [17], 40, 78-83, 86, 126, 128, 140, 153, 157
Oxalis corniculata, 153

Parker, Charles Eyre, 83
Parochetus communis, 127
Patrick, Saint: banishes snakes, 13
    biographies, 12, 47
    origins and life, 8-11
    sermon on Holy Trinity, 3, 44-47, 62, 77, 83
Patrick's crosses, 35-36, 40, 49-51
    tears, 129, 144
    thorn, 129
Pena, Pierre, 23-25
Piers, Henry (Sir), 34-35
Pluincéad, Risteard, 17
Porter, Thomas H. (Dr), 83
postage stamps, 13, 67, 73, 135-137
postcards, 5-6, 110, 113, 136
potato, 25, 39
Praeger, Robert Lloyd (Dr), 88
Prunus spinosa, 129
Pym, Susan (Miss), 98-99

Red Hand of Ulster, 117-120, 136-137
Reynolds, Joshua (Sir), 51
Robinson, Tancred (Dr), 40
Rome, 104-105
Roscommon (county), 87, 154
rose, 51, 62-64, 102, 108, 136, 137, 145
Royal Glin Hussars, 54-55, 90
Royal Irish Academy, 77, 100-101
Royal Ulster Constabulary, 137-138
Rutty, John (Dr), 82

INDEX

Saint Patrick's crosses, 35-36, 40, 49-51
Saint Patrick's tears, 129, 144
Saint Patrick's thorn, 129
Sainthill, Richard, 98-99
Scott, Walter (Sir), 74
Seaforde House (County Down), 101
seamair, 14-17, 145
seamair bhan, 93, 140
seamair bhuí, 139, 147
seamair dhearg, 92, 140
seamair óg, 3, 14
seamhair, 14
seamrach, 14, 16
seamrag nam buadh, 128-129
seamróg, 3, 14-17, 18, 26, 40, 42, 48, 145
seamróg na gceithre gcluas, 128
seamsóg, 17, 79, 96, 140
seamur, 17
Shakespeare, William, 21, 50
Shamroc, The, 60-61
shamrock: as food, 33-34, 48, 145
    black, 144
    decorating books, 110-114
    decorating churches, 83, 103-105, 132
    decorating clothes and jewellery, 106, 108
    decorating graves and monuments, 104-105, 132
    drowning of, 44, 49, 144
    etymology and meaning, 3, 14-17, 24, 42
    four-leaved, 128-129, 145
    growing outside Ireland, 143-144
    in flower, 122, 142
    selling to Americans, 124-128
    substitutes for, 127-128
    worn by regiments and soldiers, 116-117
    worn on St Patrick's Day, 36, 38, 42-44
Shamrogueshire, 39
Sharpham, Edward, 22
Shirley, James, 21, 36
Sidney, Henry (Sir), 44
Sliabh Mis (Slemish) (County Antrim), 9
Sligo (county), 140, 142, 154
sorrel, wood—see *Oxalis acetosella*
Speed, John, 28
Spenser, Edmund, 30-32, 38, 78, 130
Stafford, Earl of, 33
Stanihurst, James, 18
Stanihurst, Richard, 18, 32, 48
Steele, Richard (Dr), 40
Stokes, Whitley (Dr), 73
Swift, Jonathan (Revd Dr), 40

Tara (County Meath), 45, 47, 65, 105, 130, 133, 141
Taylor, John (The Water Poet), 30, 44
The Observer, 64
The Times, 63
thistle, 40, 51, 62-64, 74, 108, 136, 137, 145
thorn, St Patrick's, 129

Threlkeld, Caleb (Revd Dr), 41-45, 50, 80, 82, 130, 132
Tipperary (county), 104, 121, 122, 139, [144], 154
Tirechan, 13
tokens (bank and railway), 95, 134
Tone, Wolfe, 59
trefoil misinterpreted as shamrock, 130-133, 145, 150
Trifolium, 17
Trifolium dubium, 85-89, 121, 124, 130, 139, 146, 153, 155
Trifolium minus—see *T. dubium*
Trifolium pratense, 24, 86, 89, 92, 140, 153, 156
Trifolium repens, 24, 42, 79, 86, 88-89, 93, 140, 146, 153, 156
Trifolium repens 'Purpureum', 129, 144, 153
Trinity College, Dublin, 73, 102, 121
Trinity legend, 42-45, 47
Turner, Richard, 75
Tyrconnell, Earl of, 36
Tyrone (county), 154
Tyrone Volunteers, 62

Ulster Unionist Convention, 117-118
Ultan of Ardbraccan, 13
United Irishmen, 55, 58-60, 73, 81, 90, 100
United Kingdom; Royal arms of, 63-64, 137
United States of America, 124, 128
University College, Dublin, 101

Victoria, 73-75, 116-117, 136, 138, 152
Volunteer Movement, 54-58, 90

Wade, Walter (Professor), 82
Walsh, Paul (Revd), 17
Ware, James (Sir), 32
watercress—see *Nasturtium officinale*
Waterford (county), 56, 85, [141], 142, [144], 154
Westmeath (county), 34, 85, 142, 154
Wexford (county), 85, 87, [140], 154
Wicklow (county), 45, 65, 85, 87, 154
Windebank, Francis (Sir), 33
Wither, George, 28, 78, 130
wood sorrel—see *Oxalis acetosella*

Yeats, William Butler, 134
Young Citizen Volunteers, Belfast, 119-120

Zwehnsdorf (bookbinders), 1, 113

LANDSCAPE
ARCHITECTURE NOW!

IMPRINT	**PROJECT MANAGEMENT** Florian Kobler, Cologne **COLLABORATION** Harriet Graham, Turin Inga Hallsson, Cologne **PRODUCTION** Ute Wachendorf, Cologne	**DESIGN** Sense/Net Art Direction Andy Disl and Birgit Eichwede, Cologne www.sense-net.de **FINAL ARTWORK** Tanja da Silva, Cologne	**GERMAN TRANSLATION** Laila Neubert-Mader, Ettlingen **FRENCH TRANSLATION** Jacques Bosser, Paris **© VG BILD-KUNST** Bonn 2012, for the works of Catherine Mosbach and Jean Nouvel	**PRINTED IN ITALY** ISBN 978–3–8365–3676–9 © 2012 TASCHEN GMBH Hohenzollernring 53 D–50672 Cologne www.taschen.com	

LANDSCAPE
ARCHITECTURE NOW!

LANDSCHAFTS-*Architektur heute!*
PAYSAGES *contemporains!*
Philip Jodidio

TASCHEN

CONTENTS

6	INTRODUCTION	Einleitung/Introduction
50	AECOM	Pier Head and Canal Link, Liverpool, UK
		Civic Space Park, Phoenix, Arizona, USA
60	TADAO ANDO	Shiba Ryotaro Memorial Museum, Higashiosaka, Osaka, Japan
		Lee Ufan Museum, Naoshima, Kagawa, Japan
70	BALMORI ASSOCIATES	The Garden That Climbs the Stairs, Bilbao, Spain
76	PATRICK BLANC	Orchid Waltz, National Theater Concert Hall, Taipei, Taiwan
		Max Juvénal Bridge, Aix-en-Provence, France
82	M. CAFFARENA / V. COBOS / G. ALCARAZ / G. DELGADO	Centenary Park, Punta San García, Algeciras, Spain
88	JOÃO LUÍS CARRILHO DA GRAÇA	Archeological Museum of Praça Nova do Castelo de São Jorge, Costa do Castelo, Portugal
		Pedestrian Bridge, Carpinteira River, Covilhã, Portugal
98	RODRIGO CERVIÑO LOPEZ	Adriana Varejão Gallery, Inhotim Contemporary Art Center, Brumadinho, Brazil
104	DILLER SCOFIDIO + RENFRO	Hypar Pavilion Lawn, New York, New York, USA
108	VLADIMIR DJUROVIC	Hariri Memorial Garden, Beirut, Lebanon
		Salame Residence, Faqra, Lebanon
118	PETER EISENMAN	City of Culture of Galicia, Santiago de Compostela, Spain
124	GLAVOVIC STUDIO	Young Circle ArtsPark, Hollywood, Florida, USA
130	GUSTAFSON GUTHRIE NICHOL	Lurie Garden, Chicago, Illinois, USA
		Robert and Arlene Kogod Courtyard, Washington, D.C., USA
140	GUSTAFSON PORTER	Old Market Square, Nottingham, UK
		Diana, Princess of Wales Memorial Fountain, Hyde Park, London, UK
150	ZAHA HADID	Eleftheria Square Redesign, Nicosia, Cyprus
154	HERZOG & DE MEURON	Plaza de España, Santa Cruz de Tenerife, Canary Islands, Spain
160	STEVEN HOLL	Vanke Center / Horizontal Skyscraper, Shenzhen, China
		HEART: Herning Museum of Contemporary Art, Herning, Denmark
172	MARTIN HURTADO	Izaro Estate, Casablanca Valley, Chile
		Morandé Winery Productive Services, Casablanca Valley, Chile
182	JAMES CORNER FIELD OPERATIONS / DILLER SCOFIDIO + RENFRO	The High Line, New York, New York, USA
190	RAYMOND JUNGLES	1111 Lincoln Road, Miami Beach, Florida, USA
		Coconut Grove, Florida Garden, Coconut Grove, Florida, USA
202	LEDERER + RAGNARSDÓTTIR + OEI / HELMUT HORNSTEIN	Georg-Büchner-Plaza, Darmstadt, Germany
208	MAYA LIN	Storm King Wavefield, Mountainville, New York, USA
		Eleven Minute Line, Wanås, Sweden

216	GIOVANNI MACIOCCO	Anglona Paleobotanic Park, North Sardinia, Italy
222	MICHAEL MALTZAN	Playa Vista Park and Bandshell, Los Angeles, USA
228	MECANOO	TU Campus, Mekel Park, Delft, The Netherlands
234	FERNANDO MENIS	Cuchillitos de Tristán Park, Santa Cruz de Tenerife, Canary Islands, Spain
242	EDUARDO DE MIGUEL	Cabecera Park, Valencia, Spain
248	MIRALLES TAGLIABUE EMBT	HafenCity Public Spaces, Hamburg, Germany
254	CATHERINE MOSBACH	L'autre rive (The Other Bank), Quebec, Canada
258	YOSHIAKI NAKAMURA	Miho Museum Gardens, Shigaraki, Shiga, Japan
264	NBGM LANDSCAPE ARCHITECTS	The Freedom Park, Salvokop, Tshwane, Pretoria, South Africa
270	VICTOR NEVES	Reorganization of the Riverside of Esposende, Esposende, Portugal
276	NIETO SOBEJANO	Plaza de Santa Bárbara, Madrid, Spain
282	OFFICINA DEL PAESAGGIO	Community Gardens, Chiasso, Switzerland
		YTL Residence Garden, Kuala Lumpur, Malaysia
290	OKRA	Domplein, Utrecht, The Netherlands
		Afrikaanderplein, Rotterdam, The Netherlands
296	PALERM & TABARES DE NAVA	García Sanabria Park, Santa Cruz de Tenerife, Canary Islands, Spain
302	JOHN PAWSON	Sackler Crossing, Royal Botanic Gardens, Kew, London, UK
308	RENZO PIANO	Renovation and Expansion of the California Academy of Sciences, San Francisco, California, USA
314	PLASMA STUDIO	"Flowing Gardens," Xi'an World Horticultural Fair 2011, Xi'an, China
326	SHUNMYO MASUNO	SanShin-Tei, One Kowloon Office Building, Hong Kong, China
332	KEN SMITH	Kids Rock, Children's Play Environment, Orange County Great Park, Irvine, California, USA
		40 Central Park South, New York, New York, USA
		Santa Fe Railyard Park and Plaza, Santa Fe, New Mexico, USA
346	TAYLOR CULLITY LETHLEAN	Royal Botanic Gardens, Cranbourne, Melbourne, Australia
352	LUIS VALLEJO	Santander Group City Campus, Boadilla del Monte, Madrid, Spain
362	MICHAEL VAN VALKENBURGH	Connecticut Water Treatment Facility, New Haven, Connecticut, USA
		Alumnae Valley Landscape Restoration, Wellesley College, Wellesley, Massachusetts, USA
		Teardrop Park, Battery Park City, New York, New York, USA
374	WEISS/MANFREDI	Olympic Sculpture Park, Seattle, Washington, USA
380	WEST 8	Governors Island Park and Public Space Master Plan, New York, New York, USA
		Madrid RIO, Madrid, Spain
		Toronto Central Waterfront, Toronto, Canada
398	WIRTZ INTERNATIONAL	Ernsting's Family Headquarters, Coesfeld-Lette, Germany
404	WORK ARCHITECTURE COMPANY	Public Farm 1, Long Island City, New York, USA
412	INDEX	
416	CREDITS	

INTRODUCTION

DESPERATELY SEEKING REDEMPTION
The idea of forming the landscape around the centers of power, be they palaces or great squares, is an ancient one, and yet the modern definition of landscape architecture dates in a formal sense from the 19th century. The work of André Le Nôtre (1613–1700), the garden designer of Louis XIV, with the great perspectives at Versailles and Vaux-le-Vicomte or the axis formed by the Tuileries Gardens and the Champs-Elysées in Paris, must surely be cited as an example of the modification of the landscape in relation to architecture. Where cities like Paris were at the time characterized by narrow, densely built streets, Le Nôtre gave meaning to space, and willfully associated it with the grandeur of the King. The garden perspectives and great basins that radiate from Versailles are expressions of power that amplify the architecture of the palace, and impose its symbolic presence at a great distance.

Architectural realizations that take into account their natural setting may not be considered landscape architecture *per se*, and yet it is the integration of the two elements (landscape and architecture) that makes the grandeur of many of the best-known historic realizations. Those who have stood on the remnants of the Great Wall of China near Beijing, built to protect the northern borders of the Empire beginning in the fifth century BCE, can fully appreciate how the Wall snakes through hills and valleys, forcibly shaped by its location and the natural setting. Also in China, Tien An Men Square and the Forbidden City are most astonishing for their use of space, space that is largely empty but which anchors mere buildings in a context of power that has been comprehensible to all for centuries.

BEAUTIFUL AND STRANGE
For the purposes of this book, focused on very recent realizations, landscape architecture is defined not only as the formation of gardens but also of buildings that have an intimate relation to nature, that in some sense spring from the earth and give meaning to space and materials. Perhaps it would be best to speak of the architecture of landscape despite accepted usage that often excludes buildings from the definition of landscape architecture. It is the Scotsman Gilbert Laing Meason (1769–1832) who invented the term landscape architecture. His book *The Landscape Architecture of the Great Painters of Italy* (1828) had to do with how buildings are placed in their sites to produce beautiful compositions. In this sense, the very origin of the term landscape architecture does, indeed, take into account buildings.

The relation of art to nature and an evolving view of the picturesque that animated much Romantic thought dates from before the time of Meason. Joseph Addison (1672–1719), the English author and politician, wrote in *The Spectator* (no. 414, June 25, 1712):

If we consider the works of nature and art, as they are qualified to entertain the imagination, we shall find the last very defective in comparison of the former; for though they may sometimes appear as beautiful or strange, they can have nothing in them of that vastness and immensity, which afford so great an entertainment to the mind of the beholder. The one may be as polite and delicate as the other, but can never show herself so august and magnificent in the design. There is something more bold and masterly in the rough

1
Balmori Associates, *The Garden That Climbs the Stairs*, Bilbao, Spain, 2009

careless strokes of nature than in the nice touches and embellishments of art. The beauties of the most stately garden or palace lie in a narrow compass, the imagination immediately runs them over, and requires something else to gratify her; but in the wide fields of nature the sight wanders up and down without confinement, and is fed with an infinite variety of images without any certain stint or number. For this reason we always find the poet in love with a country life, where nature appears in the greatest perfection, and furnishes out all those scenes that are most apt to delight the imagination.[1]

In point of fact, the dichotomy between the "natural" and the "artificial" is a key to appreciating or understanding landscape architecture in any form. Is the goal of the landscape architect, or of the architect, to somehow mimic nature, or rather to impose a new, human order on gardens or the environment of buildings. The roots of this issue run deep in history and literature, and may in a sense be linked to the primitive longing for a time before time, when all men lived in the garden. So much landscape architecture has to do with a search for an ideal garden.

To all delight of human sense exposed,
In narrow room, Nature's whole wealth, yea more,
A Heaven on Earth: For blissful Paradise
Of God the garden was, by him in the east
Of Eden planted.
John Milton, *Paradise Lost*, Chapter 4

The first person known to have used "landscape architect" as a professional title was the American Frederick Law Olmsted (1822–1903) in 1863. Together with his partner Calvert Vaux, Olmsted designed Central Park in Manhattan and numerous parks in the United States. Olmsted and Vaux won an 1857 competition to design the 280-hectare area designated between 59th and 106th Streets. Many visitors surely take Central Park for a vision of Manhattan's natural state before the construction of the city, and yet it was formed with about ten million cartloads of soil and rocks, and over four million planted trees and shrubs, opening in 1873. With its 36 bridges designed by Vaux, and various small structures that have evolved over time, Central Park stands witness to the origins of American landscape architecture, a space that few would imagine sacrificing to the encroachment of the modern city.

BREAKING BARRIERS

The degree of integration of actual architecture with landscape architecture varies according to circumstances, and in this volume, quite intentionally; there are some cases that may cross what some see as a barrier between disciplines. When do landscape and architecture merge, making it difficult to distinguish one from the other? Each building or park has its circumstances, but might not the ultimate goal of

2
Patrick Blanc, Max Juvénal Bridge,
Aix-en-Provence, France, 2008

landscape architecture be to break the (artificial) barrier that separates it from the domain of the builders of towers and bridges? In any case, no less a figure than Frank Lloyd Wright had a clear opinion about this debate:

> Change is the one immutable circumstance found in landscape. But the changes all speak or sing in unison of cosmic law, itself a nobler form of change. These cosmic laws are the physical laws of all man-built structures as well as the laws of the landscape. Man takes a positive hand in creation whenever he puts a building upon the earth beneath the sun. If he has a birthright at all, it must consist in this: that he, too, is no less a feature of the landscape than the rocks, trees, bears, or bees, that nature to which he owes his being. Continuously nature shows him the science of her remarkable economy of structure in mineral and vegetable constructions to go with the unspoiled character everywhere apparent in her forms.
> Frank Lloyd Wright, *The Future of Architecture*[2]

SMALL, BUT PERFECTLY FORMED

The question of just when architecture and landscape become landscape architecture is a rather complicated one that this book prefers to view in an inclusive way. A first approach might be to look at small urban projects where the designers (landscape architects or architects) bring nature into the city, creating a rapport where often concrete, glass, and steel had hidden any hint of things living and green. The Garden That Climbs the Stairs (Bilbao, Spain, 2009; page 72) covers an expanse of just 80 square meters. Its creator Diana Balmori refers to a "dynamic urban space" to describe what is a garden on a stairway situated near a number of quite prestigious works of contemporary architecture. She says: "In one gesture, it narrates a story of landscape taking over and expanding over the Public Space and Architecture, therefore transforming the way that the stairs and the space is perceived and read by the user." This short description of a small garden already does something to relieve the ambiguity that nature engenders in a fundamentally artificial urban environment. Landscape is seen here as having the power to "transform" its architectural environment, and in any case to bring a counterpoint to the hard materials of construction.

GROWING UP THE WALLS

The noted French botanist Patrick Blanc has created a number of his "vertical gardens" all over the world. Noting that plants really do not need soil, but rather nutrients and water, Blanc patented a method that allows him to design and create gardens that literally cover walls, whether inside or outside. Two of his recent projects published here—Orchid Waltz, National Theater Concert Hall (Taipei, Republic of China, 2009; page 78) and the Max Juvénal Bridge (Aix-en-Provence, France, 2008; page 80)—certainly alter the perception of their environments. In the second instance, a luxuriant wall of greenery softens the otherwise harsh lines of a modern bridge that local authorities had wanted to cover with brick or stone. Thoroughly unexpected, Blanc's gardens appear to be artificial if only because they are vertical, and yet they are very much alive. Patrick Blanc has a doctorate in botany and engages in highly technical studies concerning the survival of plants in extreme environments. He does not call himself a landscape architect, and yet he has the capacity to modify the very perception of a building or a neighborhood in a city. In some sense, he redefines the word "landscape" making use of the fundamental characteristics of nature to challenge the accepted idea of what the place of a garden should be vis-à-vis architecture.

Vladimir Djurovic is a Beirut-based landscape architect who is undertaking such ambitious projects as the gardens around the new Aga Khan Museum of Islamic Art in Toronto. In Beirut, Djurovic had already created Samir Kassir Square (2004), winner of a 2007 Aga Khan Architecture Award, which measures just 850 square meters, quite small for an urban park, and yet this realization brought an unexpected feeling of peace and relief from the bustle of the city to its neighborhood. His more recent Hariri Memorial Garden (ongoing; page 110), also in Beirut, is more than twice as large, but remains limited in size as compared to other elements of the urban environment. In this instance, using bands of stone, water, and glass that step down toward the city center, the landscape architect succeeds in creating what might be called a modest monumentality, appropriate to the veneration in which many Lebanese hold the assassinated former Prime Minister Rafic Hariri (1944–2005) who is commemorated here. Where the American sculptor Richard Serra has been quoted as saying "the difference between art and architecture is that architecture serves a purpose," something similar might be said about landscape designs. Pleasing to the eye, places of rest and congregation, and in the instance of the Hariri Memorial Garden, a place to pause and remember, landscape designs may not be entirely without purpose, but at their best, these two can be called art.

A FARM IN A PLAYGROUND

Working in the context of the courtyard of the PS1 Contemporary Art Center, which is the Long Island City branch of the Museum of Modern Art, the New York architects WORK AC created an installation they called Public Farm 1 in 2008 (page 406), occupying just over one thousand square meters. PS1 is a converted school and the courtyard used for this installation was once a playground. This "fully functioning urban farm in the form of a folded plane made of structural cardboard tubes" certainly dealt with the architecture that surrounded it, and assumed a kind of architectural presence in and of itself. The point here, though, is not quite the same as for the other small incursions of nature presented here thus far. The idea that a farm, a man-made construct usually existing only far from urban areas, could take root, even temporarily in the dense city environment of Queens is a statement about just what landscape means, and how it relates to architecture. Vineyards, rice paddies, or fields of corn form the landscape in many parts of the world without usually being the object of much attention from landscape architects. And yet the process of the domestication of nature, which is at the heart of the architect's profession, is shared by the farmers who align endless rows of crops. Nature is molded and directed to a purpose, but even the underlying suppositions of farming are questioned by an installation such as that imagined by WORK AC in 2008.

The Teardrop Park (Battery Park City, New York, New York, USA, 2003–05; page 370) by Michael Van Valkenburgh Associates covers about 7000 square meters and is intended primarily as a space for children. Great care was taken to make the space environmentally responsible and sustainable. The designers point out that children's areas in parks have gradually seen the presence of nature reduced in favor of equipment, and they have sought to reverse this trend. A central feature of this park is the 8.2-meter-high, 51-meter-long stacked blue stone Ice-Water Wall, certainly of a scale to qualify as an architectural element. Like architecture itself, landscape architecture often must serve purposes beyond purely aesthetic considerations. Dealing with practicality, as is the case in a children's space, might well be within the grasp of persons not trained as landscape architects. With a firm like that of Michael Van Valkenburgh, a higher level of conception is reached, associating a sure sense of space and aesthetics with knowledge of the environment, careful attention to the actual play areas, and considerations about the general trends of park design.

The work of the New York firm Diller Scofidio + Renfro is clearly at the forefront of contemporary architecture, with such realizations as the Institute of Contemporary Art in Boston (2006). More recently DS+R has been involved in the renovation of Lincoln Center, the cultural complex located on Manhattan's Upper West Side. Their Hypar Pavilion (2011; page 106), located in Lincoln Center's North Plaza, has a very unusual roof. Its slanted surface allows visitors to profit from a thousand-square-meter lawn, which could hardly be more in the center of the urban environment. Aesthetically stunning, with its clean, angled lines, the Hypar Lawn is precisely at the intersection of architecture and landscape architecture, taking both into account, producing a result which blends the two in a seamless manner. Nature (of a tame sort) takes its place in an otherwise mineral environment, to the great appreciation of passersby.

BLUE MARBLES AND BLACK RUBBER SHARDS
Ken Smith's garden at 40 Central Park South (New York, New York, USA, 1999–2006; page 336) measures just 580 square meters and makes use of mineral or artificial elements such as crushed white marble, recycled black rubber shards, and underlit blue glass marble arrayed in bands that constitute the "strong linearity" sought by the noted landscape architect. Though he was surely not the first to dare such an approach, it was Smith who created the rooftop gardens at New York's Museum of Modern Art (2003–05) using plastic rocks and artificial trees, resolving the weight issues that plague any rooftop installation. In so doing, he also did away with almost all the maintenance that living plants might require. A private garden intended for the display of modern sculptures, the 40 Central Park South project surely does not go as far as the MoMA rooftop, but Ken Smith has shown repeatedly that nature is but one element of landscape design. Recycled black rubber shards may not appear to have much to do with nature, but contemporary landscape architecture does not exclude the use of such materials.

Architects and landscape architects continue to form the urban landscape, sometimes on a much larger scale than the thousand meters of the Hypar Lawn. Benedetta Tagliabue, the former associate of Enric Miralles (Miralles Tagliabue Arquitectes Associats – EMBT), has taken on public spaces in Hamburg's HafenCity (Germany, 2006; page 250). Using varying surface treatments and colors, taking into account the marked possibilities of flooding at the water's edge, the architect has sought to bring "people nearer to the water and its moods." Though not forcibly surprising, this declaration underlines one of the fundamental actions of landscape architecture in the context of the contemporary urban environment—to bring residents into a closer rapport with nature, where the realities of modern construction and design have long

3
AECOM, Civic Space Park, Phoenix,
Arizona, USA, 2008–09

had the opposite effect. Another group of talented Spanish architects, Nieto Sobejano, has recently completed an intervention concerning more than a hectare in Madrid. Their Plaza de Santa Bárbara (Spain, 2009; page 278) is part of a larger effort including the new Barceló market (2009–11). Lawn areas, play spaces, and a glass pavilion for books and a flower shop mark the square, and point to a subtle integration of natural and artificial elements. With this level of sophistication, there is no question of inviting "nature" in any unfettered sense into the heart of the Spanish capital. Rather, some natural elements are incorporated into a square that is very much part of the city. Landscape in any broader sense here takes a secondary role to an extension and integration of architecture into the urban milieu.

HER SECRET IS PATIENCE

Another example of fairly large-scale intervention in an urban context is the Civic Space Park in Phoenix (Arizona, USA, 2008–09; page 56). Undertaken by AECOM (Architecture, Engineering, Consulting, Operations, and Management), and in particular its landscaping unit (formerly EDAW), this 1.12-hectare public park is located in downtown Phoenix. This project dealt with historic preservation issues, but also with the insertion of works of art, including a huge work by Janet Echelman (*Her Secret Is Patience*). The designers explain: "The urban space had to serve as a community gathering space, a pedestrian passage, an urban oasis of shade and serenity, a place for learning, a place for visiting, as well as a commons for the adjacent Arizona State University campus." These complex and occasionally somewhat contradictory requirements point to the reasons that Phoenix may have called on a firm as large as AECOM, given the resources and nature of the intervention. The intense use that urban park space is often put to means that the landscape architects must provide for numerous different types of users and occupation of the space. If they manage, as AECOM did in this case, to make the park attractive, that in itself is no small accomplishment.

Another AECOM project, the Pier Head and Canal Link (Liverpool, UK, 2005–07; page 52) concerned a 2.5-hectare area located in the Liverpool Maritime Mercantile City zone which is a UNESCO World Heritage site. It was listed by UNESCO in 2004 because "the city and the port of Liverpool are an exceptional testimony to the development of maritime mercantile culture in the 18th and 19th centuries, contributing to the building up of the British Empire. It was a center for the slave trade, until its abolition in 1807, and to emigration from northern Europe to America."[3] The designers state: "The space has the flexibility and capacity to provide a safe and enjoyable venue for large festivals attracting crowds of up to 35 000, but is equally appealing as a place for quietly contemplating the handsome buildings, local war memorials, and views across the River Mersey." This intervention, as well as many other similar ones in different parts of the world, has to do with tying together an existing urban fabric, of creating coherence where time and the previous uses of space had not worked in favor of the presence of large numbers of visitors. Again, in Liverpool, as in Phoenix, the task was delicate for AECOM and they appear to have successfully navigated past the potential difficulties.

4
West 8, Madrid RIO, Madrid, Spain, 2007–11

DOUBLE DUTCH

The Dutch firm OKRA has worked on the 5.6-hectare area of Rotterdam's Afrikaanderplein (The Netherlands, 2003–05; page 294). An intriguing aspect of this intervention is that it seeks to take into account and reconcile the ethnically varied populations that live near the square. A mosque, market, playground, and a fenced "quiet area" are part of the design, though the fences are meant to be drawn back to allow easy access in the day. There is ample greenery here, but also a marked social intent and a desire to combine aesthetics with practicality and a real service to the community in the broadest definition of the term. Though architects often shy away from claiming any social benefits from their designs, it would appear that landscape architecture retains something of this utopian desire to make life better, especially in urban environments, where nature has often been reduced to its minimal expression. From the therapeutic benefits of relative quiet and the presence of nature to an aesthetic sense meant to purvey a feeling of "quality," landscape architects in the city often seem to be engaged in a kind of mission to improve living conditions.

Another Dutch group, West 8, has carried forward more than one large urban project. Their work on the 80-hectare Madrid RIO project (Spain, 2006–11; page 386) concerns the creation of park spaces above the now buried M30 motorway. Eight thousand pines in the so-called Salon de Pinos, a variety of cherry trees along the Avenida de Portugal, and a "modern interpretation of the orchard" in the Huerta de la Partida are amongst the interventions carried out in the vast spaces created when large amounts of automobile traffic were quite simply channeled underground. This type of covering of urban motorways has been carried out in numerous other cities. Here, fortunately, city authorities have had the clairvoyance to not simply open the new spaces to even denser construction. The role of landscape architecture here is, indeed, to create spaces that improve the urban environment, and make it more livable for the ordinary person, no small accomplishment in itself.

AGRI-TECTURE TO THE RESCUE

The project being carried forward in Manhattan by James Corner Field Operations and Diller Scofidio + Renfro, the High Line (New York, New York, USA, 2009 [Phase 1], 2011 [Phase 2]–ongoing; page 184) at present covers an area of 1.1 hectares in an unusual location—along a disused elevated train line. The designers state: "Inspired by the melancholic, unruly beauty of the High Line where nature has reclaimed a once vital piece of urban infrastructure, the team retools this industrial conveyance into a postindustrial instrument of leisure—a reflection about the very categories of 'nature' and 'culture' in our time. By changing the rules of engagement between plant life and pedestrians, our strategy of agri-tecture combines organic and building materials into a blend of changing proportions that accommodate the wild, the cultivated, the intimate, and the hyper-social." Section 1 of the High Line project is open to the public and provides a well-used promenade above the developing district of the Lower West Side that houses the galleries of Chelsea and stunning new buildings by the likes of Jean Nouvel, Shigeru Ban, and Frank O. Gehry. Though demolition of the High Line tracks was long considered, this project makes use of a large urban space that had served no function since train operations ceased in 1980. The remnants of industrial activity in an urban context

5
Peter Eisenman, City of Culture of Galicia, Santiago de Compostela, Spain, 2001–

often provide generous space and conditions that could hardly be reproduced today. The architects and landscape architects involved have made the High Line into a modern and attractive space despite the fact that the original function of the structure no longer exists. In this context it might be said that the role of landscape architecture is to bring urban space back to life.

The Freedom Park in Pretoria, South Africa (NBGM Landscape Architects with OCA Architects, 2002–10; page 266) is not located far from the center of the executive capital of South Africa, and yet, with its 30 hectares, it represents a sufficiently large space to be able to generate its own rules and design, without the direct influence of the urban milieu. The creation of this park, mandated by Nelson Mandela, "is structured around four key ideas: reconciliation, nation building, freedom of people, and humanity. The making of the landscape seeks to recognize the spiritual origins of these ideas, and manifest them symbolically in physical form." The idea that landscape architecture can interpret or give physical form to such lofty ideals is an intriguing one, perhaps more suited to an emerging democracy than to older nations. The Freedom Park does reveal another aspect of what landscape architecture is ultimately capable of—symbolizing no less than the struggles of a great nation.

IF THAT IS WHAT YOU MEAN BY REDEMPTION I'LL BUY IT

As a number of examples published in this book demonstrate, the matter of landscape architecture often becomes difficult to dissociate from that of architecture itself. This is undoubtedly the case of the Hypar Lawn by Diller Scofidio + Renfro already referred to here. The City of Culture of Galicia (Santiago de Compostela, Spain, 2001–; page 120) by Peter Eisenman might be the ultimate example in recent architecture of buildings that seem to spring from the earth, or that form a landscape of their own. When he is asked if he was in some sense trying to integrate these buildings into nature, or to create a new form of nature, Peter Eisenman states: "Not nature, but unnatural nature. Through advanced computation processes we have the capacity to create unnatural nature. I wanted to create something that would seem like nature, but under closer inspection one would realize that it is not nature. I call it unnatural nature. Our buildings in Santiago look like hilltop. They don't look like they have been placed there. They are made to seem and look like they have come out of the ground like giant mountains. In other words, it is like a natural process—something that would take 10 million years has happened in 10 years. So if that is what you mean by redemption I'll buy it."[4]

Renzo Piano's renovation and expansion of the California Academy of Sciences (San Francisco, California, USA, 2005–08; page 310) approaches landscape from an entirely different perspective, that of the green roof. The undulating roof of the structure, with a surface of over one hectare, is covered with 1.8 million native California plants. Careful study of the plants themselves, but also of the seismic implications of a planted roof, was part of the preparation of this aspect of the design. It is calculated that the design of the roof reduces temperatures inside the museum by about 6°C. The roof's shape and, in fact, the entire design of the museum were conceived to form a continuum with the surrounding park environment. Though ecological concerns are in fashion at the moment, the idea of this green roof, integrated into a very con-

6
Michael Van Valkenburgh, Connecticut Water Treatment Facility, New Haven, Connecticut, USA, 2001–05

temporary building, shows how "green" desires can take the form of buildings that are in a sense integrated into the landscape, or, perhaps more precisely, form a landscape of their own. Here, truly, architecture and landscape architecture are one and the same, because of the undulating roof form, but also, primarily, because of the green roof concept.

LANDSCAPE AND ARCHITECTURE BECOME ONE

Steven Holl has expressed a frequent interest in issues related to landscape. One of his recent buildings, the Vanke Center / Horizontal Skyscraper (Shenzhen, China, 2008–09; page 162) is closely integrated into a landscape that actually contains part of the program—a 500-seat auditorium and restaurants. The landscape component of the project is also one element of the environmental strategy of Steven Holl, because its ponds, fed by a gray water system, serve to cool the air. Solar panels are used, as are local materials such as bamboo. Another building by Holl, the Whitney Water Purification Facility (South Central Connecticut, USA, 1998–2005) was the object of a landscape design by Michael Van Valkenburgh Associates (2005–07; page 364). At the time of the completion of the project, Steven Holl declared: "The design fuses the architecture of the water purification plant with the landscape to form a public park. Water treatment facilities are located beneath the park, while the public and operational programs rise up in a 110-meter-long stainless-steel sliver expressing the workings of the plant below and forming a reflective horizon line in the landscape. Like an inverted drop of water, the sliver shape creates a curvilinear interior space which opens to a large window view of surrounding landscape." The excavated debris from the site was reused and recyclable materials were used wherever possible. Existing wetland and natural vegetation were preserved, while a green roof system requiring very little maintenance was designed for the facility. The area designed by Michael Van Valkenburgh covers a total of 5.67 hectares. For the landscape architect: "The use of the most elemental of landscape architectural tools—soil, water, and plants—offsets the sleek form of the facility building. The design creates topographical variety and interest through sustainable reuse of excavated soil. Swales replace a traditional engineered drainage system. The planting program, inspired by restoration ecology, is at once primal and sophisticated in its extent and complexity." The full agreement of the architect (Holl) and the landscape architect (Van Valkenburgh) on the "natural" aspects of the project make for a remarkable harmony. In this case, architecture and landscape architecture collaborate to insert a thoroughly modern facility in an environment that looks almost as though it had remained totally undisturbed. This might in some sense point in the direction of the future of landscape architecture, where the inroads of modern civilization make the very fact of revivifying nature a creative act.

IN THE MOUNTAINS OF SHIGA

Although the apparent harmony between the landscape designer and the architect was not as great, another example of the complete integration of contemporary architecture into a natural environment is the Miho Museum by I. M. Pei (Shigaraki, Shiga, Japan, 1997). The client assigned the design of the gardens (page 260) to Yoshiaki Nakamura, well-known for his work in traditional Japanese architecture. Given the wilderness setting of the museum, located an hour's drive from Kyoto, authorities had required that much of the museum be located below grade. The original design of Pei thus provided for much of the structure to "disappear" into the hilly terrain, covered by dense local veg-

7
João Luís Carrilho da Graça,
Pedestrian Bridge, Carpinteira River,
Covilhã, Portugal, 2003–09

etation. More than 10 years after the opening of the museum, it has almost been subsumed into the landscape. Nakamura intervened in such gestures as the rows of weeping cherry trees that line the entrance path leading to the access tunnel of the museum. He worked directly with Pei and the client to create an inner garden in the North Wing of the museum, with rocks, gravel, and moss. In this instance, Pei participated in the selection of the rocks and also asked Nakamura to increase the amount of moss in the garden so as to give it a less mineral appearance. Any tensions that may have existed over such choices have now become anecdotal. The Miho Museum is one of the most spectacular modern museums in the world, and perhaps one of those that is best integrated into its beautiful natural setting. The geometric forms of roofs and windows designed by the architect are still very much visible, as is the award-winning suspension bridge that leads to the museum entrance. If it might be said that Holl's Connecticut Water Purification Facility is dealt with somewhat as an alien intrusion into the landscape—Van Valkenburgh speaks of "offsetting the sleek form of the facility building…"—here landscape and architecture become one as much as is physically possible. Within and without, harmony reigns.

Another, quite different example of the juxtaposition of nature and architecture can be seen in the Carpinteira River Pedestrian Bridge (Covilhã, Portugal, 2003–09; page 94) by the Lisbon architect João Luís Carrilho da Graça. Actually formed from a total of three footbridges in the Goldra and Carpinteira valleys, this intervention aims to create more direct links between the historic center of Covilhã and outlying residential areas. The extremely simple forms of the bridges contrast in a willful manner with the city and natural environment and yet, pedestrians come to understand and appreciate this setting in a completely different way thanks to the architect's work. It might be said that the Carpinteira River Pedestrian Bridge makes the landscape appear in a completely new way. It is manifestly an artificial addition that does not seek to integrate itself, but its crisp lines are in a sense a real act of landscape architecture. Two pillars of the bridge have a spiraling stone cladding that is intended to allow vegetation to cover these supports. After a certain time, it might appear that the bridge has somehow grown out of the hillsides as a kind of celebration of the site.

A BEAUTIFUL MIND

An interesting variation on the integration of landscape design and architecture can be seen in the One Kowloon Building in Hong Kong where the Zen priest Shunmyo Masuno has created SanShin-Tei (China, 2006–07; page 328), an installation that occupies a bit more than 1500 square meters in the entrance and 554 square meters in the lobby of the building. Head Priest of the Kenkohji Temple in Yokohama since 2001, Shunmyo Masuno also heads a firm called Japan Landscape Consultants Ltd. He explains that the name of the installation is derived from the Zen word *sanshin* (literally meaning "three mind"). The association of the number three (*san*) with the word for mind (*shin*) gives a meaning that is translated as "joy mind," "mature mind," or "great mind." These concepts are imagined as a flow of energy that goes through the building, from the top and out through the lobby, marked for example by a 9.5-meter waterfall. A transition is made from an installation that Shunmyo Masuno calls "artificial" inside the building to a more "natural" arrangement on the exterior. Given the background of the designer, it is not surprising that some religious intent is integrated into SanShin-Tei, but the basic idea remains one of confronting the

8
Gustafson Guthrie Nichol, Lurie Garden, Chicago, Illinois, USA, 2004

natural and the artificial, a constant theme in the history of landscape design. Here, the modern architecture is less the object, than a presence that calls on other forces to ensure both the success of the company concerned, but also the well-being of those who come close to One Kowloon in any sense.

While Shunmyo Masuno brings a certain expression of nature to an architectural environment, the opposite might be said of John Pawson's Sackler Crossing in the Royal Botanic Gardens (Kew, London, UK, 2006; page 304). Well-known for his defense and illustration of minimalism in architecture and interior design, Pawson created a 70-meter-long bridge that he calls "a sculptural serpentine object." A bronze alloy normally employed in submarine propellers and flat planks of hard granite are the main materials employed, clearly signifying the desire to create something solid and durable. Pawson explains that the Sackler Crossing "is there to be looked at for its own beauty, but also to allow visitors to the gardens to experience the landscape from new and unexpected vantage points." The Carpinteira River Pedestrian Bridge shares with the Sackler Crossing the creation of new vantage points on the surrounding landscape, but Pawson's bridge is so low that it appears to "float across the water" while the intervention of João Luís Carrilho da Graça seems more to trace a fine line in the air, almost floating about its location. In both cases, there is a reference to minimalism in contemporary architecture, which might by definition seem contradictory with the rugged complexity of nature. Both architects obviously reject any mimicry, seeking instead to tread lightly with a vocabulary of the present, which might also be termed a classical simplicity.

Two projects of the noted Seattle firm Gustafson Guthrie Nichol present solutions to interior and exterior garden design. Their Lurie Garden (Chicago, Illinois, USA, 2004; page 132) is actually located on the roof of the Lakefront Millennium Parking Garage in Chicago's Millennium Park. The architectural context is prestigious since the site is between a band shell designed by Frank O. Gehry and the Modern Wing of the Art Institute of Chicago by Renzo Piano. Given the rooftop location, lightweight geofoam is employed to create forms in the garden. The Lurie Garden might be seen as the contemporary version of the more formal gardens of the past, and it is in perfect harmony with its very contemporary neighbors. The issue of artifice versus nature is expressed in the wholly urban setting but also in the unseen geofoam support. It is comforting and attractive to view "nature" in these circumstances, but as has always been the case in gardens, the man-made element may be more present than any uncontrolled expression of the natural world.

A second realization by Gustafson Guthrie Nichol, the Robert and Arlene Kogod Courtyard, (Washington, D.C., USA, 2005–07; page 136) is located in the former Patent Office Building which now houses the Smithsonian American Art Museum and National Portrait Gallery. The design architects working on the project were Foster + Partners and the landscape architects were working with a courtyard enclosed by a glass canopy. Large stone planters with 7.5-meter-high trees are an essential element of their design. The landscape architects state that "the temperate palette of evergreen trees and shrubs complement and distinguish the existing architecture of the courtyard with their variations in scale, form, and texture, as well as the symmetry or asymmetry of their arrangement." In an even more sustained way than their Lurie

9
Maya Lin, Storm King Wavefield,
Mountainville, New York, USA,
2007–08

Garden, this project leads the designers into an area where an artificial, controlled concept of nature becomes more architectural than natural. This is by no means a criticism, it is rather a statement of fact. Landscape architecture here is a complement to the building itself and provides a counterpoint to the mineral environment to some extent. Courtyard gardens are practically obliged to enter into the dialogue with nature in this way, and the Gustafson Guthrie Nichol project is a distinguished and contemporary expression of the genre.

Maya Lin, born in Ohio in 1959, brings an unusual background to the area of landscape architecture, art, or architecture itself, since she practices all three disciplines. She was just 22 years old when a work that remains her most famous, the Vietnam Veterans' Memorial on the Mall in Washington, D.C., was inaugurated. Lin won an open competition and her project, a black granite wall placed essentially below ground level, has continually drawn tens of thousands of visitors. With the names of the American dead in the war—some 50 000 of them inscribed by order of death on the wall—this memorial was a source of great controversy, principally in the veteran's community. It was almost as though America was ashamed of her dead, said the critics. For others, including those who understand modern architecture and design, Maya Lin's work was a fundamentally new gesture in an area long dominated by bronze casts of men boldly holding up flags. Her work was both seminal and in a sense revolutionary in its modest audacity.

WAVES, SNAKES, AND A MAP

Since that time, although no work of hers subsequently reached the same notoriety, Maya Lin has continued to create monuments, houses, and works of art that are of the greatest interest. Two of these works are published in this volume. The first, Storm King Wavefield (Mountainville, New York, USA, 2007–08; page 210), is located in a former gravel pit. The wavelike form of the grass-covered hills created by the artist–architect is intended to give visitors a sense of "total immersion." This work certainly falls into the category of landscape architecture, and yet it is also a work of art, serving no discernable purpose except to surprise and intrigue viewers. Maya Lin forms what is most familiar—the earth—in a way that catches people unaware, and makes them think again about nature.

Her Eleven Minute Line (Wanås, Sweden, 2004; page 212) is a 152-meter-long snakelike mound, also covered with grass. Inspired by Native American burial mounds in the region of the Midwest where she grew up, Lin's design started as a sketch, as most works of art or architecture still do. Fundamentally abstract and modern, this piece also calls on ancient memories, even if the transposition from Ohio to Sweden is not an obvious one. The point may be that the earth has been formed for so many centuries that every country and civilization has buried secrets like those implied by the Eleven Minute Line. There is something *fundamental* about the work of Maya Lin that places her at the juncture of the ancient and the contemporary in pieces such as these.

Although her approach is more specifically that of a landscape architect, the work of Catherine Mosbach also approaches the area of artistic expression. Her very small (150-square-meter) temporary piece, L'autre rive (The Other Bank, Quebec, Canada, 2008; page 256) was

10
Catherine Mosbach, L'autre rive (The Other Bank), Quebec, Canada, 2008

part of celebrations of the 400th anniversary of Quebec. In this instance, Mosbach used surfaces in asphalt that evoked the early maps used to conquer the new world. Her own description of the project shows just how "artistic" her intent was. "This garden is a fable that from up close confronts the materials present—soil, plants, figures—and from afar the vision of a landscape through its provocative contours… The soil, surface of recording, differentiated from the earth's crust by the presence of life, is retranscribed here as an asphalt skin—vegetation fossilized in swamp water, then mineralized… The visitors traverse the milieu, revealing it to themselves and to others by their movement, inserted into interstitial latencies." Mosbach deals here with the landscape of the new world, but perhaps somewhat less with the ancient mysteries evoked in Maya Lin's Eleven Minute Line. Nonetheless, the two women share the conviction that the landscape can be a vehicle for the most contemporary types of art—those that rely more on installation than on any museum setting, for example. Evocative without being figurative in any way (Lin's snake is only a suggestion of a form), both of these creators could just as easily be called artists as landscape architects.

The potential for ambiguity that may exist between disciplines says a good deal about the state of contemporary landscape architecture. Though symbolism was obviously present in the earliest efforts to form the landscape into great squares or the environment of royal tombs, landscape architecture retains the ability to tie together the past and present even without further reference to such earthly powers as forgotten thrones. André Le Nôtre, as I. M. Pei has noted, made early and magnificent use of planar surfaces and reflections of the sky in his basins, in a sense prefiguring the minimalism so avidly sought by many modern architects and designers. Power, in the hands of André Le Nôtre, was nothing other than perspective and infinite space, radiating from Versailles or the other symbols of French royalty. With kings long in the past, Le Nôtre's perspectives, such as that of the Champs Elysées, remain, conveying a sense of grandeur that need not be related to Louis XIV, but rather, now, to a nation and its conquests of the spirit.

LANDSCAPE, BEFORE THE FALL

If this volume seeks to make any point, it is that landscape architecture is all around us—in architecture itself, in gardens, and in recent efforts to revive the flourishing nature that was so often swept aside by the ravages of industrialization and pollution. A building can be integrated into a landscape, or surrounded by it. An architect can build a bridge that has more to do with landscape than with the detached art of the builder. Architects like Steven Holl, I. M. Pei, or Tadao Ando (also present in this book; page 62), often have viewed their buildings as part of an environment that they also seek to fashion. Ando, with works like his underground Chichu Museum (Naoshima, Japan, 2004), places major works by Claude Monet, James Turrell, and Walter De Maria underground, reforming the initial landscape of the hillside site, above the building. This is architecture as landscape and the reverse. Ando is amply aware of the call of the ancient that the earth symbolizes, and yet he is a geometric modernist. This coming together, seen too in the work of Maya Lin, for example, is the real measure of the contemporary interest of landscape architecture. At its height, landscape architecture is an art.

Underlying most initiatives that concern the landscape, there is a concern for "perfection" that may no longer animate architecture quite so much. This concern is linked to the idea of *the* garden, symbolizing a world before sin for the Judeo-Christian traditions, or heaven itself in Islam, for example. Surprisingly, even the current trend to ecological concerns emphasizes the idea of the return to a pristine state, before the fall, as it were. The regenerative power of landscape, or more precisely of nature, inspires this continually renewed hope for redemption, even in the face of massive destruction of the natural environment for reasons of need or greed. While everything man made can be questioned or subject to moods and fashions, nature has its own legitimacy, and it is a part of this legitimacy that landscape architecture surely seeks. Man-made and yet closer to nature than other forms of expression, landscape architecture, beyond the expression of individual creativity, aspires to redemption, to a return to the initial primitive state. Frank Lloyd Wright may have dreamed of architecture that was inspired by nature, but the fact of contemporary construction is that this is very rarely the case. Art is often inspired by nature, even today, but somehow cannot aspire to any real return to the perfection of what was before… before the fall.

[1] http://www.mnstate.edu/gracyk/courses/web%20publishing/addison414.htm, accessed on November 16, 2010.
[2] Horizon Press, New York, 1953.
[3] http://whc.unesco.org/en/list/1150, accessed on November 16, 2010.
[4] Peter Eisenman, quoted on http://www.curatorialproject.com/interviews/petereisenmanii.html, accessed on November 16, 2010.

EINLEITUNG

VERZWEIFELTE SUCHE NACH ERLÖSUNG

Der Gedanke, die Machtzentren – seien es Paläste oder große öffentliche Plätze – mit gestalteter Landschaft zu umgeben, ist schon alt, aber die uns vertraute Definition von Landschaftsarchitektur wurde erst im 19. Jahrhundert geprägt. Das Werk von André Le Nôtre (1613 bis 1700), des Gartenarchitekten von Ludwig XIV., mit seinen weiten Sichtachsen in Vaux-Le-Vicomte und Versailles oder die durch den Tuileriengarten und die Champs-Élysées gebildete kilometerlange Achse in Paris müssen sicherlich als Beispiele für eine geänderte Beziehung zwischen Landschaft und Architektur angeführt werden. Zur damaligen Zeit waren Städte wie Paris sehr eng, verwinkelt und dicht bebaut. Dem stellte Le Nôtre den Freiraum mit seiner besonderen Bedeutung gegenüber und verband ihn mit der Prachtentfaltung des Sonnenkönigs. Die herrlichen Perspektiven und die großen Wasserbecken, welche die Achsen akzentuieren, die von Schloss Versailles ausgehen, sind ein Ausdruck von Macht. Sie dienen dazu, die Architektur des Palastes hervorzuheben und seine symbolische Präsenz auch aus großer Entfernung wahrnehmbar zu machen.

Architekturen, die ihr natürliches Umfeld mit einbeziehen, sind nicht unbedingt per se Landschaftsarchitektur. Aber es ist doch die Verbindung beider Elemente (Landschaft und Architektur), die einige der eindrucksvollsten und bekanntesten Beispiele in der Geschichte der Baukunst hervorgebracht hat. Wer einmal in China auf den Überresten der Großen Mauer in der Nähe von Peking stand – deren Bau im 5. Jahrhundert v. Chr. zum Schutz der nördlichen Grenze des Kaiserreichs begonnen wurde – sieht ganz deutlich, wie sich die Mauer über Berge und Täler windet und erkennt, wie sich die Gestalt der Mauer aus der Umgebung und der Topografie ergibt. Als ein weiteres Beispiel, das sich ebenfalls in China befindet, ist der Tian'anmen-Platz an der Verbotenen Stadt in Peking zu nennen, der vor allem durch seine Leere beeindruckt. Im Kontext mit den Gebäuden verkörpert dieser Freiraum ein Machtsymbol, das Jahrhunderte lang allen verständlich war.

SCHÖN UND BEFREMDLICH

Das vorliegende Buch konzentriert sich vor allem auf neueste Beispiele, und die Definition von Landschaftsarchitektur bezieht sich nicht nur auf die Gestaltung von Gärten, sondern auch auf Gebäude, die eine enge Beziehung zum Freiraum und zur Natur haben, Gebäude, die gewissermaßen der Erde entsprungen sind und dem Umfeld eine besondere Bedeutung verleihen. Vielleicht sollte man den Begriff „Architektur der Landschaft" einführen statt des allgemein üblichen Begriffs „Landschaftsarchitektur", der häufig Gebäude ausschließt. Der Begriff Landschaftsarchitektur wurde von dem Schotten Gilbert Laing Meason (1769–1832) geprägt. In seinem Buch „The Landscape Architecture of the Great Painters of Italy" (1828) ging es darum, wie Gebäude so zu platzieren sind, dass sie schöne Ensembles mit ihrer Umgebung bilden. In diesem Sinne bezieht der Begriff Landschaftsarchitektur in seinem Ursprung durchaus Gebäude mit ein.

Der Bezug der Kunst zur Natur und die Vorstellung des Malerischen, die das romantische Gedankengut beflügelten, sind älter als der oben genannte Meason. Joseph Addison (1672–1719), der englische Autor und Politiker, schrieb in der Zeitschrift „The Spectator", in der Ausgabe 414 vom 25. Juni 1712:

11
Patrick Blanc, Orchid Waltz,
National Theater Concert Hall,
Taipei, Taiwan, 2009

Wenn wir die Werke der Natur und der Kunst als solche betrachten, dass sie dazu geeignet sind, die Vorstellungskraft zu nähren, werden wir Letztere im Vergleich zu Ersteren sehr mangelhaft finden; auch wenn Kunstwerke manchmal schön oder befremdlich erscheinen, so können sie doch nicht diese Weite und unermessliche Größe in sich bergen, die dem Betrachter ein so großes Vergnügen bereiten. Die Kunst kann ebenso fein und delikat sein wie die Natur, aber doch nie so erhaben und großartig. In den rohen, absichtslosen Hieben der Natur liegt mehr Kühnheit und Meisterschaft als in den hübschen Pinselstrichen und Verzierungen der Künstler. Die Schönheit der herrschaftlichen Gärten und Paläste wird von einem sehr schmalen Grat definiert. In den Weiten der Natur jedoch wandert der Blick ungehindert nach oben und nach unten, und der Betrachter wird mit einer unendlichen Fülle von Bildern versorgt. Aus diesem Grund begegnen wir auch immer dem Dichter, der das Leben auf dem Lande liebt, wo die Natur sich in ihrer größten Vollkommenheit zeigt und alle nur möglichen landschaftlichen Szenerien ausstattet, an denen unsere Fantasie sich am meisten erfreuen kann.[1]

Die Dichotomie zwischen dem „Natürlichen" und dem „Künstlichen" ist tatsächlich der wichtigste Schlüssel zum Verständnis von Landschaftsarchitektur. Ist es das Ziel der Landschaftsarchitekten und Architekten, irgendwie die Natur nachzuahmen oder dem Umfeld von Gebäuden eine menschliche Ordnung zu verleihen? Die Wurzeln dieser Überlegung lassen sich weit in der Geschichte und der Literatur zurückverfolgen. Vielleicht sprechen sie von einer uralten Gartensehnsucht des Menschen. Landschaftsarchitektur hat wohl immer auch etwas mit der Suche nach dem vollkommenen Garten zu tun.

Zur höchsten Lust des menschlichen Geschlechts,
Ja noch mehr; einen Himmel auf der Erde,
Denn Gottes Garten ward das Paradies,
Den er im Osten Edens angepflanzt.
John Milton, „Das verlorene Paradies", 4. Gesang
(Nach der Übersetzung von Adolf Böttger)

Der Erste, der nachweislich den Begriff „Landschaftsarchitekt" als Berufsbezeichnung verwendet hat, war 1863 der Amerikaner Frederick Law Olmsted (1822–1903). Gemeinsam mit seinem Büropartner Calvert Vaux plante Olmsted den Central Park in Manhattan und zahlreiche weitere Parkanlagen in den USA. Olmsted und Vaux gewannen 1857 den Wettbewerb zur Gestaltung des 280 Hektar großen Areals zwischen der 59. und der 106. Straße. Manche Besucher glauben, dass der Central Park den natürlichen Zustand dieses Teilbereichs von Manhattan darstellt, bevor um ihn herum die Stadt gebaut wurde. Aber es waren etwa zehn Millionen Wagenladungen Erde und Fels nötig, um das Gelände zu gestalten, und mehr als vier Millionen Bäume und Gehölze wurden gepflanzt, bevor der Park 1873 eingeweiht wurde. Mit den 36 von Vaux entworfenen Brücken und den vielen kleinen Konstruktionen, die im Lauf der Zeit dort entstanden, ist der Central Park eines der ersten Beispiele amerikanischer Landschaftsarchitektur. Er ist ein Freiraum, den wohl kaum jemand den Übergriffen der sich immer weiter ausbreitenden modernen Stadt opfern würde.

BARRIEREN ÜBERWINDEN
Bis zu welchem Grad aktuelle Architektur und Landschaftsarchitektur verflochten sind, hängt immer von den Umständen ab. Im vorliegenden Band werden absichtlich einige Beispiele vorgestellt, die die vermeintliche Grenze zwischen den Disziplinen überschreiten. Wann verschmelzen Landschaft und Architektur miteinander, sodass es schwierig ist, sie auseinanderzuhalten? Jedes Gebäude, jeder Park hat sein eigenes Umfeld. Aber sollte es nicht das oberste Ziel der Landschaftsarchitektur sein, die (künstliche) Barriere zu überwinden, die sie von der Sphäre der Türme- und Brückenbauer trennt? Jedenfalls hatte niemand Geringeres als Frank Lloyd Wright eine klare Meinung zu diesem Diskussionsthema:

Wandel ist der unabänderliche Umstand, den man in der Landschaft vorfindet. Aber alle Veränderungen sprechen oder singen unisono vom kosmischen Gesetz, das selbst eine noblere Form des Wandels ist. Diese kosmischen Gesetze sind die physischen Gesetze aller von Menschenhand gebauten Strukturen ebenso wie die Gesetze der Landschaft. Sooft der Mensch ein Gebäude auf der Erde errichtet, greift er positiv in die Schöpfung ein. Wenn der Mensch überhaupt ein Geburtsrecht hat, dann dies: dass er ebenso ein Teil der Landschaft ist wie die Felsen, Bäume, Bären oder Bienen, dass er Teil der Natur ist, der er sein Dasein verdankt. Die Natur zeigt ihm unaufhörlich die Wissenschaft ihrer bemerkenswerten wohldurchdachten Strukturen der mineralischen und der pflanzlichen Konstruktionen, die einhergehen mit dem unverbildeten Wesen, das in allen ihren Formen sichtbar ist.
Frank Lloyd Wright, „The Future of Architecture"[2]

KLEIN, ABER PERFEKT IN FORM
Die Frage, wann denn Architektur und Landschaft zu Landschaftsarchitektur werden, ist ziemlich schwierig zu beantworten, und im vorliegenden Buch soll diese Frage ganzheitlich betrachtet werden. Ein erster Ansatz könnte sein, dass man sich kleinen Projekten in der Stadt widmet, bei denen die Planer (Landschaftsarchitekten oder Architekten) Natur in die Stadt bringen und dort eine Wechselbeziehung herstellen, wo sonst häufig Beton, Glas und Stahl jeden Hauch Grün unterdrücken. The Garden That Climbs the Stairs, der Garten, der die Treppe hinaufwächst (Bilbao, Spanien, 2009, Seite 72), ist gerade einmal 80 Quadratmeter groß. Die Planerin dieses Gartens, Diana Balmori, spricht von einem „dynamischen urbanen Raum", um zu beschreiben, was denn ein Garten auf einer Treppe in unmittelbarer Nähe zu repräsentativen Werken zeitgenössischer Architektur ist. Sie sagt: „Mit einer einzigen Geste erzählt der Garten eine Geschichte, wie sich Landschaft ausbreitet und den öffentlichen Raum und die Architektur vereinnahmt, und dadurch ändert sich die Art und Weise, wie die Treppe und der Raum vom Nutzer wahrgenommen und gelesen werden." Diese kurze Beschreibung eines kleinen Gartens trägt dazu bei, die Vieldeutigkeit, die die Natur im eigentlich künstlichen urbanen Umfeld erzeugt, zu mildern. Der Freiraumgestaltung wird hier die Kraft zuge-

12
Vladimir Djurovic, Hariri Memorial
Garden, Beirut, Lebanon, 2005–

schrieben, die architektonische Umgebung „verändern" zu können, eine Kraft, mit der ein Kontrapunkt zu den harten Baumaterialien gesetzt werden kann.

AN DEN WÄNDEN HOCHWACHSEN

Der bekannte französische Botaniker Patrick Blanc hat überall auf der Welt seine „vertikalen Gärten" angelegt, die zu seinem Markenzeichen geworden sind. Nachdem er festgestellt hatte, dass Pflanzen eigentlich keine Erde, sondern nur Nährstoffe und Wasser brauchen, ließ sich Blanc eine Methode patentieren, mit deren Hilfe er Gärten anlegen kann, die Innen- und Außenmauern bedecken. Zwei seiner neuesten Kreationen, die im vorliegenden Buch vorgestellt werden – Orchid Waltz in der Konzerthalle des Nationaltheaters in Taipeh (Republik China, 2009; Seite 78) und die Max-Juvénal-Brücke (Aix-en-Provence, Frankreich, 2008; Seite 80) – verändern sicherlich die Wahrnehmung ihres gesamten Umfeldes. Beim zweiten Beispiel mildert eine üppig mit Grün bewachsene Wand die harten Linien einer modernen Brücke, die nach dem Willen der örtlichen Behörden ursprünglich mit Klinker oder Naturstein verkleidet werden sollte. Patrick Blancs Gärten scheinen nur deshalb künstlich zu sein, weil sie vertikal sind, aber entgegen allen Erwartungen sind sie sehr lebendig. Patrick Blanc hat in Botanik promoviert und beschäftigt sich mit hochwissenschaftlichen Studien zum Überleben von Pflanzen unter extremen Umweltbedingungen. Er selbst nennt sich nicht Landschaftsarchitekt, aber er ist in der Lage, die Wahrnehmung eines Gebäudes oder eines Wohnviertels in einer Stadt zu verändern. In gewisser Weise definiert er das Wort „Landschaft" neu, indem er elementare Eigenarten der Natur verwendet und sät so Zweifel an der üblichen Vorstellung dessen, was ein Garten im Vergleich zur Architektur zu sein hat.

Vladimir Djurovic ist ein in Beirut ansässiger Landschaftsarchitekt, der so ehrgeizige Projekte wie die Gartenanlagen um das neue Aga Khan Museum für islamische Kunst in Toronto realisiert. In Beirut hat Djurovic bereits den Samir-Kassir-Platz gestaltet (2004), für den er 2007 mit dem Aga Khan Architecture Award ausgezeichnet wurde. Die Parkanlage ist nur 850 Quadratmeter groß, also ziemlich klein für einen städtischen Park, aber diese kleine Grünanlage verleiht dem Wohnviertel eine unerwartet friedliche und entspannte Atmosphäre, die die Hektik der Stadt vergessen lässt. Sein neuestes Projekt, das er ebenfalls in Beirut baut, der Hariri Memorial Garden (laufendes Projekt; Seite 110) ist mehr als doppelt so groß, aber im Vergleich zu anderen Elementen des urbanen Umfelds von der Größe her immer noch gemäßigt. Bei diesem Projekt arbeitet er mit Bändern aus Naturstein, Wasser und Glas, die sich stufenartig hinunter zum Stadtzentrum entwickeln. Es gelingt ihm, mit diesen reduzierten Materialien und Formen etwas zu schaffen, das man als bescheidene Monumentalität bezeichnen kann. Sie ist der Verehrung angemessen, die viele Libanesen für ihren ermordeten, früheren Premierminister Rafiq Hariri (1944–2005) empfinden, dem dieses Denkmal gewidmet ist. Der amerikanische Bildhauer Richard Serra hat einmal gesagt „der Unterschied zwischen Kunst und Architektur ist der, dass Architektur einem Zweck dient". Etwas Ähnliches könnte man über Freiraumgestaltungen sagen. Es sind Orte, die das Auge erfreuen, die zum Verweilen und zum Treffen einladen, und im Fall des Hariri Memorial Garden in Beirut ist es ein Ort zum Innehalten und Erinnern. Freiraumgestaltungen mögen nicht ohne jeden Zweck sein, aber diese beiden bemerkenswerten Beispiele kann man durchaus als Kunstwerke bezeichnen.

EIN BAUERNHOF AUF DEM SPIELPLATZ

Im Rahmen der Hofgestaltung des P.S. 1 Contemporary Art Center, der Dependance des Museum of Modern Art auf Long Island, haben die New Yorker Architekten von WORK AC 2008 eine Installation geschaffen, der sie den Namen Public Farm 1 (Seite 406) gegeben haben und die gerade etwas mehr als 1000 Quadratmeter umfasst. P.S. 1 ist eine umgenutzte Schule, und der Hof war früher einmal der Spielplatz. Dieser „voll funktionierende urbane Bauernhof in Form einer abgeknickten Ebene, deren Konstruktion aus Papprören besteht" reagiert auf die umgebenden Gebäude und ist selbst zu einer Art Architektur geworden. Allerdings handelt es sich hier um etwas anderes als die anderen kleinen Vorstöße der Natur, die bisher vorgestellt wurden. Die Vorstellung, dass ein Bauernhof, ein vom Menschen geschaffenes Gebilde, das man normalerweise nur sehr weit von urbanen Bereichen entfernt findet, hier, in dem dicht besiedelten städtischen Umfeld von Queens Wurzeln schlagen könnte, und sei es nur temporär, ist eine Aussage über die Bedeutung von Landschaftsarchitektur und ihre Beziehung zur Architektur. Weinberge, Reisfelder oder Maisfelder bestimmen die Landschaft in vielen Teilen der Erde, aber Landschaftsarchitekten schenken ihnen zumeist kaum Aufmerksamkeit. An dem Prozess der Domestizierung, der Zähmung der Natur – dies ist im Grunde genommen ja der Beruf eines Architekten – sind Bauern ebenfalls beteiligt, wenn sie endlose Reihen von Feldfrüchten in der Landschaft anpflanzen. Die Natur wird für einen bestimmten Zweck geformt und geleitet. Aber selbst die grundlegenden Voraussetzungen für die Landwirtschaft werden von einer Installation wie der 2008 von WORK AC geschaffenen infrage gestellt.

Der Teardrop Park in New York (Battery Park City, 1999–2006; Seite 370) von Michael Van Valkenburgh Associates ist etwa 7000 Quadratmeter groß und ist vor allem als Bereich für Kinder gedacht. Es wurde größter Wert auf Umweltfreundlichkeit und Nachhaltigkeit der Gestaltung gelegt. Die Planer weisen darauf hin, dass die Kinderspielbereiche in den Parks immer weniger Naturnähe aufweisen, dafür aber immer besser mit Geräten ausgestattet sind und dass sie mit ihrem Entwurf versucht haben, diesen Trend umzukehren. Ein zentrales Element der Anlage ist eine 8,2 Meter hohe und 51 Meter lange, aus geschichtetem Blaustein errichtete Mauer, die „Eiswassermauer", die man aufgrund ihres Maßstabs ohne Zweifel als starkes architektonisches Element bezeichnen kann. Wie die Architektur muss auch die Landschaftsarchitektur oft Zwecke erfüllen, die nicht nur ästhetischen Gesichtspunkten folgen. Zweckmäßigkeit, wie hier im Fall eines Kinderspielplatzes, entspricht eigentlich dem Verständnis von Personen, die nicht als Landschaftsarchitekten ausgebildet sind. Die Mitwirkung eines Büros wie das von Michael Van Valkenburgh garantiert ein höheres Niveau, weil hier ein sensibles Raumempfinden mit ästhetischen Vorstellungen und Kenntnissen über Natur und Umwelt zusammenkommt. Spielbereiche werden sorgfältig geplant und mit Überlegungen zu allgemeinen Tendenzen der aktuellen Parkgestaltung verbunden.

Die Arbeiten des New Yorker Büros Diller Scofidio + Renfro, zum Beispiel beim Bau des Institute of Contemporary Art (Institut für zeitgenössische Kunst) in Boston (2006), sind eindeutig in der ersten Reihe zeitgenössischer Architektur zu sehen. Eine neuere Arbeit des Büros DS+R war die Beteiligung an der Renovierung des Lincoln Center, einem Kulturzentrum in der Upper West Side von Manhattan. Das Büro hat den Hypar Pavilion (2011; Seite 106) auf der North Plaza des Lincoln Center mit seiner ungewöhnlichen Dachkonstruktion ausgeführt. Den

13
Diller Scofidio + Renfro,
Hypar Pavilion Lawn, New York,
New York, USA, 2011

Besuchern steht auf dem geneigten Dach eine 1000 Quadratmeter große Rasenfläche zur Verfügung, und dies mitten in einem urbanen Umfeld, wie man es sich nicht lebendiger vorstellen kann. Ästhetisch beeindruckend mit seiner sauberen Linienführung, liegt der Hypar Pavilion Lawn genau an der Schnittstelle zwischen Architektur und Landschaftsarchitektur. Er hat von beidem etwas und verbindet beide Disziplinen nahtlos miteinander. Zur großen Freude der Passanten nimmt hier die Natur (in ihrer gezähmten Version) ihren Platz in einer ansonsten steinernen Umgebung ein.

BLAUE MURMELN UND SCHWARZE GUMMISTÜCKE

Der von Ken Smith gestaltete Garten in 40 Central Park South in New York (2006; Seite 336) ist nur 580 Quadratmeter groß. Der bekannte Landschaftsarchitekt hat hier Naturstein und künstliche Materialien verwendet wie zum Beispiel weißen Marmorsplitt, Stücke aus recyceltem schwarzem Gummi und von unten beleuchtete, blaue Glasmurmeln. Er hat diese Materialien in Streifen zu der von ihm beabsichtigten „strengen Linearität" angeordnet. Auch wenn Smith nicht der Erste war, der eine solche gewagte Gartengestaltung realisierte, so war er es doch, der 2003 bis 2005 den Dachgarten des Museum of Modern Art in New York konzipierte, indem er dort Kunststofffelsen und künstliche Bäume platzierte und so das Problem des Gewichts löste, das bei jeder Dachgartengestaltung auftritt. Mit dieser Lösung wurden auch die Probleme der Pflege, die alle lebendigen Pflanzen benötigen, aus der Welt geschafft. Das Projekt 40 Central Park South ist ein privater Skulpturengarten, bei dessen Gestaltung er nicht so weit geht wie bei dem Dachgarten für das MoMA, aber Ken Smith hat bereits mehrfach unter Beweis gestellt, dass Natur nur ein Element der Gartengestaltung ist. Recycelte schwarze Gummistücke haben auf den ersten Blick nicht viel mit Natur zu tun, aber in der heutigen Landschaftsarchitektur wird die Verwendung auch solcher Materialien nicht ausgeschlossen.

Architekten und Landschaftsarchitekten gestalten Stadtlandschaften und dies häufig in einem größeren Maßstab als die 1000 Quadratmeter des Hypar Lawn. Benedetta Tagliabue, die ehemalige Büropartnerin von Enric Miralles (Miralles Tagliabue Arquitectes Associats – EMBT), hat einige öffentliche Freiräume in der Hamburger HafenCity (2006; Seite 250) gestaltet. Sie verwendete hier unterschiedliche Farben und Oberflächenstrukturen, und unter Berücksichtigung der hohen Wahrscheinlichkeit von Hochwasser, war sie bestrebt, die „Leute näher an das Wasser und seine Launen" heranzuführen. Auch wenn dies nicht unbedingt überrascht, so unterstreicht diese Aussage doch ein wesentliches Bemühen der Landschaftsarchitektur im Kontext des heutigen urbanen Umfeldes – die Bewohner sollen dort, wo die Realitäten modernen Bauens und Designs schon längst ihre nachteiligen Auswirkungen zeigen, in eine engere Beziehung zur Natur gebracht werden. Das Büro der talentierten spanischen Architekten Nieto und Sobejano hat vor Kurzem in Madrid ein Projekt fertiggestellt, das mehr als einen Hektar umfasst. Die Plaza de Santa Bárbara (2009; Seite 278) ist Teil eines größeren Projekts, zu dem auch der neue Barceló-Markt (2009-11) gehört. Rasenflächen, Spielbereiche und ein gläserner Pavillon mit einem Buch- und einem Blumenladen kennzeichnen den Platz und weisen dezent auf die Verbindung von natürlichen und künstlichen Elementen hin. Bei diesem Grad der Kultiviertheit ist es keine Frage mehr, „Natur" uneingeschränkt in das Herz der spanischen Hauptstadt einzuladen. Vielmehr sind bei diesem Platz mitten in der Stadt einige natürliche Elemente eingebunden. Landschaft im weiteren Sinn ist hier zweitrangig gegenüber der Architektur, die hier in das urbane Umfeld integriert ist.

14
Nieto Sobejano, Plaza de Santa Bárbara, Madrid, Spain, 2009

14

IHR GEHEIMNIS IST DIE GEDULD

Ein weiteres Beispiel für einen großmaßstäblichen Entwurf in einem urbanen Kontext ist der Civic Space Park, in Phoenix (Arizona, USA, 2008–09; Seite 56). Das ausführende Büro ist die Abteilung für Landschaftsarchitektur (ehemals EDAW) des Planungsbüros AECOM (Architecture, Engineering, Consulting, Operations, and Management). Der 1,12 Hektar große Park befindet sich in der Innenstadt von Phoenix. Bei diesem Projekt ging es einerseits um den Erhalt von historischen Elementen und andererseits um die Einbindung von Kunstwerken, unter anderem eine großformatige Skulptur von Janet Echelman („Her Secret Is Patience" – Ihr Geheimnis ist die Geduld). Die Planer sagen dazu: „Der urbane Raum sollte ein kommunaler Versammlungsort werden, ein Fußgängerbereich, eine heitere, schattige urbane Oase, ein Ort zum Lernen, ein Ort, den man gerne aufsucht, und ein Freiraum für die benachbarte Arizona State University." Diese komplexen und auch ein wenig widersprüchlichen Anforderungen lassen erkennen, aus welchen Gründen die Stadt Phoenix bei den vorhandenen Ressourcen und der Art des Bauvorhabens ein so großes Büro wie AECOM beauftragt hat. Die intensive Nutzung, der ein städtischer Park ausgesetzt ist, bedeutet sehr oft, dass die Landschaftsarchitekten den Freiraum so gestalten müssen, dass er einer Vielzahl unterschiedlicher Nutzer und Nutzungen gerecht wird. Wenn es, wie in diesem Fall dem Büro AECOM, gelingt, den Park attraktiv zu gestalten, dann ist dies keine geringe Leistung.

Ein weiteres Projekt von AECOM ist die Neugestaltung der Pieranlagen und des Verbindungskanals von Liverpool (Großbritannien, 2005–07; Seite 52). Das Gesamtgelände umfasst 2,5 Hektar und befindet sich in der Liverpool Maritime Mercantile City. Der ehemalige Handelshafen von Liverpool wurde 2004 in die Liste des UNESCO-Weltkulturerbes aufgenommen, weil „die Stadt und der Hafen von Liverpool ein herausragendes Beispiel für die Entwicklung der Handelsmarine im 18. und 19. Jahrhundert sind, die entscheidend zum Aufbau des British Empire beigetragen hat. Bis zur Abschaffung der Sklaverei im Jahr 1807, war Liverpool ein Zentrum des Sklavenhandels, und es war der wichtigste Hafen für die nordeuropäischen Auswanderer nach Amerika."[3] Die Planer bemerken dazu: „Dieser Raum bietet ausreichende Flexibilität und Kapazität, um hier sicher und in angenehmer Umgebung Großveranstaltungen mit bis zu 35 000 Besuchern durchzuführen. Gleichzeitig lädt dieser Ort zur Betrachtung der stattlichen Gebäude und Kriegsdenkmäler ein, und er bietet schöne Ausblicke über den Fluss Mersey." Bei diesem Projekt gilt es, wie bei vielen anderen, ähnlichen Projekten überall auf der Welt, das vorhandene städtische Gefüge miteinander zu verweben und dort einen Zusammenhang herzustellen, wo frühere Zeiten und Nutzungen nicht dazu geeignet waren, eine große Anzahl von Besuchern aufzunehmen. Die Aufgabe in Liverpool war ebenso schwierig wie die in Phoenix, aber das Büro AECOM hat offensichtlich alle Schwierigkeiten erfolgreich umschifft.

DOPPELT NIEDERLÄNDISCH

Das niederländische Büro OKRA hat den 5,6 Hektar großen Park Afrikaanderplein in Rotterdam geplant (Niederlande, 2003–05; Seite 294). Ein interessanter Aspekt dieses Projektes ist der Versuch, die unterschiedlichen Ethnien angehörenden Anwohner um den Park herum zu berücksichtigen und einander anzunähern. Eine Moschee, ein Markt, ein Spielplatz und eine umzäunte „Ruhezone" sind Teile des Entwurfs. Allerdings kann der Zaun teilweise geöffnet werden, damit man tagsüber leichten Zutritt zu diesem Bereich hat. Es gibt viel Grün, aber auch einen deutlichen sozialen Anspruch und den Wunsch, Ästhetik mit Zweckmäßigkeit zu verbinden und den Anwohnern einen wirklichen Dienst im weitesten Sinn des Wortes zu erweisen. Architekten schrecken ja häufig davor zurück, mit ihren Entwürfen sozial etwas Positives bewirken zu wollen, dagegen scheinen Landschaftsarchitekten immer noch ein wenig der Utopie anzuhängen, das Leben mit ihren Projekten verbessern zu können, besonders das Leben im städtischen Umfeld, wo die Natur häufig auf ein Minimum beschränkt ist. Landschaftsarchitekten, deren Projekte in einer Stadt ausgeführt werden, scheinen häufig eine Art Mission zur Verbesserung der Lebensbedingungen erfüllen zu wollen. Dazu gehören der therapeutische Vorteil von relativer Stille und das Vorhandensein von Natur, das dem ästhetischen Empfinden ein Gefühl von „Qualität" vermitteln soll.

Ein anderes niederländisches Büro, West 8, hat mehr als nur ein großes städtebauliches Projekt ausgeführt. Das Büro hat das RIO-Projekt in Madrid (2006–11; Seite 386) realisiert, ein 80 Hektar großes Gelände mit Parkanlagen über der nun unterirdisch geführten Stadtautobahn M 30. 8000 Pinien im sogenannten Salón de Pinos, verschiedene Kirschbaumsorten auf der Avenida de Portugal und eine „moderne Interpretation eines Obstbaumgartens" in der Huerta de la Partida gehören zu den großzügigen Freiräumen, die dort entstanden sind, wo der dichte Autoverkehr schlicht in den Untergrund verbannt wurde. Städtische Autobahnen sind in vielen anderen Städten in Tunnel verlegt worden. Aber hier haben die städtischen Behörden glücklicherweise viel Weitsicht bewiesen und die neu entstandenen Flächen nicht einfach für eine noch dichtere Bebauung freigegeben. Mithilfe der Landschaftsarchitektur sind hier Räume entstanden, die das urbane Umfeld verbessern und es für den normalen Bürger lebenswerter machen. Auch das ist eine beachtliche Leistung.

AGRI-TEKTUR IST DIE RETTUNG

Das Projekt High Line der Büros James Corner Field Operations und Diller Scofidio + Renfro in Manhattan (New York, 2009, Bauabschnitt 1, 2011; Bauabschnitt 2, laufendes Projekt; Seite 184) umfasst derzeit ein fast 2,5 Kilometer langes, recht ungewöhnliches Gelände und zwar eine nicht mehr genutzte Hochbahntrasse. Die Planer kommentieren: „Wir haben uns von der melancholischen, spröden Schönheit der High Line anregen lassen, wo sich die Natur ein Stück ehemals vitaler, urbaner Infrastruktur zurückerobert hat. Unser Team verwandelt den Gleiskörper in eine postindustrielle Freizeitlandschaft, die mit Leben und Wachstum erfüllt ist. Indem wir die Regeln der Beziehungen zwischen dem Pflanzenleben und den Fußgängern verändert haben, verbindet unsere Strategie der Agri-Tektur organische und künstliche Werkstoffe, sodass eine abwechslungsreiche Mischung aus wilden, naturnahen Bereichen, Pflanzenkulturen sowie intimen und hypersozialen Bereichen entsteht." Bauabschnitt 1 des High-Line-Projekts ist der Öffentlichkeit zugänglich und wird gern als Promenade in der Lower West Side genutzt, die sich mit den Kunstgalerien in Chelsea und den fantastischen neuen Gebäuden von Jean Nouvel, Shigeru Ban und Frank O. Gehry zum neuen Szeneviertel entwickelt. Obwohl man lange erwogen hatte, die Gleisanlagen der High Line abzubrechen, führt dieses Projekt nun einen großen urbanen Freiraum einer neuen Nutzung zu, der seit der Stilllegung der Strecke 1980 keine Funktion mehr hatte. Die Überreste einer industriellen Nutzung in einem urbanen Kontext bieten oft großzügige Freiräume und Bedingungen, die man heute sonst kaum zur Verfügung stellen könnte. Die an dem Projekt beteiligten Architekten und Landschaftsarchitekten haben die Tatsache, dass die ursprünglichen

Funktionen des Bauwerks nicht mehr existieren, zu nutzen gewusst und die High Line in einen modernen und attraktiven Freiraum verwandelt. In diesem Zusammenhang kann man mit Recht sagen, dass die Rolle des Landschaftsarchitekten darin besteht, urbanen Raum mit neuem Leben zu füllen.

Der Freedom Park in Pretoria (NBGM Landscape Architects zusammen mit OCA Architects, 2002–10; Seite 266) liegt nicht weit vom Stadtzentrum der Verwaltungshauptstadt von Südafrika entfernt. Mit seinen 30 Hektar ist er groß genug, um seinen eigenen Regeln zu folgen und sich als eigenständiger Entwurf zu präsentieren, ohne dass das städtische Umfeld unmittelbaren Einfluss ausübt. Nelson Mandela selbst hatte angeordnet, diesen Park anzulegen, „dem vier Schlüsselgedanken zugrundeliegen: Versöhnung, Aufbau einer Nation, Freiheit des Volkes und Humanität. Die Landschaftsgestaltung versucht, die geistigen Ursprünge dieser Gedanken zu erkennen und sie symbolisch umzusetzen." Die Vorstellung, dass Landschaftsarchitektur solche erhabenen Ideale interpretieren und ihnen eine physische Gestalt geben kann, ist faszinierend und passt vielleicht besser zu einer sich im Aufbau befindlichen Demokratie als zu älteren Nationen. Der Freedom Park in Pretoria zeigt noch einen weiteren Aspekt dessen, was Landschaftsarchitektur leisten kann – sie vermag es sogar, die Kämpfe einer großen Nation symbolisch darzustellen.

WENN SIE DAS MIT ERLÖSUNG MEINEN, BIN ICH DAMIT EINVERSTANDEN

Eine ganze Reihe der in diesem Buch vorgestellten Beispiele zeigt, dass man Landschaftsarchitektur oft kaum von der Architektur selbst trennen kann. So ist es auch bei dem bereits erwähnten Projekt Hypar Lawn von Diller Scofidio + Renfro. Die Kulturstadt von Galicien (Santiago de Compostela, Spanien, seit 2001; Seite 120) von Peter Eisenman kann man in der aktuellen Architektur wahrscheinlich als typisches Beispiel für Gebäude bezeichnen, die förmlich aus der Erde herauszuwachsen scheinen oder für sich selbst eine Landschaft bilden. Als Peter Eisenman gefragt wurde, ob er in irgendeiner Weise versuchen würde, die Gebäude der Natur anzupassen, oder ob er eine neue Form von Natur schaffen wolle, antwortete er: „Nein, keine natürliche Natur, eher eine unnatürliche Natur. Mithilfe von hoch entwickelten Computerprogrammen sind wir heute in der Lage, eine unnatürliche Natur zu schaffen. Ich wollte etwas schaffen, das wie Natur scheint, sich aber bei näherer Betrachtung sogleich als nicht wirkliche Natur erweist. Ich nenne das unnatürliche Natur. Unsere Gebäude in Santiago sehen aus wie eine Hügelkuppe. Sie sind so gestaltet, als würden sie wie Berge aus dem Gelände herauswachsen. Mit anderen Worten, es war wie bei einem natürlichen Prozess, der normalerweise zehn Millionen Jahre dauert, und hier waren es gerade mal zehn Jahre. Wenn Sie das mit Erlösung [vom Fluch der Architektur] meinen, bin ich damit einverstanden."[4]

Bei der Sanierung und Erweiterung der California Academy of Sciences (San Francisco, Kalifornien, USA, 2005–08; Seite 310) durch Renzo Piano wird Landschaft von einer völlig anderen Perspektive aus betrachtet, nämlich der des begrünten Daches. Die mehr als ein Hektar große wellenförmige Dachlandschaft ist mit mehr als 1,8 Millionen heimischen Pflanzen aus Kalifornien bepflanzt. Im Vorfeld dieses Entwurfs wurden sowohl sorgfältige Pflanzenstudien als auch mögliche Auswirkungen von Erdbeben auf ein begrüntes Dach durchgeführt. Berechnun-

15
NBGM Landscape Architects, The
Freedom Park, Salvokop, Tshwane,
Pretoria, South Africa, 2002–10

gen haben ergeben, dass dieser Dachentwurf die Innenraumtemperaturen des Museums um etwa 6 °C senkt. Die Dachform und der gesamte Museumsentwurf sind so konzipiert, dass sie mit der umgebenden Parklandschaft eine Einheit bilden. Auch wenn ökologische Themen momentan sehr in Mode sind, so zeigt doch die Idee für dieses grüne Dach eines hochmodernen Gebäudes, wie „grüne" Ansprüche in Gebäude umgesetzt werden können, die in gewisser Hinsicht in die Landschaft eingebettet sind, oder – genauer gesagt – eine eigene Landschaft darstellen. In diesem Fall bilden Architektur und Landschaftsarchitektur sowohl durch die bewegte Form der Dachlandschaft als auch aufgrund des Gründachkonzeptes wirklich eine Einheit.

ARCHITEKTUR UND LANDSCHAFT VERSCHMELZEN ZUR EINHEIT

Steven Holl hat wiederholt sein Interesse an Themen bekundet, die mit Landschaft zu tun haben. Eines seiner neuesten Gebäude, das Vanke Center/Horizontal Skyscraper (Shenzhen, China, 2008–09; Seite 162), ist in die umgebende Landschaft eingebettet. Der Bau beinhaltet unter anderem ein Auditorium mit 500 Sitzplätzen und mehrere Restaurants. Der Aspekt der Landschaftsgestaltung bei diesem Projekt gehört ebenfalls zu der von Steven Holl verfolgten Umweltstrategie. So dienen die mit Grauwasser gespeisten Teiche zur Luftkühlung. Es werden sowohl Solarzellen eingesetzt als auch örtliche Materialien wie Bambus. Für ein anderes Gebäude von Steven Holl, die Kläranlage Whitney Water Purification Facility (South Central Connecticut, USA, 1998–2005), hat das Büro Michael Van Valkenburgh Associates (2001–05; Seite 364) die Landschaftsplanung ausgeführt. Als das Projekt fertiggestellt war, erklärte Steven Holl: „Der Entwurf verbindet die Architektur des Klärwerks mit der Landschaft, und so entsteht ein öffentlicher Park. Die eigentliche Wasseraufbereitungsanlage befindet sich unter dem Park, wohingegen die Betriebssteuerung und die öffentlichen Bereiche in einer 110 Meter langen Edelstahlröhre untergebracht sind, die durch ihre Form die Funktionen der darunterliegenden Anlage verkörpert und in der Landschaft eine spiegelnde Horizontlinie zeichnet. Der Querschnitt der Röhre hat die Form eines umgekehrten Wassertropfens, und so ist auch der Innenraum gekrümmt, der mit einem großen Fenster zur umgebenden Landschaft geöffnet ist." Der Erdaushub wurde wiederverwertet, und überall wo es möglich war, wurden recycelbare Materialien eingesetzt. Der vorhandene Feuchtbiotop und die natürliche Vegetation blieben erhalten. Das Dach der Kläranlage wurde als pflegeleichtes Gründach gestaltet. Das von Michael Van Valkenburgh geplante Gelände umfasst etwa 5,6 Hektar. Der Landschaftsarchitekt bemerkt zu dem Projekt: „Die Verwendung der grundlegenden Gestaltungselemente der Landschaftsarchitekten – Erde, Wasser und Pflanzen – dient als Ausgleich für die schlanke Form des Gebäudes. Der Entwurf schafft eine topografische Vielfalt und ist wegen der nachhaltigen Wiederverwendung des Aushubmaterials interessant. Bodensenken ersetzen das herkömmliche technische Drainagesystem. Das Bepflanzungsprogramm beruft sich auf die Ökologie der Sanierung und ist in seinem Umfang und seiner Komplexität sowohl einfach als auch sorgfältig durchdacht." Die völlige Übereinstimmung von Architekt (Holl) und Landschaftsarchitekt (Van Valkenburgh) bezüglich der Aspekte der „Natur" hat dazu beigetragen, dass dieses Projekt eine bemerkenswerte Harmonie ausstrahlt. In diesem Fall haben Architekt und Landschaftsarchitekt sehr eng zusammengearbeitet, um ein hochmodernes Industriegebäude in ein Umfeld einzubetten, das fast so aussieht, als hätte hier nie ein Eingriff stattgefunden. Das mag vielleicht in gewisser Hinsicht die Richtung aufzeigen, in der sich die Landschaftsarchitektur in Zukunft entwickeln wird, nämlich als Reaktion auf die Übergriffe der modernen Zivilisation die Natur als kreativen Akt zu neuem Leben zu erwecken.

*16
Yoshiaki Nakamura, Miho Museum Gardens, Shigaraki, Shiga, Japan, 1995–97*

16

IN DEN BERGEN DER PRÄFEKTUR SHIGA

Auch wenn beim folgenden Beispiel das Verhältnis zwischen Architekt und Landschaftsarchitekt nicht nur von Harmonie bestimmt war, gehört das Miho Museum von I. M. Pei (Shigaraki, Shiga, Japan, 1997) doch zu den Bauwerken, bei denen eine zeitgenössische Architektur vollständig in ein natürliches Umfeld eingebettet ist. Der Bauherr beauftragte Yoshiaki Nakamura mit der Gartengestaltung (Seite 260), der für seine Arbeiten im Kontext mit traditioneller japanischer Architektur bekannt ist. Das Museum liegt etwa eine Autostunde von Kioto entfernt auf einem unberührten Gelände. Aus diesem Grund hatten die Behörden verlangt, dass große Teile des Museums unterirdisch gebaut werden sollten. Der Originalentwurf von Pei sah vor, einen Großteil des Gebäudes in dem hügeligen Terrain „verschwinden" zu lassen und es mit einer dichten Vegetationsschicht zu bedecken. Mehr als zehn Jahre nach der Eröffnung ist das Museum fast Teil der Landschaft geworden. Nakamura rahmte den Zugangsweg bis zur Öffnung des Eingangstunnels zum Museum mit Reihen von Trauerzierkirschen. Er arbeitete bei der Gestaltung eines Gartens im Nordflügel mit Fels, Kies und Moos direkt mit Pei und dem Bauherren zusammen. Bei diesem Projekt war Pei an der Auswahl der Felsblöcke beteiligt und bat Nakamura auch, mehr Moos im Garten anzupflanzen, damit er nicht so steinern wirkt. Alle Spannungen, die anlässlich solcher Auswahlkriterien bestanden, haben längst nur noch anekdotischen Wert. Das Miho Museum gehört weltweit zu den spektakulärsten modernen Museumsbauten, und vielleicht ist es auch eines, das am besten in seine wunderschöne Naturumgebung eingebettet ist. Die geometrischen Dach- und Fensterformen der Architektur sind immer noch sichtbar, ebenso wie die preisgekrönte Hängebrücke, die zum Museumseingang führt. Wenn man mit Recht behaupten kann, dass Holls Kläranlage in Connecticut ein wenig wie ein Fremdkörper in der Landschaft wirkt – Van Valkenburgh spricht von dem „Ausbalancieren der schlanken eleganten Gestalt des Betriebsgebäudes..." –, dann sind bei dem Miho Museum die Landschaft und die Architektur wirklich optimal verschmolzen. Es ist eine sehr harmonische Verbindung von Innen- und Außenraum entstanden.

Ein völlig anderes Beispiel für das Nebeneinander von Natur und Architektur ist die Fußgängerbrücke über den Carpinteira (Covilhã, Portugal, 2003–09; Seite 94) des Lissaboner Architekten João Luís Carrilho da Graça. Das Projekt umfasst drei Fußgängerbrücken in den Flusstälern des Goldra und des Carpinteira, mit denen das historische Stadtzentrum von Covilhã besser als bislang an die außerhalb liegenden Wohnviertel angebunden werden soll. Die extrem einfachen, reduzierten Formen der Brückenbauwerke stehen in einem bewussten Kontrast zur Architektur der Stadt und zur umgebenden Natur. Aber die Fußgänger, die über diese Brücken gehen, sind dank der Arbeit des Architekten in der Lage, diesen Rahmen in einer völlig anderen Weise zu verstehen und wertzuschätzen. Man kann mit Recht sagen, dass man durch die Fußgängerbrücke über den Carpinteira die Landschaft in einer völlig neuen Weise wahrnimmt. Sie ist ganz offensichtlich künstlich hinzugefügt und versucht gar nicht erst, sich zu integrieren, aber ihre strengen Linien verkörpern doch in gewisser Weise eine authentische Landschaftsarchitektur. Zwei Brückenpfeiler sind spiralförmig mit Naturstein verkleidet, auf dem sich Pflanzen niederlassen und die Stützen verbergen sollen. Nach einer Weile soll es dadurch so aussehen, als ob die Brücke aus dem Gelände herausgewachsen wäre. Sie wird zu einer Hommage an den Ort.

A BEAUTIFUL MIND – EIN WUNDERBARER GEIST

Eine interessante Variante der Verbindung von Landschaftsplanung und Architektur stellt das One Kowloon Building in Hongkong dar, wo der Zen-Priester Shunmyo Masuno SanShin-Tei (China, 2006–07; Seite 328) geschaffen hat. Dieses Projekt nimmt etwas mehr als 1500 Quadratmeter vor dem Eingangsbereich des Gebäudes und etwa 550 Quadratmeter der großen Empfangshalle ein. Shunmyo Masuno ist seit 2001 Oberpriester des Tempels Kenkoh-ji in Yokohama und Direktor der Agentur Japan Landscape Consultants Ltd. Er erklärt, dass der Name der Installation von dem Zen-Wort „sanshin" (wörtlich übersetzt: „drei Geisteshaltungen") abgeleitet ist. Durch die Verbindung der Zahl drei (san) mit dem Wort für Gedanke oder Geist (shin) entsteht ein Begriff, den man mit „freudiger Geist", „reifer Geist oder Verstand" oder „großer Geist" übersetzen kann. Diese Gedankenkonzepte werden als Energiefluss begriffen, der von oben, vom Dach, bis hinunter in die Eingangshalle fließt. Er wird unter anderem durch einen 9,5 Meter hohen Wasserfall markiert. Von der Installation im Gebäudeinneren, die Shunmyo Masuno als „künstlich" bezeichnet, gibt es eine Art Übergang nach draußen zu einer „natürlicheren" Gestaltung. Wenn man den geistigen Hintergrund des Planers betrachtet, so überrascht es nicht, dass SanShin-Tei auch eine religiöse Bedeutung hat. Allerdings ist der Kontrast zwischen dem Natürlichen und dem Künstlichen ein immer wiederkehrender Grundgedanke in der Geschichte der Gartenkunst. In diesem Fall ist die moderne Architektur weniger ein Objekt als die Verkörperung einer Haltung, die andere Kräfte bemüht, um beides, den Erfolg des betreffenden Unternehmens und auch das Wohlbefinden derer, die sich in irgendeiner Weise One Kowloon nähern, darzustellen.

Während Shunmyo Masuno eine Ausdrucksform der Natur in ein architektonisches Umfeld hineinträgt, so kann man von John Pawsons Brückenprojekt der Sackler Crossing in den Royal Botanic Gardens (Kew, London, 2006; Seite 304) das Gegenteil sagen. Pawson ist als Vertreter und Verteidiger einer minimalistischen Architektur und Innenarchitektur bekannt. In den Botanischen Gärten in London schuf er eine 70 Meter lange Brücke, die er selbst als „schlangenförmiges skulpturales Objekt" bezeichnet. Als Materialien setzte er neben Granitplatten auch Metallteile aus einer Bronzelegierung ein, die normalerweise für U-Boot-Schiffsschrauben verwendet wird. Alle Materialien signalisieren, dass man etwas Solides, Dauerhaftes schaffen wollte. Pawson erklärt, dass die Sackler Crossing dazu da ist, wegen ihrer Schönheit bewundert zu werden, und auch dazu, dass die Besucher der Botanischen Gärten die umgebende Landschaft aus neuen, unerwarteten Blickwinkeln erfahren können. Die Fußgängerbrücke über den Carpinteira und auch die Sackler Crossing bieten neue Blickbeziehungen zur umgebenden Landschaft, aber die Brücke von Pawson ist so niedrig, dass sie über dem Wasser zu schweben scheint, wohingegen die Arbeit von João Luís Carrilho da Graça als zarte Linie durch die Luft über dem Gelände geführt wird. In beiden Fällen wird dem Minimalismus in der zeitgenössischen Architektur eine Referenz erwiesen, die vielleicht per definitionem im Widerspruch zu der Wildheit und Vielschichtigkeit der Natur steht. Beide Architekten lehnen offensichtlich jede Art von Mimikry ab und versuchen mit dem heute zur Verfügung stehenden architektonischen Vokabular eine Leichtigkeit im Ausdruck finden, die man als klassisch und schlicht bezeichnen kann.

Zwei Projekte des bekannten, in Seattle ansässigen Büros Gustafson Guthrie Nichol präsentieren Lösungen für Gartenplanungen im Innen- und im Außenbereich. Der Lurie Garden (Chicago, Illinois, USA, 2004; Seite 132) ist ein Garten auf dem Dach des Lakefront-Millennium-

*17
John Pawson, Sackler Crossing,
Royal Botanic Gardens, Kew, London,
UK, 2006*

Parkhauses im Millennium Park von Chicago. Die umgebende Architektur ist sehr anspruchsvoll, befindet sich doch das Parkhaus zwischen der von Frank O. Gehry entworfenen Konzertmuschel und dem Modern Wing, dem Flügel für moderne Kunst des Chicago Art Institute von Renzo Piano. Da es sich um einen Dachgarten handelt, wurde leichter EPS-Hartschaum (Geofoam) zur Modulierung des Gartens verwendet. Man kann den Lurie Garden als zeitgenössische Version formaler Gartengestaltungen bezeichnen, wie man sie aus der Vergangenheit kennt, und er harmoniert perfekt mit seinen zeitgenössischen Nachbarn. Der Gegensatz von Künstlichem und Natürlichem wird in diesem urbanen Umfeld deutlich zum Ausdruck gebracht. Dazu gehört auch der unsichtbare Unterbau aus Geofoam. Es ist einerseits beruhigend und andererseits ausgesprochen attraktiv, „Natur" unter diesen Gegebenheiten zu betrachten. Aber es ist so, wie immer bei Gärten, dass das vom Menschen geschaffene Element unter Umständen stärker präsent ist als ein unkontrollierter Ausdruck der Natur selbst.

Ein zweites Projekt von Gustafson Guthrie Nichol ist der Robert und Arlene Kogod Courtyard, (Washington, D. C., USA, 2005–07; Seite 136). Dieses Projekt befindet sich im ehemaligen Gebäude des Patentamts, das von Foster + Partners umgebaut worden ist, und in dem sich nun das Smithsonian American Art Museum und die National Portrait Gallery befinden. Die Landschaftsarchitekten hatten die Aufgabe, einen mit Glas überdachten Innenhof zu gestalten. Ein Hauptgestaltungselement sind große Pflanztröge aus Naturstein, die mit 7,5 Meter hohen Bäumen bepflanzt sind. Die Landschaftsarchitekten sagen dazu: „Die Bepflanzung des Robert and Arlene Kogod Courtyard betont das gesamte Entwurfskonzept. Die gemäßigten Farben der immergrünen Bäume und Gehölze, die unterschiedlichen Größen, Formen und Texturen sowie ihre symmetrische oder asymmetrische Anordnung vervollständigen die vorhandene Architektur und betonen sie." Dieses Projekt, das noch nachhaltiger ist als der Lurie Garden, führt die Planer in einen Bereich, in dem eine künstliche, kontrollierte Auffassung von Natur eher einen architektonischen als einen natürlichen Charakter annimmt. Dies soll keinesfalls als Kritik verstanden werden, es ist einfach nur eine Feststellung. Hier ist Landschaftsarchitektur eine Vervollständigung des Gebäudes und bildet gewissermaßen einen Kontrapunkt zu der steinernen Umgebung. Gartenhöfe müssen fast zwangsläufig auf diese Weise mit der Natur in einen Dialog treten, und das Projekt des Büros Gustafson Guthrie Nichol ist ein hervorragendes zeitgenössisches Beispiel dafür.

Maya Lin wurde 1959 in Ohio geboren. Sie ist eine ungewöhnliche Persönlichkeit und arbeitet in drei Disziplinen: Landschaftsarchitektur, Kunst und Architektur. Im Alter von nur 22 Jahren realisierte sie das Vietnam Veterans' Memorial auf der Mall in Washington, D. C. Dieses Werk ist das bisher bekannteste von ihr geplante Projekt. Lin hatte den offenen Wettbewerb für diese Gedenkstätte gewonnen, und ihr Entwurf, eine Mauer aus schwarzem Granit, die zum größten Teil in die Erde eingelassen ist, zieht immer noch jährlich Zehntausende Besucher an. In die Mauer sind die Namen der etwa 50 000 in diesem Krieg getöteten Amerikaner in der Reihenfolge ihres Todesdatums eingraviert. Die Gedenkstätte rief große Kontroversen, besonders unter den Kriegsveteranen, hervor. Kritiker sagten, es sei, als schäme sich Amerika seiner Toten. Für andere, die etwas von moderner Architektur und Design verstehen, war die Arbeit von Maya Lin eine grundlegend neue Geste in einem Bereich, wo es lange Zeit nur heroische Bronzefiguren mit wehenden Fahnen gab. Ihre Arbeit war bahnbrechend und in ihrer zurückhaltenden Kühnheit geradezu revolutionär.

18
Gustafson Guthrie Nichol, Robert and
Arlene Kogod Courtyard,
Washington, D.C., USA, 2005–07

WELLEN, SCHLANGEN UND EINE LANDKARTE

Seitdem hat Maya Lin weitere hochinteressante Denkmäler, Wohnhäuser und Kunstwerke geschaffen. Allerdings erlangten diese Arbeiten nicht den gleichen Bekanntheitsgrad wie ihr Kriegerdenkmal. Zwei ihrer Arbeiten werden im vorliegenden Band vorgestellt. Storm King Wavefield (Mountainville, New York, USA, 2007–08; Seite 210) befindet sich in einem ehemaligen Steinbruch. Die wellenförmigen, mit Rasen bedeckten Hügel, mit denen die Künstlerin und Architektin hier die Landschaft moduliert hat, sollen dem Besucher das Gefühl des „völligen Eintauchens" vermitteln. Diese Arbeit gehört eindeutig zur Kategorie Landschaftsarchitektur, ist aber zugleich auch ein Kunstwerk, das keinem besonderen Zweck dient, außer dass es den Betrachter überraschen und faszinieren soll. Maya Lin gestaltet die uns so vertraute Erde in einer völlig unerwarteten Art. Dadurch wird das Interesse des Besuchers geweckt, und man beginnt, über die Natur anders nachzudenken.

Ihre Eleven Minute Line (Wanås, Schweden, 2004; Seite 212) ist ein 152 Meter langer, schlangenlinienförmiger, ebenfalls mit Rasen bewachsener Hügel. Lin ließ sich bei dieser Arbeit von den Grabhügeln der Indianer im Mittelwesten Amerikas inspirieren, wo sie selbst aufgewachsen ist. Der Entwurf begann mit einer Handskizze, wie eigentlich immer noch die meisten Kunstwerke oder architektonischen Entwürfe. Diese Arbeit ist im Prinzip abstrakt und modern, aber zugleich ruft sie uralte Erinnerungen wach, auch wenn der Standortwechsel von Ohio nach Schweden nicht unbedingt naheliegend ist. Vielleicht will die Künstlerin damit zum Ausdruck bringen, dass die Erde so viele Jahrhunderte gebraucht hat, um ihre Gestalt anzunehmen, dass jedes Land und jede Kultur Geheimnisse in der Erde birgt, wie Eleven Minute Line sie andeutet. Das Werk von Maya Lin hat elementaren Charakter, sodass sie mit Arbeiten wie den hier vorgestellten an der Verbindungsstelle zwischen antiker und zeitgenössischer Kunst steht.

Der Ansatz von Catherine Mosbach ist eindeutig mehr der einer Landschaftsarchitektin, aber auch ihre Arbeiten sind künstlerisch geprägt. Ihr Beitrag zur 400-Jahrfeier von Quebec, war der mit 150 Quadratmetern sehr kleine temporäre Garten L'autre rive (Das andere Ufer, Quebec, Kanada, 2008; Seite 256). Mosbach gestaltete Asphaltflächen so, dass sie an Landkarten erinnern, wie man sie zur Zeit der Entdeckung der Neuen Welt benutzte. Ihre eigene Beschreibung des Projekts zeigt, wie „künstlerisch" ihr Ansatz war. „Dieser Garten ist ein Mythos, der von Nahem betrachtet die vorhandenen Materialien – Erde, Pflanzen, Formen – einander gegenüberstellt. Von Weitem hat man den Eindruck einer Landschaft mit anregenden Umrisslinien … Die Erde als Speicherfläche, die sich von der Erdkruste dadurch unterscheidet, dass sie lebendig ist, wird hier als dünne Asphalthaut dargestellt – als Vegetation, die im Wasser der Sümpfe erst versteinerte und dann mineralisierte … Die Besucher gehen durch dieses Milieu und entdecken es, aufgrund ihrer eigenen Bewegung, zwischen den verborgenen Elementen des Unbewussten." Mosbach beschäftigt sich mit der Landschaft der Neuen Welt, aber weniger mit den uralten Mysterien, die von Maya Lin mit ihrer Eleven Minute Line heraufbeschworen werden. Beide Frauen jedoch sind davon überzeugt, dass Landschaft ein Medium für die neuesten Ausdrucksformen der Kunst sein kann – Kunst in Form von Installationen und nicht unbedingt als Objekte, die für ein Museum bestimmt sind. Beide arbeiten mit dem Atmosphärischen, ohne in irgendeiner Form figurativ zu sein, so ist ja auch die Schlange von Maya Lin nur eine formale Andeutung. Man könnte beide ebenso gut als Künstlerinnen wie auch als Landschaftsarchitektinnen bezeichnen.

Der Rang, den die zeitgenössische Landschaftsarchitektur innehat, spielt mit dem Potenzial an Mehrdeutigkeit zwischen den beiden Disziplinen. Auch wenn die frühesten Beispiele der Landschaftsgestaltung mit der Anlage großer Plätze oder der Gestaltung des Umfelds von Königsgräbern immer sehr symbolträchtig waren, so ist Landschaftsarchitekur auch heute in der Lage, Vergangenheit und Gegenwart miteinander zu verbinden, ohne auf irdische Mächte Bezug nehmen zu müssen, die schon längst in Vergessenheit geraten sind. Der amerikanische Architekt Ieoh Ming Pei wies darauf hin, dass André Le Nôtre bereits sehr früh Wasserflächen mit ihren fantastischen Himmelsspiegelungen in seinen Entwürfen eingesetzt hat und damit ein Vorreiter des Minimalismus war, den heute viele Architekten und Planer anstreben. Macht drückte sich bei André Le Nôtre in den grenzenlosen Sichtachsen und dem unendlichen Raum von Versailles oder den anderen Symbolen der französischen Krone aus. Auch wenn die französischen Könige schon lange der Vergangenheit angehören, so bestehen die von Le Nôtre geschaffenen Sichtachsen, wie die Champs-Élysées immer noch. Sie vermitteln ein Gefühl von Pracht, die nicht mit Ludwig XIV. in Verbindung gebracht werden muss, sondern sie sind nun Ausdruck der Größe einer Nation und ihrer geistigen Errungenschaften.

DIE LANDSCHAFT VOR DEM FALL
Wenn das vorliegende Buch ein Ziel hat, dann dieses – es will zeigen, dass wir von Landschaftsarchitektur umgeben sind. Sie zeigt sich in der Architektur selbst, in Gärten und in den aktuellen Bemühungen, die Natur wieder zu beleben, die durch den Raubbau der Industrialisierung und die Umweltverschmutzung in der Vergangenheit vernachlässigt und ins Abseits geraten ist. Ein Gebäude kann in eine Landschaft eingebettet oder von einer Landschaft umgeben sein. Ein Architekt kann eine Brücke bauen, die die Vorzüge der umgebenden Landschaft hervorhebt, wobei die Konstruktion als solche dahinter zurücktritt. Architekten wie Steven Holl, I. M. Pei oder Tadao Ando (der im vorliegenden Buch auf Seite 60 ebenfalls mit einer Arbeit präsentiert wird) betrachten ihre Gebäude oft als Teile eines Umfeldes, das sie ebenfalls gestalten wollen. So präsentiert Ando in seinem unterirdischen Chichu Museum (Naoshima, Japan, 2004) Meisterwerke von Claude Monet, James Turrell und Walter De Maria in unterirdischen Ausstellungssälen und gestaltete die Flanke des Hügels über dem Gebäude neu. Das ist Architektur als Landschaft und umgekehrt. Tadao Ando ist sich der uralten Symbolik der Erde bewusst, bleibt aber als Modernist der Geometrie treu. Diese Verbindung, die man auch im Werk von Maya Lin beobachten kann, ist das wirkliche Maß für das heutige Interesse an der Landschaftsarchitektur. Sehr gute Landschaftsarchitektur ist immer auch Kunst.

Die meisten landschaftsarchitektonischen Entwürfe sind mit dem Streben nach „Perfektion" verbunden, die nicht unbedingt daran ausgerichtet ist, die Architektur zu beleben. Es geht vor allem um den Grundgedanken des Gartens an sich, der nach jüdisch-christlicher Tradition die Welt vor dem Sündenfall symbolisiert oder nach islamischen Vorstellungen das himmlische Paradies. Selbst die heutigen, ökologisch geprägten Tendenzen nehmen überraschenderweise den Gedanken auf, wieder zu einem Stadium der Reinheit zurückzukehren, vor dem Sündenfall. Die Erneuerungskraft der Landschaft, oder genauer gesagt der Natur, weckt die immer neue Hoffnung auf Erlösung, selbst angesichts der massiven Zerstörungen der Natur, die durch Not oder durch Gier geschehen. Alles Menschengemachte kann infrage gestellt werden, ist Launen und Moden unterworfen, wohingegen Natur ihre eigene Legitimation hat, und Ziel der Landschaftsarchitektur ist es, diese Legitimation zumindest teilweise zu erreichen. Landschaftsarchitektur ist zwar vom Menschen gemacht, aber sie ist der Natur näher als andere Ausdrucksformen. Sie ist nicht nur Ausdruck individueller Kreativität, sie sucht nach Erlösung, nach einer Rückkehr zu einem Urzustand. Frank Lloyd Wright hat von einer Architektur geträumt, die von der Natur inspiriert ist, aber bei den heutigen Bauwerken ist dies nur sehr selten der Fall. Kunst ist heute oft von der Natur inspiriert, aber sie kann nicht wirklich darauf hoffen, in einen Zustand der Perfektion zurückzukehren, den es einmal gab … vor dem Sündenfall.

[1] http://www.mnstate.edu/gracyk/courses/web%20publishing/addison414.htm, Zugriff am 16. November 2010.
[2] „Horizon Press", New York, 1953.
[3] http://whc.unesco.org/en/list/1150, Zugriff am 16. November 2010.
[4] Peter Eisenman, zitiert auf der Website http://www.curatorialproject.com/interviews/petereisenmanii.html, Zugriff am 16. November 2010.

INTRODUCTION

RECHERCHE RÉDEMPTION DÉSESPÉRÉMENT

Si l'idée de mettre en forme le paysage autour des lieux du pouvoir – palais ou grandes places – est ancienne, la définition moderne de l'architecture paysagère ne remonte qu'au XIXe siècle. Les œuvres du jardinier de Louis XIV André Le Nôtre (1613–1700), comme la grande perspective de Versailles, les jardins de Vaux-le-Vicomte ou l'axe des jardins des Tuileries et des Champs-Élysées à Paris, sont autant d'exemples de modification d'un paysage en relation avec l'architecture. Alors que des villes comme Paris se caractérisaient à l'époque par leur réseau de rues étroites bordées d'immeubles, Le Nôtre donna un nouveau sens à l'espace, qu'il associa dans ses projets à l'expression de la grandeur royale. Les perspectives et les grands bassins qui rayonnent à partir du château de Versailles expriment le pouvoir. Ils amplifient l'architecture du palais et confèrent une dimension nouvelle et impressionnante à sa présence symbolique.

Les œuvres architecturales qui prennent en compte le cadre naturel ne relèvent pas forcément de l'architecture du paysage, mais c'est précisément l'intégration de ces deux éléments (paysage et architecture) qui fait la grandeur de beaucoup de réalisations célèbres. Lorsqu'on arpente les superbes vestiges de la Grande Muraille de Chine, près de Pékin, construite à partir du Ve siècle av. J.-C. pour protéger les frontières septentrionales de l'Empire, comment ne pas comprendre le sens de l'irruption dans le paysage de ce mur qui serpente à travers les montagnes et les vallées, en se jouant de son cadre naturel ? Toujours en Chine, la place Tianan men et la Cité interdite étonnent encore par leur mise en majesté d'un espace pratiquement vide qui ancre les bâtiments dans un contexte de pouvoir et de puissance, perçu comme tel et par tous depuis des siècles.

BELLES ET ÉTRANGES

Dans cet ouvrage, axé sur des réalisations très récentes, l'architecture paysagère est définie non seulement comme la création de jardins, mais également celle de constructions entretenant une relation intime avec la nature, qui viennent en un sens de la terre et donnent une signification à l'espace et aux matériaux utilisés. Peut-être devrait-on parler plutôt d'architecture du paysage même si, selon la définition classique, le bâti est souvent exclu de ce type de projets. C'est l'Écossais Gilbert Laing Meason (1769–1832) qui a inventé en anglais le terme de *landscape architecture*. Son ouvrage sur l'architecture du paysage chez les grands peintres de l'Italie (*The Landscape Architecture of the Great Painters of Italy*, 1828) traitait des bâtiments figurant dans les sites représentés et participant à de splendides compositions. Dans cette optique, l'origine même du terme d'architecture du paysage prend en compte le bâti. Les relations entre l'art, la nature et la vision évoluée du pittoresque qui animait en grande partie la pensée romantique précèdent néanmoins de loin Meason. Joseph Addison (1672–1719), auteur et politicien anglais, écrivait ainsi dans le journal *The Spectator* (n° 414, 25 juin 1712) :

Si nous considérons les œuvres de la nature et de l'art en ce qu'elles sont capables de divertir l'imagination, nous trouvons que les dernières sont très inférieures aux premières ; car si elles peuvent parfois paraître tout aussi belles et étranges, elles ne sauraient posséder en elles cette vastitude et cette immensité qui offrent un tel enchantement à l'esprit de l'observateur. L'une peut être aussi

19

raffinée et délicate que l'autre, mais ne saura jamais se montrer aussi auguste et magnifique dans son dessein. Il y a quelque chose de plus audacieux et de plus magistral dans les traits grossiers et désordonnés de la nature que dans les touches de pinceau et les embellissements de l'art. Les beautés des jardins ou des palais les plus princiers sont contenues dans un cercle étroit. L'imagination va immédiatement au-delà et demande quelque chose d'autre pour se satisfaire, mais dans le vaste champ de la nature, le regard va de tous côtés sans restriction et se nourrit d'une infinité d'images sans se limiter. C'est pour cette raison que tous les poètes aiment la vie à la campagne, là où la nature apparaît dans sa perfection la plus grande et leur offre la multiplicité de scènes plus aptes à ravir leur imagination [1].

En fait, la dichotomie entre le « naturel » et « l'artificiel » est l'une des clés de l'appréciation ou de la compréhension de l'architecture du paysage, quelle que soit sa forme. L'objectif de l'architecte paysagiste, ou de l'architecte, serait-il, d'une certaine façon, de reproduire la nature, ou plutôt d'imposer un ordre humain aux jardins ou à l'environnement des bâtiments ? L'origine de ce débat remonte à très loin, aussi bien dans l'histoire que dans la littérature, et pourrait presque être reliée à l'aspiration primale vers des temps très anciens, quand tous les hommes vivaient dans la nature. Ainsi, une bonne part de l'architecture du paysage serait en rapport avec la quête du jardin idéal.

… dans un étroit espace, il voit renfermée pour les délices des sens de l'homme toute la richesse de la nature, ou plutôt il voit un ciel sur la terre ; car ce bienheureux Paradis était le jardin de Dieu, par lui-même planté à l'orient d'Eden.
John Milton, *Le Paradis perdu*, livre IV
(traduction de François-René de Chateaubriand)

La première personne connue à avoir fait professionnellement état de sa capacité d'architecte paysagiste a été l'Américain Frederick Law Olmsted (1822–1903), en 1863. En collaboration avec son associé Calvert Vaux, Olmsted a dessiné Central Park à Manhattan, et de nombreux autres parcs aux États-Unis. Olmsted et Vaux avaient remporté, en 1857, le concours organisé pour l'aménagement d'une zone de 280 hectares entre la 59e et la 106e Rue. De nombreux visiteurs imaginent peut-être que Central Park est une vision de l'état naturel de l'île de Manhattan avant son urbanisation et ignorent qu'il a fallu faire venir environ dix millions de charrettes de terre et de rochers et planter plus de quatre millions d'arbres et de buissons avant son ouverture en 1873. Parsemé de 36 ponts dessinés par Vaux et de diverses petites constructions qui ont évolué dans le temps, Central Park reste le témoin des origines de l'architecture paysagère aux États-Unis, un espace que personne n'imaginerait pouvoir être sacrifié aux empiètements de la ville moderne.

BRISER LES BARRIÈRES

Le degré d'intégration du bâti et de l'architecture du paysage varie selon les circonstances et dans cet ouvrage figurent volontairement quelques exemples qui bousculent ce que certains voient comme des barrières entre les deux disciplines. À quel moment le paysage et

*19
Vladimir Djurovic, Salame Residence
Faqra, Lebanon, 2009*

*20
WORK Architecture Company, Public
Farm 1, Long Island City, New York,
USA, 2008*

l'architecture fusionnent-ils au point de devenir indistinguables l'un de l'autre ? Chaque bâtiment ou parc est déterminé par les circonstances, mais le but ultime de l'architecture du paysage n'est-il pas de faire tomber ces barrières (artificielles) qui la séparent du champ d'action des constructeurs de ponts et de tours ? Un intervenant aussi important que Frank Lloyd Wright avait son opinion sur le sujet :

> *Le changement est l'une des caractéristiques immuables du paysage. Mais les changements nous parlent tous – ou chantent tous – à l'unisson de la loi cosmique, qui est elle-même une forme plus noble de changement. L'homme s'empare positivement de la création dès qu'il érige une construction sur la terre et sous le soleil. Son droit héréditaire, s'il existe, consiste en ceci : il n'appartient pas moins au paysage que les rochers, les arbres, les ours ou les abeilles, que la nature à laquelle il doit son existence. Sans cesse, la nature lui montre la science de sa remarquable économie de structure dans des constructions minérales et végétales dont le caractère original apparaît dans la multiplicité de ses formes.*
> Frank Lloyd Wright, *The Future of Architecture* [2] (traduction)

PETIT, MAIS PARFAITEMENT FORMÉ

La question de l'instant auquel l'architecture et le paysage se transforment en architecture du paysage est assez complexe et ce livre a opté pour une vision globale de ce débat. Une première approche pourrait consister à prendre en compte de petits projets urbains dans lesquels les concepteurs (architectes paysagistes ou architectes) intègrent la nature dans la cité en créant avec elle un rapport là où le béton, l'acier et le verre éliminaient souvent toute trace de vie et de verdure. Le « Jardin qui monte les escaliers » (The Garden That Climbs the Stairs, Bilbao, Espagne, 2009, page 72) couvre tout juste 80 mètres carrés. Sa créatrice, Diana Balmori, se réfère à « un espace urbain dynamique » pour décrire ce qu'est un jardin installé sur un escalier, à proximité de plusieurs réalisations architecturales contemporaines prestigieuses : « Dans son geste, il raconte une histoire du paysage prenant le pas et se développant sur l'espace public et le bâti, transformant ainsi la façon dont les escaliers et l'espace sont perçus et lus par l'usager. » Cette brève description d'un petit jardin atténue en partie l'ambiguïté qu'engendre la présence de la nature dans un environnement urbain fondamentalement artificiel. Le paysage s'attribue ici le pouvoir de « transformer » son environnement architectural et vient en contrepoint des matériaux « durs » du bâti.

GRIMPER AUX MURS

Le célèbre botaniste français Patrick Blanc a installé plusieurs de ses « jardins verticaux » dans le monde. Ayant observé que les plantes n'avaient pas réellement besoin d'un sol, mais plutôt de nutriments et d'eau, il a breveté un procédé qui lui permet de concevoir et de créer des jardins recouvrant littéralement les murs, intérieurs ou extérieurs. Deux de ses projets récents publiés dans ces pages – *Orchid Waltz* (« Valse d'orchidées »), pour la salle de concert du Théâtre national de Taipei (République de Chine, 2009, page 78) et le pont Max Juvénal (Aix-en-Provence, France, 2008, page 80) – modifient la perception de leur environnement. Dans le second cas, un mur de végétation luxuriante adoucit la raideur d'un pont moderne que les autorités locales voulaient habiller de pierre ou de brique. Totalement surprenantes et bien que très vivantes, les créations de Blanc semblent néanmoins artificielles, ne serait-ce que par leur développement vertical. Docteur en botanique, Patrick Blanc a mené des recherches avancées sur la survie des plantes en environnements extrêmes. Il ne se qualifie pas d'architecte paysagiste, mais possède néanmoins la capacité de modifier la perception d'un immeuble ou d'un quartier. En un certain sens, il propose une redéfinition du terme même de « paysage » à partir des caractéristiques fondamentales de la nature et remet en cause les conventions sur la place d'un jardin par rapport à une architecture.

Vladimir Djurovic est un architecte paysagiste basé à Beyrouth, auteur de projets aussi ambitieux que les jardins du nouveau musée de l'Aga Khan, un musée d'art islamique, à Toronto. À Beyrouth, il a créé le square Samir Kassir (2004), salué par un prix Aga Khan d'architecture en 2007. Si elle ne mesure que 850 mètres carrés, ce qui est assez petit à l'échelle urbaine, cette réalisation apporte un sentiment de paix et de détente qui contraste avec l'animation du quartier. Plus récent, son jardin-mémorial Hariri (en cours, page 110), également à Beyrouth, est plus de deux fois plus vaste, mais reste de dimensions limitées par rapport à d'autres éléments de son environnement urbain. À l'aide de bandeaux de pierre, d'eau et de verre qui descendent en escalier vers le centre de la ville, Djurovic propose une monumentalité retenue, appropriée à la vénération que de nombreux Libanais portent à l'ancien premier ministre assassiné, Rafic Hariri (1944–2005). En matière d'aménagement du paysage, on pense souvent à cette citation du sculpteur américain Richard Serra : « La différence entre l'art et l'architecture est que celle-ci est au service d'un but. » Agréables à regarder, espaces de repos et de rassemblement et, comme dans le cas du jardin-mémorial Hariri, lieu de méditation et de souvenir, ces projets ne sont pas sans objectifs, mais se rapprochent, sous leurs meilleurs aspects, d'une création artistique.

UNE FERME SUR UN TERRAIN DE JEUX

C'est dans la cour du Centre d'art contemporain P.S.1, l'annexe du MoMA de New York à Long Island, que les architectes de WORK AC ont créé une installation appelée Public Farm (« Ferme publique ») en 2008 (page 406). Elle occupe un peu plus de mille mètres carrés. Le P.S.1 est une ancienne école dont la cour servait naguère de terrain de jeux. Cette « ferme urbaine entièrement opérationnelle en forme de plan incliné plié fait de tubes de carton structurel » tient compte de l'architecture qui l'entoure et a imposé sa présence propre. L'intérêt de ce projet n'est pas exactement le même que celui des interventions de dimensions réduites présentées plus haut. L'idée qu'une ferme, un bâtiment construit n'existant qu'en dehors des zones urbaines, puisse prendre racine, même temporairement, dans un environnement aussi dense que celui du Queens est une prise de position sur le sens du paysage et ses liens avec l'architecture. Les vignes, les rizières et les champs de blé forment le paysage de nombreuses parties du monde sans que les architectes paysagistes ne s'en soient préoccupés. Cependant, le processus de domestication de la nature, qui est au centre de la pratique architecturale, concerne également les cultivateurs qui alignent à l'infini leurs rangées de plantes céréalières. La nature est alors orientée vers un objectif, mais une installation comme celle de WORK AC remet même en question les principes de base de l'agriculture.

21
Miralles Tagliabue EMBT, HafenCity Public Spaces, Hamburg, Germany, 2006

21

Destiné aux enfants, le Teardrop Park (Battery Park City, New York, 1999–2006, page 370) de Michael Van Valkenburgh Associates, en forme de larme, occupe une superficie d'environ 7000 mètres carrés. Ses créateurs se sont particulièrement attachés à en faire un lieu écologique et durable. Ils font remarquer que les zones réservées aux enfants dans les parcs ont vu peu à peu la présence de la nature se réduire en faveur de divers équipements et veulent inverser cette tendance. L'une des installations les plus impressionnantes de ce parc est un mur-cascade fait d'un empilement de schiste bleu de 8,2 mètres de haut et 51 mètres de long, dimensions qui en font évidemment un élément très architectural. De même que l'architecture du bâti, celle du paysage doit souvent remplir des fonctions qui vont au-delà de considérations purement esthétiques. Apporter des solutions pratiques, comme dans le cas d'un terrain de jeux pour enfants, n'est pas forcément l'apanage des praticiens du paysage. Une agence comme celle de Michael Van Valkenburgh atteint un niveau de conception plus élevé en associant un sens très sûr de l'espace et de l'esthétique à la connaissance de l'environnement et de ce qu'est réellement une aire de jeux dans le cadre des tendances actuelles des parcs urbains.

Le travail de l'agence new-yorkaise Diller Scofidio + Renfro est à l'avant-garde de l'architecture contemporaine, comme en témoigne son Institut d'art contemporain à Boston (2006). Plus récemment, DS+R s'est attaqué à la rénovation du Lincoln Center, le vaste complexe culturel situé à l'angle de Central Park à Manhattan. Leur «pavillon Hypar» (Hypar Pavilion, 2011, page 106), sur la place nord du Lincoln Center, présente une toiture très curieuse : une pelouse inclinée de mille mètres carrés, qui offre aux visiteurs un havre de repos au cœur du brouhaha urbain. Étonnante par la rigueur de son dessin, la «pelouse Hypar» (Hypar Lawn) se situe précisément à l'intersection de l'architecture et de l'architecture du paysage et les prend toutes deux en compte pour les fondre dans un ensemble intégré. La nature (bien que contrôlée) retrouve une place dans un environnement par ailleurs très minéral, pour le plus grand plaisir des passants.

MARBRE BLEU ET CAOUTCHOUC NOIR

Le jardin créé par Ken Smith pour le 40 Central Park South (New York, 2006, page 336) mesure tout juste 580 mètres carrés et utilise des éléments artificiels ou minéraux comme le marbre blanc broyé, des débris de caoutchouc noir recyclés et des bandeaux de verre poli bleu rétroéclairés qui provoquent cet effet de « forte linéarité » recherché par le célèbre architecte paysagiste. Bien qu'il ne soit certainement pas le premier à oser une telle approche, il a créé le jardin sur le toit du Museum of Modern Art de New York (2003–05) à l'aide de rochers en plastique et d'arbres artificiels pour résoudre le problème de poids d'une installation classique en toiture. Ce faisant, il supprimait du même coup une grande partie de l'entretien que nécessitent les plantes naturelles. Destiné à la présentation de sculptures modernes, le jardin du 40 Central Park South ne va pas aussi loin que celui du MoMA, mais il illustre bien la démarche de Smith, pour qui la nature n'est qu'un élément parmi d'autres. Des débris de caoutchouc noir recyclés peuvent paraître bien étrangers à la nature, mais l'architecture du paysage contemporaine n'exclut pas ce type de matériaux.

Les architectes et les architectes paysagistes mettent en forme le paysage urbain, parfois à des échelles bien plus importantes que les mille mètres carrés de l'Hypar Lawn. Benedetta Tagliabue, l'ancienne partenaire d'Enric Miralles (Miralles Tagliabue Arquitectes Associats – EMBT) s'est ainsi chargée de l'aménagement d'espaces publics dans la nouvelle HafenCity (« cité du Port ») à Hambourg (Allemagne, 2006, page 250). En recourant à différents traitements de surface et à des couleurs variées, tout en prenant en compte les risques d'inondation, l'architecte a cherché à « rapprocher les gens de l'eau et de ses humeurs ». Bien qu'elle ne surprenne pas vraiment, cette déclaration met en évidence le sens fondamental des interventions de l'architecture du paysage dans le contexte de l'environnement urbain contemporain : rapprocher les citadins de la nature, alors que les projets et réalisations « modernes » ont longtemps exercé l'effet inverse. Un autre groupe de talentueux architectes espagnols, Nieto Sobejano, a récemment parachevé un chantier de plus d'un hectare à Madrid. Leur Plaza de Santa Bárbara (2009, page 278) fait partie d'un vaste projet qui comprend, entre autres, le nouveau marché Barceló (2009–11). Des pelouses, des aires de jeux et un pavillon de verre faisant office de boutique de fleurs et de livres animent cette place, qui intègre avec subtilité éléments artificiels et naturels. Ce niveau de sophistication n'est pas une invite à faire entrer la « nature » au cœur de la capitale espagnole, mais il vise plutôt à incorporer quelques éléments naturels dans une place qui fait vraiment partie de la ville. Ici, le paysage au sens plus large n'occupe qu'un rôle secondaire par rapport à l'extension et l'intégration de l'architecture dans le milieu urbain.

SON SECRET EST LA PATIENCE
Autre exemple d'intervention d'échelle assez importante dans un contexte urbain : le « parc de l'Espace civique » à Phoenix (Civic Space Park, Arizona, États-Unis, 2008–09, page 56), œuvre d'AECOM (Architecture, Engineering, Consulting, Operations and Management) et en particulier de son équipe de paysagistes (anciennement EDAW). Ce parc public d'un peu plus de un hectare se situe dans le centre même de Phoenix. Il aborde les enjeux de la conservation du patrimoine, mais aussi de l'insertion d'œuvres d'art, dont une énorme pièce de Janet Echelman appelée *Her Secret Is Patience* (« Son secret est la patience »). Pour ses concepteurs : « Cet espace urbain devait servir de lieu de rassemblement, de passage pour les piétons, d'oasis urbaine offrant ombrage et sérénité, d'endroit pour apprendre, de parc à visiter, mais aussi de lieu de détente pour le campus adjacent de l'université d'Arizona. » La nature de cette intervention et ses contraintes complexes, parfois un peu contradictoires, expliquent pourquoi la ville de Phoenix a fait appel à une agence aussi importante qu'AECOM. L'usage intense des parcs urbains exige aussi que les architectes offrent des réponses à divers types d'usages et d'occupation de l'espace. Qu'AECOM ait réussi à rendre ce parc séduisant est déjà en soi une vraie réussite.

Un autre projet d'AECOM, le Pier Head and Canal Link (Liverpool, Royaume-Uni, 2005–07, page 52), concernait un terrain de 2,5 hectares situé dans la Maritime Mercantile City (« Cité marchande et maritime ») de Liverpool, inscrite au patrimoine mondial de l'UNESCO. Ce classement lui avait été accordé en 2004 parce que « la ville et le port de Liverpool constituent un témoignage exceptionnel du développement d'une culture marchande maritime aux XVIIIe et XIXe siècles, qui a contribué à l'essor de l'Empire britannique. C'était un centre du commerce d'esclaves, jusqu'à son abolition en 1807, et de l'émigration de l'Europe du Nord vers l'Amérique [3]. » Pour l'équipe d'AECOM :

22
OKRA, Afrikaanderplein, Rotterdam,
The Netherlands, 2003–05

« Cet espace possède la souplesse et la capacité nécessaires pour constituer un lieu sûr et agréable, apte à accueillir des événements attirant jusqu'à 35 000 personnes, tout en offrant également la possibilité de contempler dans le calme les superbes bâtiments anciens, les mémoriaux de guerre et les vues sur la Mersey. » Cette intervention, comme beaucoup d'autres similaires dans différents pays du monde, s'efforce de retisser un tissu urbain existant, de créer une cohérence là où le temps et des usages antérieurs ne favorisaient pas l'ouverture de ces lieux à un grand nombre de visiteurs. À Liverpool comme à Phoenix, la tâche d'AECOM était délicate et l'agence semble avoir su négocier avec succès les écueils potentiels de ces projets.

COMPLEXITÉS NÉERLANDAISES

L'agence néerlandaise OKRA est intervenue sur l'Afrikaanderplein (2003–05, page 294), une zone de 5,6 hectares à Rotterdam. L'un des aspects intriguants de cette intervention est l'effort déployé pour prendre en compte et tenter de réconcilier diverses populations d'origines ethniques différentes qui vivent à proximité de cette place. Ce projet inclut une mosquée, un marché, un terrain de jeux et une « zone de calme » entourée de grilles, conçues pour être ouvertes en journée afin de faciliter l'accès. La verdure est très présente, tout comme l'intention sociale et le désir de combiner esthétique et aspects pratiques, et d'apporter un vrai service à la communauté au sens le plus large. Si les architectes évitent souvent de se targuer des bénéfices sociaux qu'apporteraient leurs projets, l'architecture paysagère conserve en partie ce désir utopique de rendre la vie meilleure, en particulier dans un environnement urbain où la nature est souvent réduite à sa plus faible expression. Entre les bienfaits thérapeutiques d'un calme relatif, la présence de la nature et une création esthétique censée offrir un ressenti de « qualité », les architectes paysagistes œuvrant en milieu urbain se sentent souvent en charge d'une mission : améliorer les conditions de vie.

Une autre agence néerlandaise, West 8, est l'auteur de plusieurs réalisations urbaines importantes. Son travail sur le projet Madrid RIO (2006–11, page 386) concerne la création de parcs sur 80 hectares au-dessus de l'autoroute M30, désormais souterraine. Les 8000 pins plantés dans le Salón de Pinos, les diverses variétés de cerisiers qui poussent le long de l'Avenida de Portugal et « une interprétation moderne du verger » dans la Huerta de la Partida comptent parmi les interventions les plus significatives pratiquées sur ces vastes espaces, sous lesquels circulent dorénavant les voitures. La couverture des autoroutes urbaines est une problématique à laquelle sont aujourd'hui confrontées de nombreuses villes. Les autorités municipales madrilènes ont heureusement eu la clairvoyance de ne pas céder à la densification urbaine. Le rôle de l'architecture du paysage est ici de créer des espaces qui améliorent la qualité de l'environnement urbain et rendent la vie plus agréable pour les citadins, ce qui est n'est pas une mince réussite.

L'AGRI-TECTURE À LA RESCOUSSE

Projet en cours de réalisation à Manhattan signé James Corner Field Operations et Diller Scofidio + Renfro, la High Line (New York, 2009 [phase 1], 2011 [phase 2], en cours, page 184) couvre dès à présent 1,1 hectare. Sa situation est inhabituelle puisqu'il s'agissait, à l'origine, d'une voie de chemin de fer suspendue. Selon les aménageurs : « Inspirée par la beauté originale et mélancolique de la High Line, où la nature avait repris ses droits sur un élément d'infrastructure urbaine jadis vital, l'équipe a transformé cet équipement industriel en outil postindustriel de loisir, qui est aussi une réflexion sur les catégories mêmes de "nature" et de "culture" à notre époque. En modifiant les règles des rapports entre la vie végétale et les piétons, notre stratégie d'agri-tecture combine éléments organiques et matériaux de construction selon des proportions variées, adaptées au caractère sauvage de la nature, aux plantes cultivées, à l'intime et à l'hypersocial. » La section 1 de la High Line, déjà ouverte au public, propose une promenade appréciée au-dessus du quartier en cours de rénovation du Lower West Side, qui abrite les galeries d'art de Chelsea et d'étonnants bâtiments signés, entre autres, Jean Nouvel, Shigeru Ban ou Frank O. Gehry. Si la démolition de cette voie suspendue avait longtemps été envisagée, ce projet a su réutiliser un vaste espace urbain qui avait perdu son sens depuis l'arrêt des trains en 1980. Les vestiges d'une activité industrielle en contexte urbain offrent souvent de vastes espaces d'intervention dans des sites qu'il serait difficile de retrouver aujourd'hui. Les architectes et les paysagistes auteurs de ce projet ont su faire de la High Line un espace urbain moderne et séduisant en profitant de la disparition de la fonction originelle des lieux. L'un des rôles de l'architecture paysagère est ainsi de ramener à la vie certains espaces urbains.

Le « parc de la Liberté » de Pretoria, en Afrique du Sud, (Freedom Park, NBGM Landscape Architects avec OCA Architects, 2002–10, page 266) est situé à proximité du centre de la capitale administrative de l'Afrique du Sud. Ses 30 hectares représentent un espace suffisamment vaste pour qu'il puisse imposer ses propres règles et accueillir un projet original, peu influencé par son cadre urbain. La création de ce parc, demandée par Nelson Mandela, « s'est structurée autour de quatre idées principales : réconciliation, construction d'une nation, liberté du peuple et humanité. La construction de ce paysage cherche ainsi à prendre en compte les origines spirituelles de ces idées et à les exprimer symboliquement sous forme physique ». L'idée que l'architecture du paysage pourrait interpréter ou donner une forme physique à des concepts aussi nobles intrigue et répond peut-être davantage aux ambitions d'une nation émergente qu'à celles de pays plus anciens. Le Freedom Park révèle néanmoins un autre aspect de ce dont est capable l'architecture paysagère : symboliser les combats d'une grande nation.

« SI C'EST CE QUE VOUS APPELEZ RÉDEMPTION, JE SUIS D'ACCORD. »
Comme un certain nombre d'exemples publiés dans cet ouvrage le montrent, la matière même de l'architecture du paysage est devenue souvent difficile à dissocier de l'architecture tout court. C'est indéniablement le cas de l'Hypar Lawn de Diller Scofidio + Renfro, présentée plus haut. La Cité de la culture de Galice (Ciudad de la Cultura de Galicia, Saint-Jacques-de-Compostelle, Espagne, 2001–, page 120) de Peter Eisenman pourrait bien être l'exemple type de ce mouvement récent en faveur de bâtiments paraissant sortir des profondeurs de la terre ou constituer littéralement un nouveau paysage. Lorsqu'on lui demande s'il a essayé d'intégrer ces bâtiments à la nature ou même de créer une nouvelle forme de nature, Peter Eisenman répond : « Pas de nature, mais de nature non-naturelle. Grâce à divers moyens informatiques avancés, nous avons maintenant la capacité de créer une nature non-naturelle. Je voulais créer quelque chose

23
*Steven Holl, Vanke Center /
Horizontal Skyscraper,
Shenzhen, China, 2008–09*

23

qui semble être la nature, mais dont on réaliserait après un examen rapproché que ce n'est pas le cas. Je l'appelle nature non-naturelle. Nos bâtiments à Saint-Jacques-de-Compostelle font penser au sommet d'une colline. Ils ne donnent pas l'impression d'avoir été placés là, sur ce site. Ils ont été pensés pour donner l'impression d'être sortis du sol un peu comme des montagnes. En d'autres termes, ils reproduisent un processus naturel, quelque chose qui aurait pris normalement dix millions d'années et qui s'est produit ici en dix ans. Si c'est ce que vous appelez rédemption, je suis d'accord [4]. »

La rénovation et l'extension de l'Académie des sciences de Californie par Renzo Piano (Academy of Sciences, San Francisco, Californie, 2005–08, page 310) se rapprochent elles aussi d'un paysage, mais d'un point de vue totalement différent, puisqu'il s'agit surtout de son toit végétalisé. La toiture ondulée de ce bâtiment, qui mesure plus d'un hectare, est recouverte de 1,8 million de plantes originaires de Californie. Le projet a notamment débuté par une étude approfondie de ces végétaux, mais également des contraintes de poids d'un tel toit en cas de séisme. On a calculé que cette protection naturelle permettait de diminuer la température intérieure du musée d'environ 6° C. La toiture et l'aménagement tout entier du musée forment un continuum avec le parc qui l'entoure. Si les préoccupations écologiques sont actuellement à la mode, cette idée de « toit vert » intégré à un bâtiment très contemporain montre comment la volonté de « verdir » peut prendre la forme de constructions intégrées en quelque sorte au paysage ou, mieux encore, constituant un paysage en soi. Dans le cas de ce projet, architecture et architecture paysagère ne font plus qu'un grâce à la forme paysagée du toit et à sa végétalisation.

QUAND LE PAYSAGE ET L'ARCHITECTURE NE FONT QU'UN

Steven Holl a fréquemment manifesté son intérêt pour la problématique du paysage. L'une de ses réalisations récentes, le gratte-ciel horizontal du Centre Vanke (Horizontal Skyscraper – Vanke Center, Shenzhen, Chine, 2008–09, page 162) s'insère totalement dans un paysage qui contient par ailleurs d'autres parties du programme, dont un auditorium de 500 places et des restaurants. L'élément paysager du projet relève également de la stratégie environnementale de l'architecte, puisque les bassins, alimentés par les eaux grises, servent à rafraîchir l'air. Des panneaux solaires sont également utilisés, de même que des matériaux naturels locaux comme le bambou. Un autre bâtiment de Holl, la station d'épuration des eaux du Connecticut (Connecticut Water Treatment Facility, Connecticut, États-Unis, 1998–2005) a fait l'objet d'un traitement paysager par Michael Van Valkenburgh Associates (2001–05, page 364). À l'achèvement du projet, Steven Holl a précisé : « Ce projet fusionne l'architecture des installations d'épuration des eaux avec le paysage pour donner naissance à un parc public. Les installations techniques sont implantées sous le parc, tandis que la partie publique du programme et le service de contrôle des opérations occupent un fourreau d'acier inoxydable de 110 mètres de long, qui exprime la fonction des installations souterraines et marque le paysage d'une ligne d'horizon réfléchissante. Sa forme de goutte d'eau inversée génère un espace intérieur curviligne, qui s'ouvre par une grande baie vitrée sur le paysage environnant. » Les déblais du creusement du sol ont été réutilisés et des matériaux recyclables ont été employés dès que cela était possible. Les zones humides et la végétation naturelle ont été préservées et les installations protégées par une toiture végétalisée qui ne demande que très peu d'entretien. La zone aménagée par Michael Van Valkenburgh couvre une surface de 5,67 hectares.

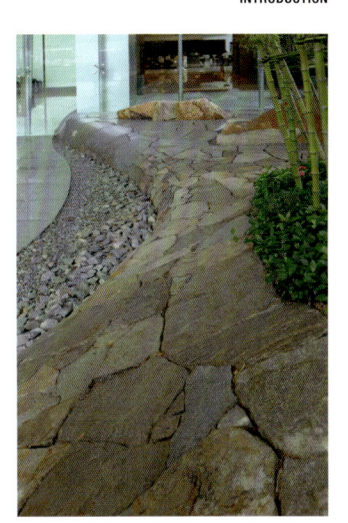

24
*Shunmyo Masuno, SanShin-Tei,
One Kowloon Office Building, Hong
Kong, China, 2006–07*

Pour lui : « Le recours aux outils les plus basiques de l'architecture paysagère – la terre, l'eau et les plantes – vient contrebalancer la forme lisse du bâtiment. Le projet apporte une diversité topographique et suscite un intérêt visuel grâce à la réutilisation de la terre extraite lors du creusement du sol. Des baissières remplacent les systèmes traditionnels de drainage. Le programme des plantations, inspiré de principes écologiques de restauration, est à la fois élémentaire et sophistiqué dans sa complexité et son étendue. » L'accord entre l'architecte – Holl – et l'architecte paysagiste – Van Valkenburgh – sur les aspects « naturels » de ce projet a permis d'obtenir une harmonie remarquable. Ici, l'architecture et l'architecture du paysage collaborent étroitement pour insérer des installations industrielles modernes dans un environnement qui donne l'impression d'avoir été presque totalement préservé. C'est peut-être l'annonce d'une nouvelle orientation pour l'architecture paysagère : faire renaître la nature dans un acte créatif de réaction aux débordements de la civilisation moderne.

DANS LES MONTAGNES DE SHIGA

Bien que l'harmonie entre le paysagiste et l'architecte n'ait pas été aussi poussée que dans l'exemple précédent, le musée Miho de I. M. Pei (Miho Museum, Shigaraki, Shiga, Japon, 1997) est un nouvel exemple d'intégration complète d'une architecture contemporaine dans un environnement naturel. Le client avait confié la conception des jardins (page 260) à Yoshiaki Nakamura, connu pour ses interventions dans le domaine de l'architecture japonaise traditionnelle. Étant donnée la nature sauvage du site, à une heure de Kyoto, les autorités avaient demandé qu'une partie importante du bâtiment soit souterraine. Le projet original de Pei en faisait ainsi « disparaître » une grande portion dans le terrain vallonné, recouvert d'une végétation dense. Plus de dix années après son inauguration, le musée s'est presque entièrement fondu dans le paysage. On reconnaît l'intervention de Nakamura dans des gestes comme les alignements de cerisiers pleureurs qui bordent le chemin menant au tunnel d'accès du musée. Il a directement travaillé avec Pei et le client pour créer le jardin intérieur de l'aile nord, à base de rochers, de graviers et de mousse. Pei a participé au choix des rochers et a demandé à Nakamura de renforcer la présence de la mousse dans le jardin pour atténuer son aspect minéral. Les tensions qui ont pu se développer autour de ces choix relèvent maintenant de l'anecdote. Le Miho Museum est l'un des plus spectaculaires musées contemporains du monde et peut-être l'un des mieux intégrés à son splendide cadre naturel. Les formes géométriques des toitures et des fenêtres restent très visibles, de même que le pont suspendu (primé) qui conduit à l'entrée du musée. S'il est possible de considérer la Connecticut Water Treatment Facility de Holl comme une intrusion dans le paysage (Van Valkenburgh parle de « contrebalancer la forme lisse du bâtiment »), ici, le paysage et l'architecture se fondent autant que cela est physiquement possible. L'harmonie règne à l'intérieur comme à l'extérieur.

Autre exemple, assez différent, de cette juxtaposition de la nature et de l'architecture : la passerelle piétonnière au-dessus de la rivière Carpinteira (Covilhã, Portugal, 2003–09, page 94) par l'architecte lisboète João Luís Carrilho da Graça. Cette intervention, qui fait partie d'un programme de trois nouvelles passerelles dans les vallées de la Goldra et de la Carpinteira, a pour objectif de créer une liaison directe entre le centre historique de Covilhã et des quartiers résidentiels périphériques. Les formes extrêmement simples de l'ouvrage contrastent volontairement avec le style architectural de la ville et son cadre naturel. Néanmoins, ceux qui l'empruntent peuvent mieux appréhender ce cadre

25
Victor Neves, Reorganization of the
Riverside of Esposende, Esposende,
Portugal, 2006–07

25

et l'apprécier de façon entièrement différente grâce au travail de l'architecte. On pourrait dire que cette passerelle présente le paysage d'une façon complètement inédite. Il s'agit manifestement d'un objet artificiel qui ne cherche pas l'intégration, mais ses lignes droites constituent, en un sens, un geste authentique d'architecture paysagère. Deux piliers de la passerelle sont parés d'un appareillage de pierre en spirale qui devrait faciliter l'accrochage de la végétation. Dans quelques années, on aura sans doute l'impression que l'ouvrage a « poussé » à partir des flancs des collines, comme pour célébrer le site.

ÉLÉVATION DE L'ESPRIT

Une variation intéressante sur l'intégration du projet paysager et de l'architecture a été réalisée à Hong-Kong : l'immeuble One Kowloon, pour lequel le prêtre zen Shunmyo Masuno a créé SanShin-Tei (Chine, 2006–07, page 328), une installation occupant un peu plus de 1500 mètres carrés devant l'entrée et 554 mètres carrés dans le hall d'accueil. Prêtre et supérieur du temple Kenko-ji de Yokohama depuis 2001, Shunmyo Masuno dirige également une agence appelée Japan Landscape Consultants Ltd. Le titre de son installation vient du terme zen *sanshin* signifiant « trois esprits ». L'association du nombre trois (*san*) au mot esprit (*shin*) prend un sens que l'on peut traduire par « joie de l'esprit », « esprit mûr » ou « esprit élevé ». Ces concepts représentent les flux d'énergie qui parcourent le bâtiment, de son sommet à son hall d'entrée, et sont matérialisés, par exemple, par une cascade de 9,5 mètres de haut. Une transition est aménagée entre l'installation intérieure, que Masuno qualifie d'« artificielle », et les arrangements plus « naturels » en extérieur. Étant donnée la personnalité du concepteur, il n'est pas surprenant d'y trouver un certain contenu religieux, mais l'idée de base reste néanmoins celle d'une confrontation entre le naturel et l'artificiel, thème constant dans l'histoire de la création de paysage. Ici, architecture moderne est moins un objet qu'une présence qui s'appuie sur d'autres forces pour assurer à la fois la réussite de l'entreprise et le bien-être de ceux qui s'approchent de One Kowloon.

Alors que Shunmyo Masuno apporte une certaine expression de la nature dans un environnement architectural, John Pawson fait presque l'inverse avec son Sackler Crossing (« pont Sackler ») dans les jardins botaniques royaux de Kew (Royal Botanic Gardens, Kew, Londres, 2006, page 304). Bien connu pour sa défense et son illustration du minimalisme en architecture et architecture intérieure, Pawson a créé là une passerelle de 70 mètres de long qu'il qualifie d'« objet sculptural de forme serpentine ». Les principaux matériaux utilisés sont un alliage de bronze habituellement réservé à la fabrication des hélices de sous-marins et des dalles de granit dur, afin de construire un ouvrage solide et durable. Pawson explique que ce petit pont « est là pour être admiré pour sa beauté intrinsèque, mais aussi pour permettre aux visiteurs des jardins de découvrir le paysage sous des angles nouveaux et inattendus ». La passerelle piétonnière de la Carpinteira et le Sackler Crossing partagent cette approche de création de points de vue nouveaux sur le paysage environnant, mais l'ouvrage de Pawson est si bas qu'il semble « flotter sur l'eau », alors que l'intervention de João Luís Carrilho da Graça n'est qu'une simple ligne tendue, suspendue au-dessus de son site. Les deux font référence au minimalisme architectural contemporain qui, par définition, semble contradictoire avec la complexité sauvage de la nature. Les deux architectes rejettent à l'évidence tout esprit d'imitation et cherchent au contraire à imaginer un vocabulaire actuel, dans lequel on peut aussi percevoir une certaine simplicité classique.

Deux projets de l'intéressante agence de Seattle Gustafson Guthrie Nichol apportent des solutions nouvelles à l'aménagement de jardins intérieurs et extérieurs. Leur « jardin Lurie » (Lurie Garden, Chicago, Illinois, 2004, page 132) est situé sur le toit du parking Lakefront Millennium Parking Garage, situé dans le Millennium Park à Chicago. Son contexte architectural est prestigieux, puisque le site se trouve entre un auditorium en plein air dessiné par Frank O. Gehry et l'aile moderne du Chicago Art Institute, signée Renzo Piano. Les formes ont été développées grâce à une mousse de faible poids, parfaite pour un toit, appelée Geofoam. Le Lurie Garden est en quelque sorte une version contemporaine des jardins plus formels du passé et se trouve en parfaite harmonie avec ses voisins, d'esprit très contemporain. L'opposition entre nature et artifice s'exprime dans ce cadre entièrement urbain, mais aussi dans le choix du support invisible qu'est la Geofoam. La présence de la « nature » est agréable et rassurante, mais comme toujours dans les jardins, l'élément créé par l'homme impose davantage sa présence que ne pourrait le faire une expression non contrôlée du monde naturel.

Une seconde réalisation de Gustafson Guthrie Nichol, la « cour Robert et Arlene Kogod » (Robert and Arlene Kogod Courtyard, Washington, DC, 2005–07, page 136), se situe à l'intérieur de l'ancien siège de l'administration des brevets, transformé par l'agence Foster + Partners en un musée, qui abrite le Smithsonian American Art Museum et la National Portrait Gallery. Les architectes paysagistes avaient été chargés d'intervenir sur une cour protégée par un auvent de verre. L'un des éléments essentiels de leur projet est une série de grandes jardinières de pierre plantées d'arbres de 7,5 mètres de haut. L'agence de Seattle explique que « la palette discrète d'arbres et de buissons à feuilles persistantes complète et anime l'architecture existante de la cour par des variations d'échelle, de forme et de texture, ainsi que par la symétrie ou l'asymétrie des implantations ». Dans une approche encore plus durable que pour le Lurie Garden, les concepteurs de ce projet explorent un domaine dans lequel le concept d'une nature artificielle et contrôlée s'exprime de façon plus architecturale que naturelle. Ce n'est en rien une critique, mais plutôt un constat. Ici, l'architecture paysagère complète un bâtiment et vient en contrepoint de l'environnement minéral. Les cours traitées en jardins sont pratiquement obligées de dialoguer avec la nature. Le projet de Gustafson Guthrie Nichol est une expression contemporaine remarquable de ce type d'approche.

Maya Lin, née dans l'Ohio en 1959, est un cas un peu à part dans le domaine de l'architecture paysagère, puisqu'elle est à la fois paysagiste, architecte et artiste. Elle n'avait que 22 ans lors de l'inauguration de son œuvre la plus fameuse, le mémorial aux Vétérans de la guerre du Vietnam (Vietnam Veterans Memorial) situé dans le parc National Mall de Washington. Lin avait remporté le concours et son projet, un mur de granit noir implanté en grande partie au-dessous du niveau du sol, attire chaque année des dizaines de milliers de visiteurs. Gravé des noms des Américains morts pendant ce conflit – 50 000 noms de soldats dans l'ordre chronologique de leur mort –, ce mémorial avait soulevé d'importantes controverses, en particulier dans la communauté des anciens combattants. C'était presque comme si l'Amérique avait honte de ses morts, disaient certains. Pour d'autres, qui comprenaient peut-être mieux l'architecture et le design de notre époque, l'œuvre de Maya Lin était un geste novateur dans un domaine longtemps dominé par les statues en bronze de soldats brandissant un drapeau. En un sens, son œuvre était à la fois fondamentale et révolutionnaire dans la modestie de son audace.

26
Herzog & de Meuron, Plaza de España, Santa Cruz de Tenerife, Canary Islands, Spain, 2006–08

26

DES VAGUES, UN SERPENT ET UNE CARTE

Depuis, même si ses autres travaux n'ont pas atteint une notoriété semblable, Maya Lin a créé d'autres monuments et ses œuvres artistiques sont d'un grand intérêt. Deux d'entre elles sont publiées dans ce volume. La première, le *Storm King Wavefield* (Mountainville, New York, 2007–08, page 210), est située dans une ancienne gravière. Ce champ de vagues gazonnées créé par l'artiste-architecte est censé donner au visiteur le sentiment d'une « immersion totale ». Ce travail entre à coup sûr dans la catégorie de l'architecture paysagère, mais est aussi une œuvre d'art qui n'a d'autre but perceptible que de créer la surprise et d'interpeller le spectateur. Maya Lin donne une forme à notre environnement le plus familier – la terre –, mais d'une manière qui capte l'attention des visiteurs et les fait réfléchir différemment sur la nature.

Son *Eleven Minute Line* (Wanås, Suède, 2004, page 212) est un monticule de forme serpentine de 152 mètres de long et recouvert d'herbe. Inspirée par les tumulus des Indiens américains du Middle West, région où elle a grandi, Lin avait commencé son projet par un croquis, comme la plupart des artistes et des architectes. Fondamentalement abstraite et moderne, cette œuvre évoque elle aussi d'anciennes réminiscences, même si leur transposition de l'Ohio à la Suède n'est pas évidente. Peut-être l'artiste veut-elle dire que la terre a été formée il y a si longtemps que chaque pays ou civilisation possède ses propres secrets enfouis. On éprouve quelque chose qui a trait au fondamental face au travail de Maya Lin, et cette approche, dans des œuvres telles que celles-ci, se place à la jonction de l'art ancien et contemporain.

Bien que sa vision relève plus spécifiquement de l'architecture paysagère, Catherine Mosbach manifeste également des préoccupations artistiques. Sa petite pièce temporaire de 150 mètres carrés intitulée *L'autre rive* (Québec, 2008, page 256) lui avait été commandée dans le cadre du 400ᵉ anniversaire de la fondation du Québec. Des surfaces d'asphalte rappelaient certaines cartes utilisées lors de la conquête du Nouveau Monde. Sa description du projet témoignait de son intention artistique : « Ce jardin est une fable qui, de près, confronte les matériaux en présence – sol, plante, figure – et de loin, la vision d'un paysage en ses contours évocateurs. [...] Le sol, surface d'enregistrement qui se différencie de la croûte terrestre par la présence de la vie, est ici retranscrit par la peau d'asphalte – végétation fossilisée dans l'eau du marais, puis minéralisée… Lorsque les visiteurs traversent ce milieu, il se découvre à eux et aux autres spectateurs au gré de leurs déplacements, qui s'insinuent dans les espaces interstitiels. » Mosbach évoque le paysage du Nouveau Monde, mais peut-être moins les mystères anciens abordés dans le *Eleven Minute Line* de Maya Lin. Néanmoins, ces deux femmes partagent la conviction que le paysage peut aussi être le véhicule de démarches artistiques des plus contemporaines, de celles qui relèvent davantage de l'installation que du cadre traditionnel du musée. Évocatrices sans être figuratives en aucune façon (le « serpent » de Lin est juste une suggestion de forme), ces deux créatrices pourraient tout aussi bien être qualifiées d'artistes que d'architectes paysagistes.

L'état de l'architecture contemporaine du paysage joue beaucoup sur le potentiel d'ambiguïté entre ces deux disciplines. Si le symbolisme était à l'évidence présent dans les tentatives anciennes de donner forme à un paysage, par exemple dans les grandes places ou dans l'environnement créé autour de tombes royales, l'architecture du paysage a conservé cette capacité à faire le lien entre le passé et le présent, sans se référer à des pouvoirs établis oubliés. Comme le fait remarquer I. M. Pei, André Le Nôtre a su très tôt faire un usage somptueux des plans d'eau et des reflets du ciel dans ses bassins, ce qui préfigurait en un sens le minimalisme recherché par de nombreux designers et architectes actuels. L'expression du pouvoir chez Le Nôtre n'était rien d'autre que l'affirmation orgueilleuse de perspectives et d'espaces infinis, rayonnant à partir du palais de Versailles et d'autres monuments symboliques de la royauté française. Les Bourbons appartiennent désormais à un passé très lointain, mais les perspectives de Le Nôtre, comme celle des Champs-Élysées, demeurent et expriment un sens de la grandeur qui n'est plus celle de Louis XIV mais d'une nation.

LE PAYSAGE D'AVANT LA CHUTE

Si cet ouvrage a bien un objectif, c'est de montrer que l'architecture paysagère est omniprésente, par le truchement de l'architecture, des jardins et des tentatives nouvelles de raviver une nature souvent mise à l'écart par les conséquences de l'industrialisation et de la pollution. Un bâtiment peut s'intégrer dans un paysage ou n'être qu'entouré par celui-ci. Un architecte peut jeter un pont qui relèvera davantage du paysagisme que de l'art du constructeur. Des architectes comme Steven Holl, I. M. Pei ou Tadao Ando (également représenté dans ce livre, page 62) considèrent souvent leurs œuvres construites comme des éléments d'un environnement plus vaste qu'ils cherchent également à modifier. Ando, dans son musée Chichu (Naoshima, Japon, 2004), présente des œuvres majeures de Claude Monet, James Turrell ou Walter de Maria sous terre et a reconstitué le paysage initial du flanc de colline au-dessus du bâtiment. C'est à la fois de l'architecture intégrée au paysage et l'inverse. Ando est simplement conscient de l'appel de la terre, tout en restant un moderniste féru de géométrie. Cette fusion, que l'on constate aussi dans le travail de Maya Lin, par exemple, donne la mesure de l'intérêt contemporain de l'architecture paysagère. Dans ses meilleures illustrations, elle devient un véritable art.

La plupart de ces interventions montrent un souci de « perfection » qui n'anime peut-être plus autant l'architecture elle-même. Il tient sans doute au concept d'un jardin qui symboliserait le monde d'avant le péché originel, selon la tradition judéo-chrétienne, ou le paradis dans l'Islam, par exemple. Même la tendance actuelle à l'écologie fait resurgir l'idée d'un retour à un état premier, avant la chute. Le pouvoir régénérateur du paysage, ou plus précisément de la nature, inspire cet espoir sans cesse renouvelé de rédemption, y compris face aux destructions massives de l'environnement naturel provoquées par la nécessité ou la cupidité. Alors que tout ce qui est fait par la main de l'homme peut être remis en question ou soumis à ses humeurs ou à ses modes, la nature possède sa légitimité propre et c'est certainement une part de cette légitimité que revendique l'architecture paysagère. Œuvre de l'homme et cependant plus proche de la nature que d'autres formes d'expression, l'architecture paysagère, au-delà des expressions individuelles de créativité, aspire à la rédemption, à un retour à un état primitif. Frank Lloyd Wright a pu rêver d'une architecture inspirée par la nature, mais les réalisations contemporaines n'en sont que très rarement l'illustration. L'art s'inspire souvent de la nature, même aujourd'hui, mais ne peut aspirer à un réel retour à la perfection de ce qui existait avant… avant la chute.

---

[1] Citation originale : http://www.mnstate.edu/gracyk/courses/web%20publishing/addison414.htm, consulté le 16 novembre 2010.
[2] Horizon Press, New York, 1953.
[3] http://whc.unesco.org/en/list/1150, consulté le 16 novembre 2010 ; http://whc.unesco.org/fr/list/1150/ pour la version française.
[4] Citation originale de Peter Eisenman : http://www.curatorialproject.com/interviews/petereisenmanii.html, consulté le 16 novembre 2010.

Pier Head and Canal Link

AECOM

AECOM, Design + Planning London
The Johnson Building
77 Hatton Garden, London EC1N 8JS, UK
Tel: +44 20 30 09 21 00 / Fax: +44 20 30 09 21 99
E-mail: warren.osborne@aecom.com / Web: www.aecom.com/landscapes

AECOM, Design + Planning Phoenix
2777 Camelback Road, Suite 200, Phoenix, AZ 85016, USA
Tel: +1 602 337 2516 / Fax: +1 602 337 2620
E-mail: chad.atterbury@aecom.com / Web: www.aecom.com/landscapes

AECOM brings together a wide variety of disciplines in design, engineering, consultancy, and project and program management. It has merged a variety of companies including EDAW, Tishman Construction, Davis Langdon, and Ellerbe Beckett. Warren Osborne, Director of Landscape Architecture in AECOM's London Design + Planning studio, has been with the firm since 1995. He has worked on the plans for Education City (Doha, Qatar, ongoing); the Pier Head (Liverpool, UK, 2005–07, published here); and the Park Hyatt Hotel (Mallorca, Spain, 2009–11). James Haig-Streeter, now Associate Principal in AECOM's San Francisco Design + Planning studio, helped lead Pier Head; he previously oversaw the implementation of Royal Victoria Square in London (2000) and was the project landscape architect for Piccadilly Gardens in Manchester (UK, 2003), as well as the project manager for a number of large landscape regeneration projects in Blackpool (UK, 2008). Chad Atterbury, Associate in AECOM's Phoenix Design + Planning studio, was the landscape designer for the Phoenix Civic Space Park (Arizona, USA, 2008–09, also published here). Other ongoing firm work includes the Dwight D. Eisenhower Memorial, Washington, D.C. (USA, 2009–15) and the Masdar City Public Realm (Abu Dhabi, UAE, 2007–20/25).

AECOM bietet eine breite Palette unterschiedlicher Disziplinen in den Bereichen Design, Hochbau, Beratung sowie Projektmanagement an. Zu der Gruppe gehören zahlreiche Unternehmen wie zum Beispiel EDAW, Tishman Construction, Davis Langdon und Ellerbe Beckett. Warren Osborne, der Leiter der Abteilung für Landschaftsarchitektur der AECOM in deren Londoner Entwurfs- und Planungsbüro (Design + Planning studio), ist seit 1995 in dem Büro tätig. Er war an der Planung für die Education City (Doha, Qatar, im Bau), dem Pier Head (Liverpool, Großbritannien, 2005–07, hier vorgestellt) und dem Park Hyatt Hotel (Mallorca, Spanien, 2009–11) beteiligt. James Haig-Streeter, derzeit leitender Mitarbeiter im Entwurfs- und Planungsbüro der AECOM in San Francisco, war an der Bauleitung des Pier-Head-Projekts beteiligt. Zuvor war er als Bauleiter bei der Ausführung des Royal Victoria Square in London (2000) tätig und war leitender Landschaftsarchitekt der Piccadilly Gardens in Manchester (Großbritannien, 2003). Außerdem war er Projektmanager bei einer Reihe größerer Projekte für Landschaftserneuerung in Blackpool (Großbritannien, 2008). Chad Atterbury ist Mitarbeiter im Entwurfs- und Planungsbüro der AECOM in Phoenix. Er war der Landschaftsarchitekt des Phoenix Civic Space Park (Arizona, USA, 2008–09, ebenfalls hier vorgestellt). Zu den derzeit im Bau befindlichen Projekten der Gruppe gehören das Dwight D. Eisenhower Memorial, Washington, D. C. (USA, 2009–15) und das Masdar City Public Realm (Abu Dhabi, Vereinigte Arabische Emirate, 2007–20/25).

L'agence **AECOM** regroupe des spécialistes dans les domaines du design, de l'ingénierie, de la consultance et de la gestion de programmes et de projets. Elle est issue de la fusion de plusieurs sociétés dont EDAW, Tishman Construction, Davis Langdon et Ellerbe Beckett. Warren Osborne, directeur de l'architecture du paysage au studio Design + Planning (conception et urbanisme) d'AECOM à Londres, fait partie de l'agence depuis 1995. Il a travaillé sur les plans de la Cité de l'éducation (Education City, Doha, Qatar, en cours) ; le quartier du Pier Head (Liverpool, GB, 2005–07, publié ici) et le Park Hyatt Hotel (Majorque, Espagne, 2009–11). James Haig-Streeter, aujourd'hui directeur associé du studio Design + Planning d'AECOM à San Francisco, a collaboré au projet du Pier Head (publié ici). Il avait auparavant supervisé le chantier du Royal Victoria Square à Londres (2000), avait été architecte-paysagiste sur le projet des Piccadilly Gardens à Manchester (GB, 2003) et directeur de projet pour un certain nombre d'importants aménagements paysagers à Blackpool (GB, 2008). Chad Atterbury, associé du studio Design + Planning d'AECOM à Phoenix, a conçu le Civic Space Park de Phoenix (Arizona, 2008–09, publié ici). Autres projets en cours de l'agence : le mémorial Dwight D. Eisenhower à Washington, DC (2009–15) et les espaces publics de Masdar City (Abou Dhabi, EAU, 2007–20/25).

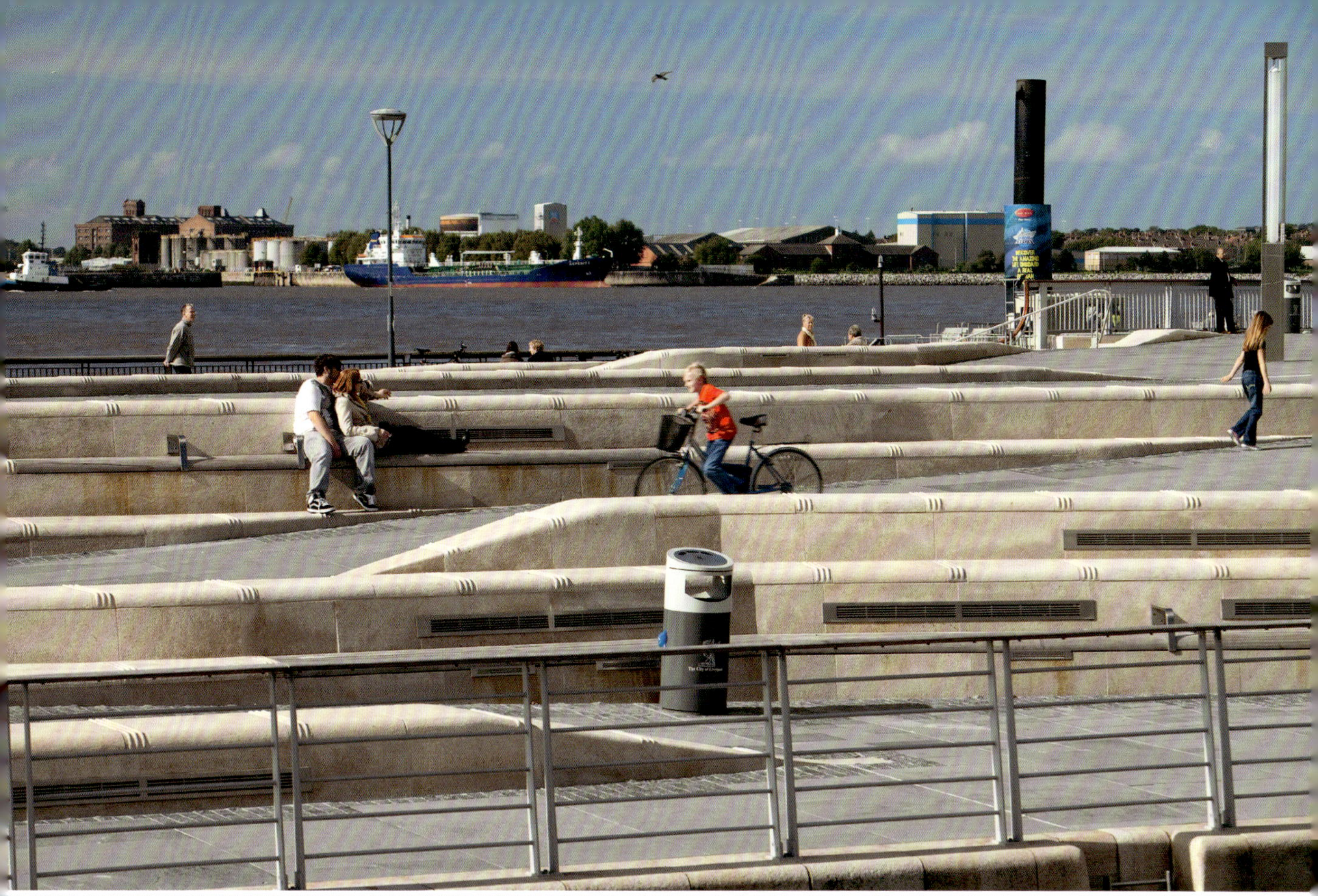

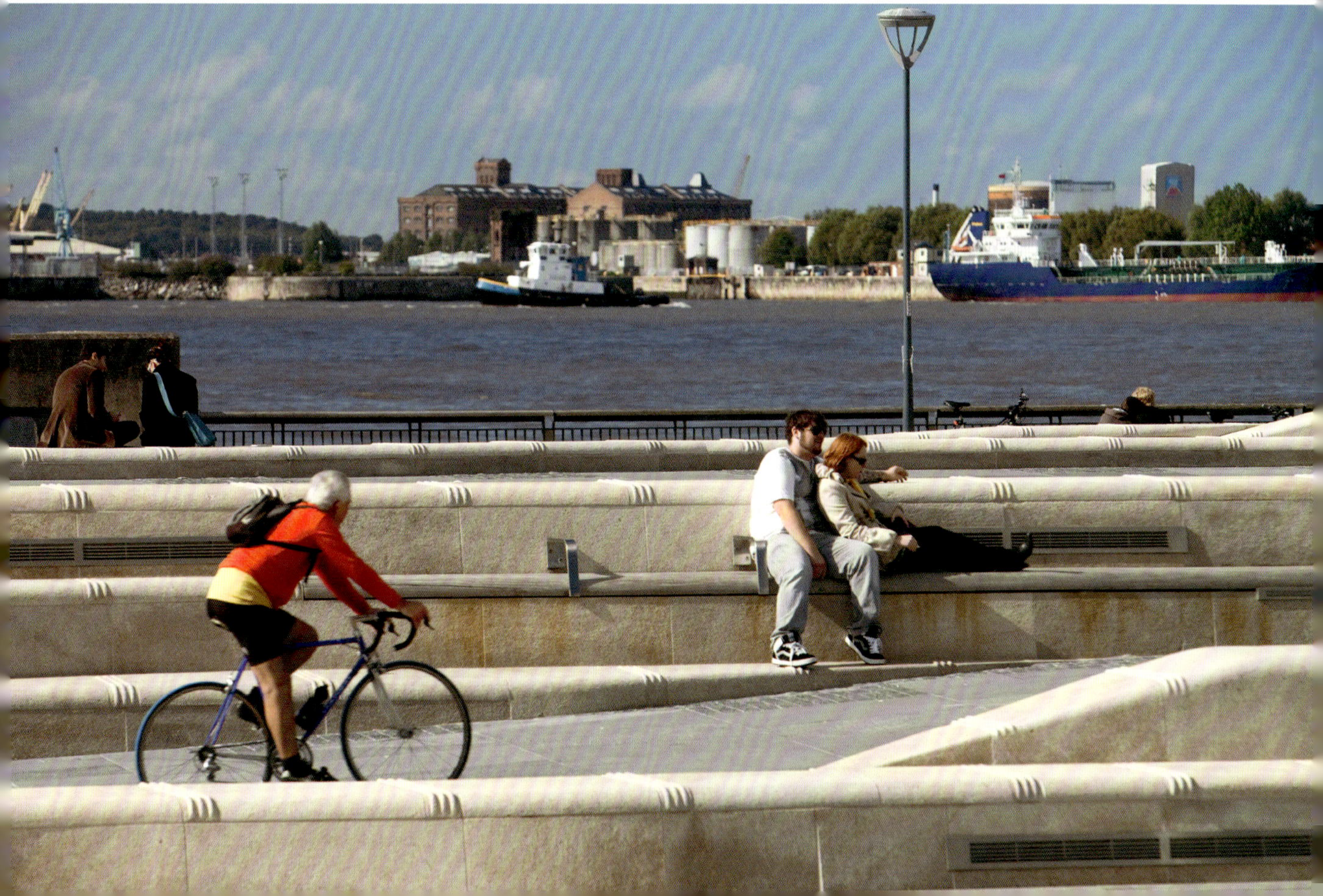

PIER HEAD AND CANAL LINK
Liverpool, UK, 2005–07

Address: Pier Head, Liverpool, UK. Area: 2.5 hectares
Client: Liverpool City Council, British Waterways—Canal Link. Cost: £22 million
Collaboration: Arup, 20/20 Liverpool, Graham Festenstein Lighting Design

With its mixture of new and older architecture, and its waterside presence, the Pier Head and Canal Link might not be thought of as "landscape design" in the most accepted definition of the term—it is, rather, an urban, mineral, park space.

Mit seiner Mischung von alter und neuer Architektur und der Lage am Wasser möchte man den Pier Head und den Canal Link nicht unbedingt als „Landschaftsgestaltung" im engeren Wortsinn bezeichnen, sondern eher als urbane Freiraumgestaltung.

En associant des bâtiments historiques ou nouveaux et la présence de l'eau, le Pier Head and Canal Link est moins un projet de « paysage » au sens courant qu'un parc urbain de nature minérale.

Sculpted and angled, the different levels of the walkway provide for convenient and easy-to-maintain seating, as well as the security implied in such an open space.

Die vielfältigen, abgetreppten, polygonalen Ebenen der Promenade bieten sich einerseits als Sitzgelegenheiten an, andererseits schaffen sie auch die notwendige Sicherheit, die in einem solchen Freiraum vorausgesetzt wird.

Sculptées en lignes brisées et étagées sur plusieurs niveaux, les promenades offrent de multiples assises faciles à entretenir et le niveau de sécurité requis dans un espace aussi ouvert.

The area near the Three Graces buildings in Liverpool is part of a UNESCO World Heritage site (Liverpool Maritime Mercantile City, a total of 136 hectares listed in 2004), and Liverpool was the 2008 European City of Culture. The Pier Head and Mann Island areas were imagined as an international gateway to the city. A canal extension linking the Leeds and Liverpool Canal and the dockland water basins was part of this project. The designers state: "The completed project has transformed a windswept square into an attractive and inviting public space for tourists, local residents, shoppers, and workers, and has provided improved connections between the docks along the famous Merseyside waterfront. The space has the flexibility and capacity to provide a safe and enjoyable venue for large festivals attracting crowds of up to 35 000, but is equally appealing as a place for quietly contemplating the handsome buildings, local war memorials, and views across the River Mersey." It is estimated that an extra 200 000 people come to the space every year as a result of the project.

Der Bereich um das Gebäudeensemble Three Graces in Liverpool gehört zum UNESCO-Weltkulturerbe (Liverpool Maritime Mercantile City – der Handelshafen von Liverpool mit einer Gesamtfläche von 136 Hektar – wurde 2004 aufgenommen), außerdem war Liverpool 2008 Europäische Kulturhauptstadt. Pier Head und Mann Island sollen das internationale Tor zur Stadt bilden. Ein Teil des Projektes war die Erweiterung des Kanals, der den Leeds- und Liverpool-Kanal mit den Hafenbecken verbindet. Die Planer bemerken dazu: „Durch diese Planung wurde aus einem zugigen Platz ein einladender öffentlicher Raum für Touristen, Anwohner, Kunden und Angestellte. Dank des Projekts wurden die Verbindungen zwischen den Hafenanlagen am Ufer des Mersey verbessert. Dieser Raum bietet ausreichende Flexibilität und Kapazität, um hier sicher und in angenehmer Umgebung Großveranstaltungen mit bis zu 35 000 Teilnehmern durchzuführen. Gleichzeitig lädt dieser Ort zur Betrachtung der stattlichen Gebäude und Kriegerdenkmäler ein, und er bietet schöne Ausblicke über den Fluss Mersey." Man schätzt, dass nach Abschluss des Projekts jährlich etwa 200 000 Besucher zusätzlich hierher kommen.

La zone située à proximité des immeubles des Three Graces à Liverpool fait partie des 136 hectares de la Maritime Mercantile City de Liverpool, inscrite en 2004 au patrimoine mondial de l'UNESCO. La ville a été Capitale européenne de la culture en 2008. Pier Head et Mann Island constituent aujourd'hui une porte d'entrée internationale de la ville. L'extension du canal qui relie la voie navigable Leeds-Liverpool aux bassins de la zone portuaire fait partie de ce projet. Pour AECOM : « Le projet final a transformé une place balayée par les vents en un espace public intéressant qui attire les touristes, les habitants, les badauds et les travailleurs, et fournit de meilleures connexions entre les docks le long du fameux front de mer de Merseyside. Cet espace possède la souplesse et la capacité nécessaires pour constituer un lieu sûr et agréable apte à accueillir des événements attirant jusqu'à 35 000 personnes, tout en offrant la possibilité de contempler dans le calme les superbes bâtiments anciens, les mémoriaux de guerre et les vues sur la Mersey. » On estime à 200 000 le nombre de visiteurs supplémentaires attirés chaque année par ces nouveaux aménagements.

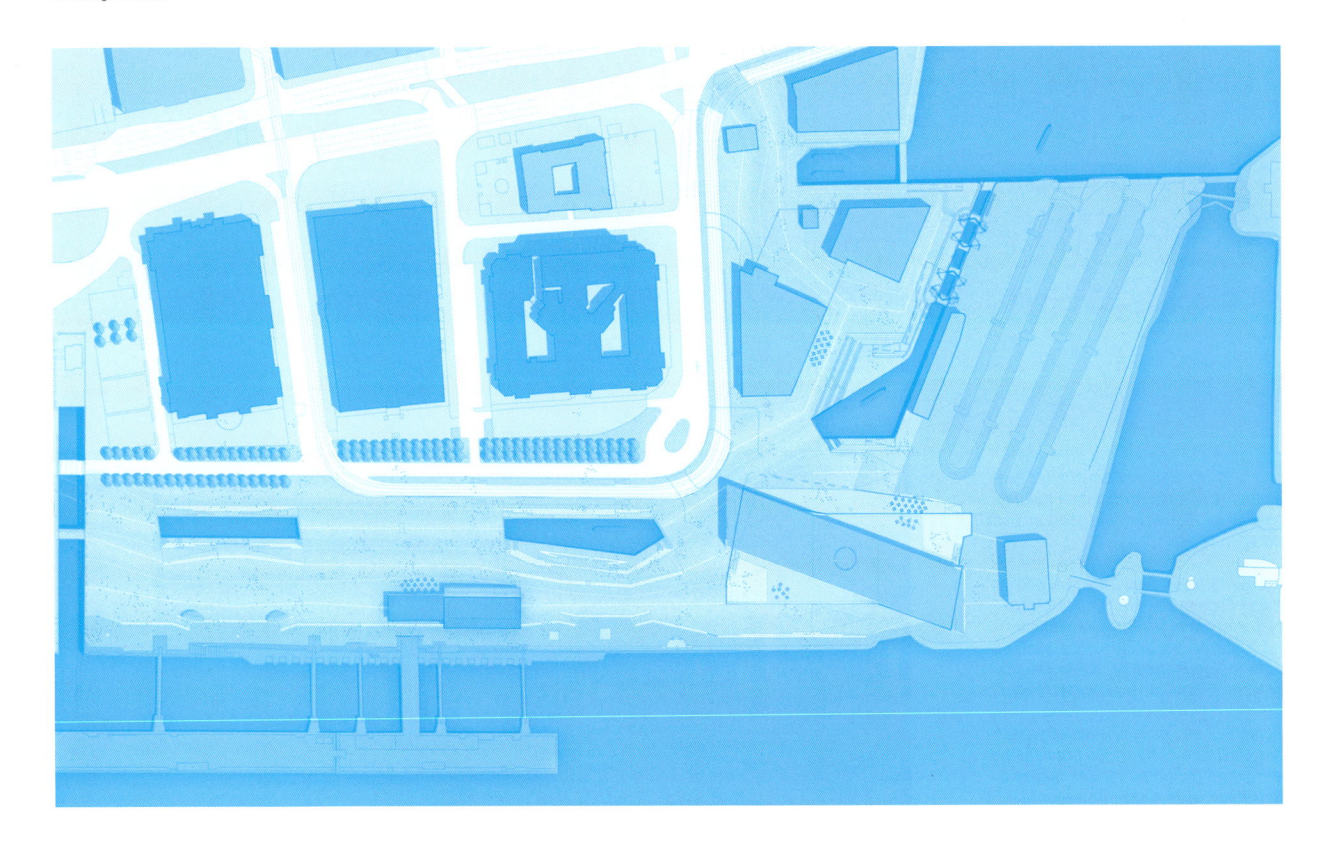

The area concerned has been referred to as the "Three Graces" because of the Royal Liver Building (1911, seen above left and below right in the photos opposite), the Cunard Building (1916), and the Port of Liverpool Building (1907). Above, a site plan.

Der gestaltete Bereich nennt sich die „Drei Grazien" wegen des Royal Liver Building (1911), oben links und unten rechts auf den gegenüberliegenden Fotos zu sehen, des Cunard Building (1916) und des Port of Liverpool Building (1907). Oben ein Lageplan.

Le nom de la zone des « Trois grâces » tire son nom de la présence des immeubles du Royal Liver (1911, sur les photos en haut à gauche et en bas à droite), de la Cunard (1916) et du port de Liverpool (1907). Ci-dessus, plan du site.

AECOM

CIVIC SPACE PARK

Phoenix, Arizona, USA, 2008–09

*Address: 424 North Central Avenue, Phoenix, AZ, USA
Area: 1.12 hectares. Client: City of Phoenix Parks Department
Cost: $11.9 million. Collaboration: Chad Atterbury, Chris Moore*

This public park located in downtown Phoenix includes a field of white columns under a canopy. This feature includes "vertical arrays of color-changing LEDs" that were "inspired by lightning touching down during Arizona's summer monsoon storms." The lighting patterns and colors vary according to visitor movement or even the seasons. A sculpture by the artist Janet Echelman entitled *Her Secret Is Patience* is also part of the park. No less than 44 meters high, 108 meters wide, and 88 meters deep, this work is made of painted galvanized steel, colored polyester braided twine netting, and lights. As for the overall project, AECOM explains: "The design team had some difficult goals to achieve including not only a revitalization of a downtown site but also incorporating historic preservation guidelines and new development demands. The urban space had to serve as a community gathering space, a pedestrian passage, an urban oasis of shade and serenity, a place for learning, a place for visiting, as well as a commons for the adjacent Arizona State University campus." Rainwater collection systems and solar panels on top of the canopies are part of an effort to make a truly "sustainable" park.

Zu diesem innerstädtischen öffentlichen Park in Phoenix gehört auch ein überdachter Bereich mit weißen Säulen. Teil dieser Installation sind „vertikal angeordnete LEDs, die ihre Farben verändern". Sie sind „von den Blitzen inspiriert, die in den sommerlichen Monsungewittern in Arizona herunterfahren". Muster und Farben der Lichter verändern sich mit den Bewegungen der Besucher und sind auch je nach Jahreszeit verschieden. Eine Skulptur der Künstlerin Janet Echelman mit dem Titel „Her Secret Is Patience" ist ebenfalls Teil des Parks. Diese Arbeit aus verzinktem Stahl und einem Netz aus farbigen, geflochtenen Polyesterschnüren und Lichtern ist 44 Meter hoch, 88 Meter tief und hat einen Durchmesser von 108 Metern. AECOM bemerkt zu dem gesamten Projekt: „Das Entwurfsteam musste einige Schwierigkeiten meistern. Nicht nur dieser Bereich der Innenstadt sollte mit neuem Leben erfüllt werden, es mussten auch Vorschriften der Denkmalpflege und die Forderungen nach Neubebauung erfüllt werden. Der urbane Freiraum sollte als Versammlungsort dienen, Fußgängerverbindungswege anbieten und eine heitere Oase mit ausreichendem Schattenangebot mitten in der Stadt sein, ein Ort, an den man sich zum Lernen zurückziehen kann und den man genauso aufsucht wie den angrenzenden Campus der Arizona State University." Ein Regenwassersammelsystem und Solarzellen auf den Dachflächen gehören zum Nachhaltigkeitskonzept, das dem Park zugrundeliegt.

Ce parc public situé au centre de Phoenix comprend un portique de colonnes blanches dressées sous un auvent, qui abrite des « barres verticales de LED aux couleurs changeantes » dont l'idée a été « inspirée des éclairs qui frappent cette région pendant les tempêtes d'été ». Les motifs et les couleurs de ces éléments lumineux varient selon les déplacements des visiteurs et même selon les saisons. Une sculpture de l'artiste Janet Echelman intitulée *Her Secret Is Patience* (son secret est la patience) s'élève au-dessus du parc. De pas moins de 44 mètres de haut, 108 de large et 88 de profondeur, cette œuvre est constituée d'acier galvanisé peint, de filets de polyester de couleur et de lumières. Pour AECOM : « L'équipe de conception était confrontée à des objectifs difficiles : non seulement la revitalisation de ce site du centre-ville, mais aussi l'incorporation d'éléments historiques à préserver et l'intégration de nouvelles constructions. Cet espace urbain devait servir de lieu de rassemblement, de passage pour les piétons, d'oasis urbaine offrant ombrage et sérénité, d'endroit pour apprendre, de parc à visiter, mais aussi de lieu de détente pour le campus adjacent de l'université d'Arizona. » Un système de collecte des eaux de pluie et des panneaux solaires installés sur l'auvent témoignent des efforts déployés pour en faire un parc « durable ».

Left, an aerial view of the space; right, an image showing part of Her Secret Is Patience, *a large-scale sculpture by Janet Echelman commissioned for the park.*

Links, Luftaufnahme des gesamten Areals; rechts, Teilansicht der Großskulptur „Her Secret Is Patience" von Janet Echelman, eine Auftragsarbeit für den Park.

À gauche, vue aérienne du parc. À droite, image partielle de Her Secret Is Patience, *énorme sculpture de Janet Echelman, spécialement commandée pour le parc.*

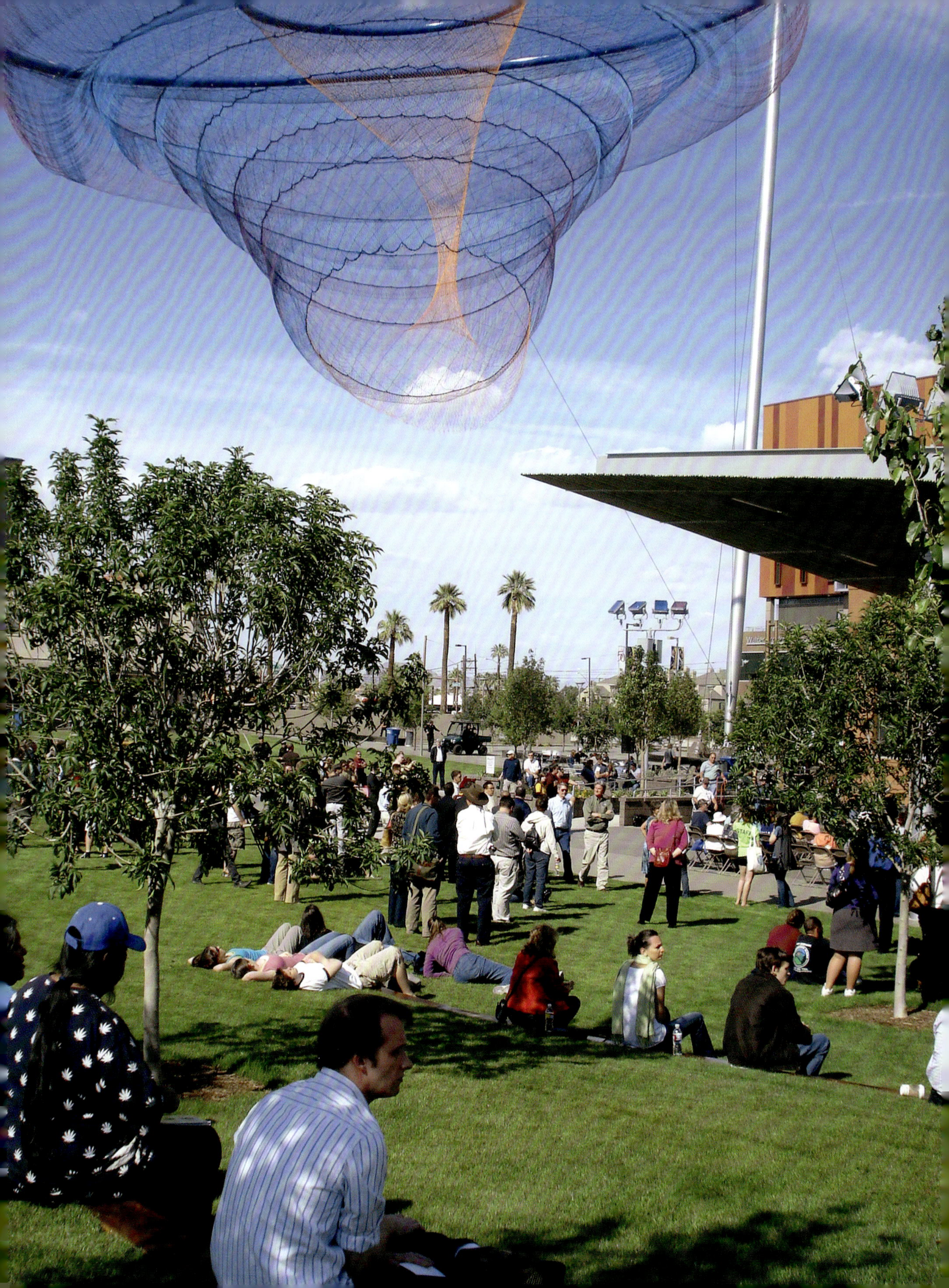

The Echelman sculpture hangs above the park space in the night images above and opposite, changing color and its degree of visibility with the darkening night.

Die Nachtansichten oben und rechts zeigen die über dem Park hängende Skulptur von Echelman. Mit der einbrechenden Nacht verändern sich ihre Farben, und sie wird damit auch immer besser sichtbar.

La sculpture d'Echelman en suspension au-dessus du parc vue la nuit, ci-dessus et page de droite. Elle change de couleurs et modifie son apparence au fur et à mesure que la nuit tombe.

AECOM

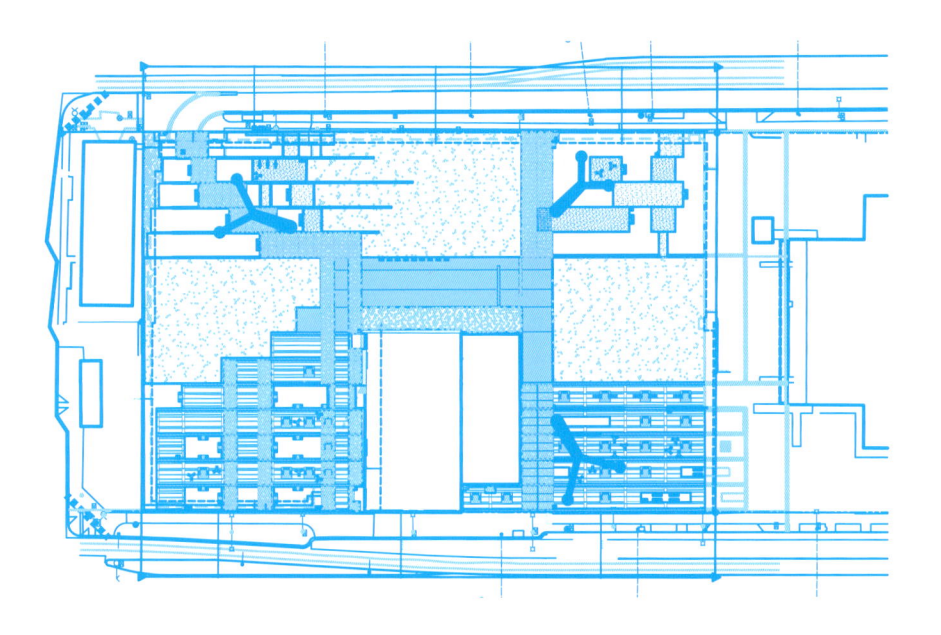

Left, a plan of the park. With its canopied spaces and open areas, the park provides a variety of different degrees of shading and protection. The architects have also carefully studied the night lighting.

Links ein Grundriss des Parks. Mit seinen überdachten und offenen Bereichen bietet der Park mehr oder weniger schattige und geschützte Bereiche. Die Architekten haben auch die nächtliche Beleuchtung sorgfältig geplant.

À gauche, plan du parc. Composé de zones ouvertes et d'autres plus abritées, le parc offre différents niveaux d'ombrage et de protection. Les architectes ont soigneusement étudié l'éclairage nocturne.

Shiba Ryotaro Memorial Museum

TADAO ANDO

*Tadao Ando Architect & Associates
Osaka
Japan*

Born in Osaka in 1941, **TADAO ANDO** was self-educated as an architect, largely through his travels in the United States, Europe, and Africa (1962–69). He founded Tadao Ando Architect & Associates in Osaka in 1969. He has received the Alvar Aalto Medal, Finnish Association of Architects (1985); the Medaille d'or, French Academy of Architecture (1989); the 1992 Carlsberg Prize; and the 1995 Pritzker Prize. Notable buildings include Church on the Water (Hokkaido, Japan, 1988); Japan Pavilion Expo '92 (Seville, Spain, 1992); Forest of Tombs Museum (Kumamoto, Japan, 1992); Rokko Housing (Kobe, Japan, 1983–93); the Suntory Museum (Osaka, Japan, 1994); Awaji Yumebutai (Awajishima, Hyogo, Japan, 1997–2000); the Pulitzer Foundation for the Arts (Saint Louis, Missouri, USA, 1997–2000); the Modern Art Museum of Fort Worth (Fort Worth, Texas, USA, 1999–2002); and the Chichu Art Museum on the Island of Naoshima in the Inland Sea (Japan, 2004). More recently he has completed the Omotesando Hills complex (Tokyo, 2006); 21_21 Design Sight (Tokyo, Japan, 2004–07); Tokyu Toyoko Line Shibuya Station (Shibuya-ku, Tokyo, Japan, 2006–08); an expansion of the Clark Art Institute (Williamstown, Massachusetts, USA, 2006–08); and the renovation of the Punta della Dogana (Venice, Italy, 2007–09). In 2010, Tadao Ando completed the Stone Sculpture Museum (Bad Münster am Stein, Germany, 1996–2010); the Shiba Ryotaro Memorial Museum (Higashiosaka, Osaka, Japan, 1998–2001/2010, published here); the WSJ-352 Building on the Novartis Campus (Basel, Switzerland, 2004–10); and the Lee Ufan Museum (Naoshima, Kagawa, Japan, 2007–10, also published here). He is working on the Abu Dhabi Maritime Museum (Abu Dhabi, UAE, 2006–); and a house for the designer Tom Ford near Santa Fe (New Mexico, USA).

TADAO ANDO wurde 1941 in Osaka geboren, er ist als Architekt Autodidakt und hat seine Eindrücke vor allem bei Reisen in die Vereinigten Staaten, Europa und Afrika in den Jahren 1962 bis 1969 gesammelt. 1969 gründete er das Büro Tadao Ando Architect & Associates in Osaka. 1985 wurde er vom Finnischen Architektenverband mit der Alvar-Aalto-Medaille ausgezeichnet, 1989 mit der Goldmedaille der Französischen Akademie für Architektur, 1992 erhielt er den Carlsberg-Preis und 1995 den Pritzker-Preis. Zu seinen bemerkenswerten Gebäuden gehören die Kirche auf dem Wasser (Hokkaido, Japan, 1988), der Japanische Pavillon auf der Expo '92 (Sevilla, Spanien, 1992), das Museum des Gräberwalds (Kumamoto, Japan, 1992), das Wohngebäude Rokkō (Kobe, Japan, 1983–93), das Suntory Museum (Osaka, Japan, 1994), Awaji Yumebutai (Awajishima, Hyogo, Japan, 1997–2000), die Pulitzer Foundation for the Arts (Saint Louis, Missouri, USA, 1997–2000), das Modern Art Museum in Fort Worth (Texas, USA, 1999–2002) und das Kunstmuseum Chichu auf der Insel Naoshima im Seto-Binnenmeer (Japan, 2004). Zu den neueren von Tadao Ando realisierten Gebäuden gehören der Einkaufs- und Apartmentkomplex Omotesando Hills (Tokio, 2006), der Kunstraum 21_21 Design Sight (Tokio, 2004–07), der Bahnhof Shibuya an der Tokyu-Toyoko-Linie (Shibuya-ku, Tokio, 2006–08), ein Erweiterungsbau des Clark Art Institute (Williamstown, Massachusetts, USA, 2006–08) und der Umbau des Museums Punta della Dogana (Venedig, 2007–09). 2010 vollendete Tadao Ando das Steinskulpturenmuseum in Bad Münster am Stein (1996–2010), das Shiba Ryotaro Memorial Museum (Higashiosaka, Osaka, Japan, 1998–2001/10, hier vorgestellt), das Gebäude WSJ-352 auf dem Novartis Campus (Basel, 2004–10) und das Lee Ufan Museum (Naoshima, Kagawa, Japan, 2007–10, ebenfalls hier vorgestellt). Zu seinen laufenden Projekten gehören das Abu Dhabi Maritime Museum (Abu Dhabi, VAE, seit 2006) und ein Wohnhaus für den Modedesigner Tom Ford in der Nähe von Santa Fe (New Mexico, USA).

Né à Osaka en 1941, **TADAO ANDO** est un architecte autodidacte formé en grande partie par ses voyages aux États-Unis, en Europe et en Afrique (1962–69). Il fonde Tadao Ando Architect & Associates à Osaka en 1969. Il est titulaire de la médaille Alvar Aalto de l'Association finlandaise des architectes (1985), de la médaille d'or de l'Académie d'architecture de Paris (1989), du prix Carlsberg (1992) et du prix Pritzker (1995). Il a notamment réalisé l'église sur l'eau (Hokkaido, Japon, 1988) ; le pavillon du Japon pour l'Exposition universelle de 1992 (Séville, Espagne) ; le Musée de la forêt des tombes (Kumamoto, Japon, 1992) ; les immeubles de logements Rokko (Kobe, Japon, 1983–93) ; le musée Suntory (Osaka, Japon, 1994) ; le Centre de conférences Awaji Yumebutai (Awajishima, Hyogo, Japon, 1997–2000) ; la Pulitzer Foundation for the Arts (Saint Louis, Missouri, 1997–2000) ; le Modern Art Museum of Fort Worth (Texas, US, 1999–2002) et le musée Chichu sur l'île de Naoshima, en mer Intérieure (Japon, 2004). Plus récemment, il a achevé le complexe Omotesando Hills (Tokyo, 2006) ; le centre de design 21_21 Design Sight (Tokyo, 2004–07) ; la gare de Shibuya de la ligne Tokyu Toyoko (Shibuya-ku, Tokyo, 2006–08) ; une extension du Clark Art Institute (Williamstown, Massachusetts, US, 2006–08) et la rénovation des entrepôts de la Punta della Dogana (Venise, 2007–09). En 2010, Tadao Ando a terminé le Steinskulpturenmuseum (Bad Münster am Stein, Allemagne, 1996–2010) ; le musée mémorial Shiba Ryotaro (Higashiosaka, Osaka, Japon, 1998–2001/2010, publié ici) ; le bâtiment WSJ-352 sur le campus Novartis (Bâle, Suisse, 2004–10) et le musée Lee Ufan (Naoshima, Kagawa, Japon, 2007–10, publié ici). Il travaille actuellement sur le Musée maritime d'Abou Dhabi (EAU, 2006–) et sur une maison pour le styliste Tom Ford, près de Santa Fe (Nouveau-Mexique, US).

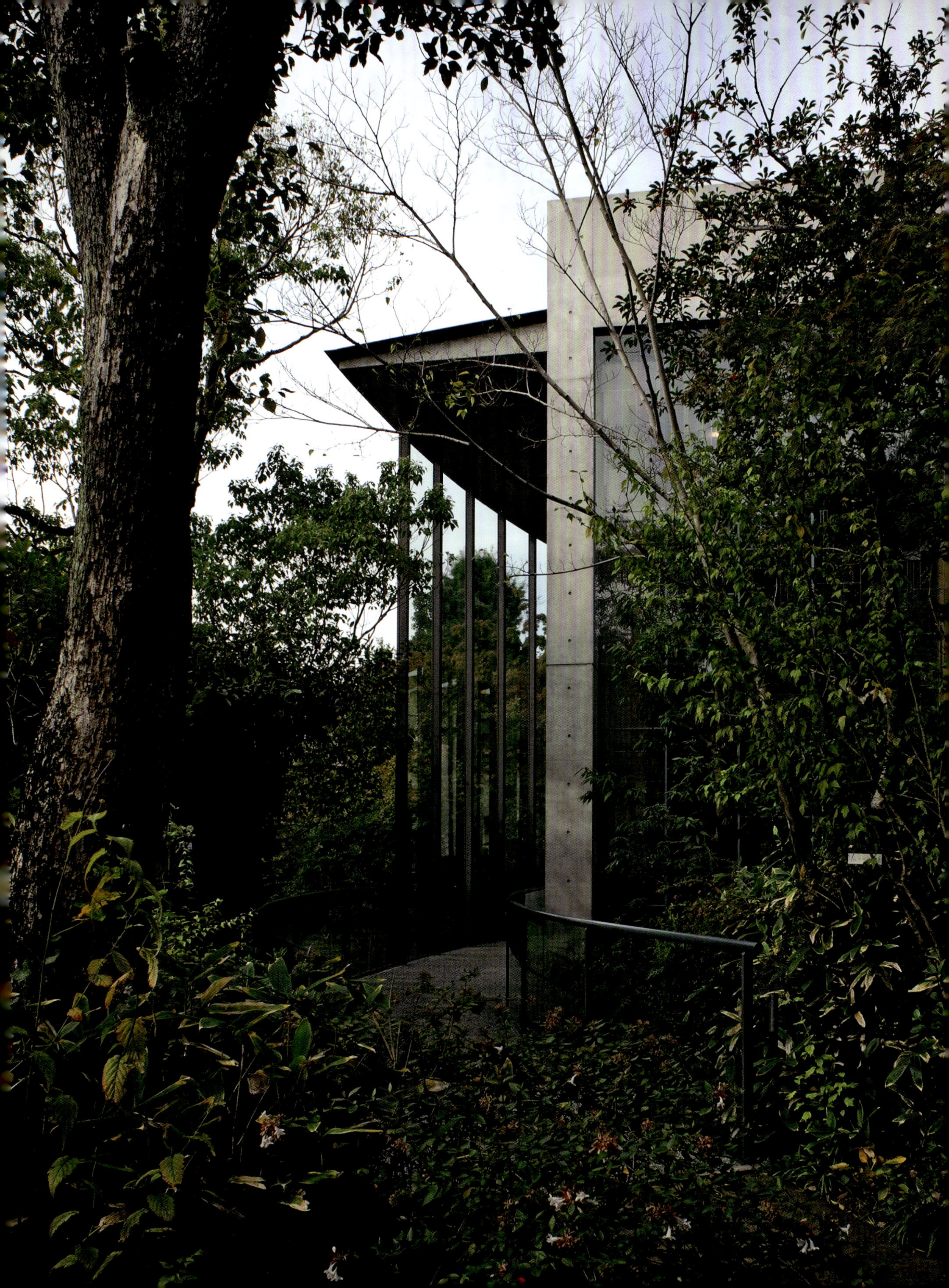

SHIBA RYOTARO MEMORIAL MUSEUM

Higashiosaka, Osaka, Japan, 1998–2001/2010

*Address: 3-11-18 Shimokosaka, Higashiosaka, Osaka Prefecture 577-0803, Japan, +81 6 6726 3860, www.shibazaidan.or.jp/index.html. Area: 1010 m² (site); 997 m² (total floor)
Client: Shiba Ryotaro Memorial Foundation. Cost: not disclosed*

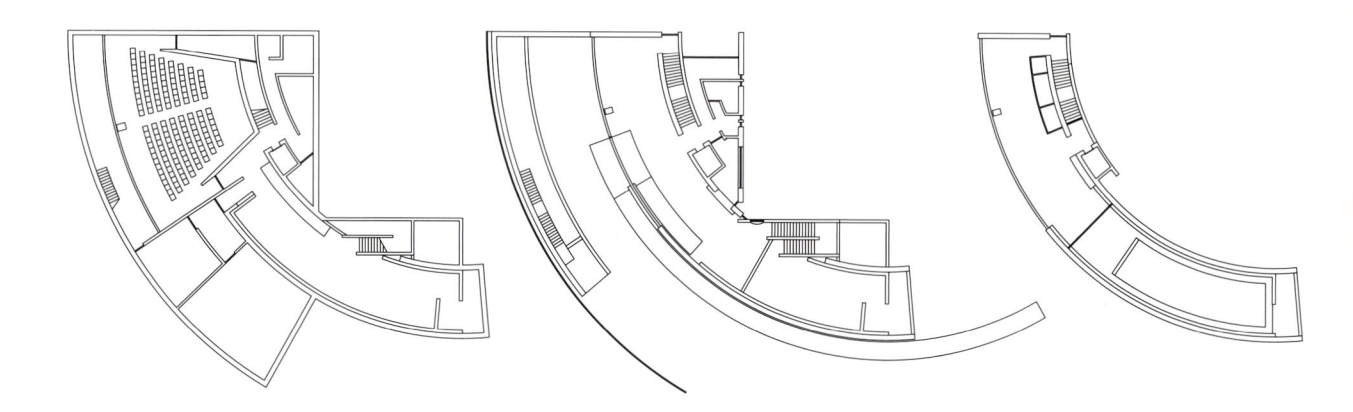

Located in a suburb of Osaka, this surprising museum is set in the former property of the writer Ryotaro Shiba. Born in 1923 in Osaka, Shiba (whose real name was Teiichi Fukuda) is considered one of the greatest modern Japanese writers. He died in 1996. Visitors to the Memorial file through the garden past the writer's old house with its large window opening onto his carefully preserved office. Tadao Ando explains: "The site of this memorial museum has expanded step by step and it now occupies the complete block; the greenery of the garden has grown around the perimeter of the building, and the complete site has virtually become a forest. Initially the project included three gardens: the garden of the original residence, the front garden that develops along the circular volume of the memorial hall, and the public garden on the north side between the residence and the memorial hall. In 2010, a fourth garden facing south, adjacent to the garden of the original residence, was added. In this way, the continuity of these four intimate gardens was accomplished. I hope the satisfaction of this small landscape will strengthen the dialogue between the hearts of visitors and Ryotaro Shiba." It is as of this point that the new structure becomes visible, with its glass and reinforced-concrete structure. Measuring some 997 square meters in area, the Memorial itself is essentially situated below grade, where a plunging library space reminds architecturally aware visitors of Tadao Ando's interest in the spaces seen in Piranesi's engravings.

Dieses ungewöhnliche Museum befindet sich in einem Vorort von Osaka auf dem ehemaligen Grundstück des Schriftstellers Ryotaro Shiba. Shiba, dessen richtiger Name Teiichi Fukuda war, wurde 1923 in Osaka geboren. Er gilt als einer der größten modernen Autoren Japans. Er starb 1996. Besucher werden hinter dem alten Wohnhaus des Schriftstellers mit seinem großen Fenster, durch das man in sein sorgfältig erhaltenes Büro blicken kann, durch den Garten zur Gedenkstätte geführt. Tadao Ando erklärt dazu: „Das Gelände für die Gedenkstätte ist Schritt für Schritt immer größer geworden und nimmt nun die Fläche eines ganzen Häuserblocks ein. Das Grün des Gartens hat sich um das Gebäude herum ausgebreitet, und so hat sich das gesamte Areal geradezu in einen Wald verwandelt. Ursprünglich gehörten zu diesem Projekt drei Gärten: der Garten des ursprünglichen Wohnhauses, der Vorgarten, der sich entlang der gekrümmten Fassade der Gedenkstätte entwickelt und der öffentliche Garten auf der Nordseite zwischen dem Wohnhaus und der Gedenkstätte. 2010 kam ein vierter Garten hinzu, der an das Wohnhaus anschließt und nach Süden orientiert ist. So wurde die Folge der vier intimen Gartenräume vervollständigt. Ich hoffe, dass die Verwirklichung dieser kleinen Gartengestaltung den Dialog zwischen den Herzen der Besucher und Ryotaro Shiba noch verstärkt." Erst jetzt wird der Neubau der Gedenkstätte, eine Stahlbeton- und Glaskonstruktion, sichtbar. Mit seinen etwa 997 Quadratmetern Grundfläche ist das Gebäude größtenteils abgesenkt. Die bogenförmige, auf beiden Seiten mit Regalsystemen ausgestattete Bibliothek erinnert architekturgeschichtlich bewanderte Besucher daran, dass sich Tadao Ando für Räume interessiert, wie man sie auf den Stichen Piranesis entdecken kann.

Situé en banlieue d'Osaka, cet étonnant musée a été aménagé dans l'ancienne propriété de l'écrivain Ryotaro Shiba. Né en 1923 à Osaka et mort en 1996, celui-ci (dont le vrai nom était Teiichi Fukuda) est considéré comme l'un des plus grands écrivains modernes japonais. Les visiteurs atteignent ce mémorial après avoir traversé un jardinet et longé la maison de l'écrivain, où une grande fenêtre révèle son bureau soigneusement conservé. Tadao Ando explique : « Le mémorial s'est agrandi par étapes et occupe maintenant la totalité de l'espace urbain qui lui était alloué. Le jardin s'est développé autour du bâtiment et le site s'est pratiquement transformé en forêt. Initialement, le projet comprenait trois jardins : celui de la résidence d'origine, le jardin de devant qui s'étend le long du volume circulaire du musée et un jardin public au nord du site, entre la résidence et le musée. En 2010, un quatrième jardin orienté au sud et adjacent au premier a été ajouté pour créer une continuité entre ces quatre espaces intimes. J'espère que la satisfaction procurée par ce petit coin de paysage renforcera le dialogue entre le cœur des visiteurs et le souvenir de Ryotaro Shiba. » C'est dans cet environnement que s'élève le nouveau bâtiment d'Ando, en verre et béton armé. La majeure partie de ses 997 m² se trouve en sous-sol. Sa bibliothèque en contrebas rappelle aux visiteurs l'intérêt de Tadao Ando pour les gravures d'architecture de Piranèse.

The Shiba Ryotaro Memorial Museum was completed by Tadao Ando in 2001 in the grounds of the former house of the author. In 2010, the architect was asked to redesign the landscape areas around the old house and the newer structure that was formed like an arc.

Tadao Ando vollendete 2001 das Shiba Ryotaro Memorial Museum, das sich auf dem Grundstück des ehemaligen Wohnhauses des Schriftstellers befindet. 2010 wurde der Architekt gebeten, die Gartenbereiche um das alte Wohnhaus und das bogenförmige Museumsgebäude herum neu zu gestalten.

Situé dans l'ancienne propriété de l'écrivain, le Musée du mémorial de Shiba Ryotaro, en forme d'arc, a été achevé par Tadao Ando en 2001. En 2010, il a été demandé à l'architecte de reconcevoir les aménagements paysagers autour de la maison ancienne et du musée.

The staircase that leads into the main library space (right) contrasts wood with Ando's favorite concrete. The leaded-glass window offers a partial view out to the garden.

Dieser Treppenlauf zum Hauptraum der Bibliothek (rechts) zeigt den Kontrast zwischen Holz und Beton, dem Lieblingsmaterial von Ando. Durch das bleigefasste Fenster hat man Ausblick auf einen Teil des Gartens.

Le sol en bois contraste avec le béton (matériau favori de l'architecte) de l'escalier qui descend vers la bibliothèque principale (à droite). Le mur en vitrail offre une vue partielle sur le jardin.

The main library space. This space, like that of the stairway on the right were part of the project completed by Ando in 2001. Then, as now, the architect had a strong interest in making the landscape of his buildings part of his projects.

Der Hauptraum der Bibliothek. Wie das Treppenhaus auf der rechten Seite gehört er zu dem Projekt, das Tadao Ando 2001 realisierte. Bereits damals war ihm sehr daran gelegen, die Landschaft zum Bestandteil seiner Projekte zu machen.

La bibliothèque principale. Comme l'escalier de la page de droite, cet espace fait partie du projet achevé par Ando en 2001. Aujourd'hui encore, l'architecte s'intéresse fortement aux aménagements paysagers, parties intégrantes de ses projets.

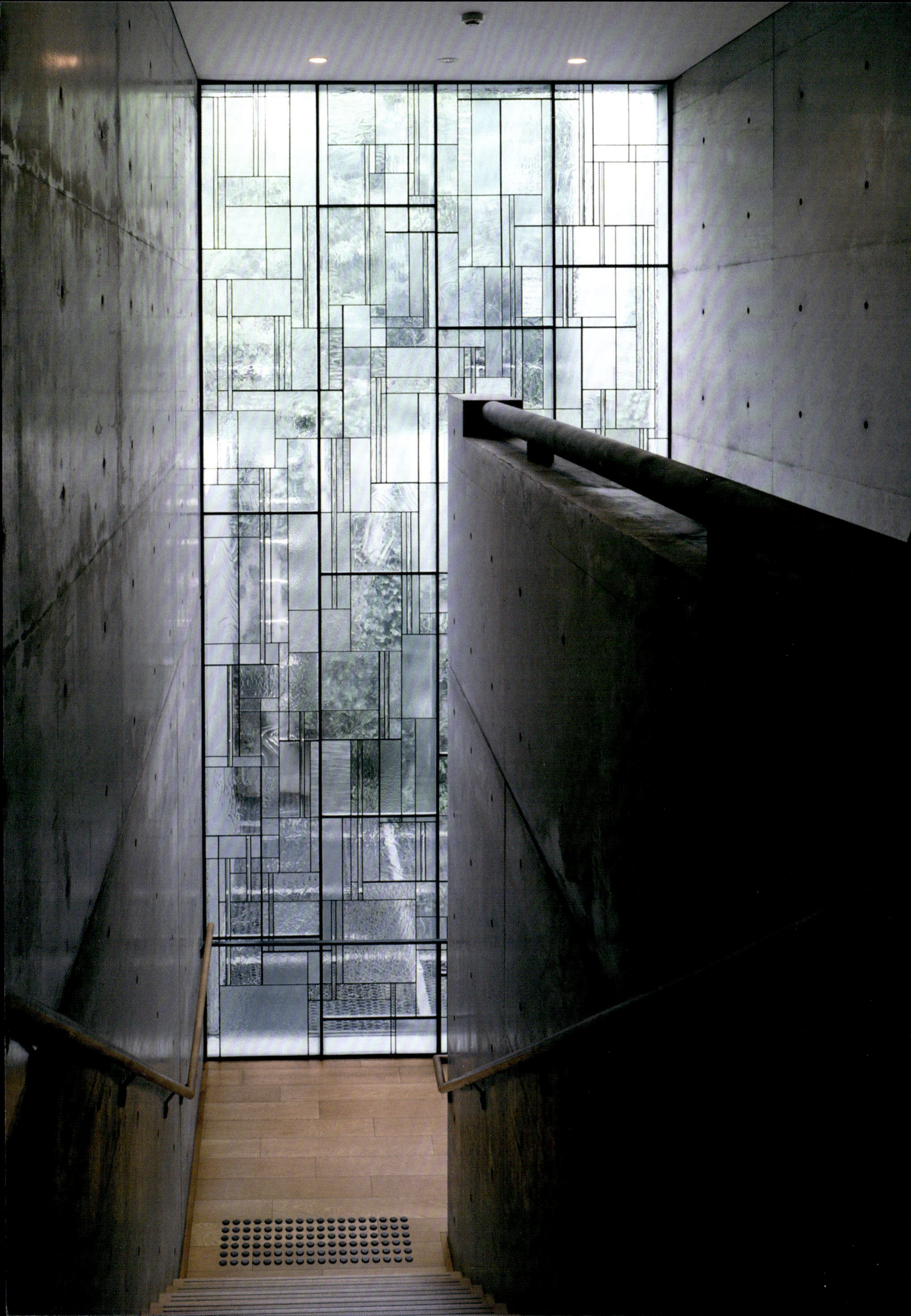

LEE UFAN MUSEUM
Naoshima, Kagawa, Japan, 2007–10

Address: 1390 Azakuraura, Naoshima, Kagawa 7613110, Japan, www.benesse-artsite.jp
Area: 9860 m^2 (site), 443 m^2 (museum). Client: Naoshima Fukutake Art Museum Foundation. Cost: not disclosed

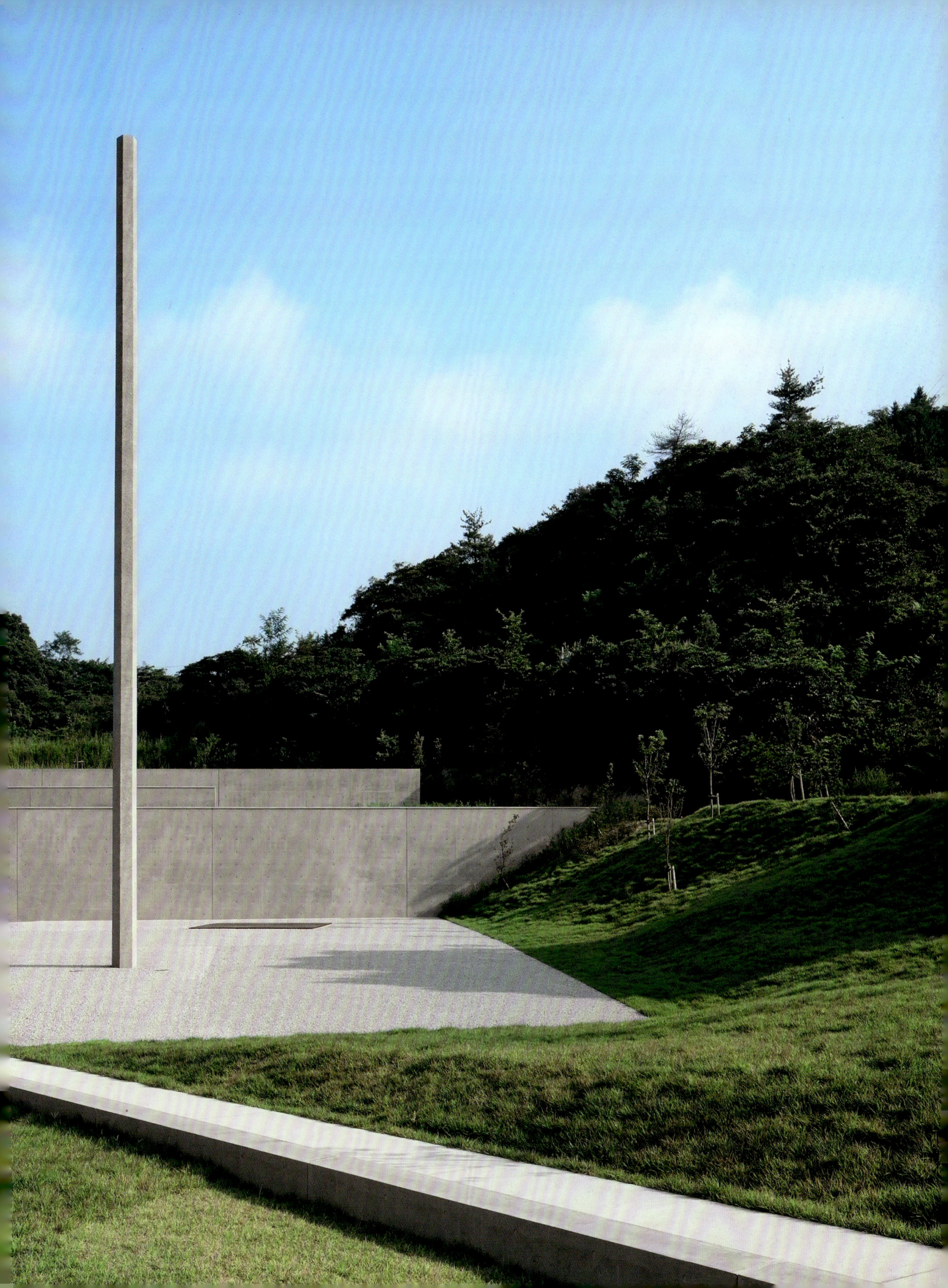

After the completion of his Benesse House Museum (formerly Naoshima Contemporary Art Museum, 1992) and Chichu Art Museum (2004), Tadao Ando's third museum on the island of Naoshima opened in June 2010. The museum, located between the two earlier structures, is exclusively dedicated to the artworks of Lee Ufan, a well-known contemporary Korean artist. "As in some of my previous architecture, the unity with nature and merging it into the landscape was the main theme of the design," states Tadao Ando. A 30 x 30-square-meter front courtyard paved with cobblestones welcomes visitors. The visitors are then led to a second, triangular entrance court through a passage made up by two straight, parallel walls appearing as a slice in the topography of the site. Three rectangular galleries can be found in the building buried underground. Their different scale, light conditions, and materiality are a result of a thorough dialogue with the artist, hence carefully adjusted to the exhibited artworks. "My intention," concludes the architect, "was to create correlation with the strict profile of the art pieces by Lee Ufan and make the visitors feel like being inhaled to the depth of the space."

Nach dem Benesse House Museum (ehemals Museum der Gegenwartskunst von Naoshima, 1992) und dem Kunstmuseum Chichu (2004) wurde das dritte Museumsgebäude von Tadao Ando auf der Insel Naoshima im Juni 2010 eröffnet. Das Museum liegt zwischen den beiden älteren Gebäuden und ist ausschließlich dem Werk von Lee Ufan, einem bekannten koreanischen Gegenwartskünstler gewidmet. Tadao Ando sagt über dieses Bauwerk: „Wie schon bei einigen meiner früheren Gebäude war das Hauptanliegen des Entwurfs die Einheit mit der Natur und die Verschmelzung mit der Landschaft." Die Besucher kommen in einem 30 x 30 Meter großen Hof mit Kopfsteinpflaster an. Von dort werden sie durch eine Passage, die von zwei geraden, parallel angeordneten Mauern begrenzt wird, zu einem dreieckigen Eingangshof weitergeleitet. Diese Mauern wirken wie ein Schnitt in die Topografie des Geländes. In dem unterirdischen Gebäudeteil befinden sich drei rechteckige Galerieräume. Ihr unterschiedlicher Maßstab, die Belichtungsbedingungen und ihre Materialität sind Ergebnis eines intensiven Dialogs mit dem Künstler und sorgfältig auf die ausgestellten Kunstwerke abgestimmt. Tadao Ando bemerkt dazu: „Meine Absicht war es, eine Beziehung zu den strengen Formen der Kunstwerke von Lee Ufan herzustellen und den Besuchern das Gefühl zu vermitteln, von der Tiefe des Raums aufgesogen zu werden."

Après le musée de la Benesse House (anciennement Musée d'art contemporain de Naoshima, 1992) et le musée Chichu (2004), le troisième musée édifié par Tadao Ando sur l'île de Naoshima a ouvert ses portes en juin 2010. Implanté entre les deux constructions existantes, il est exclusivement consacré aux œuvres de Lee Ufan, célèbre artiste coréen contemporain. « Comme dans certaines de mes réalisations antérieures, le thème principal de ce projet a été la recherche de l'unité avec la nature et la fusion avec le paysage », explique Tadao Ando. Une cour antérieure carrée de 30 x 30 m et pavée de galets accueille les visiteurs. Ils sont ensuite dirigés vers une seconde cour d'entrée triangulaire par le biais d'un passage délimité par deux murs rectilignes parallèles, formant comme une tranchée dans la topographie du site. Le bâtiment contient trois galeries de plan rectangulaire aménagées en sous-sol. Les différences d'échelle, d'éclairage et de matériaux sont l'aboutissement d'un dialogue approfondi avec l'artiste, ce qui explique leur subtile adéquation avec les œuvres présentées. « Mon intention, conclut l'architecte, était de créer une corrélation avec le caractère rigoureux des œuvres de Lee Ufan et de donner aux visiteurs l'impression d'être aspirés dans les profondeurs de cet espace. »

The powerful concrete walls designed by Tadao Ando contrast in their geometric rigor with the more "natural" form of the works of Lee Ufan. This project is part of Ando's ongoing involvement in the architecture and art on the island of Naoshima in the Inland Sea of Japan.

Die mächtigen, von Tadao Ando entworfenen Betonmauern bilden mit ihrer geometrischen Strenge einen Kontrast zu den „natürlicheren" Formen der Arbeiten von Lee Ufan. Dieses Projekt gehört zu dem Ensemble von Architektur und Kunst auf der Insel Naoshima im Seto-Binnenmeer, an dem Ando maßgeblich beteiligt ist.

Les puissants murs en béton conçus par Tadao Ando contrastent par leur rigueur géométrique avec les formes plus « naturelles » des œuvres de Lee Ufan. Ce musée fait partie des interventions de longue date de l'architecte dans les projets architecturaux et artistiques développés sur l'île de Naoshima en mer Intérieure du Japon.

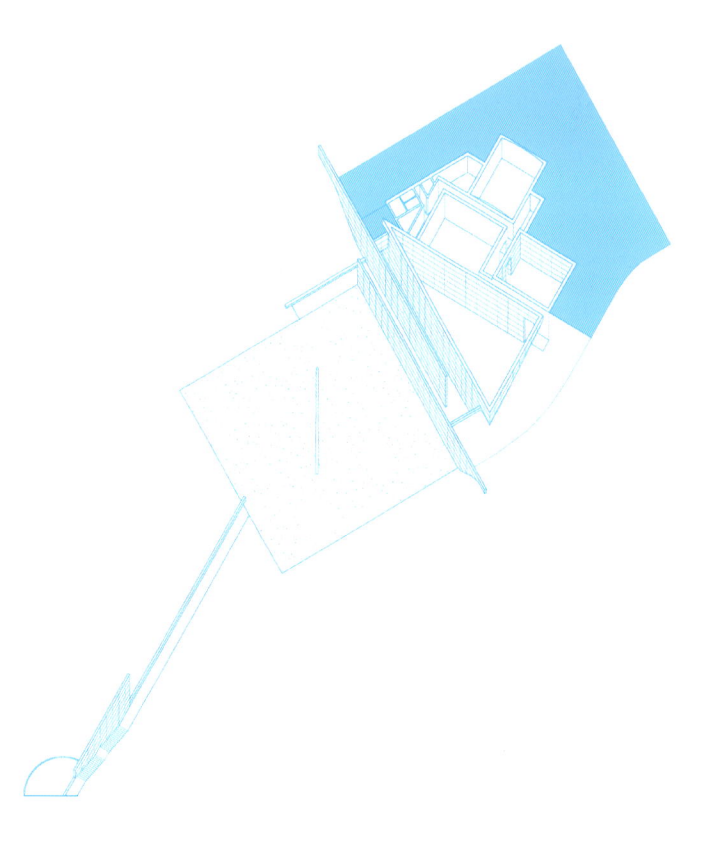

BALMORI ASSOCIATES

Balmori Associates
833 Washington Street, 2nd Floor
New York, NY 10014
USA

Tel: +1 212 431 9191
Fax: +1 212 431 8616
E-mail: info@balmori.com
Web: www.balmori.com

DIANA BALMORI was born in Gijón, Spain, and attended the Undergraduate Architecture Program at the University of Tucumán (Argentina, 1949–52), before studying at the University of California at Los Angeles (1968–70, B.A.; 1970–75, Ph.D. in Urban History). She was a Partner for Landscape and Urban Design with Cesar Pelli & Associates (New Haven, 1981–90). She is the founding Principal of Balmori Associates created in 1990 in New Haven, with an office in New York since 2001. In 2006, Diana Balmori was appointed a Senior Fellow in Garden and Landscape Studies at Dumbarton Oaks in Washington, D.C., and is serving her second term on the US Commission of Fine Arts. She has served as a jury member for the Bilbao Jardín 2009 garden festival. Her recent built work includes MPPAT (Master Plan for Public Administrative Town, Sejong, Korea, 2007); Duke University Master Plan (with Cesar Pelli & Associates, Durham, North Carolina, USA, 2008); the Garden That Climbs the Stairs (Bilbao, Spain, 2009, published here); Botanical Research Institute of Texas (BRIT, Fort Worth, USA, 2011); Campa de los Ingleses (Bilbao, 2007–); and Plaza Euskadi (Bilbao, 2011). According to the firm's own description: "Through research, collaboration, and innovation, Balmori Associates explore and expand the boundaries between nature and structure."

DIANA BALMORI ist in Gijón, Spanien, geboren und studierte von 1949 bis 1952 Architektur an der Universität von Tucumán in Argentinien. Sie setzte ihre Studien an der University of California in Los Angeles fort (1968–70, B. A., 1970–75, Promotion in Stadtgeschichte). Sie war als Partnerin im Büro Cesar Pelli & Associates (New Haven, 1981–90) für Garten- und Landschaftsplanung tätig und gründete 1990 ihr eigenes Büro Balmori Associates in New Haven, das seit 2001 auch in New York ansässig ist. 2006 wurde Diana Balmori Senior Fellow für den Studiengang Garten und Landschaft in Dumbarton Oaks in Washington, D. C., und ist zum zweiten Mal Mitglied der US Commission of Fine Arts. Beim Gartenfestival Bilbao Jardín 2009 fungierte sie als Jurymitglied. Zu ihren neuesten Arbeiten gehören der MPPAT (Masterplan für das Verwaltungsviertel von Sejong, Korea, 2007), der Masterplan für die Duke University (mit Cesar Pelli & Associates, Durham, North Carolina, USA, 2008), der Garten, der die Treppe hinaufwächst (Bilbao, Spanien, 2009, hier vorgestellt), das Botanische Forschungsinstitut von Texas (BRIT, Fort Worth, USA, 2011), der Park Campa de los Ingleses (Bilbao, seit 2007) und die Plaza Euskadi (Bilbao, 2011). Das Büro beschreibt seine Arbeitsweise wie folgt: „Aufgrund von Recherchen, Zusammenarbeit sowie innovativer Ansätze sondiert Balmori Associates die Grenzen zwischen Natur und Struktur und erweitert sie."

DIANA BALMORI, née à Gijón (Espagne), a suivi le programme de préparation aux études d'architecture de l'université de Tucumán (Argentine, 1949–52), avant d'étudier à l'université de Californie à Los Angeles (1968–70, B.A.; 1970–75, Ph.D. en histoire urbaine). Elle a été partenaire pour l'urbanisme et le paysage chez Cesar Pelli & Associates (New Haven, 1981–90). Elle dirige l'agence Balmori Associates qu'elle a fondée en 1990 à New Haven et qui possède un bureau à New York depuis 2001. En 2006, elle a été nommée *senior fellow* en études du paysage à l'institut de Dumbarton Oaks, à Washington, et siège pour la seconde fois à la Commission américaine des beaux-arts. Elle a été membre du jury du festival Bilbao Jardín 2009. Parmi ses réalisations récentes : le plan directeur de la ville administrative à Sejong (MPPAT, Sejong, Corée, 2007); le plan directeur de l'université Duke (avec Cesar Pelli & Associates, Durham, Caroline du Nord, US, 2008); The Garden That Climbs the Stairs (Bilbao, 2009, publié ici); le Botanical Research Institute of Texas (BRIT, Fort Worth, Texas, US, 2011); le parc Campa de los Ingleses (Bilbao, 2007–) et la Plaza Euskadi (Bilbao, 2011). « Par la recherche, la collaboration et l'innovation, Balmori Associates explore et repousse les limites entre nature et structure », précise le descriptif de l'agence.

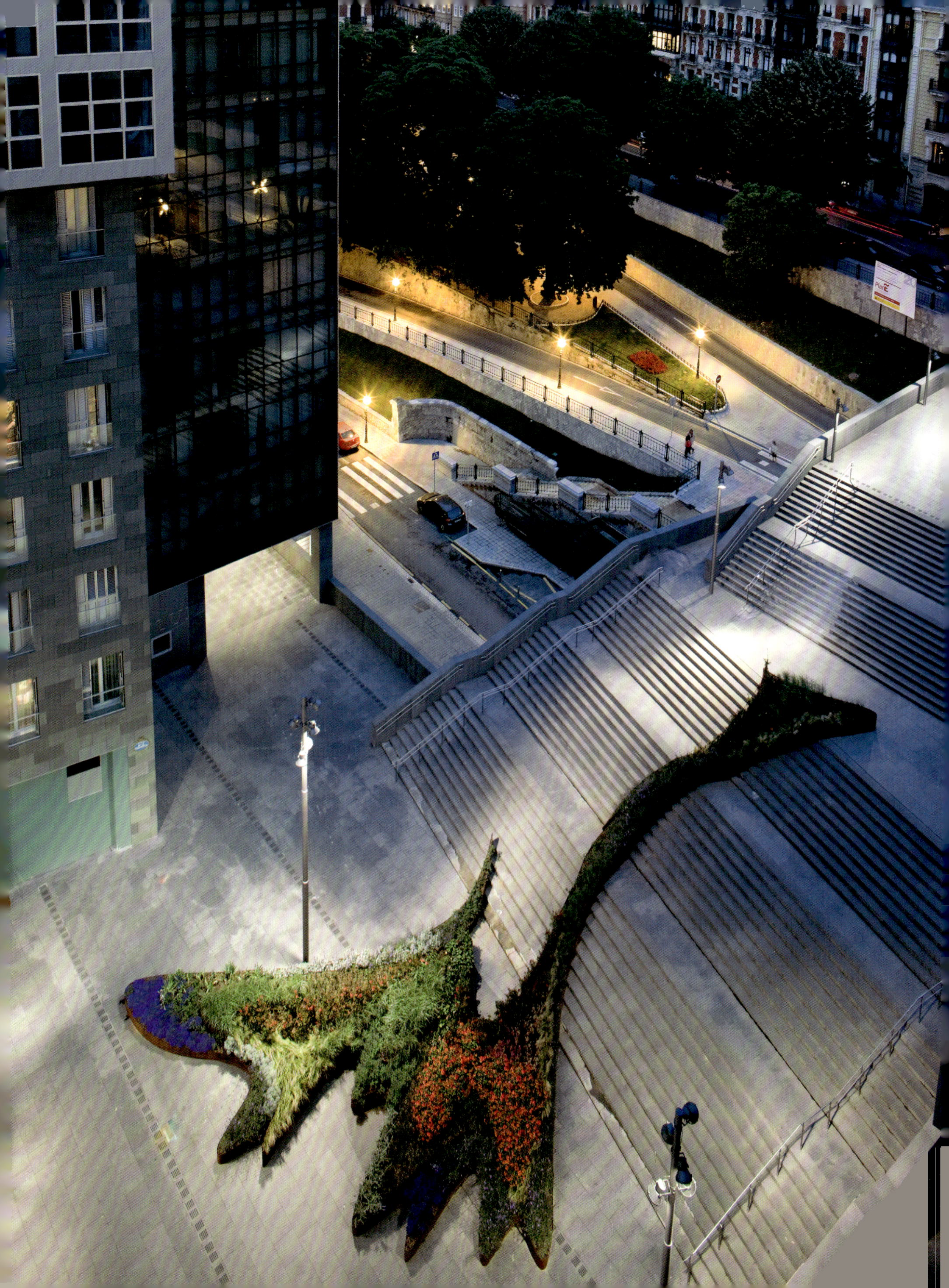

THE GARDEN THAT CLIMBS THE STAIRS
Bilbao, Spain, 2009

Address: Isozaki Atea, Ensanche, Bilbao, Spain. Area: 80 m²
Client: II International Competition: Bilbao Jardin 2009 and Fundación Bilbao 700
Cost: €12 000

With its curvilinear planters arcing across the geometric and mineral area of the square and stairs, the "Garden That Climbs the Stairs" introduces nature to an otherwise quite arid space.

Mit den geschwungenen Pflanzkübeln, die quer über den geometrisch gestalteten steinernen Platz und die Treppen führen, bringt der „Garten, der die Treppe hinaufwächst" Natur in einen sonst recht faden Freiraum.

Par ses jardinières curvilignes qui « poussent » dans l'univers géométrique et minéral de la place et des escaliers, ce « Jardin qui monte les escaliers » introduit la nature dans un environnement par ailleurs assez aride.

Making use of Corten steel, the planters assume a slightly irregular profile as they "climb" the steps, giving passersby something to look at, smell, and comment about.

Die Pflanzkübel aus Cor-Ten-Stahl verändern in dem Maß, wie sie die Treppe „emporsteigen" etwas ihr Profil, sodass die Passanten etwas zum Hinschauen, zum Riechen und natürlich auch zum Kommentieren haben.

Alors qu'elles « montent » les escaliers, ces jardinières en acier Corten au profil assez irrégulier donnent aux passants matière à regarder, sentir et commenter.

As a member of the jury for the second Bilbao Jardín festival (2009), Diana Balmori was invited to create a garden in the city. The location she chose is a stairway located between two 83-meter-high towers by Arata Isozaki (Isozaki Atea, 2004–08). These steps lead to Santiago Calatrava's bridge over the Nervión River (Campo Volantin Footbridge, 1990–97). Balmori sought to create numerous contrasts in this small garden: between red flowers, green grass, and gray paving, for example. The firm states: "The garden climbs the stairs, running in undulating lines of different textures and colors. Envisioned as a dynamic urban space, it moves in time and with the seasons. Its lush planting cascades down as though the garden was flowing or melting, bleeding the colors into each other. In one gesture, it narrates a story of landscape taking over and expanding over the Public Space and Architecture, therefore transforming the way that the stairs and the space is perceived and read by the user." The garden is one of several projects by Balmori Associates in Bilbao that include the Abandoibarra Master Plan (competition winner, 1997), Plaza Euskadi (2011), and Campa de los Ingleses (competition winner, 2007 / under construction).

Als Mitglied der Jury für das zweite Gartenfestival von Bilbao (2009) wurde Diana Balmori eingeladen, einen Garten mitten in der Stadt zu realisieren. Sie entschied sich für eine Treppenanlage zwischen zwei 83 Meter hohen Hochhäusern von Arata Isozaki (die Isozaki Atea, 2004–08). Die Treppe führt zu Santiago Calatravas Brücke über den Nervión (die Fußgängerbrücke Campo Volantin, 1990–97). Balmori hat in diesem kleinen Garten zahlreiche starke Kontraste geschaffen, zum Beispiel rote Blumen, grünes Gras und graues Pflaster. Das Büro sagt dazu: „Der Garten steigt die Treppen in geschwungenen Linien mit unterschiedlichen Texturen und Farben hinauf. Er ist als dynamischer urbaner Raum gedacht und ändert sich mit der Zeit und den Jahreszeiten. Die üppige Bepflanzung fließt wie eine Kaskade die Treppe hinunter, dabei verschmelzen die Farben miteinander. Mit einer einzigen Geste erzählt dieser Garten wie Landschaft ‚an die Macht kommt' und sich über den öffentlichen Raum und die Architektur ausbreitet. So verändert sich bei dem Nutzer die Art, wie er die Treppe und den Freiraum wahrnimmt." Dieser Garten ist eines von mehreren Projekten, die das Büro Balmori Associates in Bilbao realisiert hat. Dazu gehören auch der Masterplan für Abandoibarra (Wettbewerbssieger 1997), die Plaza Euskadi (2011) und der Park Campa de los Ingleses (Wettbewerbssieger 2007, im Bau).

Membre du jury du second festival Bilbao Jardín en 2009, Diana Balmori a été invitée à y créer un jardin. Elle a choisi un lieu particulier : un grand escalier entre deux tours de 83 m de haut dues à Arata Isozaki (Isozaki Atea, 2004–08). Il conduit au pont jeté par Santiago Calatrava sur le Nervión (passerelle de Campo Volantin, 1990–97). Balmori a cherché à multiplier les contrastes, par exemple entre les fleurs rouges, l'herbe verte et le pavement gris : « Le jardin monte les escaliers en formant des vagues de différentes textures et couleurs. Conçu comme un espace urbain dynamique, il change avec le temps et les saisons. Ses plantations luxuriantes retombent en cascade comme si le jardin coulait sur les marches ou se fondait avec elles, les couleurs débordant les unes sur les autres. Dans ce geste, il raconte l'histoire d'un paysage qui se développe et prend le dessus sur l'espace public et l'architecture, transformant ainsi la façon dont l'escalier et l'espace sont perçus par leurs usagers. » Ce jardin fait partie d'un ensemble de projets de Balmori Associates à Bilbao, dont le plan directeur d'Abandoibarra (concours remporté, 1997), la Plaza Euskadi (2011) et le parc Campa de Los Ingleses (concours remporté, 2007–).

BALMORI ASSOCIATES

The drawing of the work (below) shows its protean forms. The adaptation to the rising stair gives the installation an even more "organic" appearance and dynamism than it might have had on a uniformly flat surface.

Die Zeichnung (unten) zeigt deutlich die fließenden Formen. Die Anpassung an den Treppenlauf lässt diese Installation noch organischer und dynamischer wirken, als dies auf einer ebenen Fläche der Fall gewesen wäre.

Le dessin (ci-dessous) montre la qualité protéiforme du projet. Son adaptation aux escaliers lui donne un aspect encore plus organique et un dynamisme que cette installation n'aurait pas eu sur une surface uniformément plane.

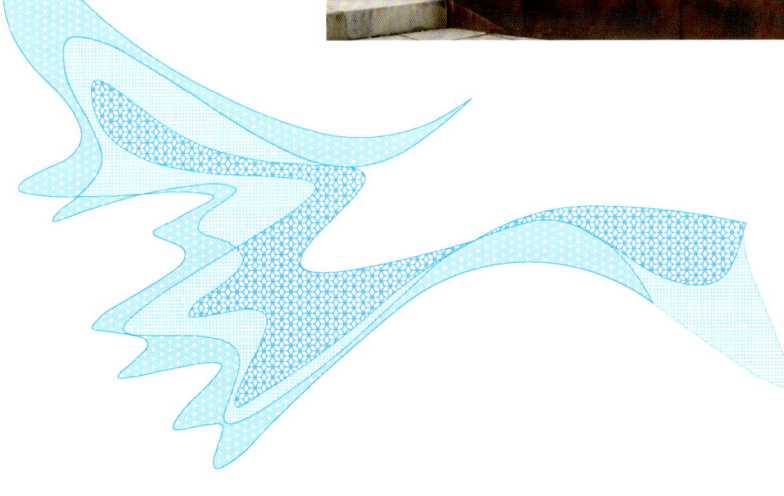

Orchid Waltz

PATRICK BLANC

Patrick Blanc

E-mail: info@murvegetalpatrickblanc.com
Web: www.verticalgardenpatrickblanc.com

Born in Paris, France, in 1953, **PATRICK BLANC** received his doctorate in Botany in 1989 (Docteur d'Etat ès Sciences, Université Pierre et Marie Curie, Paris 6). He does research on the comparative growth rates of plants and their capacity to adapt to extreme environments at the CNRS (Laboratoire d'Ecologie, Brunoy, since 1982). He has published more than 50 articles in scientific journals since 1977. He currently teaches at the University of Paris VI, and has patented his system of "mur végétal," or vertical gardens. Recent vertical gardens installed by Patrick Blanc include those at the European Parliament (Brussels, Belgium, 2006); the CaixaForum (Madrid, 2006–07); a concert hall in Taipei (Taiwan, 2007); the Plaza de España (in collaboration with Herzog & de Meuron; Santa Cruz de Tenerife, Canary Islands, Spain, 2008; page 156); and the Museum of Natural History in Toulouse (France, 2008). He also installed works at Jean Nouvel's Cartier Foundation (1998) and the Quai Branly Museum (2004), both in Paris. His work was named one of the "50 Best Inventions of the Year" by *Time* magazine (2009). Current work includes the Max Juvénal Bridge (Aix-en-Provence, France, 2008, published here); Orchid Waltz, National Theater Concert Hall (Taipei, 2009, also published here); Green Office (Meudon, France, 2011); Drew School (San Francisco, USA, 2011); and Tower One Central Park (Sydney, Australia, 2013).

PATRICK BLANC wurde 1953 in Paris geboren. 1989 promovierte er in Botanik (Docteur d'Etat ès Sciences, Université Pierre et Marie Curie, Paris 6). Seit 1982 forscht er am CNRS (Laboratoire d'Écologie, Brunoy) über die relativen Wachstumsraten von Pflanzen und ihre Fähigkeit, sich an extreme Umweltbedingungen anzupassen. Seit 1977 hat er über 50 Artikel in Fachzeitschriften veröffentlicht. Derzeit lehrt er an der Universität Paris VI. Sein System der „mur végétal", des vertikalen Gartens, hat er patentieren lassen. Seine neuesten vertikalen Gärten hat Patrick Blanc am Europäischen Parlament in Brüssel angelegt (2006), am CaixaForum (Madrid, 2006–07), in einer Konzerthalle in Taipeh (Taiwan, 2007), an der Plaza de España (in Zusammenarbeit mit Herzog & de Meuron in Santa Cruz de Tenerife, Kanarische Inseln, Spanien, 2008, Seite 156) und beim Museum für Naturgeschichte in Toulouse (Frankreich, 2008) realisiert. Er hat ebenfalls Arbeiten für die Fondation Cartier von Jean Nouvel (1998) und beim Musée du Quai Branly (2004) realisiert, die sich beide in Paris befinden. Seine Arbeiten wurden 2009 vom „Time Magazine" zu den „50 besten Erfindungen des Jahres" gewählt. Zu seinen neuesten Arbeiten zählen die Max-Juvénal-Brücke (Aix-en-Provence, Frankreich, 2008, hier vorgestellt), Orchid Waltz im Konzertsaal des Nationaltheaters von Taipeh (Taiwan, 2009, ebenfalls hier vorgestellt), das Green Office (Meudon, Frankreich, 2011), die Drew School (San Francisco, USA, 2011) und der Tower One Central Park (Sydney, Australien, 2013).

Né à Paris en 1953, **PATRICK BLANC** est docteur en botanique (doctorat d'État ès sciences de l'université Pierre et Marie Curie, Paris VI, 1989). Depuis 1982, il effectue des recherches sur les taux de croissance comparés des plantes et leur capacité à s'adapter à des environnements extrêmes, au Laboratoire d'écologie du CNRS à Brunoy. Il a publié plus de 50 articles scientifiques depuis 1977. Il enseigne actuellement à l'université de Paris VI et a breveté son système de « mur végétal », ou jardin vertical. Parmi les lieux qui ont accueilli ses récents jardins verticaux figurent le Parlement européen (Bruxelles, 2006) ; le CaixaForum (Madrid, 2006–07) ; une salle de concert à Taipei (Taiwan, 2007) ; la Plaza de España, en collaboration avec Herzog & de Meuron (Santa Cruz de Tenerife, îles Canaries, Espagne, 2008, page 156) ; le Muséum d'histoire naturelle de Toulouse (2008) et deux réalisations parisiennes de Jean Nouvel : la Fondation Cartier (1998) et le musée du quai Branly (2004). Son travail a été retenu parmi les « 50 meilleures inventions de l'année » du magazine *Time* en 2009. Ses projets actuels : le pont Max Juvénal (Aix-en-Provence, France, 2008, publié ici) ; Orchid Waltz, dans la salle de concert du Théâtre national de Taipei (2009, publié ici) ; l'immeuble de bureaux Green Office (Meudon, 2011) ; la Drew School (San Francisco, US, 2011) et la Tower One Central Park (Sydney, Australie, 2013).

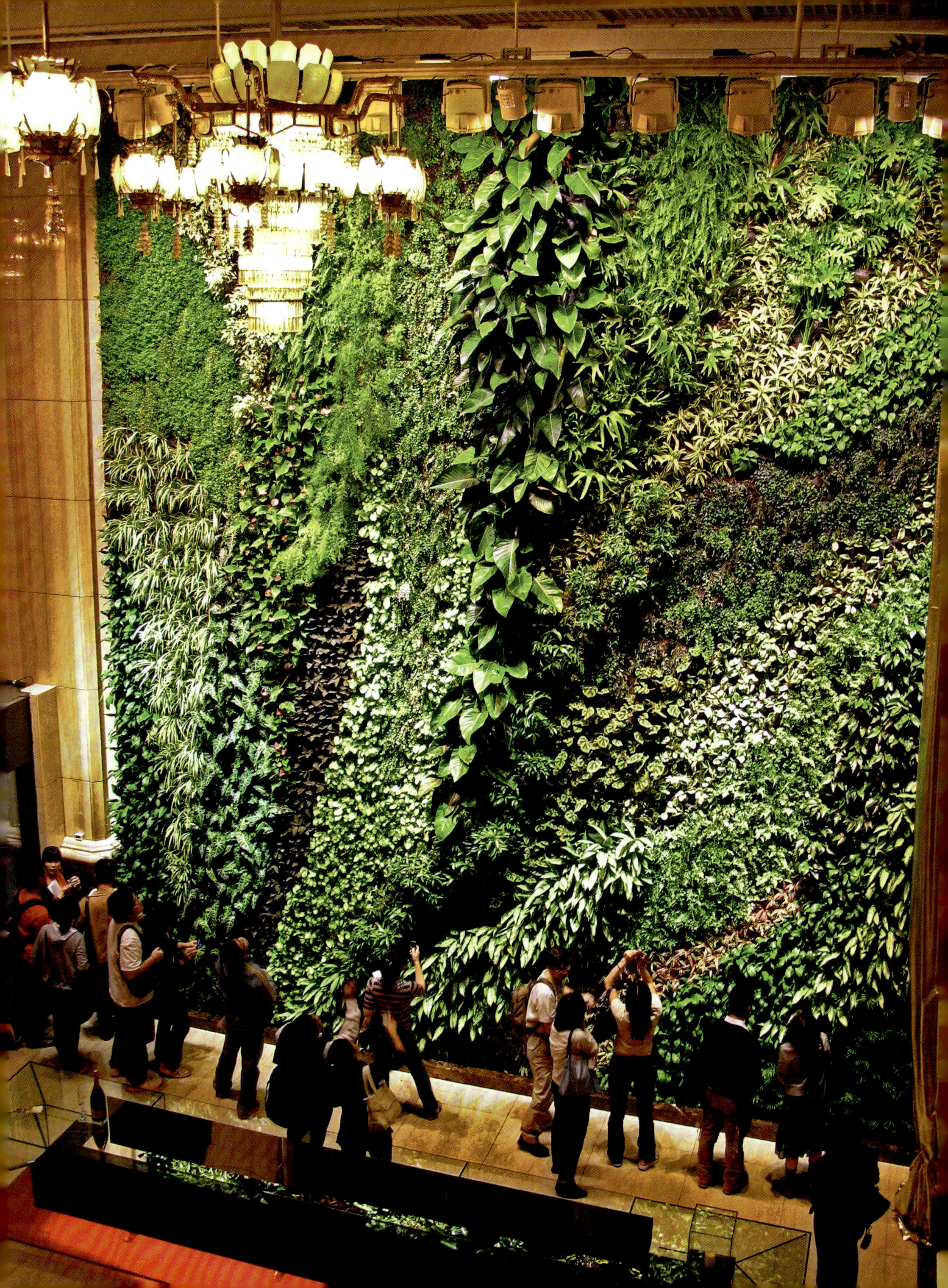

ORCHID WALTZ

National Theater Concert Hall, Taipei, Taiwan, 2009

Address: 10048 21–1 Chung-Shan South Road, Taipei, Taiwan, Republic of China, +886 2 3393 9888, www.ntch.edu.tw
Area: 64 m² (8 m x 8 m). Client: National Theater of Taiwan
Cost: not disclosed

Patrick Blanc recently created works in the National Theater in Taipei. His **ORCHID WALTZ** is made up of two separate green walls. According to Blanc, "one displays Taiwan's native species, and the other demonstrates Taiwan's technological advances." "The Butterfly Dance," one of these walls, features 250 plants including 46 hybrid species of orchids together with 2000 Adiantums. The other wall, entitled "The Wild Dance," makes use of 25 species of native Taiwan orchids mixed with nine other kinds of foliage, and an overall total of 2900 plants. Patrick Blanc has patented his "mur végétal" system which allows plants to grow in pockets of feltlike plastic irrigated with a nutrient solution through plastic pipes. Tchen Yu-chiou, the Chairperson of the National Theater Concert Hall, was inspired by Blanc's work at the Musée du Quai Branly in Paris and asked him to design these pieces for Taipei. "Mr. Blanc treats plants like human beings," she has stated. "I was moved to see him create a conversation between humans and plants." Having already commissioned Patrick Blanc to create another work, entitled "Green Symphony," for the National Concert Hall in October 2007, Tchen Yu-chiou asked Blanc to work on this second project for the National Theater.

Patrick Blanc realisierte seine neuesten Arbeiten im Nationaltheater von Taipeh. Sein **ORCHID WALTZ** besteht aus zwei unterschiedlichen begrünten Wänden. Patrick Blanc bemerkt dazu: „…auf der einen Mauer werden heimische Spezies aus Taiwan gezeigt, und auf der anderen werden die technologischen Fortschritte des Landes symbolisch dargestellt". Bei der Wandgestaltung „Tanz der Schmetterlinge" verwandte Blanc 2000 Adiantums (Frauenhaarfarn) und 250 andere Pflanzen (unter anderem 46 Orchideenhybriden). Bei der anderen Wand mit dem Titel „Der wilde Tanz" handelt es sich um eine Komposition aus 25 in Taiwan heimischen Orchideenarten und neun anderen Sorten von Blattpflanzen. Insgesamt sind hier 2900 Pflanzen zusammengestellt. Patrick Blanc hat sein System der „mur végétal" patentieren lassen. Damit ist es möglich, Pflanzen in Taschen aus einem filzartigen Kunststoff zu ziehen, die über Plastikschläuche mit einer Nährlösung versorgt werden. Tchen Yu-chiou, die Vorstandsvorsitzende der Konzerthalle des Nationaltheaters, war von Blancs Arbeit am Musée du Quai Branly in Paris begeistert und bat ihn, in Taipeh tätig zu werden. Sie sagt: „Monsieur Blanc behandelt Pflanzen als wären es menschliche Wesen. Ich war sehr gerührt von der Art, wie er ein Gespräch zwischen Menschen und Pflanzen herstellt." Nachdem Patrick Blanc bereits im Oktober 2007 den Auftrag erhalten hatte, für die Konzerthalle eine Arbeit mit dem Titel „Green Symphony" zu komponieren, bat ihn Tchen Yu-chiou, dieses zweite Projekt für das Nationaltheater auszuführen.

Patrick Blanc a récemment travaillé pour le Théâtre national de Taipei. Son **ORCHID WALTZ** se répartit sur deux murs séparés. « L'un représente des espèces végétales de Taiwan, l'autre les progrès technologiques du pays », explique-t-il. L'un de ces murs, *Butterfly Dance*, se compose de 250 plants de 46 hybrides d'orchidées et de 2000 adiantums. L'autre, *The Wild Dance*, met en scène 25 espèces d'orchidées d'origine taiwanaise associées à neuf variétés de plantes vertes, représentant au total 2900 pieds. Patrick Blanc a breveté un système de « mur végétal » dans lequel les plantes croissent dans des poches de feutre plastique alimentées en solution nutritive par des tuyaux. Tchen Yu-chiou, présidente de la salle de concert du Théâtre national, très intéressée par l'intervention de Patrick Blanc au musée du quai Branly à Paris, lui avait demandé de venir à Taipei. « Monsieur Blanc traite les plantes comme des êtres humains », a-t-elle déclaré. « J'ai été très émue de le voir créer un dialogue entre les plantes et les hommes. » C'est la seconde réalisation commandée par Tchen Yu-chiou à Patrick Blanc, qui avait déjà réalisé une œuvre intitulée *Green Symphony* pour la salle de concert en octobre 2007.

Patrick Blanc's designs are an unexpected mixture of natural exuberance and carefully planned use of his knowledge of botany. The drawing on the left and the completed work in Taipei show how he plans his effects.

Die Entwürfe von Patrick Blanc sind eine unerwartete Mischung von natürlicher Üppigkeit und der sorgfältig geplanten Umsetzung seiner botanischen Kenntnisse. Die Zeichnung links und die vollendete Arbeit in Taipeh verdeutlichen, wie er seine Effekte plant.

Les projets de Patrick Blanc sont un mélange inattendu d'exubérance naturelle et de mise en œuvre minutieuse de sa connaissance de la botanique. Le dessin de gauche et l'œuvre mise en place à Taipei montrent comment a été programmé l'effet visuel.

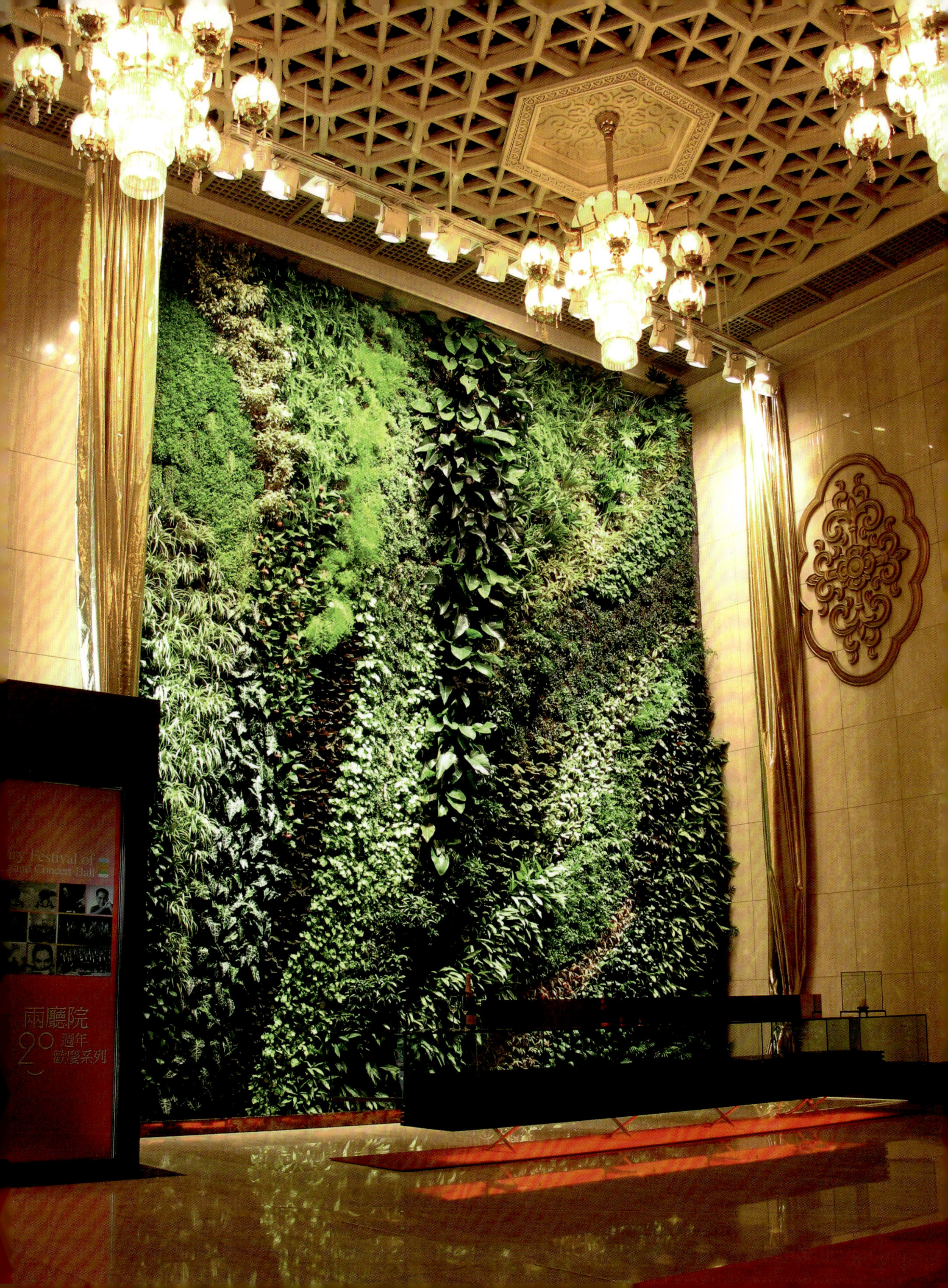
Festival of
Concert Hall
兩廳院
20 週年
歡慶系列

MAX JUVÉNAL BRIDGE
Aix-en-Provence, France, 2008

*Area: 650 m². Client: SEMEPA and the City of Aix-en-Provence
Cost: not disclosed*

"I like to reintegrate nature where one least expects it," Blanc told *The New York Times*. The newspaper called the **PONT MAX JUVÉNAL** in Aix "the most beautiful overpass you've ever seen." Blanc's garden on this modern bridge resembles an entire landscape except that it is vertical rather than horizontal. This bridge runs over railway tracks and the Avenue Max Juvénal near the Grand Théâtre de Provence. It had originally been envisaged to cover this bridge in brick or stone, but local authorities selected Blanc's work in order to soften the largely mineral aspect of the environment. Approximately 15 meters high, the installation has approximately 30 to 35 plants per square meter for a total of about 20 000 plants.

„Ich habe Freude daran, Natur dort wieder anzusiedeln, wo man es am wenigsten erwartet", sagte Patrick Blanc der „New York Times". Die Zeitung nannte den **PONT MAX JUVÉNAL** in Aix „die schönste Überführung, die man je gesehen hat". Blanc hat auf dieser modernen Brücke einen Garten geschaffen, der wie eine ganze Landschaft wirkt, nur mit dem Unterschied, dass sie sich vertikal statt horizontal entwickelt. Die Brücke überquert Gleiskörper und die Avenue Max Juvénal in der Nähe des Grand Théâtre de Provence. Ursprünglich war vorgesehen, die Brücke mit Klinker oder Naturstein zu verkleiden, aber die Stadt entschied sich für Blancs Arbeit, weil so das ansonsten sehr steinerne Umfeld optisch abgemildert wird. Die Installation ist etwa 15 Meter hoch, und pro Quadratmeter wachsen hier 30 bis 35 Pflanzen, insgesamt also etwa 20 000 Pflanzen.

« J'aime redonner ses droits à la nature là où on l'attend le moins », dit Patrick Blanc dans un article du *New York Times*. Le journal a par ailleurs qualifié le **PONT MAX JUVÉNAL** d'Aix de « plus bel ouvrage d'art jamais vu ». Le jardin aménagé par Blanc sur ce pont moderne évoque un paysage qui serait vertical et non horizontal. L'ouvrage franchit des voies ferrées et l'avenue Max Juvénal, non loin du Grand Théâtre de Provence. On avait, à l'origine, envisagé de l'habiller de brique ou de pierre, mais les autorités locales ont sélectionné le projet de Patrick Blanc pour atténuer l'aspect très minéral du lieu. D'environ 15 m de haut, l'installation compte de 30 à 35 plantes par mètre carré, soit au total près de 20 000 plants.

Patrick Blanc transforms an otherwise banal roundabout into a gentle landscape with his remarkable vertical garden—where it is "least expected," as he says.

Mit seinem bemerkenswerten vertikalen Garten verwandelt Patrick Blanc ein ansonsten banales Umfeld in eine freundliche Landschaft – dort wo man es „am wenigsten erwartet", wie er selbst sagt.

Patrick Blanc a transformé l'environnement de ce banal rond-point en un paysage délicat par ce remarquable jardin vertical, « là où il était le moins attendu », dit-il.

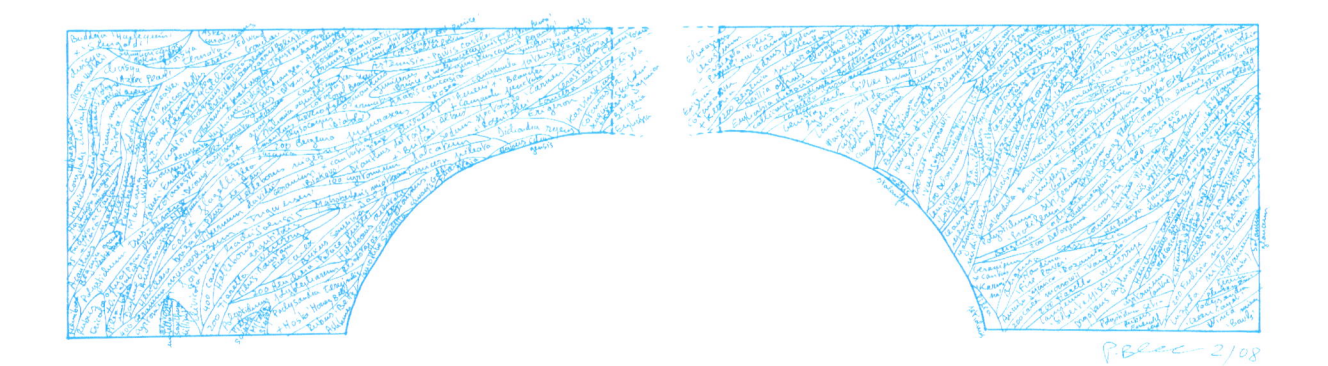

M. CAFFARENA / V. COBOS / G. ALCARAZ / G. DELGADO

María Caffarena de la Fuente
Justiniano 3, 3º, 28004 Madrid, Spain
Tel: +34 619 85 16 88 / Fax: +34 91 308 51 77
E-mail: marcaff@gmail.com

Víctor Cobos Márquez
Maximilianstr. 12, 13187 Berlin, Germany
Tel: +34 654 62 83 87
E-mail: vitorbos@gmail.com

Andrés García Alcaraz
Linaje 4, 4b–1º, 29001 Málaga, Spain
Tel: +34 653 46 82 29 / Fax: +34 95 221 41 00
E-mail: agaralc@coamalaga.es

Bernardo Gómez Delgado
Pasaje Mallol no. 22, 41003 Seville, Spain
Tel: +34 695 88 65 68 / Fax: +34 95 453 23 46
E-mail: bernardogomez303@arquihuelva.com

The architects involved in the Centenary Park published here worked together only for this project. They all received their architecture degree from the Higher Technical Architecture School of Seville. **MARÍA CAFFARENA DE LA FUENTE** was born in Algeciras in 1979, graduated in 2003, and did postgraduate studies in Madrid (DEA, 2008). She established her own office in 2003. **VÍCTOR COBOS MÁRQUEZ** was born in 1971 near Cádiz, graduated in 1999, and established his own office (E1 Architecture) the same year. **ANDRÉS GARCÍA ALCARAZ** was born in 1971 in Málaga, graduated in 2002, and established his own office the same year. **BERNARDO GÓMEZ DELGADO** was born in 1972 in Seville, graduated in 2003, and opened his own office the following year. Their work includes the restoration of D'Amato House as the head office of the International University Menéndez Pelayo (Cobos with Mario Ortiz; UIMP, La Línea de la Concepción, Spain, 2001–04); public space at Muñoz Cobos/Santísimo Street (Caffarena/Cobos with Mario Ortiz; Algeciras, 2002–04); Centenary Park at Punta de San García (Caffarena/Cobos/Alcaraz/Delgado; Algeciras, 2007, published here); Social Housing (Cobos/Alcaraz/Delgado; Aljaraque, Huelva, 2007–09); 142 dwellings for young and elderly people at San Bernardo district (Caffarena/Cobos with Alberto Nicolau; Seville, 2003–10); and the Marina "el Terrón" (Delgado with Republica-dm; Lepe, Huelva, 2010–11).

Die Architekten des hier vorgestellten Parque del Centenario (Punta San García, Algeciras, Spanien, 2007) haben nur bei diesem Projekt zusammengearbeitet und haben alle ihre Architekturausbildung in Sevilla absolviert. **MARÍA CAFFARENA DE LA FUENTE** wurde 1979 in Algeciras geboren, machte 2003 ihren Abschluss und schloss 2008 ein Postgraduiertenstudium in Madrid ab (DEA). 2003 gründete sie ihr eigenes Büro. **VÍCTOR COBOS MÁRQUEZ** wurde 1971 bei Cádiz geboren, machte 1999 seinen Abschluss und gründete im selben Jahr sein eigenes Büro, E1 Architecture. **ANDRÉS GARCÍA ALCARAZ** wurde 1971 in Málaga geboren, beendete 2002 sein Studium und gründete im selben Jahr sein eigenes Büro. **BERNARDO GÓMEZ DELGADO** wurde 1972 in Sevilla geboren und machte 2003 seinen Studienabschluss. Er gründete sein Büro im darauffolgenden Jahr. Zu den Projekten der vier gehören die Sanierung des D'Amato-Gebäudes als Hauptverwaltung der Internationalen Universität Menéndez Pelayo (Cobos mit Mario Ortiz, UIMP, La Línea de la Concepción, 2001–04), ein öffentlicher Raum zwischen den Straßen Muñoz Cobos und Santísimo (Caffarena/Cobos mit Mario Ortiz, Algeciras, 2002–04), der Parque del Centenario in Punta de San García (Caffarena/Cobos/Alcaraz/Delgado, Algeciras, 2007, hier vorgestellt), Sozialwohnungen (Cobos/Alcaraz/Delgado, Aljaraque, Huelva, 2007–09), 142 Wohnungen für junge und alte Menschen im Stadtteil San Bernardo, Sevilla (Caffarena/Cobos mit Alberto Nicolau, 2003–10), und die Marina „el Terrón" in Lepe (Delgado mit Republica-dm, Lepe, Huelva, 2010–11).

Les architectes qui ont collaboré au projet du Parque del Centenario publié ici mènent des carrières séparées. Ils sont tous diplômés de l'École supérieure technique d'architecture de Séville. **MARÍA CAFFARENA DE LA FUENTE**, née à Algeciras en 1979, a obtenu son diplôme en 2003 et a effectué des études de troisième cycle à Madrid (DEA, 2008). Elle a fondé son agence en 2003. **VÍCTOR COBOS MÁRQUEZ**, né en 1971 près de Cadix et diplômé en 1999, a fondé son agence E1 Architecture la même année. **ANDRÉS GARCÍA ALCARAZ**, né en 1971 à Málaga, a obtenu son diplôme et fondé son agence en 2002. **BERNARDO GÓMEZ DELGADO**, né en 1972 à Séville et diplômé en 2003, a ouvert son agence en 2004. Parmi leurs travaux en Espagne : la restauration et transformation du Chalet D'Amato en siège de l'Université internationale Menéndez Pelayo (Cobos et Mario Ortiz, UIMP, La Línea de la Concepción, 2001–04) ; un espace public dans le secteur des rues Muñoz Cobos et Santísimo (Caffarena/Cobos avec Mario Ortiz, Algeciras, 2002–04) ; le Parque del Centenario à la Punta de San García (Caffarena/Cobos/Alcaraz/Delgado, Algeciras, 2007, publié ici) ; des logements sociaux (Cobos/Alcaraz/Delgado, Aljaraque, Huelva, 2007–09) ; 142 logements pour jeunes gens et retraités dans le quartier de San Bernardo (Caffarena/Cobos avec Alberto Nicolau, Séville, 2003–10) et la Aula Marina El Terrón (Delgado avec Republica-dm, Lepe, Huelva, 2010–11).

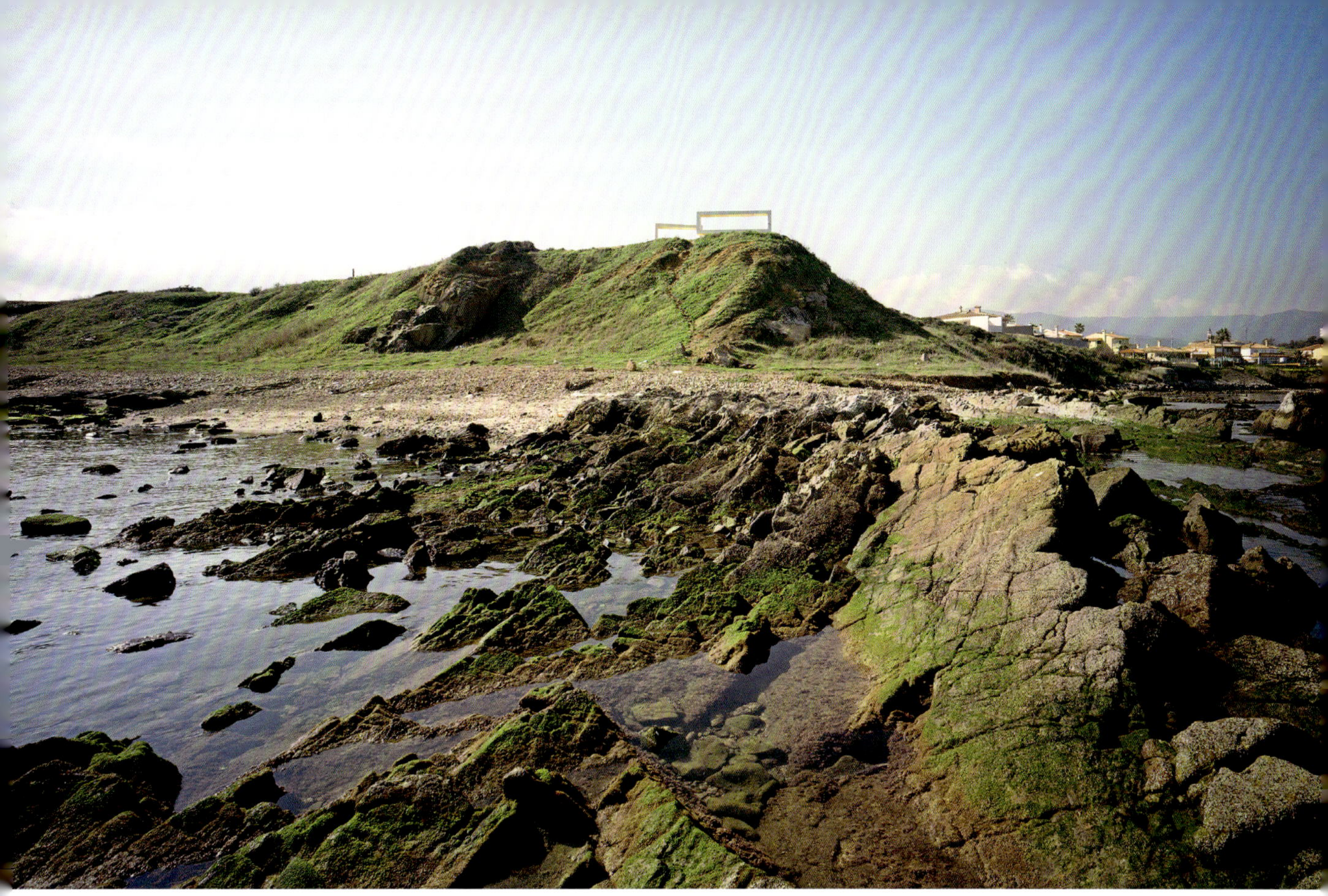
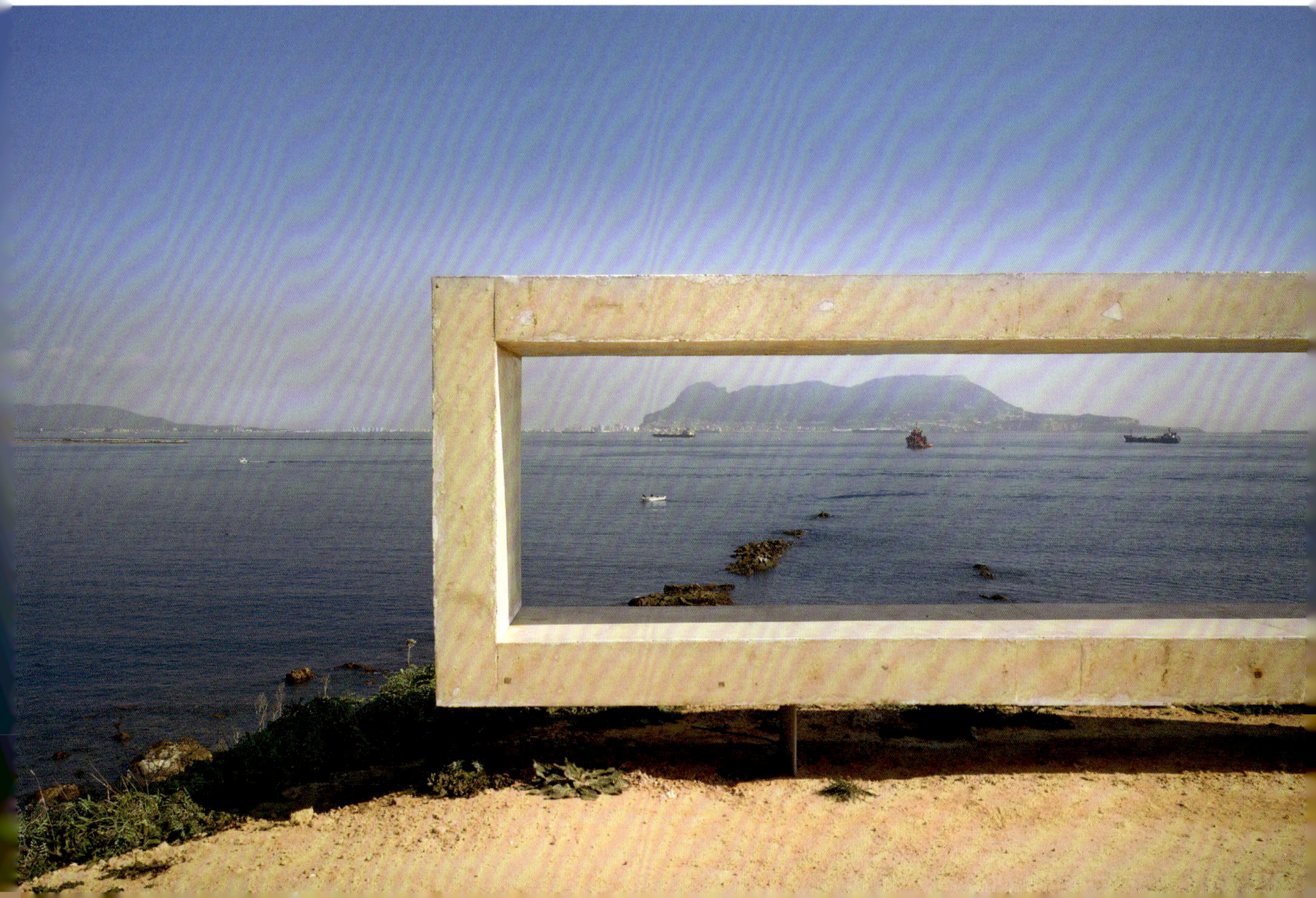

CENTENARY PARK
Punta San García, Algeciras, Spain, 2007

Address: Calle Delfin 51, 11207 Algeciras, Spain
Area: 100 700 m². Client: Bahía de Algeciras Port Authority. Cost: €842 839

The Punta de San García is a cape located at the southwest end of the bay of Algeciras, opening to the Straits of Gibraltar. It is part of a national park area. The architects sought to emphasize the history of this location at the very end of the Mediterranean and to "establish a dialogue between natural and artificial elements." They state: "The idea is not to make an urban park, but to demonstrate another experience of the landscape, with minimal equipment and infrastructure… implementing a series of elements that provide shelter and allow contemplation of the surroundings and self-understanding." A concrete slab rises from the entrance area, folding into an entrance portico and a facility for a snack bar and toilets. A path leads visitors to a viewpoint, where concrete elements "are positioned to serve as a frame that emphasizes the view" while providing benches where one can sit, protected from the strong winds. The visitor path leads to various structures on the site such as the Tower of San García, the oldest fortification, where a viewing tower is located.

Die Punta de San García ist ein Kap an der Südwestspitze der Bucht von Algeciras, die sich zur Straße von Gibraltar öffnet, und ist Teil eines Nationalparks. Die Architekten wollten die Geschichte dieses ganz am Ende des Mittelmeeres gelegenen Ortes heraufbeschwören und einen „Dialog zwischen natürlichen und künstlichen Elementen herstellen. Wir wollten keinen Stadtpark machen, sondern eine andere Erfahrung von Landschaft mit einem Minimum an Ausstattung und Infrastruktur vermitteln. Dazu haben wir eine Reihe von Elementen eingesetzt, die Schutz bieten, die Betrachtung der Umgebung und Selbstfindung erlauben." Im Eingangsbereich steht eine senkrechte Betonplatte, die dann abknickt und den Eingang ebenso überdacht wie eine Snackbar und die Toilettenanlagen. Ein Pfad führt zu einem Aussichtspunkt, wo Betonelemente „so positioniert sind, dass sie die Aussicht durch einen Rahmen betonen", zugleich dienen diese Betonelemente als Bänke, auf denen man vor dem starken Wind geschützt sitzen kann. Der Besucherpfad führt unter anderem zum Wehrturm von San García, der ältesten Festungsanlage, wo sich ein Aussichtsturm befindet.

La Punta de San García est un cap situé à l'extrémité sud-ouest de la baie d'Algeciras qui s'ouvre sur le détroit de Gibraltar. Ce site fait partie d'un parc national, se. Les architectes ont voulu mettre en scène l'histoire de ce lieu à la pointe extrême de la Méditerranée et « établir un dialogue entre des éléments naturels et artificiels ». Ils précisent : « L'idée n'est pas de réaliser un parc urbain, mais de faire la démonstration d'une autre expérience du paysage, au moyen d'infrastructures et d'équipements minimaux […] en mettant en place une série d'éléments servant d'abris pour permettre la contemplation de l'environnement et la méditation. » L'entrée se signale par une dalle de béton repliée en portique, un café et des toilettes. Un chemin conduit les visiteurs jusqu'à un point de vue, où des éléments en béton « positionnés de manière à servir de cadre à la vue » servent aussi de bancs et protègent du vent. Le chemin dessert ainsi différents éléments dont le fort de San García, la plus ancienne des fortifications présentes sur les lieux, où a été aménagée une tour d'observation.

The architects have made use of a subtle mixture of actual structures and concrete elements, such as those above, that can only be described as sculptural, framing the views beyond.

Die Architekten haben eine sehr einfühlsame Mischung von modernen Strukturen und Betonelementen ausgeführt, wie oben zu sehen ist. Man kann sie als Skulpturen bezeichnen, die die Ausblicke rahmen.

Les architectes ont pratiqué un subtil mélange de constructions et d'éléments en béton, comme ceux représentés ci-dessus, qui méritent pleinement le qualificatif de sculptures. Ils cadrent des vues sur le paysage.

M. CAFFARENA / V. COBOS / G. ALCARAZ / G. DELGADO

The architecture is minimalist, combining steel and stone with the very old olive trees planted on the site. Below, a plan of the entire park.

Die minimalistische Architektur kombiniert Stahl und Naturstein mit uralten Olivenbäumen, die auf dem Gelände gepflanzt sind. Unten, Lageplan des gesamten Parks.

L'architecture d'esprit minimaliste combine des éléments en pierre et en acier avec de très anciens oliviers plantés sur le site. Ci-dessous, plan de l'ensemble du parc.

P 86

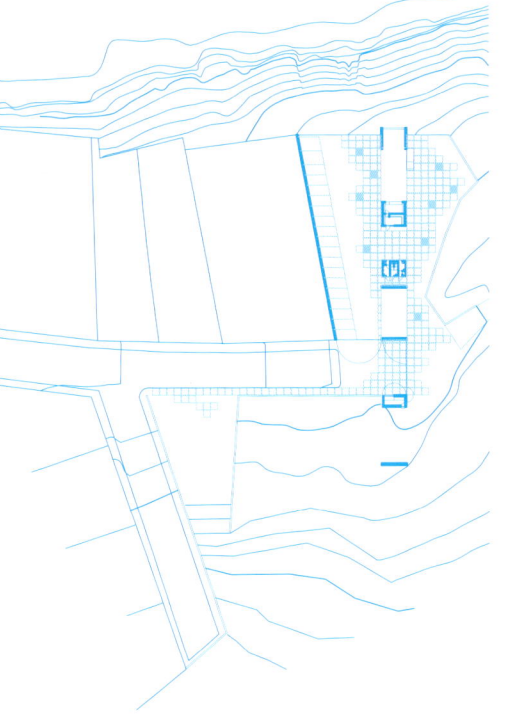

The architects have used a rather industrial vocabulary for the viewing tower in the park, seen on the left page and above. Their intervention is, in all cases, as minimal as possible.

Für den Aussichtsturm (links und oben abgebildet) haben die Architekten eine der Industriearchitektur verwandte Sprache gewählt. Ihre Eingriffe sind so gering wie möglich.

Les architectes ont choisi un langage plutôt industriel pour cette tour d'observation (à gauche et ci-dessus). Leur intervention est restée aussi minimale que possible.

São Jorge Castle Renovation

JOÃO LUÍS CARRILHO DA GRAÇA

João Luís Carrilho da Graça
JLCG Arquitectos, Lda
Calçada Marquês de Abrantes, n° 48-2°dto
1200–719 Lisbon
Portugal

Tel: +351 21 392 02 00
Fax: +351 21 395 02 32
E-mail: arquitectos@jlcg.pt
Web: www.jlcg.pt

JOÃO LUÍS CARRILHO DA GRAÇA graduated from the ESBAL (Lisbon School of Fine Arts) in 1977. He lectured at the Faculty of Architecture of the Technical University of Lisbon between 1977 and 1992, and has been a Professor at the Lisbon Autonomous University between 2001 and 2010 and at Evora University since 2005. He headed the architecture department at both institutions until 2010. His work includes the Knowledge of the Seas Pavilion (Expo '98, Lisbon, 1998); Portuguese Republic Documentation and Information Center (Lisbon, 2002); Oriente Foundation Museum (Lisbon, 2008); Music School of Lisbon Polytechnic Institute (2008); Theater and Auditorium (Poitiers, France, 2008); Pedestrian Bridge over the Carpinteira River (Covilhã, Portugal, 2003–09, published here); and the Archeological Museum of Praça Nova do Castelo de São Jorge (Costa do Castelo, 2008–10, also published here), all in Portugal unless stated otherwise. In 2010, he won the first prize in a competition to design a new Cruise Terminal in Lisbon.

JOÃO LUÍS CARRILHO DA GRAÇA schloss 1977 sein Studium an der ESBAL, der Kunsthochschule in Lissabon, ab. Von 1977 bis 1992 lehrte er an der Architekturfakultät der Technischen Universität Lissabon, und von 2001 bis 2010 war er Gastprofessor an der privaten Universidade Autónoma de Lisboa – UAL, der Autonomen Universität Lissabon, ebenso wie an der Universität Évora seit 2005. Bis 2010 war er Leiter der Abteilung für Architektur an beiden Universitäten. Zu seinen Projekten gehören der Pavillon über das „Wissen der Meere" (Expo '98, Lissabon, 1998), das Dokumentations- und Informationszentrum der Republik Portugal (Lissabon, 2002), das Museum der Stiftung Oriente (Lissabon, 2008), die Musikschule des Politechnischen Instituts Lissabon (2008), Theater und Auditorium (Poitiers, Frankreich, 2008), die Fußgängerbrücke über den Carpinteira (Covilhã, Portugal, 2003–09, hier vorgestellt) und das Archäologische Museum Praça Nova do Castelo de São Jorge (Costa do Castelo, Portugal, 2008–10, ebenfalls hier vorgestellt). 2010 gewann er bei dem Wettbewerb für den Entwurf eines neuen Kreuzfahrtterminals in Lissabon den ersten Preis.

JOÃO LUÍS CARRILHO DA GRAÇA est diplômé de l'École des beaux-arts de Lisbonne (ESBAL, 1977). Il a enseigné à la faculté d'architecture de l'Université technique de Lisbonne de 1977 à 1992, puis a été professeur à l'Université autonome de Lisbonne de 2001 à 2010 et à l'université d'Evora depuis 2005. Il a dirigé le département d'architecture de ces deux institutions jusqu'en 2010. Parmi ses réalisations : le Pavillon de la connaissance des mers (Exposition universelle de 1998, Lisbonne) ; le Centre de documentation et d'information de la République portugaise (Lisbonne, 2002) ; le musée de la Fondation de l'Orient (Lisbonne, 2008) ; l'École supérieure de musique de l'Institut polytechnique de Lisbonne (2008) ; le théâtre et auditorium de Poitiers (2008) ; une passerelle sur la Carpinteira (Covilhã, Portugal, 2003–09, publiée ici) et le Musée d'archéologie do Castelo de São Jorge (Costa do Castelo, 2008–10, publié ici). En 2010, il a remporté le concours pour la conception d'un nouveau terminal de croisières à Lisbonne.

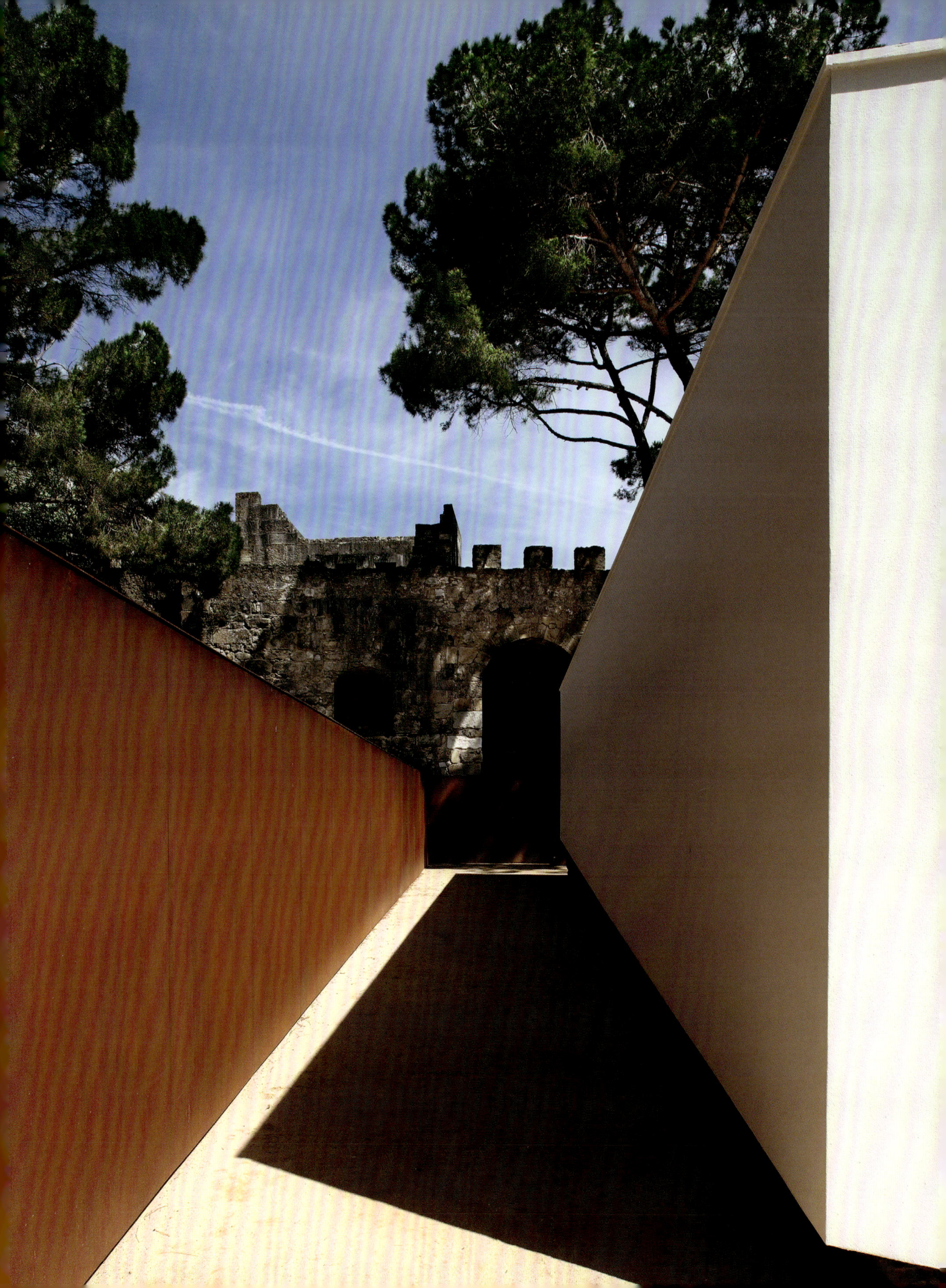

ARCHEOLOGICAL MUSEUM OF PRAÇA NOVA DO CASTELO DE SÃO JORGE

Costa do Castelo, Portugal, 2008–10

Address: Praça Nova do Castelo de São Jorge, Costa do Castelo, Lisbon 1100–179, Portugal, +351 02 11 20 50 50
Area: 3500 m². Client: EGEAC (Empresa de Gestão de Equipamentos e Animação Cultural). Cost: €1 million
Collaboration: João Gomes da Silva (Landscape Architect)

The hill on which the São Jorge Castle is located is known to be the site of the first human settlement in the area of Lisbon, as became apparent in archeological digs that were undertaken beginning in 1996. The most important artifacts were placed in the Castle Museum, but the archeological area itself was the object of protection and conversion into a site that can be visited. A Corten steel "membrane" or perimeter wall was created by the architect to allow access and broad views of the site. Limestone steps, landings, and seating guide visitors through the excavated areas. Mosaics dating from the 15th-century Palace of the Bishop of Lisbon are protected by a structure whose underside is covered in a black mirrored surface. A polycarbonate-and-wood canopy and white walls both protect and suggest the original forms of 11th-century Muslim structures on the site. The abstract walls are lightly anchored in the ground and appear to float above the ruins. Evidence of the original Iron Age settlement underlies the whole and is exposed through slits in the Corten membrane that encourage visitors to discover the deepest and most ancient level of civilization of Lisbon.

Wie man bei den 1996 begonnenen archäologischen Grabungen feststellte, befanden sich auf dem Hügel des Castelo São Jorge die frühesten Siedlungen des Gebiets von Lissabon. Die bedeutendsten Fundstücke werden im Burgmuseum ausgestellt, und die archäologische Grabungsstätte wurde unter Denkmalschutz gestellt. Sie wurde umgestaltet und kann nun besichtigt werden. Der Architekt errichtete als umlaufende Trennwand eine „Membrane" aus Cor-Ten-Stahl, über die hinweg man die archäologische Stätte überblickt. Durch eine Öffnung kann man hinuntersteigen. Kalksteinstufen, Podeste und Sitzgelegenheiten führen die Besucher durch die Ausgrabungsstätte. Bodenmosaike, die man in dem im 15. Jahrhundert errichteten Palast des Bischofs von Lissabon gefunden hat, werden durch eine Konstruktion geschützt, deren Unterseite schwarz verspiegelt ist. Ein Schutzdach aus Polykarbonat und Holz, das auf weißen Wänden ruht, schützt die original erhaltenen Strukturen der maurischen Siedlung aus dem 11. Jahrhundert und zeichnet sie nach. Die abstrakten Mauern sind nur leicht im Boden verankert und scheinen über den Ruinen zu schweben. Unter dem gesamten Bereich liegen Zeugnisse der eisenzeitlichen Siedlung, die durch Schlitze in der Cor-Ten-Stahl-Membrane sichtbar gemacht werden. Diese Schlitze fordern die Besucher dazu auf, die tiefsten und ältesten Schichten der Kultur von Lissabon zu entdecken.

La colline sur laquelle se dresse le château de São Jorge serait le site des premières implantations humaines dans la région de Lisbonne, comme l'ont montré des fouilles archéologiques pratiquées à partir de 1996. Les objets découverts les plus importants ont été déposés au musée du Château, mais la zone archéologique elle-même a été protégée et aménagée pour les visiteurs. Une « membrane » en acier Corten – un mur périmétrique – a été érigée pour permettre l'accès et rendre le site plus lisible. Des marches en pierre calcaire, des paliers et des bancs permettent au visiteur de parcourir confortablement les zones dégagées par les fouilles. Des mosaïques du XV siècle appartenant au palais de l'évêque de Lisbonne sont protégées par une construction dont le plafond est recouvert d'une matière noire réfléchissante. L'auvent en bois et polycarbonate et les murs blancs qui protègent les fouilles rappellent la forme des constructions musulmanes du XIe siècle. Les murs abstraits, légèrement ancrés dans le sol, donnent l'impression de flotter au-dessus des ruines. Les vestiges des anciennes constructions datant de l'âge de fer servent de base et certaines sont visibles par des fentes découpées dans la membrane en Corten, pour encourager les visiteurs à découvrir les anciennes traces de civilisation à Lisbonne.

An elevation drawing shows the new structures in the space above grade and the dark outlines of the architect's intervention.

Der Aufriss zeigt die neuen, oberirdischen Gebäude und, in Grau, die Umrisslinien des architektonischen Eingriffs.

Un dessin d'élévation montre les nouvelles structures à l'air libre et, en sombre, les contours de l'intervention de l'architecte.

JOÃO LUÍS CARRILHO DA GRAÇA

Using a thin wall made of Corten steel, whose weathered appearance makes it immediately look "old," the architect has skillfully defined the limits of the site and carefully inserted the small, white, minimalist evocation of an original 11th-century Muslim building.

Mit einer dünnen Cor-Ten-Stahlwand, deren Patina sie recht schnell „alt" wirken lässt, definiert der Architekt sehr geschickt die Raumkanten und gliedert sorgfältig das kleine, weiße, minimalistische Gebäude ein, das einem originalen maurischen Gebäude aus dem 11. Jahrhundert nachempfunden ist.

À l'aide d'un fin muret en acier Corten dont l'aspect patiné exprime le passage du temps, l'architecte a habilement précisé les limites du terrain et inséré avec délicatesse des évocations minimalistes de constructions musulmanes du XIe siècle.

Though it is difficult to harmonize a contemporary intervention with an ancient site, the architect succeeds because of the modesty of his concept and his real respect for this place. Even those who do not like contemporary architecture can find little to complain about here.

Es ist schwierig, einen modernen Entwurf mit einem alten Gelände harmonisch zu verbinden. Dem Architekten gelingt dies hier, weil er in seiner Grundhaltung sehr bescheiden ist und dem Ort wirklich seinen Respekt erweist. Selbst Menschen, denen zeitgenössische Architektur nicht gefällt, haben an diesem Projekt kaum etwas auszusetzen.

L'architecte a réussi l'intégration harmonieuse toujours délicate d'une intervention sur un site ancien grâce à la modestie de son concept et à son respect du lieu. Même les adversaires de l'architecture contemporaine y trouvent peu matière à critique.

JOÃO LUÍS CARRILHO DA GRAÇA

PEDESTRIAN BRIDGE
Carpinteira River, Covilhã, Portugal, 2003–09

Address: Covilhã, Portugal
Area: 956 m² Client: Poliscovilhã, S.A. Cost: €1.6 million
Collaboration: AFA, Consult Lda. (António Adão da Fonseca and Carlos Quinaz);
ARPAS, Luís Cabral and Lucile Dubroca (Landscape Architecture)

This project includes three footbridges in the Goldra and Carpinteira valleys and aims to create more direct links between the historic center of Covilhã and outlying residential areas. The architect has used a steel structure that is reduced to its simplest expression. Stone applied in a spiral pattern has been used to clad two pillars. When they eventually become covered in greenery, these pillars will appear to blend into the landscape, or rather the stunning form of the bridge itself will seem to emerge from the landscape. The bridge forms a distinctive landscape element, tying together parts of the landscape that otherwise appear to be distinct or separate. The use of such modern forms in this context is surprising but as the images published here demonstrate, the project is both convincing and visually spectacular.

Diese Brücke ist eine von drei Fußgängerbrücken in den Flusstälern des Goldra und des Carpinteira, mit denen das historische Stadtzentrum von Covilhã besser an die außerhalb liegenden Wohnviertel angebunden werden soll. Der Architekt hat sich für eine auf die einfachste Formensprache reduzierte Stahlkonstruktion entschieden. Zwei Brückenpfeiler sind spiralförmig mit Naturstein verkleidet, auf dem sich Pflanzen niederlassen und die Stützen verbergen sollen. Nach einer Weile werden die beiden Stützen optisch mit der Landschaft verschmelzen, und die spektakuläre Gestalt der Brücke wird dann so aussehen, als würde sie aus der Landschaft herauswachsen. Die Brücke ist ein eigenständiges Element, das Teile der Landschaft miteinander verbindet, die sonst klar voneinander getrennt wären. Die Verwendung so hochmoderner Formen in diesem Kontext überrascht, aber wie die hier veröffentlichten Bilder zeigen, ist das Projekt überzeugend und optisch ausgesprochen spektakulär.

L'objectif de cette passerelle piétonnière en trois sections qui franchit les vallées de la Goldra et de la Carpinteira est d'offrir un accès direct entre le centre historique de la ville de Covilhã et ses quartiers résidentiels extérieurs. L'architecte a conçu un ouvrage en acier réduit à sa plus simple expression. Les deux piliers parés de pierres selon un motif en spirale se fondront dans l'environnement lorsque la végétation grimpante les aura recouverts. La forme étonnante de la passerelle paraîtra encore plus surgir du paysage. Le pont, qui relie des zones jusqu'alors bien séparées, devient un élément distinctif de ce paysage. Si une forme aussi moderne peut surprendre dans un tel contexte, le projet reste néanmoins convaincant et très spectaculaire.

Most cities, even relatively small ones, are plagued by roads and highways, with little space given to pedestrians, who are, nonetheless, the real "users" of a city. This project aims to give back to the people who walk some of their own space.

Die meisten Städte, selbst relativ kleine, leiden unter dem Straßen- und Autobahnnetz, das Fußgängern nur wenig Raum lässt. Dabei sind gerade sie die wirklichen „Nutzer" der Stadt. Dieses Projekt soll den Fußgängern ihren Raum zurückgeben.

La plupart des villes, même relativement petites, sont enlaidies par les voies d'accès et les autoroutes qui ne laissent guère de place aux piétons, pourtant vrais « usagers » de l'espace public. Ce projet veut leur rendre un peu de l'espace qui leur revient.

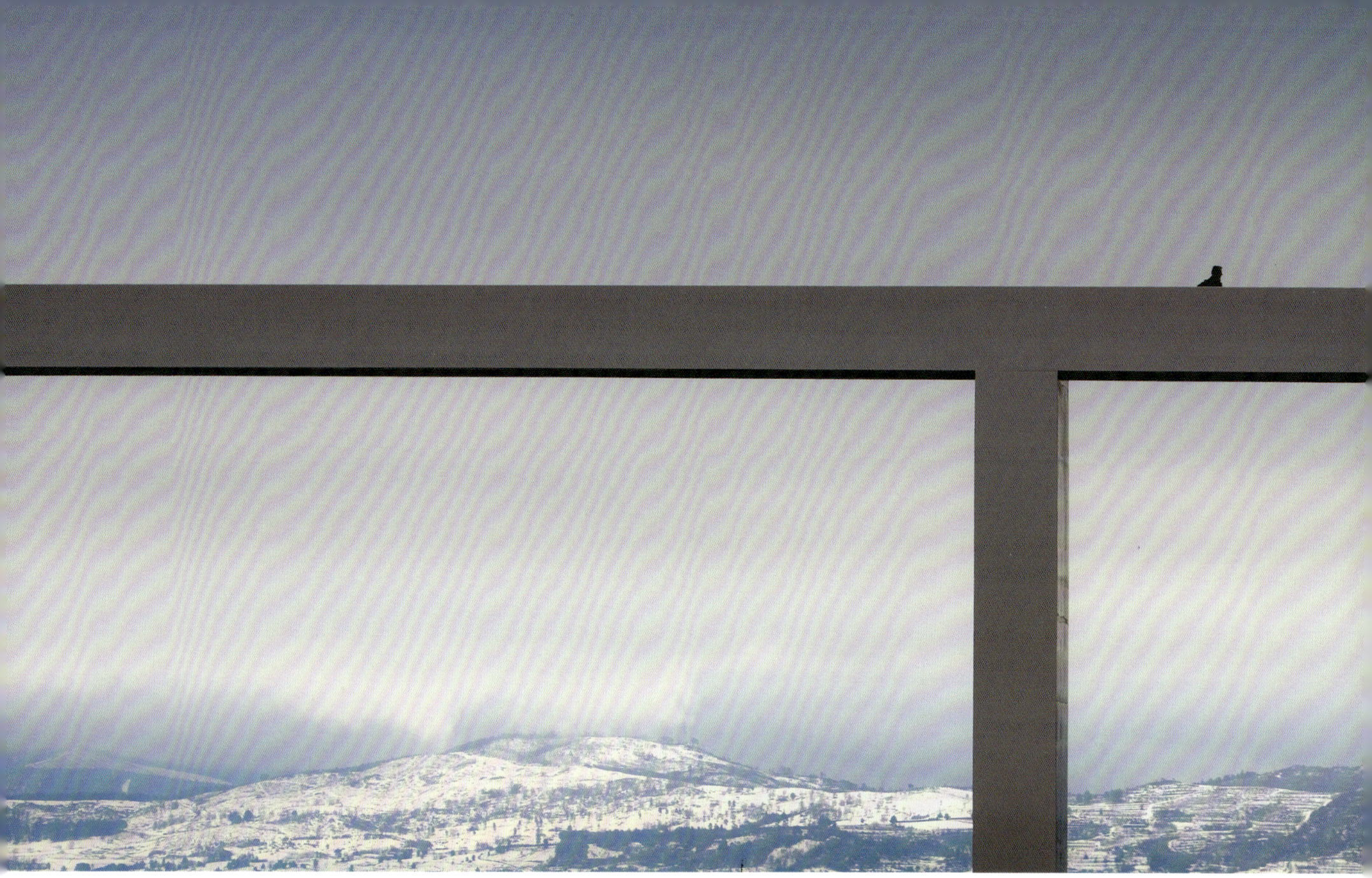

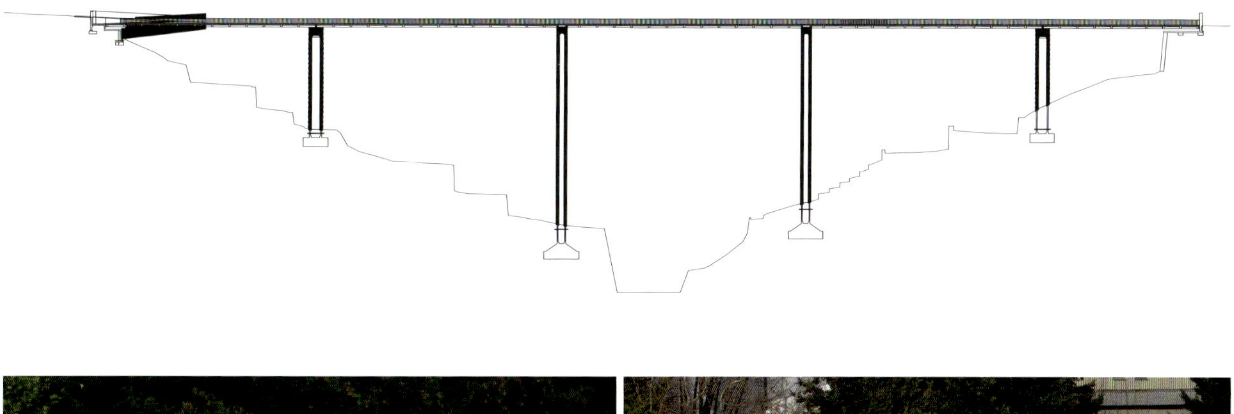

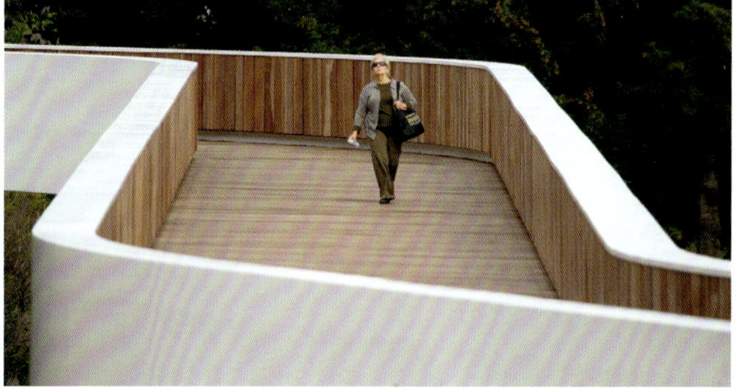

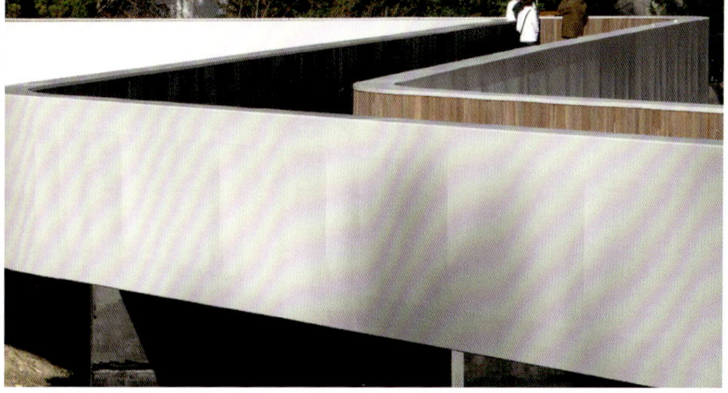

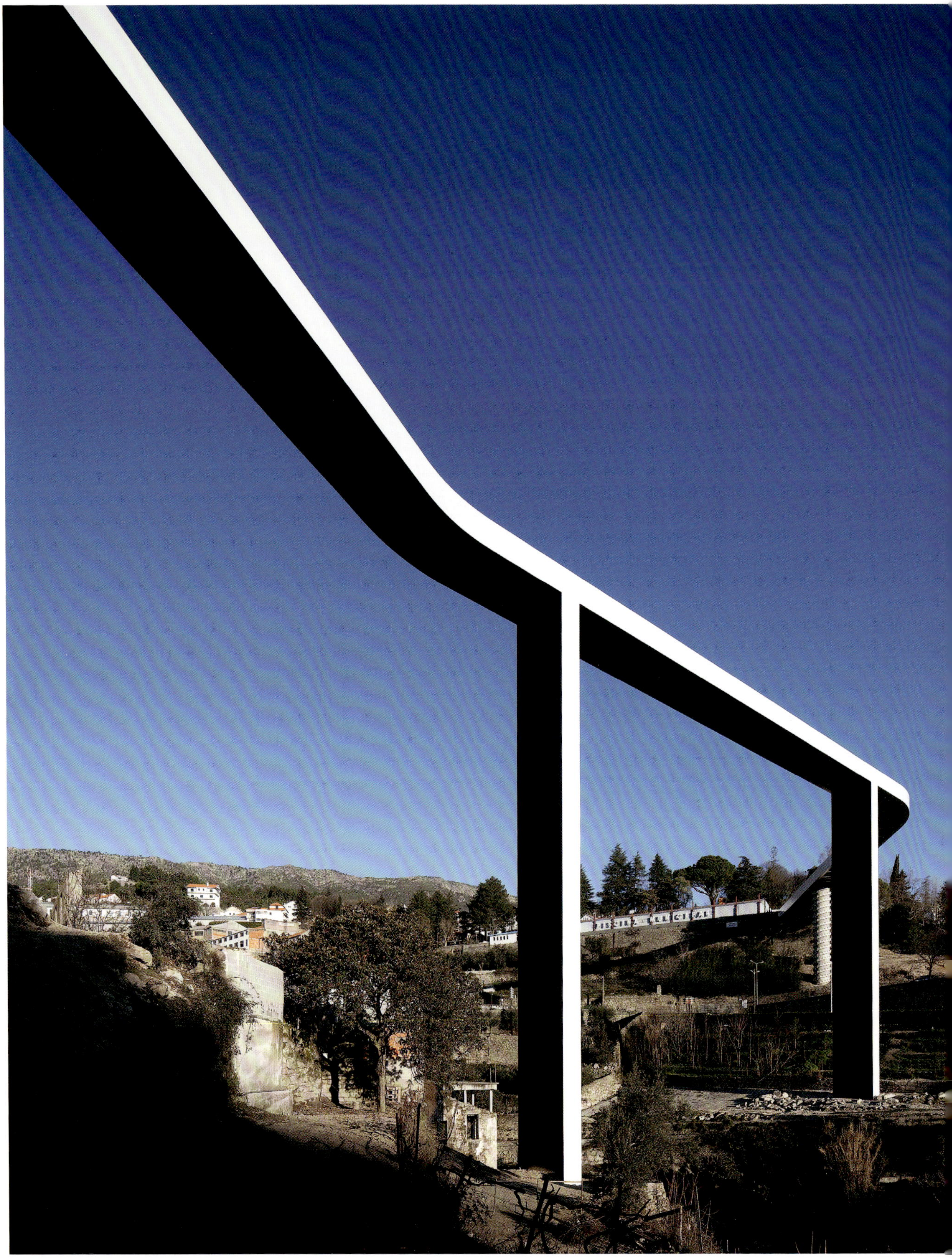

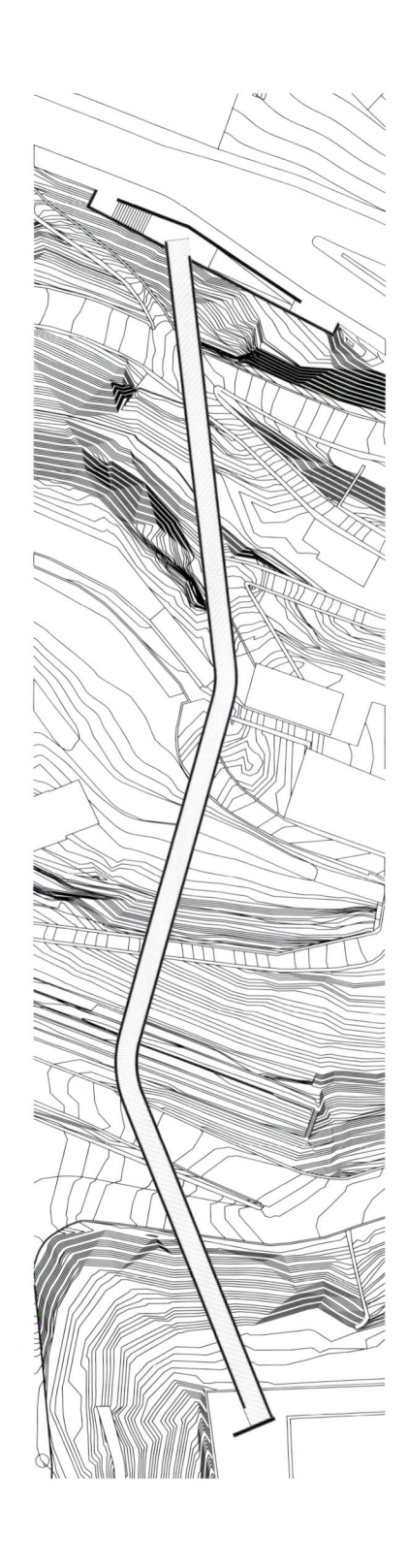

The simple elegance of the design and its very thin supporting columns make it a sculptural presence in the landscape, whose use is obvious nonetheless.

Die schlichte Eleganz des Entwurfs und die sehr schlanken Stützen verwandeln diese Brücke in eine Skulptur, die in der Landschaft steht, deren Funktion als Brücke aber trotzdem offensichtlich ist.

Sa simple élégance et la grande finesse de ses piles confèrent à cette passerelle une forte présence sculpturale, sans rien masquer de sa fonction.

Adriana Varejão Galle

RODRIGO CERVIÑO LOPEZ

Rodrigo Cerviño Lopez
Tacoa Arquitetos Associados
Avenida Ipiranga 200 D 162
01046–010 São Paulo, SP
Brazil

Tel: +55 11 3159 3045
E-mail: tacoa@tacoa.com.br
Web: www.tacoa.com.br

RODRIGO CERVIÑO LOPEZ was born in São Paulo, Brazil, in 1972. He graduated from the FAU-USP (University of São Paulo, Faculty of Architecture and Urbanism) in 2003. He lives and works in São Paulo, where he has been a Principal of Tacoa Arquitetos Associados, together with Fernando Falcon, since 2005. Their recent work includes Galpão Fortes Vilaça (São Paulo, 2008); Adriana Varejão Gallery, Inhotim Contemporary Art Center (Brumadinho, 2008, published here); Museu da Imagem e do Som (Rio de Janeiro, competition, 2009); Plot 40 House (São Roque, 2009); Loja Alexandre Herchcovitch: Fashion Mall, São Conrado (Rio de Janeiro, 2010); and the Vila na vila (São Paulo, 2011), all in Brazil.

RODRIGO CERVIÑO LOPEZ wurde 1972 São Paulo, Brasilien, geboren. 2003 schloss er sein Studium an der Fakultät für Architektur und Städtebau an der Universität São Paulo (FAU-USP) ab. Er lebt und arbeitet in São Paulo, wo er seit 2005 zusammen mit Fernando Falcon das Architekturbüro Tacoa Arquitetos Associados leitet. Zu ihren neuesten Arbeiten gehören Galpão Fortes Vilaça (São Paulo, 2008), die Galerie Adriana Varejão, das Zentrum für zeitgenössische Kunst Inhotim (Brumadinho, 2008, hier vorgestellt), das Museum für Bild und Ton (Rio de Janeiro, Wettbewerb, 2009), das Wohnhaus Plot 40 (São Roque, 2009), die Loja Alexandre Herchcovitch: Fashion Mall in São Conrado (Rio de Janeiro, 2010) und die Vila na vila (São Paulo, 2011). Alle Projekte befinden sich in Brasilien.

RODRIGO CERVIÑO LOPEZ, né à São Paulo au Brésil en 1972, est diplômé de la FAU-USP (université de São Paulo, faculté d'architecture et d'urbanisme, 2003). Il vit et travaille à São Paulo, où il dirige Tacoa Arquitetos Associados avec Fernando Falcon depuis 2005. Parmi leurs réalisations récentes au Brésil : la galerie Galpão Fortes Vilaça (São Paulo, 2008) ; la galerie Adriana Varejão du Centre d'art contemporain Inhotim (Brumadinho, 2008, publié ici) ; le Musée de l'image et du son (Rio de Janeiro, concours, 2009) ; la maison Plot 40 (São Roque, 2009) ; la boutique Alexandre Herchcovitch du Fashion Mall de São Conrado (Rio de Janeiro, 2010) et la Vila na vila (São Paulo, 2011).

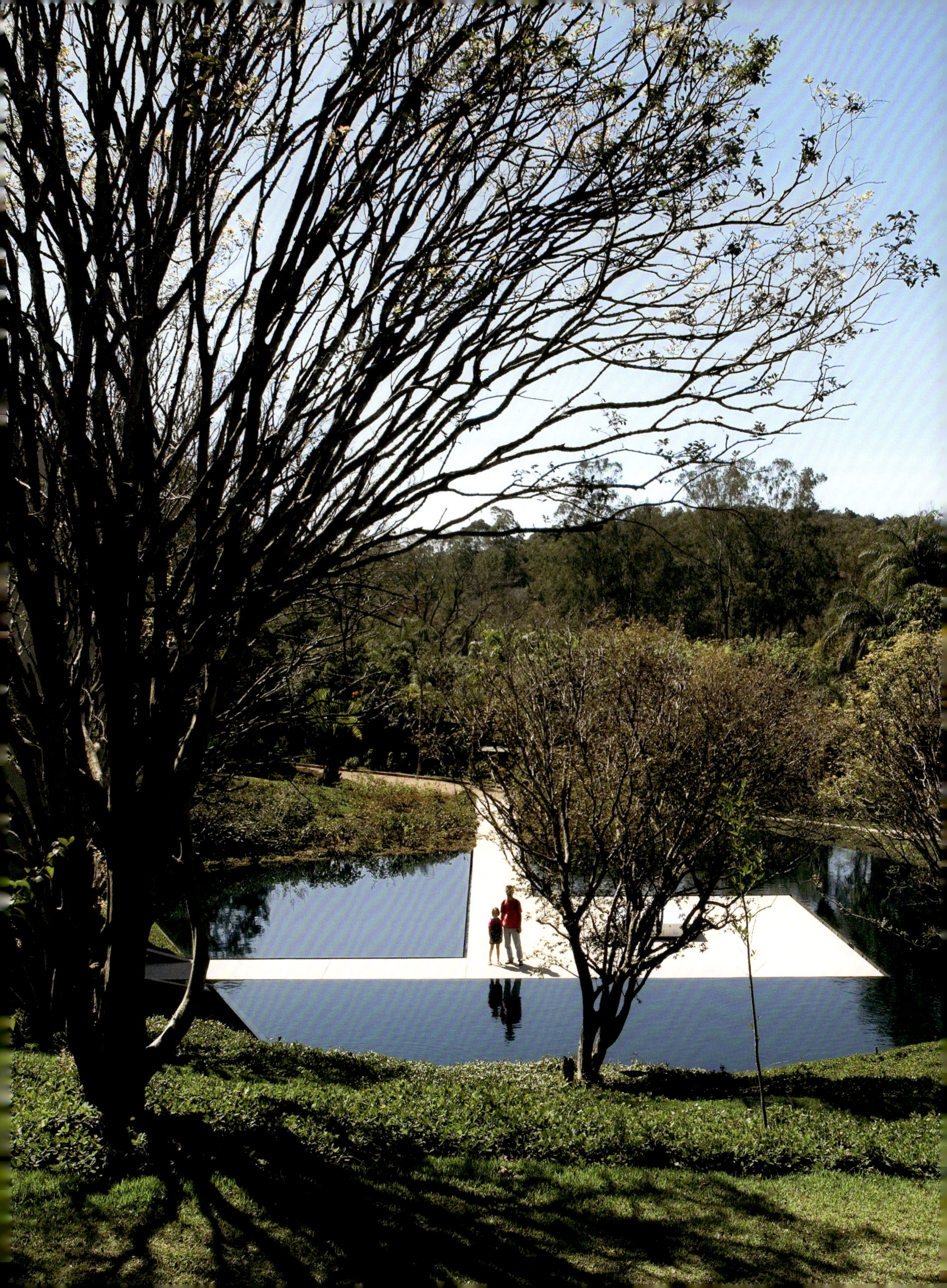

ADRIANA VAREJÃO GALLERY

Inhotim Contemporary Art Center, Brumadinho, Brazil, 2008

Address: Inhotim Centro de Arte Contemporânea, Rua B20, Inhotim,
Brumadinho, Minas Gerais 35460–000, Brazil, +55 31 3227 0001, www.inhotim.org.br
Area: 823 m². Client: Inhotim Centro de Arte Contemporânea. Cost: €500 000

The Inhotim Contemporary Art Center is located in the village of Brumadinho near Belo Horizonte, the capital of Minas Gerais. A personal initiative of the businessman Bernardo Paz, the museum is composed of a number of pavilions located in a 35-hectare park. The **ADRIANA VAREJÃO GALLERY** is a structure for two works by the artist acquired by the museum and exhibited previously at the Cartier Foundation in Paris: the sculpture *Linda do Rosário* and the polyptych *Celacanto Provoca Maremoto*. Adriana Varejão also created four other works for the building. The architect explains that the project "occupies a hillside with a small slope, typical of the topography of Minas Gerais that is composed of old and smooth hills, partially surrounded by the native forest." Formerly used to store containers the site had its topography modified to create a large flat surface. The architect sought to recompose the original topography of the site while inserting the gallery in the form of a block of reinforced concrete into the hillside.

Das Zentrum für zeitgenössische Kunst Inhotim befindet sich im Dorf Brumadinho, in der Nähe der Hauptstadt Belo Horizonte des Bundesstaates Minas Gerais. Das Museum geht auf eine Initiative des Geschäftsmanns Bernardo Paz zurück. Es besteht aus einer Reihe von Pavillons, die in dem 35 Hektar großen Park verteilt sind. Die **GALERIE ADRIANA VAREJÃO** ist ein Gebäude, das eigens für zwei Werke der Künstlerin errichtet wurde – die Skulptur „Linda do Rosário" und das Polyptychon „Celacanto Provoca Maremoto". Das Museum hat sie von der Fondation Cartier in Paris erworben, wo sie vorher ausgestellt waren. Adriana Varejão schuf weitere vier Arbeiten für das Gebäude. Der Architekt erklärt, dass „das Projekt auf einem leicht abfallenden Hügel liegt. Diese Topografie mit alten, sanften Hügeln, die zum Teil vom Urwald umgeben sind, ist typisch für den Bundesstaat Minas Gerais". Da auf dem Gelände früher Container gelagert wurden, hatte man die Topografie verändert und eine große ebene Fläche geschaffen. Ziel des Architekten war es, die ursprüngliche Topografie wiederherzustellen, und so schob er den als Stahlbetonblock ausgeführten Baukörper in den Hang hinein.

Le Centre d'art contemporain Inhotim est situé dans le village de Brumadinho, près de Belo Horizonte, capitale de l'État de Minas Gerais. Initiative de l'homme d'affaires Bernardo Paz, ce musée se compose de plusieurs pavillons implantés à l'intérieur d'un parc de 35 ha. La **GALERIE ADRIANA VAREJÃO** a été conçue pour accueillir deux œuvres de cette artiste acquises par le musée et qui étaient précédemment exposées à la Fondation Cartier à Paris : la sculpture *Linda do Rosário* et le polyptyque *Celacanto Provoca Maremoto*. Adriana Varejão a également créé quatre œuvres supplémentaires pour la galerie. L'architecte explique que le projet « occupe un terrain présentant une légère dénivellation, typique de la topographie du Minas Gerais, composé d'anciennes collines partiellement recouvertes par la forêt ». Servant jadis de dépôt de conteneurs, le site avait été transformé en vaste plateau. L'architecte a cherché à recomposer la topographie d'origine et a inséré le « bloc » que forme cette galerie en béton armé à flanc de colline.

RODRIGO CERVIÑO LOPEZ

The simple, blocklike form of the building is placed in the landscaped site, with the pond and platform on the water reflecting and echoing the architecture.

Der einfache massive Block ist in die Landschaft hineingeschoben. Die davor gelegene Wasserfläche mit der darauf „schwimmenden" Plattform spiegeln die Architektur.

Le bâtiment, simple boîte posée dans le paysage, fait face à un étang où il se reflète et sur lequel a été posée une plate-forme de plan carré.

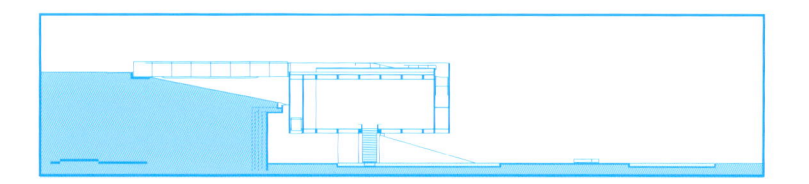

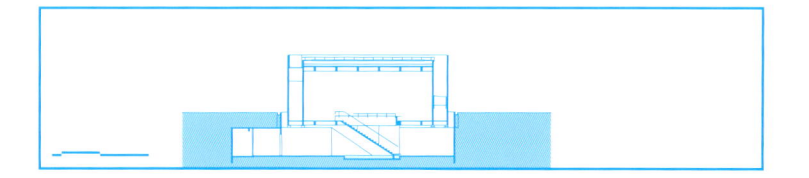

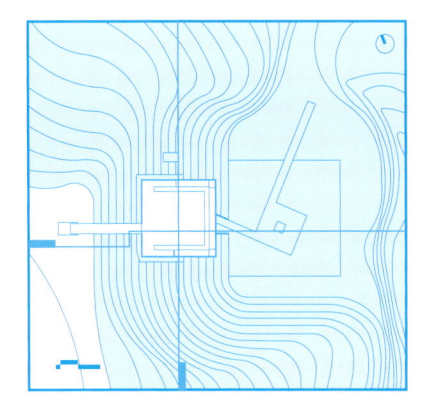

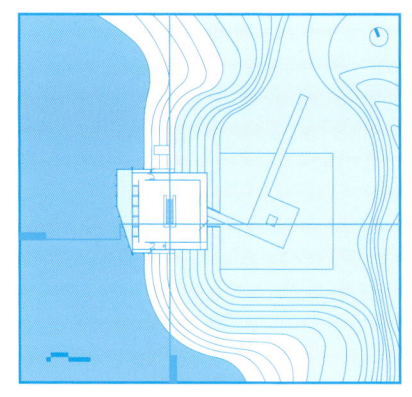

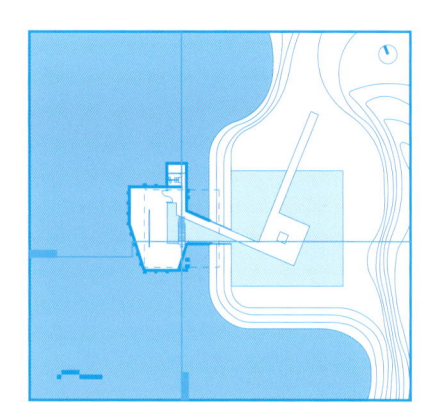

The pavilion itself and the platform in the water are square in form, with the water platform rotated vis-à-vis the building.

Der Pavillon und auch die Plattform auf dem Wasser sind rechteckig, wobei die Plattform in Bezug auf das Gebäude leicht gedreht ist.

Le pavillon et la plate-forme sont de plan carré, cette dernière étant légèrement pivotée devant le bâtiment.

Left, the approach bridge leading to the pavilion. The strict lines of the architecture find a constant contrast in the gentle landscape that surrounds it.

Links, die Zugangsbrücke zum Pavillon. Die strenge Linienführung der Architektur steht im Gegensatz zu der sanften Landschaft der Umgebung.

À gauche, la passerelle d'accès au pavillon. La rigueur géométrique de l'architecture contraste avec le paysage vallonné.

DILLER SCOFIDIO + RENFRO

Diller Scofidio + Renfro
601 West 26th Street, Suite 1815
New York, NY 10001
USA

Tel: +1 212 260 7971 / Fax: +1 212 260 7924
E-mail: disco@dsrny.com / Web: www.dsrny.com

ELIZABETH DILLER was born in Łódź, Poland, in 1954. She received her B.Arch degree from Cooper Union School of Architecture in 1979 and is a Professor of Architecture at Princeton University. **RICARDO SCOFIDIO** was born in New York in 1935. He graduated from Cooper Union School of Architecture and Columbia University, and is now Professor Emeritus of Architecture at Cooper Union. **CHARLES RENFRO** was born in Baytown, Texas, in 1964. He graduated from Rice University and Columbia University. Diller + Scofidio was founded in 1979 and Renfro became a Partner in 2004. According to their own description: "DS+R is a collaborative, interdisciplinary studio involved in architecture, the visual arts, and the performing arts." They completed the Brasserie Restaurant in the Seagram Building (New York, 2000); the Viewing Platforms at Ground Zero in Manhattan (New York, 2001); the Blur Building (Expo 02, Yverdon-les-Bains, Switzerland, 2002); and the Institute of Contemporary Art in Boston (Massachusetts, 2006). Recently completed projects include the Lincoln Center Redevelopment Project in New York, including the expansion of the Juilliard School of Music (2009), the renovation of Alice Tully Hall (2009), public spaces throughout the campus, and the Hypar Pavilion Lawn (2011, published here); the conversion of the High Line, a 2.4-kilometer stretch of elevated railroad, into a New York City park (Phase 1, 2009; Phase 2, 2011; page 184); and the Creative Arts Center at Brown University (Providence, Rhode Island, 2011), all in the USA unless stated otherwise.

ELIZABETH DILLER wurde 1954 in Łódź, Polen, geboren. 1979 machte sie ihren Bachelor in Architektur an der Cooper Union School of Architecture und ist derzeit Professorin für Architektur an der Princeton University. **RICARDO SCOFIDIO** wurde 1935 in New York geboren. Er schloss seine Studien an der Cooper Union School of Architecture und an der Columbia University ab und ist heute emeritierter Professor für Architektur an der Cooper Union. **CHARLES RENFRO** wurde 1964 in Baytown, Texas, geboren. Er studierte an der Rice und der Columbia University. Das Büro Diller + Scofidio wurde 1979 gegründet, Renfro kam als Partner 2004 hinzu. Das Büro beschreibt seine Tätigkeit wie folgt: „DS+R ist ein interdisziplinär zusammenarbeitendes Büro, das sich mit Architektur, bildender und darstellender Kunst beschäftigt." Zu den von DS+R ausgeführten Projekten gehören das Brasserie-Restaurant im Seagram Building (New York, 2000), die Aussichtsplattformen am Ground Zero in Manhattan (2001), das Blur Building (Expo 02, Yverdon-les-Bains, Schweiz, 2002) und das Institute of Contemporary Art in Boston (Massachusetts, USA, 2006). Zu den neueren abgeschlossenen Projekten zählen die Sanierung des Lincoln Center for the Performing Arts in New York, zu dem auch der Erweiterungsbau der Juilliard School of Music (2009) gehört, und die Renovierung der Alice Tully Hall (2009). Hinzu kamen öffentliche Freiräume auf dem gesamten Campus des Lincoln Center for the Performing Arts und die Rasenfläche auf dem Hypar Pavilion (2011, hier vorgestellt), außerdem die Umwandlung der High Line, einer 2,4 Kilometer langen ehemaligen Hochbahnstrecke, in einen Park mitten in New York City (Bauabschnitt 1, 2009, Bauabschnitt 2, 2011, Seite 184) und das Creative Arts Center an der Brown University (Providence, Rhode Island, 2011), alle in den USA, sofern nicht anders angegeben.

Née à Łódź (Pologne) en 1954, **ELIZABETH DILLER** est titulaire d'un B.Arch. de l'École d'architecture de la Cooper Union (1979). Elle est professeur d'architecture à l'université de Princeton. **RICARDO SCOFIDIO**, né à New York en 1935, est diplômé de l'École d'architecture de la Cooper Union et de l'université Columbia. Il est aujourd'hui professeur émérite d'architecture à la Cooper Union. **CHARLES RENFRO**, né à Baytown (Texas, US) en 1964, est diplômé de l'université Rice et de l'université Columbia. Renfro a rejoint en 2004 l'agence Diller + Scofidio, fondée en 1979. Ils se décrivent ainsi : « DS+R est une agence interdisciplinaire collaborative qui se consacre à l'architecture, aux arts visuels et aux arts du spectacle. » Elle a réalisé, entre autres, le restaurant Brasserie du Seagram Building (New York, 2000) ; la plateforme d'observation de Ground Zero à Manhattan (2001) ; le Blur Building de l'Exposition nationale suisse de 2002 (Yverdon-les-Bains) et l'Institute of Contemporary Art de Boston (Massachusetts, US, 2006). Parmi ses projets récents aux États-Unis figurent : le projet de redéveloppement du Lincoln Center à New York, comprenant l'extension de la Juilliard School of Music (2009), la rénovation de l'Alice Tully Hall (2009), divers espaces publics et la Hypar Pavilion Lawn (2011, publiée ici) ; la transformation en parc de la High Line à New York, une section de voie ferrée suspendue de 2,4 km (Phase 1, 2009 ; Phase 2, 2011 ; page 184) et le Creative Arts Center de l'université Brown (Providence, Rhode Island, 2011).

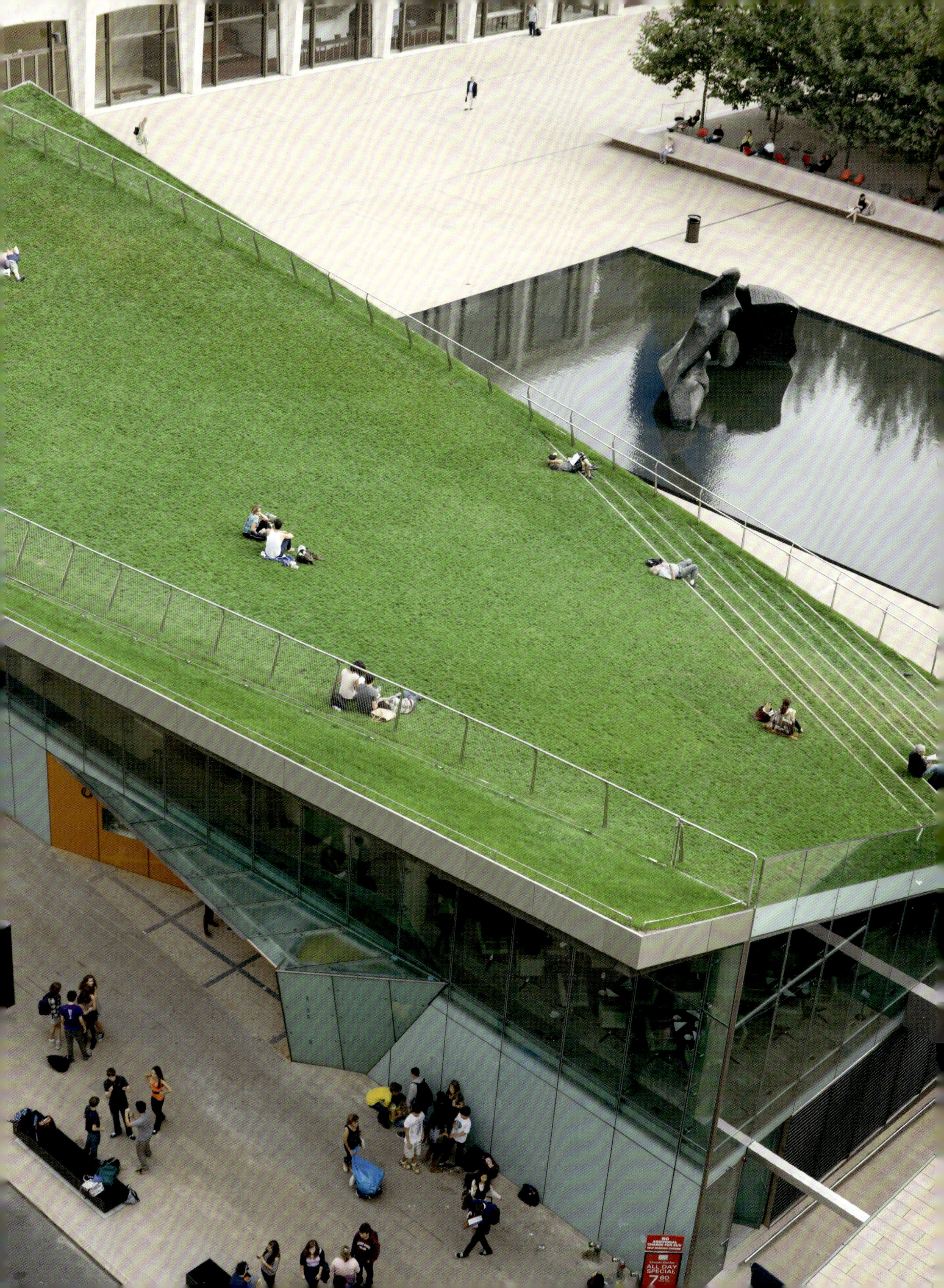
NO
ADDITIONAL
CHARGE FOR SUV
ALL DAY
SPECIAL
7 60

HYPAR PAVILION LAWN

New York, New York, USA, 2011

Address: 142 West 65th Street, New York, NY 10023, USA
Area: 670 m². Client: Lincoln Center for the Performing Arts. Cost: not disclosed
Collaboration: FXFowle (Executive Architect), Ove Arup & Partners (Structural & MEP Engineers),
Mathews Nielsen Landscape Architects (Landscape Architect)

As part of their ongoing Lincoln Center for the Performing Arts Redevelopment project, Diller Scofidio + Renfro designed "a twisting lawn that acts as an occupiable grass roof over a glass pavilion restaurant." The **HYPAR LAWN** is located in Lincoln Center's North Plaza and is oriented away from the noise of the city "to create a bucolic urbanism." The geometry of the lawn directly corresponds to the contoured wood ceiling of the restaurant below, framing views to the plaza and the street. A tall fescue mixed with Kentucky bluegrass was chosen for the lawn because of its durability. The increased thermal mass of the grass roof dramatically reduces the mechanical loads of the restaurant below. Water is drained through the structural columns underneath the lawn surface.

Für das noch nicht abgeschlossene Sanierungsprojekt des Lincoln Center for the Performing Arts hat das Büro Diller Scofidio + Renfro eine „in sich gewundene, schräge Rasenfläche entworfen, die als benutzbares Rasendach auf einem verglasten Pavillon mit Restaurant dient". Der **HYPAR LAWN** befindet sich an der North Plaza des Lincoln Centers. Er liegt vom Lärm der Stadt abgewendet, um eine Atmosphäre des „bukolischen Urbanismus" zu schaffen. Die Gestalt der Rasenfläche folgt den Konturen der Holzdecke des darunter befindlichen Restaurants und bietet Ausblicke auf die Plaza und die Straße. Aus Gründen der Widerstandsfähigkeit wurde als Rasen eine Mischung aus hohem Schwingelgras (Festuca) und Wiesen-Rispengras (Poa pratensis) gewählt. Dank der thermischen Eigenschaften des Rasendachs wird im darunterliegenden Restaurant der Energiebedarf deutlich reduziert. Das Wasser wird durch die tragenden Stützen unterhalb der Rasenfläche abgeleitet.

Dans le cadre du projet (en cours) de redéveloppement du Lincoln Center, l'agence Diller Scofidio + Renfro a conçu une « pelouse inclinée praticable servant de toiture à un pavillon de verre qui abrite un restaurant ». L'**HYPAR LAWN** est située sur la place nord du Lincoln Center, à l'écart du bruit de la ville, pour « créer un urbanisme bucolique ». Le plan de la pelouse correspond exactement à celui du plafond en bois du restaurant qu'elle couvre et offre différentes vues sur la place et la rue. L'herbe est un mélange de fétuque élevée et de pâturin des prés, tous deux choisis pour leur résistance. La masse thermique élevée du toit végétalisé réduit de façon spectaculaire les besoins mécaniques du restaurant en matière d'énergie. L'eau de pluie est drainée par des colonnes structurelles installées sous la surface de la pelouse.

Above, the tilted, grass-covered roof is placed above the restaurant and opposite a reflecting pond with a work by Henry Moore placed at its center.

Oben, das geneigte Rasendach auf dem Restaurant und davor eine spiegelnde Wasserfläche mit einer Skulptur von Henry Moore in der Mitte.

Ci-dessus, la toiture inclinée semée de gazon recouvre le restaurant. Devant, le bassin est animé en son centre par une œuvre d'Henry Moore.

The lawn adds readily usable leisure space to an otherwise rather strict and largely mineral square. The grass roof also, of course, serves to reduce energy requirements for the restaurant.

Die Rasenfläche bietet einen weiteren, gut nutzbaren Erholungsraum in einer sonst mit ihrem Plattenbelag ziemlich strengen Platzanlage. Das Rasendach reduziert zudem den Energiebedarf des Restaurants.

La pelouse offre un espace de détente parfaitement utilisable à l'intérieur d'une place aux contours assez stricts et à l'esprit minéral. La toiture végétalisée permet de réduire la consommation énergétique du restaurant.

Hariri Memorial Garden

VLADIMIR DJUROVIC

Vladimir Djurovic Landscape Architecture (VDLA)
Villa Rizk
Broumana
Lebanon

Tel: +961 4 86 2444/555
Fax: +961 4 86 2462
E-mail: info@vladimirdjurovic.com
Web: www.vladimirdjurovic.com

VLADIMIR DJUROVIC was born to a Serb father and a Lebanese mother in 1967. He received a degree in Horticulture from Reading University in England in 1989 and his M.A. in Landscape Architecture from the University of Georgia in 1992, after having worked at EDAW in Atlanta. Vladimir Djurovic Landscape Architecture (VDLA) was created in 1995 in Beirut, Lebanon. The office has participated in and won several international competitions, such as Freedom Park South Africa (2003). They have completed numerous private residences in Lebanon, including the F House (with Nabil Gholam; Dahr El Sawan, 2000–04). The firm won a 2008 Award of Honor in the residential design category from the American Society of Landscape Architects (ASLA) for its Bassil Mountain Escape project in Faqra (Lebanon). After the work on the award-winning Samir Kassir Square (Beirut, 2004), current work includes the landscaping of the Wynford Drive site in Toronto (Canada) to accommodate the Aga Khan Museum by Fumihiko Maki and the Ismaili Centre by Charles Correa. VDLA won the international competition for this design in 2006. Other current work includes the Salame Residence (Faqra, Lebanon, 2009, published here); Beirut Marina (Solidere, BCD, Lebanon; architect: Steven Holl); Beirut Terraces (Solidere, BCD, Lebanon; architect: Herzog & de Meuron); 3 Beirut (Solidere, BCD, Lebanon; architect: Foster + Partners); and the Hariri Memorial Garden (Beirut, 2005–11, also published here).

VLADIMIR DJUROVIC wurde 1967 als Sohn eines serbischen Vaters und einer libanesischen Mutter geboren. 1989 schloss er sein Studium mit einem Hochschulabschluss in Gartenbau an der Universität Reading in England ab. 1992, nachdem er im Büro EDAW in Atlanta gearbeitet hatte, machte er seinen Master in Landschaftsarchitektur an der Universität von Georgia. 1995 gründete er in Beirut, Libanon, sein Büro Vladimir Djurovic Landscape Architecture (VDLA). Das Büro hat an verschiedenen internationalen Wettbewerben teilgenommen und sie gewonnen, wie zum Beispiel 2003 beim Freedom Park in Pretoria, Südafrika. Zu den Projekten des Büros gehören zahlreiche Aufträge für den privaten Wohnungsbau im Libanon, unter anderem das Haus F (mit Nabil Gholam, Dahr El Sawan, 2000–04). 2008 gewann das Büro für sein Projekt Bassil Mountain Escape in Faqra, Libanon, den Ehrenpreis der American Society of Landscape Architects (ASLA) in der Kategorie Wohnhausdesign. Nach dem preisgekrönten Entwurf für den Samir-Kassir-Platz (Beirut, 2004) arbeitet Djurovic derzeit an der Landschaftsgestaltung des Wynford-Drive-Geländes in Toronto (Kanada), auf dem das von Fumihiko Maki entworfene Aga Khan Museum sowie das Ismaili Centre von Charles Correa errichtet werden. Das Büro VDLA gewann 2006 den internationalen Wettbewerb für diesen Entwurf. Weitere aktuelle Projekte im Libanon sind das Wohnhaus Salame (Faqra, 2009, hier vorgestellt), der Jachthafen von Beirut (Solidere, BCD, Architekt: Steven Holl), die Beirut Terraces (Solidere, BCD, Architekt: Herzog & de Meuron), 3 Beirut (Solidere, BCD, Architekt: Foster + Partners) und die Gedenkstätte Hariri Memorial Garden (Beirut, 2005–11, ebenfalls hier vorgestellt).

VLADIMIR DJUROVIC, né d'un père serbe et d'une mère libanaise en 1967, est titulaire d'un diplôme d'horticulture de l'université de Reading (GB, 1989). Il a obtenu un M.A. en architecture du paysage à l'université de Géorgie (1992) après avoir travaillé dans l'agence EDAW à Atlanta. L'agence Vladimir Djurovic Landscape Architecture (VDLA) a été fondée en 1995 à Beyrouth. Elle a remporté plusieurs concours internationaux, dont celui du Freedom Park en Afrique du Sud (2003), et a réalisé de nombreuses résidences privées au Liban, dont la F House (Dahr El Sawan, 2000–04, avec Nabil Gholam). En 2008, elle a reçu le prix d'honneur dans la catégorie « projets résidentiels » de l'American Society of Landscape Architects (ASLA) pour la retraite de montagne Bassil à Faqra, au Liban. Après avoir remporté un prix pour la place Samir Kassir (Beyrouth, 2004), l'agence travaille actuellement sur les aménagements paysagers du site de Wynford Drive à Toronto (Canada), qui regroupe le musée de l'Aga Khan de Fumihiko Maki et l'Ismaili Centre de Charles Correa. VDLA en avait remporté le concours international en 2006. Parmi ses autres projets figurent la Salame Residence (Faqra, Liban, 2009, publiée ici) ; la marina de Beyrouth (Solidere, BCD, Liban, architecte : Steven Holl) ; Beirut Terraces (Solidere, BCD, Liban, architectes : Herzog & de Meuron) ; 3 Beirut (Solidere, BCD, Liban, architectes : Foster + Partners) et le jardin-mémorial Hariri (Beyrouth, 2005–11, publié ici).

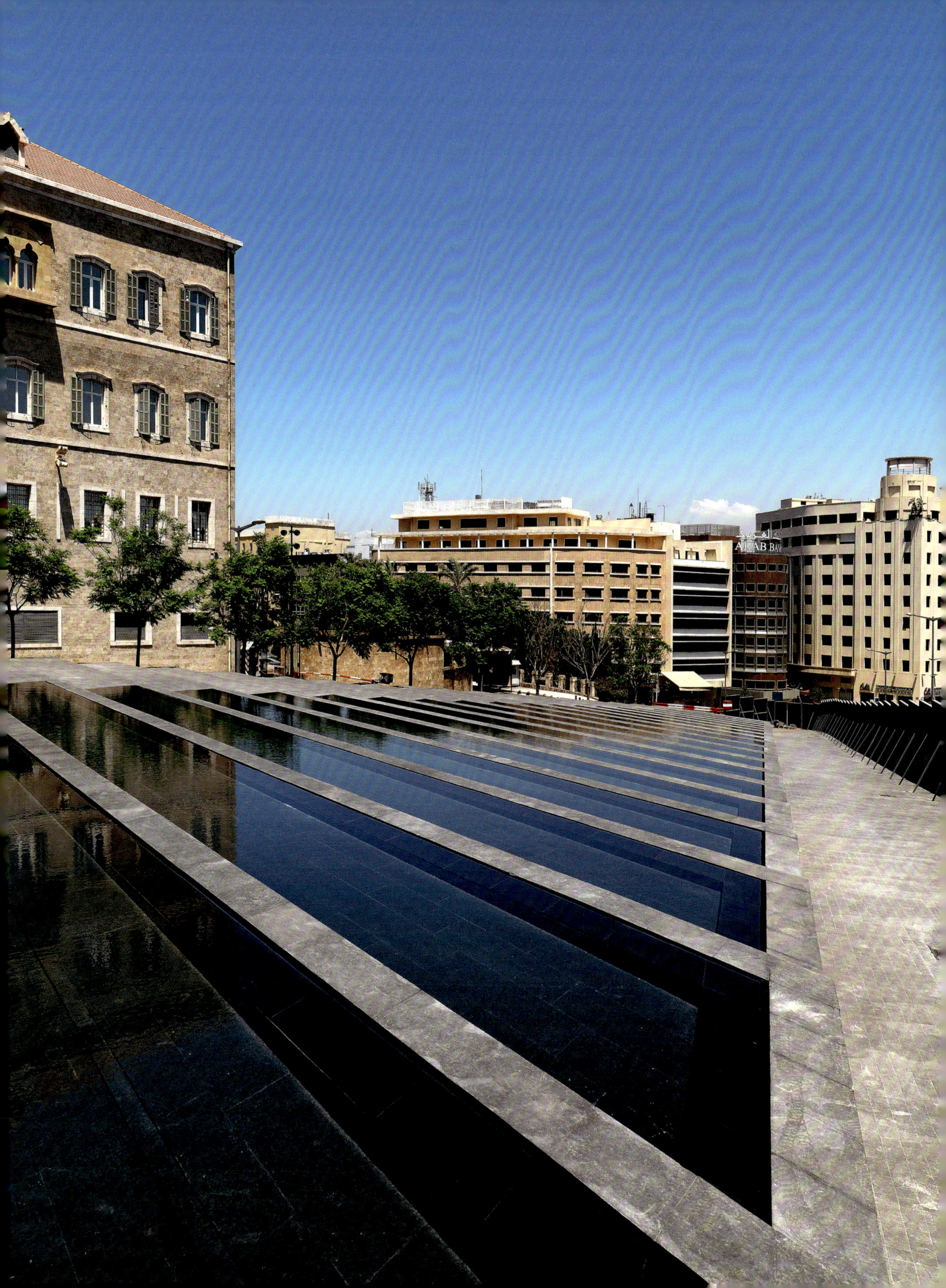

HARIRI MEMORIAL GARDEN
Beirut, Lebanon, 2005–11

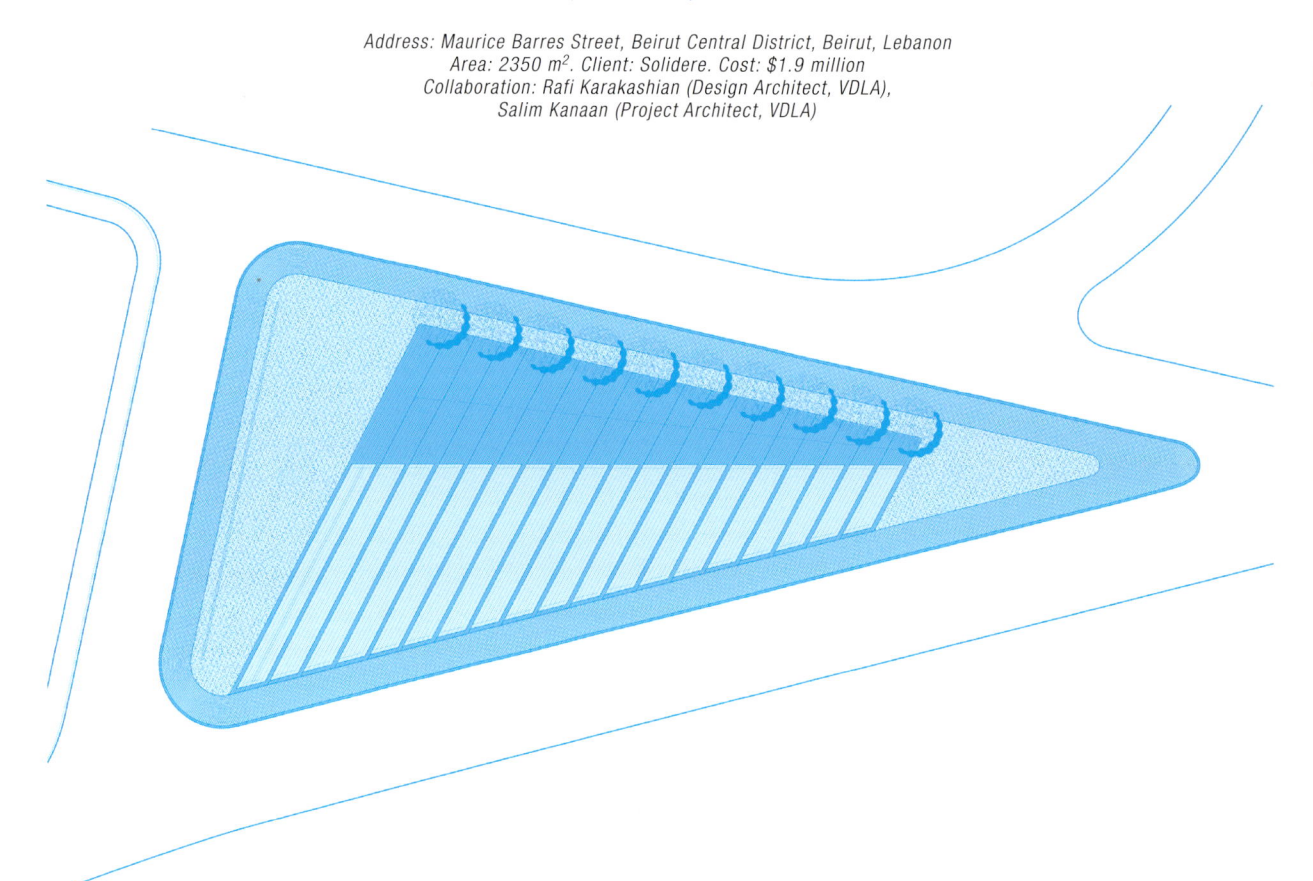

*Address: Maurice Barres Street, Beirut Central District, Beirut, Lebanon
Area: 2350 m^2. Client: Solidere. Cost: $1.9 million
Collaboration: Rafi Karakashian (Design Architect, VDLA),
Salim Kanaan (Project Architect, VDLA)*

Located in front of the Grand Serail (the Government headquarters), on a hill near one of the main gateways to the city center, the **HARIRI MEMORIAL GARDEN** is a homage to Lebanon's ex-Prime Minister, who was assassinated in 2005. The garden was conceived as "a monument rather than a public space" and is made up of long planar surfaces in gray stone, water, and grass, stepping down in the direction of the city center. The triangular form of the site emphasizes the movement in the direction of the downtown area. A row of jacaranda trees marks the boundary between the steps and the façade of the Serail. Vladimir Djurovic states: "The project presents a limited palette of elements and materials, charged with symbolic significance. The steps symbolize the gradual rebuilding of Beirut and an open invitation to the city. The basalt stone planes symbolize grief, sobriety, and perseverance. The water mirrors symbolize life, purity, peace, and the immaterial. Grass symbolizes tenderness and compassion. The jacaranda trees symbolize joy, sorrow, hope, and through their cycle of birth and death, life's constant renewal."

Der **HARIRI MEMORIAL GARDEN** liegt gegenüber dem Grand Serail, dem Regierungspalast der libanesischen Regierung, auf einer Anhöhe in unmittelbarer Nähe einer der Hauptzufahrtsstraßen zum Stadtzentrum. Die Parkanlage ist eine Hommage an den früheren libanesischen Ministerpräsidenten Hariri, der 2005 ermordet wurde. Die Konzeption der Anlage ist „mehr ein Denkmal als ein öffentlicher Freiraum". Lange, ebene Stufen aus grauem Naturstein, Wasser und Rasen führen hinunter in Richtung des Stadtzentrums. Die Dreiecksform des Geländes betont die Bewegung in Richtung Innenstadt. Eine Reihe von Jakarandabäumen markiert die Grenze zwischen den Stufen und der Fassade des Serail. Vladimir Djurovic bemerkt dazu: „In diesem Projekt werden nur sehr wenige Elemente und Materialien eingesetzt, die alle einen großen Symbolwert haben. Die Stufen stehen für den schrittweisen Wiederaufbau von Beirut und sind eine offene Einladung, in die Stadt zu kommen. Die Basaltflächen drücken Trauer und Ernsthaftigkeit aus. Die Wasserflächen symbolisieren Leben, Reinheit, Frieden und das Immaterielle. Rasen steht für Zärtlichkeit und Mitleid. Die Jakarandabäume bringen Freude, Schmerz und Hoffnung zum Ausdruck und stehen mit ihrem Vegetationszyklus für die ständige Erneuerung des Lebens."

Situé face au Grand Sérail (le siège du gouvernement), sur la pente d'une colline à proximité de l'une des principales entrées du centre de la capitale, le **HARIRI MEMORIAL GARDEN** a été créé en hommage au premier ministre du Liban assassiné en 2005. Conçu davantage comme « un monument plutôt qu'un lieu public », le jardin se compose de plusieurs plans allongés de pierre grise, d'eau et de gazon descendant en escalier vers le centre-ville. La forme triangulaire du terrain renforce ce mouvement descendant. Un alignement de jacarandas marque la limite entre l'emmarchement et la façade du Sérail. Vladimir Djurovic précise : « Ce projet utilise une palette limitée d'éléments et de matériaux chargés de sens. Les marches symbolisent la reconstruction graduelle de Beyrouth et sont une invitation à aller vers la ville. Les plans en basalte représentent la peine, la sobriété et la persévérance. Les miroirs d'eau sont l'image de la vie, de la pureté, de la paix et de l'immatérialité. L'herbe est un symbole de tendresse et de compassion. Les jacarandas représentent la joie, la tristesse, l'espoir et, à travers leur cycle biologique, le renouveau permanent de la vie. »

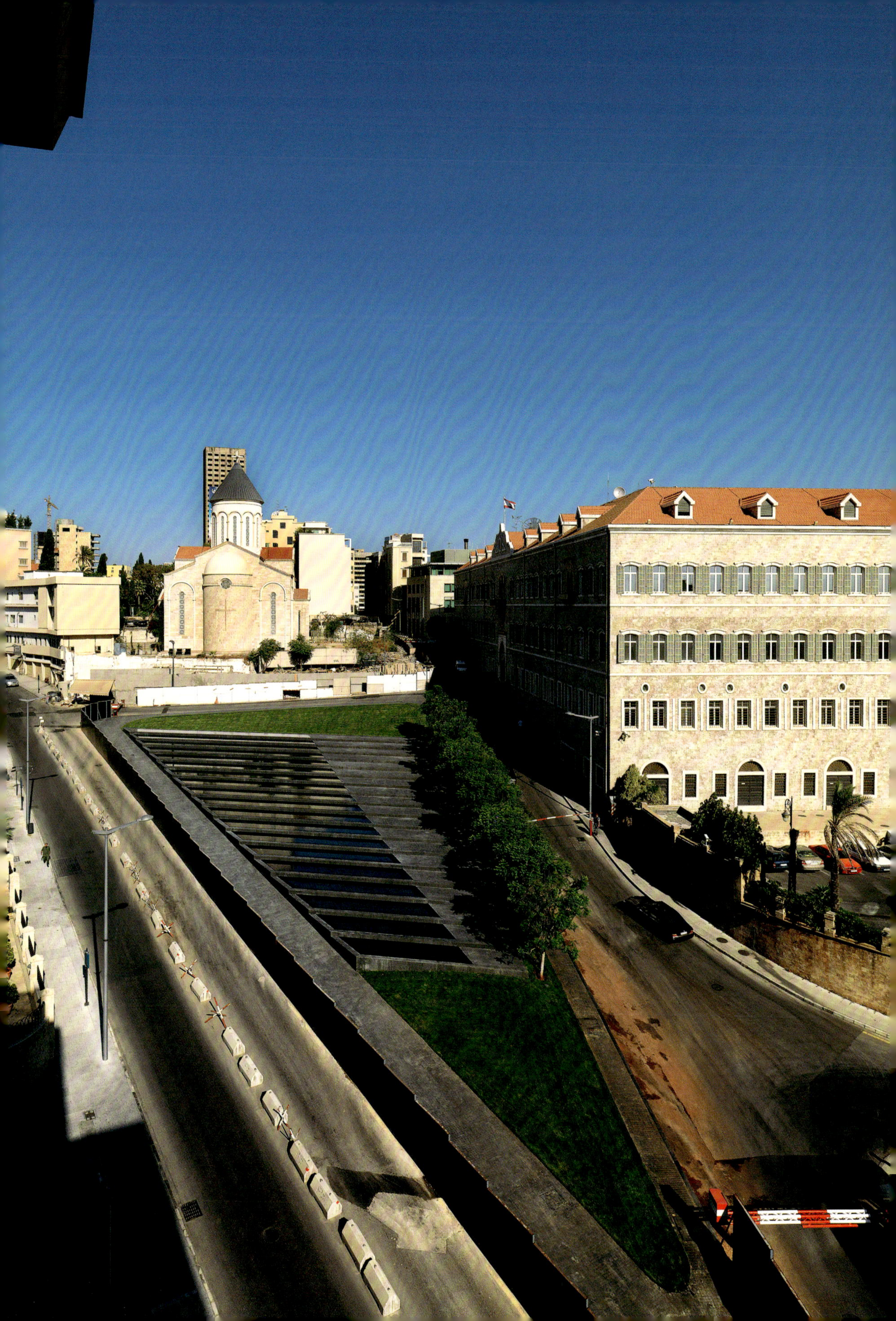

P 112

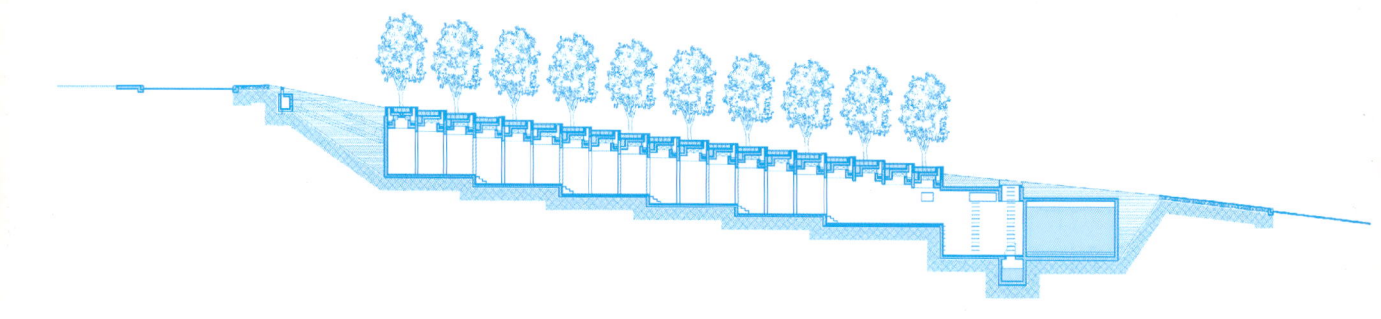

With its sequence of rectangular pools, the Memorial is both subtle and evocative without in any sense being figurative, or controversial, an important element in this city of delicate inter-community relations.

The landscape architect Vladimir Djurovic has found an elegant solution that combines mineral elements with water, trees, and grass, all in a restrained space near the seat of government.

Durch die Abfolge der Wasserbecken wirkt diese Gedenkstätte subtil und zugleich bewegend, ohne figurativ oder provokant zu sein. Das ist ein wichtiges Element in dieser Stadt mit ihren schwierigen Bevölkerungsbeziehungen.

Vladimir Djurovic hat an diesem räumlich sehr eingezwängten Platz in der Nähe des Regierungssitzes eine elegante Lösung gefunden, indem er Stein, Wasser, Bäume und Rasenflächen miteinander kombiniert.

Séquence de bassins rectangulaires, le mémorial est à la fois évocateur et subtil sans être figuratif ni provocant, élément qui compte dans cette ville aux relations intercommunautaires troublées.

L'architecte paysagiste Vladimir Djurovic a proposé une solution élégante qui combine l'eau, les arbres, le gazon et des éléments minéraux dans un espace restreint face au siège du gouvernement.

SALAME RESIDENCE
Faqra, Lebanon, 2009

Area: 9130 m². Client: Tony Salame
Cost: not disclosed. Collaboration: Rafi Karakashian (Design Architect, VDLA)

As is his habit, Djurovic has created a minimal, essentially geometric design for this private residence. Stone elements surrounded by water form a sculptural composition (left page).

Djurovic hat, seiner Entwurfssprache getreu, einen minimalistischen, geometrischen Entwurf für dieses Privathaus vorgelegt. Natursteinelemente in einer Wasserfläche verdichten sich zu einer Komposition mit skulpturalem Charakter (linke Seite).

Pour cette résidence privée, Djurovic a imaginé selon son habitude un projet minimaliste, essentiellement géométrique. Des éléments en pierre entourés d'eau forment une composition sculpturale (page de gauche).

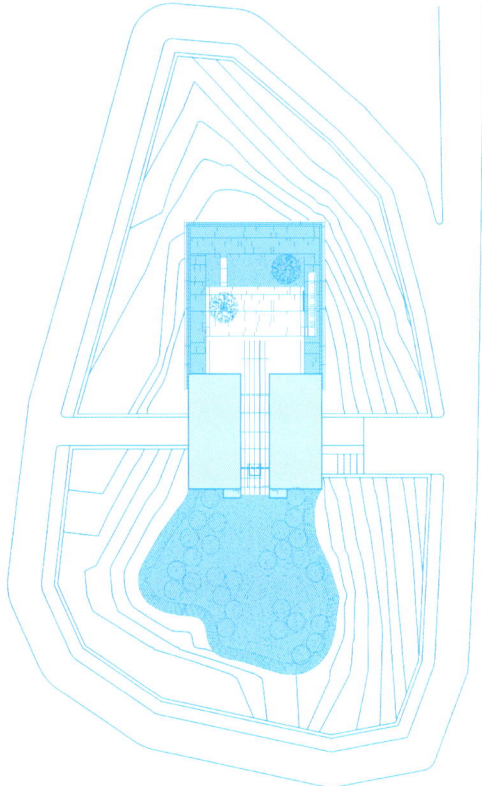

Designed for the founder of a Lebanese luxury department store chain, this property is located on a remote site 2700 meters above sea level. The designer states: "The landscape program was to create a serene environment that captures the beauty of the mountain ranges beyond and provides ample space for contemplation, yet allowing for occasional gatherings and entertainment to occur." At the front of the garden, the granite base of the house was extended to provide the required outdoor space. A pool creates a boundary on the front side, while in the rear "the hilltop was carved out, almost as a crater / sunken garden, providing a safe, informal, green, and organic space in contrast to the formal expression in the front."

Dieses Anwesen, das für den Gründer einer libanesischen Luxusladenkette entworfen wurde, liegt abgelegen auf einer Höhe von 2700 Metern über dem Meeresspiegel. Der Planer erklärt dazu: „Der Auftrag für die Landschaftsgestalter war, ein heiteres Umfeld zu schaffen, das die Schönheit der Bergketten im Hintergrund in den Entwurf einbezieht und viel Raum für Kontemplation lässt. Der Garten sollte aber auch für gelegentliche Feste und Einladungen geeignet sein." Vor dem Wohnhaus wurde die Granitbodenplatte verlängert, sodass der geforderte Außenraum entstand. Ein großes Wasserbecken vor dem Haus markiert eine Grenze, wohingegen an der Rückseite des Gebäudes „die Hügelkuppe fast wie ein Krater ausgehöhlt wurde und hier eine Art Senkgarten entstanden ist, ein lockerer, grüner, organischer Raum, der einen schönen Kontrast zu der strengen Formalität des Gartens vor dem Wohnhaus bildet."

Conçue pour le fondateur d'une chaîne de grands magasins de luxe libanais, cette propriété se trouve dans un site retiré, à 2700 m d'altitude. « Le programme d'aménagement paysager consistait à créer un environnement serein qui sache capter la beauté des montagnes environnantes et procure des espaces généreux, favorables à la contemplation, mais qui se prêtent aussi à des réunions et réceptions occasionnelles », explique le designer. À l'avant du jardin, le socle en granit de la maison se prolonge pour offrir un espace de vie à l'extérieur. Un grand bassin crée une limite visuelle devant la maison, tandis qu'à l'arrière « le sommet de la colline a été creusé, presque comme un cratère ou un jardin creux, pour aménager un espace vert de forme libre, protégé et naturel, qui contraste avec le parti pris formel de l'avant de la résidence ».

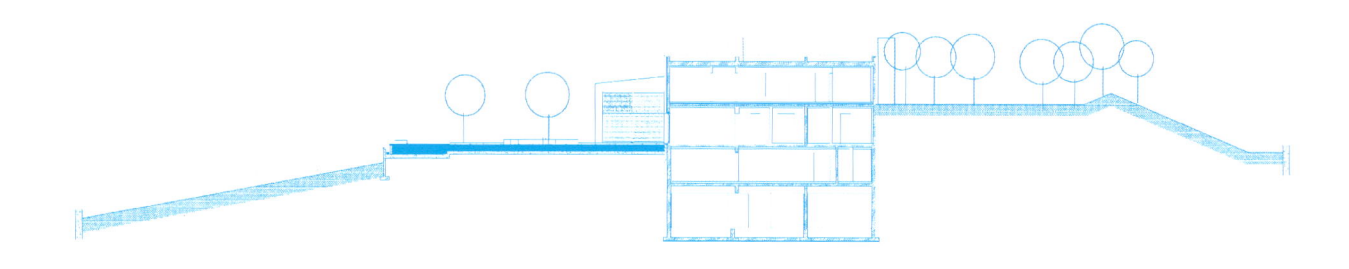

In the dramatic mountainous landscape, the designer has made use of long reflecting pools that seem to bring earth and sky together.

In der dramatischen Berglandschaft hat der Planer lange, spiegelnde Wasserflächen entworfen, die Erde und Himmel zu vereinen scheinen.

Au milieu d'un spectaculaire paysage de montagnes, le designer a créé de longs bassins réfléchissants où le ciel et la terre semblent se rejoindre.

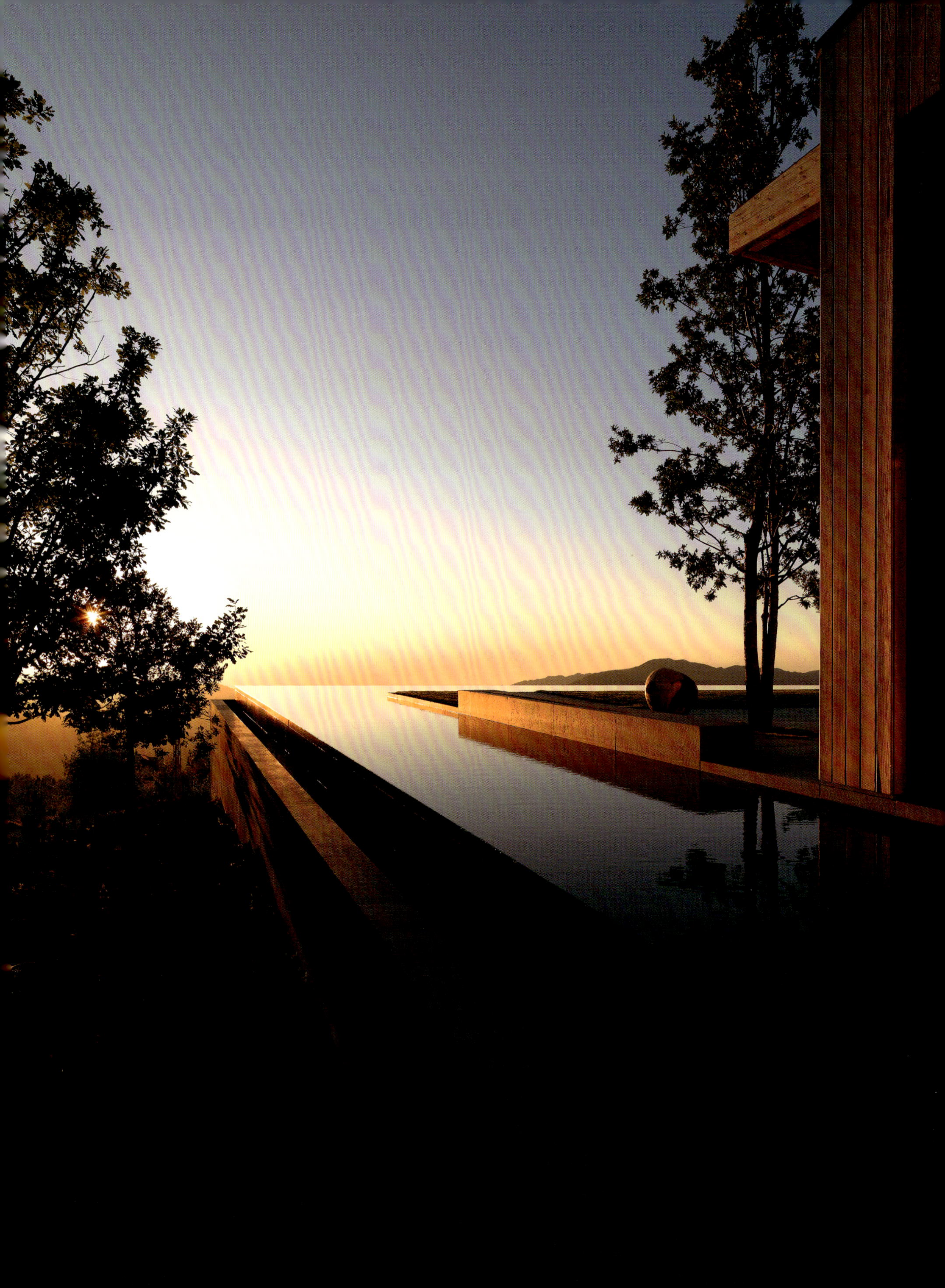

City of Culture of Galicia

PETER EISENMAN

*Eisenman Architects P.C.
41 West 25th Street
New York, NY 10010
USA*

*Tel: +1 212 645 1400
Fax: +1 212 645 0726
E-mail: info@eisenmanarchitects.com
Web: www.eisenmanarchitects.com*

Born in Newark, New Jersey, in 1932, **PETER EISENMAN** received an architecture degree from Columbia, and a Ph.D. at the University of Cambridge, UK. Having established his reputation as a theorist, he came into view first as a member of the "New York Five" with Richard Meier, John Hejduk, Charles Gwathmey, and Michael Graves. He set up his practice in 1980, and came to the attention of a wider public with the Wexner Center for the Visual Arts (Ohio State University, Columbus, Ohio, USA, 1982–89). His major projects include the Koizumi Sangyo Building (Tokyo, 1987–89); Greater Columbus Convention Center (Columbus, Ohio, USA, 1989–93); and the Aronoff Center for Design and Art (University of Cincinnati, Cincinnati, Ohio, USA, 1988–96). He participated in the competition for the new World Trade Center in 2002 with Richard Meier, Charles Gwathmey, and Steven Holl. Recent projects include the Memorial to the Murdered Jews of Europe (Berlin, 1998–2005); the Pozzuoli Waterfront Master Plan (with AZ Studio, Interplan Seconda, Pozzuoli, Italy, 2007–); and the City of Culture of Galicia (Santiago de Compostela, Spain, 2001–, published here). The Project Director for the City of Culture of Galicia Sandra Hemingway, who studied at Indiana University and received her M.Arch from Carnegie Mellon University, joined the firm in 1999.

PETER EISENMAN wurde 1937 in Newark, New Jersey, geboren. Er schloss sein Architekturstudium an der Columbia University ab und promovierte dann in Cambridge in Großbritannien. Er war bereits ein renommierter Architekturtheoretiker, als er zusammen mit Richard Meier, John Hejduk, Charles Gwathmey und Michael Graves Mitglied der „New York Five" wurde. Sein eigenes Büro gründete er 1980. Einem breiteren Publikum wurde er durch seinen Entwurf für das Wexner Center for the Visual Arts (Ohio State University, Columbus, Ohio, USA, 1982–89) bekannt. Zu seinen wichtigsten Projekten zählen das Koizumi Sangyo Building (Tokio, 1987–89), das Kongresszentrum Greater Columbus Convention Center (Columbus, Ohio, USA, 1989–93) und das Aronoff Center for Design and Art – Zentrum für Design und Kunst (University of Cincinnati, Cincinnati, Ohio, USA, 1988–96). 2002 nahm er mit Richard Meier, Charles Gwathmey und Steven Holl an dem Wettbewerb für die Neubebauung des World Trade Center teil. Zu seinen neuesten Projekten gehören das Holocaust-Mahnmal, Denkmal für die ermordeten Juden in Europa (Berlin, 1998–2005), der Masterplan für das Hafenviertel von Pozzuoli (mit AZ Studio, Interplan Seconda, Pozzuoli, Italien, seit 2007) und die Cidade da Cultura de Galicia – die Kulturstadt von Galicien (Santiago de Compostela, Spanien, seit 2001, hier vorgestellt). Die Projektleiterin der Kulturstadt von Galicien, Sandra Hemingway, trat 1999 nach ihrem Studium an der Indiana University und dem Abschluss als Master der Architektur an der Carnegie Mellon University in das Büro ein.

Né à Newark (New Jersey) en 1932, **PETER EISENMAN** a obtenu son diplôme d'architecture à l'université Columbia (US) et son Ph.D. à l'université de Cambridge (GB). Surtout connu au début de sa carrière comme théoricien, il a fait partie du groupe des « New York Five » avec Richard Meier, John Hejduk, Charles Gwathmey et Michael Graves. Il fonde son agence en 1980 et atteint la notoriété avec le Wexner Center for the Visual Arts (université de l'Ohio, Columbus, Ohio, US, 1982–89). Parmi ses réalisations majeures : l'immeuble Koizumi Sangyo (Tokyo, 1987–89) ; le Greater Columbus Convention Center (Columbus, Ohio, US, 1989–93) et l'Aronoff Center for Design and Art (université de Cincinnati, Cincinnati, Ohio, US, 1988–96). Il a participé au concours pour le nouveau World Trade Center en 2002 avec Richard Meier, Charles Gwathmey et Steven Holl. Plus récemment, il a réalisé le mémorial aux Juifs d'Europe assassinés (Berlin, 1998–2005) ; le plan directeur du front de mer de Pouzzoles (avec AZ Studio, Interplan Seconda, Pouzzoles, Italie, 2007–) et la Cité de la culture de Galice (Saint-Jacques-de-Compostelle, Espagne, 2001–, publié ici). La directrice de ce projet de la Cité de la culture de Galice, Sandra Hemingway, qui a étudié à l'université d'Indiana et obtenu son M.Arch. à l'université Carnegie Mellon, a rejoint l'agence en 1999.

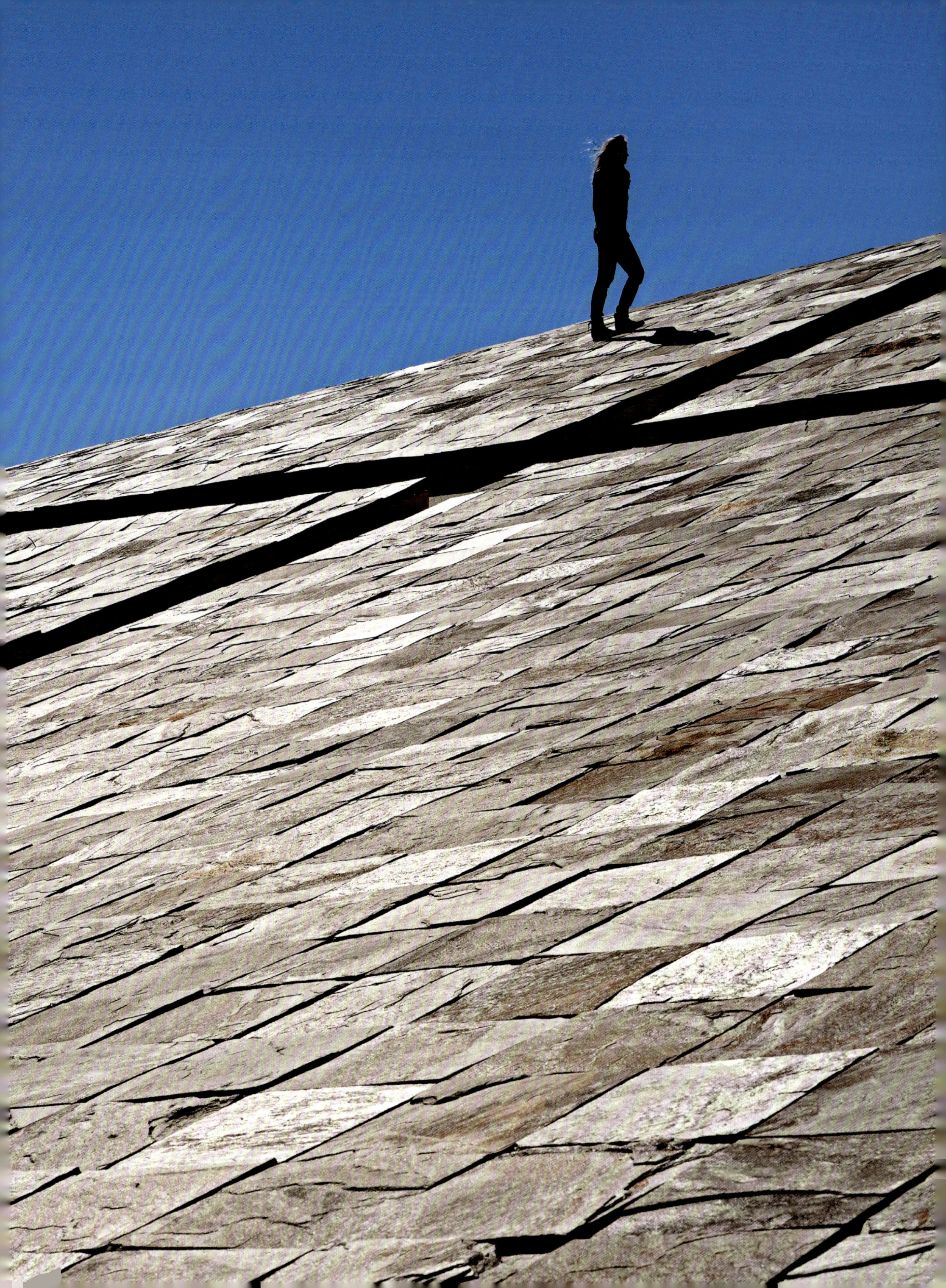

CITY OF CULTURE OF GALICIA
Santiago de Compostela, Spain, 2001–

Address: Galician City of Culture Foundation, Hospital de San Roque, Rúa de San Roque no. 2, 15704 Santiago de Compostela, Galicia, Spain, +34 881 99 75 65, www.cidadedacultura.org
Area: 150 000 m². Client: Fundación Cidade da Cultura de Galicia. Cost: not disclosed
Collaboration: Sandra Hemingway, Andrés Perea

The Parliament of Galicia hosted a competition organized by the Department of Culture, Social Communications, and Tourism of the Xunta de Galicia for the **CITY OF CULTURE OF GALICIA** (Cidade da Cultura de Galicia or CCG) in February 1999. A short list of architects, including Ricardo Bofill, José Manuel Gallego, Annette Gigon and Mike Guyer, Steven Holl, Rem Koolhaas, Daniel Liebeskind, Juan Navarro, Jean Nouvel, Dominique Perrault, and César Portela submitted proposals, and the jury selected Peter Eisenman in August 1999. The cornerstone for the 93 000-square-meter project, set in a 70-hectare site, was laid on February 15, 2001. The six buildings of the project "are conceived as three pairs: the Museum of Galicia and the International Art Center; the Center for Music and Performing Arts and the Central Services building; and the Library of Galicia and Galician Archives." According to the architects: "The heights of all of the buildings rise in gentle curves that seem to reconstruct the shape of the hilltop with their collective rooflines." The zones around the buildings feature landscape and water elements. The Arboretum of Galicia, an area of gardens and native woodland, "conceived as both a recreational and an educational facility," is located beyond the built areas of the complex.

Im Februar 1999 veranstaltete das Parlament von Galicien einen Wettbewerb für die Errichtung der **KULTURSTADT VON GALICIEN** (Cidade da Cultura de Galicia oder CCG), der vom Ministerium für Kultur, Soziale Kommunikation und Tourismus der Xunta de Galicia organisiert worden war. Ein kleiner Kreis von Architekten reichte Vorschläge ein, unter ihnen Ricardo Bofill, José Manuel Gallego, Annette Gigon und Mike Guyer, Steven Holl, Rem Koolhaas, Daniel Liebeskind, Juan Navarro, Jean Nouvel, Dominique Perrault und César Portela. Im August 1999 entschied sich die Jury für den Entwurf von Peter Eisenman. Der Grundstein für das 93 000 Quadratmeter umfassende Projekt auf einem 70 Hektar großen Gelände wurde am 15. Februar 2001 gelegt. Die sechs Gebäude des Gesamtprojektes sind als „drei Paare konzipiert: das Museum von Galicien und das Internationale Kunstzentrum, das Zentrum für Musik und darstellende Künste und das Zentrale Dienstleistungsgebäude sowie die Bibliothek von Galicien und das Galicische Archiv". Die Architekten beschreiben ihr Projekt folgendermaßen: „Alle Gebäude erheben sich in einem sanften Schwung und scheinen mit ihrer einheitlichen Dachform die Konturen der umgebenden Landschaft nachzuzeichnen." Die Bereiche um die Gebäude herum sind mit landschaftsarchitektonischen Elementen und Wasserspielen gestaltet. Jenseits des bebauten Areals entsteht das Arboretum von Galicien mit Gartenanlagen, die mit heimischen Bäumen und Sträuchern bepflanzt sind. Es soll „sowohl zur Erholung als auch zur Information dienen".

Le département de la Culture, de la Communication sociale et du Tourisme de la junte de Galice avait organisé en 1999 un concours pour la **CITÉ DE LA CULTURE DE GALICE** (CCG). Un nombre limité d'architectes y avait participé : Ricardo Bofill, José Manuel Gallego, Annette Gigon et Mike Guyer, Steven Holl, Rem Koolhaas, Daniel Liebeskind, Juan Navarro, Jean Nouvel, Dominique Perrault, César Portela et Peter Eisenman, finalement retenu par le jury en août 1999. La première pierre de ce projet de 93 000 m² édifié sur un terrain de 70 ha a été posée le 15 février 2001. Les six bâtiments « sont conçus en trois paires : le musée de Galice et le Centre d'art international ; le Centre de la musique et des arts du spectacle et le bâtiment central des services ; la bibliothèque de Galice et les archives de Galice ». « La couverture de chaque bâtiment s'élève en suivant une courbe douce qui semble reprendre la forme des collines à travers l'ensemble des toitures », explique l'architecte. Tout autour, le terrain a été paysagé et des bassins ont été aménagés. L'Arboretum de Galice, « pensé comme une installation à la fois éducative et de loisirs », est situé à l'arrière de la partie bâtie du complexe.

PETER EISENMAN

More than 10 years in the works, this complex inspires itself from the very land it is built in, making architecture into landscape in many senses.

Seit mehr als zehn Jahren wird an diesem Gebäudeensemble gearbeitet, das seine Anregungen aus der umgebenden Landschaft bezieht, um dann die Architektur wieder in vielfältiger Hinsicht in Landschaft zu verwandeln.

Aboutissement de plus de dix ans de travaux, ce complexe s'inspire de la nature même du sol sur lequel il est édifié, faisant à de nombreux égards de l'architecture un paysage.

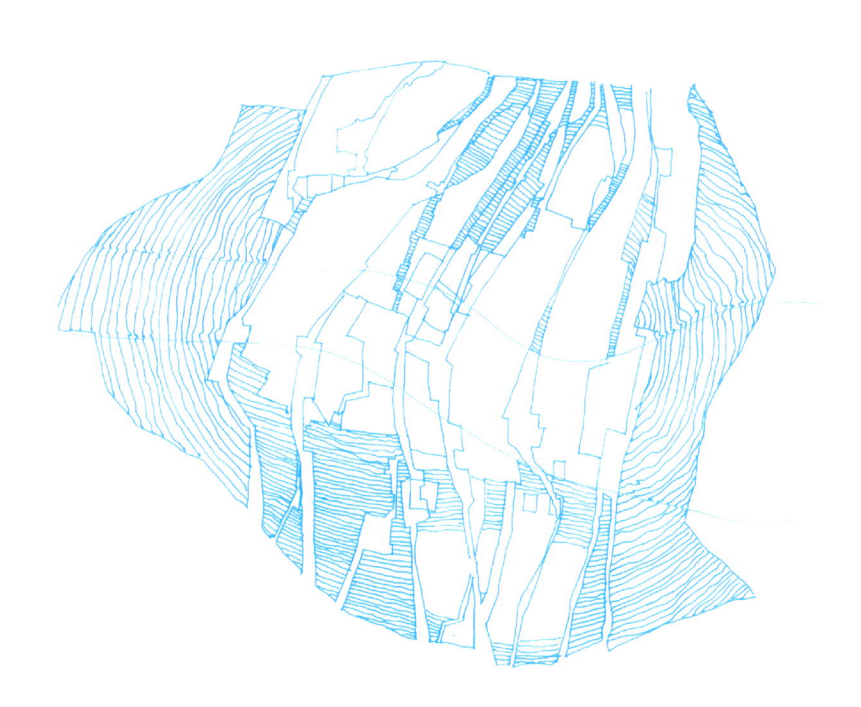

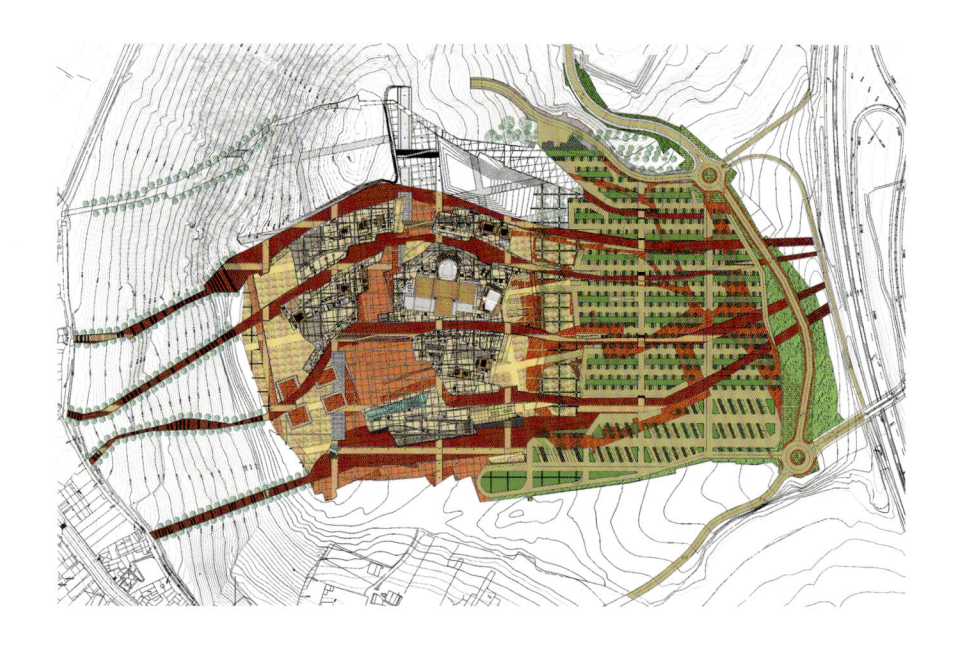

A plan of the complex (left) shows how the overall forms blend into the landscape in an almost natural way.

Der Grundriss des Geländes (links) zeigt, wie die Gesamtformen fast natürlich mit der Landschaft verschmelzen.

Le plan du complexe (à gauche) montre à quel point ses formes se fondent dans le paysage de manière presque naturelle.

Beneath the stone-covered roof, the lines and volumes of the building continue the hilly composition.

Unter dem Natursteindach setzen die Linien und Volumen der Gebäude die sanft gewellte Komposition fort.

Sous le toit dallé de pierre, les lignes et les volumes semblent prolonger les vallonnements.

Young Circle ArtsPark

GLAVOVIC STUDIO

Glavovic Studio Inc.
724 NE 3rd Avenue
Fort Lauderdale, FL 33304
USA

Tel: +1 954 524 5728
Fax: +1 954 524 5729
E-mail: info@glavovicstudio.com
Web: www.glavovicstudio.com

MARGI NOTHARD was born in 1963 in Harare, Zimbabwe. She attended the University of Kwa Zulu-Natal in Durban as an undergraduate. She received her M.Arch degree from the Southern California Institute of Architecture (SCI-Arc) in 1992. Margi Nothard is the President and Design Principal of Glavovic Studio Inc. The firm's work includes the Girls' Club (Fort Lauderdale, 2007); Young Circle ArtsPark (Hollywood, Florida, 2003–07, published here); the Museum of Art|NSU: Art Plaza (Fort Lauderdale, 2011); and Young @ Art / Broward County Children's Museum and Reading Center Project (Davie, Florida, 2012), all in the USA.

MARGI NOTHARD wurde 1963 in Harare, Zimbabwe, geboren. Sie studierte zuerst an der University of Kwa Zulu-Natal in Durban und machte 1992 ihren Master in Architektur am Southern California Institute of Architecture (SCI-Arc). Margi Nothard ist Firmenchefin und leitende Entwurfsarchitektin des Büros Glavovic Studio Inc. Zu den Arbeiten des Büros gehören die Sammlung Girls' Club (Fort Lauderdale, 2007), der ArtsPark mit Kunstinstallationen im Stadtteil Young Circle (Hollywood, Florida, 2003–07, hier vorgestellt), das Museum of Art|NSU: Art Plaza (Fort Lauderdale, 2011) und Young @ Art/Broward County Children's Museum and Reading Center Project – Kindermuseum und Lesezentrum (Davie, Florida, 2012), alle in den USA.

MARGI NOTHARD, née en 1963 à Harare au Zimbabwe, a étudié à l'université de Kwa Zulu-Natal à Durban. Elle a obtenu son M.Arch. à l'Institut d'architecture de Californie du Sud (SCI-Arc, 1992). Elle est présidente et responsable de la conception de Glavovic Studio Inc. Aux États-Unis, l'agence a réalisé, entre autres, le Girls' Club (Fort Lauderdale, Floride, 2007) ; le Young Circle ArtsPark (Hollywood, Floride, 2003–07, publié ici) ; le Musée d'art|NSU: Art Plaza (Fort Lauderdale, Floride, 2011) et le Young @ Art Children's Museum/Broward County Library (Davie, Floride, 2012).

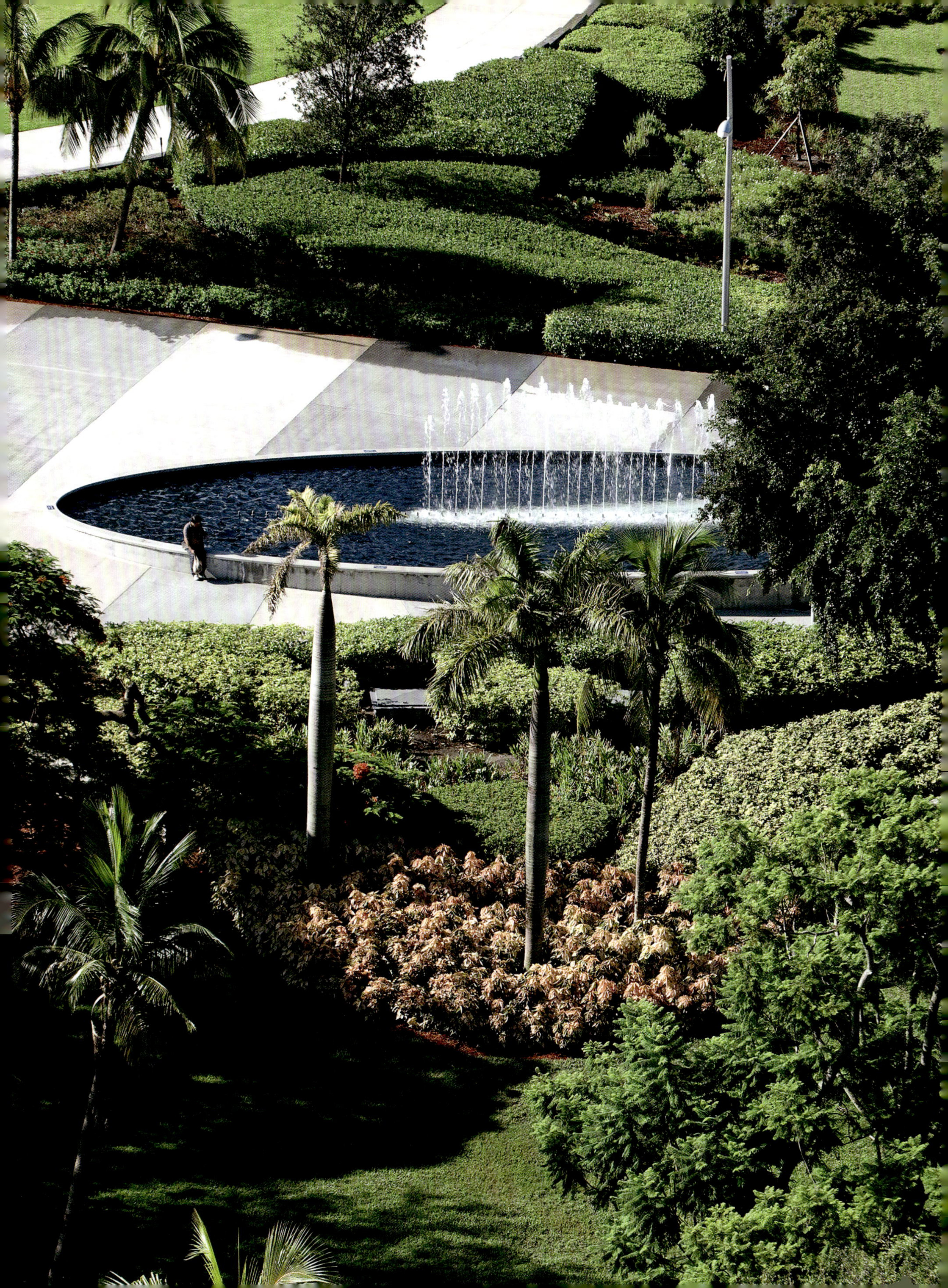

YOUNG CIRCLE ARTSPARK
Hollywood, Florida, USA, 2003–07

*Address: One Young Circle, Hollywood Boulevard at US1, Hollywood, FL 33020, USA, +1 954 921 3520, www.hollywoodfl.org/artspark
Area: 4 hectares (park); 1300 m² (pavilion). Client: City of Hollywood. Cost: $13 million (Phase I); $26 million overall
Collaboration: Marvin Scharf, IBI Group (Executive Architect)*

One challenge of this site is that it is located in the center of Hollywood and is surrounded by a 4-5 lane one-way federal highway (US 1) with multiple cross entries. The architectural program was divided into two main buildings (Performing Arts and Visual Arts). The landscaping is based on an emerging spiral that becomes a cantilevered canopy. The flat center of the park was raised to provide variety, and to accommodate trees. The architect states: "Materials such as poured-in-place concrete connect the buildings to the earth and reinforce their emergence. These built structures then pierce the sky with metal and wood panels providing a counterpoint to the landscaped earth. The tree and building canopy are one with the blue sky. Art as environment, art as light, art as sound, and art as architecture."

Eine Herausforderung bei diesem Gelände ist seine Lage: mitten im Stadtzentrum von Hollywood und von der vier- bis fünfspurigen Bundesstraße US 1 mit zahlreichen Einmündungen umgeben. Das Raumprogramm wurde auf zwei Hauptgebäude verteilt (darstellende und bildende Künste). Die Landschaftsgestaltung basiert auf einer sich in die Höhe schraubenden Spirale, die zu einem ausladenden Baldachin wird. Das flache Zentrum des Parks wurde mit Erde aufgeschüttet, um das Gelände lebendiger zu gestalten und um Bäume pflanzen zu können. Die Architektin sagt dazu: „Mit Materialien wie Ortbeton werden die Gebäude mit der Erde verbunden und ihr Erscheinen betont. Diese gebauten Strukturen durchdringen dann mit Metall- und Holzpaneelen den Himmel und bilden einen Kontrapunkt zu der landschaftlich gestalteten Erdoberfläche. Der Baum- und Gebäudebaldachin vereinigt sich mit dem blauen Himmel. Kunst als Umfeld, Kunst als Licht, Kunst als Klang und Kunst als Architektur."

L'un des défis posés par ce terrain du centre d'Hollywood était d'être un immense rond-point encerclé par l'autoroute fédérale US 1 de quatre à cinq voies. Le programme architectural a été réparti en deux bâtiments principaux (arts du spectacle et arts visuels). L'aménagement paysager repose sur une spirale qui se termine en un auvent en porte-à-faux. La partie centrale a été surélevée pour donner davantage de variété au relief, puis plantée d'arbres. « Des matériaux comme le béton coulé sur site ancrent les bâtiments dans le sol et renforcent l'impression d'émergence. Ces constructions pointent ensuite vers le ciel leurs panneaux de métal et de bois qui viennent en contrepoint du sol aménagé. Les arbres et l'auvent ne font qu'un avec le ciel. L'art comme environnement, l'art comme lumière, l'art comme son et l'art comme architecture. »

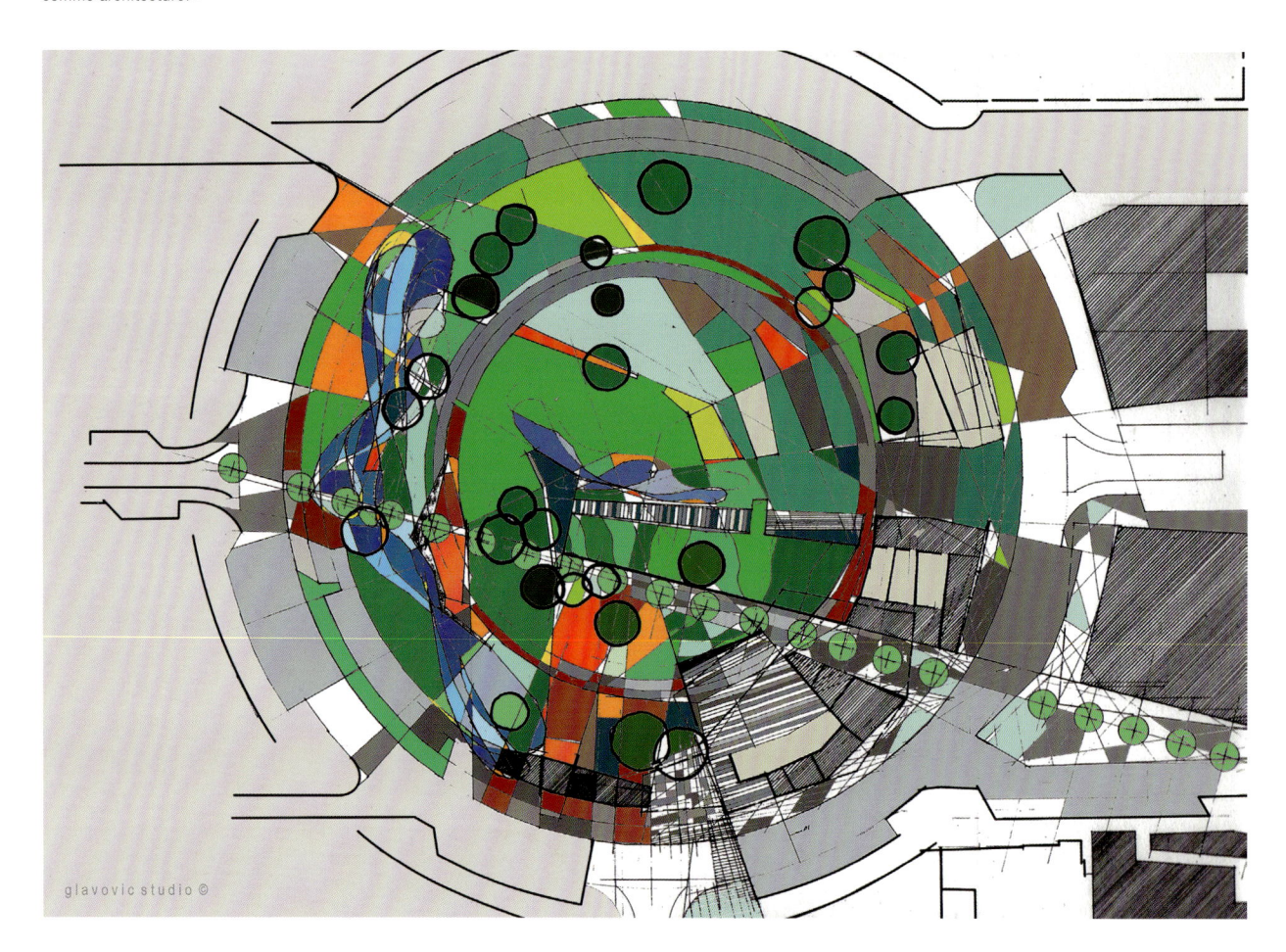

GLAVOVIC STUDIO

The unusual round form of the park is visible both in the overall image and the plan on the left, which also gives an indication of the planting and the use of the space. A building in the park is seen on the right.

Die ungewöhnliche Kreisform des Parks kann man sowohl auf der Luftaufnahme als auch auf dem Grundriss links erkennen, der auch Angaben zur Bepflanzung und Nutzung der verschiedenen Freiräume enthält. Rechts ein Gebäude in diesem Park.

L'étonnant plan circulaire du parc, vu dans le plan à gauche et l'image ci-dessus, précise les plantations et l'utilisation de l'espace. À droite, un des bâtiments édifiés à l'intérieur du parc.

The design is uncluttered and colorful. The designs on the ground echo the angled architecture of the arts centers that are the main architectural feature of the park.

Das Design ist schlicht und farbenfroh. Die Muster auf dem Boden nehmen die Linienführung der winkelförmigen Architektur der Kunstzentren auf, der größten Gebäude im Park.

La composition est dégagée et colorée. Les parterres font écho à l'architecture des centres d'art, principaux éléments architecturals du parc.

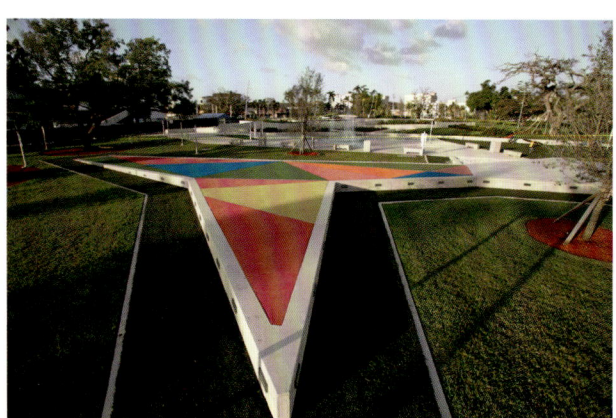

Night lighting allows the park spaces to be used beyond the more usual day periods. On the whole, the outlines of the park forms and the planting are simple and efficient.

Wegen der nächtlichen Beleuchtung kann der Park auch bei Dunkelheit genutzt werden. Insgesamt sind die Konturen der Parkgestaltung und die Bepflanzung einfach und rationell.

L'éclairage nocturne permet d'utiliser les lieux au-delà des horaires habituels. Les formes construites et les plantations se répondent de façon simple et efficace.

GUSTAFSON GUTHRIE NICHOL

Gustafson Guthrie Nichol Ltd
Pier 55, Floor 3
1101 Alaskan Way
Seattle, WA 98101
USA

Tel: +1 206 903 6802
Fax: +1 206 903 6804
E-mail: contact@ggnltd.com
Web: www.ggnltd.com

KATHRYN GUSTAFSON was born in 1951 in the US. She attended the University of Washington (Seattle, 1970), the Fashion Institute of Technology (New York, 1971), and the École Nationale Supérieure du Paysage (Versailles, France, 1979). Gustafson Porter Ltd was founded in 1997 by Kathryn Gustafson and Neil Porter. In 1999, Gustafson, with Partners **JENNIFER GUTHRIE** and **SHANNON NICHOL**, established the practice Gustafson Guthrie Nichol Ltd in Seattle. Aside from Towards Paradise (Venice Architecture Biennale, Venice, Italy, 2008), the firm has completed the Lurie Garden in the Millennium Park (Chicago, Illinois, 2004, published here); and the Kogod Courtyard at the Smithsonian Institution (Washington, D.C., 2005–07, also published here). Current work includes CityCenterDC, a mixed-use project being developed with Foster + Partners (Washington, D.C., 2010–); the Bill and Melinda Gates Foundation Campus (Seattle, 2011); and the Smithsonian National Museum of African American History and Culture gardens (Washington, D.C., 2015).

KATHRYN GUSTAFSON wurde 1951 in den Vereinigten Staaten geboren. 1970 studierte sie an der University of Washington (Seattle), ab 1971 am Fashion Institute of Technology an der State University of New York, wo sie ihren Abschluss im Fach Modedesign machte. Danach ging sie nach Paris und studierte Landschaftsarchitektur an der École nationale supérieure du paysage (Versailles) und machte dort 1979 ihr Diplom. Das Büro Gustafson Porter Ltd wurde 1997 von Kathryn Gustafson und Neil Porter gegründet. 1999 gründete Gustafson gemeinsam mit ihren Büropartnerinnen **JENNIFER GUTHRIE** und **SHANNON NICHOL** das Büro Gustafson Guthrie Nichol Ltd in Seattle. Neben dem Projekt „Towards Paradise" für die Architekturbiennale Venedig 2008 hat das Büro 2004 den Lurie Garden im Millennium Park in Chicago, Illinois ausgeführt (hier vorgestellt) sowie den Kogod Courtyard, einen Innenhof im Gebäude der Smithsonian Institution (Washington, D. C., 2005–07, ebenfalls hier vorgestellt). Zu den im Bau befindlichen Projekten gehören das CityCenterDC, ein Projekt für eine Mischnutzung, das gemeinsam mit dem Büro Foster + Partners (Washington, D. C., seit 2010) geplant wird, sowie der Bill and Melinda Gates Foundation Campus (Seattle, 2011) und die Gartenanlage des Smithsonian National Museum of African American History and Culture (Washington, D. C.), die 2015 fertiggestellt werden soll.

KATHRYN GUSTAFSON, née aux États-Unis en 1951, a étudié à l'université de Washington (Seattle, 1970), à l'Institut des technologies de la mode (New York, 1971) et à l'École nationale supérieure du paysage (Versailles, 1979). L'agence Gustafson Porter Ltd a été fondée en 1997 par Kathryn Gustafson et Neil Porter. En 1999, Gustafson et ses associés, **JENNIFER GUTHRIE** et **SHANNON NICHOL**, ont fondé l'agence Gustafson Guthrie Nichol Ltd à Seattle. En dehors de *Towards Paradise* (Biennale d'architecture de Venise, 2008, publié ici), l'agence a réalisé le Lurie Garden dans le Millennium Park (Chicago, Illinois, US, 2004, publié ici) et la Kogod Courtyard à la Smithsonian Institution (Washington, 2005–07, publiée ici). Elle travaille actuellement sur les projets suivants aux États-Unis : CityCenterDC, projet immobilier mixte en collaboration avec Foster + Partners (Washington, 2010–) ; le campus de la Bill and Melinda Gates Foundation (Seattle, 2011) et les jardins du Smithsonian National Museum of African American History and Culture (Washington, 2015).

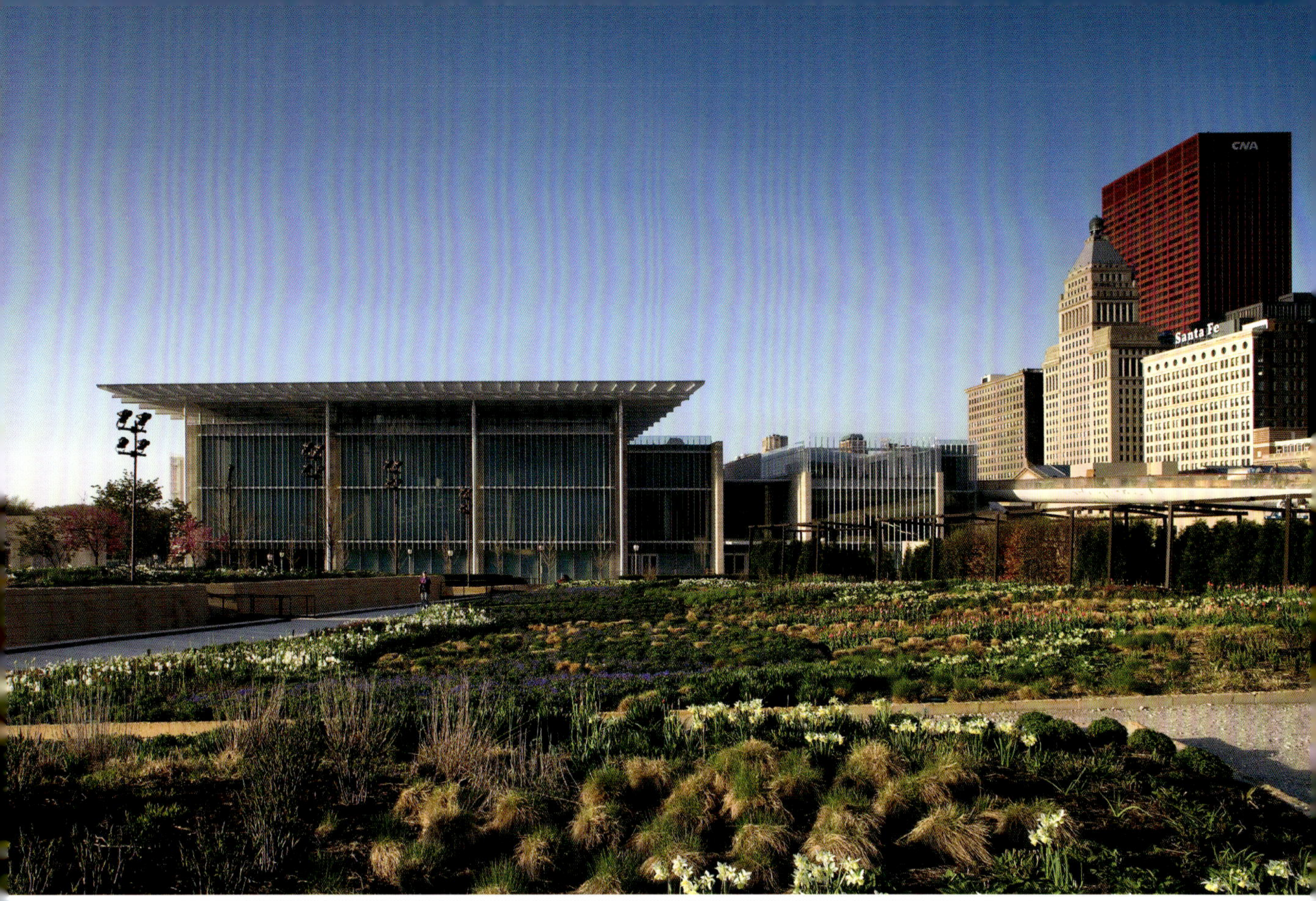
CNA
Santa Fe

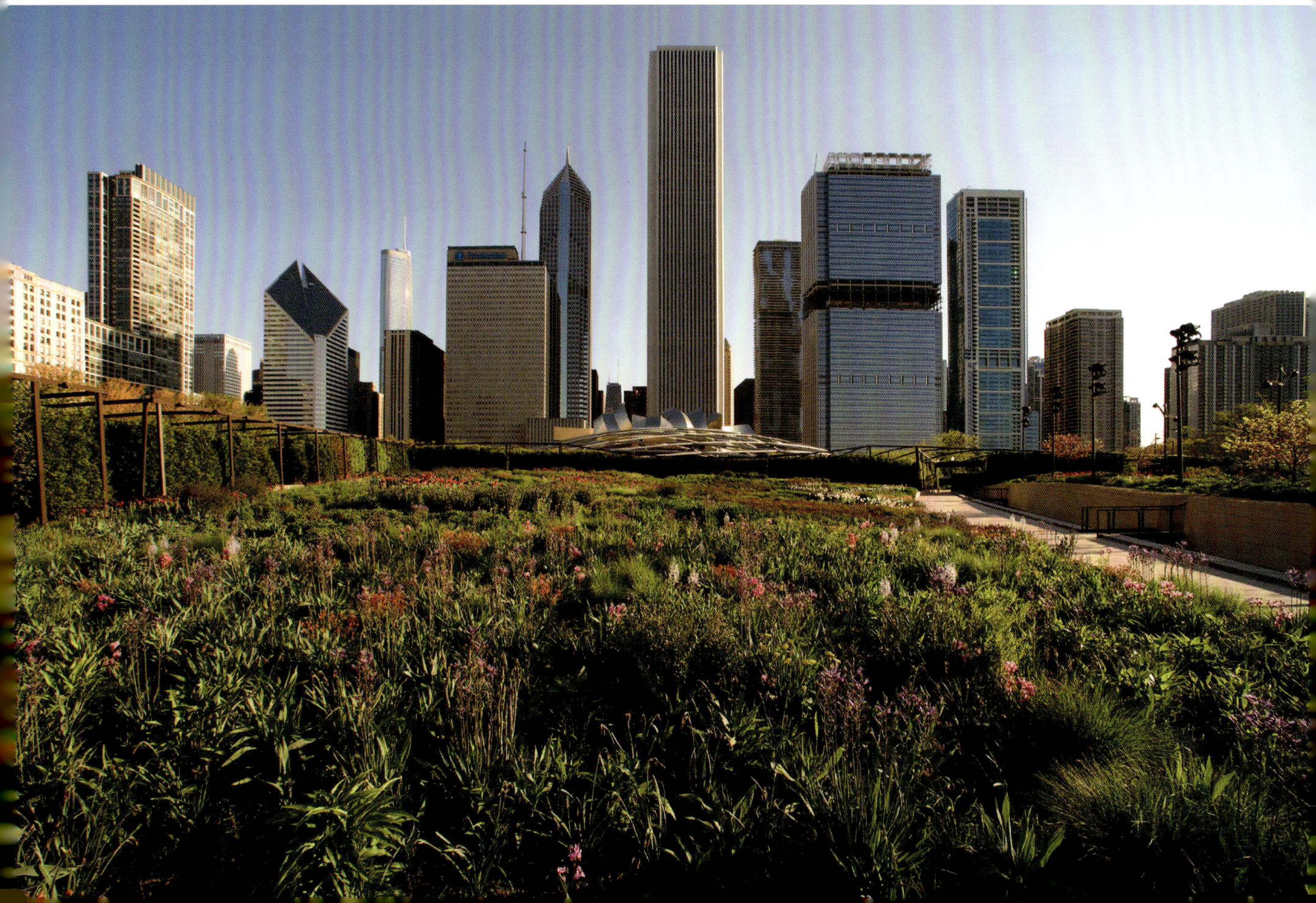

LURIE GARDEN
Chicago, Illinois, USA, 2004

Address: Monroe Street and Columbus Drive, Chicago, IL 60601, USA, www.luriegarden.org
Area: 12 629 m^2, Client: Millennium Park. Cost: not disclosed
Collaboration: Piet Oudolf (Plantsman), Robert Israel (Conceptual Reviewer)

The **LURIE GARDEN** is situated on the roof of the Lakefront Millennium Parking Garage in Chicago's Millennium Park. The site is between the band shell designed by Frank O. Gehry and a renovation of the Art Institute of Chicago by Renzo Piano. Millennium Park is part of Grant Park and the Lurie Garden "continues the precedent of Grant Park's 'rooms' with treed enclosures, perimeter circulation, and axial views; it expresses these qualities in forms that are distinct to the Garden's special site and context." A large hedge encloses the garden from the north and west. The land forms within the Garden were created using lightweight geofoam due to weight restrictions on the garage roof. Although the garden is manifestly urban in its site and surroundings, its artificial aspects are by no means visible. It is, rather, an attractive public space that has been designed to accommodate large crowds exiting nearby concerts. Its modernity sits well with the architectural environment and particularly in the presence of works by Frank Gehry and Renzo Piano. Limestone from a local midwestern quarry is used for curbing, stone stairs, stair landings, wall coping, and wall cladding in the interior of the Garden. Flamed granite is used as paving and wall veneer in the water feature and the Dark Plate area.

Der **LURIE GARDEN** ist auf dem Dach des Parkhauses Lakefront Millenium im Millenium Park von Chicago angelegt worden. Das Gelände befindet sich zwischen der von Frank O. Gehry entworfenen Konzertmuschel und dem von Renzo Piano renovierten Flügel des Art Institute of Chicago. Der Millennium Park gehört zum Grant Park, und im Lurie Garden „finden die Gartenräume des Grant Park mit den von Bäumen umschlossenen Bereichen, den am Rand geführten Wegeverbindungen und den Sichtachsen ihre Fortsetzung. Die neue Gartenanlage drückt diese Qualitäten in einer Formensprache aus, die das besondere Gelände und den Kontext des Gartens hervorhebt." Eine große Hecke begrenzt den Garten an der Nord- und Westseite. Die Geländemodulation im Garten ist mit leichtem EPS-Hartschaum (Geofoam) gestaltet, um den Gewichtsbeschränkungen auf dem Parkhausdach Rechnung zu tragen. Auch wenn es sich aufgrund der Lage und der Umgebung um einen ausgesprochen urbanen Garten handelt, so bleibt doch seine Künstlichkeit verborgen. Der Garten ist ein attraktiver öffentlicher Freiraum, der so geplant ist, dass er große Menschenmengen aufnehmen kann, die von den benachbarten Konzerten kommen. Seine Modernität passt gut zu dem architektonischen Umfeld, besonders zu den Arbeiten von Frank O. Gehry und Renzo Piano. Kalkstein aus einem örtlichen Steinbruch im Mittelwesten wurde für die Randsteine, Natursteintreppen, Podeste, die Verkleidung der Mauern und die Mauerkronen im Garten verwendet. Geflammter Granit wurde als Bodenpflaster und Mauerverblendung beim Wasserspiel und im Bereich der Dark Plate verwendet.

Le **LURIE GARDEN** a été aménagé sur la toiture du parking Lakefront Millennium Parking Garage, dans le Millennium Park à Chicago, entre l'auditorium en plein air signé Frank O. Gehry et l'aile du Chicago Art Institute rénovée par Renzo Piano. Le Millennium Park fait partie du Grant Park et le Lurie Garden « reprend le précédent créé par ce parc de "pièces" entourées d'arbres, de circulation périmétrique et de vues en perspective. Il exprime cette spécificité par des formes adaptées aux caractéristiques du site et du contexte ». Une importante haie protège le jardin au nord et à l'ouest. De légers vallonnements ont été créés en Géofoam légère pour éviter de trop alourdir la toiture du parking. Même si le site et l'environnement de ce jardin sont résolument urbains, ses aspects artificiels restent invisibles. C'est un lieu public attractif, conçu pour accueillir des visiteurs en grand nombre, comme les spectateurs des concerts de plein air, par exemple. Sa modernité s'intègre bien à son environnement architectural et en particulier aux réalisations de Gehry et de Piano. Les bordures, les escaliers, les paliers, les couronnements et le parement des murs sont en pierre calcaire du Midwest. Le pavement et l'habillement des fontaines et du Dark Plate sont en granit flammé.

As seen against the Chicago skyline (left page) or in the plan and "aerial" view on this page, the garden quite literally brings a breath of fresh air and natural color to the otherwise metallic and mineral urban environment.

Vor dem Hintergrund der Skyline von Chicago (linke Seite), aber auch auf dem Grundriss und der "Luftaufnahme" auf dieser Seite erkennt man, dass der Garten fast im wörtlichen Sinn eine frische Brise und natürliche Farben in ein ansonsten von Stahl und Stein geprägtes urbanes Umfeld bringt.

Devant le panorama urbain de Chicago (page de gauche) ou dans le plan et la vue « aérienne » de cette page, le jardin apporte un peu de fraîcheur et de couleurs naturelles dans un environnement minéral et métallique.

The convivial atmosphere of the park is visible in these images, which give little hint that the park was created on the rooftop of a parking area. Frank O. Gehry's band shell is visible in the image above.

Auf diesen Bildern spürt man die heitere und gesellige Atmosphäre im Park, und man denkt nicht, dass er auf dem Dach eines Parkhauses angelegt wurde. Auf dem Bild oben ist im Hintergrund die Konzertmuschel von Frank O. Gehry zu erkennen.

L'atmosphère conviviale du parc, qui se sent dans ces images, ne laisse rien deviner de son aménagement sur la toiture d'un parking. En haut à gauche, la scène d'orchestre créée par Frank O. Gehry.

Though the landscape created by the designers in this place is essentially "artificial" as opposed to being "natural," its dense planting and flowers, as seen in the image above, give a very definite impression of allowing the natural world to take its rightful place in the city.

Auch wenn der von den Planern geschaffene Gartenraum durch und durch „künstlich" ist, als Gegensatz zu „natürlich", so vermittelt die dichte Bepflanzung den Eindruck, als würde die Natur den ihr rechtmäßig zustehenden Platz in der Stadt einnehmen, wie man auf dem Bild oben erkennt.

Si le paysage créé est essentiellement « artificiel » – par opposition à « naturel » –, la densité des plantations, comme le montre l'image ci-dessus, donne l'impression que le monde naturel a retrouvé la place qui lui revenait.

ROBERT AND ARLENE KOGOD COURTYARD

Washington, D.C., USA, 2005–07

Address: 8th and F Streets, NW, Donald W. Reynolds Center, Washington, D.C. 20001, USA
Area: 2601 m². Client: The Smithsonian Institution. Cost: not disclosed
Collaboration: Foster + Partners (Design Architects)

The **KOGOD COURTYARD** is a new public space in the center of the historic United States Patent Office building. Called the Donald Reynolds Center, the structure houses the Smithsonian American Art Museum and National Portrait Gallery. Enclosed by a glass canopy, the courtyard receives museum goers or formal events for up to 1000 people. A central water feature runs the length of the space. The landscaping work is intended as a "link between the historic building and the new roof." Large stone planters with 7.5-meter-high canopy trees are used to define a central space in the courtyard. The designers state: "The planting concept for the Robert and Arlene Kogod Courtyard reinforces the overall design concept. The temperate palette of evergreen trees and shrubs complement and distinguish the existing architecture of the courtyard with their variations in scale, form, and texture as well as the symmetry or asymmetry of their arrangement." Architectural lighting is part of the project, allowing the space to be equally attractive after nightfall. Granite pavers with a flamed finish contrasts with honed white marble cladding on the planters. The design architects for the overall project are Foster + Partners.

Der **KOGOD COURTYARD** ist ein neuer öffentlicher Innenhof in dem historischen Gebäude des US-Patentamts. Das heutige Donald Reynolds Center beherbergt das Smithsonian Museum für amerikanische Kunst und die Nationale Porträtgalerie. Der Innenhof wird von einem Glasdach geschlossen, sodass sich hier die Museumsbesucher bewegen können oder auch Veranstaltungen für bis zu 1000 Personen stattfinden können. Eine zentrale Wasserinstallation wird durch die gesamte Länge des Raums geführt. Der landschaftsgestalterische Entwurf soll als „Verbindung zwischen dem historischen Gebäude und dem neuen Dach wirken". Steintröge, die mit 7,5 Meter hohen großkronigen Bäumen bepflanzt sind, definieren den zentralen Raum in diesem Innenhof. Die Planer bemerken dazu: „Die Bepflanzung des Robert and Arlene Kogod Courtyard betont das gesamte Entwurfskonzept. Die gemäßigten Farben der immergrünen Bäume und Gehölze, ihre unterschiedlichen Größen, Formen und Texturen sowie ihre symmetrische oder asymmetrische Anordnung vervollständigen die vorhandene Architektur und betonen sie." Die Beleuchtung, mit der die Architektur in Szene gesetzt wird, ist Teil des Projekts und trägt zur Attraktivität dieses Raums nach Einbruch der Dunkelheit bei. Geflammte Granitplatten als Bodenbelag kontrastieren mit der geschliffenen weißen Marmorverkleidung der Pflanztröge. Entwurfsplaner für das Gesamtprojekt ist das Büro Foster + Partners.

La **KOGOD COURTYARD** est un nouvel espace public créé au centre du siège historique de l'administration des brevets des États-Unis. Appelé Donald Reynolds Center, ce bâtiment abrite aujourd'hui le Smithsonian American Art Museum et la National Portrait Gallery. Abritée par un auvent de verre, la cour accueille les visiteurs des musées et le public de diverses manifestations (jusqu'à 1000 participants). Un bassin central occupe toute la longueur de l'espace. L'aménagement paysager est conçu pour faire le « lien entre le bâtiment historique et la nouvelle verrière ». D'importantes jardinières de pierre plantées d'arbres de plus de 7 m de haut entourent l'espace central. Les paysagistes précisent : « Le concept des plantations renforce celui de l'ensemble. La palette discrète d'arbres et de buissons à feuilles persistantes complète et anime l'architecture existante de la cour par des variations d'échelle, de forme et de texture, ainsi que par la symétrie ou l'asymétrie des implantations. » L'éclairage architectural fait partie intégrante de ce projet, afin de rendre le lieu tout aussi attractif en nocturne. Un pavement de granit flammé contraste avec le parement en marbre blanc veiné des jardinières. L'agence d'architecture responsable de l'ensemble du projet était Foster + Partners.

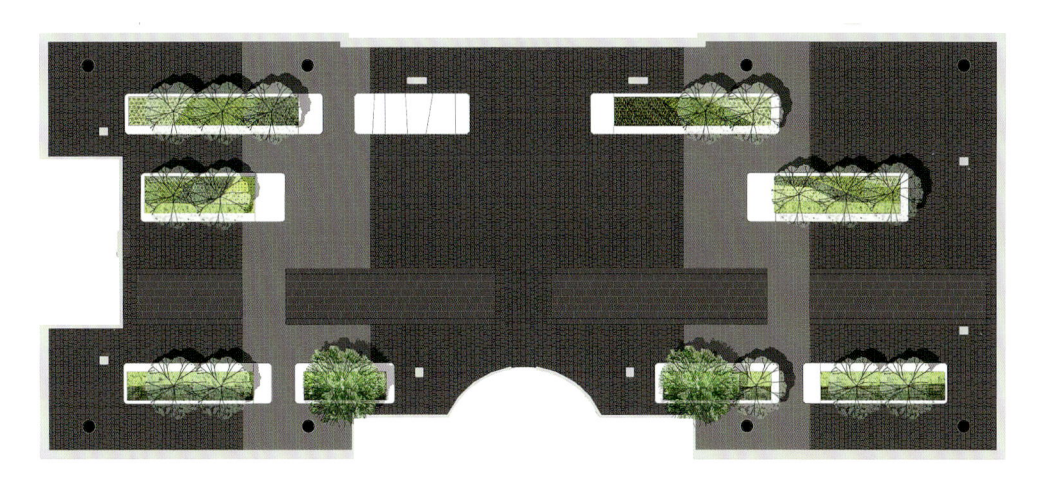

Within the walls of the Neoclassical Smithsonian American Art Museum and National Portrait Gallery, Gustafson Guthrie Nichol have created garden spaces that are placed, in the image below, under a light, curving skylight structure.

Das Büro Gustafson Guthrie Nichol hat im Innenhof des klassizistischen Gebäudekomplexes des Smithsonian Museum für amerikanische Kunst und der Nationalen Porträtgalerie unter einer leichten, gewellten Oberlichtkonstruktion Gartenräume geschaffen.

C'est sous une immense verrière ondulée tendue entre les murs du Smithsonian American Art Museum et de la National Portrait Gallery de style néoclassique que l'agence Gustafson Guthrie Nichol a créé ce jardin.

The natural presence in the building is, of course, circumscribed by the need for visitors to be able to circulate freely, but the designers have found an appropriate balance between the presence of trees and shrubs with the mineral passageways required.

Die Natur innerhalb des Gebäudes ist natürlich eingeschränkt, weil sich die Besucher frei bewegen können müssen, aber die Planer haben ein ausgewogenes Verhältnis von Bäumen und Kleingehölzen zu den geforderten Wegeverbindungen mit ihren Plattenbelägen gefunden.

La présence de la nature dans le bâtiment est conditionnée par la liberté de déplacement des visiteurs, mais les architectes ont trouvé un équilibre harmonieux dans l'implantation des jardinières d'arbres et de fleurs et les allées dallées de pierre.

Whether in daylight or after sunset, the spaces designed by Gustafson Guthrie Nichol make full use of lighting, as seen in the image of the stone planter and bench (right). A central water feature animates the space (below).

Ob bei Tag oder nach Sonnenuntergang, die von Gustafson Guthrie Nichol entworfenen Gartenräume werden durch das Licht in Szene gesetzt, wie man auf dem Bild mit dem Pflanztrog und der Bank (rechts) sehen kann. Ein Wasserbecken belebt die Mitte des Raumes (unten).

De jour comme de nuit, les espaces conçus par Gustafson Guthrie Nichol utilisent toutes les ressources de l'éclairage naturel et artificiel, comme le montrent ces images d'une jardinière de pierre et de son banc (à droite). Un bassin anime une partie de l'espace (ci-dessous).

GUSTAFSON PORTER

Gustafson Porter Ltd
Linton House
39–51 Highgate Road
London NW5 1RS
UK

Tel: +44 20 72 67 20 05
Fax: +44 20 74 85 92 03
E-mail: enquiries@gustafson-porter.com
Web: www.gustafson-porter.com

KATHRYN GUSTAFSON was born in 1951 in the US. She attended the University of Washington (Seattle, 1970), the Fashion Institute of Technology (New York, 1971), and the École National Supérieure du Paysage (Versailles, France, 1979). Gustafson Porter Ltd was founded in 1997 by Kathryn Gustafson and Neil Porter. **NEIL PORTER** was born in the UK in 1958. He attended Newcastle University School of Architecture (1977–80) and the Architectural Association (London, 1981–83). He is a joint Director and designer for Gustafson Porter. **MARY BOWMAN** was also born in the US in 1958. She attended the University of Virginia (Charlottesville, Virginia, 1976–80), and the Architectural Association (London, 1984–88). She has been a Director of the firm with Gustafson and Porter since 2002. Work of the firm includes the Diana, Princess of Wales Memorial Fountain in Hyde Park (London, UK, 2003–04, published here); the Cultuurpark Westergasfabriek (Amsterdam, The Netherlands, 2004); Old Market Square (Nottingham, UK, 2005–07, also published here); Towards Paradise (11th Venice Architecture Biennale, Venice, Italy, 2008); the Woolwich Squares (London, UK, planned completion 2011); Bay East, Gardens by the Bay (Singapore, planned completion 2012); and Citylife Fiera Milano Park (Milan, Italy, planned completion 2014).

KATHRYN GUSTAFSON wurde 1951 in den USA geboren. 1970 studierte sie an der University of Washington (Seattle) und dann ab 1971 am Fashion Institute of Technology an der State University of New York, wo sie ihren Abschluss im Fach Modedesign machte. Danach ging sie nach Paris und studierte Landschaftsarchitektur an der École nationale supérieure du paysage (Versailles) und machte dort 1979 ihr Diplom. Das Büro Gustafson Porter Ltd wurde 1997 von Kathryn Gustafson und Neil Porter gegründet. **NEIL PORTER** wurde 1958 in Großbritannien geboren. Er studierte an der Newcastle University an der School of Architecture (1977–80) und später an der Architectural Association (London, 1981–83). Er ist Mitglied der Geschäftsführung von Gustafson Porter und Entwurfsarchitekt. **MARY BOWMAN** wurde 1958 in den USA geboren. Sie studierte an der University of Virginia (Charlottesville, Virginia, 1976–80) und an der Architectural Association (London, 1984–88). Sie ist seit 2002 gemeinsam mit Gustafson und Porter Mitglied der Geschäftsführung. Zu den Projekten des Büros gehören der Memorial Fountain für Prinzessin Diana im Hyde Park (London, 2003–04, hier vorgestellt), der Cultuurpark Westergasfabriek (Amsterdam, 2004), der Old Market Square (Nottingham, GB, 2005–07, ebenfalls hier vorgestellt), die Garteninstallation „Towards Paradise" (auf der XI. Architekturbiennale in Venedig, 2008), die Woolwich Squares (London, geplante Fertigstellung 2011), Bay East, Gardens by the Bay (Singapur, geplante Fertigstellung 2012), und der CityLife Fiera Milano Park (Mailand, geplante Fertigstellung 2014).

KATHRYN GUSTAFSON, née aux États-Unis en 1951, a étudié à l'université de Washington (Seattle, 1970), à l'Institut des technologies de la mode (New York, 1971) et à l'École nationale supérieure du paysage (Versailles, 1979). L'agence Gustafson Porter Ltd a été fondée en 1997 par Kathryn Gustafson et Neil Porter. **NEIL PORTER**, né au Royaume-Uni en 1958, a étudié à l'École d'architecture de l'université de Newcastle (1977–80) et à l'Architectural Association (Londres, 1981–83). Il est directeur associé et concepteur chez Gustafson Porter. **MARY BOWMAN**, née aux États-Unis en 1958, a étudié à l'université de Virginie (Charlottesville, Virginie, 1976–80) et à l'Architectural Association (Londres, 1984–88). Elle est l'une des directrices de l'agence Gustafson and Porter depuis 2002. Parmi les réalisations de l'agence : la Diana, Princess of Wales Memorial Fountain, dans Hyde Park (Londres, 2003–04, publiée ici) ; la Cultuurpark Westergasfabriek (Amsterdam, 2004) ; la Old Market Square (Nottingham, GB, 2005–07, publié ici) ; *Towards Paradise* (XIe Biennale d'architecture de Venise, 2008) ; les Woolwich Squares (Londres, achèvement prévu en 2011) ; Bay East, Gardens by the Bay (Singapour, achèvement prévu en 2012) et le parc Fiera Milano de Citylife (Milan, achèvement prévu en 2014).

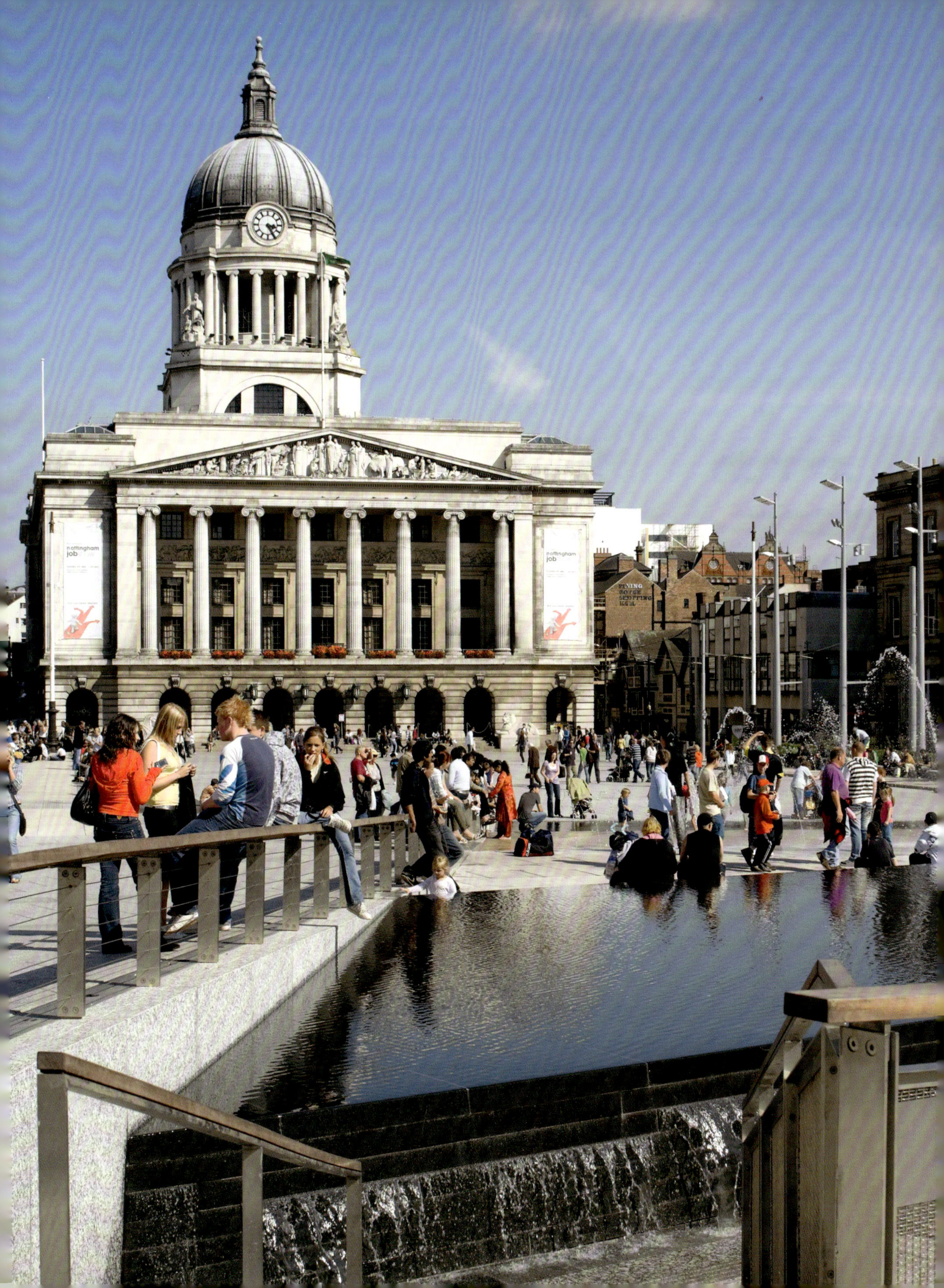
nottingham
job
nottingham
job

OLD MARKET SQUARE
Nottingham, UK, 2005–07

Address: Old Market Square, Nottingham, UK, www.oldmarketsquare.org.uk
Area: 11.5 hectares. Client: Nottingham City Council. Cost: €5.7 million

One of Britain's oldest public squares, the **OLD MARKET SQUARE** in Nottingham is the second largest such space in England after Trafalgar Square in London. In a 2004 international design competition, Gustafson Porter was selected to rethink the existing 1929 design. The designers explain: "Terraces of colored granite blocks delineate level changes and hint at the geological strata below the Square's surface. Their tapering forms accommodate rows of benches, planters, and water events set around a large flat and unobstructed surface used for markets and city events." Gustafson Porter created new diagonal routes between Chapel Bar and Smithy Row, Long Row and Friar Land leading to Nottingham Castle, giving pedestrians easier access to the middle of the square and the rest of the city. The designers conclude: "Since opening, the square has hosted some of the largest and best attended events ever staged in the city. These have included free concerts, firework displays, an ice rink, fine food fair, and a bulb and flower market. The day to day impact of the new square has been immediate, and it has already become a well-used space at lunchtimes and early evenings."

Eine der ältesten Platzanlagen in Großbritannien, der **OLD MARKET SQUARE** in Nottingham, ist der zweitgrößte Platz in England nach dem Trafalgar Square in London. In einem internationalen Wettbewerb wurden Gustafson Porter 2004 ausgewählt, die von 1929 stammende Konzeption zu überarbeiten. Die Planer erklären dazu: „Terrassen aus farbigen Granitblöcken zeichnen das Gefälle nach und weisen auf die geologischen Schichten unter dem Platz hin. Die sich verjüngenden Formen bieten Platz für Sitzbänke, Pflanztröge und Wasserspiele, die um eine große ebene Freifläche angeordnet sind, die für den Markt und Stadtfeste genutzt werden kann." Das Büro legte neue, diagonale Wege zwischen Chapel Bar und Smithy Row sowie Long Row und Friar Land an, die zum Nottingham Castle führen, sodass Fußgänger einen einfacheren Zugang zur Platzmitte und zur übrigen Stadt erhalten. Die Planer sagen abschließend: „Seit seiner Einweihung haben auf dem Platz große Veranstaltungen mit einem noch nie dagewesenen Besucherandrang stattgefunden, unter anderem Konzerte mit freiem Eintritt, Feuerwerke, eine Eisbahn, ein Delikatessenmarkt und ein Blumenmarkt. Der neue Platz hat sich unmittelbar auf den Alltag ausgewirkt und ist ein beliebter und viel genutzter Freiraum in der Mittagspause und am frühen Abend."

L'une des plus anciennes places de Grande-Bretagne, la **OLD MARKET SQUARE** à Nottingham est aussi la seconde en surface après Trafalgar Square à Londres. En 2004, Gustafson Porter a remporté le concours organisé pour repenser les aménagements de 1929. Les architectes paysagistes expliquent : « Des terrasses en pavés de granit de couleur soulignent les changements de niveaux et rappellent les strates géologiques présentes en sous-sol. Leur forme en biseau s'accompagne d'alignements de bancs, de jardinières et de bassins autour d'un vaste plan vide qui accueille des marchés et divers événements municipaux. » L'agence a également créé des allées en diagonale entre Chapel Bar et Smithy Row, entre Long Row et Friar Land, qui conduisent au château de Nottingham, pour un accès piétonnier plus aisé au cœur de la place et de la ville. Elle conclut : « Depuis son inauguration, la place a accueilli quelques-unes des plus importantes manifestations jamais organisées dans cette ville : concerts gratuits, feux d'artifice, patinoire, foire gastronomique et marché aux fleurs. L'impact de ce projet sur la vie quotidienne a été immédiat et la Old Market Square commence à être très fréquentée à l'heure du déjeuner et en début de soirée. »

As seen in the photos, this page and opposite, taken from a high vantage point or in the plan (left), the Square is essentially mineral, with a relatively small number of trees arrayed toward the periphery of the space.

Wie man auf den Fotos oben und links sowie auf dem Grundriss links sehen kann, ist das bei dieser Platzgestaltung verwendete Material vorwiegend Naturstein, und es wurden nur relativ wenige Bäume am Rand des Platzes gepflanzt.

Comme le montrent ces photos prises d'un point de vue élevé (cette page et page de gauche) ou le plan de gauche, le style de la place est essentiellement minéral. Un nombre d'arbres réduit a été planté en périphérie.

DIANA, PRINCESS OF WALES MEMORIAL FOUNTAIN

Hyde Park, London, UK, 2003–04

*Address: Hyde Park, London W2 2UH, UK, www.royalparks.org.uk/hyde_park/diana_memorial.cfm
Area: 5600 m². Client: Department for Culture Media and Sport. Cost: not disclosed*

Digitally designed texture effects mark the course of the water through the fountain, as seen on the right.

Digital entworfene Oberflächentexturen bestimmen den Lauf des Wassers durch die Brunnenanlage (rechts).

Des effets de texture obtenus par CAO animent l'écoulement de l'eau dans la fontaine (à droite).

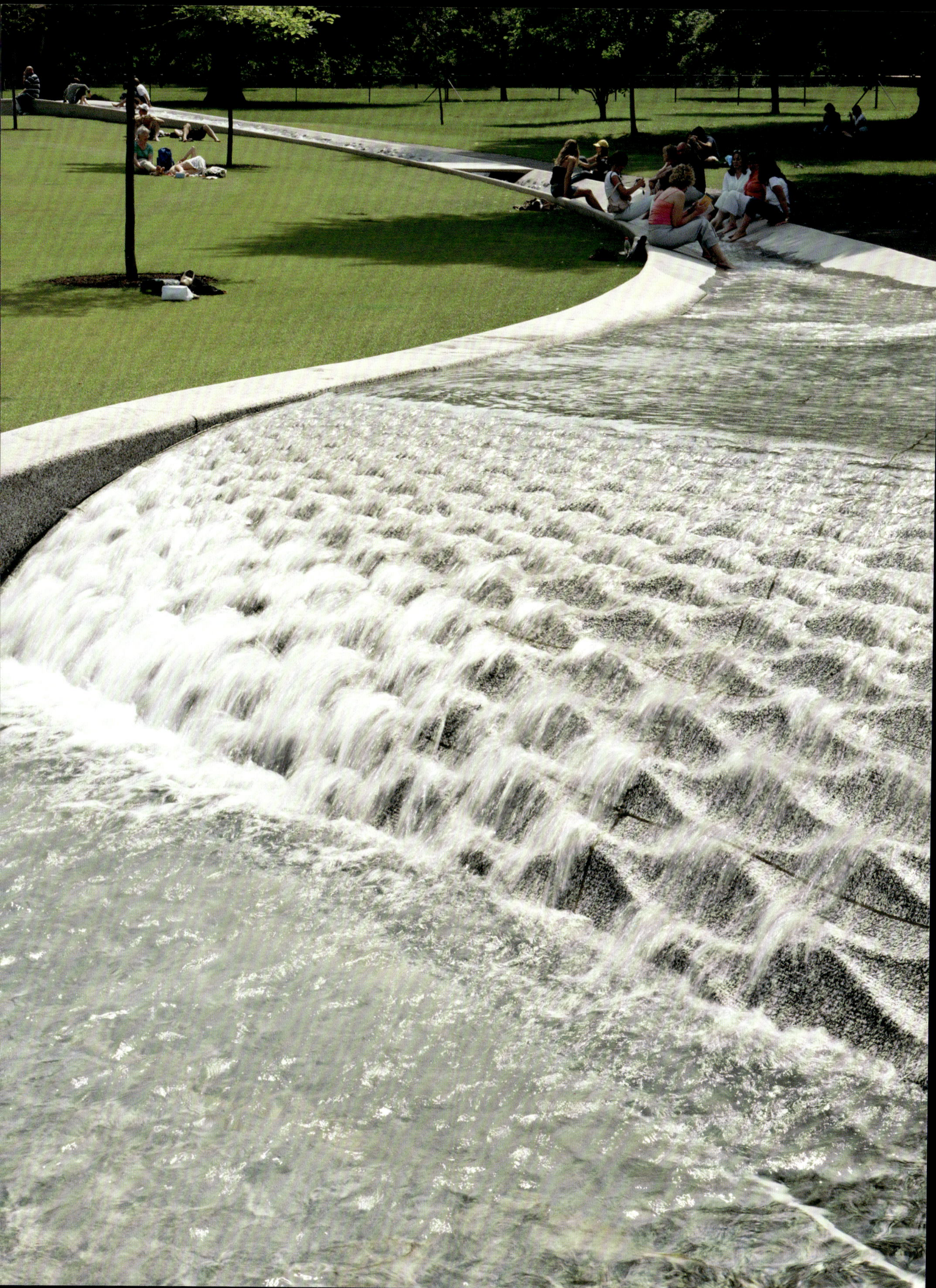

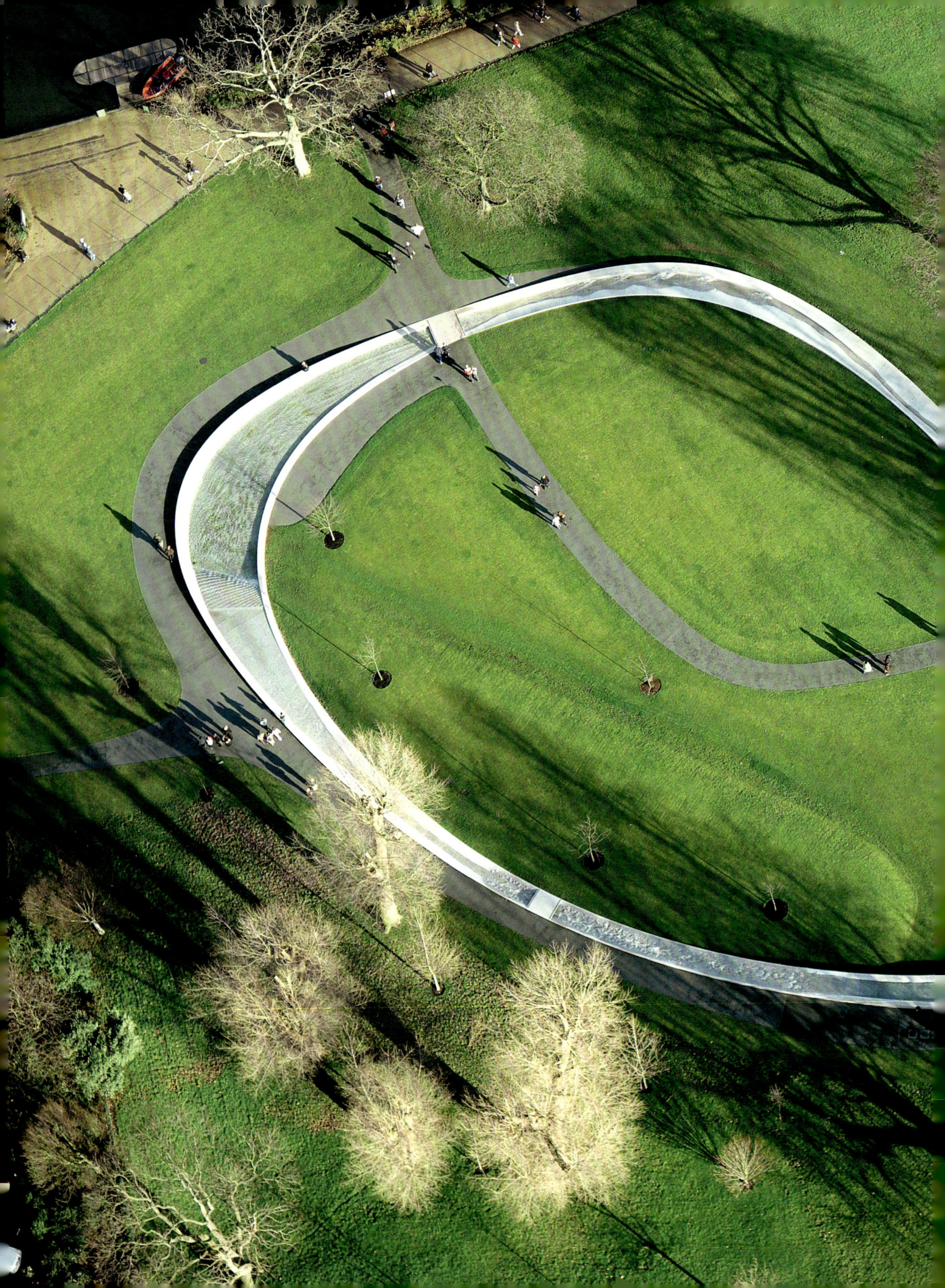

In this aerial image, the gentle, curving form of the fountain with a walkway arcing through the central space is reminiscent of a stylized heart.

Auf dieser Luftaufnahme erkennt man, dass die sanft geschwungene Form der Brunnenanlage, durch deren Mitte ein bogenförmiger Fußweg führt, an ein stilisiertes Herz erinnert.

Sur cette image aérienne, la courbe sophistiquée de la fontaine traversée par une allée rappelle un cœur stylisé.

More than two million visitors have come to the **PRINCESS OF WALES MEMORIAL FOUNTAIN** since it was inaugurated by HM the Queen in 2005. The fountain was integrated into the natural slope of the land in Hyde Park. Computer-generated models developed for the automotive industry were used on the basis of an original clay design. The 600-ton Memorial is made up of 545 unique stones. Digital design was also used for the 230 meters of texture effects seen on the fountain. The designers state: "We believe that this project has pushed the boundaries of landscape design in the United Kingdom and would like to recognize the incredible collaborative effort required to deliver one of the most high-profile landscape projects in the world."

Seit seiner Einweihung durch Königin Elisabeth II. im Jahr 2005 sind mehr als zwei Millionen Besucher zu dem **PRINCESS OF WALES MEMORIAL FOUNTAIN** gekommen. Das Wasserspiel ist in den natürlichen Geländeverlauf des Hyde Park integriert. Auf der Grundlage eines Arbeitsmodells aus Ton wurden verschiedene Modelle mithilfe von Computerprogrammen, die ursprünglich für die Automobilindustrie entwickelt worden sind, erstellt. Das Projekt besteht aus 545 einzeln gefertigten Granitwerksteinen, die zusammen über 600 Tonnen wiegen. Auch die unterschiedlichen Texturen des 230 Meter langen Wasserlaufs wurden digital entworfen. Die Planer bemerken dazu: „Wir glauben, dass mit diesem Projekt die Grenzen der Landschaftsplanung in Großbritannien gesprengt worden sind, und wir möchten auch die unglaubliche, gemeinsame Leistung würdigen, die notwendig war, um eines der weltweit anspruchsvollsten landschaftsarchitektonischen Projekte zu realisieren."

Depuis son inauguration par la reine en 2005, plus de deux millions de visiteurs se sont déjà rendus à la **FONTAINE MÉMORIAL DE LA PRINCESSE DE GALLES**. Le mémorial est inséré dans une légère déclivité naturelle de Hyde Park. À partir d'une maquette originale en terre glaise, des logiciels conçus pour l'industrie automobile ont permis la mise au point technique du projet. Le monument de 600 tonnes a été construit avec 545 pierres, toutes différentes. La CAO a également servi à créer des effets de texture sur les 230 m de l'anneau. « Nous pensons que ce projet a fait bouger les limites de l'aménagement paysager au Royaume-Uni et sommes heureux de saluer l'incroyable travail de collaboration qui a permis de réaliser l'un des plus remarquables projets paysagers au monde », ont déclaré les auteurs de ce mémorial.

On a warm day, visitors sit along the edges of the fountain, set in the midst of a carefully manicured lawn.

An warmen Tagen sitzen die Besucher auf dem Rand des Brunnens, der mitten in eine gepflegten Rasenfläche eingelassen ist.

À la belle saison, les visiteurs s'assoient sur le rebord de la fontaine implantée au centre d'une pelouse soigneusement entretenue.

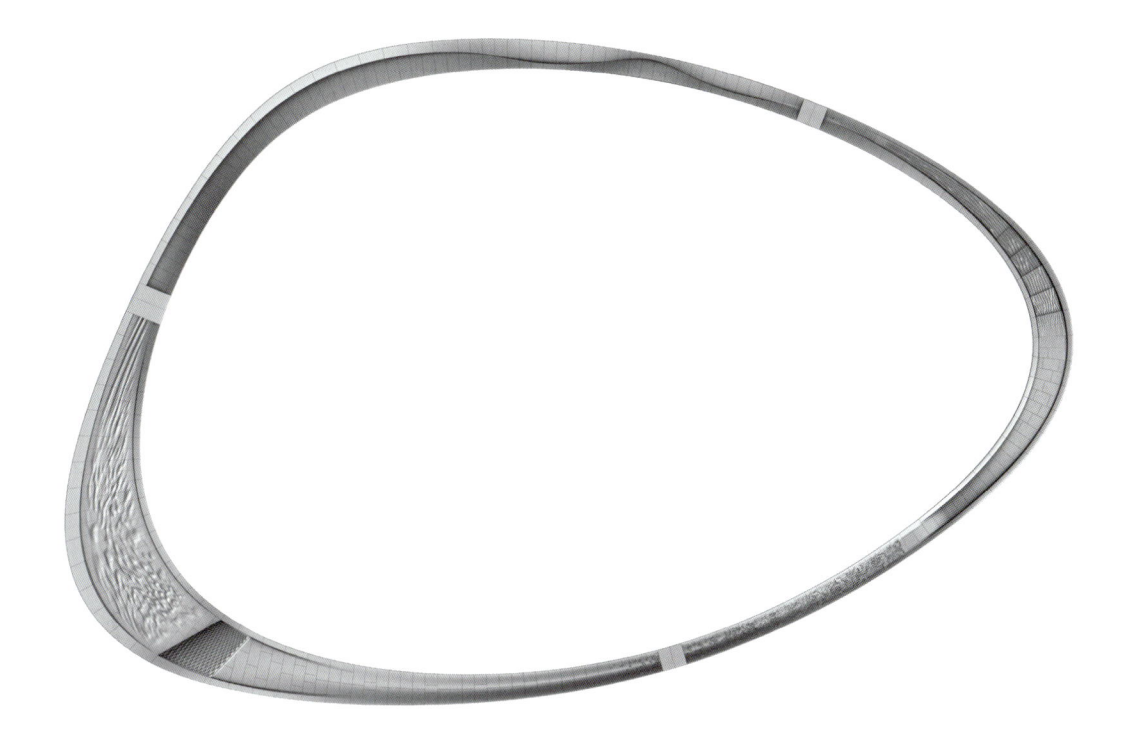

Above a "three-dimensional" drawing of the fountain, seen as a textured loop, the computer-designed texturing is visible both in the image with water and without (below).

Oben eine „dreidimensionale" Zeichnung der Brunnenanlage, die als Schleife mit unterschiedlichen Oberflächen entworfen ist. Die digital entworfenen Texturen kann man unten mit und ohne Wasser sehen.

Ci-dessus, dessin en trois dimensions de la fontaine conçue comme une boucle texturée. Cette texturation conçue par ordinateur est visible dans les deux images ci-dessous, à travers l'eau ou à sec.

Eleftheria Square Redesign

ZAHA HADID

*Zaha Hadid
Studio 9
10 Bowling Green Lane
London EC1R OBQ, UK*

*Tel: +44 207 253 51 47 / Fax: +44 207 251 83 22
E-mail: mail@zaha-hadid.com / Web: www.zaha-hadid.com*

ZAHA HADID studied architecture at the Architectural Association (AA) in London beginning in 1972 and was awarded the Diploma Prize in 1977. She then became a Partner of Rem Koolhaas in OMA and taught at the AA. In 2004, Zaha Hadid became the first woman to win the coveted Pritzker Prize. Well-known for her paintings and drawings, she had a substantial influence, even before beginning to build very much. She completed the Vitra Fire Station (Weil am Rhein, Germany, 1990–94); and exhibition designs such as that for "The Great Utopia" (Solomon R. Guggenheim Museum, New York, 1992). More recently, Zaha Hadid has entered a phase of active construction with such projects as the Bergisel Ski Jump (Innsbruck, Austria, 2001–02); Lois & Richard Rosenthal Center for Contemporary Art (Cincinnati, Ohio, USA, 1999–2003); Phæno Science Center (Wolfsburg, Germany, 2001–05); the Central Building of the new BMW Assembly Plant in Leipzig (Germany, 2005); Ordrupgaard Museum Extension (Copenhagen, Denmark, 2001–05); Lopez de Heredia Wine Pavilion (Haro, Spain, 2001–06); Mobile Art, Chanel Contemporary Art Container (various locations, 2007–); Home House (London, UK, 2007–08); Atelier Notify (Paris, France, 2007–08); and the Neil Barrett Flagship Store (Tokyo, Japan, 2008). She recently completed the MAXXI, the National Museum of 21st Century Arts (Rome, Italy, 1998–2009); Burnham Pavilion (Chicago, Illinois, USA, 2009); and the JS Bach / Zaha Hadid Architects Music Hall (Manchester, UK, 2009). Current projects include the Eleftheria Square Redesign published here (Nicosia, Cyprus, 2005–); the Guangzhou Opera House (Guangzhou, China, 2006–10); and the Sheik Zayed Bridge (Abu Dhabi, UAE, 2005–11).

ZAHA HADID begann 1972 ihr Studium an der Architectural Association (AA) in London und machte 1977 ihr Diplom mit Auszeichnung. Danach arbeitete sie als Partnerin von Rem Koolhaas bei OMA und lehrte an der AA. 2004 erhielt Zaha Hadid als erste Frau den begehrten Pritzker-Preis. Sie war wegen ihrer Malereien und Zeichnungen bekannt und beeinflusste die Architekturszene, noch bevor sie viel bauen konnte. Sie plante die Vitra-Feuerwehrstation (Weil am Rhein, 1990–94) und entwarf Ausstellungsdesign wie zum Beispiel für „The Great Utopia" (Solomon R. Guggenheim Museum, New York, 1992). In neuerer Zeit begann für Zaha Hadid eine Phase aktiven Bauens mit Projekten wie der Bergisel-Schanze (Innsbruck, Österreich, 2001–02), dem Lois & Richard Rosenthal Center for Contemporary Art (Cincinnati, Ohio, USA, 1999–2003), dem Phæno-Wissenschaftszentrum (Wolfsburg, 2001–05), dem Zentralgebäude des BMW-Werks Leipzig (2005), der Erweiterung des Ordrupgaard-Museums (Kopenhagen, 2001–05), dem Lopez-de-Heredia-Weinpavillon (Haro, Spanien, 2001–06), dem Mobile-Art-Pavillon für Chanel, einem mobilen Ausstellungspavillon (unterschiedliche Ausstellungsorte, seit 2007), Home House (London, 2007–08), dem Atelier Notify (Paris, 2007–08) und dem Neil Barrett Flagshipstore (Tokio, 2008). Die neuesten von ihr realisierten Gebäude sind das MAXXI, das Nationalmuseum für Kunst des 21. Jahrhunderts (Rom, 1998–2009), der Burnham Pavilion (Chicago, Illinois, USA, 2009) und die JS Bach/Zaha Hadid Architects Music Hall (Manchester, GB, 2009). Zu den laufenden Projekten gehören die Neugestaltung des Eleftheria-Platzes (Nicosia, Zypern, seit 2005, hier vorgestellt), das Opernhaus von Guangzhou (China, 2006–10) und die Sheik-Zayed-Brücke (Abu Dhabi, VAE, 2005–11).

ZAHA HADID a fait ses études d'architecture à l'Architectural Association (AA) de Londres de 1972 à 1977, année d'obtention de son diplôme, salué par le Prix du diplôme. Elle est ensuite devenue partenaire de Rem Koolhaas à l'OMA et a enseigné à l'AA. En 2004, elle a été la première femme à remporter le prix Pritzker. Connue pour ses tableaux et ses dessins, elle avait déjà une influence considérable avant de se mettre à construire. Parmi ses réalisations figurent le poste d'incendie de l'usine Vitra (Weil-am-Rhein, Allemagne, 1990–94) et des projets pour des expositions comme « The Great Utopia » au Solomon R. Guggenheim Museum à New York (1992). Elle a également à son actif le tremplin de saut à ski de Bergisel (Innsbruck, Autriche, 2001–02) ; le Centre d'art contemporain Lois & Richard Rosenthal (Cincinnati, Ohio, US, 1999–2003) ; le Centre des sciences Phæno (Wolfsburg, Allemagne, 2001–05) ; le bâtiment central de la nouvelle usine BMW de Leipzig (Allemagne, 2005) ; l'extension du musée Ordrupgaard (Copenhague, Danemark, 2001–05) ; la boutique des caves López de Heredia (Haro, Espagne, 2001–06) ; le Mobile Art, Chanel Contemporary Art Container (divers lieux depuis 2007, aujourd'hui à Paris) ; le club Home House (Londres, 2007–08) ; l'Atelier Notify (Paris, 2007–08) et le magasin amiral Neil Barrett (Tokyo, 2008). Elle a récemment achevé le MAXXI, Musée national des arts du XXIe siècle (Rome, 1998–2009) ; son Burnham Pavilion (Chicago, Illinois, US, 2009) et le JS Bach/Zaha Hadid Architects Music Hall (Manchester, GB, 2009). Ses projets en cours incluent la restructuration de la place Eleftheria (Nicosie, Chypre, 2005–, publiée ici) ; l'Opéra de Canton (Canton, Chine, 2006–10) et le pont Sheik Zayed (Abou Dhabi, EAU, 2005–11).

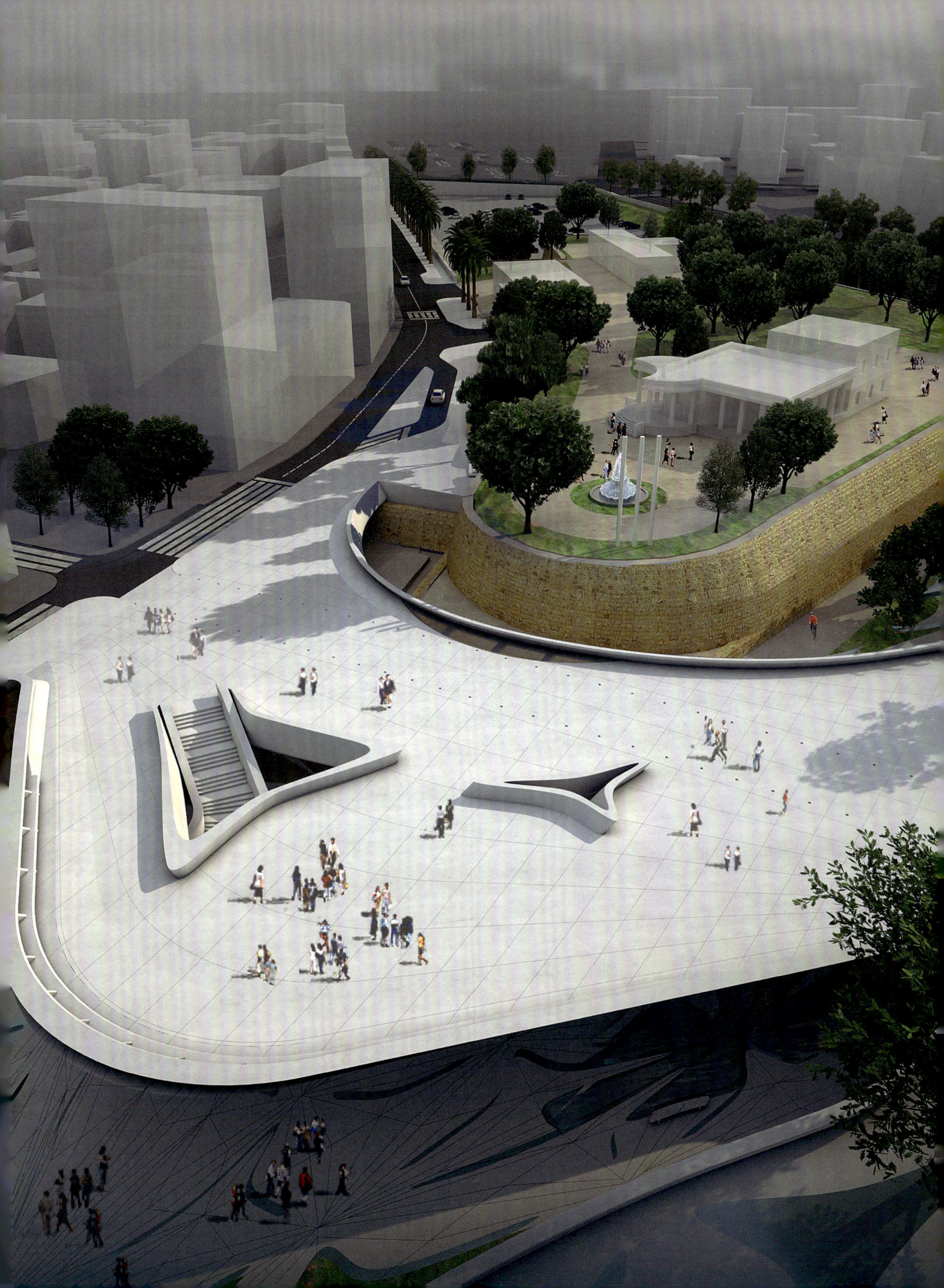

ELEFTHERIA SQUARE REDESIGN
Nicosia, Cyprus, 2005–

Area: 20 000 m². Client: City of Nicosia. Cost: not disclosed
Collaboration: Christos Passas (Project Director)

In this project, the architect takes on not only a complex site, but one that is charged with history, old and new. As they explain the situation: "On one hand, the presence of the massive Venetian fortification, which was designed to protect the city from invaders, de facto defines the extent of the ancient city and separates it from the modern city outside the walls. On the other hand, the 'Green Line' separates the city's two communities making it the last divided capital of Europe. The significance of the intervention at **ELEFTHERIA SQUARE** lies in the fact that it can become a catalyst, an opportunity for the urban unification of the whole of Nicosia. An urban intervention that would be architecturally coherent and continuous and that would reinstate the Venetian Wall as a main part of the identity of the capital." Making use of a moat that becomes a "green belt" and a main park for the city, Zaha Hadid proposed to reduce car traffic, placing parking lots beneath the green areas. A continuous line of palm trees is proposed to delineate a walking path near the old Venetian walls of the city. On hold for some time for financial reasons, the project looked set to advance as this book went to print.

Bei diesem Projekt geht es nicht nur um ein komplexes Gelände, auf das sich die Architektin einlassen musste, sondern auch um einen geschichtsträchtigen Ort, der wie folgt beschrieben wird: „Auf der einen Seite ist da die starke Präsenz der massiven venezianischen Festungsanlage, die zum Schutz der Stadt gegen Invasoren errichtet wurde und die de facto die Grenze der Altstadt bildet und sie von der modernen Stadt außerhalb der Mauern trennt. Auf der anderen Seite ist da die ‚Green Line', die die beiden Bevölkerungsgruppen trennt und Nikosia zur letzten geteilten Hauptstadt Europas macht. Die Neugestaltung des **ELEFTHERIA-PLATZES** ist deshalb so bedeutend, weil er zum Katalysator werden kann und damit eine Möglichkeit der Wiedervereinigung der Stadt darstellt. Ein städtebaulicher Eingriff, der architektonisch kohärent und durchgängig ist und der venezianischen Stadtmauer wieder ihre Rolle als Hauptidentitätsmerkmal der Hauptstadt zurückgeben kann." In ihrem Entwurf schlägt Zaha Hadid vor, den alten Stadtgraben zu einem Grüngürtel umzugestalten und ihn zu einem wichtigen Park für die Stadt zu machen. Um den Verkehr zu reduzieren, werden unter den Grünflächen Tiefgaragen gebaut. Eine Palmenreihe soll einen Spazierweg an der alten venezianischen Stadtmauer entlang begleiten. Das Projekt musste aus finanziellen Gründen eine Weile gestoppt werden, aber als das vorliegende Buch in den Druck ging, schien es, als könnten die Arbeiten fortgesetzt werden.

Pour ce projet, l'architecte était confrontée non seulement à un site complexe, mais aussi à un lieu chargé d'histoire, ancienne et récente : « D'un côté, les fortifications vénitiennes massives, édifiées pour protéger la ville des envahisseurs, délimitent *de facto* la cité ancienne et la séparent de la ville moderne hors les murs ; de l'autre, la "ligne verte" divise la ville et en fait la dernière capitale d'Europe coupée en deux. Le sens de notre intervention repose sur l'idée que la **PLACE ELEFTHERIA** peut devenir un catalyseur, offrir l'opportunité d'une réunification urbaine de l'ensemble de Nicosie. Cette intervention doit être cohérente et continue sur le plan architectural et redonner à la muraille vénitienne un rôle essentiel dans l'identité de la capitale », explique l'agence. Zaha Hadid a proposé de transformer une douve en « ceinture verte » et en parc, de réduire la circulation automobile et de construire des parkings sous les espaces plantés. Un alignement continu de palmiers soulignera la promenade le long des remparts vénitiens. Retardé un certain temps pour des raisons financières, le projet semble avoir progressé au moment de l'impression de cet ouvrage.

Zaha Hadid has had a career-long interest in the integration of her architecture with the landscaped environment and this is clearly the case in this instance, with paths and lines of force flowing near the old Venetian wall.

Zaha Hadid hat sich immer darum bemüht, ihre Architektur in die umgebende Landschaft zu integrieren. Das erkennt man auch deutlich an diesem Entwurf mit seinen Wegen und Freiflächen, die sich nahe der alten venezianischen Stadtmauer befinden.

Zaha Hadid s'intéresse depuis ses débuts à l'intégration de l'architecture dans l'environnement paysager. C'est clairement le cas ici, dans les cheminements et les lignes de force qui partent de l'ancienne muraille vénitienne.

Plaza de España

HERZOG & DE MEURON

Herzog & de Meuron, Rheinschanze 6, 4056 Basel, Switzerland
Tel: +41 61 385 57 57 / Fax: +41 61 385 57 58
E-mail: info@herzogdemeuron.com

JACQUES HERZOG and **PIERRE DE MEURON** were both born in Basel, Switzerland, in 1950. They received degrees in architecture from the ETH, Zurich, in 1975, after studying with Aldo Rossi, and founded their partnership in Basel in 1978. The partnership has grown over the years: Christine Binswanger joined the practice in 1994, successively followed by Ascan Mergenthaler in 2004 and Stefan Marbach in 2006. Herzog & de Meuron won the 2001 Pritzker Prize, and both the RIBA Gold Medal and Praemium Imperiale in 2007. They were chosen to design Tate Modern in London (1994–2000), and, in 2005, Herzog & de Meuron were commissioned by the Tate to develop a scheme for the extension of the gallery and its surrounding areas: the Tate Modern Project. More recently, they have built the Fórum 2004 Building and Plaza (Barcelona, 2000–04); the de Young Museum (San Francisco, California, USA, 1999–2005); the Walker Art Center, Expansion of the Museum and Cultural Center (Minneapolis, Minnesota, USA, 1999–2005); Allianz Arena (Munich, Germany, 2001–05); the CaixaForum (Madrid, 2001–08); TEA, Tenerife Espacio de las Artes (Santa Cruz de Tenerife, Canary Islands, Spain, 1999–2008); the National Stadium for the 2008 Olympic Games in Beijing (2002–08); and Plaza de España (Santa Cruz de Tenerife, Canary Islands, Spain, 2006–08, published here). Among their latest works are VitraHaus, a new building to present Vitra's "Home Collection" on the Vitra Campus in Weil am Rhein (Germany, 2006–09); and 1111 Lincoln Road, a mixed-use parking facility in Miami (Florida, USA, 2005–10; page 192). Projects currently under construction include the Elbphilharmonie in Hamburg (Germany); a new building for Roche in Basel, Roche Building 1 (Basel, Switzerland); and the new Miami Art Museum (Florida, USA).

JACQUES HERZOG und **PIERRE DE MEURON** wurden beide 1950 in Basel geboren. Sie studierten bei Aldo Rossi und machten 1975 ihr Diplom in Architektur an der ETH Zürich. 1978 gründeten sie ein Büro in Basel. Über die Jahre sind neue Partner in das Büro eingestiegen, so kam 1994 Christine Binswanger hinzu, danach folgten 2004 Ascan Mergenthaler und 2006 Stefan Marbach. Herzog & de Meuron erhielten 2001 den Pritzker-Preis und 2007 sowohl die RIBA-Goldmedaille als auch den Praemium Imperiale. Sie entwarfen die Tate Modern in London (1994–2000), und 2005 wurden sie von der Tate beauftragt einen Entwurf für eine Erweiterung des Museums und der umliegenden Bereiche vorzulegen: das Tate Modern Extension Project. Zu ihren neueren Arbeiten gehören das Gebäude und die Platzanlage Fórum 2004 (Barcelona, 2000–04), das de Young Museum (San Francisco, Kalifornien, USA, 1999–2005), das Walker Art Center, Erweiterungsbau des Musuems und des Kulturzentrums (Minneapolis, Minnesota, USA, 1999–2005), die Allianz Arena (München, 2001–05), das CaixaForum (Madrid, 2001–08), TEA, Tenerife Espacio de las Artes (Santa Cruz de Tenerife, Kanarische Inseln, Spanien, 1999–2008), das Nationalstadion für die Olympischen Sommerspiele 2008 in Peking (2002–08) und die Plaza de España (Santa Cruz de Tenerife, 2006–08, hier vorgestellt). Zu ihren neuesten ausgeführten Projekten gehören das VitraHaus, ein neues Gebäude zur Präsentation der „Home Collection" von Vitra in Weil am Rhein (2006–09), und 1111 Lincoln Road, ein Parkhaus mit Mischnutzung in Miami (Florida, USA, 2005–10, Seite 192). Zu den derzeit im Bau befindlichen Projekten gehören die Elbphilharmonie in Hamburg, ein Neubau für das Pharmaunternehmen Roche in Basel, Roche Bau 1, und das neue Miami Art Museum (Florida, USA).

JACQUES HERZOG et **PIERRE DE MEURON**, tous deux nés à Bâle en 1950, sont diplômés en architecture de l'ETH de Zurich (1975), où ils ont étudié auprès d'Aldo Rossi. En 1978, ils fondent leur agence à Bâle. Christine Binswanger les rejoint en 1994, suivie d'Ascan Mergenthaler en 2004 et de Stefan Marbach en 2006. Ils ont remporté le prix Pritzker en 2001, la médaille d'or du RIBA ainsi que le Praemium Imperiale en 2007. Ils ont conçu la Tate Modern de Londres (1994–2000) pour laquelle, en 2005, ils ont présenté un projet d'extension du musée et d'aménagement de son environnement : le Tate Modern Project. Parmi leurs autres réalisations : l'immeuble et la place du Fórum 2004 (Barcelone, 2000–04) ; le de Young Museum (San Francisco, Californie, 1999–2005) ; l'extension du musée et centre culturel Walker Art Center (Minneapolis, Minnesota, US, 1999–2005) ; l'Allianz Arena (Munich, Allemagne, 2001–05) ; le CaixaForum (Madrid, 2001–08) ; le TEA, Tenerife Espacio de las Artes (Santa Cruz de Tenerife, îles Canaries, Espagne, 1999–2008) ; le stade national pour les Jeux olympiques de Pékin 2008 (2002–08) et la Plaza de España (Santa Cruz de Tenerife, îles Canaries, Espagne, 2006–08, publiée ici). Plus récemment, ils ont achevé la VitraHaus, un nouveau bâtiment pour présenter la collection « Home » de Vitra sur le Vitra Campus (Weil-am-Rhein, Allemagne, 2006–09) et le 1111 Lincoln Road, un immeuble mixte à Miami (Floride, US, 2005–10, page 192). Parmi leurs projets actuels figurent l'Elbphilharmonie à Hambourg (Allemagne) ; un nouvel immeuble pour le groupe Roche à Bâle, Roche Building 1 et le nouveau Miami Art Museum (Floride, US).

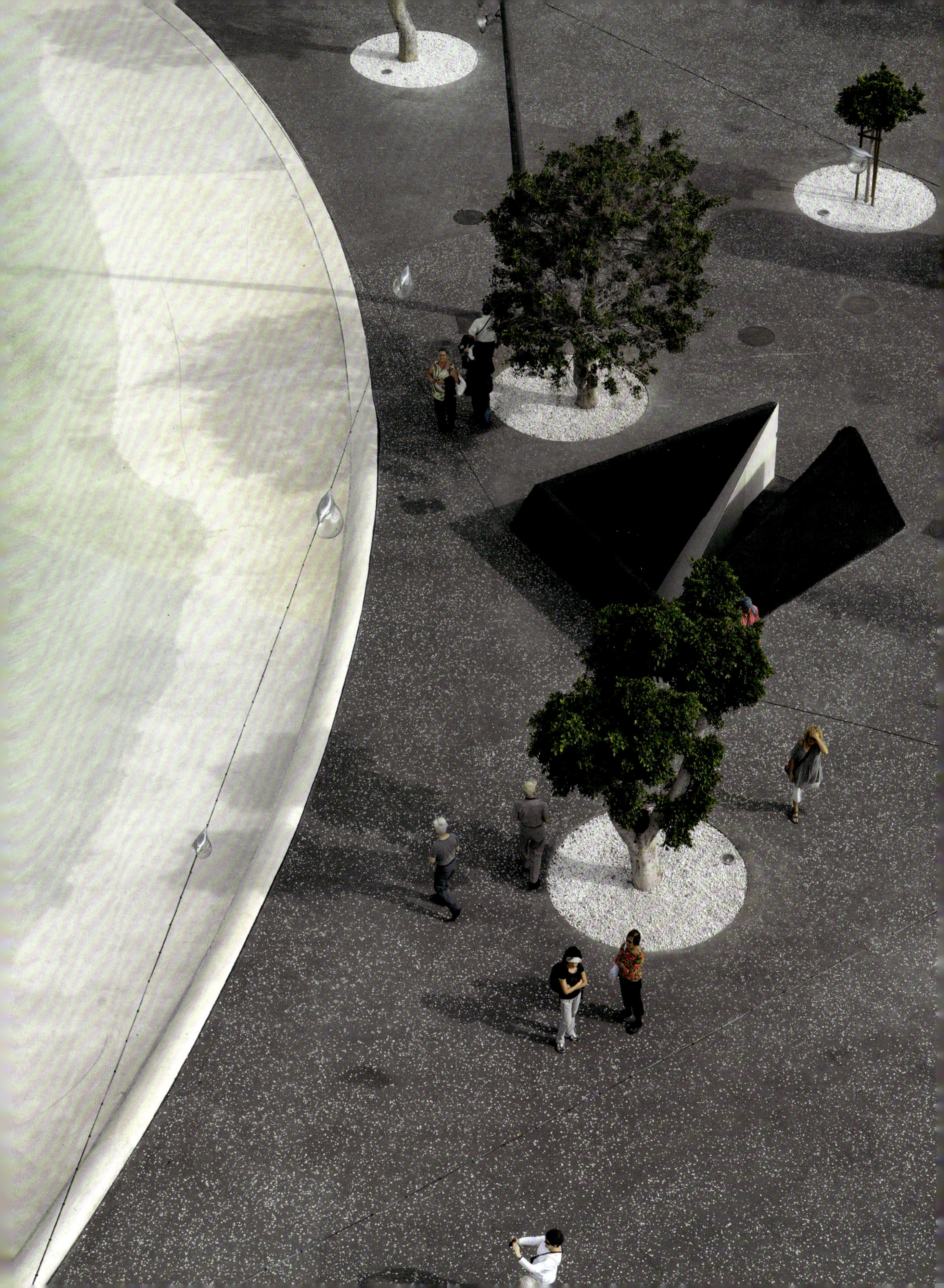

PLAZA DE ESPAÑA

Santa Cruz de Tenerife, Canary Islands, Spain, 2006–08

Area: 38 080 m² (Plaza). Client: Cabildo Insular de Tenerife
Cost: not disclosed

The architects conceived the remodeled **PLAZA DE ESPAÑA** and the new Plaza de las Islas Canarias "as a single connected public space that functions on two superposed levels: a lower level for heavy traffic, the projected railroad, and car parks; and an upper level that links the pedestrian zone of the city with the new marina and the sea." Above all, they have imagined this square as an integral part of the particular geological and historic realities of Tenerife. They state: "The new Plaza de España and the Plaza de las Islas Canarias will form a crust of land on top of the subterranean flow of traffic, extending to the marina and the sea. Above ground this crust will have the appearance of artificial nature, resembling a lava flow or a sprawling beach: the new beach of Santa Cruz with many urban facilities and amenities, including cafés, bars, kiosks, bus shelters, meeting places, as well as storage areas for the large celebrations in the city such as Carnival or the Christmas concert." The large round basin in the plaza has a fountain that operates four times a day in conjunction with low and high tide.

Die Architekten haben die neu gestaltete **PLAZA DE ESPAÑA** und die neue Plaza de las Islas Canarias „als einen einzigen, zusammenhängenden öffentlichen Freiraum konzipiert, dessen Funktionen sich auf zwei übereinandergelagerten Ebenen verteilen: die untere Ebene ist für Verkehrsbauwerke reserviert, unter anderem für die geplante Eisenbahnlinie und für die Tiefgarage, die obere Ebene bildet die Verbindung zwischen der innerstädtischen Fußgängerzone und dem neuen Jachthafen und dem Meeresufer." Herzog & de Meuron haben sich diesen Platz vor allem als Teil der geologischen und historischen Gegebenheiten von Teneriffa gedacht. Sie kommentieren: „Die neue Plaza de España und die Plaza de las Islas Canarias sollen ein Stück Erde oberhalb des unterirdisch fließenden Verkehrs bilden, das sich in Richtung Jachthafen und Meer fortsetzt. Da dieses Stück Land vom Untergrund abgehoben ist, hat es etwas Künstliches, es erinnert an fließende Lava oder einen breiten Strand: Der neue Strand von Santa Cruz mit zahlreichen städtischen Einrichtungen und Annehmlichkeiten, wie zum Beispiel Cafés, Bars, Kiosken, Bushaltestellen und Versammlungsorten bietet Raum für große Veranstaltungen, wie zum Beispiel den Karneval oder das Weihnachtskonzert." Der Springbrunnen in dem großen runden Wasserbassin auf dem Platz ist an die Gezeiten von Ebbe und Flut gebunden und sprudelt viermal am Tag.

Les architectes ont conçu cette restructuration de la **PLAZA DE ESPAÑA** et la nouvelle Plaza de las Islas Canarias « comme un espace connecté unique fonctionnant sur deux niveaux superposés : un niveau inférieur pour la circulation lourde, la voie ferrée projetée et des parkings, et un niveau supérieur qui relie la zone piétonnière à la mer et à la nouvelle marina ». Mais par dessus tout, ils ont pensé cette place en tant que partie intégrante des réalités historiques et géologiques particulières de Tenerife : « La nouvelle Plaza de España et la Plaza de las Islas Canarias formeront une croûte de terre au-dessus du flux souterrain de la circulation jusqu'à la marina et à l'océan. Au niveau du sol, cette croûte aura l'aspect d'une nature artificielle ressemblant à une coulée de lave ou à une vaste plage : la nouvelle plage de Santa Cruz, qui offrira de nombreux équipement urbains et lieux d'accueil, dont des cafés, des bars, des kiosques, des abris de bus, des lieux de rencontre, mais aussi des aires d'entreposage pour les grandes fêtes organisées par la ville, comme le carnaval ou le concert de Noël. » Le vaste bassin central est doté d'une fontaine qui fonctionne quatre fois par jour, lorsque la marée est haute ou basse.

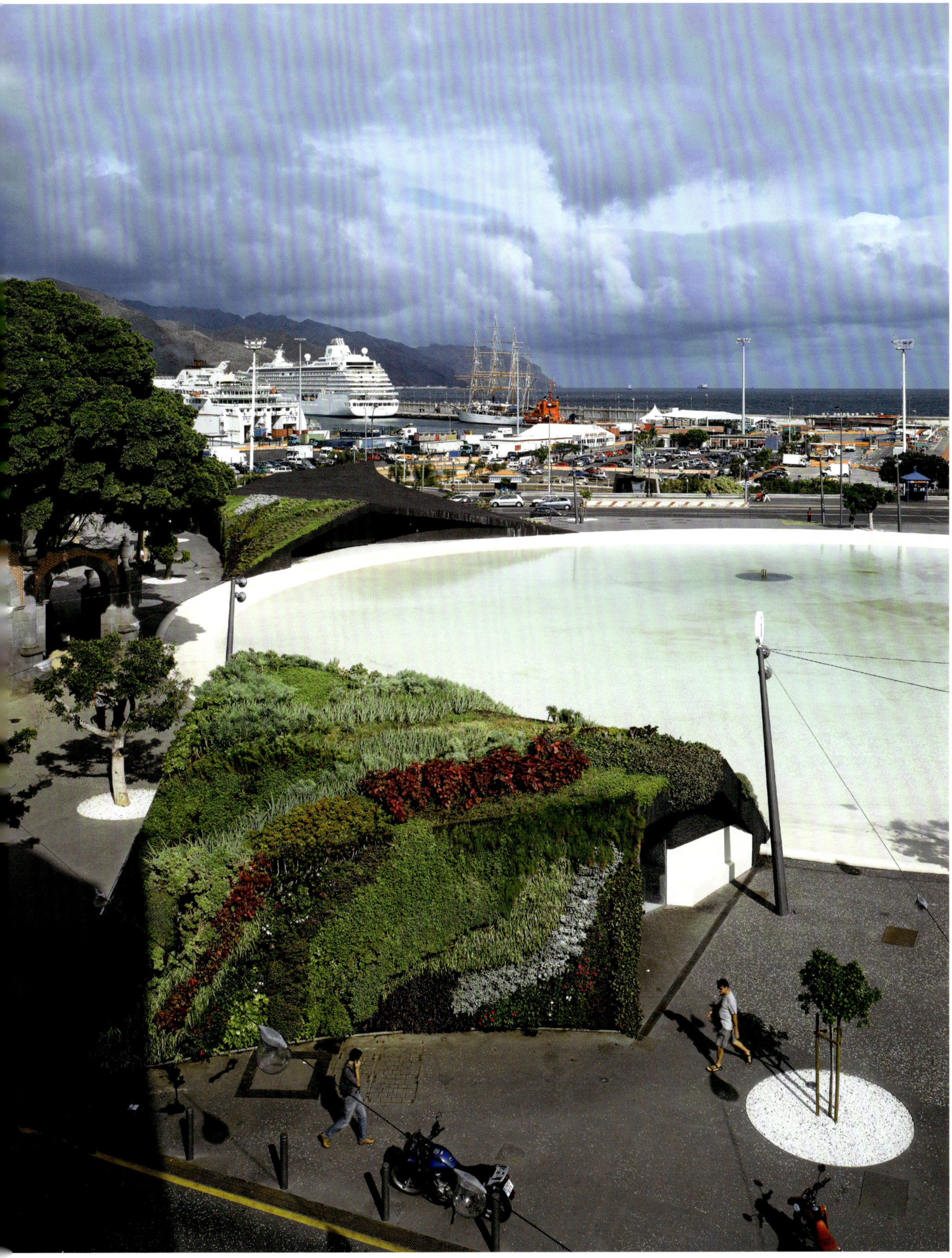

In plan (above) and in an image, the square appears to rotate around the round basin, with an almost cosmological pattern of planting and lighting dispersed in an irregular pattern around this organizing element.

Im Grundriss (oben) und auf dem Foto scheint sich der Platz mit seinem fast kosmologischen Muster aus Pflanzen und unregelmäßig verteilten Beleuchtungskörpern um das kreisrunde Wasserbecken zu drehen, dem organisierenden Grundelement dieses Platzes.

En plan (ci-dessus) comme en image, la place semble s'articuler autour de l'immense bassin rond en un motif quasi cosmique de plantations et de luminaires disposés sur un mode aléatoire autour d'un élément central organisateur.

The French botanist Patrick Blanc collaborated with the Swiss architects for the vegetal walls and roofs of the pavilions arrayed near the round basin.

Der französische Botaniker Patrick Blanc arbeitete mit den Schweizer Architekten bei der Gestaltung der bewachsenen Mauern und Dächer der Pavillons zusammen, die in der Nähe des runden Wasserbeckens stehen.

Le botaniste français Patrick Blanc a collaboré avec les architectes suisses pour la végétalisation des murs et des toitures des pavillons répartis autour du bassin rond.

Vanke Center / Horizontal Skyscraper

STEVEN HOLL

*Steven Holl Architects, P.C.
450 West 31st Street, 11th floor
New York, NY 10001, USA*

*Tel: +1 212 629 7262 / Fax: +1 212 629 7312
E-mail: nyc@stevenholl.com / Web: www.stevenholl.com*

Born in 1947 in Bremerton, Washington, **STEVEN HOLL** obtained his B.Arch degree from the University of Washington (1970). He studied in Rome and at the Architectural Association in London (1976). He opened his own office in New York in 1976. Holl has taught at the University of Washington, Syracuse University, and, since 1981, at Columbia University. His notable buildings include Void Space / Hinged Space, Housing (Nexus World, Fukuoka, Japan, 1991); Stretto House (Dallas, Texas, 1992); Chapel of Saint Ignatius, Seattle University (Seattle, Washington, 1997); and an extension to the Cranbrook Institute of Science (Bloomfield Hills, Michigan, 1999). Winner of the 1998 Alvar Aalto Medal, Steven Holl's more recent work includes the Turbulence House in New Mexico for the artist Richard Tuttle (2005); the Pratt Institute Higgins Hall Center Insertion (Brooklyn, New York, 2005); and the New Residence at the Swiss Embassy (Washington, D.C., 2006) all in the USA unless stated otherwise. He recently won the competition (2009) for the Glasgow School of Art (Glasgow, UK), and completed an expansion and renovation of the Nelson-Atkins Museum of Art (Kansas City, Missouri, USA, 1999–2007); Linked Hybrid (Beijing, 2003–09); the Knut Hamsun Center (Hamarøy, Norway, 2006–09); HEART: Herning Museum of Contemporary Art (Herning, Denmark, 2007–09, published here); and the Vanke Center / Horizontal Skyscraper (Shenzhen, China, 2008–09, also published here), which won the 2011 AIA Honor Award for Architecture. Current projects include Cité de l'Océan et du Surf (Biarritz, France, 2005–10, with Solange Fabião); the Nanjing Museum of Art and Architecture (China, 2008–10); Shan-Shui Hangzhou (master plan, Hangzhou, China, 2010–); and the Hangzhou Music Museum (Hangzhou, China, 2010–).

STEVEN HOLL wurde 1947 in Bremerton, Washington, geboren. 1970 schloss er sein Studium an der University of Washington mit dem B. Arch. ab. Danach studierte er in Rom und an der Architectural Association in London (1976). 1976 gründete er sein Büro in New York. Holl lehrte an der University of Washington und der Syracuse University sowie seit 1981 an der Columbia University. Zu seinen Projekten zählen die Wohnanlage Void Space/Hinged Space (Nexus World, Fukuoka, Japan, 1991), das Stretto House (Dallas, Texas, 1992), die Kapelle St. Ignatius, Seattle University (Seattle, Washington, 1997), und ein Erweiterungsbau des Cranbrook Institute of Science (Bloomfield Hills, Michigan, 1999). 1998 wurde Holl mit der Alvar-Aalto-Medaille ausgezeichnet. Zu seinen neueren Bauwerken gehören das Turbulence House in New Mexico für den Künstler Richard Tuttle (2005), die Erweiterung des Pratt Institute Higgins Hall Center (Brooklyn, New York, 2005) und die neue Residenz der Schweizer Botschaft (Washington, D. C., 2006). 2009 gewann er den Wettbewerb für die Erweiterung der Glasgow School of Art (Glasgow, Schottland), und er vollendete einen Anbau und die Renovierung des Nelson-Atkins Museum of Art (Kansas City, Missouri, USA, 1999–2007), außerdem die Wohnanlage Linked Hybrid (Peking, 2003 bis 2009), das Knut Hamsun Center (Hamarøy, Norwegen, 2006–09), das HEART: Herning Museum of Contemporary Art (Herning, Dänemark, 2007–09, hier vorgestellt) und das Vanke Center/Horizontal Skyscraper (Shenzhen, China, 2008–09, ebenfalls hier vorgestellt), ein Projekt, für das er 2011 den Ehrenpreis für Architektur des American Institute of Architects (AIA) erhielt. Zu seinen laufenden Projekten zählen die Cité de l'Océan et du Surf (Biarritz, Frankreich, 2005–10, mit Solange Fabião), das Nanjing Museum of Art and Architecture (China, 2008–10), der Masterplan für Shan-Shui Hangzhou (China, seit 2010) und das Hangzhou Music Museum (seit 2010).

Né en 1947 à Bremerton (Washington), **STEVEN HOLL** a obtenu son B.Arch. à l'université de Washington (1970) et étudié à Rome et à l'AA à Londres (1976). Il ouvre une agence à New York en 1976. Il a enseigné à l'université de Washington, à l'université de Syracuse et, depuis 1981, à l'université Columbia. Parmi ses réalisations les plus notables : les immeubles de logements Void Space/Hinged Space (Nexus World, Fukuoka, Japon, 1991) ; la Stretto House (Dallas, Texas, US, 1992) ; la chapelle de Saint-Ignace, université de Seattle (Seattle, Washington, 1997) ; et une extension du Cranbrook Institute of Science (Bloomfield Hills, Michigan, US, 1999). Il est titulaire de la médaille Alvar Aalto 1998. Parmi ses réalisations plus récentes : la Turbulence House pour l'artiste Richard Tuttle (Nouveau-Mexique, US, 2005) ; le Higgins Hall du Pratt Institute (Brooklyn, New York, 2005) et la nouvelle résidence de l'ambassade de Suisse (Washington, 2006). Il a récemment remporté le concours (2009) pour la Glasgow School of Art (Glasgow, GB) et a achevé une extension et la rénovation du Nelson-Atkins Museum of Art (Kansas City, Missouri, US, 1999–2007). Il a aussi réalisé le complexe résidentiel Linked Hybrid (Pékin, 2003–09) ; le Centre Knut Hamsun (Hamarøy, Norvège, 2006–09) ; le HEART Herning Museum of Contemporary Art (Herning, Danemark, 2007–09, publié ici) et le Vanke Center–Horizontal Skyscraper (Shenzhen, Chine, 2008–09, publié ici), qui lui a valu le prix d'honneur d'architecture de l'AIA en 2011. Plus récemment : la Cité de l'océan et du surf (Biarritz, 2005–10, avec Solange Fabião) ; Musée d'art et d'architecture de Nankin (Chine, 2008–10) ; le plan directeur de Shan-Shui Hangzhou (Hangzhou, Chine, 2010–) et le Musée de la musique de Hangzhou (2010–).

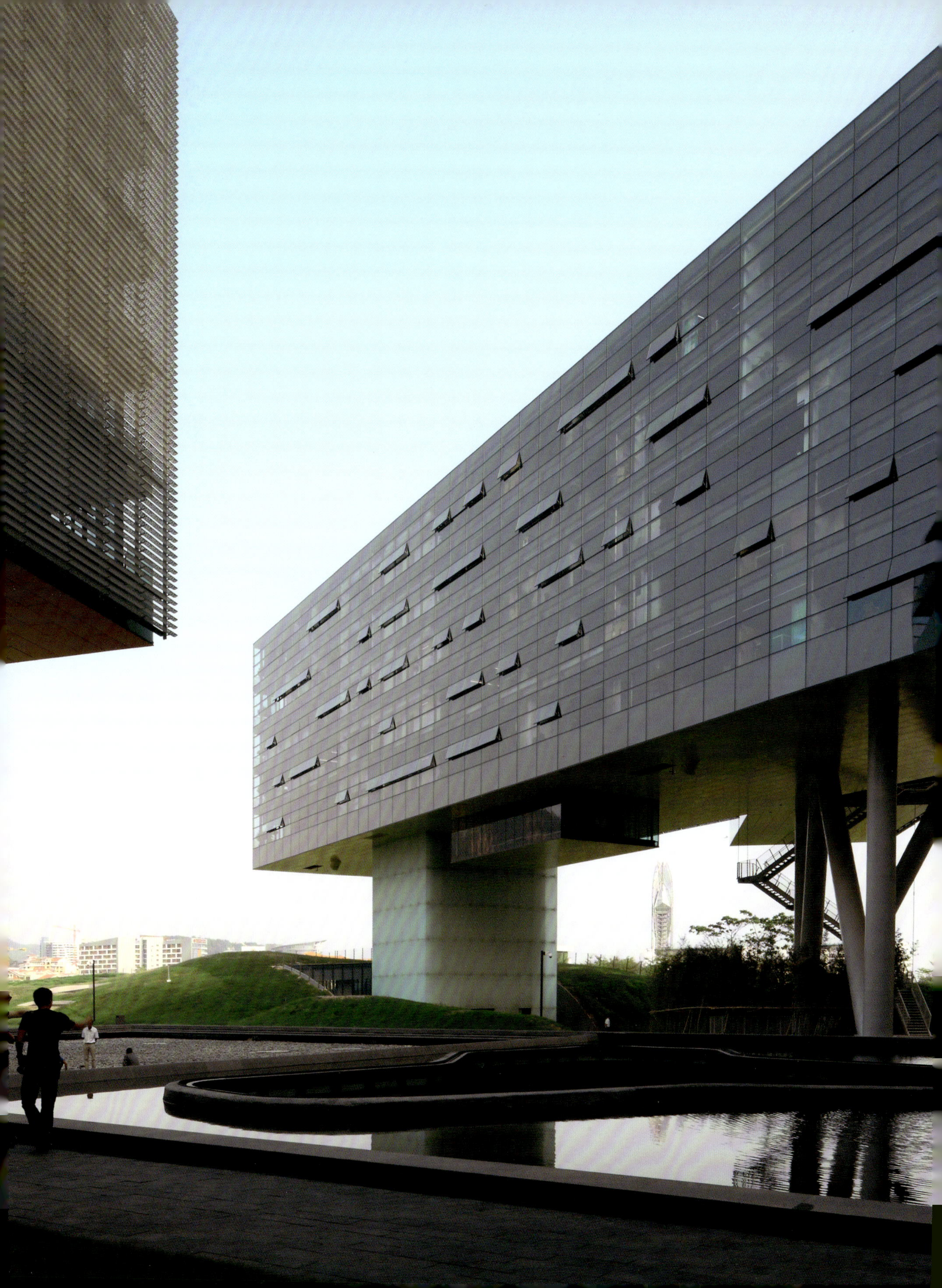

VANKE CENTER / HORIZONTAL SKYSCRAPER
Shenzhen, China, 2008–09

Address: Neihuan Road, Vanke Center, Dameisha, Yantian District, Shenzhen 518083, China
Area: 120 445 m² (building), 52 000 m² (landscape area)
Client: Shenzhen Vanke Real Estate Co. Cost: not disclosed

As seen from certain angles, the complex appears to be an assembly of different buildings. Though cladding varies, however, the entire very large building is part of a single complex.

Aus bestimmten Blickwinkeln wirkt der Komplex wie ein Ensemble aus verschiedenen Gebäuden. Trotz der unterschiedlichen Fassadenverkleidung ist es aber ein einziger Gesamtkomplex.

Vu sous certains angles, le complexe donne l'impression d'un assemblage de différentes constructions d'habillages variés, qui constituent en réalité un seul bâtiment.

In the drawing to the right, the idea of the "horizontal skyscraper" is made clear. The office, residential, and hotel spaces are grouped in different parts of the building, which is supported on large pillars.

Auf der Zeichnung rechts wird der Gedanke des „horizontalen Hochhauses" deutlich. Büros, Wohnungen und das Hotel sind in unterschiedlichen Teilen des auf großen Stützen ruhenden Gebäudes zusammengefasst.

Le dessin de droite, illustre le concept de « gratte-ciel horizontal ». Les bureaux, les logements et l'hôtel sont installés dans différentes parties de l'immeuble qui repose sur d'énormes piliers.

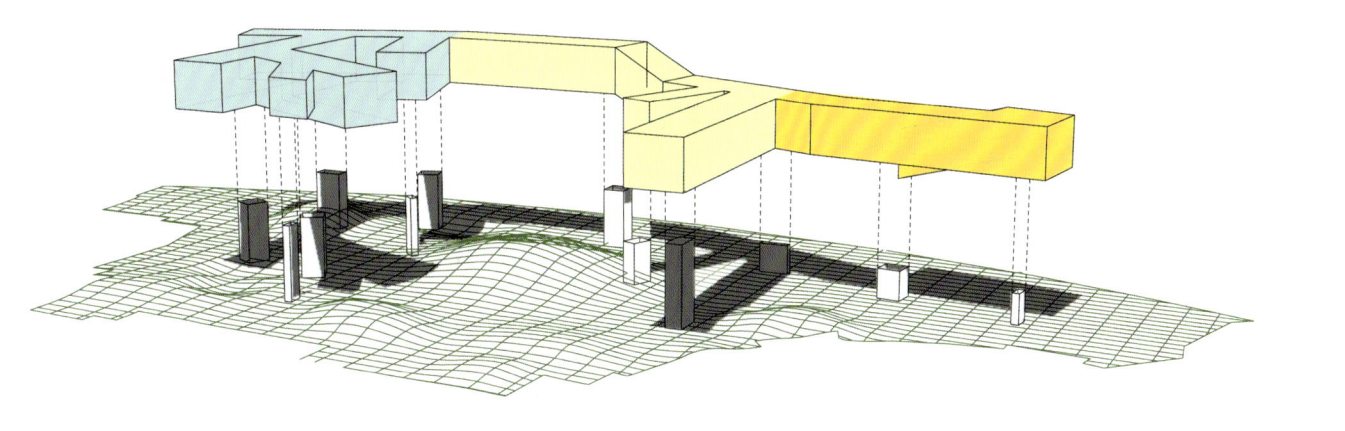

STEVEN HOLL

It is with some pride that Steven Holl describes this mixed-use complex as a **"HORIZONTAL SKYSCRAPER"** that is as long as the Empire State Building is tall. The Vanke Center includes hotel space, offices for the Vanke Company, serviced apartments, and a public park. While Holl's Linked Hybrid (Beijing, 2003–09) emphasizes a certain verticality, the Vanke Center is mainly about horizontal lines inserted into a tropical landscape. In fact, the landscape contains a conference center, and, under mounds, a 500-seat auditorium and restaurants. The office explains: "The building appears as if it were once floating on a higher sea that has now subsided, leaving the structure propped up high on eight legs. The decision to float one large structure right under the 35-meter height limit, instead of several smaller structures each catering to a specific program, generates the largest possible green space open to the public on the ground level." The landscape component of the project is also one element of the environmental strategy of Steven Holl because its ponds, fed by a gray-water system, serve to cool the air. Solar panels are used as well as local materials such as bamboo.

Steven Holl beschreibt diesen Gebäudekomplex mit einem gewissen Stolz als **„HORIZONTALEN WOLKENKRATZER"**, der so lang ist, wie das Empire State Building hoch ist. Zum Vanke Center gehören ein Hotel, Büros für die Firma Vanke Company, ein Aparthotel und ein öffentlicher Park. Holls Projekt Linked Hybrid hat noch eine gewisse Vertikalität. Beim Vanke Center herrschen horizontale Linien vor, die in eine tropische Landschaft eingefügt sind. Auf dem Gelände befinden sich ein Konferenzzentrum und unter künstlichen Hügeln verborgen ein Auditorium mit 500 Sitzplätzen sowie Restaurants. Das Büro erklärt: „Das Gebäude wirkt, als wäre es früher einmal auf einem höheren Wasserspiegel geschwommen, und nun sieht man den auf acht Stützen gesetzten Gebäudekomplex. Die Entscheidung, ein großes Gebäude über dem Erdboden schweben zu lassen und dabei knapp unter der vorgeschriebenen Grenze von 35 Meter Höhe zu bleiben, anstatt mehrere kleine Gebäude zu errichten, hat zur Folge, dass ein größtmöglicher öffentlicher Freiraum entsteht." Die Freiraumgestaltung gehört zur Umweltstrategie von Steven Holl, denn die dort angelegten Teiche werden mit Grauwasser gespeist und dienen dazu, die Luft zu kühlen. Solarzellen kommen ebenso zum Einsatz wie ortstypische Baumaterialien, zum Beispiel Bambus.

Steven Holl présente avec une certaine fierté ce complexe immobilier mixte comme un **« GRATTE-CIEL HORIZONTAL »** aussi long que l'Empire State Building est haut. Le Vanke Center regroupe un hôtel, des bureaux de la Vanke Company, des appartements en résidence et un parc public. Si le Linked Hybrid de Holl (Pékin, 2003–09) mettait l'accent sur une certaine verticalité, le Vanke Center joue essentiellement de son horizontalité, insérée dans un paysage de type tropical. Les aménagements paysagers abritent également un centre de conférences, des restaurants et un auditorium de 500 places. Selon le descriptif : « On pourrait imaginer que l'immeuble a longtemps flotté sur une mer aujourd'hui disparue, qui, en se retirant, l'aurait laissé reposer sur le fond sur huit grands pieds. La décision de faire "flotter" l'immeuble juste sous la limite de hauteur autorisée de 35 m, plutôt que d'opter pour plusieurs constructions autonomes plus basses consacrées chacune à un programme différent, a permis d'accroître au maximum la surface des espaces verts ouverts au public. » La partie paysagère du projet relève de la stratégie environnementale de l'architecte, les bassins, alimentés par les eaux grises, servant à rafraîchir l'air. Des panneaux solaires et des matériaux naturels locaux comme le bambou sont également utilisés.

Steven Holl has shown a consistent interest in the landscaping around his buildings. Planting adds to the continuity of the whole and relieves the appearance of density that would otherwise have dominated.

Steven Holl hat sich schon immer für die Landschaftsgestaltung um seine Gebäude interessiert. Pflanzen tragen zur Kontinuität des Ganzen bei und mildern die Massigkeit der Baukörper, die sonst dominieren würden.

Steven Holl manifeste un intérêt de longue date pour les aménagements paysagers de ses réalisations. Les plantations contribuent à la continuité de l'ensemble et allègent l'impression de densité qu'un tel projet aurait pu susciter.

At the same time as the architecture appears to rise from the earth or to enter it, the planting rises up on the roof of the building, as seen in the image above, right. Below, an overall plan of the complex.

So wie die Architektur sich von der Erde zu lösen scheint oder in sie eindringt, erobern sich die Pflanzen das Dach des Gebäudes, wie man auf dem Bild oben rechts erkennen kann. Unten: Gesamtplan des Komplexes.

L'architecture semble sortir du sol ou y plonger, d'autant plus que les plantations recouvrent la toiture (en haut à droite). Ci-dessous, plan général du complexe.

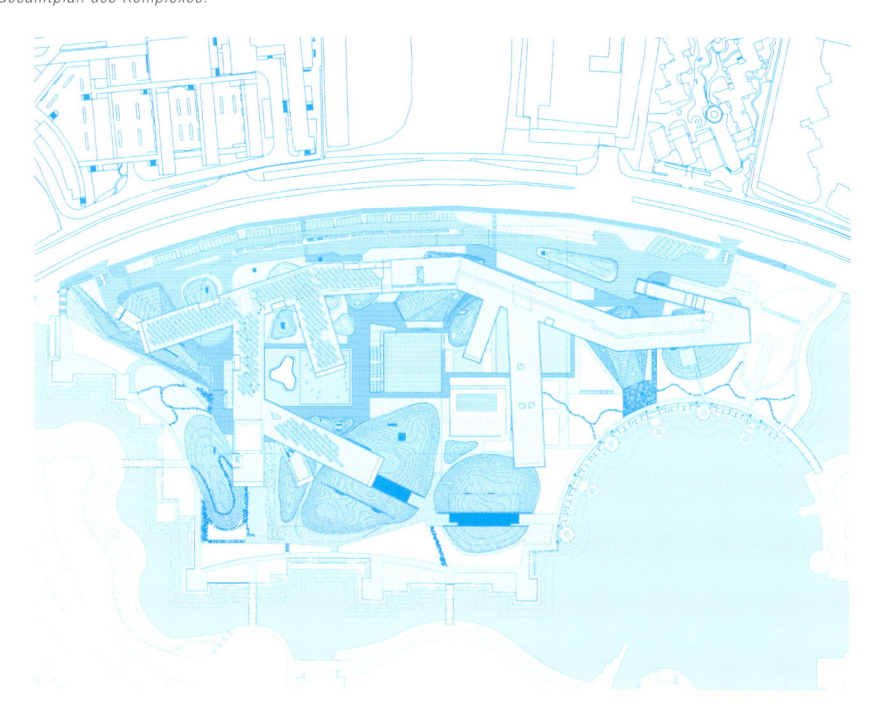

Though major elements of the design, such as the suspended blocks seen above, are perfectly rectilinear, the architect uses color and natural forms to soften the mass of the building.

Die prägnanten Entwurfselemente, wie die aufgeständerten Gebäudeblöcke oben, sind alle streng rechteckig. Um der Masse der Gebäude ihre Dominanz zu nehmen, verwendet der Architekt Farbe und eine aus der Natur entlehnte Formensprache.

Alors que les principaux éléments du projet, comme les blocs suspendus vus ci-dessus, sont de plan parfaitement rectiligne, l'architecte a utilisé des formes et des couleurs naturelles pour adoucir la masse de l'ensemble.

Suspended stairs and passageways, and some irregular forms, connect and enliven the spaces within the "horizontal skyscraper."

Frei schwebende Treppen und Stege sowie unregelmäßige Formen verbinden und beleben die Freiräume in diesem „horizontalen Wolkenkratzer".

Des escaliers, des passages suspendus et quelques formes libres, relient et animent les espaces interstitiels de ce « gratte-ciel horizontal ».

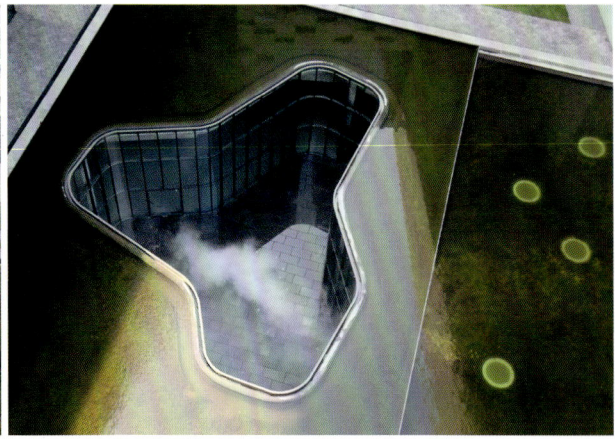

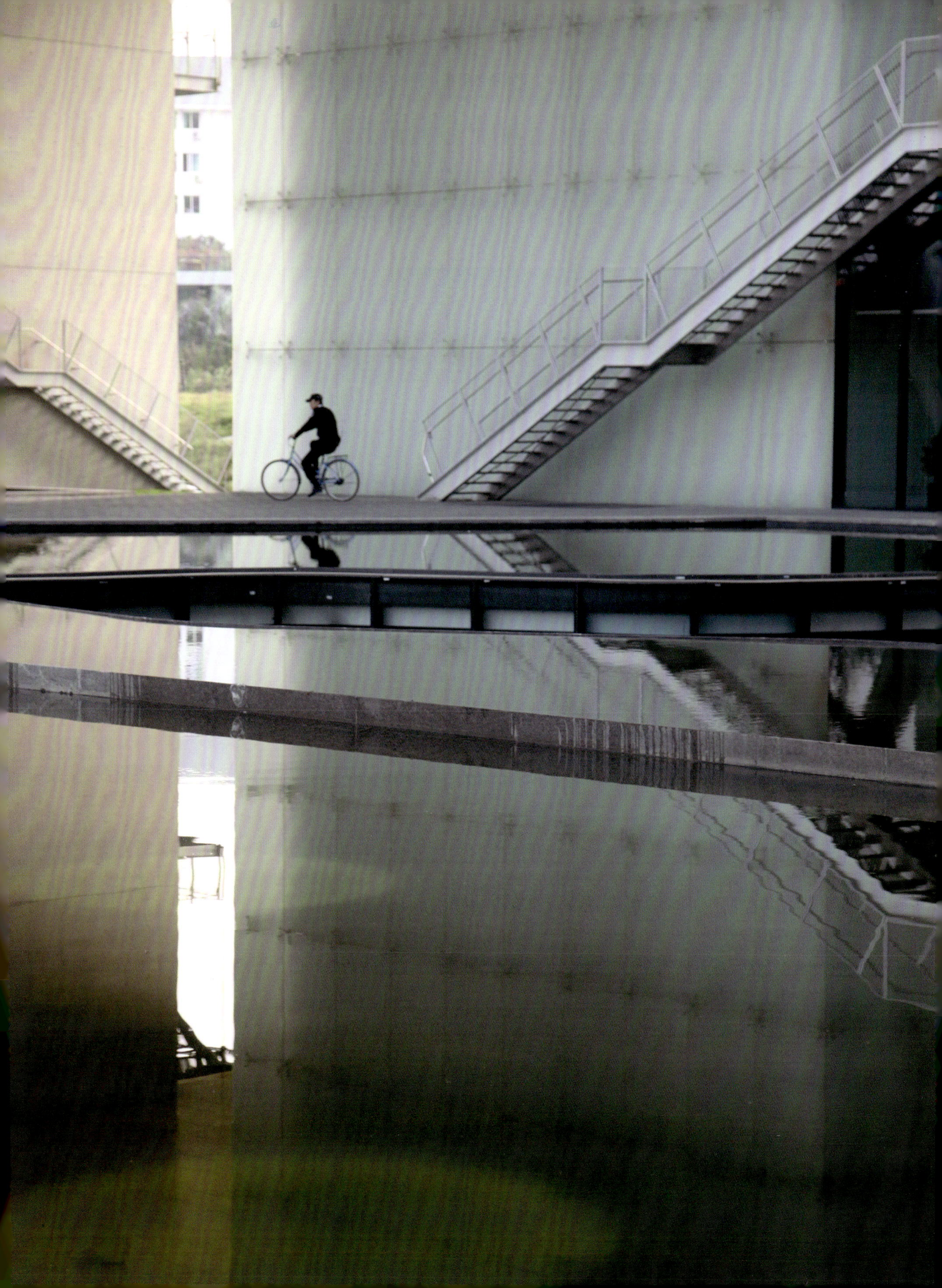

HEART: HERNING MUSEUM OF CONTEMPORARY ART

Herning, Denmark, 2007–09

Address: Birk Centerpark 8, 7400 Herning, Denmark, +45 9712 1033, www.heartmus.com
Area: 5600 m². Client: Herning Center of the Arts. Cost: not disclosed

Here, as in other projects, Steven Holl contrasts the crisp forms of the architecture with the presence of rounded lines in the earth and a basin set in the grass around the museum.

Steven Holl setzt hier, wie auch bei anderen Projekten, auf den Kontrast zwischen der spröden, streng linearen Architektur des Museums und den umgebenden gerundeten Erdformen sowie dem in die Rasenfläche eingelassenen Wasserbecken.

Ici, comme dans d'autres projets, Steven Holl fait contraster les formes tendues caractéristiques de son architecture avec des éléments plus doux comme les terrassements ou le bassin entouré de pelouses devant le musée.

This center combines visual arts and music through the **HERNING MUSEUM OF CONTEMPORARY ART**, the MidWest Ensemble, and the Socle du Monde. The intention of Steven Holl was to "fuse landscape and architecture" in a single-story structure that includes temporary exhibition galleries, a 150-seat auditorium, museum, rehearsal rooms, restaurant, library, and offices. The Museum features a collection of 46 works by Piero Manzoni and has an ongoing interest in textiles, given the fact that the corporate sponsor (Herning) is a shirt manufacturer. Steven Holl states that the "roof geometry resembles a collection of shirt sleeves laid over the gallery spaces: the curved roofs bring balanced natural lights to the galleries. The loose edges of the plan offer spaces for the café, auditorium, lobby, and offices." Orthogonal gallery spaces are conceived as "treasure boxes". The unusual surface of the building's exterior adds interest, as the architect explains: "Truck tarps were inserted into the white concrete formwork to yield a fabric texture to the building's exterior walls." The scheme features a 3700-square-meter "bermed landscape of grass mounds and pools" that conceals parking and service areas.

Das **HERNING MUSEUM OF CONTEMPORARY ART** verbindet mit dem MidWest Ensemble und dem Socle du Monde die visuellen Künste mit der Musik. Steven Holls Ziel war es, durch einen eingeschossigen Baukörper, in dem Räume für Wechselausstellungen, ein Auditorium mit 150 Sitzplätzen, ein Museum, Übungsräume, ein Restaurant, eine Bibliothek und Büros untergebracht sind, „die Landschaft und die Architektur miteinander zu verschmelzen". Das Museum besitzt 46 Werke von Piero Manzoni und ist besonders an Textilien interessiert, zumal der Hauptsponsor Herning Hemden produziert. Holl merkt an, dass „die Geometrie des Dachs an Hemdenärmel erinnert, die über die Ausstellungsräume gehängt sind: Durch die geschwungenen Dächer gelangt diffuses Tageslicht in die Ausstellungsräume. Die freien Ecken im Grundriss bieten Platz für das Café, das Auditorium, die Eingangshalle und die Büros." Die rechtwinkligen Ausstellungsräume sind als „Schatztruhen" gedacht. Die ungewöhnliche Fassade macht den Bau besonders interessant. Holl erklärt: „In die weiße Betonschalung sind Fahrzeugplanen eingelegt worden, um den Außenfassaden eine textile Struktur zu verleihen." Eine 3700 Quadratmeter große Gartenanlage mit Wasserbecken und Rasenwellen verbirgt die Parkplätze und Dienstleistungsbereiche.

Ce centre allie arts visuels et musique grâce au **HERNING MUSEUM OF CONTEMPORARY ART**, au Ensemble MidWest et au Socle du Monde. L'intention de Steven Holl était ici de « fusionner le paysage et l'architecture » dans une construction d'un seul niveau avec galeries d'expositions temporaires, auditorium de 150 places, musée, salles de répétition, restaurant, bibliothèque et bureaux. Le musée présente notamment une collection de 46 œuvres de Piero Manzoni et cultive un intérêt pour les textiles (le mécène, Herning, est un fabricant de chemises). Steven Holl explique que « la géométrie des toitures fait penser à des manches de chemises repliées sur les volumes des galeries : les toits incurvés permettent de capter un éclairage naturel équilibré, orienté vers les galeries. Les parties en projection accueillent un café, l'auditorium, le hall d'entrée et des bureaux ». Les galeries orthogonales sont conçues comme des « coffrets à bijoux ». Le traitement de surface des murs extérieurs est étonnant : « Des bâches de camion glissées dans les coffrages du béton blanc ont donné aux murs un aspect textile. » Le projet comprend aussi un « paysage de bermes, de monticules gazonnés et de bassins » qui s'étend sur 3700 m² et dissimule parkings et installations techniques.

In an aerial view of the building, it is evident that its own lines prolong and reinforce the landscaping, making it appear to be at the nexus of the forces that surround it.

Auf der Luftaufnahme des Gebäudes wird deutlich, dass seine Linienführung in der Landschaft eine Antwort findet und die Freiraumgestaltung zu einer wichtigen Verbindung wird.

Vue aérienne du bâtiment montrant comment ses axes prolongent et renforcent les aménagements paysagers, semblant les placer ainsi au cœur du réseau de forces qui l'entoure.

The rounded forms that project from the museum into the landscape can be seen in these images, together with the rectangular basin.

Auf diesen Bildern kann man die gerundeten Formen, die von dem Museum in die Landschaft streben, gut erkennen, ebenso das rechteckige Wasserbecken.

Des avancées en partie arrondies se projettent du musée vers le paysage en bordure d'un grand bassin rectangulaire.

MARTIN HURTADO

Martin Hurtado Arquitectos
Av. Los Conquistadores 1700, Of. 302
Santiago, RM 7520282
Chile

Tel: +56 2 371 3010
Fax: +56 2 489 2119
E-mail: contacto@martinhurtado.cl
Web: www.martinhurtado.cl

MARTIN HURTADO COVARRUBIAS was born in Santiago, Chile, in 1965. He created his architectural firm there in 1993, having received his degree from the Catholic University of Chile (1989). He worked in the office of Enrique Browne and Borja Huidobro from 1990 to 1993. His projects include Greenvic Organic Apple Packing Plant (Colchagua Valley, 2006); master plan and infrastructure design for the Izaro Estate (Casablanca Valley, 2006, published here); Morandé Winery Productive Services (Casablanca Valley, 2006, also published here); the Las Camelias Stables (San Bernardo, Metropolitana Region, 2008); San Francisco Javier School (Puerto Montt, 2009); Marine Research Laboratory (Las Cruces, Valparaiso, 2010); and the Colegio María Auxiliadora School (Puerto Montt, 2010), all in Chile.

MARTIN HURTADO COVARRUBIAS wurde 1965 in Santiago de Chile geboren. Er gründete 1993 dort auch sein eigenes Architekturbüro, nachdem er 1989 an der Katholischen Universität von Chile sein Diplom in Architektur gemacht hatte. Von 1990 bis 1993 arbeitete er im Büro Enrique Browne und Borja Huidobro. Zu seinen Projekten gehören die Verpackungsfabrik für Greenvic Bioäpfel (Valle Colchagua, 2006), der Masterplan und der Entwurf für die Infrastruktur der Finca Izaro (Valle Casablanca, 2006, hier vorgestellt), der Masterplan und die Wirtschafts- und Verwaltungsgebäude der Kellerei Viña Morandé (Valle Casablanca, 2006, ebenfalls hier vorgestellt), die Stallungen des Gestüts Las Camelias (San Bernardo, Región Metropolitana, 2008), die Schule San Francisco Javier (Puerto Montt, 2009), ein Meeresforschungslabor in Las Cruces (Valparaiso, 2010) und die Schule Colegio María Auxiliadora (Puerto Montt, 2010), alle in Chile.

MARTIN HURTADO COVARRUBIAS, né à Santiago du Chili en 1965, a fondé son agence dans cette ville en 1993, après avoir obtenu son diplôme d'architecte à l'Université catholique du Chili (1989). Il avait auparavant travaillé dans l'agence d'Enrique Browne et de Borja Huidobro de 1990 à 1993. Parmi ses réalisations figurent les suivantes (toutes situées au Chili) : l'usine de conditionnement de pommes biologiques de Greenvic (vallée de Colchagua, 2006) ; le plan directeur et la conception des infrastructures du domaine Izaro (vallée de Casablanca, 2006, publié ici) ; les installations des services de production viticole de Morandé (Vallée de Casablanca, 2006, publié ici) ; les écuries du haras Las Camelias (San Bernardo, Région Metropolitana, 2008) ; l'établissement scolaire Colegio San Francisco Javier (Puerto Montt, 2009); le laboratoire de recherches marines (Las Cruces, Valparaíso, 2010) et l'établissement scolaire Colegio María Auxiliadora (Puerto Montt, 2010).

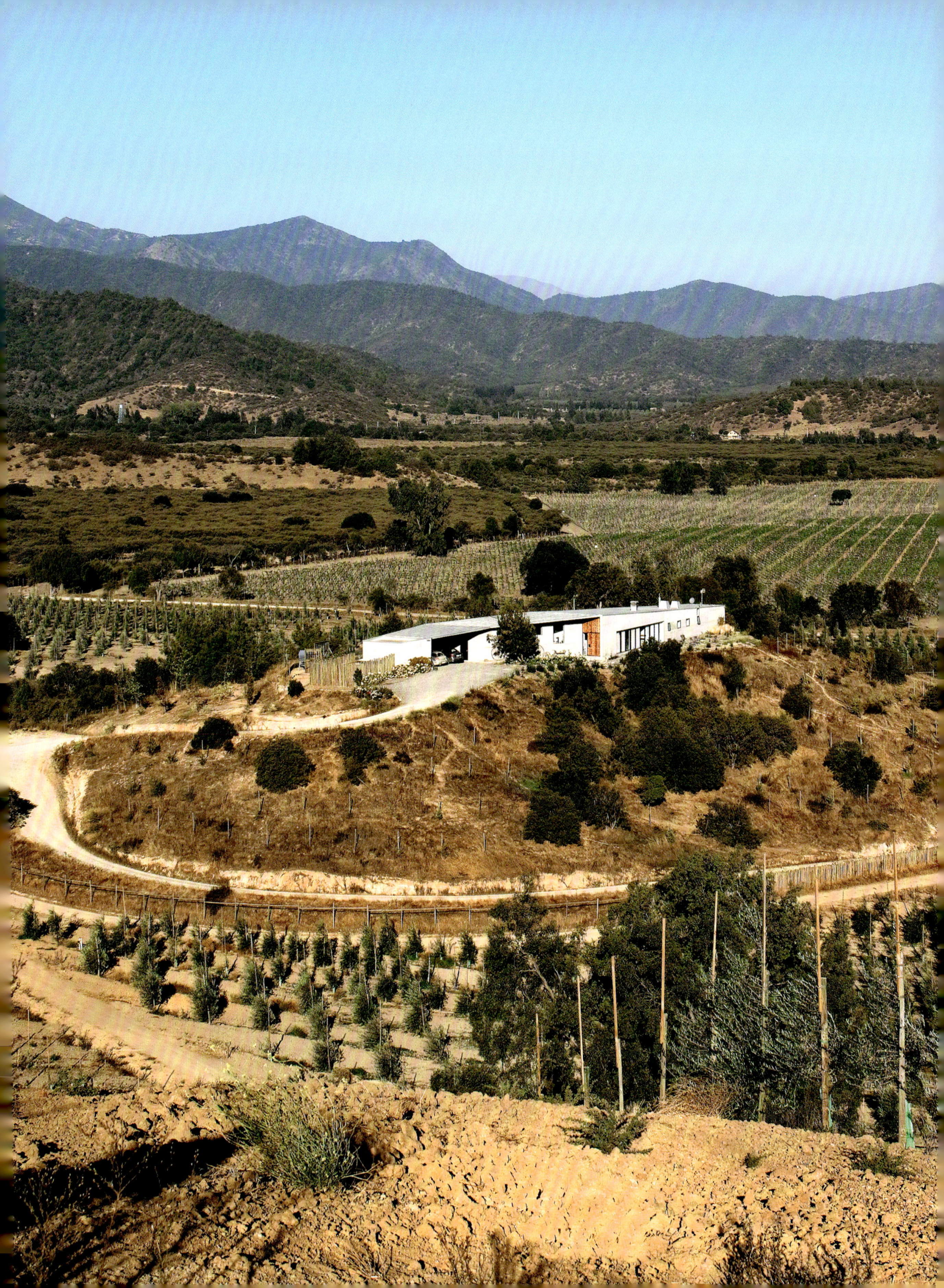

IZARO ESTATE
Casablanca Valley, Chile, 2006

Area: 1000 hectares (site), 5000 m² (various buildings). Client: Izaro Fund
Cost: $900 per square meter. Collaboration: Ignacio Correa, Ivan Salas, Andrés Suarez

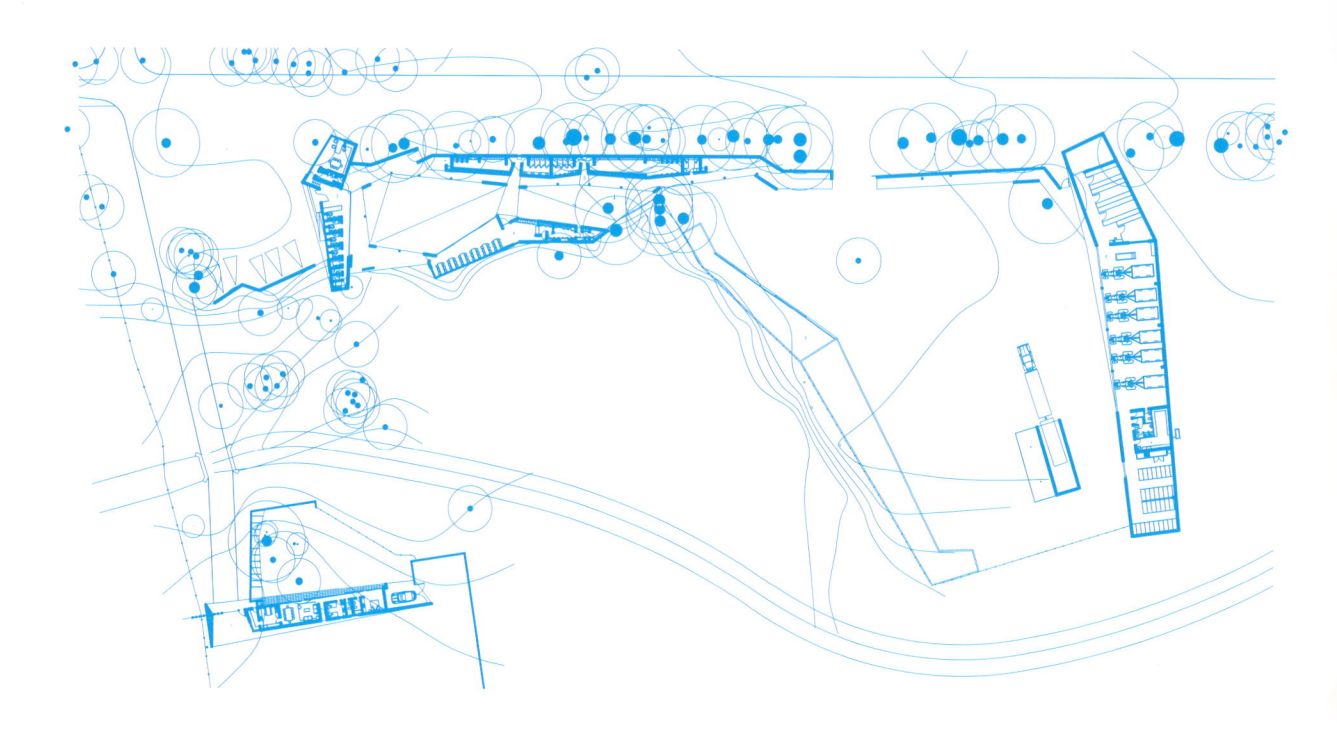

    The project involved the creation of a master plan for a new agricultural development situated in previously abandoned fields. The architect explains: "This implied defining entrances, circulation, program distribution, and a general order of construction that would give coherence and unity to the whole territory, in such a way it could become a landscape." Concrete was the main building material "given its versatility and noble austerity with no need for maintenance." Concrete was also chosen to avoid having to call on outside technology. The final program involved a porter's house, administrative services, dressing and dinning rooms, machinery warehouse, resident personnel housing, stables, irrigation installations, tailings, roads, paths, and lookout points. Each structure was designed with respect to the precise topography of its site. The architect concludes on a modest note: "The materials used were those available at the local hardware store."

    Zu diesem Projekt gehörte die Ausarbeitung eines Masterplans für ein neues Landgut, das auf einer landwirtschaftlichen Brache errichtet werden sollte. Der Architekt erklärt dazu: „Das bedeutete, wir mussten Zufahrten, Verkehrswege, ein Bebauungsprogramm und die Verteilung der Gebäude festlegen sowie Vorgaben für die einheitliche Gestaltung der Gebäude machen, damit sich alles kohärent zu einer landschaftsgestalterischen Einheit fügte." Hauptbaumaterial war Beton, „weil er so vielseitig ist und eine noble Zurückhaltung ausstrahlt. Außerdem ist er pflegeleicht." Man entschied sich auch für Beton, weil dafür keine Technologie von außerhalb in Anspruch genommen werden musste. Zum endgültigen Bauprogramm gehören ein Pförtnerhaus, Verwaltung, Umkleiden und Speisesaal, Lagerräume für Maschinen und Produkte, Personalwohnungen, Stallungen, Bewässerungsanlagen, Absetzbecken, Straßen, Wege und Aussichtspunkte. Jedes Bauwerk ist so konstruiert, dass es sich an die Topografie des Geländes anpasst. Der Architekt merkt bescheiden an: „Wie haben nur Materialien verwendet, die man im örtlichen Baumarkt kaufen konnte."

    Le projet comprenait la création du plan directeur d'un nouveau domaine agricole aménagé sur des terres en friche. Comme l'explique l'architecte : « Cette recherche impliquait de définir des entrées, des circulations, la distribution du programme et de trouver un ordre constructif donnant de la cohérence et de l'unité à l'ensemble de ce territoire, de telle façon qu'il devienne réellement un paysage. » Le béton est le principal matériau de construction choisi « pour sa souplesse d'utilisation, la noblesse de son austérité et son absence d'entretien ». Il a également été retenu pour éviter de faire appel à des technologies extérieures. Le programme final contenait une maison de gardien, des bureaux administratifs, une salle à manger, un vestiaire, un entrepôt pour les machines agricoles, des logements pour les employés résidents, des étables, des installations d'irrigation, des routes carrossables, des chemins et des points d'observation. Chaque bâtiment a été conçu dans le respect précis de la topographie du site. L'architecte conclut en faisant remarquer que « les matériaux utilisés ont été trouvés dans le commerce local ».

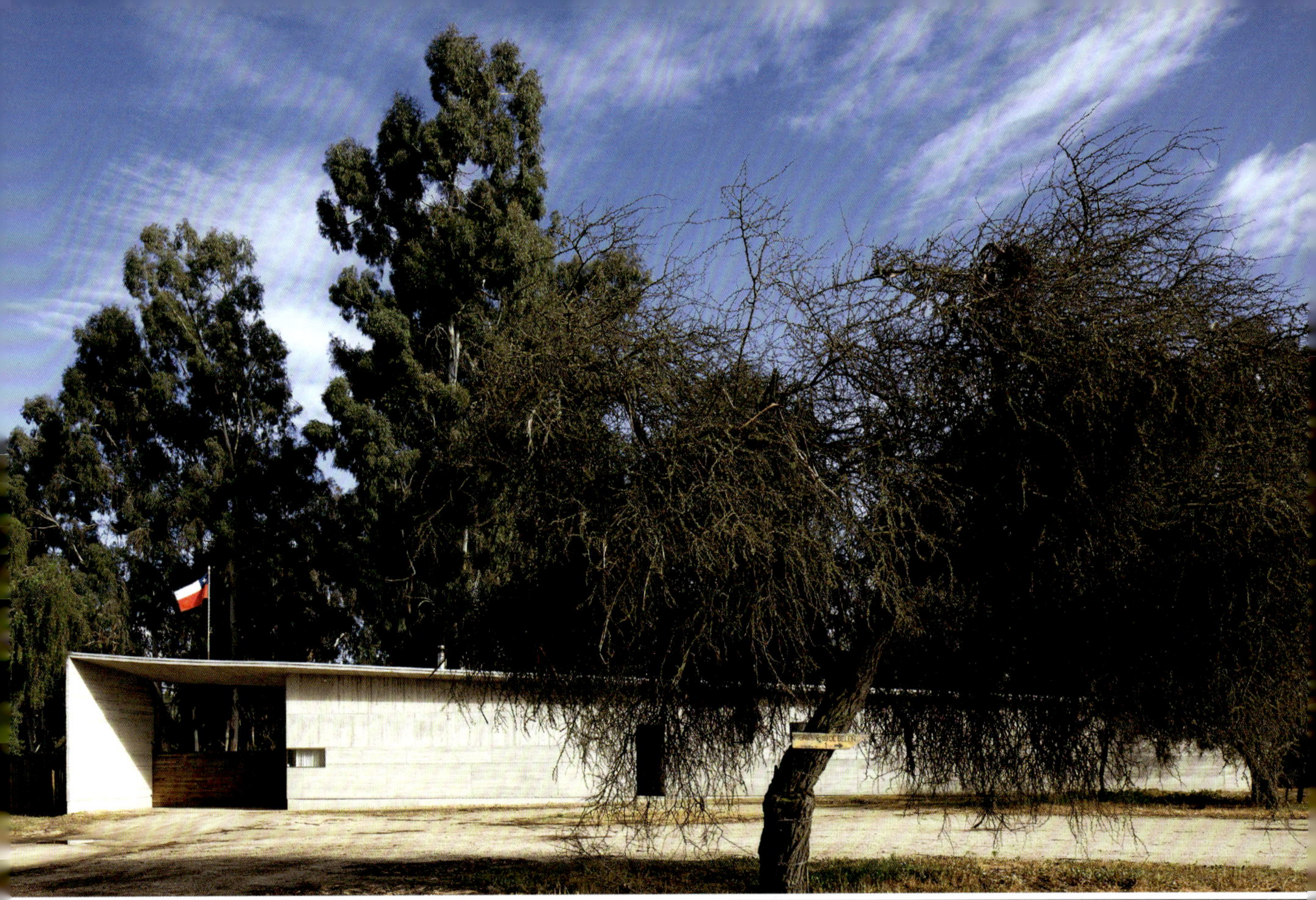

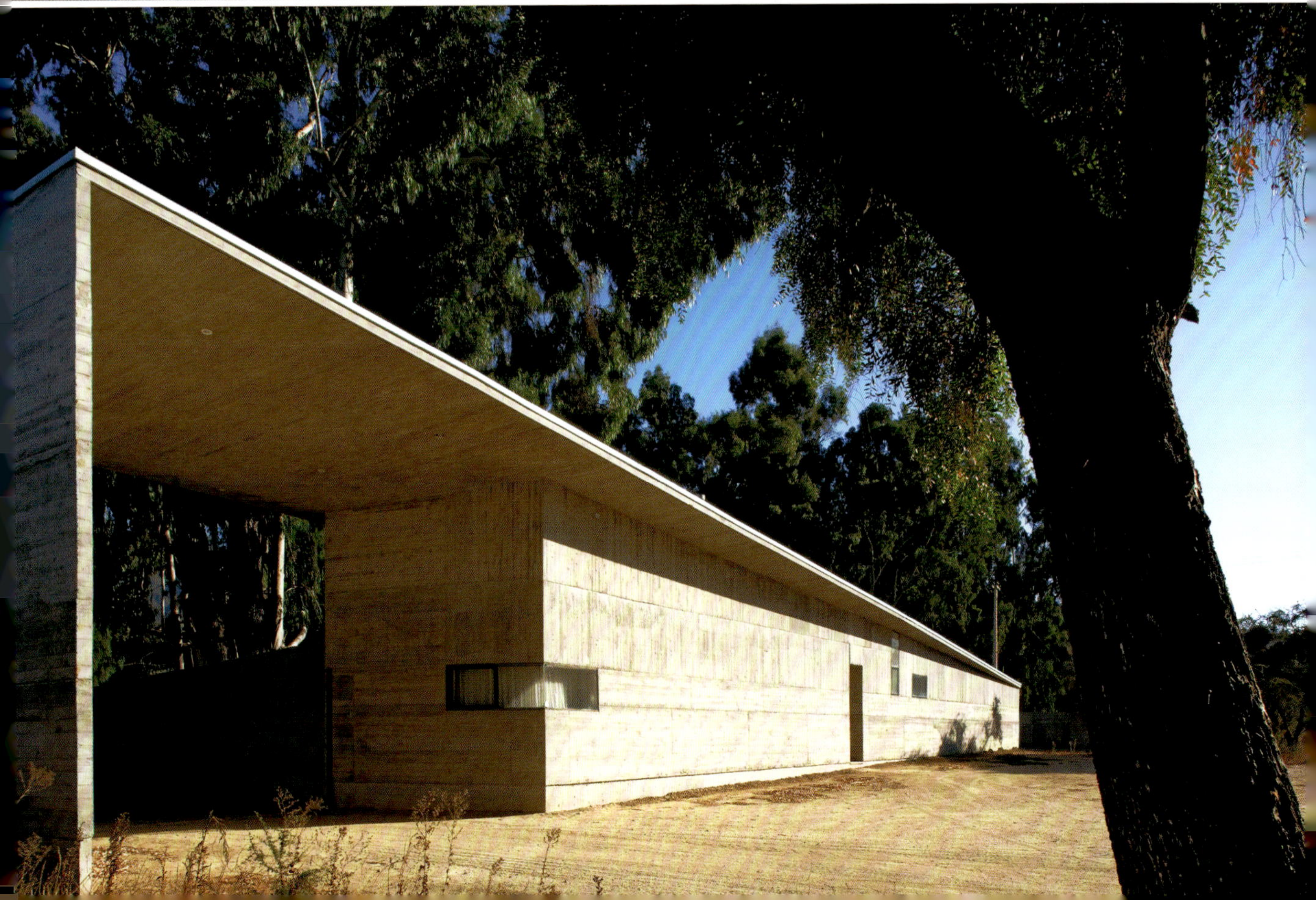

The disposition of the architectural elements with a gentle and almost natural use of planting creates a landscape, as the architect himself puts it.

Aus der Anordnung der Architekturelemente im Zusammenspiel mit einer zurückhaltenden, fast natürlichen Bepflanzung entsteht ein Landschaftsraum, wie der Architekt es formuliert.

Comme l'explique l'architecte, c'est le positionnement des éléments architecturaux et le recours discret, presque naturel, aux plantations, qui créent le paysage.

The placement of the buildings in the site, as seen in the plan to the right or in the image above where stone, earth, and wood come together, creates an unusual degree of symbiosis between architecture and nature.

Wie man auf dem Plan rechts oder auch auf dem Bild oben erkennen kann, schafft die Anordnung der Gebäude auf diesem Gelände, wo es Stein, Erde und Wald gibt, eine ungewöhnliche Symbiose zwischen Architektur und Natur.

L'implantation des bâtiments sur le terrain (plan de droite ou image ci-dessus) et l'association de la pierre, de la terre et du bois, créent un degré de symbiose inhabituel entre l'architecture et la nature.

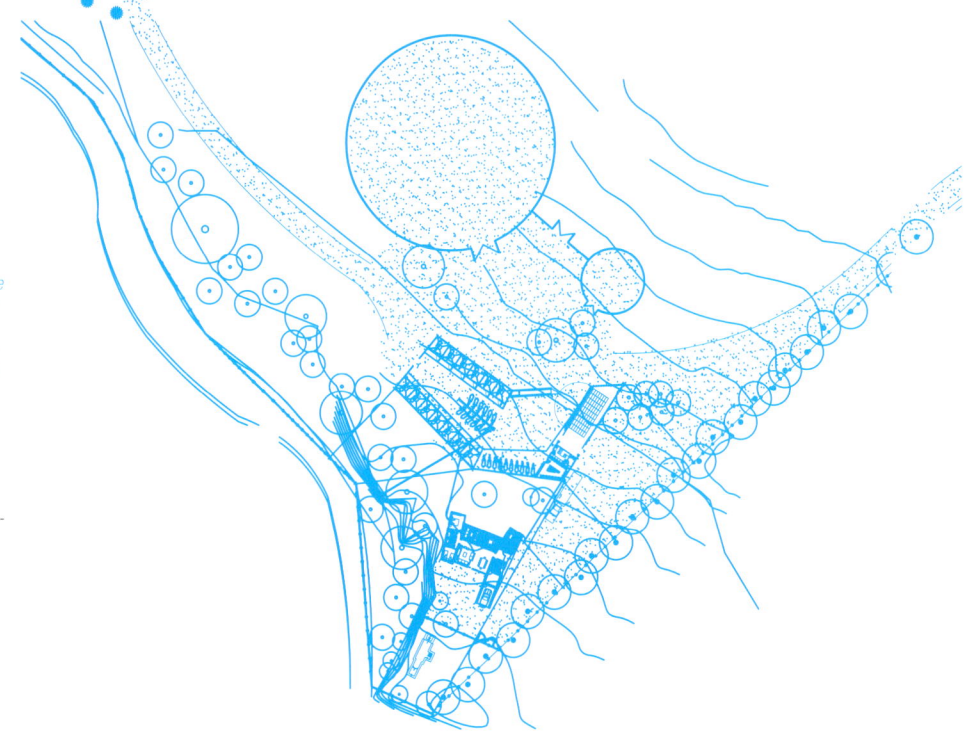

MORANDÉ WINERY PRODUCTIVE SERVICES
Casablanca Valley, Chile, 2006

Area: 552 hectares, 1718 m² (four buildings)
Client: Morandé Winery. Cost: $650 per square meter
Collaboration: Ignacio Correa, Ivan Salas, Andrés Suarez

This large site is marked by ravines, water reservoirs, hills, and vineyards. As the architect views his role: "The objective was to accomplish a coherent order to all actions that are to be considered in a wine-producing project for a leading winery in Chile." There are some 300 hectares of planted fields and approximately 12 000 square meters of structures on the site. The Productive Services buildings designed by Hurtado establish "a counterpoint between nature and architecture with geometric and rigorous order, as opposed to the winding forms of existing topography." Here, too, Martin Hurtado has sought to make use of local building materials and technology.

Dieses weitläufige Gelände ist durch Schluchten, Stauseen, Hügel und Weinberge gekennzeichnet. Der Architekt sieht seine Rolle wie folgt: „Ziel war es, eine gemeinsame Ordnung für alle Aktivitäten im Zusammenhang mit der Weinproduktion einer der bedeutendsten Kellereien in Chile zu schaffen." Es gibt etwa 300 Hektar Anbaufläche und nahezu 12 000 Quadratmeter bebauter Fläche auf dem Gelände. Die von Hurtado entworfenen Wirtschafts- und Verwaltungsgebäude bilden „mit ihrer strikten geometrischen Ordnung einen Kontrapunkt zwischen Natur und Architektur, weil sie einen Gegensatz zu den geschwungenen Formen der bestehenden Topografie bilden." Auch in diesem Fall hat Martin Hurtado vor Ort vorhandene Baumaterialien verwendet und die heimische Bautechnik angewandt.

Ce vaste domaine se caractérise par un paysage de collines et de vignobles, traversé de ravins et ponctué de réserves d'eau. « L'objectif était de conférer un ordre cohérent à toutes les étapes de la production du vin de l'un des principaux producteurs chiliens », explique l'architecte. Le domaine regroupe quelque 300 ha de plantations et environ 12 000 m² de bâti. Les bâtiments des services de production dessinés par Hurtado instaurent « un équilibre entre la nature et l'architecture grâce à un ordre géométrique rigoureux, qui contraste avec la douceur des formes naturelles de la topographie ». Là encore, Hurtado a tenu à utiliser le plus possible les techniques et les matériaux locaux.

The Productive Services buildings designed by the architect are placed in a natural setting that is marked by hills and different types of vegetation, as seen in the plan below and the image to the right.

Die von Hurtado entworfenen Wirtschafts- und Verwaltungsgebäude liegen in einer hügeligen Landschaft mit abwechslungsreicher Vegetation, wie man unten auf dem Plan und auf dem Bild rechts gut erkennen kann.

Les bâtiments de la production conçus par l'architecte ont été implantés dans un cadre naturel de collines et de végétation variée, comme le montrent le plan ci-dessous et l'image de droite.

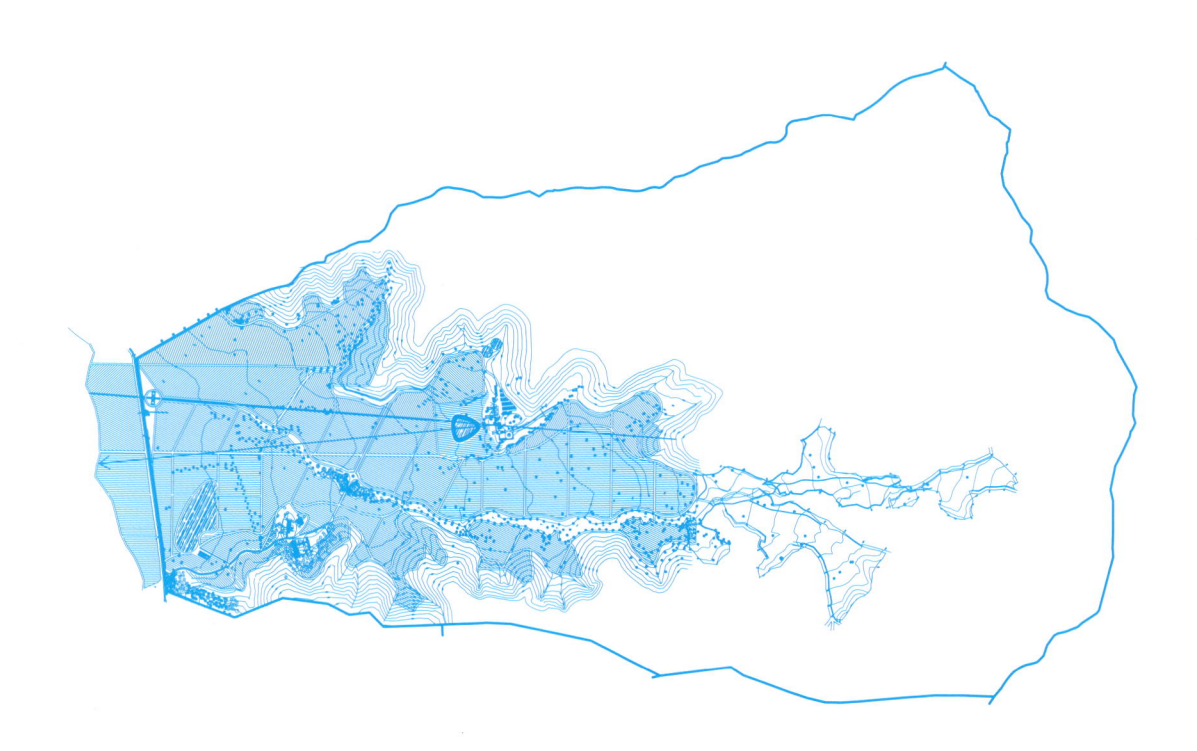

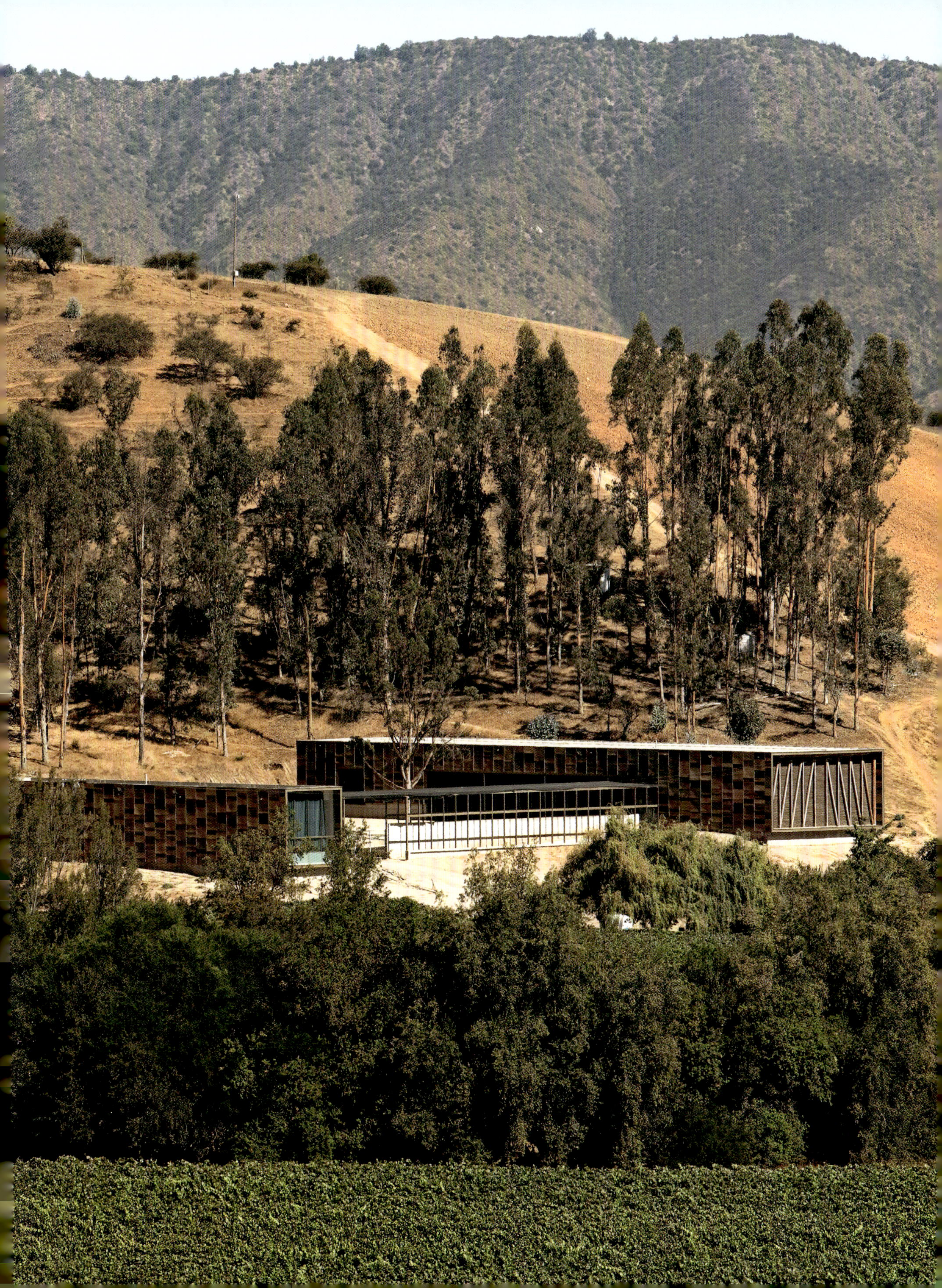

The color pattern of the building above, seen from a different angle below, relates it to the earth of the setting, but its forms are willfully rectilinear.

Die Farbpalette des oben abgebildeten Gebäudes (unten aus einem anderen Blickwinkel) harmoniert mit den Erdfarben der Umgebung. Die Konturen des Gebäudes hingegen sind bewusst geradlinig.

Si la couleur et le motif de l'habillage du bâtiment ci-dessus (vu sous un autre angle ci-dessous), rappellent la nature du terrain sur lequel il s'élève, ses formes restent volontairement rectilignes.

The architect uses concrete and clearly geometric forms, as can be seen in the plan to the right and the image above.

Der Architekt setzt Beton und klare geometrische Formen ein, wie man auf dem Plan rechts und auf dem Bild oben sehen kann.

L'architecte a fait appel à des formes très géométriques qu'il a traitées en béton, comme le montrent le plan de droite et l'image ci-dessus.

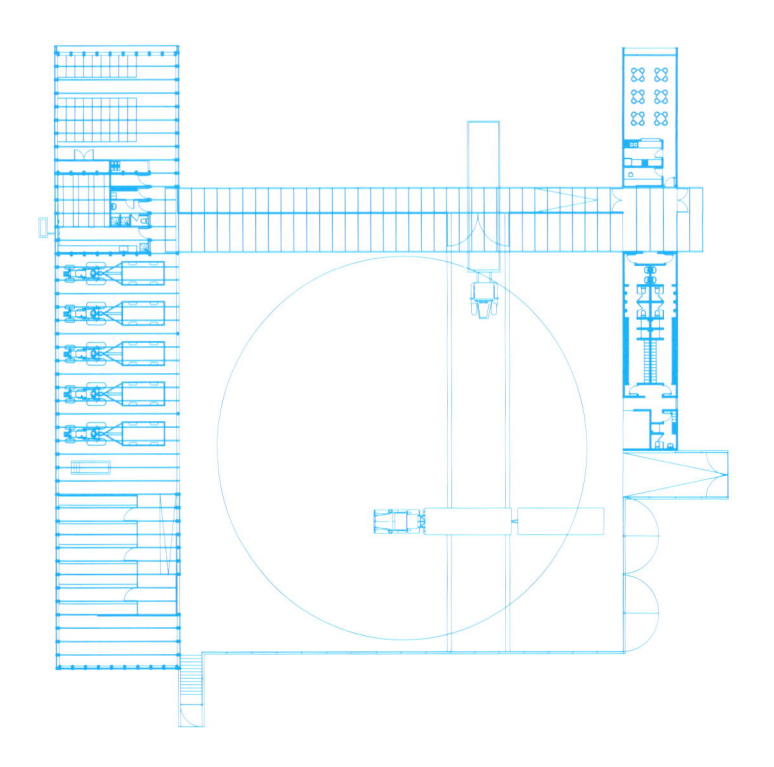

JAMES CORNER FIELD OPERATIONS / DILLER SCOFIDIO + RENFRO

James Corner Field Operations
475 10th Avenue, 10th floor, New York, NY 10018, USA
Tel: +1 212 433 1450 / Fax: +1 212 433 1451
E-mail: info@fieldoperations.net / Web: www.fieldoperations.net

Diller Scofidio + Renfro
601 West 26th Street, Suite 1815, New York, NY 10001, USA
Tel: +1 212 260 7971 / Fax: +1 212 260 7924
E-mail: disco@dsrny.com / Web: www.dsrny.com

JAMES CORNER FIELD OPERATIONS is an interdisciplinary design firm based in New York City. Corner is founder and Director of JCFO and is also Chair of the Department of Landscape Architecture at the University of Pennsylvania School of Design. He was educated at Manchester Metropolitan University, UK (B.A. in Landscape Architecture with first class honors), and the University of Pennsylvania (M.L.A/U.D.). Lisa Tziona Switkin is an Associate Partner and project leader for the High Line at JCFO with a B.A. in Urban Planning from the University of Illinois and an M.L.A. from the University of Pennsylvania. **DILLER SCOFIDIO + RENFRO** is an interdisciplinary firm based in New York City. Elizabeth Diller is Professor of Architecture at Princeton University; Ricardo Scofidio is Professor Emeritus of Architecture at the Cooper Union in New York; and Charles Renfro has served as a Visiting Professor at Rice, Columbia University, and Parsons the New School for Design. Matthew Johnson, a Senior Associate at DS+R and project leader for the High Line, is a native of Michigan. He is a graduate of the University of Michigan and holds an M.Arch from Princeton University. The team is primarily involved in thematically driven experimental works that take the form of architectural commissions, temporary installations and permanent site-specific installations, multimedia theater, electronic media, and print.

JAMES CORNER FIELD OPERATIONS ist ein interdisziplinär arbeitendes Büro in New York City. James Corner ist Gründer und Direktor des Büros und Inhaber des Lehrstuhls für Landschaftsarchitektur an der University of Pennsylvania School of Design. Er hat an der Manchester Metropolitan University in Großbritannien seinen Bachelor in Landschaftsarchitektur mit Auszeichnung bestanden, dann hat er an der University of Pennsylvania sein Studium als Master in Landschaftsarchitektur und Stadtplanung abgeschlossen. Lisa Tziona Switkin ist Büropartnerin und Projektleiterin für das Projekt High Line im Büro JCFO. Sie hat ihren Bachelor in Stadtplanung an der University of Illinois gemacht und ihren Master in Landschaftsarchitektur an der University of Pennsylvania. **DILLER SCOFIDIO + RENFRO** ist ein ebenfalls interdisziplinär arbeitendes Büro in New York City. Elizabeth Diller ist Professorin für Architektur an der Princeton University, Ricardo Scofidio ist Professor emeritus für Architektur an der Cooper Union in New York, und Charles Renfro war Gastprofessor an der Rice und der Columbia University sowie Parsons the New School for Design. Matthew Johnson, einer der Teilhaber des Büros DS+R und Projektleiter von High Line, stammt aus Michigan. Er hat seinen Studienabschluss an der University of Michigan gemacht und seinen Master an der Princeton University. Das Team beschäftigt sich vorwiegend mit experimentellen Aufträgen aus den Bereichen Architektur, temporäre und ständige ortsspezifische Installationen, Multimediatheater, elektronische und Druckmedien.

JAMES CORNER FIELD OPERATIONS est une agence interdisciplinaire new-yorkaise fondée et dirigée par James Corner, architecte paysagiste et urbaniste. Également président et professeur du département d'architecture du paysage à l'École de design de l'université de Pennsylvanie, il a fait ses études à la Manchester Metropolitan University (B.A. d'architecture du paysage, avec les honneurs de première classe), Royaume-Uni, et à l'université de Pennsylvanie (M.L.A/U.D.). Lisa Tziona Switkin, directrice associée de JCFO et directrice de projet pour la High Line, est titulaire d'un B.A. d'urbanisme de l'université de l'Illinois et d'un M.L.A. de l'université de Pennsylvanie. **DILLER SCOFIDIO + RENFRO** est une agence interdisciplinaire basée à New York. Elizabeth Diller est professeur d'architecture à l'université de Princeton, Ricardo Scofidio professeur émérite d'architecture à la Cooper Union à New York et Charles Renfro professeur invité aux universités Rice et Columbia, ainsi qu'à la Parsons New School for Design. Originaire du Michigan, Matthew Johnson, associé senior chez DS+R et directeur de projet pour la High Line, est diplômé de l'université du Michigan et titulaire d'un M.Arch. de l'université de Princeton. Leurs recherches essentiellement thématiques et expérimentales prennent la forme de commandes architecturales, d'installations temporaires, d'installations permanentes adaptées au site, de théâtre multimédia, de médias électroniques et de publications.

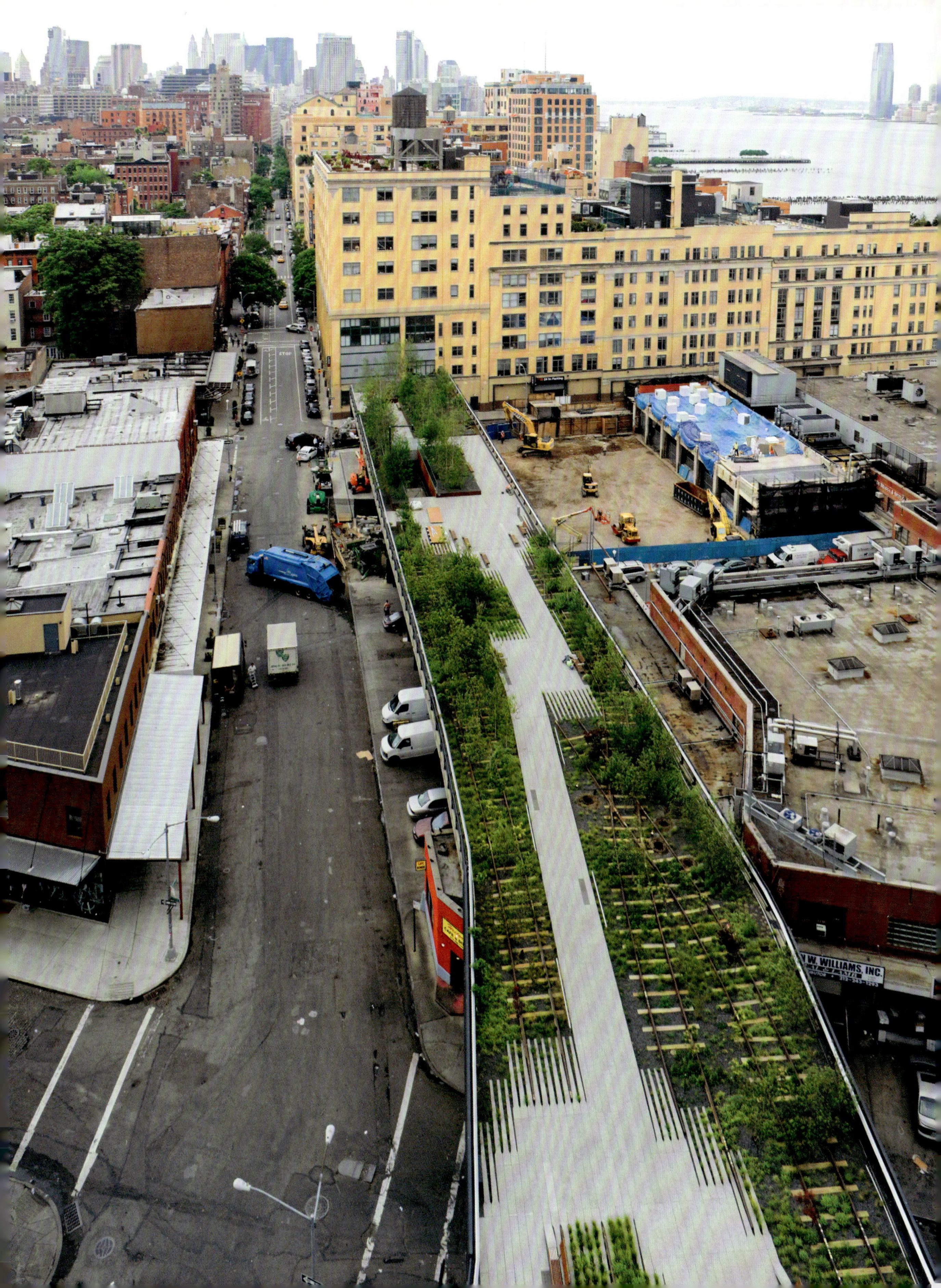
W. WILLIAMS, INC.

THE HIGH LINE
New York, New York, USA, 2009 (Phase I), 2011 (Phase II)–ongoing

Address: West Side of Manhattan, from Gansevoort Street to West 34th Street, between 10th and 11th Avenues, New York City, USA, +1 212 206 9922, www.thehighline.org . Length: 2.4 km. Client: City of New York, Friends of the High Line Cost: $156 million (Phases I and II). Collaboration: Piet Oudolf, Craig Schwitter (Principal, Buro Happold), Joseph F. Tortorella (Vice President, Robert Silman Associates), Herve Descottes (Principal, L'Observatoire International)

JAMES CORNER FIELD OPERATIONS / DILLER SCOFIDIO + RENFRO

The designers and architects who worked on the High Line had the clear challenge to make creative use of an abandoned elevated rail line in lower Manhattan. Their solution has revitalized the area and created a new park space.

Die besondere Herausforderung für die Planer und Architekten bestand darin, für diesen brachliegenden Hochbahngleiskörper mitten in Lower Manhattan eine kreative Nutzung zu finden. Ihre Lösung hat das gesamte Stadtviertel neu belebt, ein neuer Park ist entstanden.

Les designers et les architectes ayant travaillé sur le projet de la High Line dans le bas de Manhattan devaient répondre au défi de trouver une réutilisation créative de cette voie ferrée suspendue abandonnée. Leur projet a revitalisé le quartier et offert à New York un nouvel espace vert.

The High Line has offered new green areas to both residents and tourists, but also served to incite a good deal of construction and renovation along the path of the rail spur.

Die High Line bietet neue Grünräume sowohl für die Anwohner als auch für Touristen. Infolge des neuen Parks wurden entlang der ehemaligen Bahnstrecke viele Sanierungen vorgenommen, und es entstanden Neubauten.

La High Line qui offre de nouveaux espaces verts aux résidents et aux touristes, a entraîné le développement de nombreux projets de construction et de rénovation le long de son tracé.

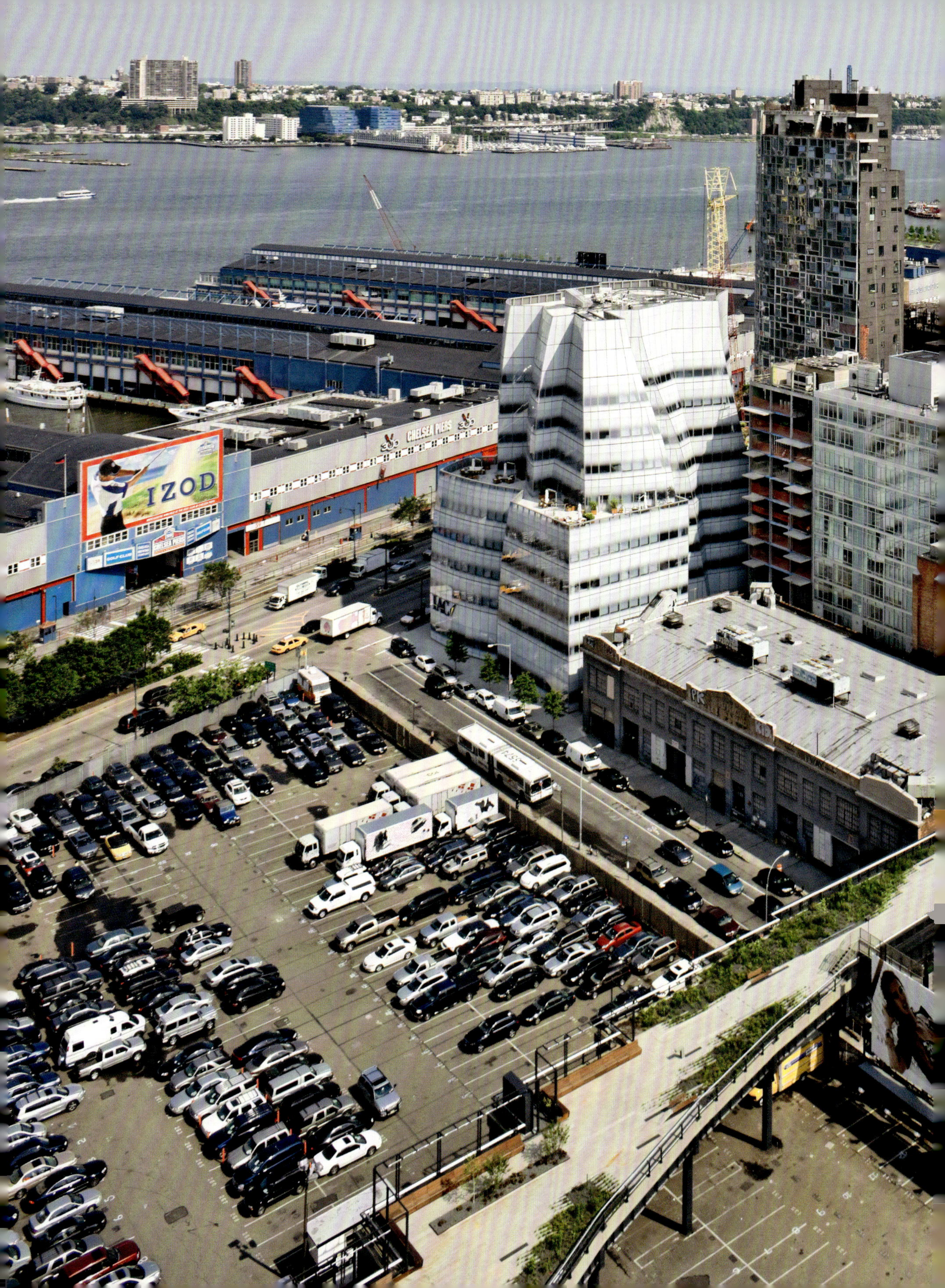
IZOD
CHELSEA PIERS

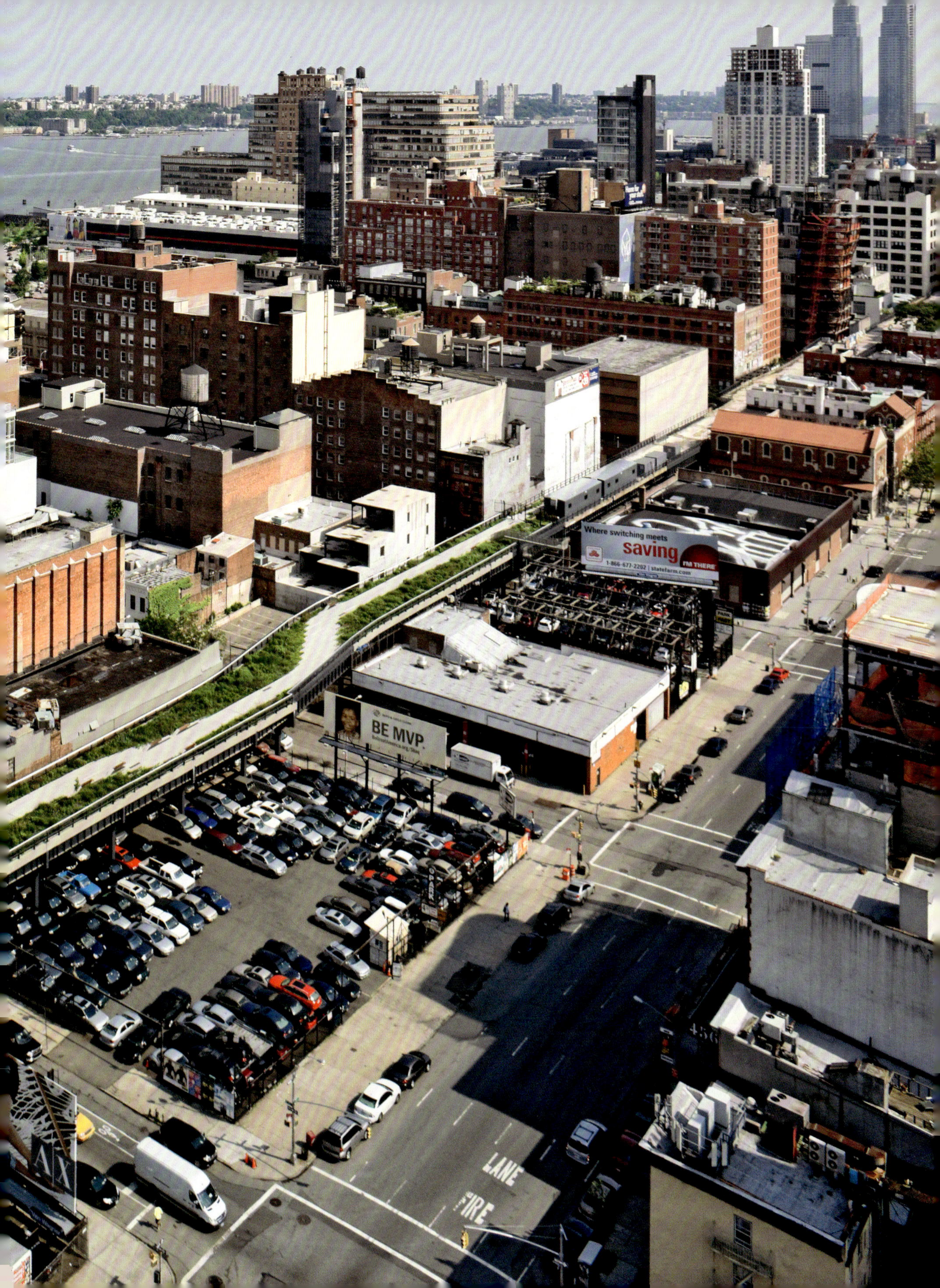
Where switching meets
saving
BE MVP
LANE

THE HIGH LINE is a public park constructed on an elevated railway running about 2.4 kilometers from Gansevoort Street in the Meatpacking District to West 34th Street on Manhattan's West Side. The West Side Improvement Project, including the High Line, was built because of the number of accidents involving the street-level railroad crossing of trains, and was put into effect in 1929. Intended to avoid the negative effects of subway lines over crowded streets, the High Line cuts through the center of city blocks. Increasing truck traffic led to the demolition of parts of the High Line in the1960s and a halt to train operations in 1980. Despite efforts to demolish the remaining structure to allow new construction, a good part of the High Line survived and a group called Friends of the High Line, created in 1999, eventually convinced authorities to renovate it rather than to allow its destruction. James Corner Field Operations, Diller Scofidio + Renfro, Piet Oudolf, and a number of other parties have participated in a collective project for the High Line to renovate and bring new life to a disused part of the city. As they say: "Inspired by the melancholic, unruly beauty of the High Line where nature has reclaimed a once vital piece of urban infrastructure, the team retools this industrial conveyance into a postindustrial instrument of leisure, life, and growth. By changing the rules of engagement between plant life and pedestrians, our strategy of agri-tecture combines organic and building materials into a blend of changing proportions that accommodate the wild, the cultivated, the intimate, and the hyper-social."

THE HIGH LINE ist ein öffentlicher, circa 2,4 Kilometer langer Park, der auf einer ehemaligen Hochbahntrasse zwischen der Gansevoort Street im Stadtviertel Meatpacking District und der West 34th Street auf der West Side von Manhattan angelegt ist. Das Projekt zur Verbesserung der Infrastruktur der West Side, zu dem auch der Bau der High Line gehörte, entstand im Jahr 1929. Die High Line wurde damals gebaut, da es zahlreiche Unfälle gegeben hatte, weil die Züge die Straßen niveaugleich querten. Mit der High Line sollte die negative Wirkung von Schienenverkehr auf beengten Straßenräumen vermieden werden, und so wurde die Schneise für die High Line mitten durch die Blockbebauung des Viertels geschlagen. Aufgrund des wachsenden Lkw-Verkehrs wurde in den 1960er-Jahren die High Line teilweise abgebrochen und der Zugverkehr 1980 ganz eingestellt. Trotz aller Versuche, die Reste der Bahntrasse abzureißen und etwas Neues zu bauen, hat die High Line überlebt. Einer 1999 gegründeten Initiative, den Friends of the High Line, gelang es, die Behörden davon zu überzeugen, die Trasse zu sanieren statt sie abzubrechen. James Corner Field Operations, Diller Scofidio + Renfro, Piet Oudolf und andere Teams haben an dem Gemeinschaftsprojekt zur Sanierung der High Line mitgearbeitet, damit diesem vernachlässigten Teil der Stadt zu neuem Leben verholfen wird. Die Planer kommentieren: „Wir haben uns von der melancholischen, spröden Schönheit der High Line anregen lassen, wo sich die Natur ehemals vitaler, urbaner Infrastruktur zurückerobert hat. Unser Team verwandelt den Gleiskörper in eine postindustrielle Freizeitlandschaft, die mit Leben und mit Wachstum erfüllt ist. Indem wir die Regeln der Beziehungen zwischen dem Pflanzenleben und den Fußgängern verändert haben, verbindet unsere Strategie der Agri-Tektur organische und künstliche Werkstoffe, sodass eine abwechslungsreiche Mischung aus wilden, naturnahen Bereichen, Pflanzenkulturen sowie intimen und hypersozialen Bereichen entsteht."

Le parc public de **LA HIGH LINE** a été aménagé sur une ancienne voie ferrée surélevée d'environ 2,4 km de long entre la rue Gansevoort, dans le quartier de Meatpacking, et la 34ᵉ rue Ouest, dans le West Side de Manhattan. Le projet d'amélioration du West Side, comprenant la High Line, avait été mis en œuvre en 1929 pour réduire le nombre d'accidents aux passages à niveau. La ligne traversait même des « blocs » du centre-ville pour limiter certains effets négatifs des voies suspendues au-dessus de rues encombrées. L'augmentation de la circulation des camions a entraîné la démolition de certaines parties de la High Line dans les années 1960 et la fin de l'exploitation de la voie en 1980. Malgré plusieurs tentatives de la détruire totalement pour faire place à de nouvelles constructions, une bonne partie des voies subsistait encore et, en 1999, une association des amis de la High Line a été créée et a fini par convaincre les autorités municipales de la rénover plutôt que de la détruire. James Corner Field Operations, Diller Scofidio + Renfro, Piet Oudolf et un certain nombre d'autres participants ont préparé un projet collectif de rénovation de la ligne et de revitalisation de ce quartier en déshérence : « Inspirée par la beauté originale et mélancolique de la High Line, où la nature avait repris ses droits sur un élément d'infrastructure urbaine jadis vital, l'équipe a transformé cet équipement industriel en outil post-industriel de loisirs, de vie et de croissance. En modifiant les règles des rapports entre la vie végétale et les piétons, notre stratégie d'agri-tecture combine éléments organiques et matériaux de construction selon des proportions variées, adaptées au caractère sauvage de la nature, aux plantes cultivées, à l'intime et à l'hypersocial. »

In some places, the original rails are left in place to give an indication of the original function of the elevated platform. The High Line connects frequently to the street level (above).

An einigen Orten hat man die Originalgleise einfach belassen, um auf die ursprüngliche Funktion der erhöhten Plattform hinzuweisen. Die High Line ist an vielen Stellen mit dem Straßenniveau verbunden (oben).

Des rails ont été laissés en place à certains endroits pour rappeler la fonction d'origine de ce jardin suspendu. La High line est fréquemment reliée au niveau de la rue (ci-dessus).

Walkways and benches are, of course, part of the scheme that allows walkers to view the essentially industrial architecture of the area from a new vantage point.

Fußwege und Bänke gehören natürlich auch zu dem Entwurf. So können Fußgänger die Industriearchitektur, die dieses Viertel prägt, aus einem neuen Blickwinkel betrachten.

L'aménagement des allées et les bancs font partie du projet, permettant aux promeneurs d'apercevoir l'architecture essentiellement industrielle du quartier sous un nouvel angle.

RAYMOND JUNGLES

Raymond Jungles, Inc.
242 SW 5th Street
Miami, FL 33130
USA

Tel/Fax: +1 305 858 6777
E-mail: raymond@raymondjungles.com
Web: www.raymondjungles.com

RAYMOND JUNGLES was born in Omaha, Nebraska, in 1956. He received a Bachelor of Landscape Architecture degree from the University of Florida (1981) and founded Raymond Jungles, Inc. in 1982. Raymond Jungles cites Roberto Burle Marx, the famous Brazilian garden designer, as his "favorite artist." The two men met in 1979 when Burle Marx came to lecture at the University of Florida. His recent work includes Coconut Grove, Florida Garden (Coconut Grove, Florida, 2003–09, published here); the Key West Botanical Garden (Key West, Florida, 2009); the Brazilian Garden at Naples Botanical Garden (Naples, Florida, 2009); the Golden Rock Inn (Nevis, West Indies, 2009); 1111 Lincoln Road (Miami Beach, Florida, 2009, also published here); and Brazilian Modern, New York Botanical Garden Orchid Show (Bronx, New York, 2009). Current work includes the Miami Beach Botanical Garden Redesign (Miami Beach, Florida, 2011); and the New World Symphony Campus Expansion (Miami Beach, 2011), all in the USA.

RAYMOND JUNGLES wurde 1956 in Omaha, Nebraska, geboren. Er machte 1981 seinen Bachelor in Landschaftsarchitektur an der University of Florida und gründete 1982 sein Büro Raymond Jungles, Inc. Raymond Jungles bezeichnet Roberto Burle Marx, den berühmten brasilianischen Gartenplaner, als seinen „Lieblingskünstler". Die beiden Männer trafen sich 1979, als Burle Marx an der University of Florida einen Vortrag hielt. Zu den neueren Projekten von Jungles gehören der Garten des Wohnhauses Coconut Grove in Florida Garden (Coconut Grove, Florida, 2003–09, hier vorgestellt), der Botanische Garten von Key West (Key West, Florida, 2009), der Brasilianische Garten im Botanischen Garten von Naples (Naples, Florida, 2009), der Garten des Golden Rock Inn auf der zu den Kleinen Antillen gehörenden Insel Nevis (2009), die Gartengestaltung 1111 Lincoln Road (Miami Beach, Florida, 2009, ebenfalls hier vorgestellt) und die Orchideenschau Brasilian Modern im Botanischen Garten von New York (Bronx, New York, 2009). Zu seinen aktuellen Arbeiten in den USA zählen die Neuplanung des Botanischen Gartens in Miami Beach (Miami Beach, Florida, 2011) und die Gartenanlagen des Erweiterungsbaus des Campus der New World Symphony (Miami Beach, 2011).

RAYMOND JUNGLES, né à Omaha (Nebraska) en 1956, est titulaire d'un B.A. d'architecture du paysage de l'université de Floride (1981) et a fondé Raymond Jungles, Inc. en 1982. Roberto Burle Marx, le célèbre concepteur brésilien de jardins, est son « artiste favori », qu'il avait rencontré en 1979, lorsque le paysagiste était venu donner une conférence à l'université de Floride. Parmi ses réalisations récentes (aux États-Unis, sauf mention contraire) : Coconut Grove, Florida Garden (Coconut Grove, Floride, 2003–09, publié ici) ; le jardin botanique de Key West (Key West, Floride, 2009) ; le jardin brésilien du jardin botanique de Naples (Naples, Floride, 2009) ; la Golden Rock Inn (Niévès, Saint-Christophe-et-Niévès, 2009) ; 1111 Lincoln Road (Miami Beach, Floride, 2009, publié ici) et le « Brazilian Modern », exposition d'orchidées du jardin botanique de New York (Bronx, New York, 2009). Parmi ses dernières réalisations aux États-Unis : la restructuration du jardin botanique de Miami Beach (Miami Beach, Floride, 2011) et l'extension du campus du New World Symphony (Miami Beach, 2011).

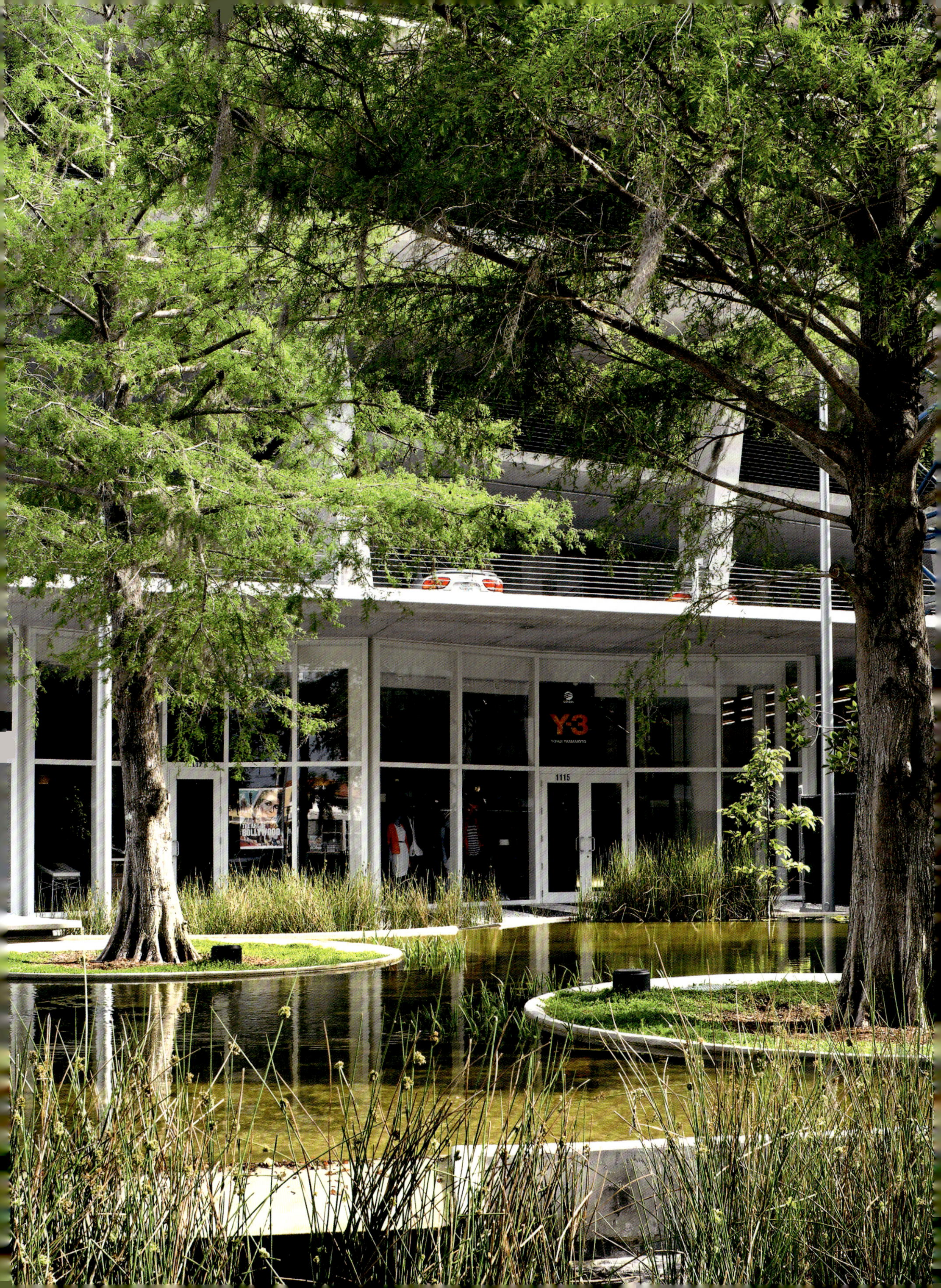
Y-3
1115

1111 LINCOLN ROAD

Miami Beach, Florida, USA, 2009

Address: 1111 Lincoln Road, Miami Beach, FL 33139, USA
Area: 4047 m². Client: Robert Wennett, UIA Management, LLC. Cost: not disclosed

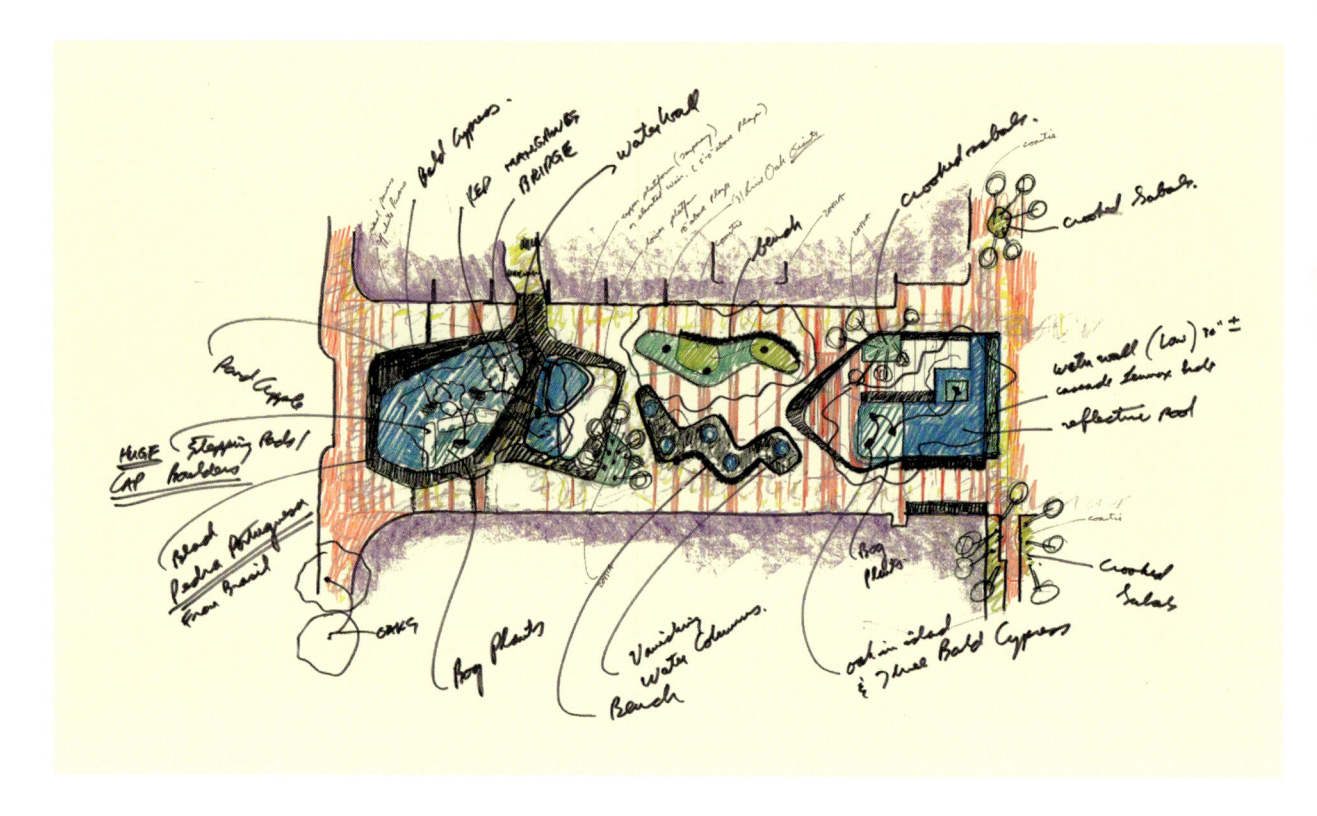

A drawing by the designer (above) shows the freely formed areas of planting and the water features in the garden. Right page, photos of the space.

Eine Zeichnung des Planers (oben) zeigt die intuitiv gestalteten Pflanzbereiche und die Wasserflächen in diesem Garten. Rechte Seite: Fotos dieses Freiraums.

Un croquis du designer (ci-dessus) montre les aires de plantations de formes libres et les bassins prévus dans le jardin. Page de droite : photos de la réalisation.

This project was carried out in collaboration with Herzog & de Meuron, architects of the mixed-use building at **1111 LINCOLN ROAD**. Raymond Jungles states: "It maintains compatibility with Morris Lapidus' original vision of an outdoor 'tropical' setting for shopping, dining, and public gathering." Water gardens, planted areas, and stripes of "Pedra Portuguesa" stone are the main elements of the design. A raised central platform allows for public events. The designer states: "A variety of Florida native trees, palms, and grasses recreate a sense of nature while providing a smooth transition between the building and human scale."

Das Projekt für dieses Gebäude mit einer gemischten Nutzung in der **1111 LINCOLN ROAD** führte Jungles in Zusammenarbeit mit dem Architekturbüro Herzog & de Meuron aus. Raymond Jungles bemerkt dazu: „Der Entwurf ist mit der ursprünglichen Vision von Morris Lapidus kompatibel, der für das Einkaufen, Essengehen und für öffentliche Versammlungen immer einen ‚tropischen' Rahmen schaffen wollte." Wassergärten, bepflanzte Bereiche und Pflasterstreifen, die in Mustern aus schwarzem und weißem Natursteinmaterial ausgeführt sind und nach ihren portugiesischen Vorbildern „Pedra Portuguesa" genannt werden, sind die wesentlichen Elemente dieses Entwurfs. Der Planer bemerkt dazu: „Verschiedene in Florida heimische Bäume, Palmen und Gräser vermitteln hier den Endruck von Natur, und zugleich schaffen sie einen sanften Übergang zwischen dem Gebäude und dem Außenraum."

Ce projet a été réalisé en collaboration avec Herzog & de Meuron, architectes de l'immeuble mixte du **1111 LINCOLN ROAD** à Miami Beach. Raymond Jungles précise : « Il [le projet] reste compatible avec la vision d'origine de Morris Lapidus d'un cadre extérieur "tropical" conçu pour faire ses courses, dîner et se rencontrer. » Des jardins d'eau, des zones plantées et des bandeaux de pierre (*pedra portuguesa*) constituent les principaux éléments de la proposition. Une plate-forme centrale surélevée est prévue pour des manifestations publiques. « Une grande variété d'arbres originaires de Floride, de palmiers et de graminées recrée un sentiment de nature tout en offrant une transition douce entre l'immeuble et l'échelle humaine », explique l'architecte.

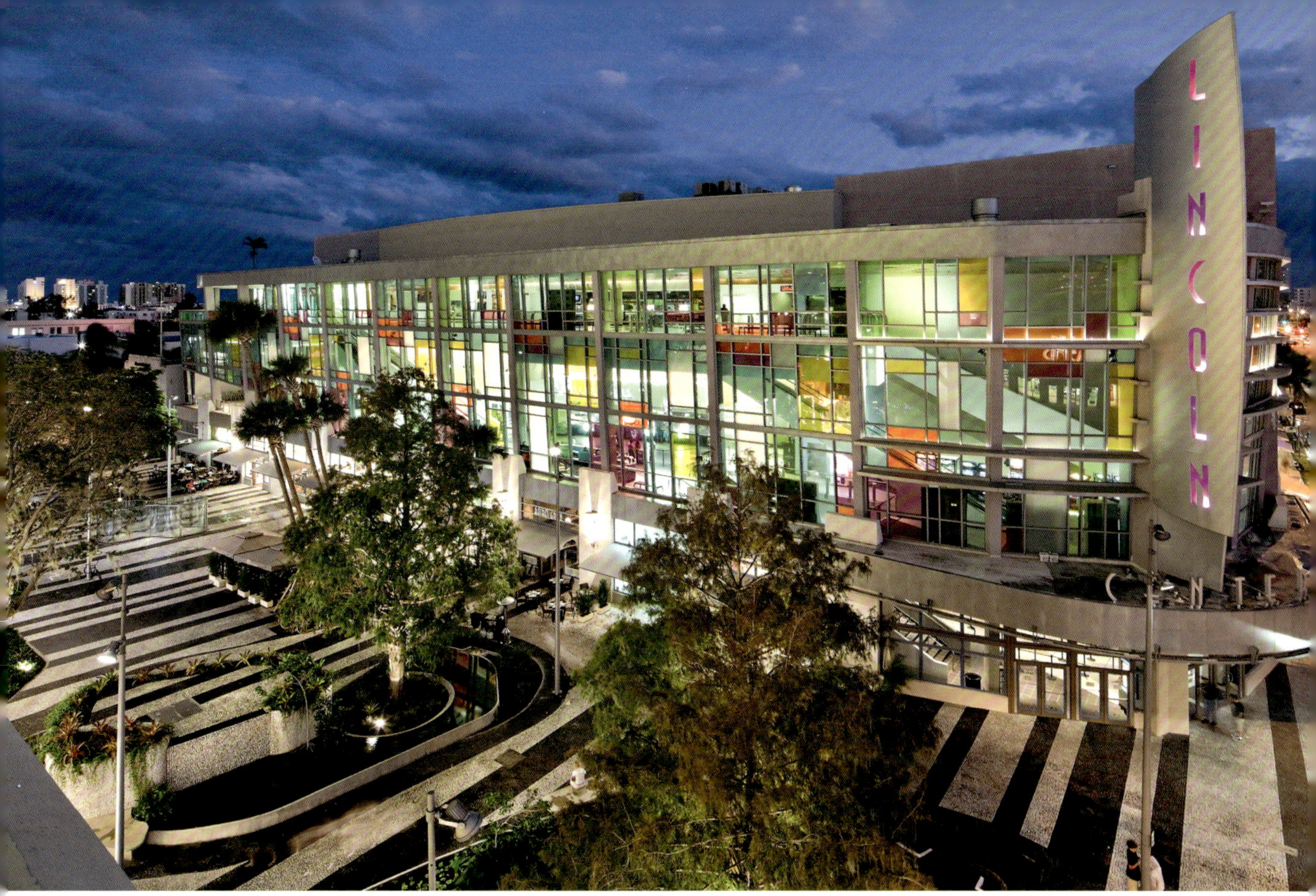
LINCOLN

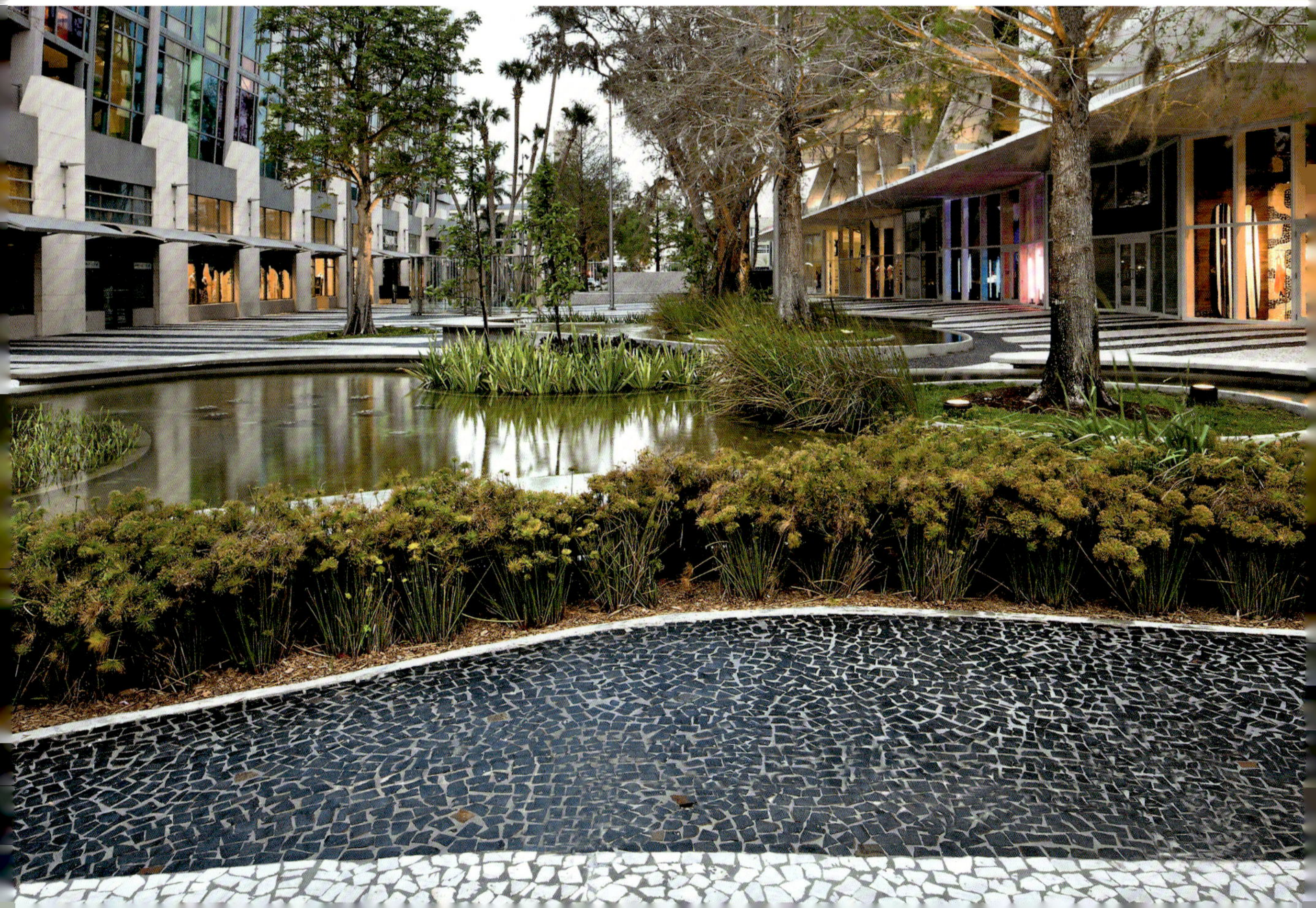

The luxuriant planting of course takes into account the climate of Miami and the fact that these outdoor spaces will be used much more frequently than gardens located further north.

Das Klima in Miami ermöglicht eine üppige Bepflanzung, und deshalb wird dieser Freiraum sicherlich auch intensiver und häufiger genutzt als weiter nördlich gelegene Gartenanlagen.

La luxuriance des plantations reflète le climat de Miami. Grâce à lui, ces espaces verts sont utilisés toute l'année, beaucoup plus fréquemment que ceux du nord des États-Unis.

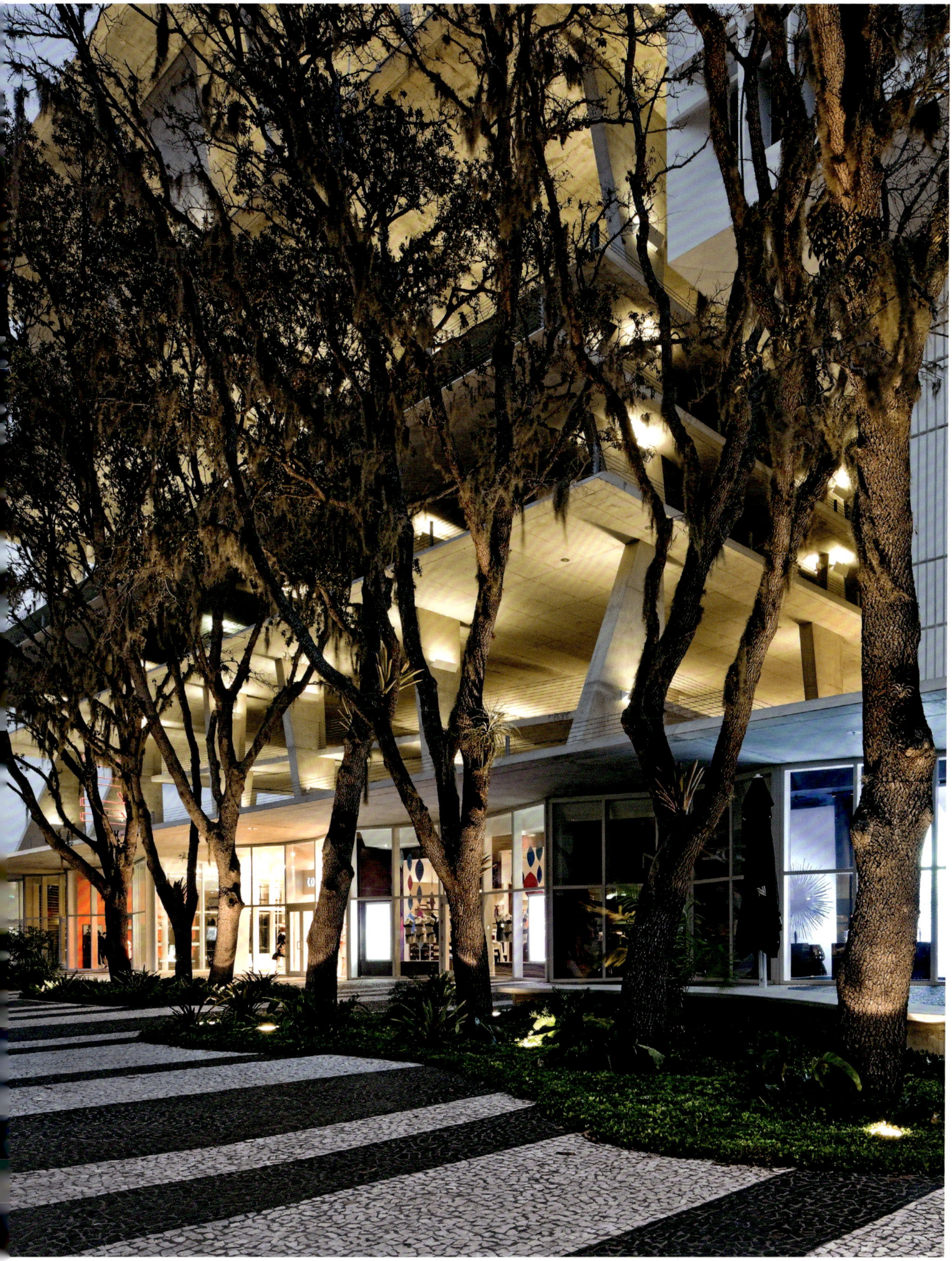

COCONUT GROVE

Florida Garden, Coconut Grove, Florida, USA, 2003–09

Area: 1.36 hectares. Client: not disclosed. Cost: not disclosed
Architecture: Alison Spear, Proun Space Studio, John Bennett and Gustavo Bonevardi

Raymond Jungles worked for six years on the grounds of this family compound. A large 80-year-old East Indian Banyan tree marks a ridge on the property, straddling three residential lots. The owner acquired 11 other properties and razed four residential structures in the course of the project, allowing for extra garden areas. The designers state: "From their first meeting, the design group decided the unifying elements for the entire site would be landforms and sculptural spaces, defined with monolithic slabs of oolite limestone. More than 500 cubic yards (382 cubic meters), which represents 80% of the stone used, was excavated on site. More than 800 indigenous, habitat-producing trees, including more than 500 palms, cycads from around the world, and fragrant flowering trees and shrubs enhance the well-balanced spaces, imparting a provocative sense of variety."

Raymond Jungles hat sechs Jahre auf dem Grundstück dieses Familienbesitzes gearbeitet. Ein großer, 80 Jahre alter, ursprünglich aus Ostindien stammender Banyanbaum wächst auf dem höchsten Punkt des Grundstücks, das in drei Parzellen unterteilt ist. Der Besitzer kaufte elf weitere Grundstücke und riss im Lauf der Bauarbeiten vier Wohnhäuser ab, um Platz für weitere Gartenbereiche zu erhalten. Die Planer merken an: „Schon beim ersten Treffen hat das Entwurfsteam beschlossen, das gesamte Gelände durch eine Modulierung der Landschaft und durch skulptural gestaltete Freiräume zu vereinheitlichen, die mit monolithischen Platten aus Kalkoolith gestaltet werden. Mehr als 382 Kubikmeter Naturstein, das sind 80 Prozent des verwendeten Materials, sind vor Ort gebrochen worden. Mehr als 800 heimische Bäume haben hier ihren neuen Standort gefunden, unter anderen mehr als 500 Palmen. Palmfarne (Cycadales) aus allen Regionen der Welt, Bäume und Gehölze mit duftenden Blüten sind die Anziehungspunkte der ausgewogenen Freiräume und vermitteln ein Gefühl von Vielfalt."

Raymond Jungles a travaillé six ans sur les plans de ce domaine familial. Un important banian d'Inde orientale, vieux de 80 ans, marque une butte sur la propriété et détermine trois parcelles résidentielles. Au cours du projet, le propriétaire a acquis onze autres propriétés et rasé quatre maisons pour agrandir ses jardins. « Dès notre première réunion, le groupe chargé de la conception a décidé que les éléments unificateurs du site seraient des mouvements de terrain et des espaces sculpturaux définis par des dalles monolithiques de calcaire oolithique. Plus de 382 m³ de pierre – soit 80 % du total utilisé – ont été trouvés sur place. Plus de 800 arbres de la région, dont plus de 500 palmiers, fougères du monde entier, arbres et buissons à fleurs odorantes, animent ces espaces bien équilibrés en imposant un sentiment provoquant de variété », explique l'architecte.

Above, an elevation drawing showing the residence and a plan of the grounds located near the house. To the right, the house and swimming pool, surrounded by the essentially tropical vegetation.

Oben, Ansicht des Wohnhauses und Lageplan der unmittelbar an das Haus angrenzenden Gartenbereiche. Das Wohnhaus mit dem Swimmingpool (rechts) ist vorwiegend von tropischer Vegetation umgeben.

Ci-dessus, dessin d'élévation de la résidence et plan des jardins. À droite, la maison et sa piscine, noyées dans une végétation de nature essentiellement tropicale.

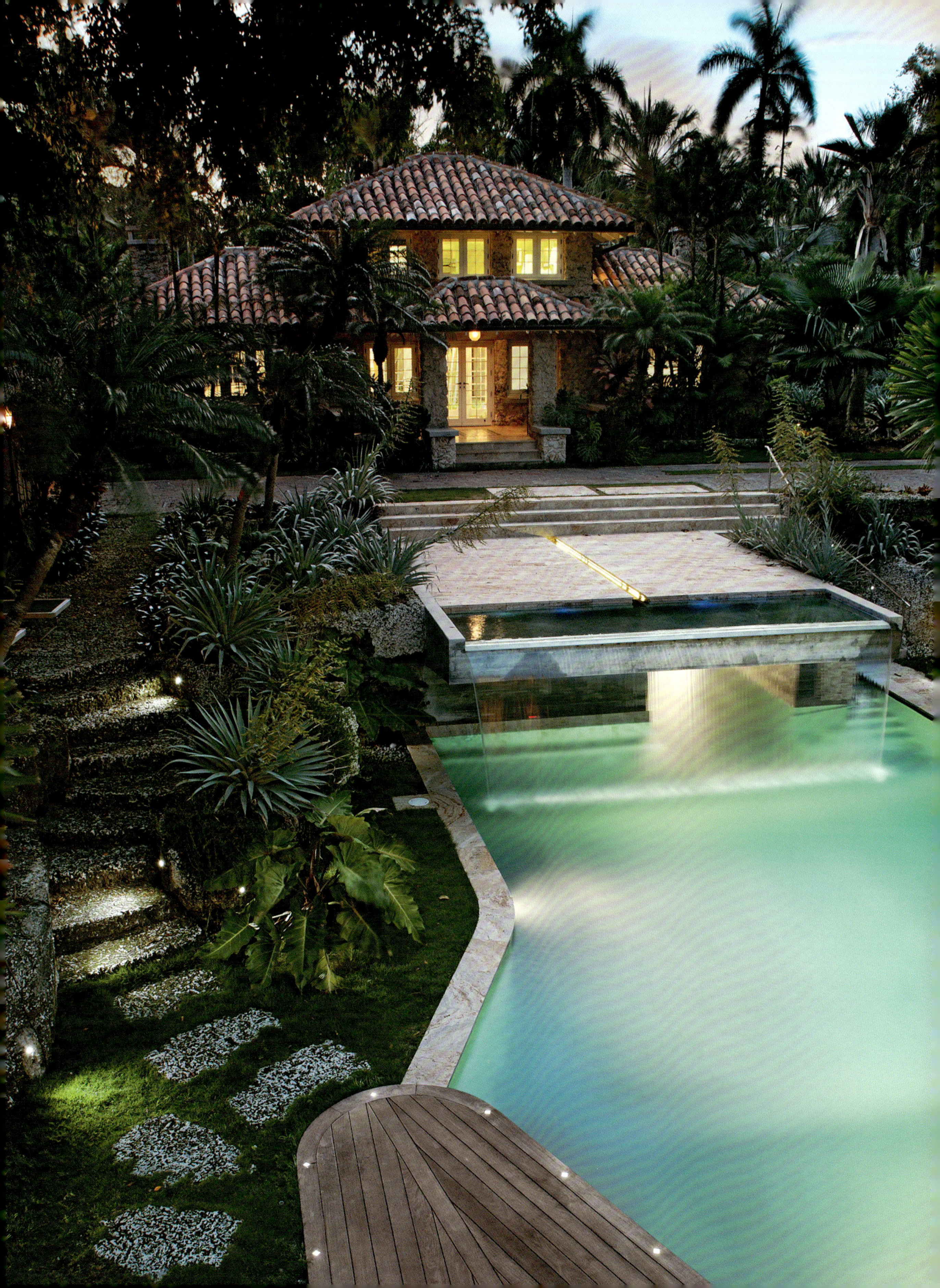

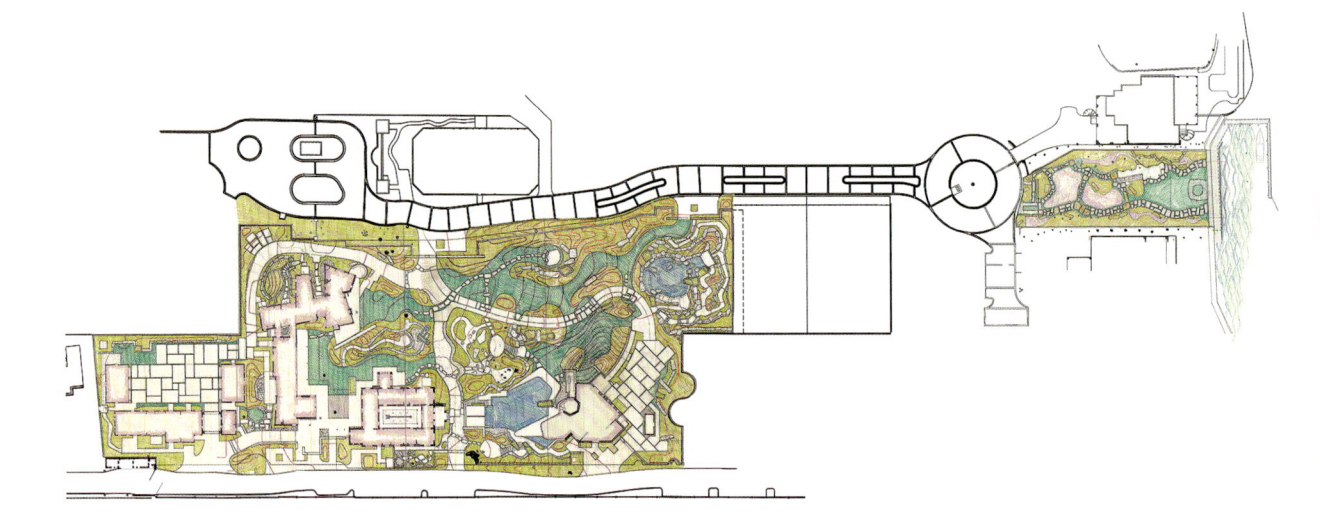

Above, a plan of the entire compound showing both the various structures and the paths that lead residents through the garden. Below, a more modern house and pool in the grounds.

Der Gesamtplan des Anwesens, oben, zeigt neben den Gebäuden auch die Wege, auf denen die Bewohner durch den Garten geleitet werden. Unten, ein modernes Haus mit Pool auf dem Gelände des Anwesens.

Ci-dessus, plan d'ensemble de la propriété montrant à la fois les différentes constructions et les allées qui sillonnent le jardin. Ci-dessous, toujours dans la même propriété, une maison moderne et sa piscine.

The landscape architect varies both the nature of the vegetation and its height, giving great variety to the treatment of the outside areas.

Der Landschaftsarchitekt spielt mit der Art und mit der Höhe der Vegetation, sodass die Außenbereiche sehr vielfältig gestaltet sind.

Les variations dans la nature des plantations et leur hauteur ont permis une grande variété de traitements.

A broad, stone path leads to a house with a waterfall. Mature trees, in good part palms in this image, alternate with lower shrubs and tropical plants near the grass.

Ein breiter, mit Natursteinplatten befestigter Weg führt zu einem Haus mit einem Wasserfall. Alte Bäume, auf diesem Bild vor allem Palmen, wechseln sich mit niedrigen Gehölzen und tropischen Pflanzen ab.

Une large allée dallée de pierre conduit à une maison près d'une cascade. De grands arbres adultes, surtout des palmiers, alternent avec des arbustes et des plantes tropicales basses.

The extent of the property and its different types of architecture allow Raymond Jungles to create an entire environment of plants and basins that make the domain into a place apart from its surroundings.

Angesichts der Größe des Anwesens und der unterschiedlichen Architekturen konnte Raymond Jungles hier eine ganze Welt von Pflanzen und Wasserbecken schaffen, die das Anwesen in einen von seiner Umgebung völlig abgeschiedenen Ort verwandelt.

L'étendue de la propriété et la variété des types d'architecture ont permis à Raymond Jungles de créer un environnement de plantes et de bassins qui singularise ce domaine dans son environnement.

LEDERER + RAGNARSDÓTTIR + OEI / HELMUT HORNSTEIN

Lederer + Ragnarsdóttir + Oei
Kornbergstr. 36
70176 Stuttgart
Germany

Tel: +49 711 225 50 60
Fax: +49 711 22 55 06 22
E-mail: mail@archlro.de
Web: www. archlro.de

ARNO LEDERER was born in 1947 in Stuttgart, Germany, and studied architecture in Stuttgart and Vienna, graduating in 1976. He worked for Ernst Gisel in Zurich (1977) and BHO (Berger Hauser Oed) in Tübingen (1978). He was a freelance architect beginning in 1979. Since 1985 he has been running a joint studio together with Jórunn Ragnarsdóttir, and since 1992 also with Marc Oei. In 1997, he was appointed Chair of Building Sciences at the Faculty of Architecture of the University of Karlsruhe. Since 2005 he has been the head of the Institute for Public Buildings and Design at the University of Stuttgart. **JÓRUNN RAGNARSDÓTTIR** was born in 1957 in Akureyri, Iceland. She studied architecture in Stuttgart, graduating in 1982. **MARC OEI** was born in 1962 in Stuttgart, where he graduated with an architecture degree in 1988. Their work includes the Hessian State Theater (Darmstadt, 2006); University of Cooperative Education (Loerrach, 2008); Georg-Büchner-Plaza (Darmstadt, 2009–10, published here); and the Bestehornpark Educational Center (Aschersleben, 2010). Their current work is focused on three museums: Art Museum (Ravensburg, 2012); City Museum (Stuttgart, 2013); and the Historical Museum (Frankfurt, 2014), all in Germany.

ARNO LEDERER wurde 1947 in Stuttgart geboren und studierte Architektur in Stuttgart und Wien. 1976 schloss er sein Studium ab. Er arbeitete für Ernst Gisel in Zürich (1977) und im Büro BHO (Berger Hauser Oed) in Tübingen (1978). Seit 1979 arbeitet er als freiberuflicher Architekt, seit 1985 in Bürogemeinschaft mit Jórunn Ragnarsdóttir, 1992 ist Marc Oei als dritter Partner hinzugekommen. 1997 übernahm Lederer den Lehrstuhl für Gebäudelehre an der Universität Karlsruhe. Seit 2005 ist er Leiter des Instituts für öffentliche Bauten und Entwerfen an der Universität Stuttgart. **JÓRUNN RAGNARSDÓTTIR** wurde 1957 in Akureyri, Island, geboren. Sie studierte in Stuttgart Architektur und diplomierte 1982. **MARC OEI** wurde 1962 in Stuttgart geboren, studierte dort Architektur und machte 1988 sein Diplom. Zu den Arbeiten des Büros gehören das Hessische Staatstheater (Darmstadt, 2006), die Berufsakademie (Lörrach, 2008), der Georg-Büchner-Platz in Darmstadt (2009–10, hier vorgestellt) und das Bildungszentrum Bestehornpark (Aschersleben, 2010). Zu ihren laufenden Planungen gehören vor allem drei Museumsbauten: das Kunstmuseum in Ravensburg (2012), das Stadtmuseum in Stuttgart (2013) und das Historische Museum in Frankfurt am Main (2014).

ARNO LEDERER, né en 1947 à Stuttgart, en Allemagne, a étudié l'architecture à Stuttgart et Vienne, et a été diplômé en 1976. Il a travaillé pour Ernst Gisel à Zurich (1977), pour BHO (Berger Hauser Oed) à Tübingen (1978) et a été architecte freelance à partir de 1979. Depuis 1985, il dirige une agence en association avec Jórunn Ragnarsdóttir et Marc Oei, qui les a rejoints en 1992. En 1997, il a été nommé titulaire de la chaire de sciences de la construction à la faculté d'architecture de l'université de Karlsruhe. Depuis 2005, il dirige l'Institut des bâtiments publics et de la conception à l'université de Stuttgart. **JÓRUNN RAGNARSDÓTTIR,** né en 1957 à Akureyri en Islande, a étudié l'architecture à Stuttgart (diplômé en 1982). **MARC OEI** est né en 1962 à Stuttgart, où il a également obtenu son diplôme d'architecture (1988). Parmi leurs réalisations en Allemagne : le Théâtre d'État de la Hesse (Darmstadt, 2006) ; l'institut technologique (Lörrach, 2008) ; la Georg-Büchner-Platz (Darmstadt, 2009–10, publiée ici) et le centre éducatif Bestehornpark (Aschersleben, 2010). Ils travaillent actuellement beaucoup sur des projets de musées : le Musée d'art de Ravensburg (2012) ; le musée de la Ville de Stuttgart (2013) et le Musée historique de Francfort (2014).

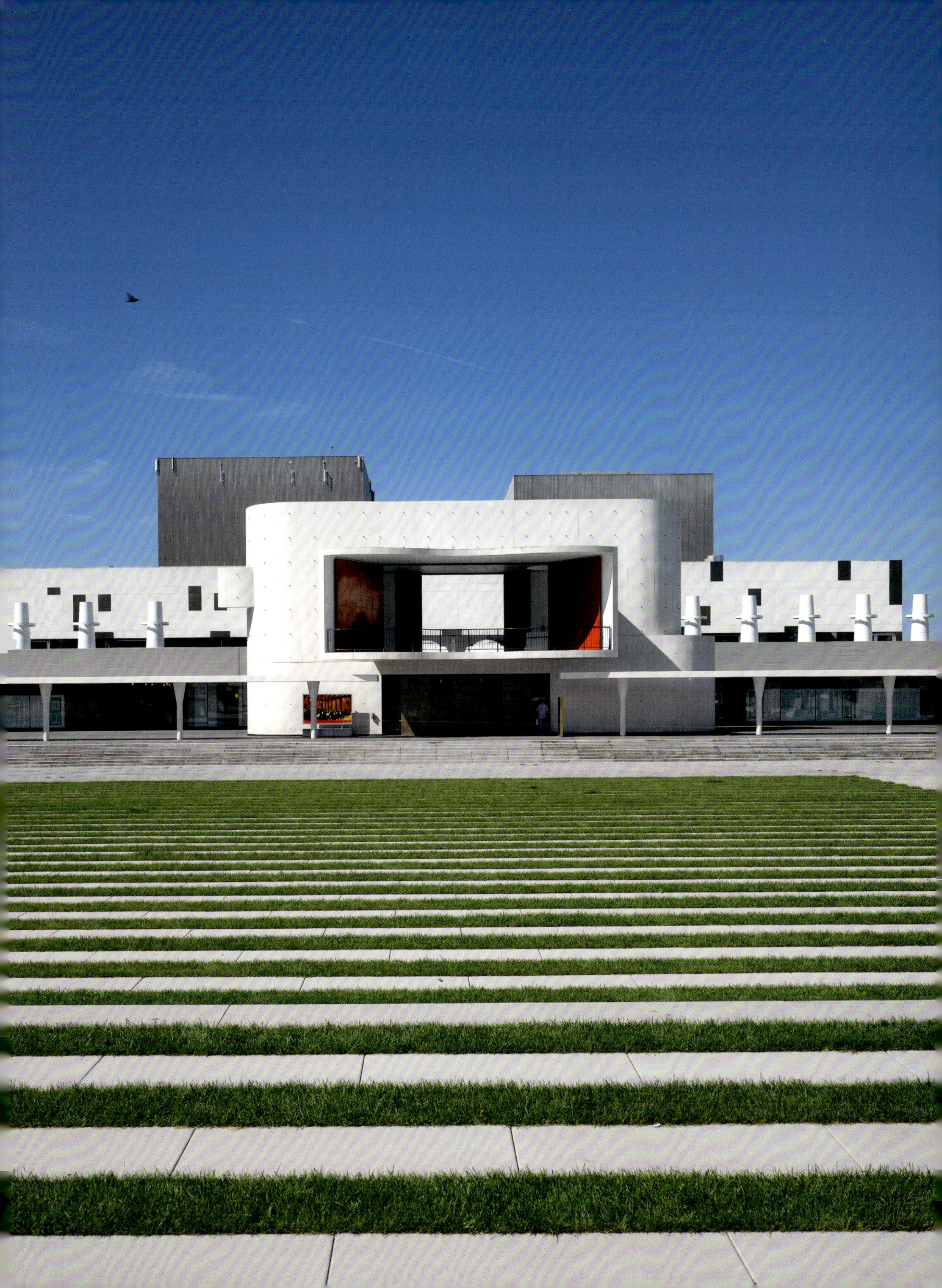

GEORG-BÜCHNER-PLAZA
Darmstadt, Germany, 2009–10

Area: 10 000 m². Client: State of Hesse (Hessisches Baumanagement, Regionalniederlassung Süd)
Cost: €4.8 million (plaza), €7.2 million (car park refurbishment)
Collaboration: Wolfram Sponer, Andrea Stahl, Christian Horle

Both the planting and the architectural treatment of the space are extremely rigorous, with the dominant theme being rows of trees, or rows of parasol seating arrangements.

Sowohl die Bepflanzung als auch die Architektur an diesem Platz sind sehr streng. Dominierendes Thema sind die Baumreihen und die parallel angeordneten Sitzgelegenheiten unter den feststehenden Sonnenschirmen.

Les plantations comme le traitement architectural de l'espace sont extrêmement rigoureux. Le thème dominant est donné par l'alignement des arbres, des parasols et des banquettes.

Located next to the Hessian State Theater in Darmstadt, designed by the architects (2006), this new plaza has a middle zone with strips of lawn that acts as a forecourt to the theater. The architects state: "It is oriented to the theater's impressive new portal, which is also used as on outdoor stage, and enhances the spatial relation of the theater to the city." On either side, rows of Japanese pagoda trees encourage pedestrians to walk or sit. Stairways lead from the plaza to the refurbished underground parking area. These stairways bring light into the garage zone which has space for 342 cars. A basin with a work by the sculptor Arnaldo Pomodoro (*Grande disco*) is located at the eastern side of the plaza. The architects conclude: "The closely related design vocabulary in the signature elements of the plaza and the theater and the use of the same materials create a coherent overall impression."

Der neue Platz ist dem 2006 ebenfalls von den Architekten entworfenen Hessischen Staatstheater in Darmstadt vorgelagert. Die leicht terrassierte Mitte des Platzes mit ihren Rasenstreifen dient als Vorplatz für das Theater. Die Architekten kommentieren dazu: „Dieser Bereich steht in engem Dialog mit dem neuen, repräsentativen Portal, das auch als Bühne nutzbar ist, dadurch wird der räumliche Bezug zwischen dem Theater und der Stadt hervorgehoben." Auf beiden Seiten sind Reihen mit japanischen Schnurbäumen gepflanzt, unter deren Kronen die Fußgänger flanieren oder sich hinsetzen können. Von der Platzanlage führen Treppen in die sanierte Tiefgarage hinunter. Über diese Treppen fällt Tageslicht in die Tiefgarage mit 342 Stellplätzen. Nach Osten wird der Platz durch eine Wasserfläche mit einer Skulptur des Bildhauers Arnaldo Pomodoro („Grande disco") begrenzt. Die Architekten bemerken abschließend: „Die eng miteinander verwandte Formensprache der Platzelemente und des Theaters sowie die Verwendung der gleichen Materialien erzeugen ein einheitliches Gesamtbild."

Située face au Théâtre d'État de la Hesse de Darmstadt conçu par l'agence en 2006, cette nouvelle place se caractérise par sa zone centrale, striée de bandes de gazon, qui sert d'avant-cour au théâtre. « Le projet orienté vers l'impressionnant nouveau balcon du théâtre, qui peut également servir de scène de plein air, améliore la relation spatiale entre le théâtre et la ville », expliquent les architectes. De chaque côté, des alignements de sophoras du Japon invitent les passants à la promenade ou à s'asseoir sous leur ombrage. Des escaliers mènent de la place aux parkings souterrains rénovés. Un bassin orné d'une œuvre du sculpteur italien Arnaldo Pomodoro (*Grande disco*) a été installé à l'extrémité de la place. « Un vocabulaire conceptuel étroitement lié aux éléments forts de la place et au théâtre, ainsi que le recours à des matériaux identiques, crée une impression d'ensemble cohérente », concluent les architectes.

The design of the park fits well with the nearby Hessian State Theater, also designed by the architects.

Der Entwurf des Parks passt gut zur Architektur des Hessischen Staatstheaters, das vom selben Büro entworfen wurde.

La conception du parc est parfaitement adaptée à l'esthétique architecturale du Théâtre d'État de la Hesse, conçu par les mêmes architectes.

A circular sculpture is set on a shallow pool of water. Below, an elevation drawing shows how the slightly sloped park is set above the underground garage of the theater.

Eine kreisförmige Skulptur steht in einem flachen Wasserbecken. Die Ansichtszeichnung unten zeigt, wie das leicht abfallende Gelände des Parks auf der Tiefgarage des Theaters angelegt ist.

Une sculpture circulaire se dresse au centre d'un bassin plat. Ci-dessous, dessin de coupe montrant comment les parkings ont été insérés sous le parc légèrement incliné vers le théâtre.

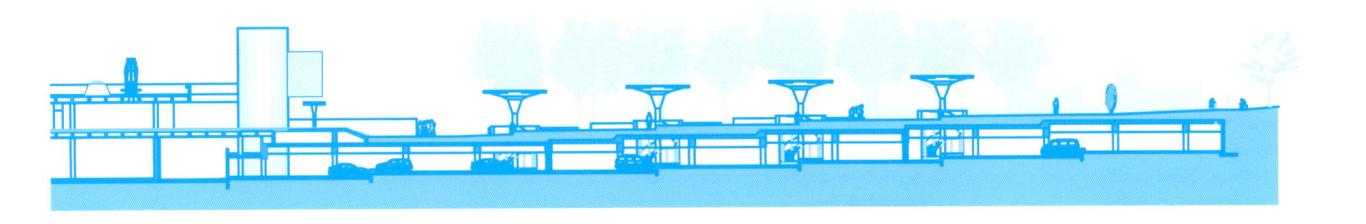

Storm King Wavefiel

MAYA LIN

Maya Lin Studio
112 Prince Street, no. 4
New York, NY 10012
USA

Tel: +1 212 941 6463 / Fax: +1 212 941 6464
E-mail: studio@mlinstudio.com
Web: www.mayalin.com

MAYA LIN was born in Athens, Ohio, in 1959. She attended Yale College and the Yale School of Architecture, receiving her M.Arch in 1986. She created her office, Maya Lin Studio, in New York the same year. By that time, she had already created what remains her most famous work, the Vietnam Veterans' Memorial on the Mall in Washington, D.C. (1981). Other sculptural work includes her Civil Rights Memorial, in Montgomery (Alabama, 1989); and Groundswell, at the Wexner Center of the Arts (Columbus, Ohio, 1993). She completed the design for the Museum of African Art in New York (with David Hotson, 1993); the Weber Residence (Williamstown, Massachusetts, 1994); the Asia/Pacific/American Studies Department (New York University, New York, 1997); and the Langston Hughes Library, Children's Defense Fund (with Martella Associates, Architects; Clinton, Tennessee, 1999). She has also completed the Greyston Bakery, Greyston Foundation (Yonkers, New York, 2003); the Riggio-Lynch Chapel, Children's Defense Fund (Clinton, Tennessee, 2004); Eleven Minute Line (Wanås, Sweden, 2004, published here); and Storm King Wavefield (Mountainville, New York, 2007–08, also published here). Work currently in progress includes the Confluence Project, a seven-part art installation (Columbia River Basin, 2006–); Folding the Field (The Farm, Kaipara Bay, New Zealand, 2010); and The Meeting Room (Queen Anne Square, Newport, Rhode Island), all in the USA unless stated otherwise.

MAYA LIN wurde 1959 in Athens, Ohio, geboren. Sie studierte am Yale College sowie der Yale School of Architecture und machte 1986 ihren Master. Im gleichen Jahr gründete sie ihr Büro Maya Lin Studio in New York. Damals hatte sie bereits ihr bislang bekanntestes Werk geschaffen, das Vietnam Veterans' Memorial auf der Mall in Washington, D. C. (1981). Weitere skulpturale Arbeiten sind das Civil Rights Memorial in Montgomery (Alabama, 1989) und „Groundswell" (Dünung) im Wexner Center of the Arts (Columbus, Ohio, 1993). Sie entwarf das Museum of African Art in New York (mit David Hotson, 1993), das Wohnhaus Weber (Williamstown, Massachusetts, 1994), die Fakultät für Asien-/Pazifik-/Amerika-Studien (New York University, New York, 1997) und die Langston Hughes Library des Children's Defense Fund (mit Martella Associates Architects, Clinton, Tennessee, 1999). Zu ihren Projekten gehören auch die Bäckerei Greyston der Stiftung Greyston (Yonkers, New York, 2003), die Riggio-Lynch Chapel des Children's Defense Fund (Clinton, Tennessee, 2004), die Erdinstallation „Eleven Minute Line" (Wanås, Schweden, 2004, hier vorgestellt) und ihre Arbeit „Wavefield" für das Storm King Art Center (Mountainville, New York, 2007–08, ebenfalls hier vorgestellt). Zu den derzeit im Bau befindlichen Arbeiten von Maya Lin gehören „Confluence Project", eine Installation von sieben Kunstwerken (Columbia River Basin, seit 2006), die Installation „Folding the Field" (The Farm, Kaipara Bay, Neuseeland, 2010) und „The Meeting Room" (Queen Anne Square, Newport, Rhode Island), alle in den USA, sofern nicht anders angegeben.

MAYA LIN, née à Athens (Ohio) en 1959, a étudié l'architecture au Collège Yale et à l'École d'architecture Yale, où elle a reçu son M.Arch. en 1986. Elle a fondé son agence, Maya Lin Studio, à New York la même année. Elle avait déjà créé ce qui reste son œuvre la plus connue : le mémorial aux Vétérans de la guerre du Vietnam, sur le Mall à Washington (1981). Elle a réalisé d'autres œuvres d'esprit sculptural comme le mémorial des Droits civiques à Montgomery (Alabama, US, 1989) et *Groundswell*, au Wexner Center of the Arts (Columbus, Ohio, US, 1993). Elle a conçu le Musée de l'art africain de New York (avec David Hotson, 1993) ; la Weber Residence (Williamstown, Massachusetts, 1994) ; le bâtiment du département des études Asie/Pacifique/Amérique (université de New York, New York, 1997) ; la bibliothèque Langston Hughes, Children's Defense Fund (avec Martella Associates, architectes, Clinton, Tennessee, 1999) ; la boulangerie Greyston, Greyston Foundation (Yonkers, New York, 2003) ; la chapelle Riggio-Lynch, Children's Defense Fund (Clinton, Tennessee, 2004) ; la *Eleven Minute Line* (Wanås, Suède, 2004, publiée ici) et le *Storm King Wavefield* (Mountainville, New York, 2007–08, publié ici). Elle travaille actuellement sur le *Confluence Project*, une installation artistique en sept parties (bassin du fleuve Columbia, 2006–) ; *Folding the Field* (The Farm, Kaipara Bay, Nouvelle-Zélande, 2010) et *The Meeting Room* (Queen Anne Square, Newport, Rhode Island). Toutes ces réalisations sont situées aux États-Unis, sauf mention contraire.

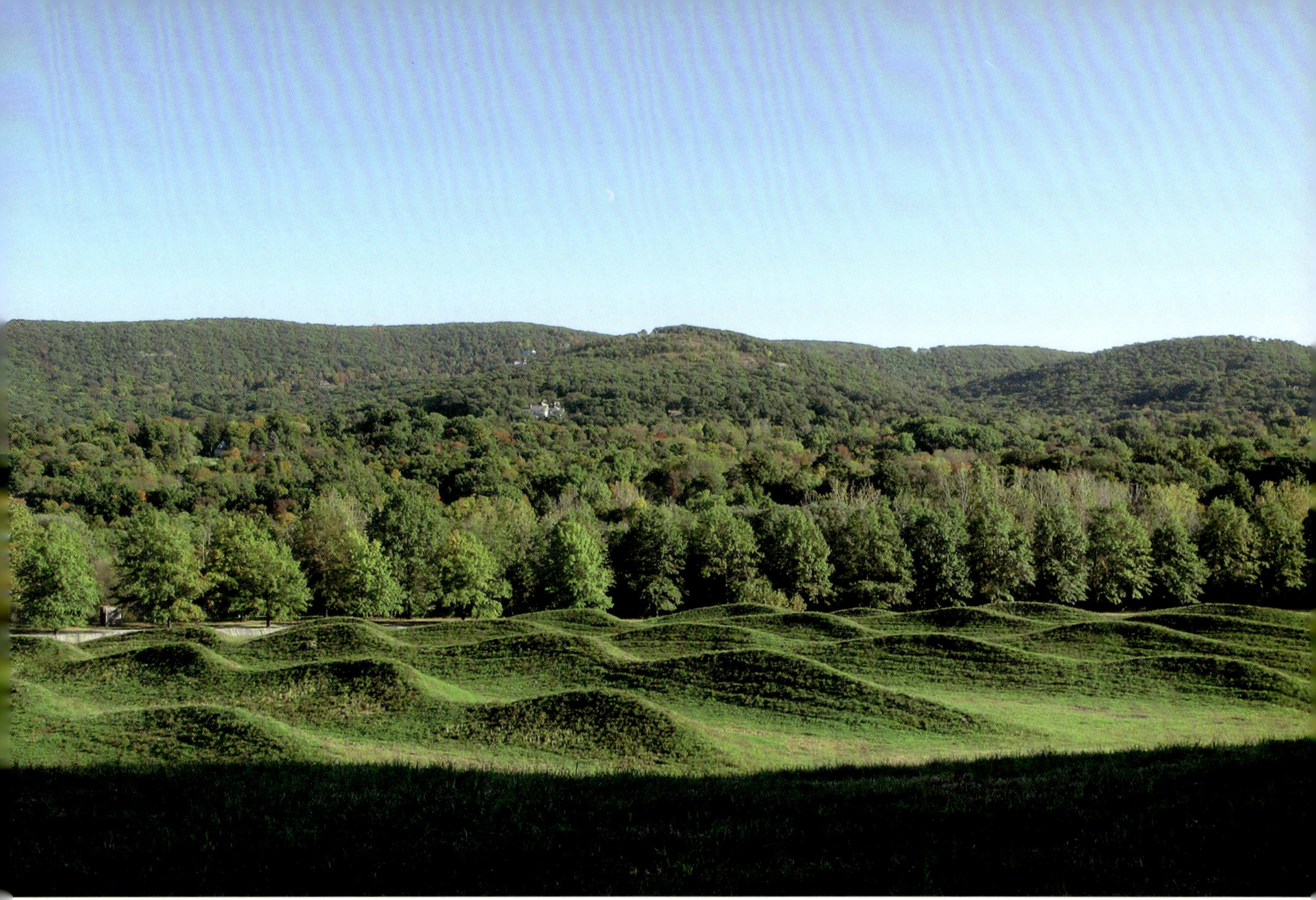

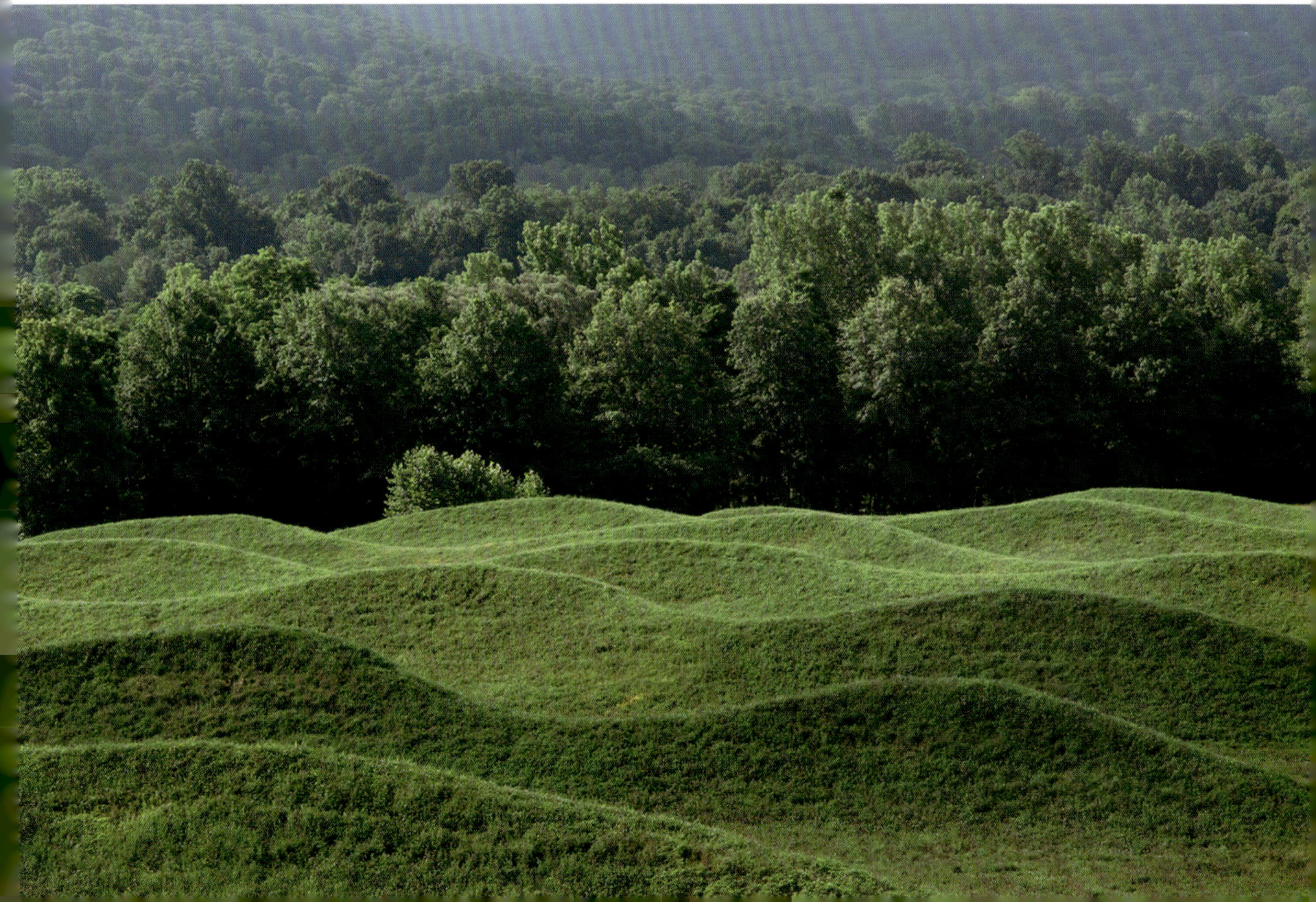

STORM KING WAVEFIELD

Mountainville, New York, USA, 2007–08

Address: Storm King Art Center, PO Box 280, Old Pleasant Hill Road, Mountainville, NY 10953, USA, +1 845 534 3115, www.stormking.org. Area: 1.6 hectares
Client: Storm King Art Center. Cost: not disclosed

The site of this work occupies a total of 4.5 hectares, but the actual earthworks cover 1.6 hectares. Formerly a gravel pit, the site is an environmental reclamation project that complies with the regulations of the New York State Department of Environmental Conservation. The **WAVEFIELD** used a maximum amount of soil from the site and has a naturally made drainage system beneath the soil. Native grasses with minimal watering needs were used, emphasizing the sustainable nature of the work. The carbon footprint of the construction of the piece was calculated and is to be offset by the planting of 260 indigenous trees. The work is made up of seven rows of earth and grass with a trough-to-trough distance of about 12 meters. According to the artist: "Because it is executed in the same scale as an actual set of waves, the viewer's experience is similar to that of being at sea, where one loses visual contact with adjacent waves. Compound curves allow for a complex and subtle reading of the space in the form of an environment that pulls the viewer into its interior and creates a sense of total immersion."

Das Gelände dieses Projekts nimmt insgesamt 4,5 Hektar ein, die derzeitigen Erdarbeiten umfassen 1,6 Hektar. Das Gelände, eine ehemalige Kiesgrube, gehört zu einem Renaturierungsprojekt, das den Bestimmungen des Ministeriums für Umweltschutz des Staates New York entspricht. Für das **„WELLENFELD"** wurde zum Großteil Erde von dem Gelände verwendet, und unter dem Erdreich befindet sich ein natürliches Drainagesystem. Es wurden nur heimische Gräser verwendet, die sehr wenig Wasser benötigen. Dadurch wird die Nachhaltigkeit der Arbeit hervorgehoben. Die CO_2-Bilanz für den Bau dieses Kunstwerks ist berechnet worden und muss durch die Pflanzung von 260 heimischen Bäumen ausgeglichen werden. Das Projekt besteht aus sieben Reihen von Erd- und Graswellen mit einem Abstand von zwölf Metern. Die Künstlerin bemerkt dazu: „Da die Arbeit im gleichen Maßstab wie echte Wellen ausgeführt ist, hat der Betrachter den Eindruck, er befände sich auf dem Meer, wo man die nächste Welle aus den Augen verliert. Dicht aufeinanderfolgende Wellenlinien ermöglichen es, den Freiraum in einer komplexen und subtilen Weise zu lesen, sodass der Betrachter in das Innere des Umfelds hineingezogen wird und das Gefühl hat, gänzlich in das Objekt eingetaucht zu sein."

Si le site de cette œuvre couvre 4,5 ha, elle n'occupe elle-même que 1,6 ha. Cette ancienne gravière devait être remise en état dans le cadre de la réglementation du département de la Conservation de l'environnement de l'État de New York. Ce **« CHAMP DE VAGUES »** utilise au maximum la terre trouvée sur place et bénéficie d'un système de drainage autonome. L'herbe semée, d'origine locale et qui n'a pas besoin d'être beaucoup arrosée, est un exemple supplémentaire de la nature durable de cette intervention. L'empreinte carbone du chantier a été calculée et compensée par la plantation de 260 arbres locaux. Le projet consiste en sept « vagues » de terre semées d'herbe, espacées d'environ 12 m. Pour l'artiste : « Parce que le projet est à la même échelle que les vagues naturelles, le spectateur découvre des sensations semblables à ce qu'il pourrait ressentir en mer, lorsque l'on perd le contact visuel avec les vagues qui se suivent. La concentration des courbes permet une lecture complexe et subtile de l'espace par l'intermédiaire d'un environnement qui engloutit le spectateur et crée un sentiment d'immersion totale. »

Above, in the early fall, the grasses covering the rolling hill forms designed by Maya Lin give it an entirely different aspect than in the winter. The work is as much sculptural as it is a piece of landscape design.

Im Frühherbst (oben), wenn Gras die von Maya Lin entworfenen, wellenförmigen Hügel bedeckt, bietet sich ein völlig anderer Anblick als im Winter. Diese Arbeit ist eher eine Skulptur als eine Landschaftsgestaltung.

Ci-dessus, en début d'automne, l'herbe qui recouvre les vagues de terre dessinées par Maya Lin leur donne un aspect entièrement différent de celui de l'hiver. L'œuvre relève autant de la sculpture que du paysagisme.

Seen in the winter (left), the Wavefield takes on an entirely abstract presence, devoid of obvious natural references related to the site.

Das Wellenfeld im Winter (links) wirkt wie ein abstraktes Kunstwerk, weil es keine sichtbaren natürlichen Bezüge zu dem Gelände gibt.

L'hiver (à gauche), le « champ de vagues » prend un aspect entièrement abstrait, dénué de toute référence naturelle au site.

ELEVEN MINUTE LINE
Wanås, Sweden, 2004

Address: Wanås Foundation, Box 67, 289 21 Knislinge, Sweden,
+46 44 660 71 (office), +46 44 661 58 (konsthall), www.wanas.se
Area: 152 meters (length); 4 meters (maximum height). Client: Wanås Foundation. Cost: not disclosed

Above, the work in the course of construction in 2004. Although the relation between this Swedish site and the American Indian burial mounds that inspired Maya Lin may be obscure, the work surely stands on its own.

Oben, während der Arbeiten an der Erdskulptur 2004. Auch wenn die Beziehung zwischen dem Standort in Schweden und den Grabhügeln der Indianer, die Maya Lin angeregt hatten, etwas im Dunklen liegt, so spricht die Arbeit doch für sich.

Ci-dessus, l'œuvre en cours de construction en 2004. Bien que la relation entre ce site suédois et les tumulus amérindiens dont s'est inspirée Maya Lin reste assez obscure, l'œuvre n'en possède pas moins une forte présence.

Inspired by burial mounds created by the Hopewell and Adena Indians in Ohio between 1000 B.C. and A.D. 700, one of which is called the Serpent Mound, Maya Lin has sought to explore the ways in which a drawing can be interpreted as a three-dimensional form. In this instance, her first "sketch" was a gravel drawing made outside Wanås Castle. Maya Lin's studio explains that "a topographic model of the site was then created in order to translate that first sketch into a drawing to fit the pasture's sloping landscape. It was 'drawn' with an understanding that both reading it from the road and walking upon it would have to be equally balanced experiences." The serpentine work, which overlaps itself twice, certainly echoes ancient forms but also reflects Maya Lin's interest in abstraction.

Maya Lin ließ sich bei dieser Arbeit von den Grabhügeln inspirieren, die von den Hopewell- und Adena-Indianern in Ohio zwischen 1000 v. Chr. und 700 n. Chr. angelegt worden waren. Einer dieser Grabhügel heißt Serpent Mound – Schlangenhügel. Maya Lin suchte bei diesem Projekt nach Wegen, wie eine Zeichnung dreidimensional umgesetzt werden kann. In diesem Fall hat sie ihre erste „Skizze" in den Kies bei Schloss Wanås gezeichnet. Das Büro von Maya Lin erklärt, dass daraufhin „ein Geländemodell erstellt wurde, um diese erste Skizze in eine Zeichnung umzusetzen, damit alles in die sanft hügelige Graslandschaft passt. Die Installation wurde so gestaltet, dass die Erfahrungen sowohl beim Betrachten von der Straße aus als auch beim Begehen ausgewogen sind". Die schlangenlinienförmige Arbeit, die sich zweimal überschneidet, erinnert zweifellos an uralte Formen, aber sie zeigt ebenso Maya Lins Interesse an der Abstraktion.

Inspirée par les tumulus funéraires des Indiens Hopewell et Adena en Ohio entre 1000 av. J.-C. et 700 apr. J.-C., dont l'un est appelé Serpent Mound, Maya Lin a cherché ici à explorer la façon dont un dessin peut s'interpréter sous forme tridimensionnelle. Son premier « croquis » est un dessin en gravier réalisé devant le château de Wanås. Le studio de Maya Lin explique qu'« une maquette topographique du site a ensuite été réalisée pour transposer ce premier croquis en dessin adapté au paysage d'un pâturage en pente. Il a été dessiné de manière à ce que sa lecture à la fois depuis la route et depuis le site lui-même offre des expériences équilibrées ». La courbe serpentine, qui se recoupe à deux reprises, fait écho à des formes anciennes, mais aussi à l'intérêt porté par Maya Lin à l'abstraction.

In the winter, the sculptural form of the line is made evident by the snow. Obviously the line has inspired at least one person to walk along its ridge.

Im Winter wird die skulpturale Linie durch den Schnee sichtbar gemacht. Man kann erkennen, dass die Linie mindestens eine Person dazu angeregt hat, auf ihr entlangzulaufen.

En hiver, la forme sculpturale de l'œuvre est mise en valeur par la neige. Un promeneur a visiblement eu envie de se promener sur sa crête.

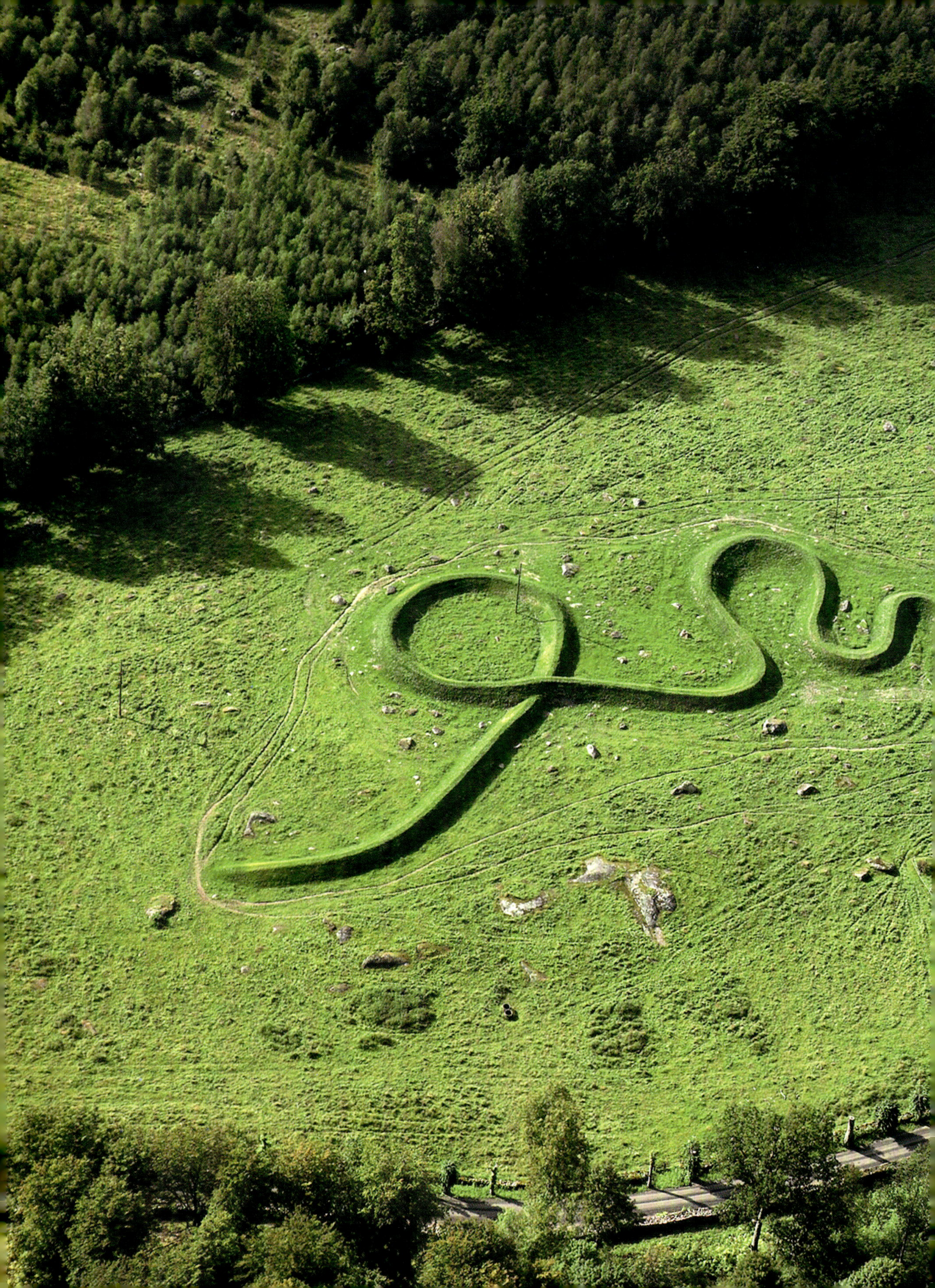

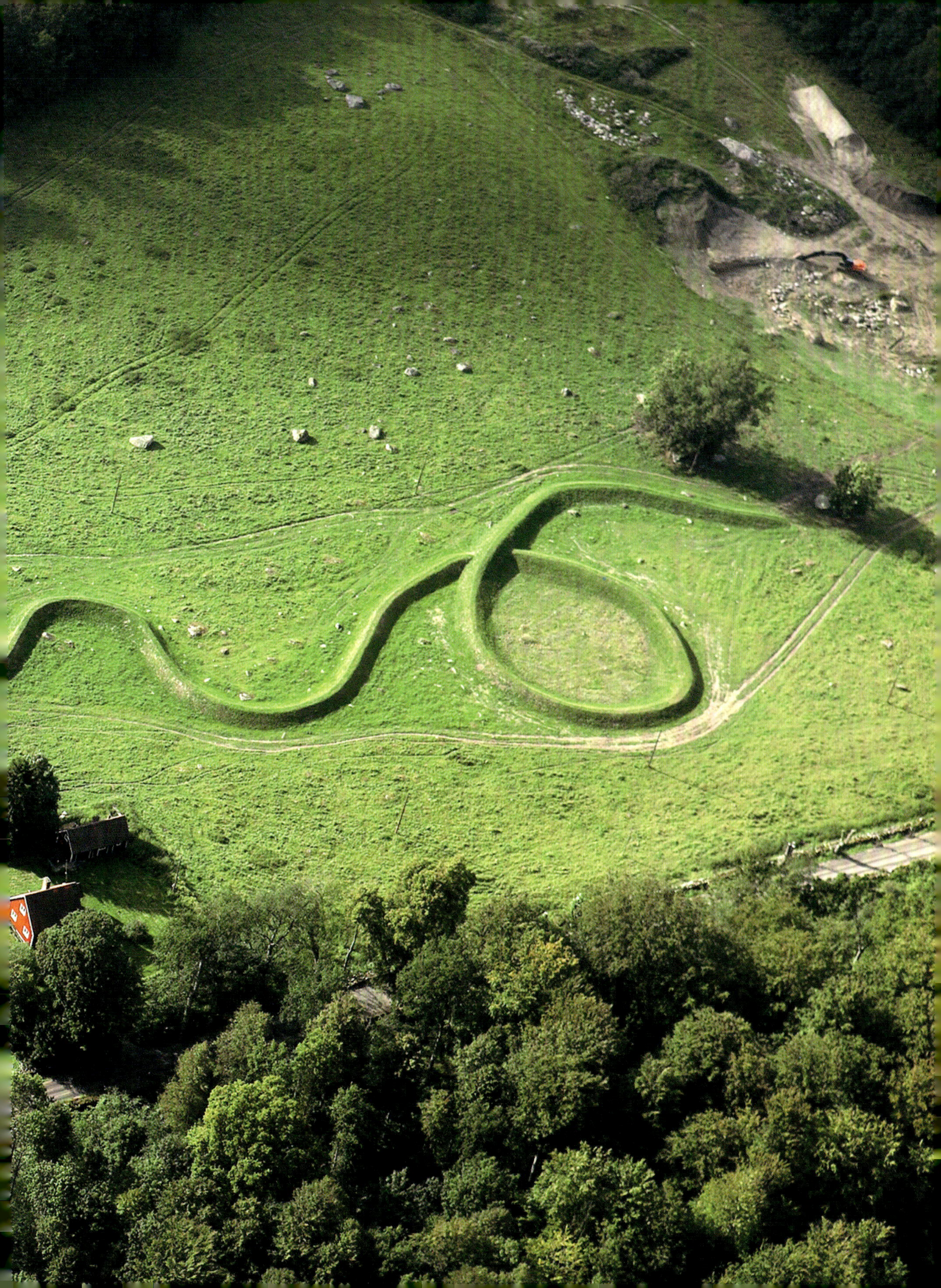

GIOVANNI MACIOCCO

Giovanni Maciocco
Piazza Duomo 6
Sassari, SS 07100
Italy

Tel: +39 79 201 50 90
Fax: +39 79 207 58 50
E-mail: vannimaciocco@gmail.com
Web: www.giovannimaciocco.it

GIOVANNI MACIOCCO was born in Olbia (Sardinia, Italy) in 1946 and holds degrees in engineering and in architecture. His research seeks to brings together the fields of architecture and urban and landscape planning. He teaches urban and territorial space planning at the Faculty of Alghero, Sardinia. He is the editor of the book series *Urban and Landscape Perspectives*. Some of his recent projects are the Museum of Restoration (Sassari, 2005); Open Theater Spaces (Alghero, 2006); the National Archeological Museum (Olbia, 2007); the Anglona Paleobotanical Park (with Domenico Bianco and Salvatore Altana, 2005–08, published here); and the Harbor Museum Porto Torres (2010). At present he is completing the restoration of the former hospital and convent of Santa Chiara in Alghero. Domenico Bianco was born in Bortigiadas (Sardinia) in 1962 and holds a degree in architecture. Among his projects are the restoration of the church of Santa Lucia (Bortigiadas, 2001); structural littoral plan in the Province of Sassari (2005); restoration of the rural church of Santa Reparata (Buddusò, 2005); and the redevelopment of the urban structure of the rural villages of Chiaramonti, Erula, and Perfugas (2008). Salvatore Altana was born in Tempio Pausania (Sardinia) in 1958 and is a surveyor. Among his projects are Fiuminaltu Lake (2002); Fluvial Park in the Coghina River Delta (2004); and an environmental project at the Conca Manna Lake (2008). All the projects are on the island of Sardinia in Italy.

GIOVANNI MACIOCCO wurde 1946 in Olbia (Sardinien, Italien) geboren und hat Universitätsdiplome in den Fachbereichen Ingenieurwesen und Architektur. Bei seinen Forschungen vereint er Architektur, Stadt- und Landschaftsplanung. Er lehrt Stadt- und Raumplanung an der Fakultät in Alghero auf Sardinien und ist Herausgeber der Buchreihe „Urban and Landscape Perspectives". Zu seinen neueren Projekten gehören das Museum für Restaurierung (Sassari, 2005), Freilichttheater in Alghero (2006), das Nationale Archäologische Museum von Olbia (Olbia, 2007), der paläobotanische Park von Anglona (mit Domenico Bianco und Salvatore Altana, 2005–08, hier vorgestellt) und das Hafenmuseum Porto Torres (2010). Derzeit restauriert er das ehemalige Krankenhaus und Kloster Santa Chiara in Alghero. Domenico Bianco wurde 1962 in Bortigiadas (Sardinien) geboren und hat sein Diplom in Architektur gemacht. Zu seinen Projekten gehören die Restaurierung der Kirche Santa Lucia (Bortigiadas, 2001), der Strukturplan für die Küstenzone der Provinz Sassari (2005), die Restaurierung der Dorfkirche Santa Reparata (Buddusò, 2005) sowie die Sanierung der dörflichen Strukturen von Chiaramonti, Erula und Perfugas (2008). Salvatore Altana wurde 1958 in Tempio Pausania (Sardinien) geboren und ist Landvermesser. Zu seinen Arbeiten gehören der Lago Fiuminaltu (2002), der Fluvialpark im Delta des Coghina-Flusses (2004) sowie ein Umwelschutzprojekt am Lago Conca Manna (2008). Alle Projekte befinden sich auf Sardinien, Italien.

GIOVANNI MACCIOCO, né à Olbia (Sardaigne, Italie) en 1946, est diplômé en ingénierie et architecture. Ses recherches veulent rapprocher architecture, urbanisme et architecture du paysage. Il enseigne l'espace urbain et les territoires à la faculté d'Alghero (Sardaigne) et dirige une collection d'ouvrages intitulée *Urban and Landscape Perspectives*. Parmi ses projets récents figurent le Musée de la restauration (Sassari, 2005); des théâtres ouverts (Alghero, 2006); le Musée national d'archéologie d'Olbia (2007); le parc paléobotanique d'Anglona (avec Domenico Bianco et Salvatore Altana, 2005–08, publié ici) et le Musée du port de Porto Torres (2010). Actuellement, il achève la restauration de l'ancien hôpital et couvent de Santa Chiara à Alghero. Domenico Bianco, né en 1962 à Bortigiadas (Sardaigne), est diplômé en architecture. Il est l'auteur de la restauration de l'église de Santa Lucia (Bortigiadas, 2001); du plan structurel du littoral de la province de Sassari (2005); de la restauration de l'église rurale de Santa Reparata (Buddusò, 2005) et de la rénovation de la structure urbaine des villages ruraux de Chiaramonti, Erula et Perfugas (2008). Salvatore Altana, né à Tempio Pausania (Sardaigne) en 1958, est géomètre. Parmi ses références figurent le lac Fiuminaltu (2002); le parc fluvial du delta du fleuve Coghina (2004) et un projet environnemental en bordure du lac Conca Manna (2008). Tous les projets cités sont situés en Sardaigne, en Italie.

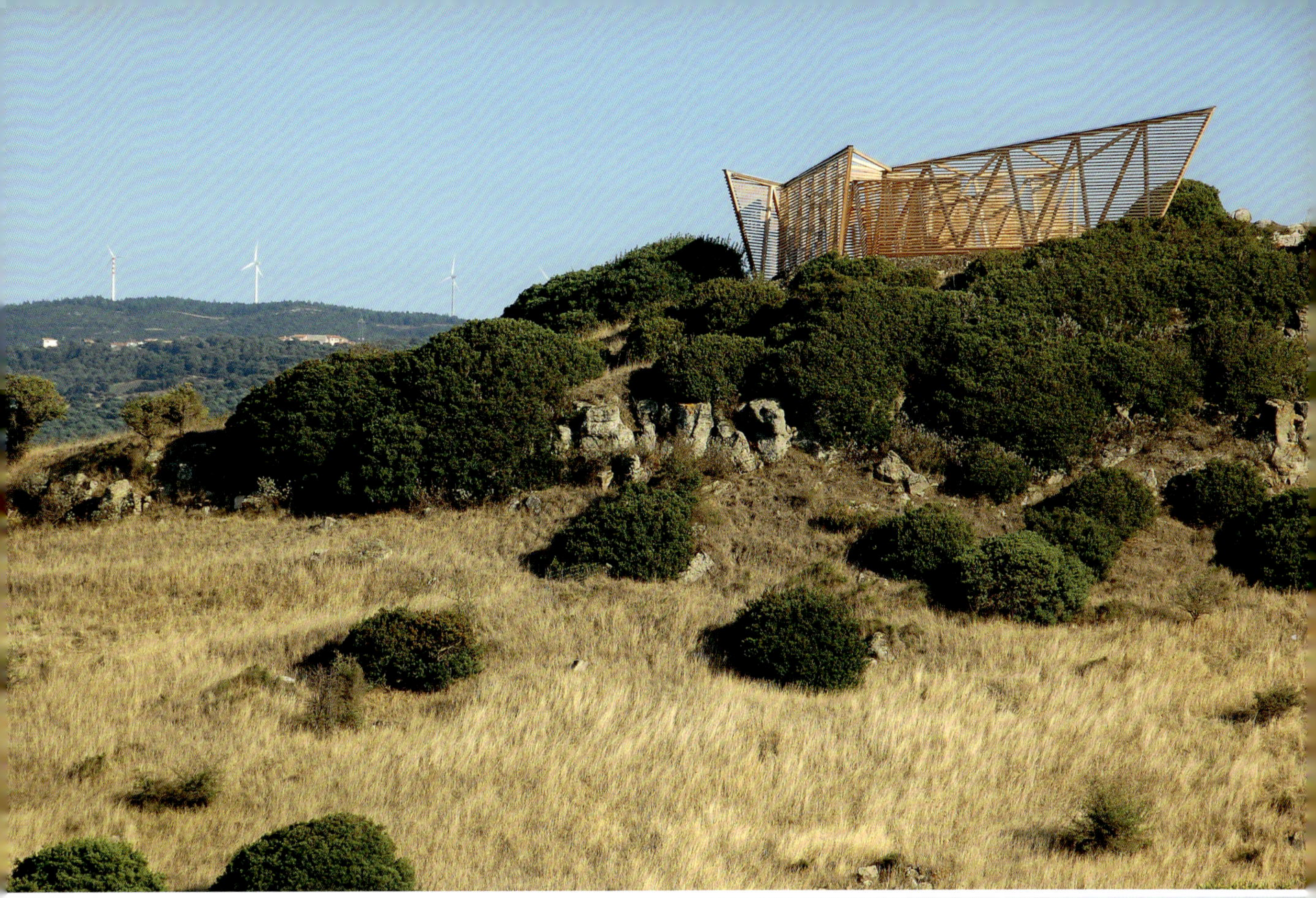
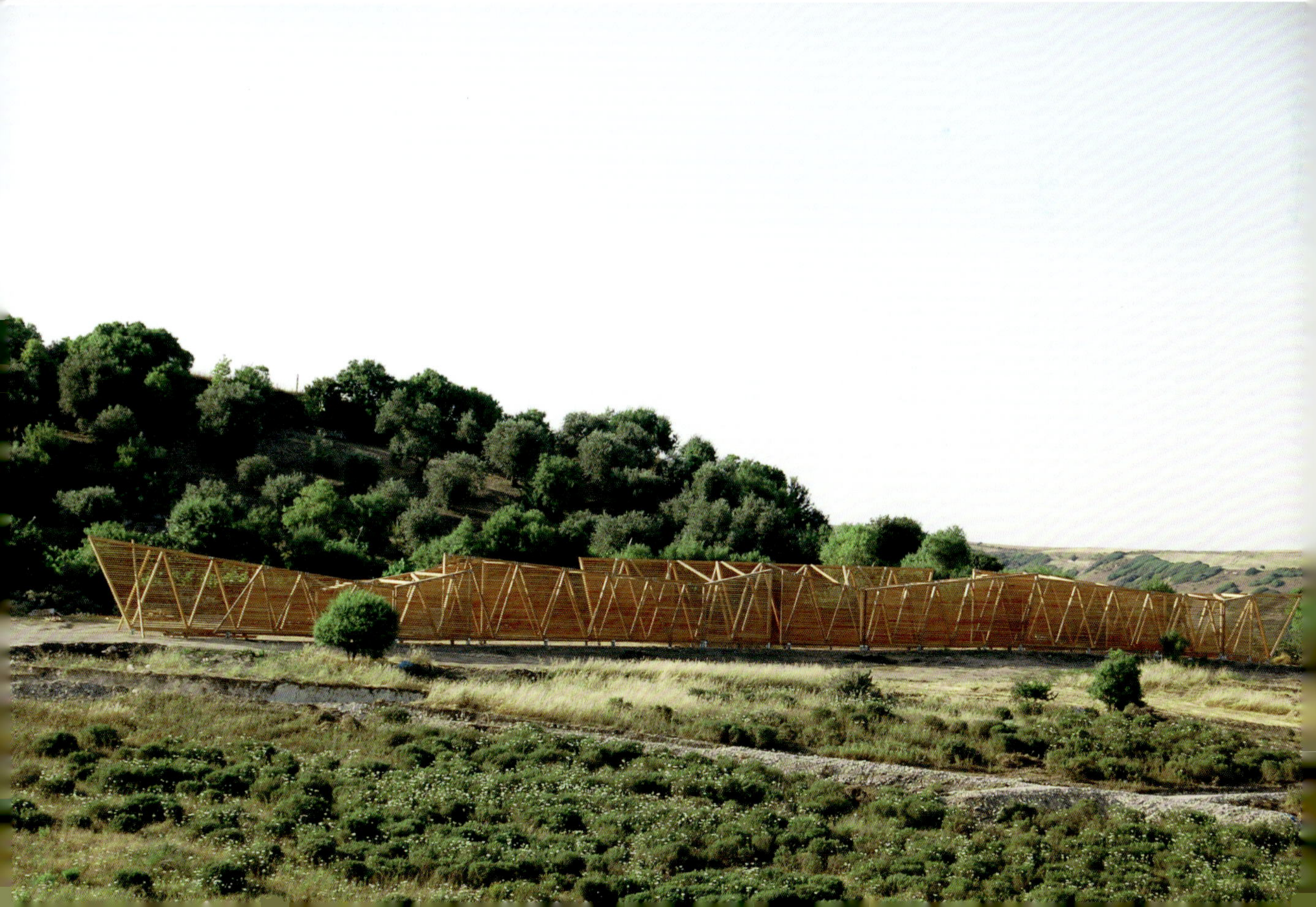

ANGLONA PALEOBOTANIC PARK
North Sardinia, Italy, 2005–08

Address: Martis, Bulzi, Laerru and Perfugas villages, Anglona region, north Sardinia, Italy
Area: 28.4 hectares (park), 3930 m² (pavilions). Client: Martis, Bulzi, Laerru and Perfugas Municipalities
Cost: €1.503 million. Collaboration: Domenico Bianco, Salvatore Altana

Giovanni Maciocco explains that "the paleobotanic area of Anglona, with its petrified forest of wood-fossil and fossil finds that go back to the Tertiary Era and are spread over a surface area of approximately 100 square kilometers, is one of the most significant paleo-environmental and geological resources of Sardinia." The broad area involved encouraged the creation of the **PALEOBOTANIC PARK** that the architects have sought to unify through their shaded pavilion structures. These are strategically placed at high points in the area. The cooperation of the four municipalities involved springs from their awareness that it is in the common good to preserve the exceptional fossil remains. The architect continues: "The shaded galleries referred to consist of light wooden structures which adapt as necessary to the topographic and hydrographic conditions of the sites, giving shape to a rich repertory of architectural variations on the principal theme of the project. It is, indeed, these adaptations that mark out favorable conditions for the creation of small stretches of water—a direct reference to the succession of lakes to which the fossil formations owe their origin." Routes through the park were also created in conjunction with the new structures.

Giovanni Maciocco erklärt, dass das „paläobotanische Gebiet von Anglona mit seinem versteinerten Wald und den Fossilien aus dem Tertiär datiert und sich über ein Gelände von ungefähr 100 Quadratkilometer erstreckt. Damit gehört es zu den bedeutendsten erhaltenen paläobotanischen Arealen und den reichsten geologischen Ressourcen von Sardinien." Da es sich um ein sehr großes Gebiet handelt, lag es nahe, einen **PALÄOBOTANISCHEN PARK** einzurichten. Die Architekten haben versucht, mit den schattenspendenden Konstruktionen eine optische Einheit zu schaffen. Die „Pavillons" sind strategisch auf den Geländeerhebungen platziert. Die Zusammenarbeit der vier betroffenen Gemeinden zeigt, dass man sich bewusst war, dass die Erhaltung der außergewöhnlichen fossilen Funde von allgemeiner Bedeutung ist. Der Architekt führt dazu aus: „Die schattenspendenden Gerüste sind leichte Holzkonstruktionen, die sich an die Topografie und an die hydrografischen Bedingungen des Geländes anpassen. Sie stellen das zentrale Thema dieses Projektes mit variantenreichen Architekturelementen dar. Es war vor allem diese Anpassung, die es ermöglichte, gute Bedingungen für kleine Wasserläufe zu schaffen – eine unmittelbare Referenz an die Seenkette, der wir die fossilen Formationen zu verdanken haben." Im Zusammenhang mit den neuen Konstruktionen wurde auch ein Wegesystem geschaffen, das den Besucher durch den Park leitet.

Giovanni Maciocco explique que « la zone paléobotanique d'Anglona – site d'une forêt pétrifiée et de bois et animaux fossiles – qui remonte à l'ère tertiaire s'étend sur une surface d'environ 100 km². C'est l'un des trésors géologiques et paléoenvironnementaux les plus importants de Sardaigne ». Les quatre communes impliquées dans le projet ont pris conscience de leur intérêt commun de préserver ces vestiges fossiles et ont créé un **PARC PALÉOBOTANIQUE** que les architectes ont cherché à unifier par des constructions en forme de pavillons-galeries, implantés en des lieux stratégiques élevés. « Ces galeries ombragées sont des structures légères en bois, adaptées aux conditions topographiques et hydrographiques des sites. Elles sont la mise en forme architecturale de multiples variantes sur le thème principal du projet et créent des conditions favorables à la mise en place de petites retenues d'eau, en référence directe à la succession de lacs auxquels les formations fossiles doivent leur origine », explique l'architecte. Des chemins ont été créés pour relier ces nouvelles constructions.

Left page, a plan of the site. This page, the architect has used a light wooden design for this shaded pavilion structure intended to give unity to the vast area of the park.

Linke Seite, Plan des Geländes. Diese Seite: Der Architekt hat die schattenspendenden Gerüste als leichte Holzkonstruktionen ausgeführt, um dem ausgedehnten Park Einheitlichkeit zu verleihen.

Page de gauche, plan du site. Photos de cette page : l'architecte a utilisé un principe de construction en lattis de bois pour ce pavillon qui veut donner une unité à ce vaste parc.

Further views of the "pavilion" design used by Giovanni Maciocco as signals within the park that underline its continuity. Below, an elevation drawing of the structure.

Weitere Ansichten der von Giovanni Maciocco entworfenen „Pavillons", die wie Signale im Park stehen und seine Einheit betonen. Unten eine Ansichtszeichnung der Konstruktion.

Autre vues du « pavillon » traité par Giovanni Maciocco comme signal de la continuité du parc. Ci-dessous, élévation du pavillon.

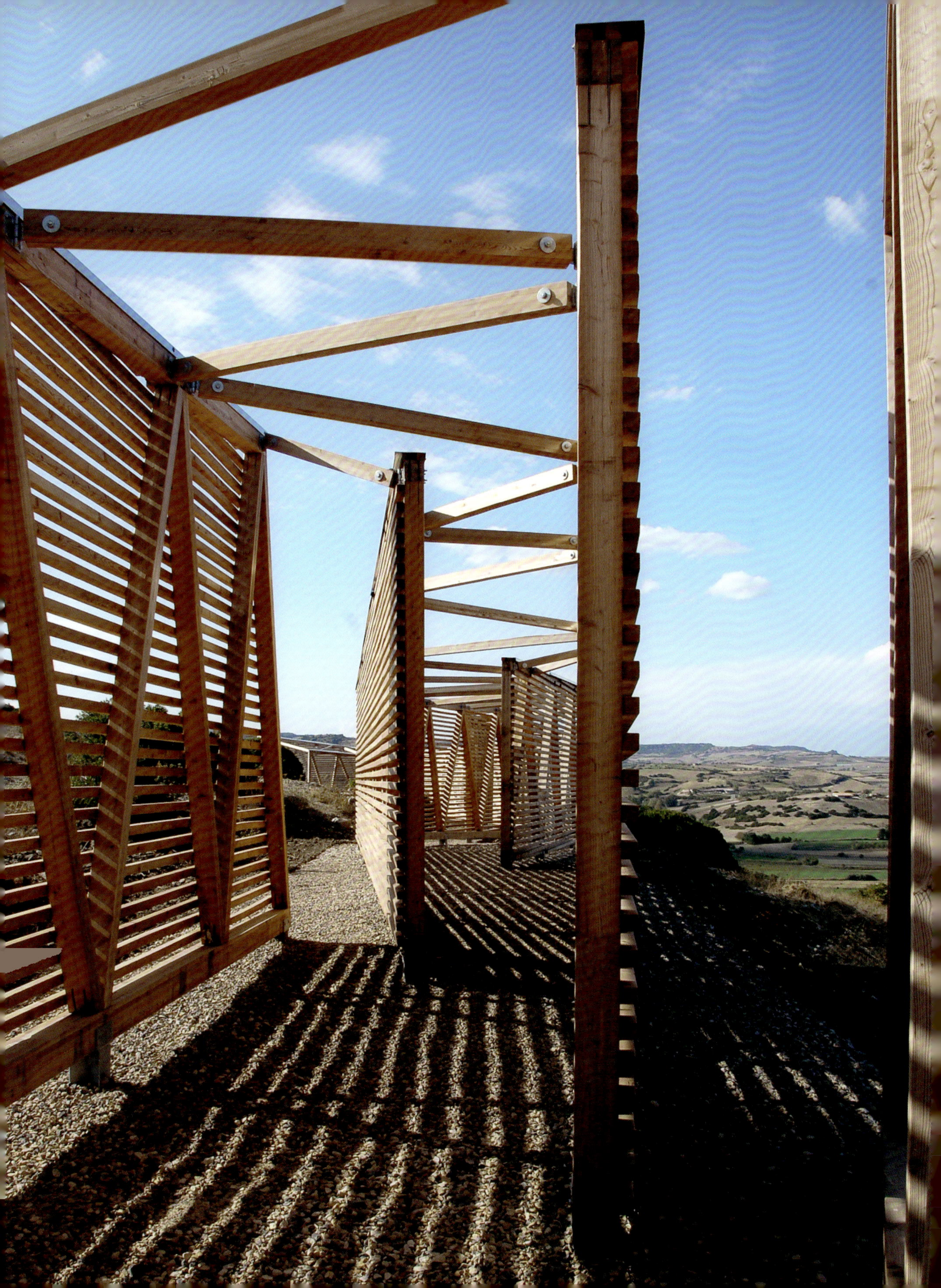

MICHAEL MALTZAN

Michael Maltzan Architecture, Inc.
2801 Hyperion Avenue, Studio 107
Los Angeles, CA 90027
USA

Tel: +1 323 913 3098 / Fax: +1 323 913 5932
E-mail: info@mmaltzan.com / Web: www.mmaltzan.com

MICHAEL MALTZAN was born in 1959 in Levittown, New York. He holds both a B.F.A. and a B.Arch from Rhode Island School of Design (1984, 1985) and an M.Arch degree from Harvard (1988). Since establishing his own firm in 1995, Michael Maltzan has been responsible for the design of a wide range of arts, educational, commercial, institutional, and residential projects, including the Mark Taper Center / Inner-City Arts (Phase 1, Los Angeles, California, 1994); Harvard-Westlake School's Feldman-Horn Center for the Arts (North Hollywood, California, 1995); the Kidspace Children's Museum in Pasadena (California, 1998); and MoMA QNS in Long Island City (New York, New York, 2002). Recent work includes Ministructure No. 16 / Book Store for the Architecture Park in Jinhua (China, 2004–06); the UCLA Hammer Museum Billy Wilder Theater in Los Angeles (California, 2006); Rainbow Apartments (Los Angeles, California, 2006); and Inner-City Arts (Phase 3, Los Angeles, California, 2008). More recently he completed the New Carver Apartments (Los Angeles, California, 2009); the Pittman Dowell Residence (La Crescenta, California, 2009); and Playa Vista Park and Bandshell (Los Angeles, 2009–10, published here). Current work includes One Santa Fe (Los Angeles, California, 2006–); the Jet Propulsion Laboratory Administration Complex (Pasadena, California, 2006–); San Francisco State University Mashouf Performing Arts Center (San Francisco, California, 2008–); Star Apartments (Los Angeles, California, 2008–); and the New Orleans Riverfront Mandeville Crossing (New Orleans, Louisiana, 2008–), all in the USA unless stated otherwise.

MICHAEL MALTZAN wurde 1959 in Levittown, New York, geboren. Er hat den Bachelor of Fine Arts und den Bachelor der Architektur an der Rhode Island School of Design (1984, 1985) gemacht und seinen Master für Architektur in Harvard (1988). Seit Gründung seines Büros 1995 hat Michael Maltzan die unterschiedlichsten Projekte realisiert: Kunstprojekte, Ausbildungsstätten, öffentliche und Geschäftsgebäude sowie Wohnhäuser. Unter anderem gehören dazu das Mark Taper Center/Inner-City Arts (Bauabschnitt 1, Los Angeles, Kalifornien, 1994), das Feldman-Horn Center for the Arts der Harvard-Westlake School (North Hollywood, Kalifornien, 1995), das Kidspace Children's Museum in Pasadena (Kalifornien, 1998) und das MoMA QNS in Long Island City (New York, 2002). Zu den neueren Projekten zählen die Ministructure No. 16, ein Buchladen für den Architekturpark in Jinhua (China, 2004–06), das UCLA Hammer Museum Billy Wilder Theater in Los Angeles (2006), die Rainbow Apartments (Los Angeles, 2006) und Inner-City Arts (Bauabschnitt 3, Los Angeles, 2008). Erst kürzlich wurden die von ihm entworfenen New Carver Apartments (Los Angeles, 2009) fertiggestellt, das Wohnhaus Pittman Dowell (La Crescenta, Kalifornien, 2009) sowie der Playa Vista Park mit Konzertmuschel (Los Angeles, 2009–10, hier vorgestellt). Derzeit arbeitet er an dem Projekt One Santa Fe (Los Angeles, seit 2006), dem Jet Propulsion Laboratory Administration Complex (Pasadena, Kalifornien, seit 2006), am Mashouf Performing Arts Center der San Francisco State University (San Francisco, Kalifornien, seit 2008), den Star Apartments (Los Angeles, seit 2008) und an der New Orleans Riverfront Mandeville Crossing (New Orleans, Louisiana, seit 2008). Alle Projekte befinden sich in den USA.

MICHAEL MALTZAN, né en 1959 à Levittown (New York), est titulaire d'un B.F.A. et d'un B.Arch. de l'École de design de Rhode Island (1984, 1985), et d'un M.Arch de Harvard (1988). Il a fondé son agence en 1995 et a réalisé de multiples projets dans les secteurs des arts, de l'éducation, du commerce, des bâtiments publics et résidentiels, dont le Mark Taper Center/Inner-City Arts (Phase 1, Los Angeles, 1994); le Feldman-Horn Center for the Arts de l'École Harvard-Westlake (North Hollywood, Californie, 1995); le Kidspace Children's Museum de Pasadena (Californie, 1998) et le MoMA QNS à Long Island City (New York, 2002). Parmi ses travaux récents : Ministructure No. 16/librairie pour le parc d'architecture de Jinhua (Chine, 2004–06); le Billy Wilder Theater du musée Hammer de l'UCLA (Los Angeles, 2006); les Rainbow Apartments (Los Angeles, 2006) et le centre éducatif Inner-City Arts (Phase 3, Los Angeles, 2008). Plus récemment : les New Carver Apartments (Los Angeles, 2009); la Pittman Dowell Residence (La Crescenta, Californie, 2009) et le Playa Vista Park avec sa scène couverte (Los Angeles, 2009–10, publié ici); l'immeuble One Santa Fe (Los Angeles, 2006–); le complexe administratif du Jet Propulsion Laboratory (Pasadena, Californie, 2006–); le Mashouf Performing Arts Center de l'université de San Francisco (2008–); l'immeuble Star Apartments (Los Angeles, 2008–) et l'aménagement du front de mer de Mandeville à La Nouvelle-Orléans (Louisiane, 2008–). Tous aux États-Unis, sauf mention contraire.

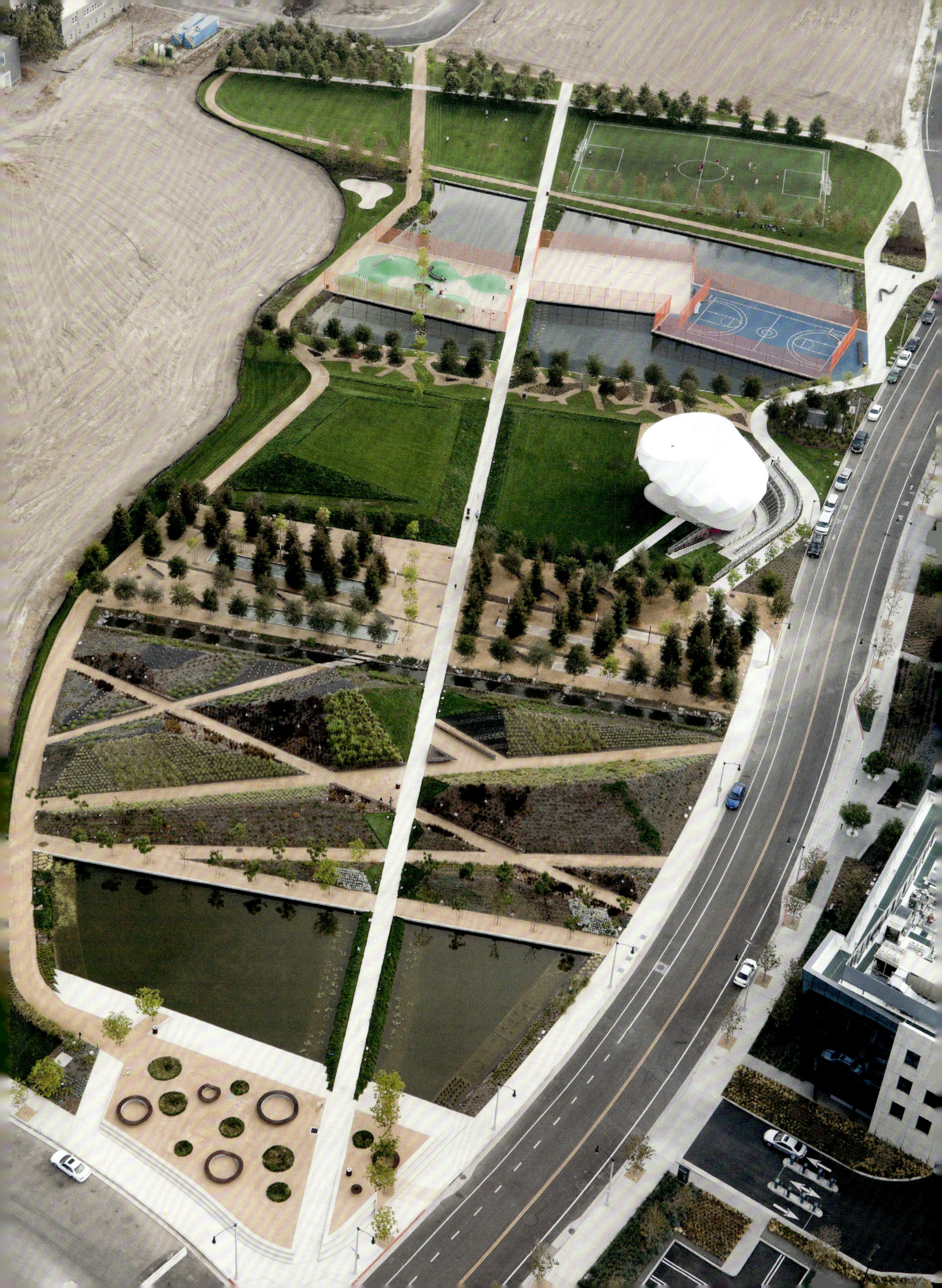

PLAYA VISTA PARK AND BANDSHELL
Los Angeles, USA, 2009–10

Address: 12045 Waterfront Drive, Playa Vista, Los Angeles, CA 90094, USA;
Bandshell at Playa Vista, 12040 Waterfront Drive, 90045 Los Angeles, CA, USA
Area: 3.2 hectares (park). Client: Playa Capital Commercial Land LLC. Cost: $9.4 million

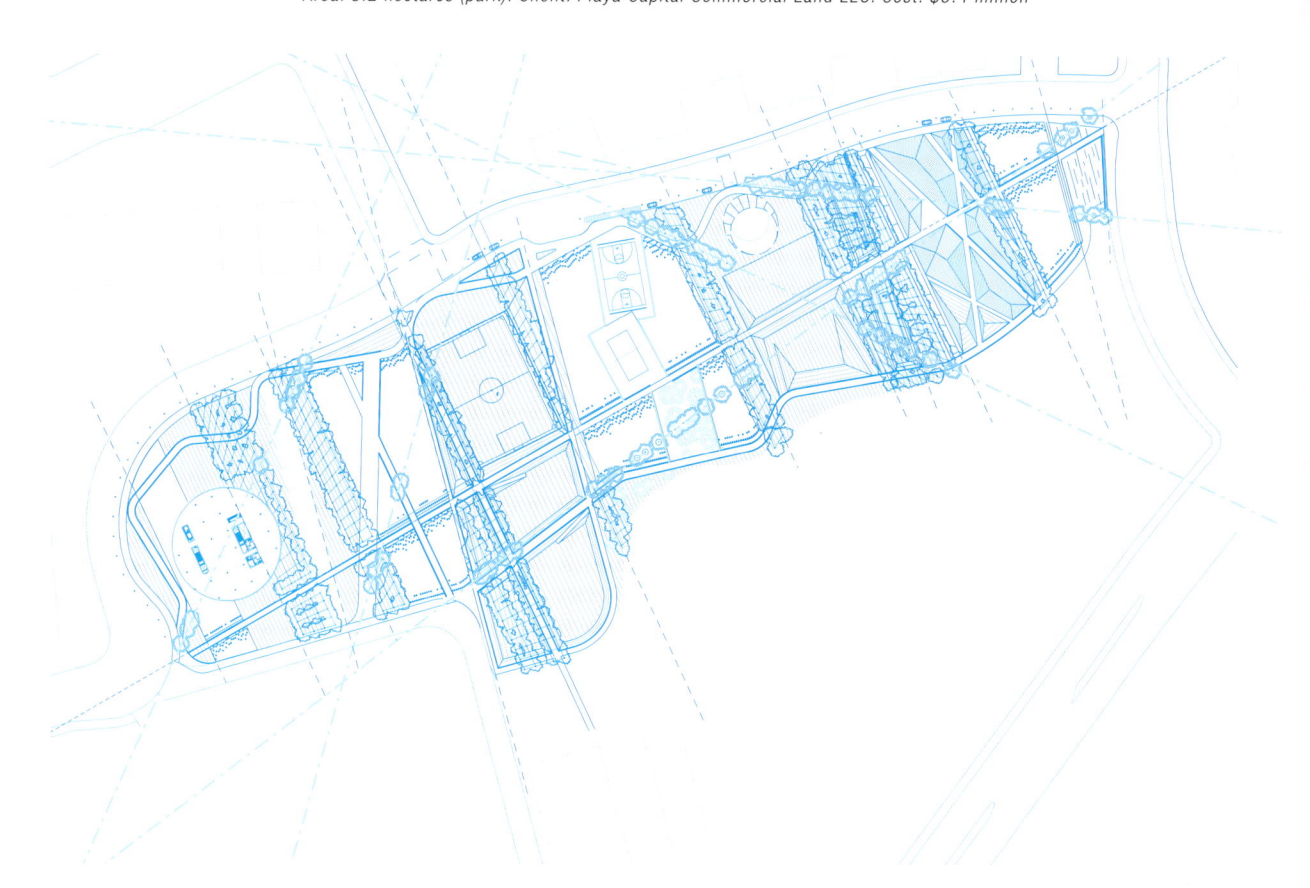

Located at the eastern end of a large urban development, the **PLAYA VISTA PARK** "links and expands the surrounding context of offices to include the Park as a whole." Designed as a series of bands formed with alders, elms, and pines that cross the site, the architect states: "The Park is not a mimetic representation of nature, but is instead defined by its activity, its utility, its performance." Soccer fields and basketball courts, and an amphitheater lawn encourage different activities and forms of meeting, as does the **BANDSHELL** with its 168-square-meter stage, also designed by Michael Maltzan. The central axis of the park running from east to west corresponds to the former Hughes runway and has been "defined by elongated stripes of color and light designed with the artist Daniel Buren." Alleys of flowering jacarandas cross the park from east to west "blooming in the spring and early summer to create strong bands of color linking the park as a whole."

Der **PLAYA VISTA PARK** liegt am östlichen Rand eines großen städtischen Siedlungsgebiets und „verbindet und erweitert das Umfeld der umliegenden Bürogebäude so, dass der Park als Ganzes miteinbezogen wird". Der Entwurf besteht aus Streifen, die mit Erlen, Ulmen und Kiefern bepflanzt sind und das Gelände queren. Der Architekt bemerkt dazu: „Der Park ist keine Nachahmung der Natur, sondern er wird durch die Aktivitäten definiert, die hier stattfinden, durch seine Nutzung und seine Funktionen." Fußballplätze und Basketballfelder sowie ein Rasenamphitheater laden zu den unterschiedlichsten Aktivitäten und Formen des Zusammentreffens ein. Auch die von Michael Maltzan entworfene **KONZERTMUSCHEL** mit ihrer 168 Quadratmeter großen Bühne gehört zu diesem Angebot. Die Hauptachse des Parks verläuft von Ost nach West und zeichnet die ehemalige Start- und Landebahn des Flughafens Hughes nach. Sie wurde mit „langen Streifen aus Farbe und Licht definiert, die zusammen mit dem Künstler Daniel Buren entworfen wurden". Alleen blühender Jakarandabäume queren den Park von Ost nach West. Sie „blühen im Frühjahr und im Frühsommer. Ihre stark farbigen Streifen wirken als Bindeglied für den gesamten Park."

Situé à l'extrémité est d'une vaste opération d'urbanisation, le **PLAYA VISTA PARK** « relie et prolonge un contexte urbain de bureaux en s'intégrant à son ensemble ». Il est divisé en plusieurs zones par une succession d'alignements d'aunes, d'ormes et de pins. « Le parc n'est pas une représentation mimétique de la nature, mais se définit plutôt par son activité, son utilité et sa performance. » Ont ainsi été installés des terrains de football et de basketball, une pelouse en amphithéâtre pour des activités et manifestations diverses, et une spectaculaire **SCÈNE COUVERTE** de 168 m², également dessinée par Michael Maltzan. L'axe central est-ouest du parc, qui correspond à l'ancienne piste de l'aérodrome Hughes, « se caractérise par de longues bandes de couleur et de lumière conçues en collaboration avec l'artiste Daniel Buren ». Des allées de jacarandas qui traversent le parc d'est en ouest « fleurissent au printemps et au début de l'été pour donner naissance à de grands bandeaux de couleur qui unifient l'ensemble ».

The irregular shape of the park is laid out by the architect with a central walkway, seen in the plan to the left and above. Other, smaller paths cut through the space, which is planted with a variety of different species.

Der Architekt hat den Park mit seinem unregelmäßigen Grundriss durch einen zentralen Weg gegliedert, wie auf dem Plan links und oben zu erkennen. Andere, schmalere Wege durchschneiden den Freiraum, in dem zahlreiche Pflanzenarten gedeihen.

Le parc de forme irrégulière est traversé par une large allée centrale (plan de gauche et ci-dessus). De petites allées et des sentiers redécoupent l'espace, planté d'espèces végétales variées.

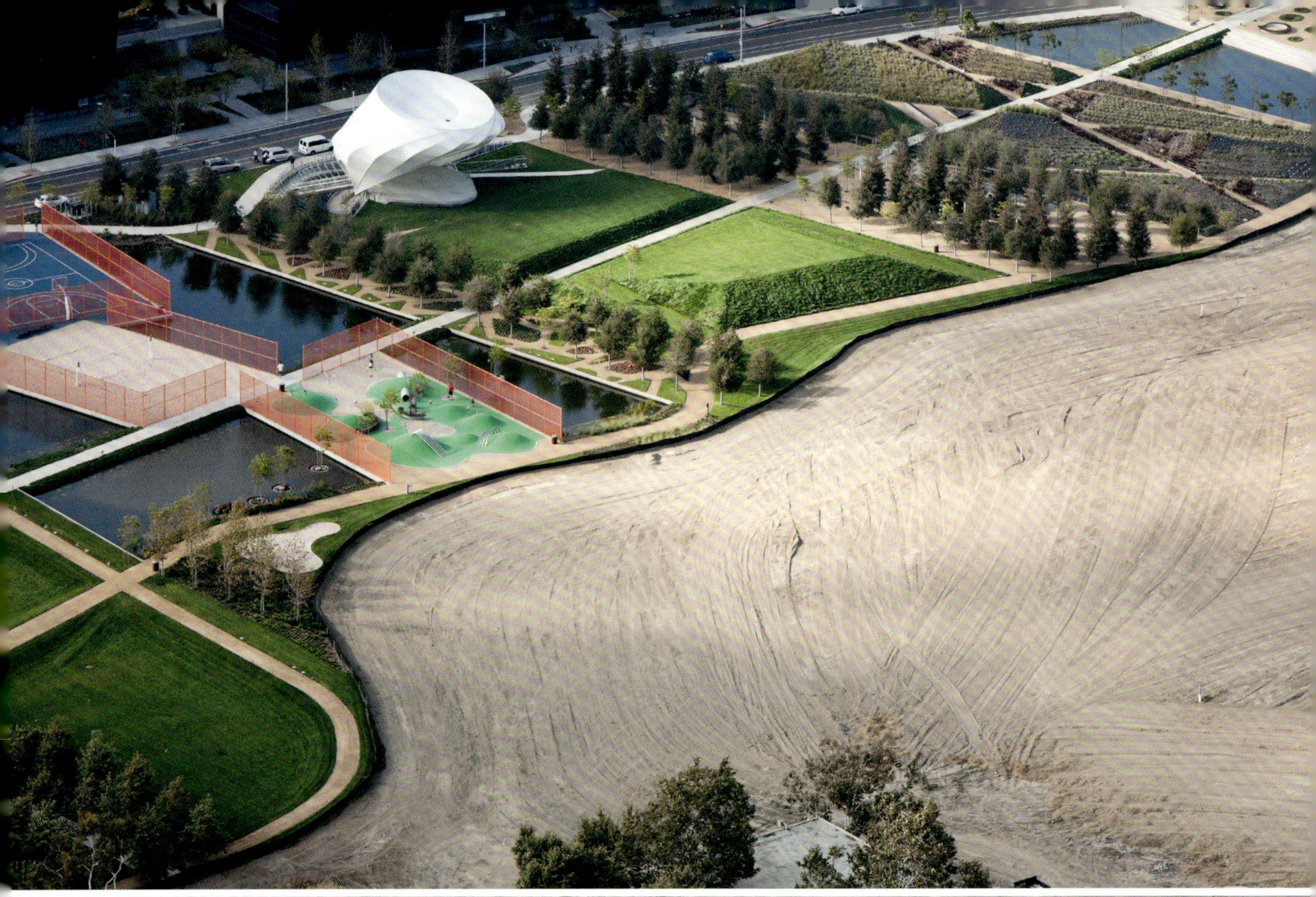

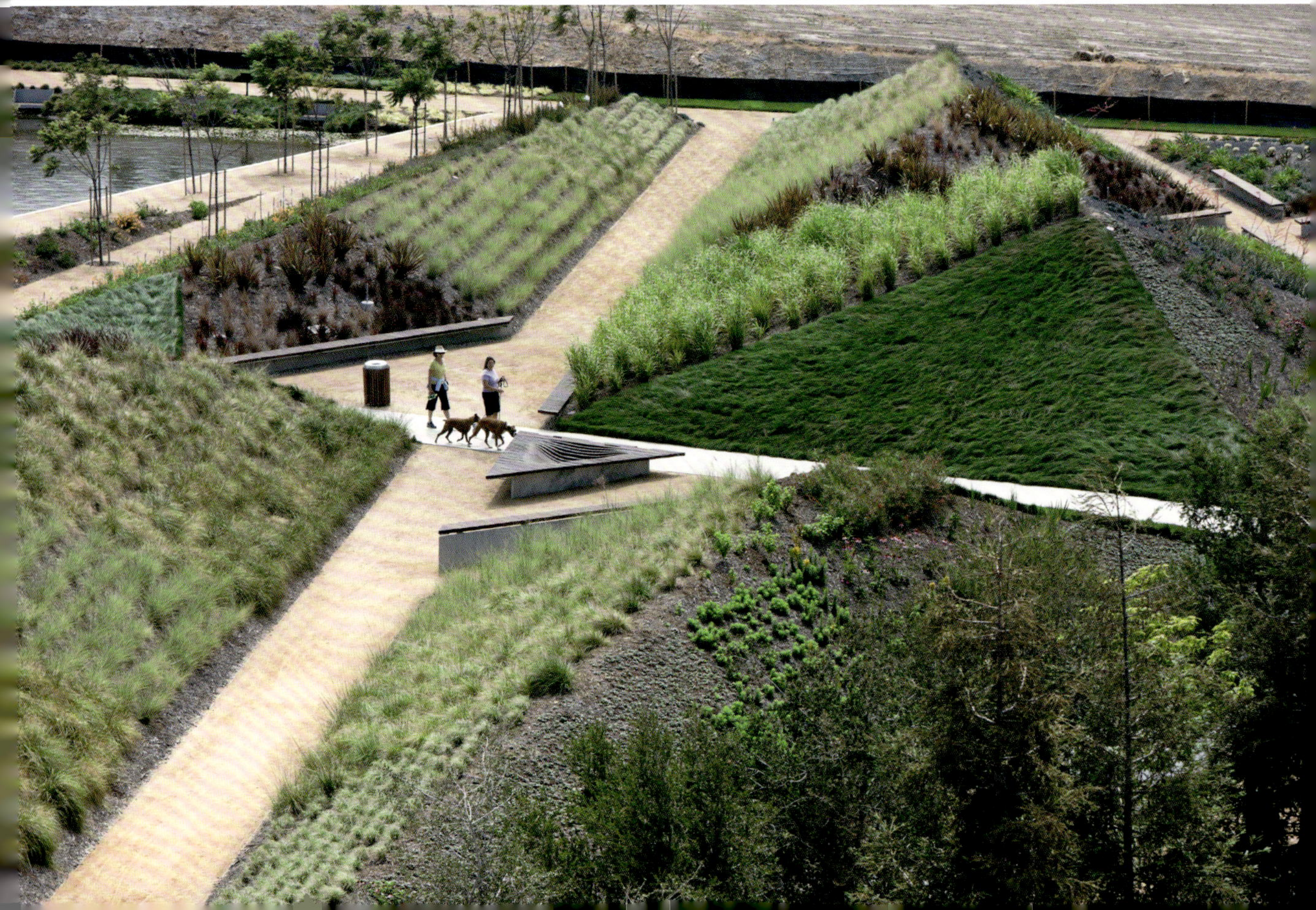

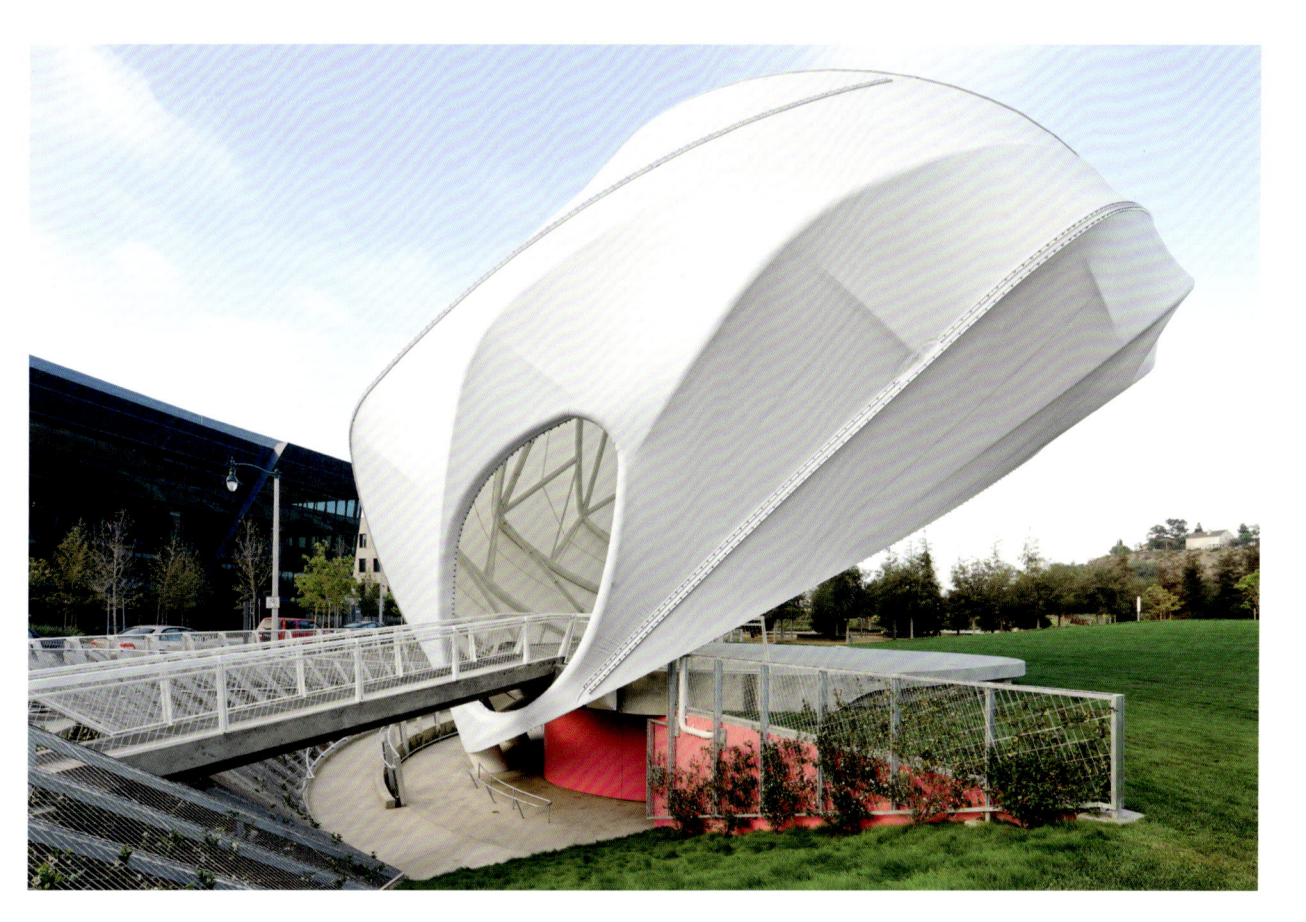

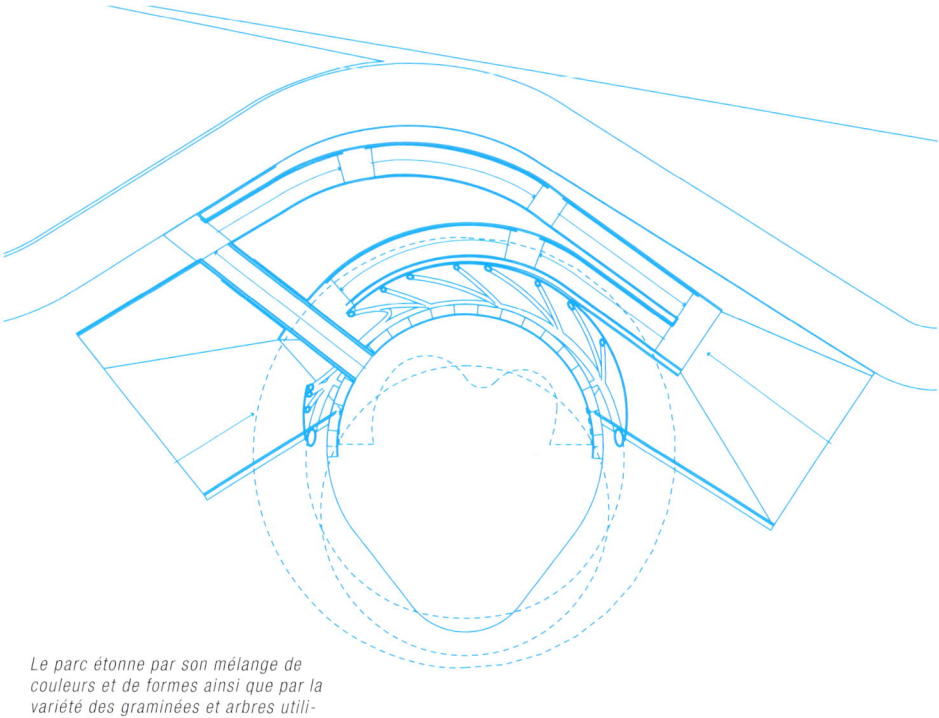

The park is unusual because of its mixture of colors and forms, as well as the variety of grasses and trees employed. Water is also part of the scheme, as is the band shell, seen above and in the plan on this page.

Dieser Park ist in seiner Mischung aus Farben und Formen und auch wegen der Vielfalt der verwendeten Gräser und Bäume außergewöhnlich. Wasser ist genauso ein Entwurfselement wie die Konzertmuschel (Foto oben und Grundriss).

Le parc étonne par son mélange de couleurs et de formes ainsi que par la variété des graminées et arbres utilisés. Les bassins et la scène couverte font également partie du projet (ci-dessus et en plan).

MECANOO

Mecanoo architecten
Oude Delft 203
2611 HD Delft
The Netherlands

Tel: +31 15 279 81 00
Fax: +31 15 279 81 11
E-mail: info@mecanoo.nl
Web: www.mecanoo.nl

FRANCINE HOUBEN, Henk Döll, and Roelf Steenhuis met in Delft and started Mecanoo architecten in 1980. Together with Chris de Weijer and Erick van Egeraat (the latter leaving in 1995) they officially founded Mecanoo architecten in 1984, directed since 2003 by Francine Houben. All attended the Technical University of Delft, graduating in 1983 and 1984. Their work includes a private house in Rotterdam (1989–91), large housing projects such as the Herdenkingsplein in Maastricht (1990–92), and smaller-scale projects such as their 1990 Boompjes Pavilion, a cantilevered structure overlooking the harbor of Rotterdam, close to the new Erasmus Bridge. Signature features of these projects include unexpected use of materials, as in the Rotterdam house, where bamboo and steel are placed in juxtaposition with concrete, for example, or an apparent disequilibrium, as in the Boompjes Pavilion. Other projects are the TU Delft Library (1998); Montevideo residential tower in Rotterdam's harbor (2006); Philips Business Innovation Center FiftyTwoDegrees (Nijmegen, 2007); TU Campus, Mekel Park (Delft, 2007–09, published here); and the La Llotja Theater and Congress Center in Lleida (Spain, 2010). Ongoing work includes the Library of Birmingham integrated with the REP Theatre (Birmingham, UK, 2013); the Wei-Wu-Ying Center for the Arts (Kaohsiung, Taiwan, 2013); and the Delft City Hall and Train Station (Delft, 2015), all in the Netherlands unless stated otherwise.

FRANCINE HOUBEN, Henk Döll und Roelf Steenhuis lernten sich in Delft kennen und gründeten 1980 das Büro Mecanoo architecten. Offizielles Gründungsjahr von Mecanoo architecten war allerdings 1984, als Chris de Weijer und Erick van Egeraat hinzukamen. Letztgenannter verließ das Büro 1995. Seit 2003 ist Francine Houben die Chefin. Alle Partner haben ihr Diplom 1983 bzw. 1984 an der Technischen Universität Delft gemacht. Zu ihren Arbeiten gehören ein privates Wohnhaus in Rotterdam (1989–91), große Wohnanlagen wie die Herdenkingsplein in Maastricht (1990–92) sowie Projekte in kleinerem Maßstab wie der 1990 entworfene Boompjes Pavillon, eine auskragende Konstruktion in der Nähe der neuen Erasmus-Brücke, von der aus man den Rotterdamer Hafen überblickt. Wesentliche Merkmale dieser Projekte sind ungewöhnliche Materialien, wie bei dem privaten Wohnhaus in Rotterdam, das aus Bambus und Stahl zusammen mit Beton errichtet wurde, oder man steht, wie im Fall des Boompjes Pavillon, vor einem scheinbar unausgewogenen Entwurf. Andere Projekte sind die Bibliothek der TU Delft (1998), der Montevideo Tower, ein Wohnhochhaus am Rotterdamer Hafen (2006), das Philips Business Innovation Center FiftyTwoDegrees (Nijmegen, 2007), der TU Campus, Mekel Park (Delft, 2007–09, hier vorgestellt), und das Theater und Kongresszentrum La Llotja in Lleida (Spanien, 2010). Zu den derzeit im Bau befindlichen Projekten gehören die Bibliothek von Birmingham mit einem gemeinsamen Zugang zum REP Theatre (Birmingham, GB, 2013), das Wei-Wu-Ying Center for the Arts (Kaohsiung, Taiwan, 2013) sowie das neue Rathaus und der Bahnhof von Delft (Delft, 2015). Sofern nicht anders angegeben, befinden sich alle Projekte in den Niederlanden.

FRANCINE HOUBEN, Henk Döll et Roelf Steenhuis se sont rencontrés à Delft et ont lancé l'agence Mecanoo architecten en 1980. Cependant, ce n'est qu'en 1984, avec Chris de Weijer et Erick van Egeraat (celui-ci les quittera en 1995) qu'ils ont officiellement fondé l'agence Mecanoo architecten, dirigée depuis 2003 par Francine Houben. Tous ont fait leurs études d'architecture à l'Université de technologie (TU) de Delft, dont ils sont sortis diplômés en 1983 et 1984. Parmi leurs réalisations : une maison à Rotterdam (1989–91) ; d'importants projets d'immeubles de logements, comme Herdenkingsplein à Maastricht (1990–92), et des interventions plus modestes comme leur pavillon sur la Boompjes (1990), construction en porte-à-faux sur le port de Rotterdam, près du pont Erasmus. Ces projets sont remarquables par leur utilisation peu conventionnelle des matériaux (la maison à Rotterdam allie béton, acier et bambou) ou par la mise en place de déséquilibres apparents (pavillon sur la Boompjes). Ils ont également réalisé la bibliothèque de la TU Delft (1998) ; la tour résidentielle Montevideo dans le port de Rotterdam (2006) ; le Philips Business Innovation Center FiftyTwoDegrees (Nimègue, 2007) ; le Mekelpark sur le campus de la TU (Delft, 2007–09, publié ici) et le Théâtre La Llotja et le Centre de congrès à Lleida (Espagne, 2010). L'agence travaille actuellement sur les projets de la bibliothèque de Birmingham qui viendra s'intégrer au REP Theatre (Birmingham, GB, 2013) ; du Centre pour les arts Wei-Wu-Ying (Kaohsiung, Taiwan, 2013) et de l'hôtel de ville et de la gare de Delft (2015). Tous les projets cités sont situés au Pays-Bas, sauf mention contraire.

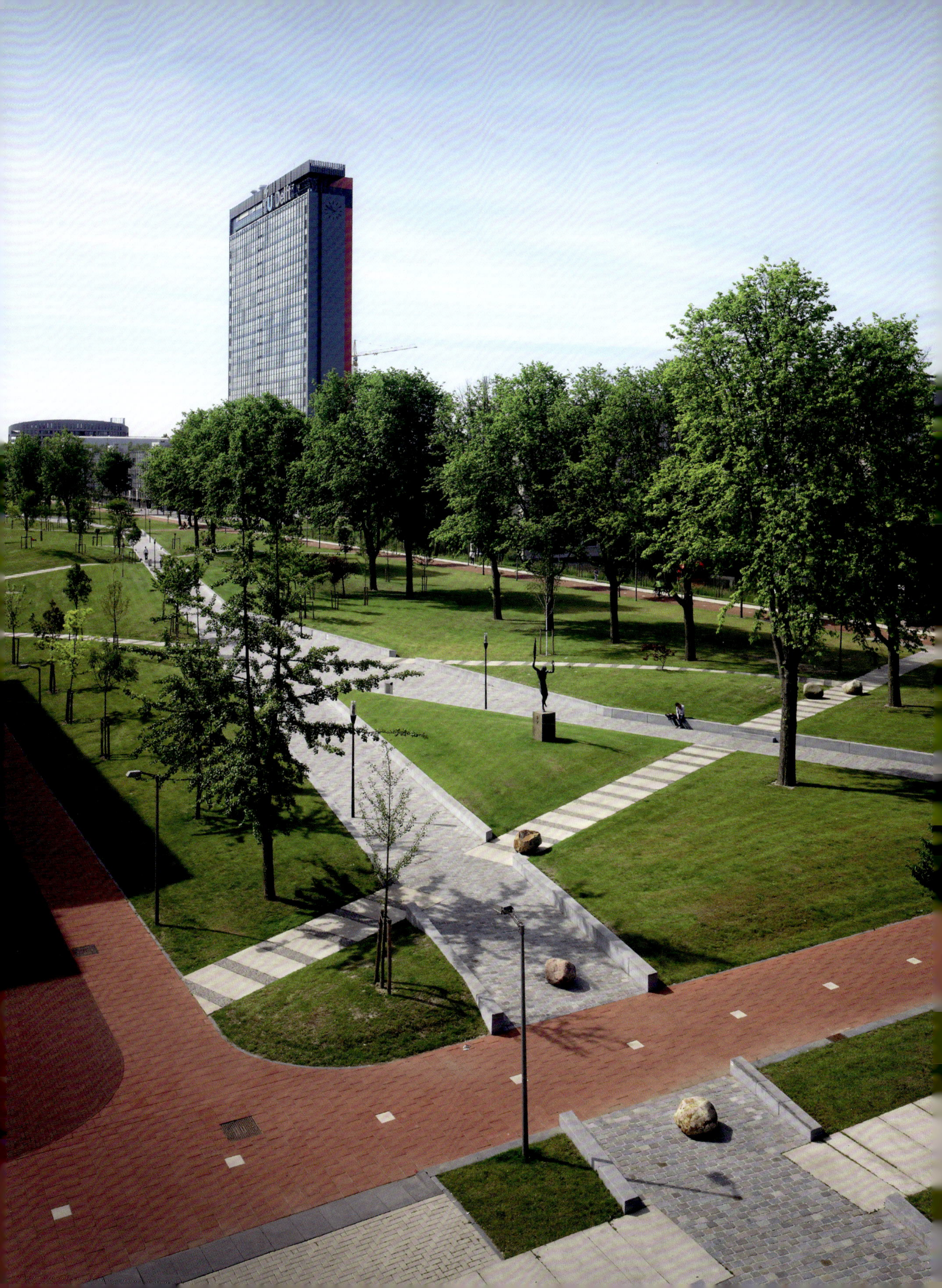
TU Delft

TU CAMPUS, MEKEL PARK
Delft, The Netherlands, 2007–09

*Address: Mekelweg, Mekel Park, 2628 BX Delft, The Netherlands, +31 15 278 46 70, www.tudelft.nl
Area: 4 hectares. Client: TU Vastgoed, Delft. Cost: €6 million*

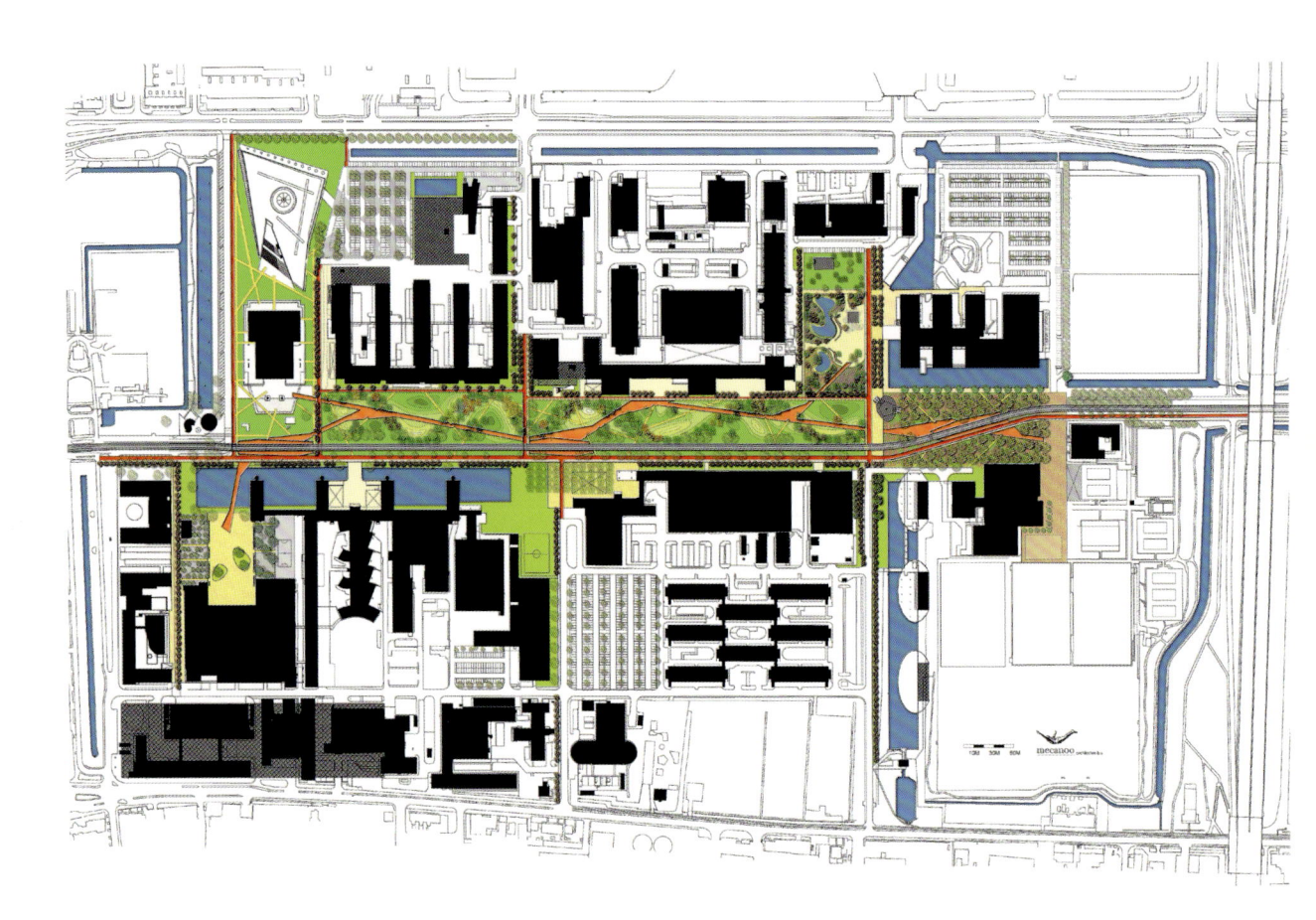

MEKEL PARK is envisaged as an informal meeting place for the students and staff of the Technical University. Three tram or bus stations were designed in the same formal language as the Nieuwe Delft promenade, which is marked by sawed granite stone and a 1547-meter-long granite bench. The promenade is "shaped like a bolt of lightning"; it links faculty buildings with one another and "symbolizes the interdisciplinary character of the university." An apparently random grid of footpaths crisscrosses the promenade while existing trees were saved or moved as much as possible. The architects write: "Flower fields and prunus trees gently announce the spring. The introduction of tram line 19, connecting Leidschendam with the Technical University via Delft Central Station, establishes the campus as a car-free zone."

Der **MEKELPARK** ist als Treffpunkt für Studenten und Mitarbeiter der TU gedacht. Es wurden drei Straßenbahnhaltestellen in der gleichen Formensprache wie die Nieuwe Delft Promenade entworfen, die durch gesägten Granit und eine 1547 Meter lange Granitbank gekennzeichnet ist. Die Fußgängerpromenade ist im Grundriss wie ein „Blitz" gestaltet. Sie verbindet die Fakultätsgebäude miteinander und „symbolisiert den interdisziplinären Charakter der Universität". Ein scheinbar zufälliges Fußwegeraster quert die Hauptpromenade mehrfach, wobei vorhandene Bäume so weit wie möglich erhalten blieben oder ggf. versetzt wurden. Die Architekten schreiben dazu: „Blumenwiesen und Zierkirschen künden zart den Frühling an. Die Straßenbahnlinie 19, die Leidschendam mit der Technischen Universität verbindet und dabei am Hauptbahnhof von Delft vorbeifährt, trägt dazu bei, dass der Universitätscampus eine autofreie Zone ist."

Le **MEKELPARK** a été pensé comme un lieu informel mis à la disposition des étudiants et du personnel de l'Université de technologie de Delft pour se détendre ou se rencontrer. Les trois arrêts de tramway et de bus créés utilisent le même langage formel que la promenade de Nieuwe Delft, caractérisée par un dallage en granit scié et un banc de granit de 1547 m de long. Ce cheminement « en forme d'éclair » relie les bâtiments universitaires entre eux et « symbolise le caractère interdisciplinaire de l'université ». Une trame d'allées d'implantation apparemment aléatoire strie le parc. Les arbres existants ont été conservés ou légèrement déplacés. « Des champs de fleurs et des prunus annoncent discrètement l'arrivée du printemps. La mise en place de la ligne de tramway n° 19, qui relie Leidschendam à l'Université technologique en passant par la gare centrale de Delft, a permis de transformer le campus en zone sans voitures. »

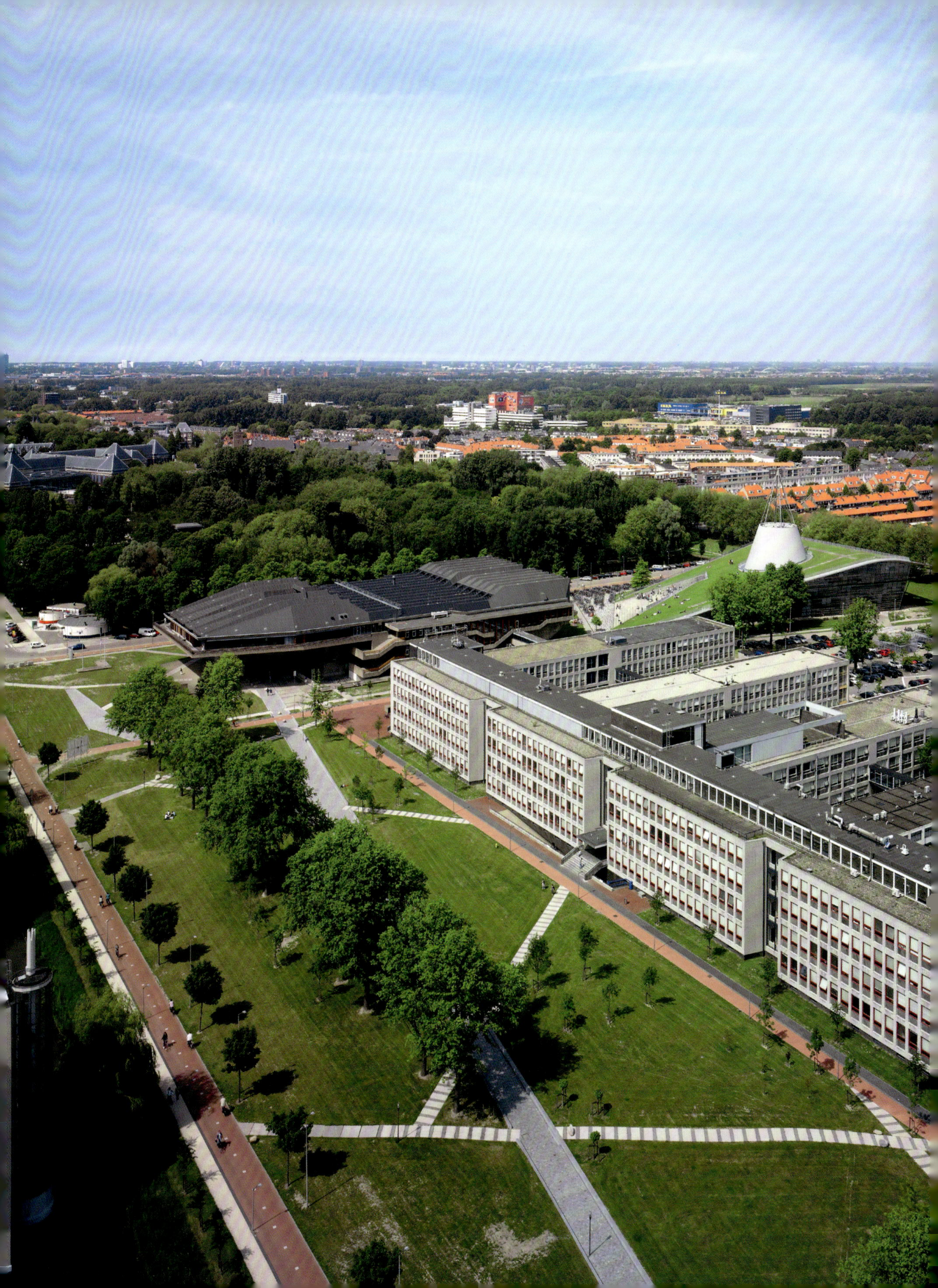

The architects have designed the central band of green space on the campus as a simple, lightly wooded grass area. There is a certain minimalism in this landscape solution.

Zentrales Gestaltungselement des Campus ist ein grünes Band – eine einfache Rasenfläche, auf der einige Bäume verteilt sind. In dieser landschaftsgestalterischen Lösung ist ein gewisser Minimalismus zu erkennen.

Les architectes ont conçu cette bande verte au milieu du campus comme un simple espace vert planté d'arbres. C'est une solution d'esprit minimaliste.

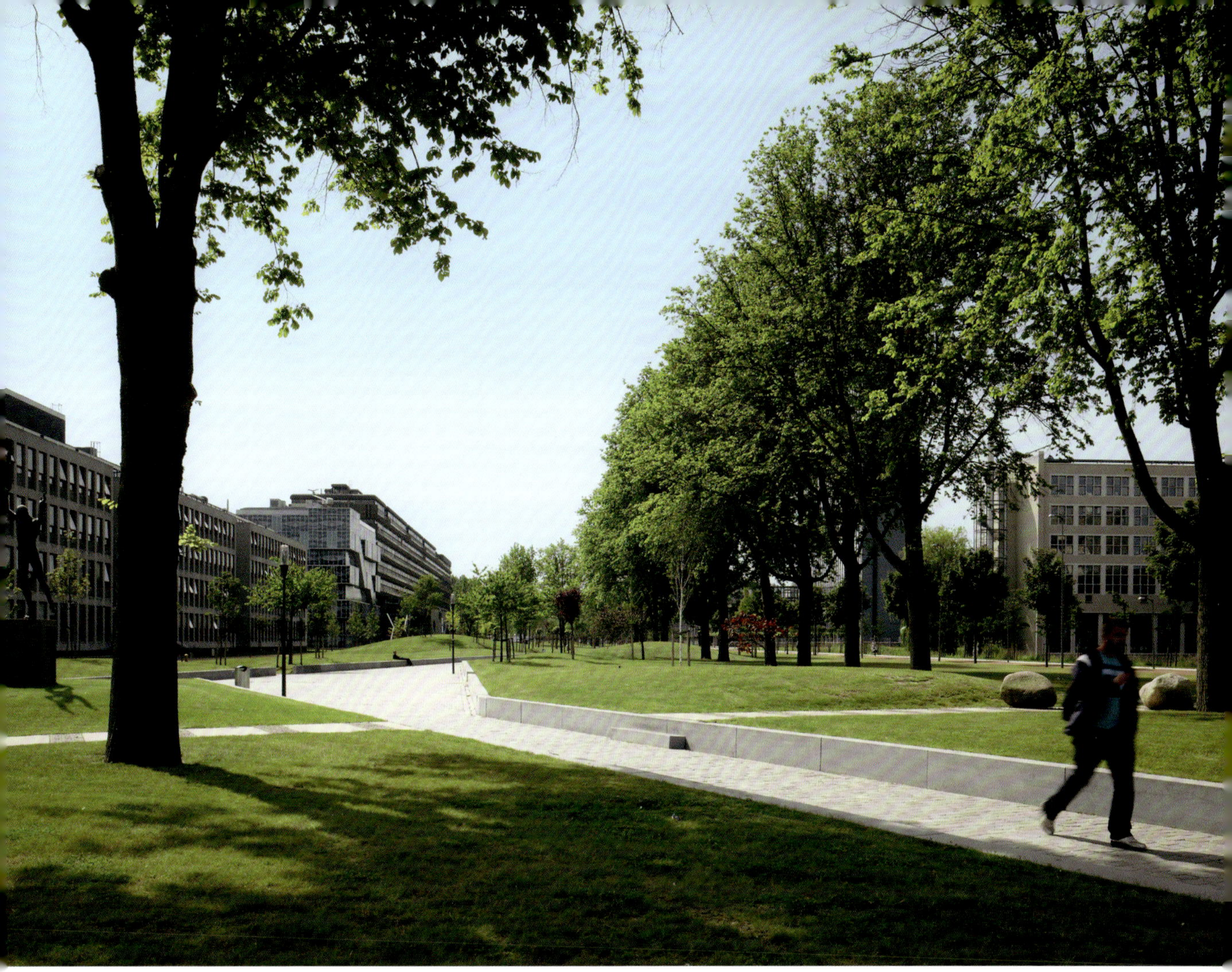

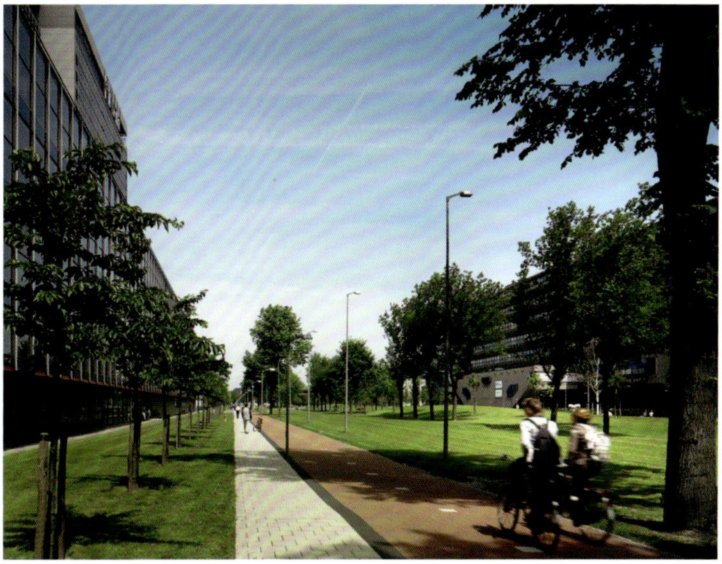

FERNANDO MENIS

Fernando Menis / Menis Aquitectos
Puerta Canseco, 35, 2º B
38003 Santa Cruz de Tenerife
Canary Islands
Spain

Tel: +34 922 28 88 38 / Fax: +34 922 15 19 25
E-mail: info@menis.es / Web: www.menis.es

Born in Santa Cruz de Tenerife (Canary Islands, Spain) in 1951, **FERNANDO MENIS** studied architecture in Barcelona, obtaining his degree in 1975 from the Escuela de Arquitectura Superior there. He went on to study urban planning in the same school. He worked from 1976 to 1980 on projects such as the Jardin des Halles in Paris, and housing in Marne-la-Vallée (both in France). He also worked with the architect Ricardo Fayos in Barcelona (1977). In 1981, he created a firm with Felipe Artengo Rufino and José María Rodriguez-Pastrana Malagón under the name Artengo, Menis, Pastrana. In 1992 the team turned into a company called Artengo, Menis, Pastrana Arquitectos (AMP Arquitectos, S.L.). In 2004 and 2005, Menis, with his new independent studio Menis Arquitectos, won first prize in competitions for the Puerto de la Cruz Harbor (Tenerife); the Cuchillitos de Tristán Park (Tenerife; published here); 55 social housing units, La Laguna (Tenerife, 2010); and for the rehabilitation of the historic center of Agulo and Vallehermoso (La Gomera). His work includes the Offices of the President of the Government (Santa Cruz de Tenerife, 1997–99); 11 bungalows (El Guincho, Tenerife, 2000–02); a Swimming Pool on the Spree (Berlin, 2003–04); and Cuchillitos de Tristán Park (Santa Cruz de Tenerife, Canary Islands, 2006–07, published here), all in Spain unless stated otherwise. The first phase of his Church of the Holy Redeemer in Tenerife was completed in 2009 and the second phase is currently in progress. The Plaza and Sacred Museum (Adeje, Tenerife South), and Jordanek Concert Hall (Torun, Poland) are both currently under construction.

FERNANDO MENIS wurde 1951 in Santa Cruz de Tenerife (Kanarische Inseln, Spanien) geboren. Er studierte Architektur in Barcelona und schloss sein Studium 1975 an der Escuela de Arquitectura Superior ab. Danach studierte er Stadtplanung an derselben Universität. Von 1976 bis 1980 arbeitete er an Projekten wie dem Jardin des Halles in Paris und Wohnungsbauprojekten in Marne-la-Vallée. 1977 war er Mitarbeiter des Architekten Ricardo Fayos in Barcelona. 1981 gründete er gemeinsam mit Felipe Artengo Rufino und José María Rodriguez-Pastrana Malagón ein Büro mit dem Namen Artengo, Menis, Pastrana. 1992 nahm das Büro die Rechtsform einer GmbH an und nannte sich nun Artengo, Menis, Pastrana Arquitectos (AMP Arquitectos, S. L.). 2004 und 2005 gewann Menis, der nun das unabhängige Büro Menis Arquitectos gegründet hatte, mehrere erste Preise bei Wettbewerben: für den Hafen von Puerto de la Cruz (Teneriffa), den Park Cuchillitos de Tristán (Teneriffa, hier vorgestellt) und 55 Sozialwohnungen in La Laguna (Teneriffa, 2010). Ebenfalls preisgekrönt wurde sein Entwurf für die Sanierung der historischen Altstadt von Agulo und Vallehermoso (La Gomera, Kanarische Inseln). Zu seinen Projekten gehören das Bürogebäude des Präsidenten der Regierung der Kanarischen Inseln (Santa Cruz de Tenerife, 1997–99), elf Bungalows (El Guincho, Teneriffa, 2000–02), ein Badeschiff in der Spree (Berlin, 2003–04) und der Park Cuchillitos de Tristán (Santa Cruz de Tenerife, 2006–07, hier vorgestellt). Alle Projekte befinden sich in Spanien, sofern nicht anders angegeben. Der erste Bauabschnitt der Iglesia del Santísimo Redentor auf Teneriffa wurde 2009 fertiggestellt, der zweite Bauabschnitt wird gerade ausgeführt. Das Museo Sacro und die Plaza de España in Adeje im Süden von Teneriffa sowie die Konzert- und Kongresshalle Jordanek in Thorn, Polen, werden zurzeit realisiert.

Né à Santa Cruz de Tenerife (îles Canaries, Espagne) en 1951, **FERNANDO MENIS** a étudié l'architecture à l'École supérieure d'architecture de Barcelone (ETSAB) dont il est sorti diplômé en 1975, puis l'urbanisme dans la même école et a travaillé de 1976 à 1980 sur des projets comme le jardin des Halles à Paris et des logements à Marne-la-Vallée. Il a également collaboré avec l'architecte Ricardo Fayos à Barcelone (1977). En 1981, il crée l'agence Artengo, Menis, Pastrana en association avec Felipe Artengo Rufino et José María Rodríguez-Pastrana Malagón. En 1992, l'agence prend le nom d'Artengo, Menis, Pastrana Arquitectos (AMP Arquitectos, S.L.). En 2004 et 2005, Menis et sa nouvelle agence indépendante, Menis Arquitectos, remportent le premier prix des concours organisés aux Canaries pour le Puerto de la Cruz (Tenerife) ; le parc Cuchillitos de Tristán (Tenerife, publié ici) ; 55 logements sociaux à La Laguna (Tenerife, 2010) et la réhabilitation du centre historique d'Agulo et de Vallehermoso (La Gomera). Parmi ses réalisations (toutes en Espagne, sauf mention contraire) : les bureaux du président du gouvernement (Santa Cruz de Tenerife, 1997–99) ; 11 bungalows (El Guincho, Tenerife, 2000–02) ; une piscine sur la Spree (Berlin, 2003–2004) et le parc Cuchillitos de Tristán (Santa Cruz de Tenerife, 2006–07, publié ici). La première phase de construction de l'église du Saint-Rédempteur à Tenerife a été achevée en 2009 et la seconde phase est en cours. La Plaza de España et le Museo Sacro (Adeje, Tenerife Sud), ainsi que la salle de concert Jordanek (Toruń, Pologne) sont actuellement en cours de construction.

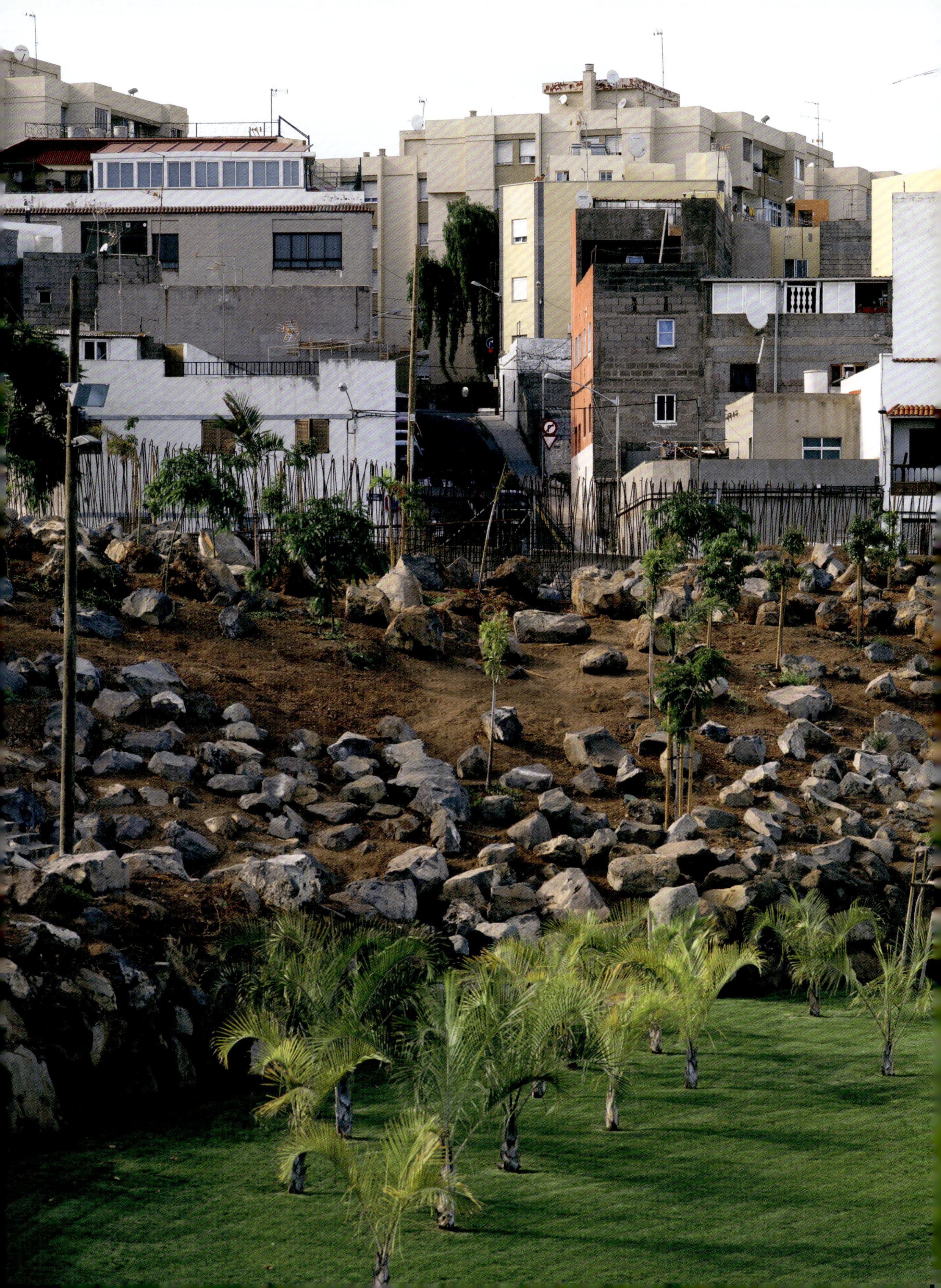

CUCHILLITOS DE TRISTÁN PARK
Santa Cruz de Tenerife, Canary Islands, Spain, 2006–07

*Address: Avenida de Los Principes S/N, neighborhood of Tristán, Santa Cruz de Tenerife, Canary Islands, Spain
Area: 32 500 m². Client: Cabildo Insular de Tenerife (Government of Tenerife)
Cost: €2.6 million. Collaboration: Bruno Rodriguez (Assistant Architect),
Arnoldo Stantos (Biology Consultancy), Zona Verde (Gardening)*

The meandering paths of the park allow visitors to go up or down the hillside in very pleasant surroundings where local vegetation is privileged.

Die mäandrierenden Fußwege laden den Besucher dazu ein, in der angenehmen Parklandschaft, in der vorwiegend heimische Pflanzen wachsen, hügelauf und hügelab zu spazieren.

Les méandres des allées du parc permettent au visiteur d'escalader ou de descendre le flanc de la colline dans un environnement agréable qui privilégie la végétation locale.

This park forms a boundary between densely populated areas. There are differences of as much as 30 meters in the natural levels of the site, a fact that influenced the design of Fernando Menis. He states: "Rather than principal and secondary paths, there is a network of walks that reconcile the different levels. The park is delimited by a footpath that shakes off the impression of being a pavement, simplifying and easing the transition between park and city. Like those inside the park, this path varies in breadth, becoming wider near the main entrances. This network of paths, with its free geometry, adapts to the topography to reinforce the image of lava flows, creating routes through the recreated landscape that offer views from different perspectives." Two thirds of the park area are covered by vegetation including 3000 square meters of grass surrounded by an amphitheater for games and events. Flowering plants were chosen to provide different colors, while a fence was designed to be completely covered with creeping plants. The topography was used to create playgrounds, a skateboard area, a swimming pool, and quieter places for the elderly. "The park," says Menis, "becomes a landscape which seems to have been unearthed from its urban surroundings to retrieve a pocket of frozen nature that spills over into the city, the Anaga mountains, and the sea."

Dieser Park liegt am Rand dicht besiedelter Wohngebiete. Das Terrain ist durch große Höhenunterschiede von bis zu 30 Metern charakterisiert. Dies hat wesentlichen Einfluss auf den Entwurf von Fernando Menis gehabt. Er bemerkt dazu: „Es gibt hier keine Haupt- und Nebenwege. Wir haben ein Wegenetz geschaffen, das die unterschiedlichen Geländeniveaus miteinander in Einklang bringt. Ein umlaufender Fußweg bildet die Grenze des Parks, der nicht einfach nur eine Pflasterfläche ist, sondern den Übergang zwischen Park und Stadt herstellt. Er variiert in seiner Breite, wie übrigens auch alle anderen Wege in diesem Park. Nähert man sich den Haupteingängen, verbreitert er sich. Dieses Wegenetz mit seiner freien Geometrie passt sich der Topografie des Geländes an und verstärkt das Bild von Lavaströmen, die sich ihren Weg durch die neue Parklandschaft bahnen und von denen man Aussichen aus unterschiedlichen Perspektiven hat." Zwei Drittel der Parkfläche sind mit Vegetation bedeckt, dazu gehört auch eine 3000 Quadratmeter große Rasenfläche, die von einer halbrunden Mauer umschlossen ist und sich für Rasenspiele und Veranstaltungen eignet. Blütenpflanzen wurden nach ihren Farben ausgewählt, und ein Zaun soll völlig mit Rankpflanzen überwuchert werden. Die Topografie wurde genutzt, um Spielplätze anzulegen, einen Skaterpark, ein Schwimmbecken und Ruhebereiche für ältere Leute. Menis sagt: „Der Park wird zur Landschaft, die mitten aus diesem urbanen Umfeld heraus zutage tritt, eine Art tiefgefrorene Natur, die man hier wiederfindet und die sich nun über die gesamte Stadt, die Anaga-Berge und das Meer ausbreitet."

Ce parc forme une limite entre des quartiers très peuplés. La dénivellation naturelle de près de 30 m a influencé le projet de Fernando Menis qui explique : « Plutôt qu'un réseau classique de cheminements principaux et secondaires, on trouve un réseau d'allées qui desservent et relient les différents niveaux. Le parc est délimité par une allée piétonnière qui fait oublier sa nature d'élément de voirie en simplifiant et facilitant la transition entre le parc et la ville. Comme pour les autres cheminements intérieurs du parc, sa largeur varie et s'accroît à proximité des entrées principales. Ce réseau d'allées de géométrie libre s'adapte à la topographie, ce qui renforce une idée de coulée de lave créant des cheminements à travers un paysage remodelé, qui offre des vues sous des perspectives différentes. » Deux tiers du parc sont recouvert de végétation, dont 3000 m² de pelouse prévue pour des jeux ou des manifestations et entourée d'un amphithéâtre. Les plantes à fleurs ont été sélectionnées pour la variété de leurs couleurs et la clôture mise en place sera entièrement recouverte de plantes grimpantes. La topographie a permis d'aménager des aires de jeux, une zone pour le skateboard, une piscine et des endroits plus calmes pour les personnes âgées. « Le parc, conclut Menis, devient ainsi un paysage qui semble avoir été exhumé de son environnement urbain pour relâcher une poche de nature, jusqu'alors figée, qui se répandrait sur la ville, jusqu'aux montagnes de l'Anaga et jusqu'à la mer. »

Above, an overall view of the park and the city of Santa Cruz. Right page, a field of rocks placed in an almost natural way, and, on the lower right, an elevation drawing showing the fairly steep slope of the site.

Oben, Blick über den Park und Santa Cruz. Rechte Seite, eine Freifläche, auf der, fast wie in der Natur, Felsbrocken verteilt sind. Auf der Zeichnung unten rechts kann man erkennen, wie steil das Gelände ist.

Ci-dessus, vue d'ensemble du parc et de la ville de Santa Cruz. Page de droite : un champ de rochers disposés de façon presque naturelle et, en bas à droite, élévation montrant l'assez forte dénivellation du terrain.

FERNANDO MENIS

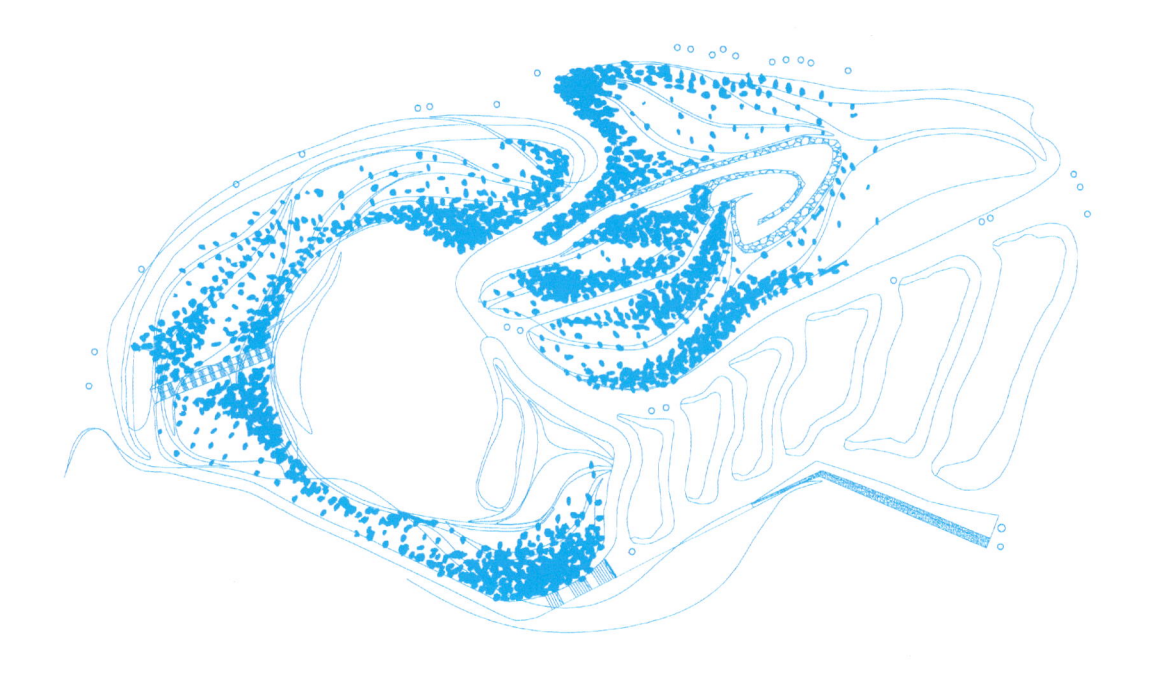

Earth and rock make up a good part of the surface of the park areas with the meandering paths covering a fairly large space as well. Stones, an ample resource in this volcanic region, form the real décor.

Ein Großteil der Parkfläche besteht aus Erde und Fels, durch die gewundene Wege führen. Steine gibt es in dieser Vulkangegend mehr als genug, hier sind sie die wichtigsten Gestaltungselemente.

Le parc se compose en grande partie de terre et de rochers, mais son réseau d'allées occupe une importante surface. Les pierres, qui font partie du paysage de cette île volcanique, constituent le vrai décor.

FERNANDO MENIS

EDUARDO DE MIGUEL

Eduardo de Miguel
C/ Alfonso de Córdoba, 8, bajo
46010 Valencia
Spain

Tel: +34 96 394 13 63
E-mail: info@landar.net

EDUARDO DE MIGUEL ARBONÉS was born in Pamplona, Spain, in 1959. In 1994, he moved to Valencia where he now works and teaches as Professor of Architectural Design at the Universidad Politécnica de Valencia. He also teaches Architectural Design at the Universidad de Navarra, Universidad Politécnica de Madrid, and at the Universidad Iberoamericana in Mexico D. F. He worked on Cabecera Park (published here) with Arancha Muñoz Criado and Vicente Corell Farinós. Born in 1962 in Valencia, **ARANCHA MUÑOZ** was educated at the University of Navarra and the Harvard GSD. She has been Director of the Territorial and Landscape Department of the Region of Valencia since 2007. **VICENTE CORELL** was born in Valencia in 1954 and graduated as an architect from the Universidad Politécnica of Valencia in 1978. He created his own firm, Corell y Monfort, arquitectos, the same year. The work of Eduardo de Miguel includes the Palau de la Música Extensión (Valencia, 2001–02); El Musical Cultural Center (Valencia, 1999–2003); Cabecera Park (Valencia, 2002–04, published here); the Benicarló Fish Market (Benicarló, 1997–2006); Landscaping of Tram Line 1 (Alicante, 2005–09); and the Serra Grossa Seafront Promenade (Alicante, in progress), all in Spain.

EDUARDO DE MIGUEL ARBONÉS wurde 1959 in Pamplona, Spanien, geboren. 1994 zog er nach Valencia, wo er heute arbeitet und als Professor für Entwerfen an der Universidad Politécnica de Valencia lehrt. Er hat auch Lehraufträge an der Universidad de Navarra, an der Universidad Politécnica de Madrid und an der Universidad Iberoamericana in Mexico D. F. Das Projekt des hier vorgestellten Parque de Cabecera bearbeitete er gemeinsam mit Arancha Muñoz Criado und Vicente Corell Farinós. **ARANCHA MUÑOZ** wurde 1962 in Valencia geboren. Sie erhielt ihre akademische Ausbildung an der Universidad de Navarra und an der Harvard Graduate School of Design. Seit 2007 ist sie Leiterin der Generaldirektion für Territorium und Landschaft der Region Valencia. **VICENTE CORELL** wurde 1954 in Valencia geboren und hat 1978 sein Diplom als Architekt an der Universidad Politécnica Valencia gemacht. Im selben Jahr gründete er sein Büro Corell y Monfort, arquitectos. Zu den Projekten von Eduardo de Miguel gehören der Erweiterungsbau des Palau de la Música (Valencia, 2001–02), das Kulturzentrum El Musical (Valencia, 1999–2003), der Parque de Cabecera (Valencia, 2002–04, hier vorgestellt), der Fischmarkt von Benicarló (Benicarló, 1997–2006), die Landschaftsgestaltung der Straßenbahnlinie 1 in Alicante (2005–09) und die derzeit im Bau befindliche Uferpromenade Serra Grossa in Alicante, alle in Spanien.

EDUARDO DE MIGUEL ARBONÉS, né à Pampelune (Espagne) en 1959, s'installe en 1994 à Valence où il travaille et enseigne en tant que professeur de conception architecturale à l'Université polytechnique de Valence, matière qu'il enseigne également à l'université de Navarre, à l'Université polytechnique de Madrid et à l'Université ibéro-américaine de Mexico. Il a travaillé sur le projet du Parque de Cabecera (publié ici) avec Arancha Muñoz Criado et Vicente Corell Farinós. Née en 1962 à Valence, **ARANCHA MUÑOZ** a étudié à l'université de Navarre et à la Harvard GSD. Elle dirige le département du Territoire et du Paysage de la région de Valence depuis 2007. **VICENTE CORELL**, né à Valence en 1954, est diplômé d'architecture de l'Université polytechnique de Valence (1978) et a fondé son agence, Corell y Monfort, arquitectos, la même année. Les réalisations d'Eduardo de Miguel comprennent l'extension du Palau de la Música (Valence, 2001–02) ; le Centre culturel El Musical (Valence, 1999–2003) ; le Parque de Cabecera (Valence, 2002–04, publié ici) ; le marché aux poissons de Benicarló (Benicarló, 1997–2006) ; le traitement paysager de la ligne de tramway nº 1 (Alicante, 2005–09) et la promenade en front de mer de Serra Grossa (Alicante, en cours), tous en Espagne.

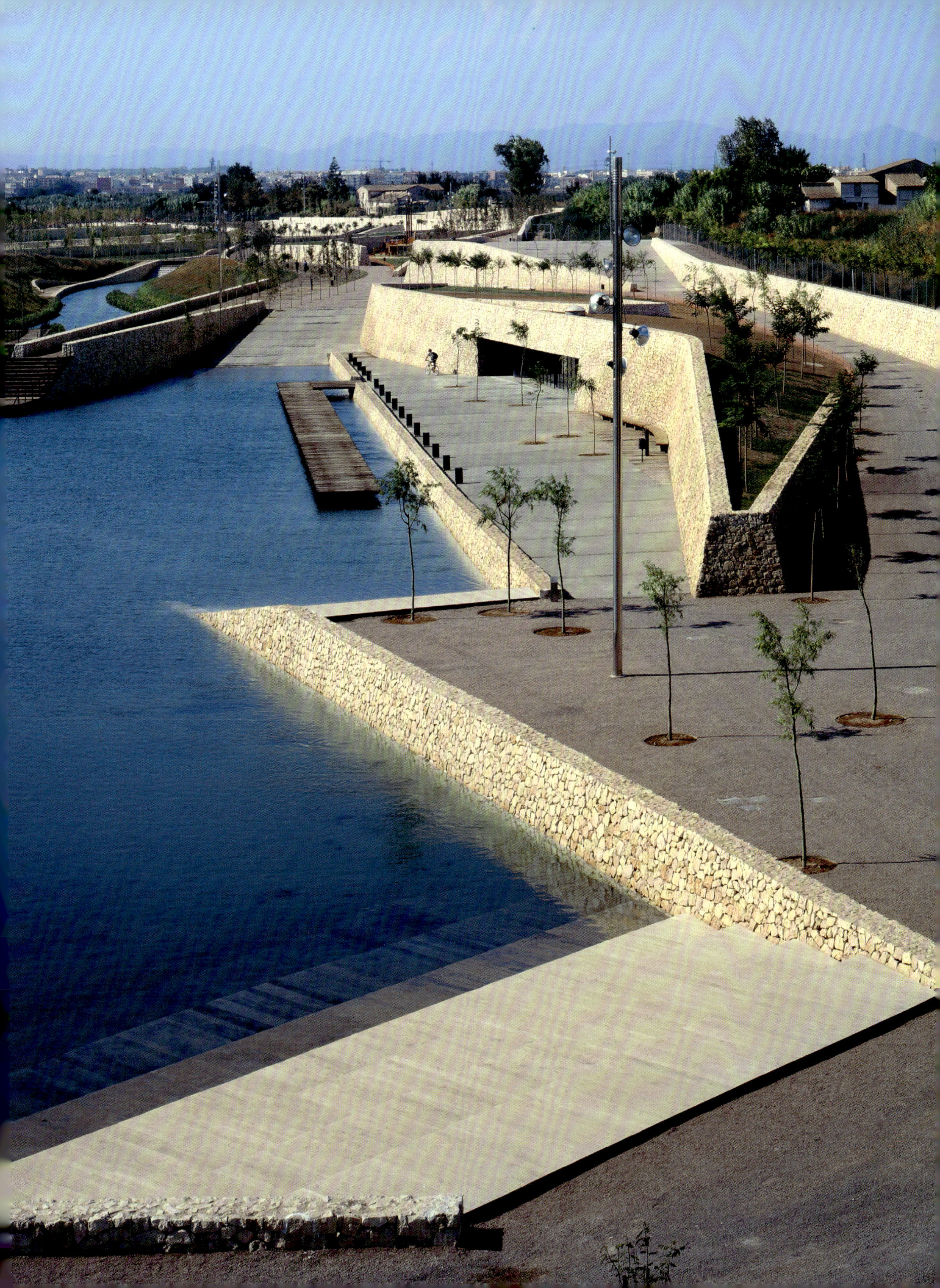

CABECERA PARK
Valencia, Spain, 2002–04

Address: Avenida 9 de Octubre, Valencia, Spain. Area: 169 256 m². Client: Ayuntamiento de Valencia
Cost: €17.3 million. Collaboration: Manel Colominas Golobardes (Agricultural Engineer)

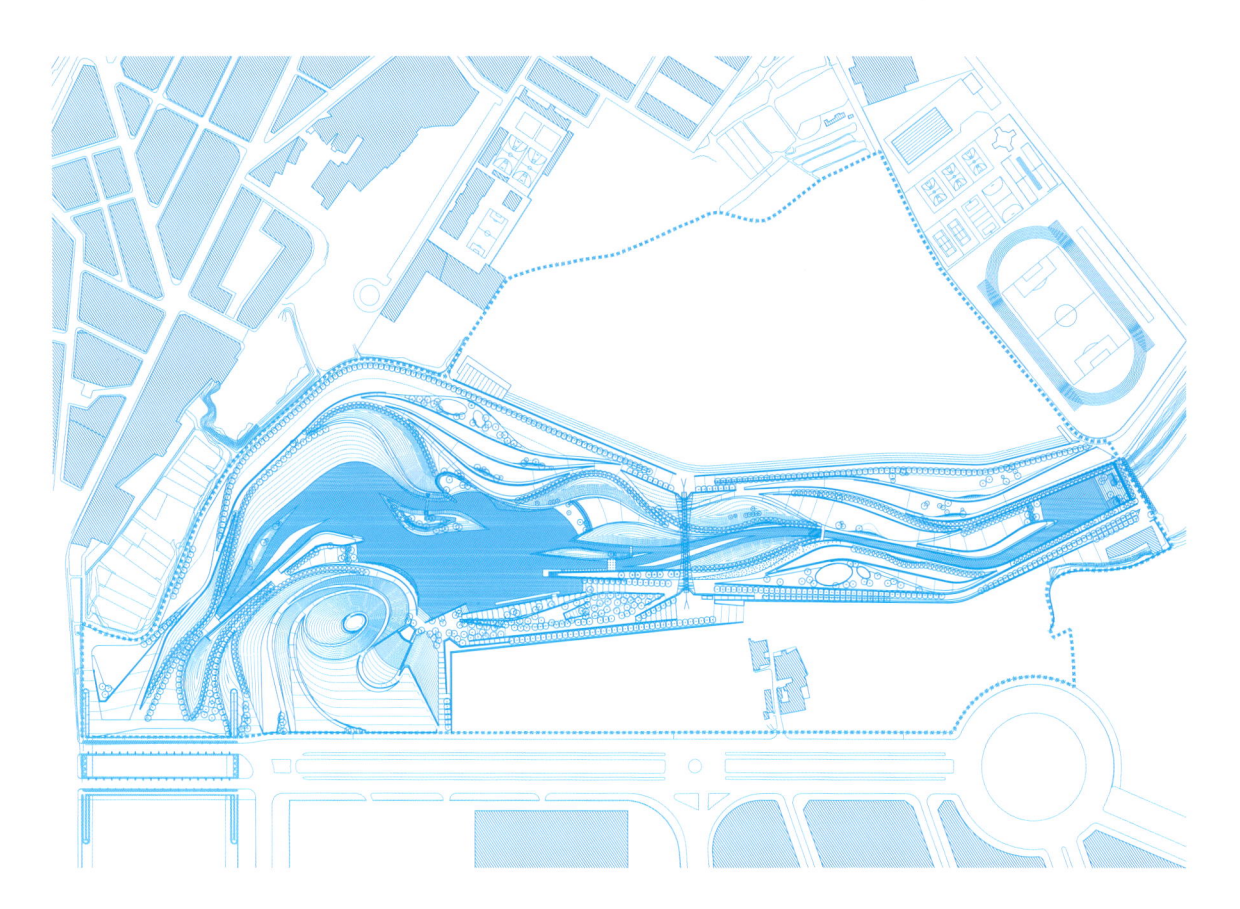

This park is located in an area about one kilometer long in the former bed of the Turia River between the edge of the city and nearby market gardens. The architect explains: "The idea for the park was to turn this space into a riverside wood, typical of the environs of Mediterranean rivers, blending vegetation, topography, and water. This strategy helped to solve the transition between the rigid embankments of the River Turia, built in the 18th century to protect the city from flash floods, and the natural riverbed upstream. The water, the vegetation, the topography, and the dry stone walls are the structuring features on which the concept of the park is based." An oval lookout hill marks the highest point of the park where the riverbed marks a 90° turn before running through the city. Island-like forms that recall the natural topography of the river bed mark the park, which has ipe wood benches, washed concrete paving, rounded gravel, and recycled railway ties constituting the paths.

Dieser etwa einen Kilometer lange Park liegt im ehemaligen Flussbett des Turia zwischen dem Rand der Stadt und den nahegelegenen Gemüseanbauflächen. Der Architekt erklärt hierzu: „Die Grundidee für diesen Park war, den vorhandenen Freiraum in einen uferbegleitenden Wald zu verwandeln, wie er für Flüsse im Mittelmeerraum typisch ist, und dabei die Vegetation, die Topografie und das Wasser zu einer Einheit zu verbinden. Diese Strategie half uns, den Übergang zwischen den starren Ufermauern des Turia, die im 18. Jahrhundert zum Schutz vor Überschwemmungen errichtet worden waren, und dem natürlichen Flußbett weiter flussaufwärts aufzufangen. Das Wasser, die Vegetation, die Topografie und die Natursteinmauern sind die strukturbildenden Elemente, auf denen die Konzeption des Parks basiert." Ein ovaler Aussichtshügel bildet den höchsten Punkt des Parks, dort, wo das Flussbett eine Biegung von 90 Grad macht, bevor es mitten durch die Stadt verläuft. Inselförmige Gestaltungen erinnern an die natürliche Topografie des Flussbettes und sind Markierungspunkte im Park. Die Parkbänke sind aus tropischem Ipé-Holz, die Bodenbeläge aus Waschbeton, Splitt und recycelten Eisenbahnschwellen.

Ce parc occupe une section d'environ un kilomètre de long de l'ancien lit du Turia, entre la délimitation de la ville et des champs de cultures maraîchères. Comme l'explique l'architecte : « L'idée du parc a été de transformer cet espace en bois au bord de l'eau – disposition typique de l'environnement des cours d'eau méditerranéens – en fusionnant la végétation, l'eau et la topographie. Cette stratégie a permis de résoudre les problèmes de transition entre la rigidité architecturale des quais du Turia, construits au XVIIIe siècle pour protéger la ville des inondations, et le lit naturel du fleuve en amont. L'eau, la végétation, la topographie et les murs en pierre sèche sont donc les éléments structurants sur lesquels repose le concept de ce parc. » Une colline-belvédère de forme ovale a été aménagée au point le plus élevé, au-dessus d'un méandre du fleuve à 90°, juste avant qu'il n'atteigne la ville. Des formes en îlots rappellent la topographie naturelle du fleuve ponctuent le parc, dont les aménagements comprennent des bancs en ipé, des sols en béton lavé, des allées en graviers ronds et des cheminements en traverses de chemin de fer recyclées.

EDUARDO DE MIGUEL

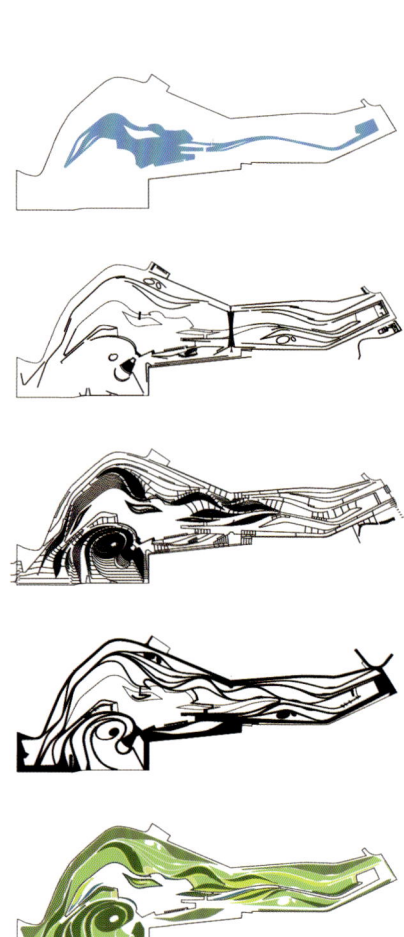

The architects have sculpted the space and the water, using the built areas to create the limits and walkways that enclose the water and greenery.

Die Architekten haben den Freiraum und das Wasser zu einer Gesamtskulptur gestaltet. Bebaute Flächen bilden die Grenzen des Parks, und Fußwege führen um das Wasser und die Grünflächen herum.

Les architectes ont sculpté l'eau et l'espace. Les parties construites et les allées sont autant de limites qui retiennent l'eau et la végétation.

EDUARDO DE MIGUEL

MIRALLES TAGLIABUE EMBT

Miralles Tagliabue Arquitectes Associats – EMBT
Passatge de la Pau, 10 bis. pral., 08002 Barcelona, Spain
Tel: +34 934 12 53 42 / Fax: +34 934 12 37 18
E-mail: info@mirallestagliabue.com / Web: www.mirallestagliabue.com

Born in Barcelona in 1955, Enric Miralles received his degree from the ETSA in that city in 1978. He died in 2000. He formed a partnership with Carme Pinós in 1983, and won a competition for the Igualada Cemetery Park on the outskirts of Barcelona in 1985 (completed in 1992). Contrary to the minimalism of other local architects like Viaplana and Piñón, with whom he worked from 1974 to 1984, or Estève Bonnel, Miralles was known for the exuberance of his style. His work includes the Olympic Archery Ranges (Barcelona, 1989–91); the La Mina Civic Center (Barcelona, 1987–92); the Morella Boarding School (Castelló, 1986–94); and the Huesca Sports Hall (Huesca, 1988–94), all in Spain. His most visible project, completed after his death, was the Scottish Parliament (Edinburgh, Scotland, UK, 1998–2004). **BENEDETTA TAGLIABUE** was born in Milan, Italy, and graduated from the IUAV in Venice in 1989. She studied and worked in New York (with Agrest & Gandelsonas) from 1987 to 1989. She worked for Enric Miralles, beginning in 1992, first becoming a Partner, then leading the studio after his death. Miralles Tagliabue EMBT completed the Rehabilitation of the Santa Caterina Market (Barcelona, 1997–2005); the Principal Building for the University Campus (Vigo, 2006); Public Spaces, HafenCity (Hamburg, Germany, 2006, published here); headquarters for Gas Natural (Barcelona, 2007); the Public Library (Palafolls, 1997–2007); and the Spanish Pavilion for the International Exhibition in Shanghai (China, 2010), all in Spain unless stated otherwise. The firm is currently working on a new metro station in Naples (Italy, 2011); Social Housing buildings and public spaces in Madrid (Spain, 2011); an Extension to the Music School in Hamburg (Germany, 2011); a Museum for the Chinese Painter Zhang Da Qian (Neijiang, China, 2010–13); and the Excellence Houhai Project in Houhai Center District (Shenzhen, China, 2011–13).

Enric Miralles wurde 1955 in Barcelona geboren und machte 1978 sein Diplom an der dortigen ETSA. Er starb 2000. Mit Carme Pinós gründete er 1983 ein Büro, das 1985 den Wettbewerb für den Friedhof von Igualada in Barcelona gewann (Fertigstellung 1992). Im Gegensatz zu dem Minimalismus lokaler Architekten wie Viaplana und Piñón, mit denen er 1974 bis 1984 zusammenarbeitete, oder Estève Bonnel war Miralles für seinen üppigen Stil bekannt. Er entwarf die Olympische Bogenschießanlage (Barcelona, 1989–91), das Stadtteilzentrum La Mina (Barcelona, 1987–92), das Internat in Morella (Castelló, 1986 bis 1994) und das Sportzentrum in Huesca (1988–94), alle in Spanien. Sein spektakulärstes Projekt ist das Schottische Parlament (Edinburgh, GB, 1998–2004). **BENEDETTA TAGLIABUE** wurde in Mailand, Italien, geboren und machte 1989 ihr Diplom in Venedig. 1987 bis 1989 studierte und arbeitete sie bei Agrest & Gandelsonas in New York. Ab 1992 arbeitete sie für Enric Miralles, nach seinem Tod wurde sie Leiterin des Büros. Zu den Projekten von Miralles Tagliabue EMBT gehören die Sanierung der Markthalle Santa Caterina (Barcelona, 1997–2005), das Hauptgebäude für den Universitätscampus von Vigo (2006), Platzgestaltungen in der HafenCity Hamburg (2006, hier vorgestellt), das Verwaltungsgebäude von Gas Natural (Barcelona, 2007), die Öffentliche Bibliothek von Palafolls (1997–2007) und der Spanische Pavillon auf der Expo 2010 in Shanghai (China, 2010), alle in Spanien, sofern nicht anders vermerkt. Derzeit ist das Büro mit der Planung einer neuen U-Bahn-Station in Neapel befasst (2011) sowie mit der Planung von Sozialwohnungen und öffentlichen Freiräumen in Madrid (2011), einem neuen Konzertsaal für die Jugendmusikschule in Hamburg (2011), einem Museum für den chinesischen Maler Zhang Da Qian (Neijiang, China, 2010 –13) und mit dem Hochhaus Tower Excellence Houhai in der Innenstadt von Houhai (Shenzhen, China, 2011–13).

Né à Barcelone en 1955, Enric Miralles, diplômé de l'ETSA en 1978, est décédé en 2000. Associé à Carme Pinós en 1983, il remporta le concours pour le cimetière d'Igualada, dans la banlieue de Barcelone, en 1985 (achevé en 1992). À l'opposé du minimalisme d'autres architectes locaux comme Viaplana et Piñón, avec lesquels il a travaillé de 1974 à 1984, ou Estève Bonnel, Miralles était connu pour l'exubérance de son style. Parmi ses réalisations en Espagne : le terrain de tir à l'arc olympique (Barcelone, 1989–91) ; le Centre social La Mina (Barcelone, 1987–92) ; l'internat de Morella (Castelló, 1986–94) et le palais des sports municipal de Huesca (1988–94). Son plus fameux projet, achevé après sa mort, a été le Parlement écossais (Édimbourg, 1998–2004). **BENEDETTA TAGLIABUE**, née à Milan est diplômée de l'IUAV de Venise (1989). Elle a étudié et travaillé à New York pour Agrest & Gandelsonas de 1987 à 1989, puis avec Enric Miralles à partir de 1992, devenant sa partenaire et prenant la direction de l'agence après la mort de celui-ci. Miralles Tagliabue EMBT a réalisé (en Espagne, sauf mention contraire) la réhabilitation du marché de Santa Caterina (Barcelone, 1997–2005) ; le bâtiment principal du campus universitaire de Vigo (2006) ; des espaces publics pour HafenCity (Hambourg, Allemagne, 2006, publiés ici) ; le siège de Gas Natural (Barcelone, 2007) ; la bibliothèque publique de Palafolls (1997–2007) et le pavillon de l'Espagne à l'Exposition universelle de Shanghai (Chine) en 2010. L'agence travaille actuellement sur une nouvelle station de métro à Naples ; des logements sociaux et des espaces publics à Madrid (2011), l'extension d'une école de musique de Hambourg (2011) ; un musée pour le peintre chinois Zhang Da Qian (Neijiang, Chine, 2010–13) et la tour Excellence Houhai, dans le quartier central de Houhai (Shenzen, Chine, 2011–13).

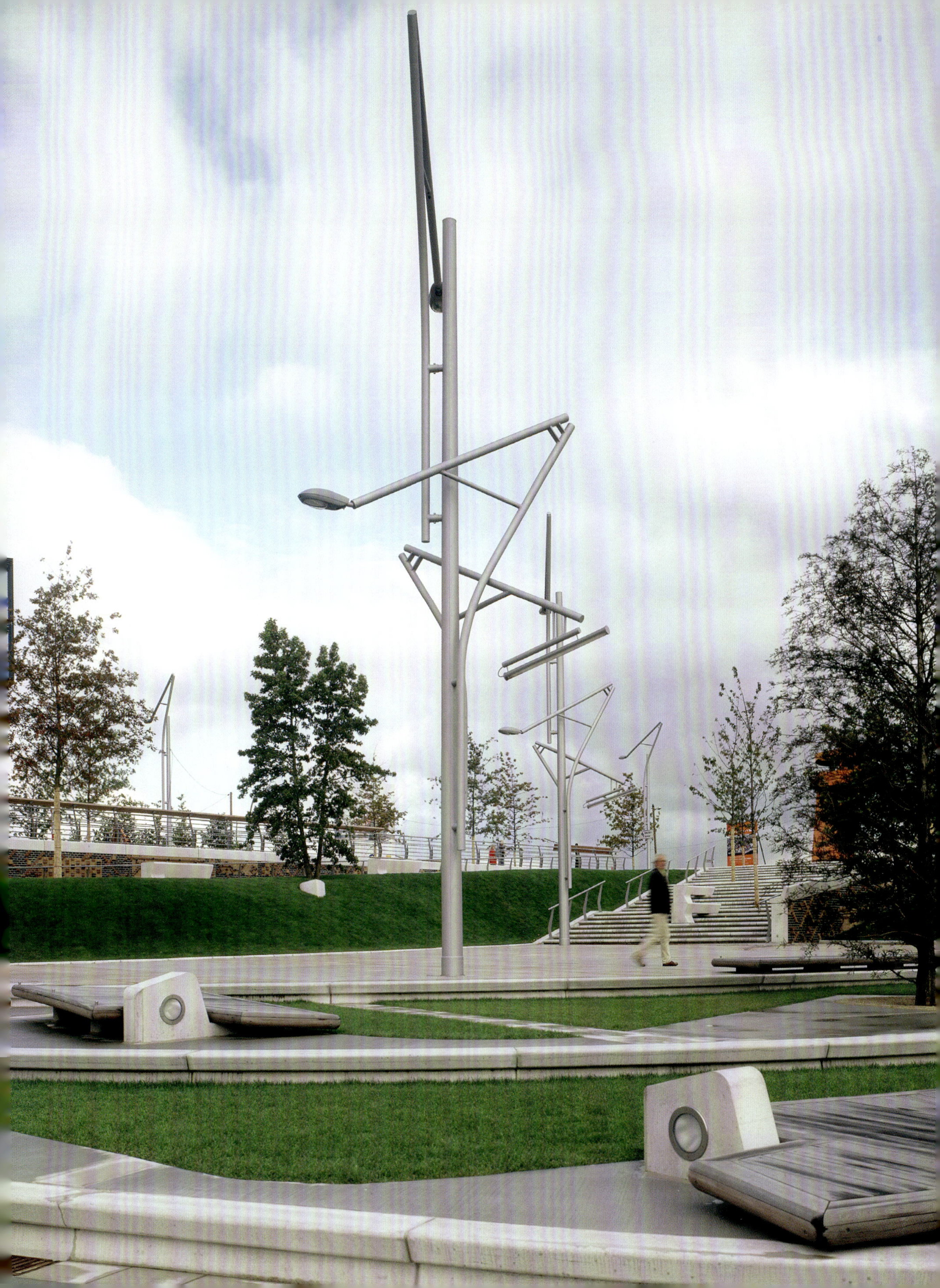

HAFENCITY PUBLIC SPACES
Hamburg, Germany, 2006

*Address: Großer Grasbrook, Hamburg, Germany
Area: 150 000 m². Client: HafenCity Hamburg GmbH. Cost: €8.6 million*

Such elements as the bent light poles and layers of stone that step down to the water enliven the space and allow walkers to enjoy the site from various different angles.

Elemente wie die abgeknickten Leuchten und die Betontreppen, die zum Wasser hinunterführen, beleben den Raum, und Spaziergänger können diesen Ort aus unterschiedlichen Perspektiven genießen.

Des éléments comme les lampadaires recourbés et les marches de pierre qui descendent vers l'eau animent l'espace et permettent aux promeneurs de découvrir le site sous différents angles.

The **HAFENCITY** project, one of the largest construction areas in Europe, has involved the conversion of former port areas of Hamburg into residential and business districts. Miralles Tagliabue was the winner of the 2002 international competition to create a new public space in this district. Different surface treatments and colors are used to "mediate" the large space, such as stairs and inclines clad in stone pavement and prefabricated concrete decorated with bird and fish patterns. The architects state: "Our intervention is dynamic and flexible. A changing landscape on a human scale, moving partially with the floods, bringing people nearer to the water and its moods." At water level a large floating platform provides access to sport and ferry boats as well as leisure areas. At the "low promenade level" (4.5 meters above the water), there are cafés and a promenade. Storm surges were calculated to inundate this area about three times a year.

Das Projekt **HAFENCITY,** eine der größten Baustellen Europas, umfasst die Umwandlung von Teilen des Hamburger Hafens in Wohn- und Geschäftsviertel. 2002 gewann das Büro Miralles Tagliabue den internationalen Wettbewerb für die Gestaltung eines öffentlichen Freiraums. Unterschiedliche Oberflächen und Farben wurden eingesetzt, um den großen Freiraum gestalterisch in den Griff zu bekommen, zum Beispiel Treppenanlagen und Rampen mit Steinverkleidung und Betonfertigteile mit dekorativen Vogel- und Fischmustern. Die Architekten stellen fest: „Unser Entwurf ist dynamisch und flexibel. Eine sich verändernde Landschaft im menschlichen Maßstab, die sich zum Teil mit den Gezeiten verändert und die Menschen dem Wasser und seinen Stimmungen näherbringt." Eine Plattform auf dem Wasser bietet Zugang zu den Fähren, den Privatbooten und den Freizeitgeländen. Auf der unteren Ebene, 4,5 Meter über dem normalen Pegelstand, befinden sich Cafés und eine Uferpromenade. Es wurde ermittelt, dass dieser Bereich etwa dreimal im Jahr durch Hochwasser überflutet wird.

Le projet de **HAFENCITY**, l'un des plus grands chantiers d'Europe, consiste en la reconversion de zones du port de Hambourg en quartier résidentiel et d'affaires. L'agence Miralles Tagliabue a remporté en 2002 le concours international organisé pour y créer un nouvel espace public. Emmarchements, plans inclinés pavés de pierre, éléments préfabriqués en béton décorés de motifs d'oiseaux et de poissons : divers traitements de surfaces et colorations servent à « médiatiser » ce vaste lieu. Comme l'expliquent les architectes : « Notre intervention est à la fois dynamique et souple. C'est un paysage changeant à échelle humaine, qui se déplace en partie avec les flots, rapproche les passants de l'eau et de ses humeurs. » Au niveau du fleuve, une importante plate-forme flottante donne accès aux ferries, aux bateaux de plaisance et à des zones de détente. Au « niveau bas » (4,5 m au-dessus de l'eau), on trouve des cafés et une promenade. Il est prévu que les tempêtes et les montées naturelles des eaux inondent cette zone environ trois fois par an.

The swirling tubular structure seen above, with fluorescent lighting attached to it, further heightens the dynamic impression given to the park by the architects.

Diese Röhrenkonstruktion mit dem fluoreszierenden Licht (oben), die wie ein Band über dem Platz schwingt, betont den dynamischen Charakter dieser Parkanlage.

Les formes brisées des lampadaires tubulaires (ci-dessus), à éclairage fluorescent renforcent l'impression de dynamisme donnée au parc par les architectes.

Above, mixing limited amounts of greenery with large stone surfaces, the architects recall something of the industrial history of the site, witnessed by the shipping cranes in the background. Right, overall plan of the site.

Die Mischung von streng abgegrenzten Grünarealen und großen, mit Betonplatten ausgelegten Flächen (oben) erinnert ein wenig an die industrielle Geschichte dieses Ortes. Die Schiffskräne im Hintergrund werden zu Zeugen. Rechts ein Gesamtplan des Geländes.

Ci-dessus, en associant des plantations de surface réduite à de grands plans dallés de pierre, les architectes évoquent en partie le passé historique du site dont témoignent les grues visibles dans le lointain. À droite, plan d'ensemble du site.

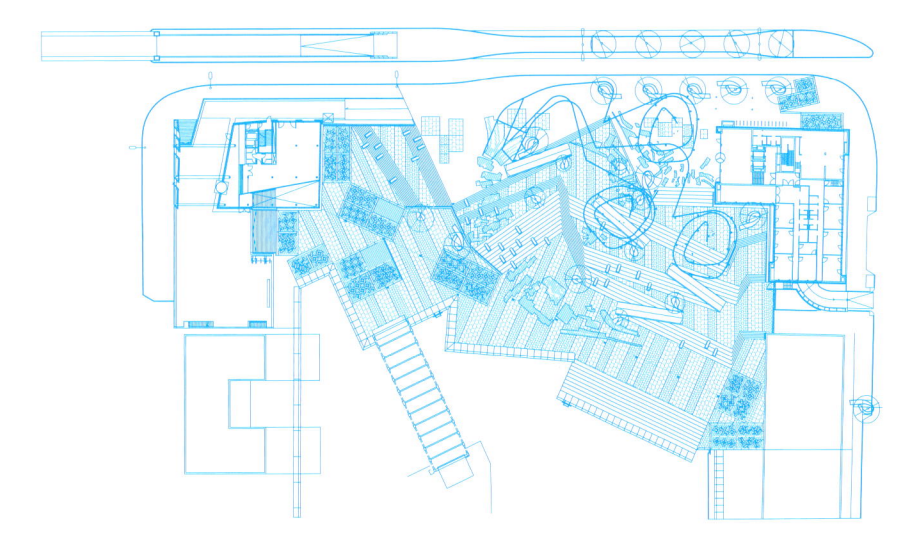

CATHERINE MOSBACH

Mosbach Paysagistes
81 Rue des Poissonniers
75018 Paris
France

Tel: +33 1 53 38 49 99
Fax: +33 1 42 41 22 10
E-mail: mp.mosbach@mosbach.fr
Web: www.mosbach.fr

After obtaining a degree in natural and life sciences from the Louis Pasteur University in Strasbourg, France, **CATHERINE MOSBACH** graduated from the École nationale supérieure du Paysage in Versailles in 1986. She has had her own landscape architecture practice in Paris since 1987. Working in the context of a master plan drawn up by Dominique Perrault (1992–97), Mosbach transformed 4.6 hectares of former industrial land on Bordeaux's right bank into a spectacular Botanical Garden (1999–2007). She was awarded the Rosa Barba European Landscape Prize at the 3rd European Biennale of Landscape in Barcelona, in 2003, for this project, and it featured in the exhibition "Groundswell, Constructing the Contemporary Landscape" at the Museum of Modern Art (MoMA, New York) in 2005. Other work includes the botanical gardens in Monaco (2003); the archeological and botanical park in Solutré (Saône-et-Loire, 1998–2006); the walk along the Saint-Denis canal (1998–2007); and L'autre rive (The Other Bank) (Quebec, Canada, 2008, published here). She is currently working on the gardens of the Louvre-Lens museum (with SANAA; Lens, ongoing), all in France unless stated otherwise.

Nach ihrem Universitätsabschluss in Natur- und Biowissenschaften an der Universität Louis Pasteur in Straßburg, Frankreich, machte **CATHERINE MOSBACH** 1986 ihr Diplom an der École nationale supérieure du paysage in Versailles. Seit 1987 hat sie ihr eigenes Büro für Landschaftsarchitektur in Paris. Auf der Grundlage des Masterplans von Dominique Perrault (1992–97) verwandelte Mosbach 4,6 Hektar einer Industriebrache am rechten Garonneufer von Bordeaux in einen spektakulären Botanischen Garten (1999–2007). 2003 wurde sie für dieses Projekt auf der 3. Biennale für Landschaftsarchitektur in Barcelona mit dem Rosa Barba European Landscape Prize ausgezeichnet. Der Entwurf wurde 2005 in der Ausstellung „Groundswell, Constructing the Contemporary Landscape" im Museum of Modern Art (MoMA, New York) gezeigt. Zu ihren weiteren Projekten gehören die Botanischen Gärten in Monaco (2003), der Archäologiepark und Botanische Garten in Solutré (Saône-et-Loire, 1998–2006), der Wanderweg entlang des Kanals Saint-Denis (1998–2007) und das Projekt L'autre rive (Das andere Ufer, Quebec, Kanada, 2008, hier vorgestellt). Derzeit arbeitet sie am Entwurf für die Gartenanlagen des Museums Louvre-Lens (zusammen mit dem Büro SANAA, Lens, im Bau). Alle Projekte befinden sich in Frankreich, sofern nicht anders vermerkt.

Après avoir obtenu un diplôme de sciences naturelles et de la vie à l'université Louis Pasteur à Strasbourg, **CATHERINE MOSBACH** a étudié à l'École nationale supérieure du paysage de Versailles, dont elle est sortie diplômée en 1986. Elle a créé et dirige son agence d'architecture paysagère à Paris depuis 1987. Dans le contexte d'un plan directeur établi par Dominique Perrault (1992–97), elle a transformé 4,6 ha de friches industrielles sur la rive droite de Bordeaux en un spectaculaire jardin botanique (1999–2007). Elle a reçu le prix européen du paysage Rosa Barba à la IIIe Biennale européenne du paysage à Barcelone en 2003 pour ce projet, qui a également été présenté dans l'exposition « Groundswell, Constructing the Contemporary Landscape » au Musée d'art moderne de New York (MoMA, 2005). Parmi ses autres réalisations figurent le jardin botanique de Monaco (2003) ; le parc botanique et archéologique de Solutré (Saône-et-Loire, 1998–2006) ; une promenade le long du canal Saint-Denis (France, 1998–2007) et le jardin L'autre rive (Québec, 2008, publié ici). Elle travaille actuellement sur les jardins du musée du Louvre-Lens (avec SANAA, Lens, en cours).

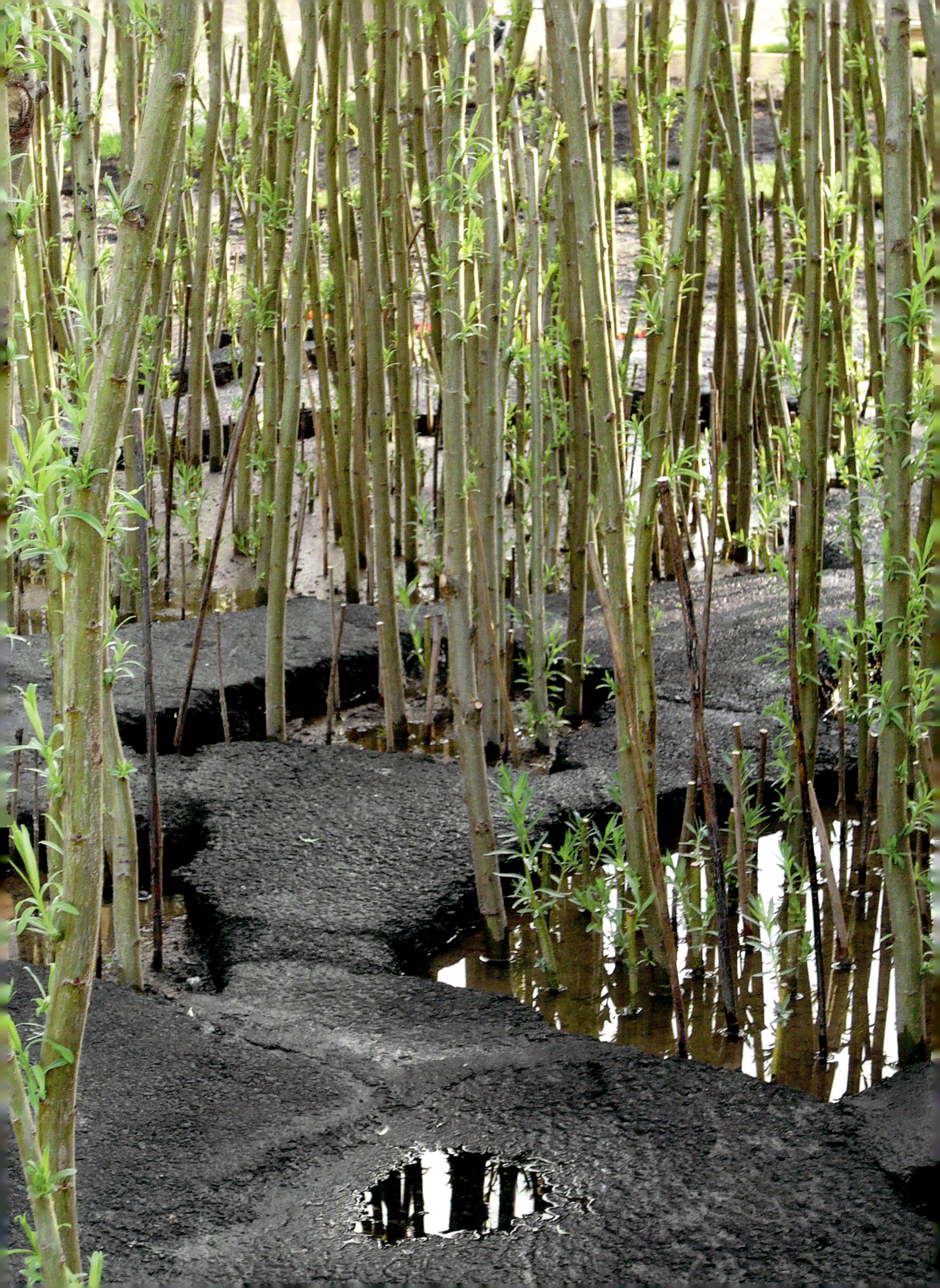

L'AUTRE RIVE (THE OTHER BANK)

Quebec, Canada, 2008

Address: Bassin Louise of the Vieux-Port, Quebec, Canada. Area: 150 m²
Client: Espace 400. Cost: not disclosed. Collaboration: Delphine Elie, Jessica Gramcko, Eiko Tomura

This project was selected as part of a competition for contemporary ephemeral gardens related to the 400th anniversary of Quebec. The jury cited it because of its references to the discovery of Canada. "The site is an area with asphalt where history is inscribed. The garden is a kind of planted cartography with a reference to nature. Very natural, but also conceptual…" The project was created in the Bassin Louise of the Vieux-Port in Quebec. The designers wrote: "This garden is a fable that from up close confronts the materials present—soil, plants, figures—and from afar the vision of a landscape through its provocative contours… The soil, surface of recording, differentiated from the earth's crust by the presence of life, is retranscribed here as an asphalt skin—vegetation fossilized in swamp water, then mineralized… The visitors traverse the milieu, revealing it to themselves and to others by their movement, inserted into interstitial latencies."

Dieses Projekt gehört zu den ausgewählten Entwürfen eines Wettbewerbs für temporäre Gärten, anlässlich des 400. Jubiläums von Quebec. Das Preisgericht entschied sich für diesen Entwurf wegen seiner Bezüge zur Entdeckung von Kanada. „Das Gelände ist eine Asphaltfläche, auf der die Geschichte eingeschrieben ist. Der Garten ist eine Art gepflanzter Landkarte mit einem starken Bezug zur Natur – sehr naturnah, aber ebenso Konzeptkunst…" Der Entwurf wurde im Bassin Louise im Alten Hafen von Quebec angelegt. Die Planer schrieben dazu: „Dieser Garten ist ein Mythos, der von Nahem betrachtet die vorhandenen Materialien – Erde, Pflanzen, Formen – einander gegenüberstellt. Von Weitem hat man den Eindruck einer Landschaft… Die Erde, als Speicherfläche, die sich von der Erdkruste dadurch unterscheidet, dass sie lebendig ist, wird hier als dünne Asphalthaut dargestellt – als Vegetation, die im Wasser der Sümpfe erst versteinerte und dann mineralisierte… Die Besucher gehen durch dieses Milieu und entdecken es, aufgrund ihrer eigenen Bewegung, zwischen den verborgenen Elementen des Unbewussten."

Ce projet a été sélectionné lors du concours de jardins contemporains éphémères organisé pour le 400ᵉ anniversaire du Québec. Créé dans le bassin Louise du Vieux-Port, le jury l'a apprécié pour ses références à la découverte du Canada. « Le site se présente comme un espace d'asphalte sur lequel l'histoire vient s'inscrire. Le jardin est une base cartographique plantée, avec référence à la nature. Très naturel, mais aussi conceptuel. » Les créatrices le décrivent ainsi : « Ce jardin est une fable qui, de près, confronte les matériaux en présence – sol, plante, figure – et de loin, la vision d'un paysage en ses contours évocateurs. […] Le sol, surface d'enregistrement qui se différencie de la croûte terrestre par la présence de la vie, est ici retranscrit par la peau d'asphalte, végétation fossilisée dans l'eau du marais, puis minéralisée… Lorsque les visiteurs traversent ce milieu, il se découvre à eux et aux autres spectateurs au gré de leurs déplacements, qui s'insinuent dans les espaces interstitiels. »

Measuring just 150 square meters, this is one of the smaller spaces selected for this book, and one of the least "durable" as well because it was designed to be ephemeral. The irregular asphalt surfaces contrast with the dense green vegetation.

Dieser Garten ist mit 150 Quadratmetern eines der kleineren und am wenigsten auf Dauer angelegten hier vorgestellten Projekte, da als temporärer Garten konzipiert. Die unregelmäßigen Asphaltflächen bilden einen Kontrast zur dichten Vegetation.

De 150 m² seulement, ce jardin est l'un des plus petits sélectionnés pour ce livre et l'un des moins « durables » puisque conçu pour être éphémère. Les surfaces en asphalte de forme irrégulière contrastent avec la densité de la végétation.

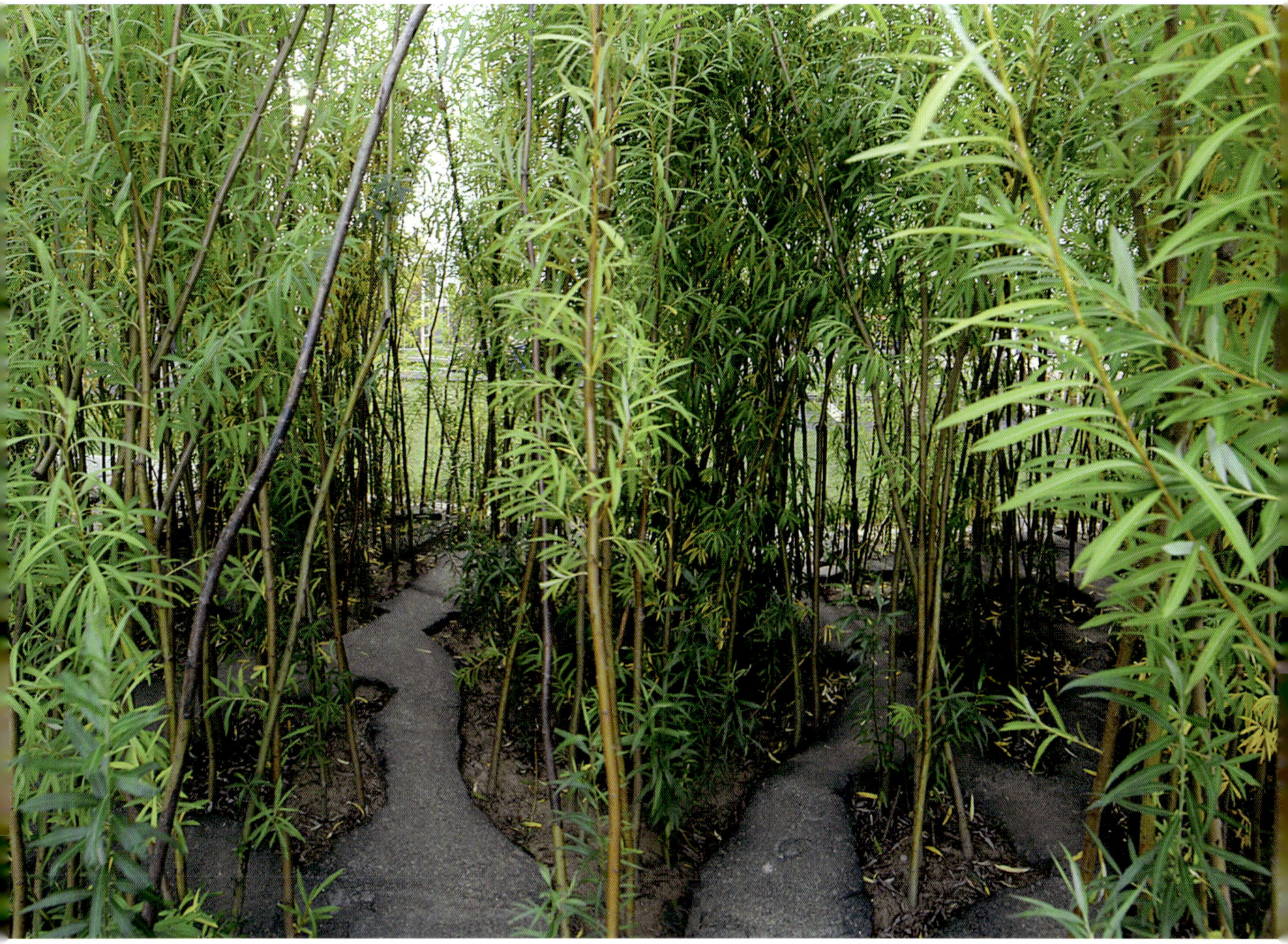

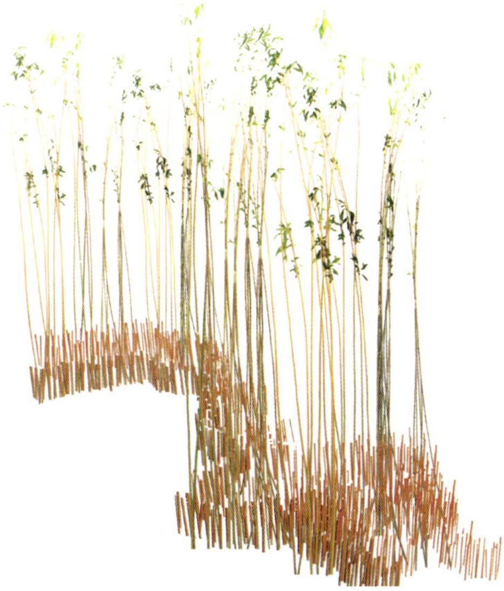

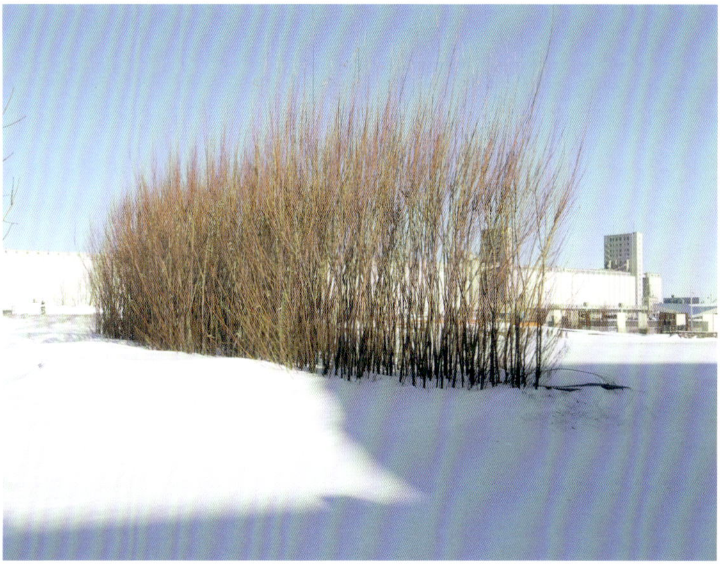

YOSHIAKI NAKAMURA

Yoshiaki Nakamura
Nakamura Sotoji Komuten
15 Nishigoshoden-cho, Murasakino
Kita-ku, Kyoto 603–8165
Japan

Tel: +81 75 451 8012
E-mail: nakamura.kohseki@aw.wakwak.com

Born in 1946, **YOSHIAKI NAKAMURA** is a son of the famous Japanese master carpenter Sotoji Nakamura. He graduated in 1968 from the Faculty of Business Administration at Ritsumeikan University in Kyoto. In 1968, he went to work for Nakamura Sotoji Komuten, Japan's leading *sukiya* building company. In 1984, he created the firm Kohseki in Kyoto as a design and construction company of *sukiya*-style buildings and wooden furniture (tea-ceremony houses) for Nakamura Sotoji Komuten. He was head of construction for the Villa Miho (architect Junzo Yoshimura with Sotoji Nakamura; Shigaraki, Shiga, Japan, 1982–85); project head and contractor for the landscaping of the Miho Museum (architect I. M. Pei; Shigaraki, Shiga, Japan, 1995–97, published here); project head for the reconstruction of the Nihon Shoin (Philadelphia, 2000); builder of the main guest room of the Kyoto State Guest House (Kyoto, 2005); and is working on Edo Avenue at Haneda Airport (Tokyo, 2010), all in Japan.

YOSHIAKI NAKAMURA wurde 1946 als Sohn des berühmten Kunsttischlers Sotoji Nakamura geboren. Er schloss 1968 sein Studium der Betriebswirtschaft an der Universität Ritsumeikan in Kioto ab. 1968 trat er in die Firma Nakamura Sotoji Komuten ein, Japans führendem Unternehmen zum Bau von „sukiya" – japanischen Teehäusern für die Teezeremonie. 1984 gründete er die Firma Kohseki in Kioto, ein Unternehmen, das sich auf den Bau von Gebäuden im Stil der „sukiya" sowie auf Holzmöbel für Nakamura Sotoji Komuten spezialisiert hat. Er war Bauleiter bei dem Projekt der Villa Miho (Architekt Junzo Yoshimura mit Sotoji Nakamura, Shigaraki, Shiga, Japan, 1982–85), Projektleiter und Garten- und Landschaftsbauer für die Gartengestaltung des Miho Museum (Architekt I. M. Pei, Shigaraki, 1995–97, hier vorgestellt), Projektleiter für die Rekonstruktion des Nihon Shoin (Philadelphia, USA, 2000) und Baumeister des Hauptraums im Gästehaus des japanischen Staates in Kioto (2005). Derzeit arbeitet er am Projekt der Edo Avenue beim Flughafen Haneda (Tokio, 2010).

Né en 1946, **YOSHIAKI NAKAMURA** est le fils du fameux maître charpentier Sotoji Nakamura. Une fois diplômé de la faculté de gestion de l'université Ritsumeikan à Kyoto en 1968, il commence à travailler pour Nakamura Sotoji Komuten, première entreprise japonaise de constructions *sukiya*. En 1984, il crée à Kyoto la société Kohseki, spécialisée dans la conception et la construction de bâtiments *sukiya* et de mobilier en bois (dans le style des maisons de cérémonie du thé), pour Nakamura Stotoji Komuten. Il est responsable de la construction de la villa Miho (architecte Junzo Yoshimura avec Sotoji Nakamura; Shigaraki, Shiga, Japon, 1982–85) ; directeur de projet et entreprise générale pour l'aménagement paysager du Miho Museum (architecte I. M. Pei, Shigaraki, Shiga, Japon, 1995–97, publié ici) ; chef de projet pour la reconstruction du Nihon Shoin (Philadelphie, US, 2000) ; constructeur de la suite principale de la résidence d'accueil des invités officiels de Kyoto (2005) et il travaille actuellement sur le projet de l'avenue Edo à l'aéroport de Haneda (Tokyo, 2010).

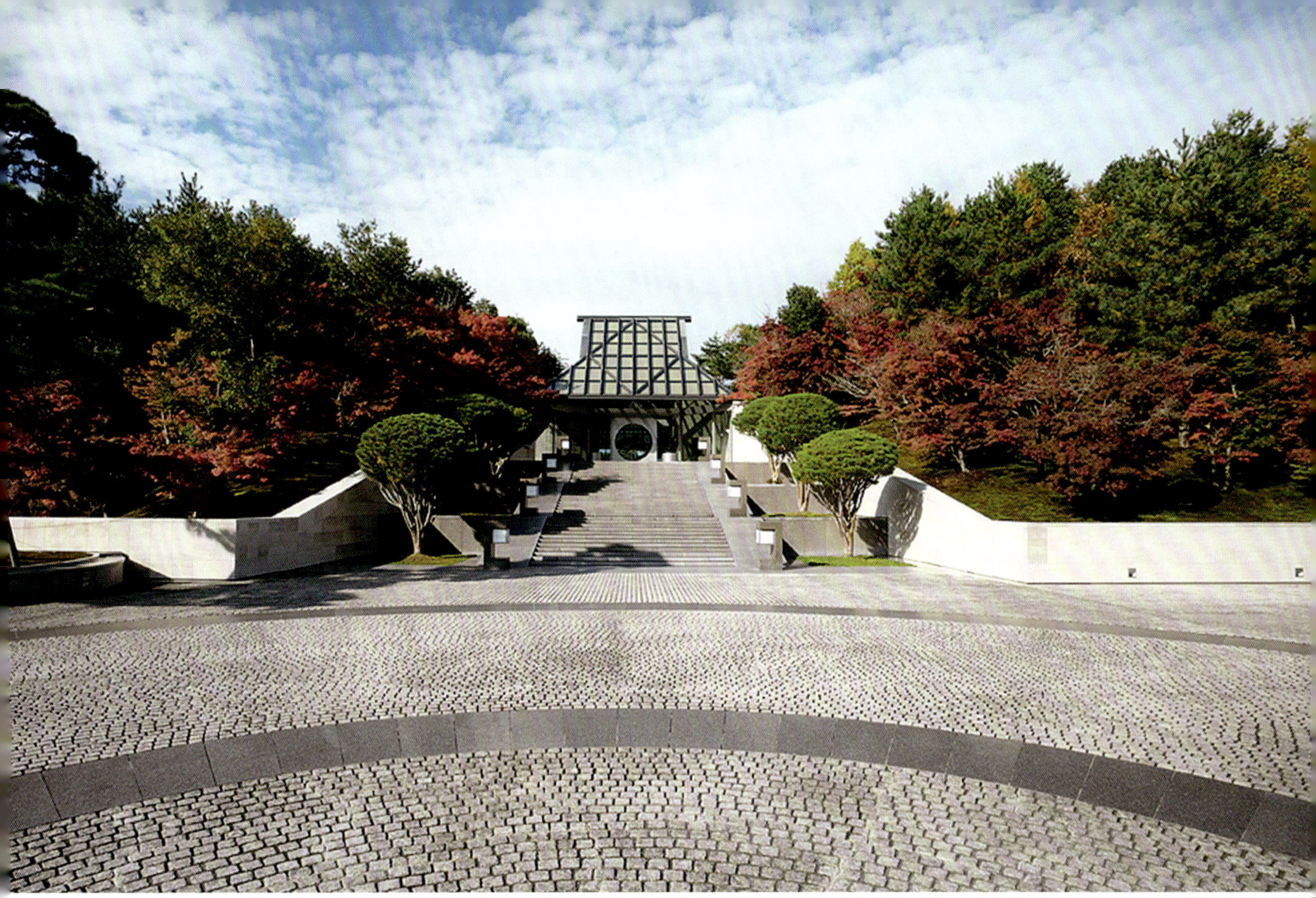

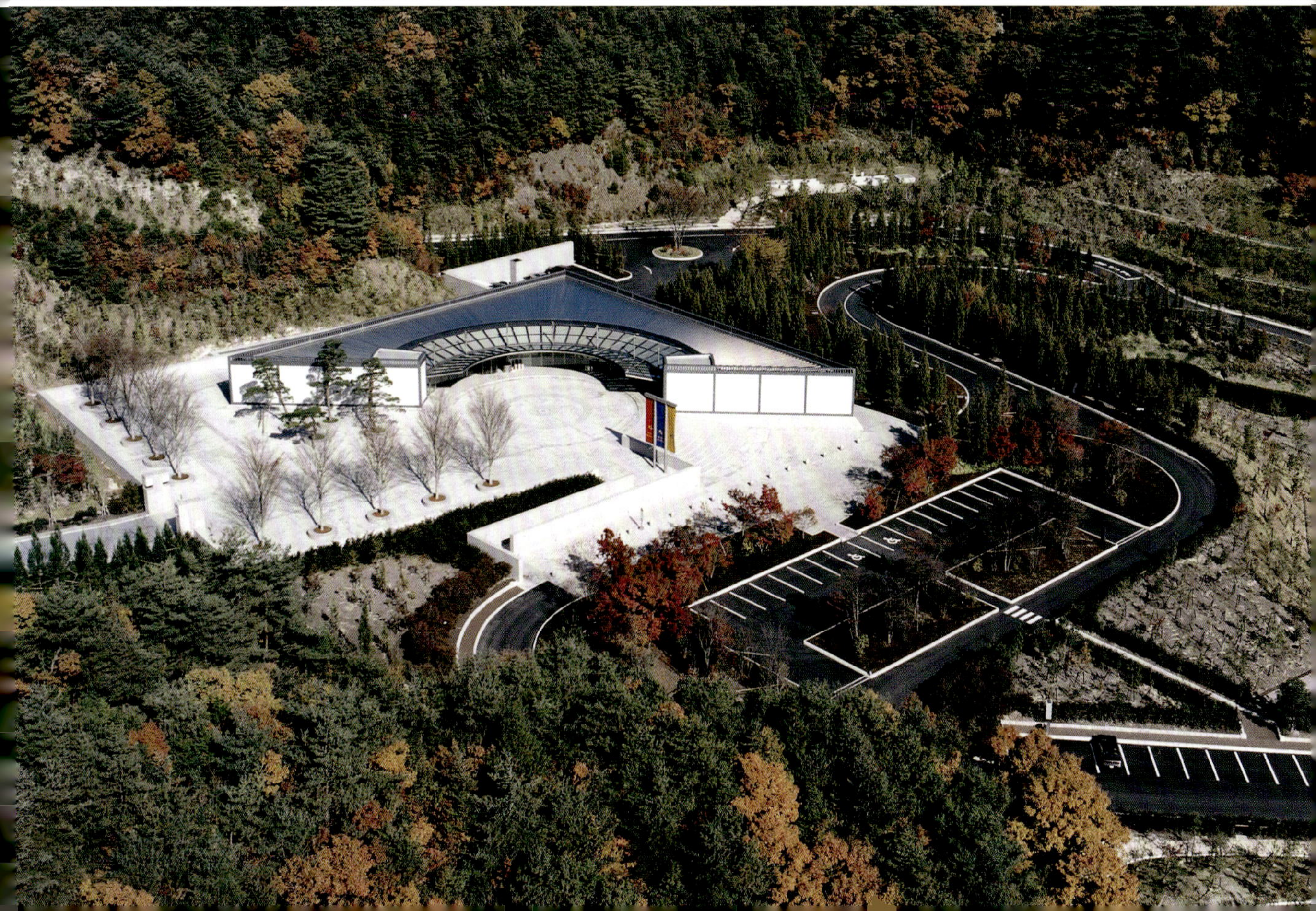

MIHO MUSEUM GARDENS
Shigaraki, Shiga, Japan, 1995–97

Address: 300 Momodani Tashiro, Shigaraki Koka, Shiga 529–1814, Japan, +81 748 82 3411, www.miho.jp
Area: 1 million m². Client: Shinji Shumeikai. Cost: not disclosed

Above, an overall view of the museum with its approach tunnel and bridge seen above the main buildings.

Oben, Gesamtansicht des Museums mit Zugangstunnel und Brücke, oberhalb der Hauptgebäude zu sehen.

Ci-dessus, vue générale du musée, du tunnel et du pont d'accès par-delà les bâtiments principaux.

Above, the suspension bridge that leads from the tunnel toward the museum. Right, the opening of the tunnel seen at the entrance side with weeping cherry trees in blossom.

Oben, die Hängebrücke, die vom Tunnel zum Museum führt. Rechts, die Tunnelöffnung mit den blühenden Trauerzierkirschen, vom Museumseingang gesehen.

Ci-dessus, le pont suspendu entre le tunnel et le musée. À droite, l'ouverture du tunnel côté entrée, donnant sur des cerisiers pleureurs en fleurs.

Designed by I. M. Pei, the **MIHO MUSEUM** is located in a hilly, wooded region near the town of Shigaraki and Lake Biwa. Because of local regulations, the museum is to a good extent buried in the site, and now, in good part, intentionally overgrown with local vegetation. Yoshiaki Nakamura was at the origin of the planting of the rows of weeping cherry trees that today line the entrance path to the museum, leading to the access tunnel and bridge. He worked closely with I. M. Pei, who was involved in the planting in and around the museum, on such areas as the rock garden enclosed in the North Wing courtyard of the museum. In collaboration with the client and Pei, Nakamura chose stones from the Saji River (Tottori) which are reputed to be amongst the best garden stones in Japan due to their attractive form and hardness. The rocks were placed with moss and pebbles in the courtyard as an allegory of the ocean and islands. I. M. Pei asked Yoshiaki Nakamura to increase the moss area of the garden because he felt that otherwise the space would be too "dry." The largest stone in this garden weighs seven tons and necessitated the placement of an extra pillar in the building, beneath the stone.

Das von Ieoh Ming Pei entworfene **MIHO MUSEUM** liegt in einer hügeligen Waldlandschaft nahe der Stadt Shigaraki und des Biwa-Sees. Aufgrund örtlicher Bauvorschriften musste das Museum zu einem großen Teil in das Gelände hineingebaut werden, und nun ist es mit heimischer Vegetation überwuchert. Nakamura hat die Trauerzierkirschen gepflanzt, die den Weg zu dem Tunnel und der Brücke des Eingangs säumen. Nakamura arbeitete sowohl bei der Bepflanzung der Innenräume als auch bei den Außenbereichen eng mit Pei zusammen, so auch beim Felsengarten im Innenhof des Nordflügels des Museums. Gemeinsam mit dem Bauherrn und Pei suchte Nakamura Steine aus dem Fluss Saji (Tottori) aus, die in Japan wegen ihrer attraktiven Form und ihrer Härte als mit die besten Steine für Gärten gelten. Die Felsbrocken wurden mit Moos und Kies als Allegorie des Ozeans und einer Inselgruppe in den Innenhof gesetzt. Ieoh Ming Pei bat Nakamura, die Moosfläche dieses Gartens zu vergrößern, weil er das Gefühl hatte, er wäre sonst zu „trocken". Der größte Fels wiegt sieben Tonnen, und man musste für ihn eine zusätzliche Stütze einbauen.

Conçu par I. M. Pei, le **MIHO MUSEUM** est situé dans une région de collines boisées, non loin de Shigaraki et du lac Biwa. Par respect de la réglementation locale, l'institution est en partie enterrée et volontairement recouverte de végétation. Yoshiaki Nakamura a été à l'origine de la plantation des alignements de cerisiers pleureurs qui bordent l'allée de l'entrée au musée menant au tunnel d'accès et au pont. Il a travaillé en étroite collaboration avec I. M. Pei, qui s'est impliqué dans les plantations du site, en particulier pour le jardin clos en rocaille de la cour nord du musée. En accord avec le client et Pei, Nakamura a choisi des pierres de la rivière Saji (Tottori), réputées comme les plus beaux rochers utilisés dans les jardins minéraux japonais pour leur forme et leur dureté. Elles ont été disposées parmi les mousses et les graviers de la cour, telle une allégorie de l'océan et des îles. I. M. Pei a demandé à Nakamura d'accroître la présence de la mousse pour éviter que l'espace ne paraisse trop « sec ». La pierre la plus volumineuse (7 t) a nécessité la pose d'un pilier de soutien supplémentaire dans le bâtiment souterrain sur lequel elle repose.

Above, an internal courtyard of the museum designed in the Japanese style in collaboration between the architect I. M. Pei and Yoshiaki Nakamura.

Ein Innenhof des Museums (oben) wurde gemeinsam von dem Architekten Ieoh Ming Pei und Yoshiaki Nakamura im japanischen Stil gestaltet.

Ci-dessus, une cour intérieure du musée aménagée en style japonais, fruit de la collaboration entre l'architecte I. M. Pei et Yoshiaki Nakamura.

Interior views of the museum, where some planters allow nature into the space, along with ample views of the lush vegetation around the structure seen through windows.

Innenraumansichten des Museums. Mit den Pflanzkübeln kommt Natur in den Innenraum. Durch die großen Fensterflächen hat man Ausblicke auf die üppige Vegetation, in die das Gebäude eingebettet ist.

Vues intérieures du musée : des jardinières permettent à la nature de pénétrer dans les volumes couverts. Les baies offrent d'amples vues sur la végétation luxuriante qui entoure les constructions.

NBGM LANDSCAPE ARCHITECTS

NBGM Landscape Architects
369 Government Road
Johannesburg North
Gauteng 2055
South Africa

Tel: +27 11 462 6967
Fax: +27 11 462 9284
E-mail: graham@newla.co.za / anton@greeninc.co.za
Web: www.newla.co.za / www.greeninc.co.za

Graham Young, born in 1953 in Johannesburg, South Africa, received a Bachelor of Landscape Architecture degree from the University of Toronto (1978) and is a Senior Lecturer in the Department of Architecture at the University of Pretoria. He was a founding member of Newtown Landscape Architects (NLA) in 1994 in Johannesburg. Johan Barnard was born in 1966 in Johannesburg. He holds a degree in Landscape Architecture from the University of Pretoria (1988) and a Master's in Urban Design from the University of the Witwatersrand (1993). He is a founding member and managing director of Newtown Landscape Architects and specializes in construction detailing, contract documentation, site supervision, and contract management. He is a past President of the Institute of Landscape Architects of South Africa and a past President and Treasurer of the South African Council for the Landscape Architectural Profession. **NBGM** is a group formed from Newtown Landscape Architects, GREENinc, and Gallery Momo. Founded in 1995 by Anton Comrie and Stuart Glen, GREENinc is a landscape architecture firm based in Johannesburg. NBGM realized the Freedom Park, published here (Salvokop, Tshwane, Pretoria, South Africa, 2002–10).

Graham Young wurde 1953 in Johannesburg, Südafrika, geboren. Er machte seinen Bachelor in Landschaftsarchitektur an der Universität Toronto, Kanada (1978), und ist Dozent an der Architekturfakultät der Universität Pretoria, Südafrika. 1994 gehörte er zu den Gründungsmitgliedern des Büros Newtown Landscape Architects (NLA) in Johannesburg. Johan Barnard wurde 1966 in Johannesburg geboren. Er hat einen Universitätsabschluss in Landschaftsarchitektur der Universität Pretoria (1988) und schloss seinen Master in Urban Design an der Universität Witwatersrand (1993) ab. Er ist ebenfalls Gründungsmitglied und Geschäftsführer des Büros Newtown Landscape Architects und auf Werkplanung, Vertragsunterlagen, Bauleitung und Vertragsdurchführung spezialisiert. Er ist ehemaliger Präsident des Institute of Landscape Architects of South Africa, eines Interessenverbandes südafrikanischer Landschaftsarchitekten, und ehemaliger Präsident und Schatzmeister des South African Council for the Landscape Architectural Profession, der südafrikanischen Landschaftsarchitektenkammer. **NBGM** ist eine Gruppe, die sich aus dem Büro Newtown Landscape Architects, GREENinc und der Galerie Momo konstituiert hat. Sie wurde 1995 von Anton Comrie und Stuart Glen gegründet, GREENinc ist ein Büro für Landschaftsarchitektur in Johannesburg. NBGM hat den hier vorgestellten Freedom Park realisiert (Salvokop, Tshwane, Pretoria, Südafrika, 2002–10).

Graham Young, né en 1953 à Johannesburg (Afrique du Sud), est titulaire d'un B.A. d'architecture du paysage de l'université de Toronto (1978) et est assistant senior au département d'architecture de l'université de Pretoria. Il a été l'un des cofondateurs de l'agence Newtown Landscape Architects (NLA) en 1994 à Johannesburg. Johan Barnard, né en 1966 à Johannesburg, est diplômé en architecture du paysage de l'université de Pretoria (1988) et titulaire d'un M.A. d'urbanisme de l'université de Witwatersrand (1993). Il est cofondateur et directeur gérant de Newtown Landscape Architects et s'est spécialisé dans le second œuvre, la documentation technique, la supervision des chantiers et la gestion des sous-traitants. Il a été président de l'Institut des architectes paysagistes d'Afrique du Sud ainsi que président et trésorier du Conseil d'Afrique du Sud pour la profession d'architecte paysagiste. **NBGM** est un groupe formé par Newtown Landscape Architects, GREENinc et Gallery Momo. Fondée en 1995 par Anton Comrie et Stuart Glen, GREENinc est une agence d'architecture paysagère basée à Johannesburg. NBGM a réalisé le Freedom Park publié ici (Salvokop, Tshwane, Pretoria, Afrique du Sud, 2002–10).

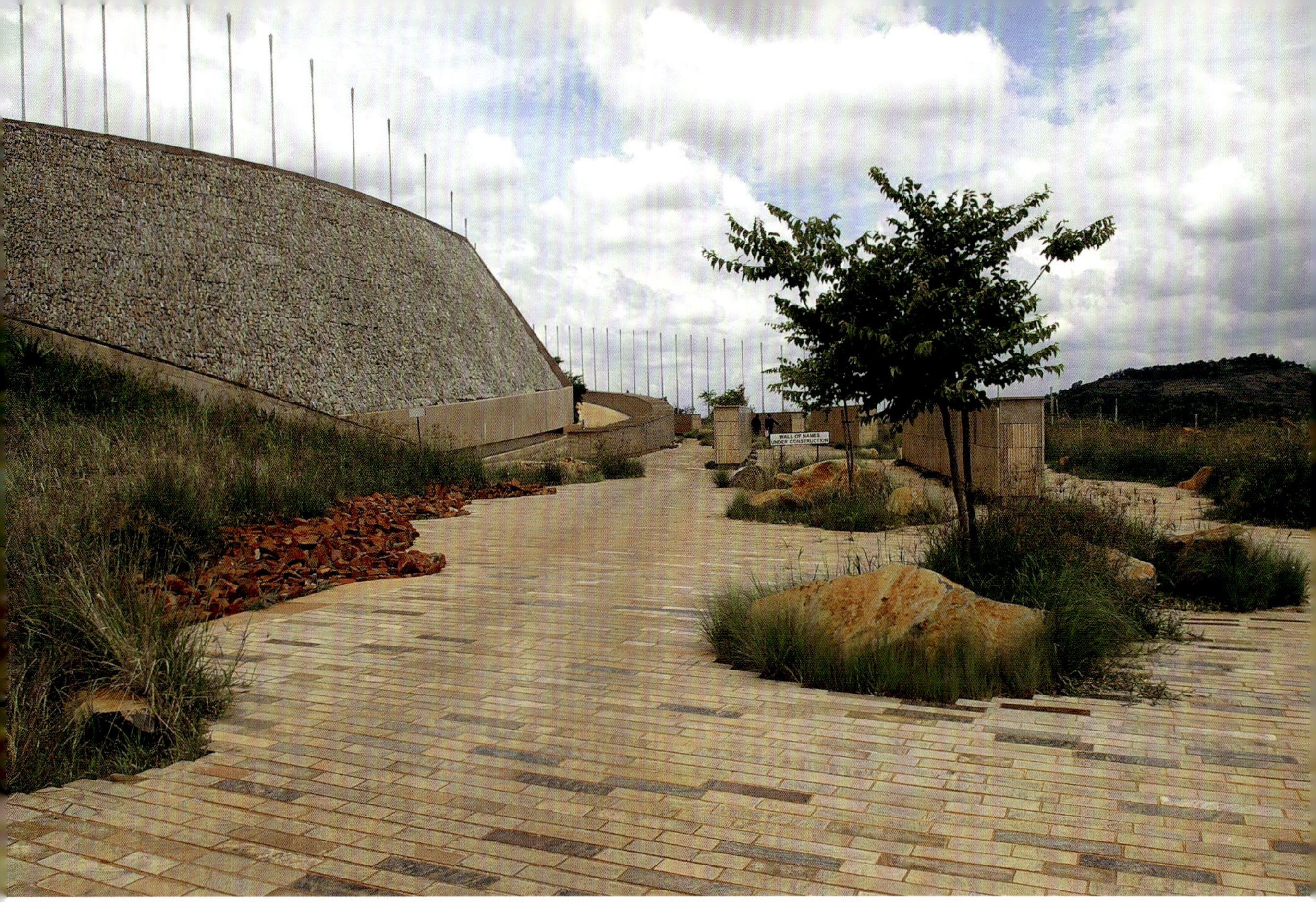
WALL OF NAMES
UNDER CONSTRUCTION

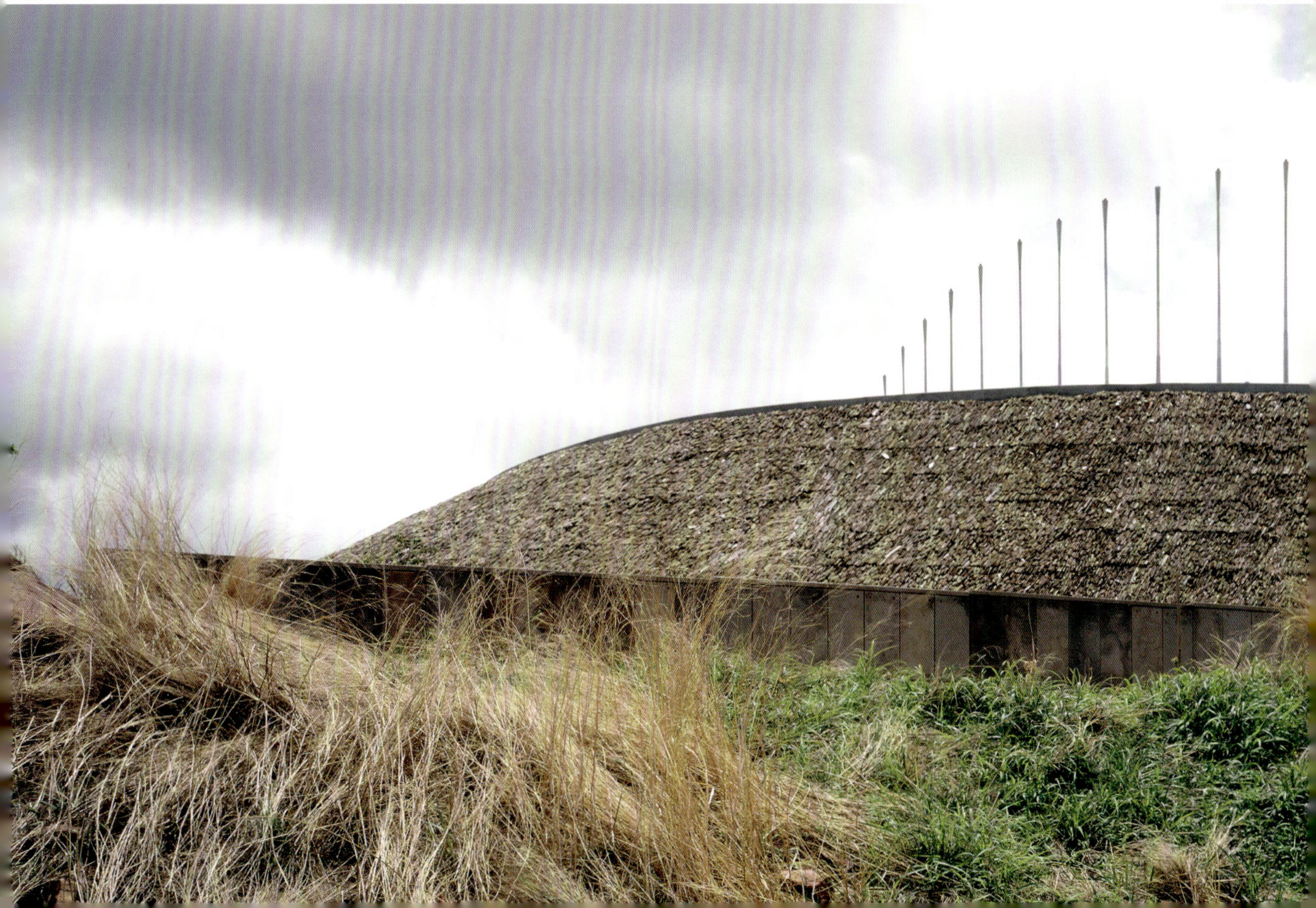

THE FREEDOM PARK
Salvokop, Tshwane, Pretoria, South Africa, 2002–10

Area: 30 hectares. Client: The Freedom Park Trust. Cost: not disclosed
Architects: OCA Architects

Mveledzo, seen above, is a stone wall–lined spiral path, which links all of the elements of Freedom Park together.

Der Mveledzo, oben, ist ein spiralförmiger Weg, der von Natursteinmauern begrenzt wird und der alle Elemente des Freedom Park verbindet.

Le Mveledzo (ci-dessus) est une allée en spirale bordée de murs, qui relie entre eux les divers éléments du Freedom Park.

Each of the nine provinces of South Africa contributed a boulder to Lesaka, a symbolic, mist-shrouded burial ground for those who died in the struggle for freedom—seen in the image below.

Jede der neun Provinzen von Südafrika hat einen Felsen für den Lesaka gestiftet, eine symbolische, hinter einem Nebelschleier verborgene Grabstätte für diejenigen, die im Kampf um die Freiheit gestorben sind (Bild unten).

Chacune des neuf provinces de l'Afrique du Sud a envoyé un rocher pour Lesaka, tombeau symbolique noyé de brume, dédié à ceux qui sont morts dans les combats pour la liberté (ci-dessous).

The architects explain that "the **FREEDOM PARK** is a project mandated by President Nelson Mandela as the natural outcome of the Truth and Reconciliation Commission process that occurred after the fall of apartheid. Its vision is structured around four key ideas: reconciliation, nation building, freedom of people, and humanity. The making of the landscape seeks to recognize the spiritual origins of these ideas, and manifest them symbolically in physical form." They refer to this project as a "garden of remembrance, a natural indigenous garden telling the story of South Africa's progression to freedom…" Landscaping and architectural elements were integrated into the site to create a space that "is unapologetically founded in African cultural expression." The park contains areas called Tiva (a large body of water); Moshate (a hospitality suite); //hapo (an interactive exhibition space); S'khumbuto (a memorial commemorating the conflicts that shaped South Africa); and Isivivane (a symbolic resting place for those who died in the struggle for freedom). These different elements are joined together by a spiral stone-walled path called Mveledzo.

Die Architekten erklären, dass „der **FREEDOM PARK** von Präsident Nelson Mandela in Auftrag gegeben wurde als natürliches Ergebnis der Arbeit der Wahrheits- und Versöhnungskommission, die nach dem Fall der Apartheid eingesetzt worden war. Dem Park liegen vier Schlüsselgedanken zugrunde: Versöhnung, Aufbau einer Nation, Freiheit des Volkes und Humanität. Die Landschaftsgestaltung versucht, die geistigen Ursprünge dieser Gedanken zu erkennen und sie symbolisch umzusetzen." Die Planer bezeichnen dieses Projekt als „Garten der Erinnerung, als natürlichen, heimischen Garten, der die Geschichte Südafrikas auf dem Weg zur Freiheit erzählt…" Landschaftsgestalterische und architektonische Elemente wurden in das Gelände eingegliedert, um einen Raum zu schaffen, „der sich auf den kulturellen Ausdruck Afrikas beruft und darauf gründet." In diesem Park gibt es Areale wie Tiva (eine große Wasserfläche), Moshate (ein Gästehaus), //hapo (ein interaktiver Ausstellungsbereich), S'khumbuto (eine Gedenkstätte, die an die Konflikte erinnert, die Südafrika seine Gestalt gegeben haben) und Isivivane (eine symbolische Ruhestätte für alle, die im Kampf um die Freiheit gestorben sind). Diese unterschiedlichen Elemente werden durch einen in Spiralform verlaufenden Weg, den Mveledzo, der von Natursteinmauern begrenzt wird, miteinander verbunden.

Les architectes expliquent que « le **FREEDOM PARK** est un projet initié par le président Nelson Mandela pour saluer le travail de la Commission Vérité et Réconciliation mise en place après la fin de l'apartheid. Sa vision s'est structurée autour de quatre idées principales : réconciliation, construction d'une nation, liberté du peuple et humanité. La construction de ce paysage prend en compte les origines spirituelles de ces idées et les exprime symboliquement sous forme physique ». Ce projet constitue ainsi « un jardin du souvenir, un jardin naturel indigène qui raconte l'histoire de la progression de l'Afrique du Sud vers la liberté… ». Le travail du paysage et les éléments architecturaux s'intègrent dans le site pour créer un espace « résolument fondé sur une expression culturelle africaine ». Le parc contient diverses sections : Tiva (un grand bassin), Moshate (une maison d'hôtes), //hapo (un lieu d'expositions interactives), S'khumbuto (un mémorial des conflits ayant marqué l'histoire du pays) et Isivivane (un lieu de repos symbolique des morts pour la liberté). Ces éléments sont réunis par une allée en spirale bordée de murs de pierres, appelée Mveledzo.

The Waterfall in the Freedom Park is part of Isivivane, near the resting place of the spirits of those who died in the struggle for freedom.

Der Wasserfall im Freedom Park ist Teil von Isivivane, nahe der Ruhestätte für die Geister derer, die im Kampf um die Freiheit gestorben sind.

La cascade du Freedom Park fait partie d'Isivivane, près du lieu de repos des esprits des combattants morts pour la liberté.

The Wall of Names, seen below, is a 697-meter-long structure inscribed with the names of those who died during the eight major conflicts that have marked the history of South Africa.

Die Mauer der Namen auf dem Bild unten ist 697 Meter lang. Auf ihr sind die Namen aller verzeichnet, die während der acht großen Aufstände gestorben sind, die die Geschichte Südafrikas kennzeichnen.

Le Wall of Names (ci-dessous) est une muraille de 697 mètres de long sur laquelle sont inscrits les noms des victimes des huit conflits majeurs qui ont marqué l'histoire de l'Afrique du Sud.

VICTOR NEVES

Victor Neves – Arquitectura e Urbanismo, Lda
Rua Das Trinas, Nº 48–RC/ e 2º
1200–859 Lisbon
Portugal

Tel: +351 21 395 16 97
Fax: +351 21 395 59 61
E-mail: victneves@sapo.pt
Web: www.victorneves.com

VICTOR NEVES was born in 1956 in Lisbon, Portugal. He obtained his doctorate at the ETSAB-UPC (Escuela Técnica Superior de Arquitectura de Barcelona Univ. – Politécnica de Cataluña, Spain). He is a Professor at the Faculty of Architecture of the Lusíada University (Lisbon). He created his own office Victor Neves – Arquitectura e Urbanismo, Lda in 1985. His work includes the Vale Flores Primary School (Feijó, Almada, 2000); a house in Cacela-a-Velha (2003); and the reorganization of the Esposende riverside (Esposende, 2006–07, published here). More recently, he has participated in competitions for a new Psychiatric Hospital (with Proconsultores and Cândido Gomes; Bejaia, Algeria, 2008); the redesign of the sea front of Pedra Alta (in collaboration with the architect Victor Mogadouro and landscape designer Francisca Pinto da Costa; northern Portugal, 2009); and a river front design for Caminha (also in collaboration with Francisca Pinto da Costa, Portugal, 2010).

VICTOR NEVES wurde 1956 in Lissabon, Portugal, geboren. Er promovierte an der ETSAB-UPC (Escuela Técnica Superior de Arquitectura de Barcelona/Universidad Politécnica de Cataluña, Spanien). Er ist Professor an der Architekturfakultät der Universität Lusíada in Lissabon. 1985 gründete er sein eigenes Büro Victor Neves – Arquitectura e Urbanismo, Lda. Zu seinen Arbeiten zählen die Grundschule Vale Flores (Feijó, Almada, 2000), ein privates Wohnhaus in Cacela-a-Velha (2003) und die Neugestaltung des Flussufers bei Esposende (Esposende, 2006–07, hier vorgestellt). Außerdem hat er an Wettbewerben teilgenommen, wie zum Beispiel für den Neubau eines Psychiatrischen Krankenhauses (mit dem Büro Proconsultores und Cândido Gomes, in Bejaia, Algerien, 2008), die Neugestaltung der Meeresküste von Pedra Alta (mit dem Architekten Victor Mogadouro und der Landschaftsplanerin Francisca Pinto da Costa, Nordportugal, 2009) sowie den Entwurf für das Flussufer in Caminha (ebenfalls mit Francisca Pinto da Costa, 2010). Alle Projekte befinden sich in Portugal, sofern nicht anders vermerkt.

VICTOR NEVES, né en 1956 à Lisbonne, est titulaire d'un doctorat en architecture de l'ETSAB-UPC (École supérieure technique d'architecture de l'université de Barcelone – Politécnica de Cataluña, Espagne). Il est professeur à la faculté d'architecture de l'université Lusíada (Lisbonne) et a créé l'agence Victor Neves, Arquitectura e Urbanismo, Lda, en 1985. Parmi ses réalisations : l'école primaire de Vale Flores (Feijó, Almada, 2000) ; une maison à Cacela-a-Velha (2003) ; la restructuration des berges à Esposende (2006–07, publié ici). Plus récemment, il a participé à des concours pour un nouvel hôpital psychiatrique (avec Proconsultores et Cândido Gomes, Bejaia, Algérie, 2008) ; la rénovation du front de mer de Pedra Alta (en collaboration avec l'architecte Victor Mogadouro et l'architecte paysagiste Francisca Pinto da Costa (nord du Portugal, 2009) et un projet de front de mer pour Caminha (également en collaboration avec Francisca Pinto da Costa, 2010), tous ces projets étant situés au Portugal, sauf mention contraire.

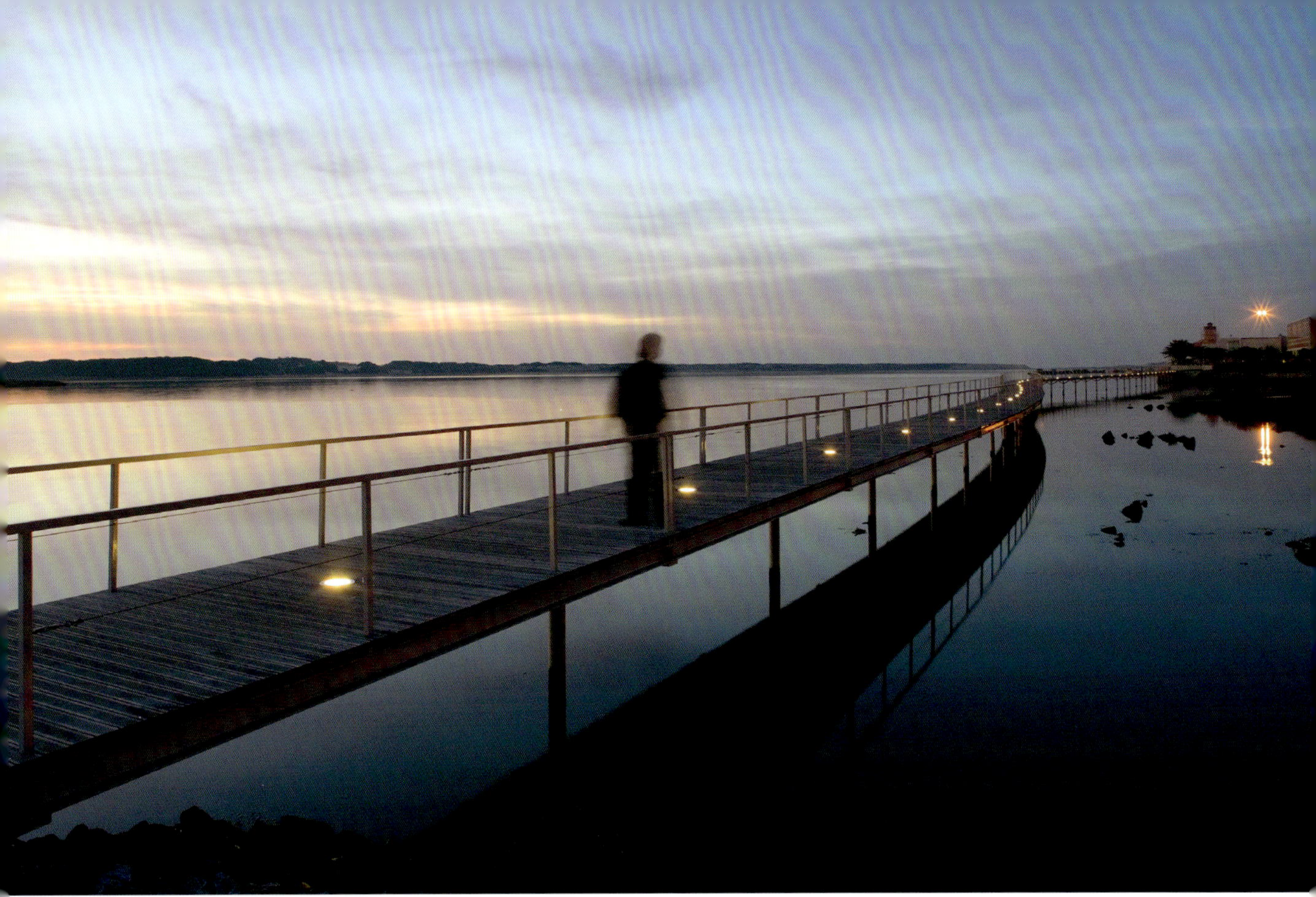

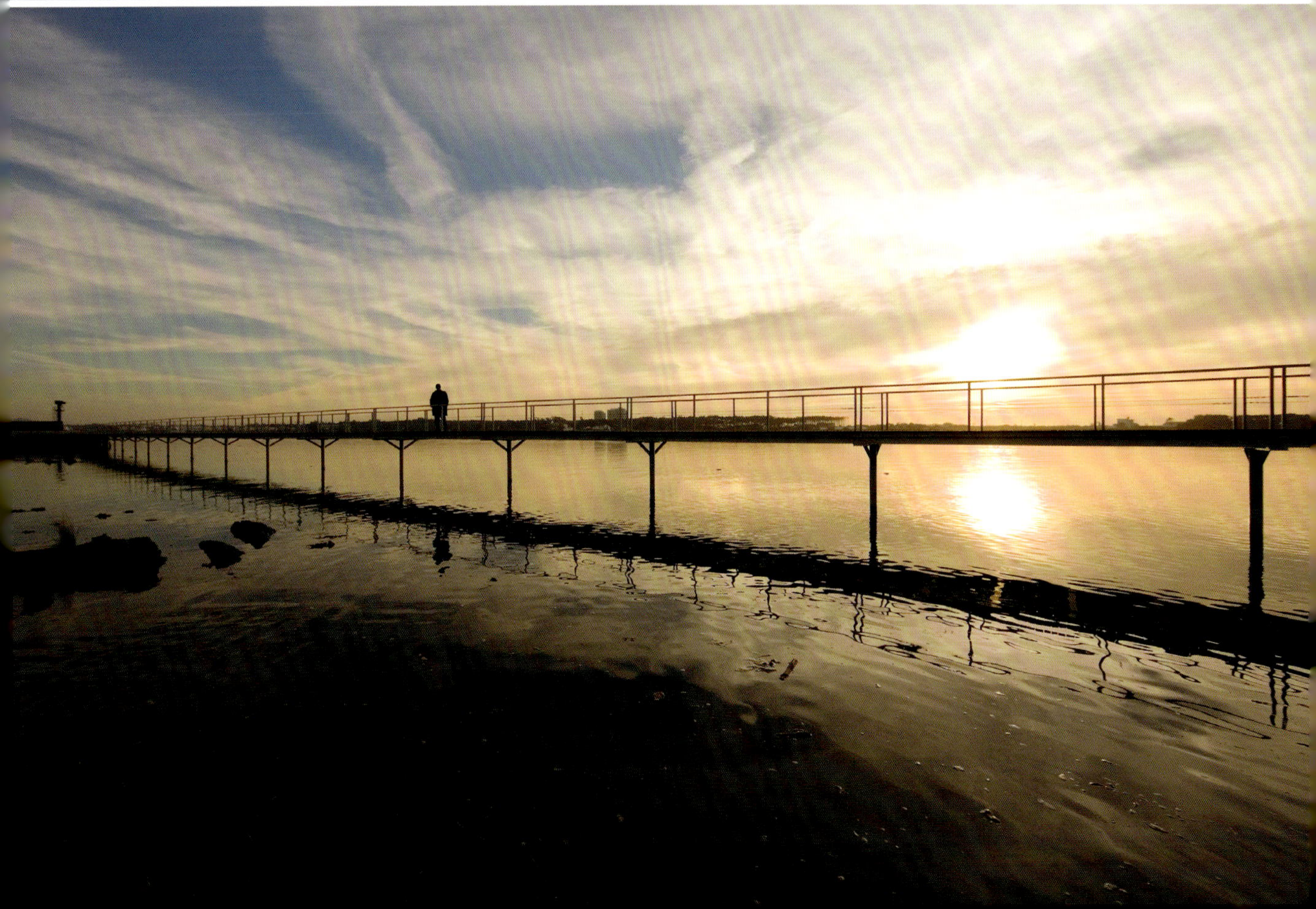

REORGANIZATION OF THE RIVERSIDE OF ESPOSENDE

Esposende, Portugal, 2006–07

Address: Av. Eduardo Arantes, Esposende, Portugal. Area: 25 500 m²
Client: Câmara Municipal de Esposende and Instituto Marítimo-Portuário
Cost: €1.7 million. Collaboration: João Nunes (PROAP), Nuno Mota,
Joana Barreto, and Mafalda Meirinho (Landscape Architecture)

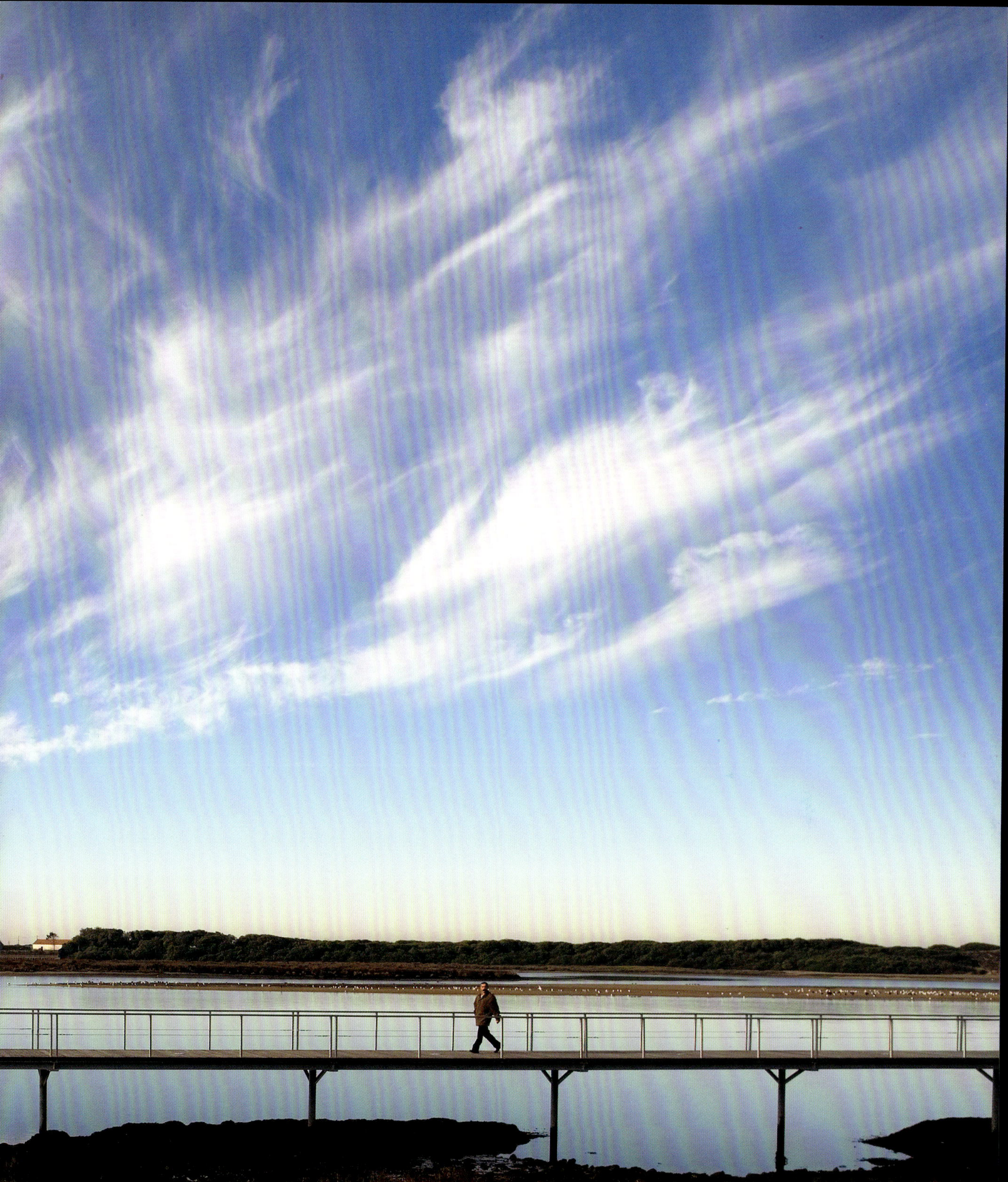

Esposende is a small village located in the north of Portugal. The project consisted essentially in the reorganization of the urban area, bordered by a riverside, in order to create a public leisure area, surrounded by several buildings related to nautical activities. The project originally included a number of five-meter glass cubes housing visitor information facilities, not erected for financial and regulatory reasons. The architect states: "The concept behind the whole proposal is to give priority to the natural landscape, denying the artificiality of formal excesses of the 'design' and also denying the profusion of materials often exhibited in recent urban interventions;" and to increase awareness and knowledge of local flora and fauna.

Esposende ist ein kleines Dorf im Norden Portugals. Bei dem Projekt ging es vor allem um die Neugestaltung des Siedlungsbereiches, der an den Fluss grenzt. Hier sollte ein öffentliches Erholungsgebiet entstehen, das von Gebäuden gerahmt wird, die mit nautischen Aktivitäten verbunden sind. Ursprünglich war vorgesehen, hier auch mehrere 5 x 5 Meter große Glaskuben aufzustellen, in denen unter anderem eine Touristeninformation untergebracht werden sollte. Dies wurde aus Kostengründen und aufgrund der geltenden Bauvorschriften nicht realisiert. Der Architekt kommentiert sein Projekt wie folgt: „Das Konzept des Gesamtentwurfs gründet sich auf den Anspruch, der Naturlandschaft Priorität einzuräumen und sich jeder Form der Künstlichkeit zu verweigern, die die formalen Exzesse des ‚Design' mit sich bringen, ebenso die Materialschlacht zu vermeiden, wie sie leider so häufig bei neueren städtebaulichen Maßnahmen zu beobachten ist." Ziel ist es, die lokale Flora und Fauna ins Bewusstsein der Menschen zu rücken und Wissen darüber zu vermitteln.

Esposende est un petit village du nord du Portugal. Le projet consistait essentiellement à restructurer la zone urbaine en bordure de rivière pour créer une aire publique de loisirs entourée de quelques bâtiments affectés aux activités nautiques. Il comportait à l'origine un certain nombre de constructions en forme de cubes de verre de 5 m, abritant divers services d'information pour les visiteurs, qui n'ont pas été réalisées pour des raisons financières et réglementaires. «Le concept de l'ensemble de cette proposition est de donner la priorité au paysage naturel, en rejetant l'aspect artificiel des excès de "design" et la profusion de matériaux que l'on trouve souvent dans les interventions urbaines récentes», explique l'architecte. Le projet encourage les passants à prendre conscience de la flore et de la faune locales, et à mieux les connaître.

The architect has intentionally made his interventions minimal, not only for cost reasons, but also in a spirit of rejection of what he sees as the excessive nature of other similar projects in other locations.

Der Architekt hat seine Eingriffe mit Absicht auf ein Minimum beschränkt. Dies geschah nicht nur aus Kostengründen, sondern auch aus einer gewissen Ablehnung von übertriebenem Design, wie man es oft bei ähnlichen Projekten findet.

L'intervention de l'architecte est restée volontairement minimale, non seulement pour des raisons de coût, mais aussi pour marquer sa désapprobation des excès d'aménagements que provoque parfois ce type d'intervention.

With its light walkway set just off the river banks, the design makes the economy of means into a visible virtue, allowing walkers to discover their own town from an entirely different angle.

Mit diesem leichten Steg, der nah am Flussufer entlang führt, wird der sparsame Einsatz der Mittel zu einer sichtbaren Tugend, können doch die Fußgänger ihre Stadt von einem völlig neuen Blickwinkel aus betrachten.

Grâce à cette allée suspendue à quelques mètres de la rive, ce projet qui fait de l'économie de moyens une vertu permet aux promeneurs de découvrir leur ville sous un angle entièrement différent.

Plaza de Santa Bárbar

NIETO SOBEJANO

Nieto Sobejano Arquitectos S. L. P.
Talavera, 4 L–5
28016 Madrid
Spain

Tel: +34 91 564 38 30 / Fax: +34 91 564 38 36
E-mail: nietosobejano@nietosobejano.com
Web: www.nietosobejano.com

FUENSANTA NIETO and **ENRIQUE SOBEJANO** graduated as architects from the ETSA Madrid (ETSAM) and the Graduate School of Architecture at Columbia University in New York. They are currently teaching at the Universidad Europea de Madrid (UEM) and at the University of the Arts (UdK) in Berlin, and are the managing partners of Nieto Sobejano Arquitectos S. L. Both have been visiting critics and/or teachers at various Spanish and international universities and institutions, such as the GSD at Harvard University, University of Arizona, Technical University (Munich), ETSA Barcelona, University of Torino, University of Stuttgart, University of Cottbus (Germany), Columbia University, and the University of Texas (Austin). From 1986 to 1991, they were the editors of the architectural journal *Arquitectura* edited by the Architectural Association of Madrid (Colegio Oficial de Arquitectos de Madrid). Their work has been exhibited, amongst other locations, at the Venice Biennale (Italy, 2000 and 2002); Bienal Española de Arquitectura (2003); Extreme Eurasia (Tokyo, 2005); and "On Site: New Architecture in Spain" (MoMA, New York, 2006). They won the National Prize for Conservation and Restoration of Cultural Patrimony for their extension of the National Sculpture Museum (Valladolid, Spain, 2007); and have recently completed the Madinat al-Zahra Museum and Research Center (Córdoba, Spain, 2005–08); the Moritzburg Museum Extension (Halle, Saale, Germany, 2006–08); and Plaza de Santa Bárbara (Madrid, Spain, 2009, published here).

FUENSANTA NIETO und **ENRIQUE SOBEJANO** diplomierten als Architekten an der ETSA Madrid und an der Graduate School of Architecture an der Columbia University in New York. Beide arbeiten sowohl als Partner in ihrem Büro Nieto Sobejano Arquitectos S. L. als auch als Dozenten an der Universidad Europea de Madrid (UEM) und der Universität der Künste (UdK) in Berlin. Beide sind als Referenten und/oder Dozenten an mehreren spanischen und ausländischen Universitäten und Institutionen tätig gewesen, wie zum Beispiel an der Graduate School of Design (GSD) der Harvard University, an der University of Arizona, an der Technischen Universität München, an der ETSA in Barcelona, der Universität Turin, der Universität Stuttgart, der Universität Cottbus, der Columbia University und der University of Texas in Austin. Von 1986 bis 1991 waren sie Herausgeber der spanischen Architekturzeitschrift „Arquitectura", die von der Architektenkammer Madrid, dem Colegio Oficial de Arquitectos de Madrid, veröffentlicht wird. Ihre Arbeiten wurden unter anderem auf der Biennale von Venedig ausgestellt (2000 und 2002), auf der Bienal Española de Arquitectura (2003), auf der Extreme Eurasia (Tokio, 2005) und in der Ausstellung „On Site: New Architecture in Spain" (MoMA, New York, 2006). Für den Erweiterungsbau des Nationalmuseums für Skulptur in Valladolid (Spanien, 2007) wurden sie mit dem Nationalpreis für Restaurierung und Erhalt von Kulturgütern ausgezeichnet. Sie haben das Museum und Forschungszentrum Madinat al-Zahra bei Córdoba (Spanien, 2005–08) gebaut, den Anbau an das Museum Moritzburg (Halle an der Saale, 2006–08) und die Neugestaltung der Plaza de Santa Bárbara in Madrid (2009, hier vorgestellt) ausgeführt.

FUENSANTA NIETO et **ENRIQUE SOBEJANO** sont architectes diplômés de l'ETSA Madrid (ETSAM) et de la Graduate School of Architecture de l'université Columbia (New York). Ils enseignent actuellement à l'Université européenne de Madrid (UEM) et à l'Université des arts (UdK) de Berlin, tout en dirigeant Nieto Sobejano Arquitectos S.L. Tous deux ont été critiques et/ou professeurs invités dans diverses universités et institutions espagnoles et internationales, dont la GSD de l'université Harvard, l'université d'Arizona, l'Université technique de Munich, l'ETSA de Barcelone, l'université de Turin, l'université de Stuttgart, l'université de Cottbus (Allemagne), l'université Columbia et l'université du Texas, à Austin. De 1986 à 1991, ils ont été rédacteurs en chef du magazine spécialisé *Arquitectura*, publié par l'Association des architectes de Madrid (Colegio Oficial de Arquitectos de Madrid). Leurs travaux ont été exposés, entre autres, à la Biennale de Venise (2000 et 2002); la Biennale espagnole d'architecture (2003); l'exposition « Extreme Eurasia » (Tokyo, 2005) et l'exposition « On Site: New Architecture in Spain » (MoMA, New York, 2006). Ils ont remporté le Prix national pour la conservation et la restauration du patrimoine culturel pour leur extension du Musée national de sculpture (Valladolid, Espagne, 2007) et ont récemment achevé le musée et centre de recherches Madinat al-Zahra (Cordoue, Espagne, 2005–08); l'extension du musée de Moritzburg (Halle, Saale, Allemagne, 2006–08) et la Plaza de Santa Bárbara (Madrid, 2009, publiée ici).

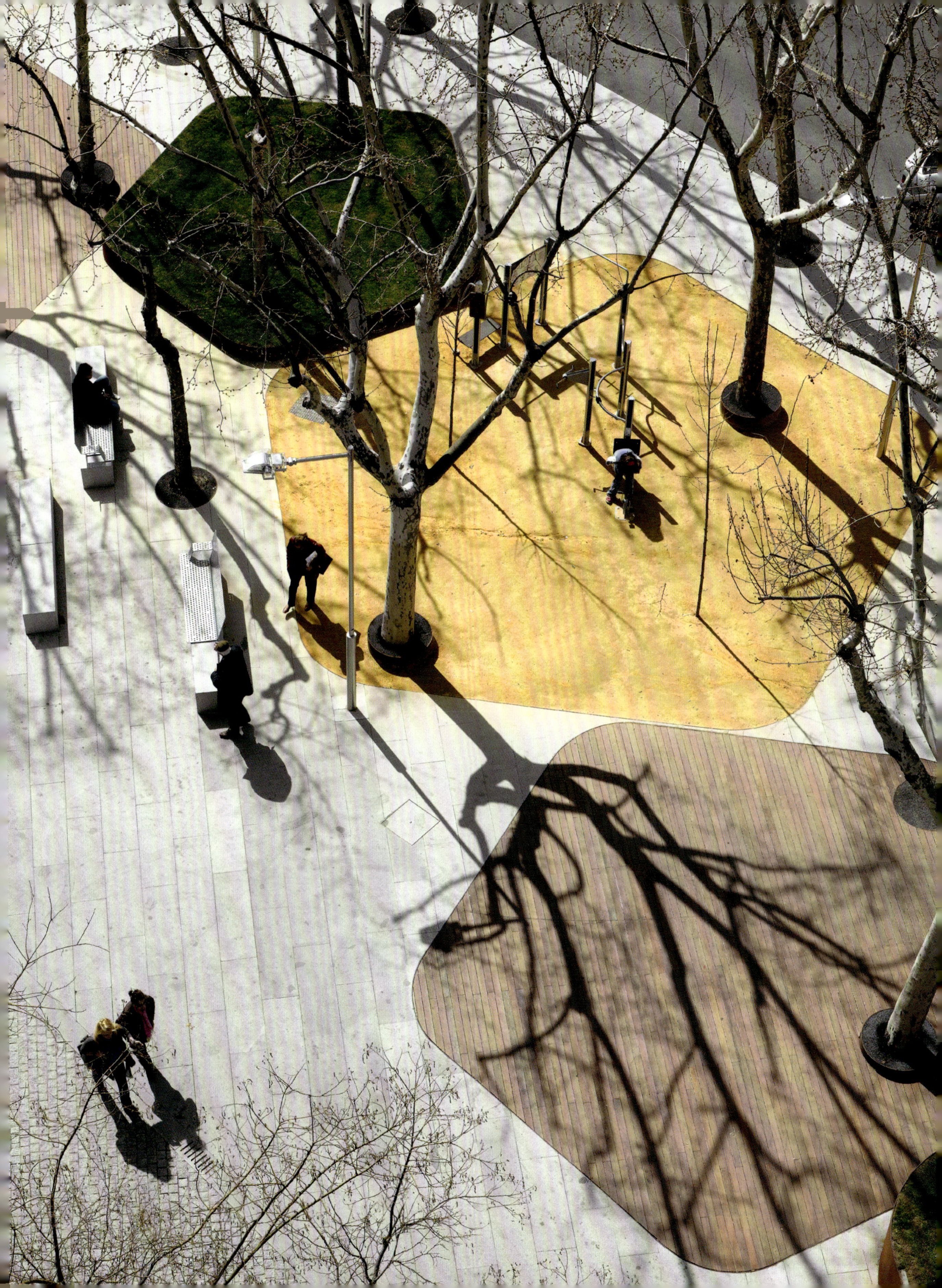

PLAZA DE SANTA BÁRBARA

Madrid, Spain, 2009

Address: Plaza de Santa Bárbara, 28004 Madrid, Spain
Area: 10 800 m². Client: Madrid City Council
Cost: not disclosed

The architects won the competition to renovate the **PLAZA DE SANTA BÁRBARA** as part of a larger project that will include a new 38 000-square-meter mixed-use building integrating the Barceló market (2009–11). In the square, the architects managed to eliminate one street, reducing vehicular traffic. A large green pedestrian area with recreational zones was defined. The architects explain that "large granite paving stones alternate with other surfaces in a pentagonal geometric pattern, similar to the forms of the newly built Barceló Temporary Market, which incorporate lawn-covered areas, games and leisure precincts, boardwalks, and a small glass pavilion serving as a book and flower shop. The new street furniture forms part of an integrated urban design and landscape concept that will link Plaza de Santa Bárbara with the neighboring streets around the future mixed-use building in construction."

Das Büro Nieto Sobejano gewann den Wettbewerb zur Neugestaltung der **PLAZA DE SANTA BÁRBARA,** die Teil eines größeren Projekts ist, zu dem auch eine 38 000 Quadratmeter umfassende Bebauung mit gemischter Nutzung unter Einbeziehung der Markthalle Barceló gehört (2009–11). Es gelang den Architekten, eine Straße, die den Platz querte, aufzuheben und so den Autoverkehr zu reduzieren. Es entstand eine große, grüne Fußgängerzone mit Ruhebereichen. Die Architekten erklären, dass „große Granitbodenplatten sich mit anderen Oberflächenmaterialien abwechseln und ein geometrisches fünfeckiges Bodenmuster bilden, das den temporären Konstruktionen auf dem Gelände der Barceló-Markthalle ähnlich ist. Es gibt auch Rasenflächen, Spiel- und Erholungsbereiche, Fußwege und einen kleinen gläsernen Blumen- und Bücherpavillon. Die neue Straßenmöblierung ist Teil des städtebaulichen Gesamtkonzepts und der Freiraumgestaltung, die die Plaza de Santa Bárbara mit den benachbarten Straßen verbindet, wo gerade Neubauten für eine gemischte Nutzung entstehen."

L'agence a remporté le concours pour la rénovation de la **PLAZA DE SANTA BÁRBARA** dans le cadre d'un projet plus important qui prévoit la construction d'un ensemble immobilier de 38 000 m² comprenant le marché Barceló (2009–11). Pour la place, les architectes ont supprimé une artère afin de réduire l'impact de la circulation automobile. Une vaste zone piétonnière plantée parsemée de zones récréatives a été créée. « D'importantes parties dallées de granit alternent avec d'autres types de surfaces de forme pentagonale, similaires aux formes du nouveau marché temporaire de Barceló, comprenant des jardinières de gazon, des aires de jeux, des allées de planches et un petit pavillon de verre faisant office de boutique de fleurs et de livres. Le nouveau mobilier urbain mis en place fait partie du concept d'urbanisme et de travail du paysage qui organise le lien entre la Plaza de Santa Bárbara et les rues avoisinantes, autour du bâtiment mixte en construction », précisent les architectes.

Islands of greenery delineated with low Corten-steel walls enliven and brighten the space of this essentially mineral square.

Grüne Inseln, die von niedrigen Cor-Ten-Stahlwänden umschlossen werden, beleben diesen Platz, auf dem das Element Stein dominiert.

Des îlots de verdure délimités par des murets en acier Corten animent et égayent ce square essentiellement minéral.

As seen in this aerial view, a number of trees planted in a slightly irregular way dot the square, whose main features are brightly colored pentagonal areas.

Wie man auf dieser Luftaufnahme erkennt, sind die Bäume unregelmäßig über den Platz verteilt. Das auffallendste Gestaltungselement sind die stark farbigen Fünfecke.

Comme le montre cette vue aérienne, un certain nombre d'arbres plantés de façon irrégulière ponctuent la place dont les éléments les plus forts restent néanmoins les plans colorés pentagonaux.

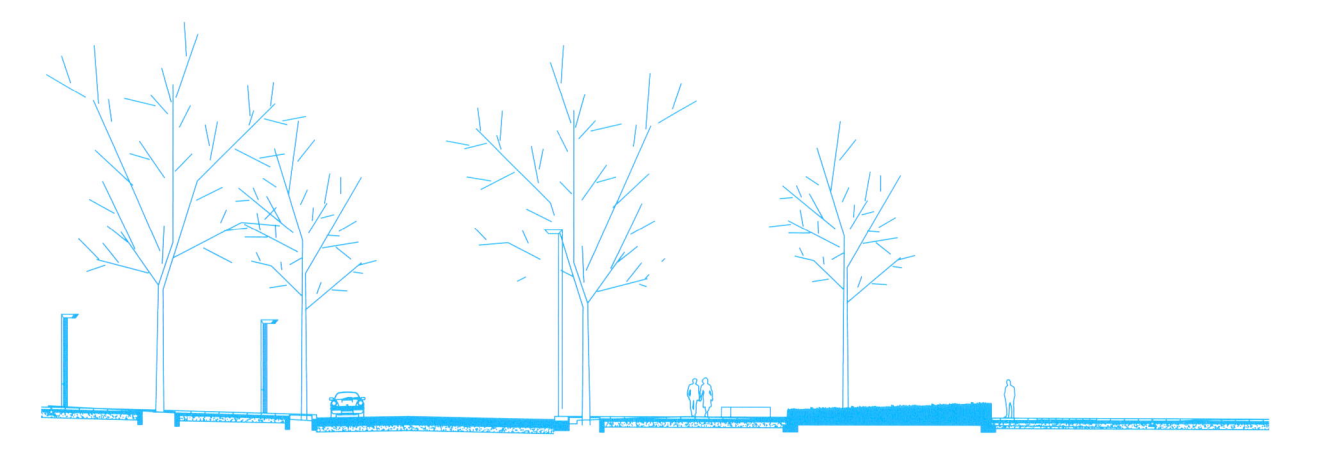

An overall plan (above) shows not only the shape of the square but also the location of the trees and the pentagonal colored zones.

Der Gesamtplan (oben) zeigt nicht nur die Umrisse des Platzes, sondern auch die Anordnung der Bäume und der farbigen Fünfecke.

Le plan général (ci-dessus) montre la forme du square mais aussi l'implantation des arbres et des plans colorés pentagonaux.

On this page and opposite (bottom), the sophisticated alternation of elevated grass-covered volumes and granite blocks gives an impression that is modern and yet unobtrusive.

Auf dieser und der gegenüberliegenden Seite unten: Das raffinierte Wechselspiel von angehobenen Rasenflächen und Granitblöcken wirkt modern und gleichzeitig dezent.

Sur cette page et page ci-contre (en bas), l'alternance sophistiquée de banquettes de verdure et de blocs de granit crée une discrète impression de modernité.

OFFICINA DEL PAESAGGIO

Sophie Agata Ambroise
Officina del Paesaggio
Post Office Box 6192
6901 Lugano
Switzerland

Tel: +41 91 922 84 24
Fax: +41 91 922 84 25
E-mail: rdv@officinadelpaesaggio.com
Web: www.officinadelpaesaggio.com

SOPHIE AGATA AMBROISE was born in Lugano, Switzerland, in 1969. She received a degree in architecture from the Politecnico of Milan, Italy. She then studied at the EHESS (École des hautes études en sciences sociales) with Augustin Berque in Paris and then at the ENSP (École nationale supérieure du paysage, Versailles), where she was influenced by Gilles Clément. In 2000, she returned to Ticino and worked as an assistant to the landscape architect Michel Desvigne at the Architecture Academy in Mendrisio. The same year she created her own office in Lugano, the Officina del Paesaggio. Her work includes the gardens of the Bulgari Hotel (Milan, Italy, 2004); Monterotondo Resort (Gavi, Italy, 2005); landscaping and terracing (Sondrio, Italy, 2007); the gardens of the YTL Residence (with Jouin Manku architects; Kuala Lumpur, Malaysia, 2008, published here); Community Gardens (Chiasso, Switzerland, 2010, also published here); and the Muse resort (Saint Tropez-Ramatuelle, France, 2010).

SOPHIE AGATA AMBROISE wurde 1969 in Lugano, Schweiz, geboren. Sie machte einen Universitätsabschluss in Architektur am Polytechnikum in Mailand, Italien. Danach studierte sie an der EHESS (École des hautes études en sciences sociales) bei Augustin Berque in Paris und später an der ENSP (École nationale supérieure du paysage) in Versailles. Hier wurde sie besonders von Gilles Clément beeinflusst. Im Jahr 2000 kehrte sie ins Tessin zurück und arbeitete als Assistentin des Landschaftsarchitekten Michel Desvigne an der Architekturakademie in Mendrisio. Im gleichen Jahr gründete sie ihr eigenes Büro in Lugano, die Officina del Paesaggio. Zu ihren Arbeiten gehören die Gartenanlagen des Hotels Bulgari in Mailand (2004) und des Monterotondo Resort (Gavi, Italien, 2005), Landschaftsgestaltung und Terrassierungsarbeiten in Sondrio, Italien (2007), die Gartenanlagen des Wohnhauses YTL (mit Patrick Jouin und Sanjit Manku Architekten, Kuala Lumpur, Malaysia, 2008, hier vorgestellt), die Kleingartenanlage in Chiasso, Schweiz (2010, ebenfalls hier vorgestellt), und das Luxushotel Muse in Saint Tropez-Ramatuelle, Frankreich (2010).

SOPHIE AGATA AMBROISE, née à Lugano (Suisse) en 1969, est diplômée en architecture du Politecnico de Milan. Elle a ensuite étudié à l'EHESS (École des hautes études en sciences sociales) auprès d'Augustin Berque à Paris, puis à l'ENSP (École nationale supérieure du paysage, Versailles), où elle a été marquée par l'influence de Gilles Clément. En 2000, elle est revenue dans le Tessin et a travaillé comme assistante de l'architecte paysagiste Michel Desvigne à l'Académie d'architecture de Mendrisio. La même année, elle a créé son agence à Lugano, l'Officina del Paesaggio. Parmi ses réalisations : les jardins de l'hôtel Bulgari (Milan, Italie, 2004) ; Monterotondo Resort (Gavi, Italie, 2005) ; des travaux de paysagisme et d'aménagement de terrasses (Sondrio, Italie, 2007) ; les jardins de la YTL Residence (avec Jouin Manku architects ; Kuala Lumpur, Malaisie, 2008, publiés ici) ; les jardins communautaires de Chiasso (Suisse, 2010, publiés ici) et ceux de l'hôtel Muse (Saint Tropez-Ramatuelle, France, 2010).

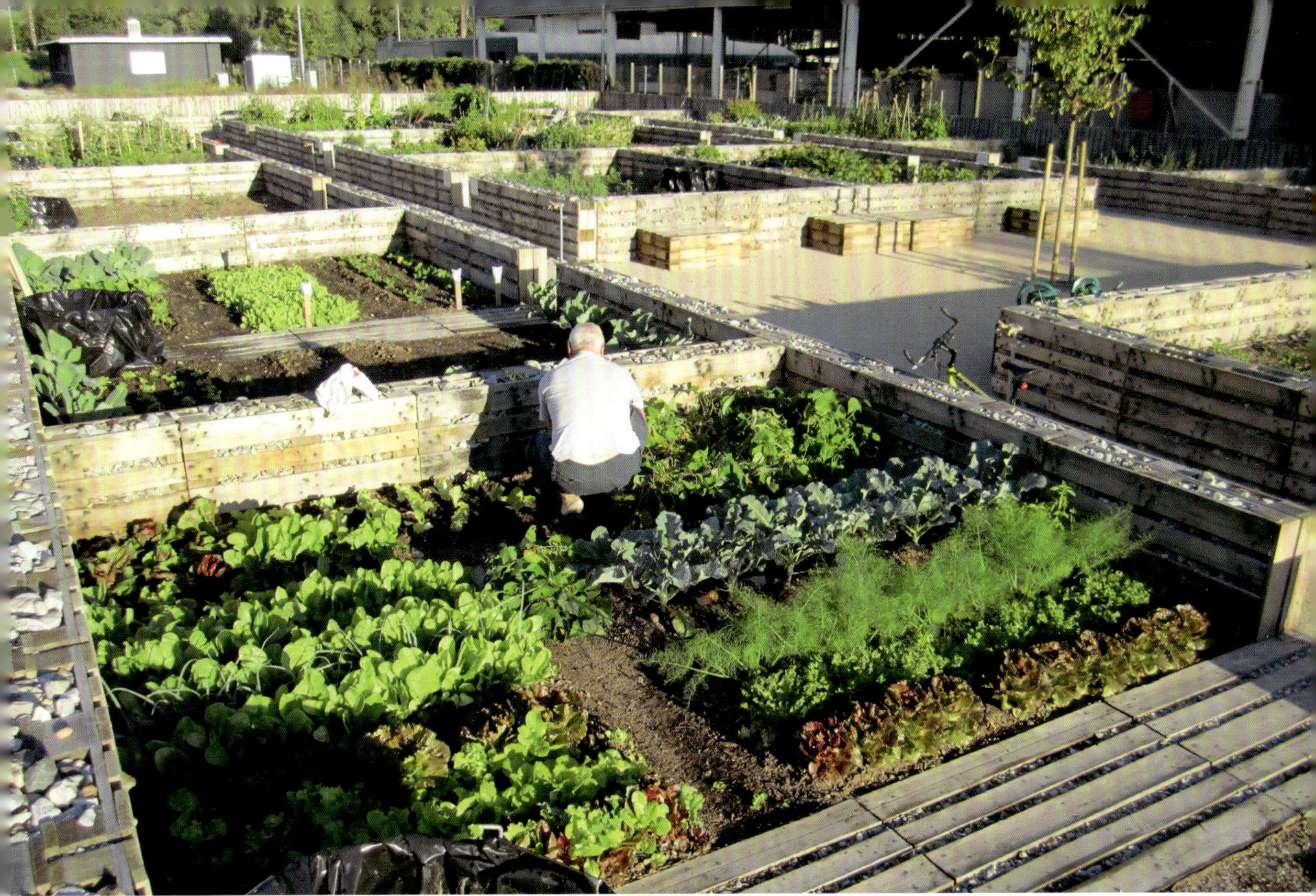

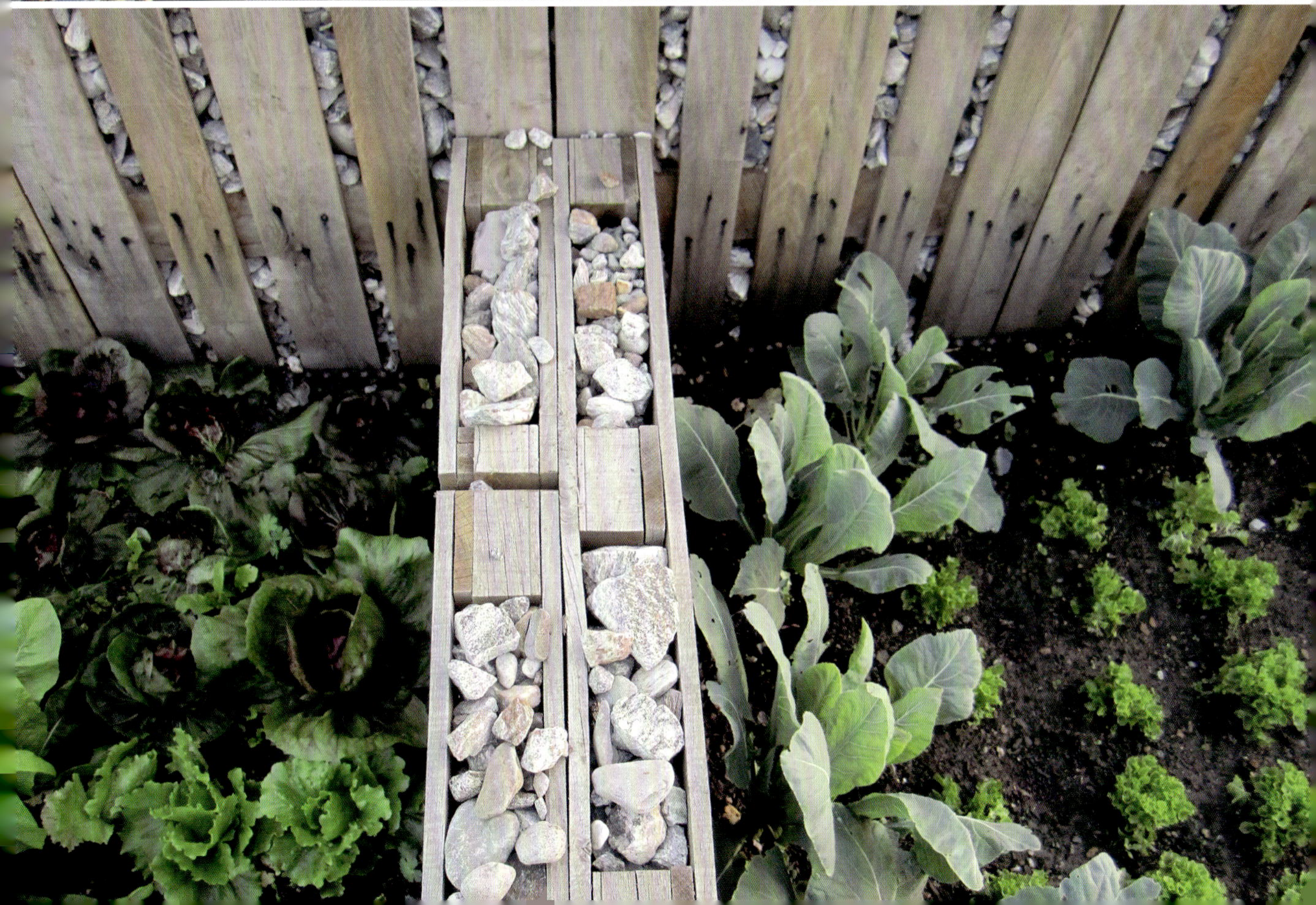

COMMUNITY GARDENS
Chiasso, Switzerland, 2010

Address: Palapenz, Chiasso, Switzerland. Area: 2400 m². Client: City of Chiasso
Cost: €200 000. Collaboration: Radix Svizzera Italiana

Sophie Agata Ambroise points out that these **COMMUNITY GARDENS** are not only a space used for cultivation but also a place to meet and relax. Intended as shared space, the project was encouraged by a group of citizens and Radix Svizzera Italiana, an association dedicated to the promotion of the quality of life in Switzerland. The city immediately approved and financed the project. Eighteen parcels of 30 square meters and 40 parcels of 15 square meters, together with a space accessible to the handicapped, form the garden. Calling on the history of the city as a transport hub, the architect used wooden pallets normally employed in the shipment of merchandise, and gravel from the Alp Transit tunnel construction zone. These oak dividers are used to signify the limits of the parcels, and Sophie Agata Ambroise also created spaces for tools and a pergola with a large wooden table, a grill, and a fountain, emphasizing the social function of the space.

Sophie Agata Ambroise betont, dass diese **KLEINGARTENANLAGE** nicht nur dazu da ist, Blumen und Gemüse anzubauen, sondern dass sie ein Ort ist, an dem man sich trifft und entspannt. Das Projekt wurde von einer Bürgerinitiative und von der Gruppe Radix Svizzera Italiana unterstützt, die sich für die Verbesserung der Lebensqualität in der Schweiz engagiert. Die Stadt Chiasso genehmigte das Projekt sofort und finanzierte es auch. Es entstanden 18 Parzellen mit 30 m² und 40 Parzellen mit 15 m², außerdem ein barrierefreier Bereich. In Anspielung daran, dass Chiasso eine Durchgangsstation für den Lastverkehr ist, verwendete die Planerin Holzpaletten, wie man sie üblicherweise für den Warentransport verwendet, und Kies von der Baustelle des Alpentransittunnels. Diese Trennelemente aus Eichenholz markieren auch die Grenzen der einzelnen Parzellen. Zu dem Entwurf gehören auch ein Bereich für die Gartengeräte sowie eine Pergola mit einem langen Holztisch, einem Grill und einem Brunnen, die die soziale Funktion dieses Freiraums betonen.

Sophie Agata Ambroise précise que ces **JARDINS COMMUNAUTAIRES** ne sont pas seulement faits pour être cultivés, mais aussi pour être un lieu de détente et de vie sociale. Mené dans un esprit coopératif, ce projet a été réalisé à l'initiative d'un groupe de citoyens et de Radix Svizzera Italiana, association pour la promotion de la qualité de la vie en Suisse. La ville a très vite approuvé et financé le projet. Le jardin comprend 18 parcelles de 30 m² et 40 de 15 m², ainsi qu'un espace accessible aux handicapés. Rappelant l'histoire de la ville, ancien nœud de communication, l'architecte s'est servie de palettes de bois généralement utilisées dans le transport de marchandises et de pierres récupérées sur le chantier du projet Alp Transit. Ces petites clôtures de chêne marquent les limites des parcelles. Des endroits pour ranger les outils ont été également créés, ainsi qu'une pergola dotée d'une grande table de bois, d'un grill et d'une fontaine, confirmant la fonction sociale de ces jardins.

OFFICINA DEL PAESAGGIO

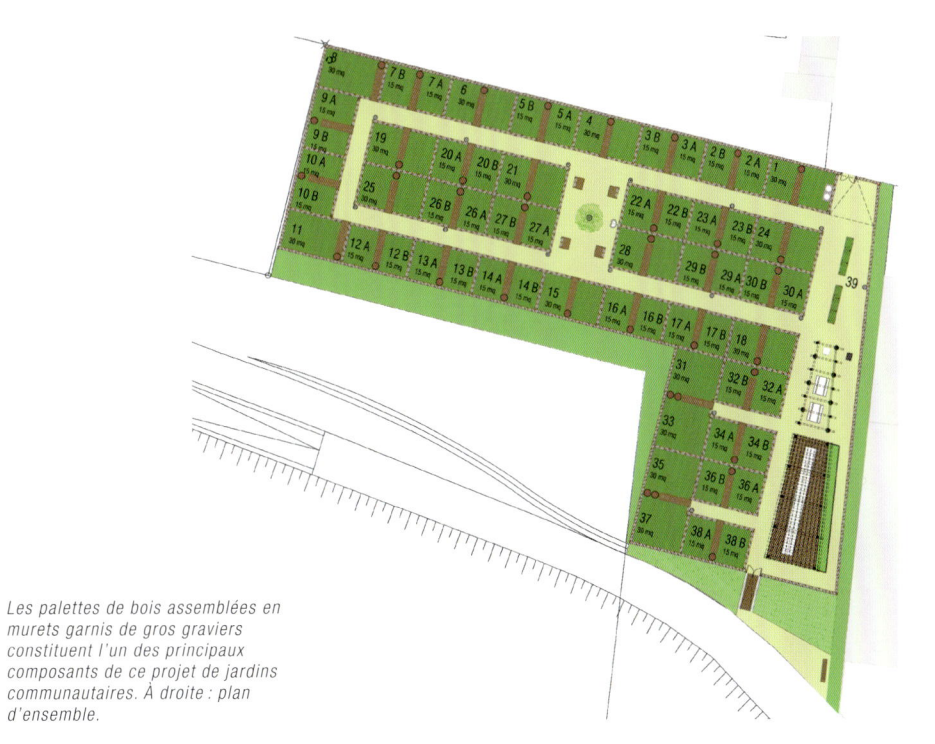

Wooden pallets assembled into low walls and gravel are amongst the main elements used in this design for a shared gardening space, seen as a whole in the plan to the right.

Niedrige Mauern aus Holzpaletten und Kies sind die Hauptgestaltungselemente für diese Gemeinschaftsgärten. Auf dem Plan rechts ist die gesamte Anlage abgebildet.

Les palettes de bois assemblées en murets garnis de gros graviers constituent l'un des principaux composants de ce projet de jardins communautaires. À droite : plan d'ensemble.

YTL RESIDENCE GARDEN
Kuala Lumpur, Malaysia, 2008

Address: Damansara Heights, Kuala Lumpur, Malaysia
Area: 10 000 m². Client: Dr Tan Sri Dato'Seri Yeoh Tiong Lay. Cost: not disclosed
Collaboration: Francesca Bellabona, Nikolaus Schoenenberger

These luxuriant gardens were created by Sophie Agata Ambroise on the site of the large **YTL RESIDENCE** designed by the Paris-based firm Jouin Manku for a prominent Chinese Malaysian family. She began her work with the family by asking: "What is the form of nature that is in the process of escaping you?" Her own response to this question was quite simply the biodiversity of the tropics, threatened by rapid urban expansion and the intensive cultivation of palm trees for their oil. Indeed, in Kuala Lumpur, tropical vegetation is still very much in evidence, but so is urban sprawl. Sophie Agata Ambroise decided, in agreement with the clients, to create a laboratory of biodiversity in the hectare of land at her disposition, inside and outside the walls of the residence. "The basis for the project," she states, "was to bring the 'wild' nature of the rain forest into the heart of the house. What a joy it is, in the process of bringing the jungle to the heart of the polluted city, to see birds and butterflies return."

Diese üppige Gartenanlage schuf Sophie Agata Ambroise auf dem Gelände des großen, von dem Pariser Architekturbüro Jouin Manku entworfenen **ANWESENS YTL** für eine bekannte chinesisch-malayische Familie. Sie begann ihre Arbeit, indem sie die Familie fragte: „Welche Form der Natur, glauben Sie, geht verloren?" Ihre eigene Antwort auf diese Frage war ziemlich einfach – es ist die Biodiversität der Tropen, die durch die rasante Ausbreitung der Städte und die intensive Bewirtschaftung mit Ölpalmen bedroht ist. In Kuala Lumpur sieht man zwar noch sehr viel tropische Vegetation, aber ebenso fällt auf, wie sehr die Stadt wuchert. Sophie Agata Ambroise entschied sich in Absprache mit den Bauherren, auf dem ihr zur Verfügung stehenden ein Hektar großen Terrain innerhalb und außerhalb des Wohngebäudes ein Labor für Biodiversität einzurichten. „Grundidee für dieses Projekt war, die ‚wilde' Natur des Regenwaldes in das Herz des Hauses zu transportieren. Es ist eine Freude, den Urwald mitten in die verschmutzte Stadt zu bringen und zu sehen, dass Vögel und Schmetterlinge zurückkehren."

Ces jardins luxuriants ont été créés par Sophie Agata Ambroise autour de la vaste **YTL RESIDENCE**, conçue par l'agence parisienne Jouin Manku pour une éminente famille sino-malaise. Elle a commencé son travail en demandant à la famille : « Quelle forme de nature est aujourd'hui sur le point de vous échapper ? » Sa propre réponse à la question fut simplement : la biodiversité tropicale, menacée par l'expansion urbaine rapide et la culture intensive du palmier à huile. La végétation tropicale est en fait encore très présente à Kuala Lumpur, mais la ville se développe très rapidement. La paysagiste a décidé, en accord avec son client, de créer un laboratoire de biodiversité sur l'hectare de terrain mis à sa disposition, aussi bien à l'extérieur qu'à l'intérieur de la maison. « La base de ce projet, explique-t-elle, a été de faire entrer la nature sauvage de la forêt équatoriale au cœur de la maison. Quelle joie de constater qu'en faisant pénétrer la jungle au cœur de la ville polluée, les oiseaux et les papillons réapparaissent. »

OFFICINA DEL PAESAGGIO

The large YTL Residence occupies almost its entire site, but Sophie Agata Ambroise has nonetheless created narrow green spaces around the outer perimeter, as seen in the plans below and the image on the left page. Above, inside the walls, approaching the main part of the house.

Das große Wohnhaus YTL besetzt fast das gesamte Grundstück. Sophie Agata Ambroise ist es dennoch gelungen, schmale grüne Räume am Rand zu schaffen, wie man auf den Grundrissen unten und dem Bild auf der linken Seite sehen kann. Oben der Zugang zum Hauptflügel des Hauses.

Si la vaste résidence YTL occupe pratiquement la totalité de son terrain, Sophie Agata Ambroise n'en a pas moins réussi à créer de petits espaces de verdure en périmètre, comme le montrent les plans ci-dessous et l'image de la page de gauche. Ci-dessus, un jardin ceint de murs à proximité de l'aile principale.

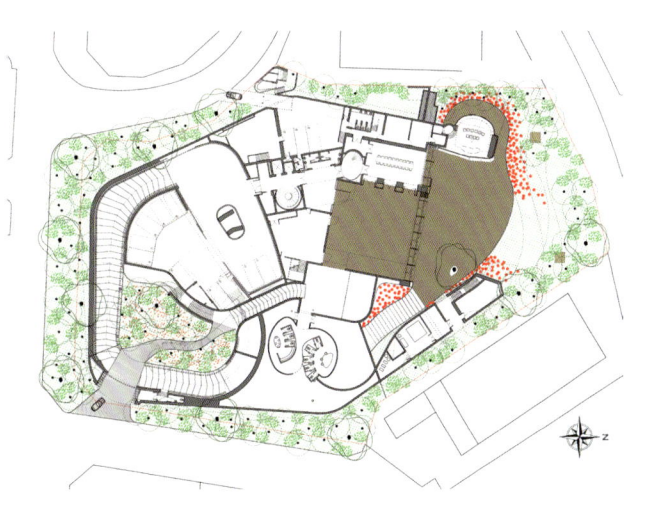

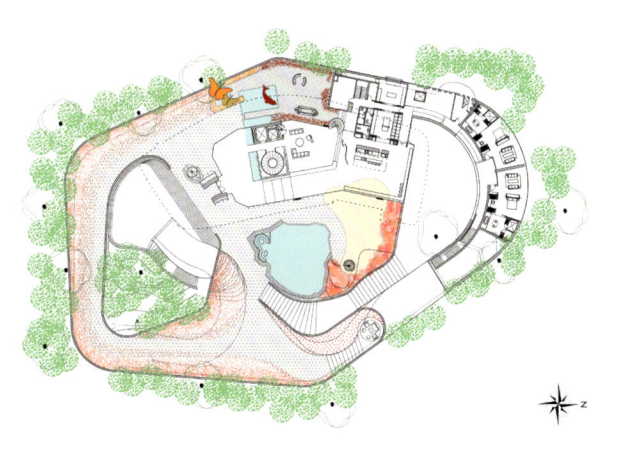

Above, the main entrance gate of the house and the densely planted interior space. Above (this page), on the main level, orchids hang freely from a wooden structure. Right page, the main volume of the house seen from the lower garden area.

Oben, der Haupteingang des Wohnhauses mit üppig bepflanztem Innenbereich. Auf der Hauptebene hängen Orchideen frei an einer Holzkonstruktion (ganz oben). Rechte Seite, der Hauptbaukörper vom unteren Gartenbereich aus gesehen.

Ci-dessus, le portail de l'entrée principale et un luxuriant jardin intérieur. En haut : au niveau principal, des orchidées retombent en suspension d'une pergola. Page de droite : la partie principale de la maison vue du jardin inférieur.

OFFICINA DEL PAESAGGIO

OKRA

OKRA Landscape Architects
Oudegracht 23
3511 AB Utrecht
The Netherlands

Tel: +31 30 273 42 49
Fax: +31 30 273 51 28
E-mail: mail@okra.nl
Web: www.okra.nl

Boudewijn Almekinders, one of the founders of **OKRA**, was born in 1966 and studied Landscape Architecture at the Agricultural University in Wageningen (1984–91). Before creating OKRA, he worked as a designer for the Floriade World exhibition (1992). Martin Knuijt, born in 1966, and Christ-Jan van Rooij, born in 1963, are two other founders of OKRA. Wim Voogt graduated in 1998 from the Van Hall Larenstein University of Applied Sciences, where he studied Gardening and Landscaping and specialized in Design. He then went to work at OKRA, initially as a designer/illustrator and subsequently, from 2002, as a project manager. Since 2010, he has been a Director and Partner. Bart Dijk, born in 1980, worked as a project manager on the Utrecht Domplein project (Utrecht, 2010, published here). The work of OKRA includes New Wood Living, a new concept for housing in forests (2001); the Nieuwveense Landen, a master plan for a 700-hectare new city development in Meppel (2003–04); and the Afrikaanderplein (also published here), a park in Rotterdam awarded the national prize for Best Public Space 2003–05 (OAP), all in the Netherlands.

Boudewijn Almekinders, einer der Gründer von **OKRA,** wurde 1966 geboren und studierte Landschaftsarchitektur an der Landwirtschaftlichen Universität in Wageningen, Niederlande (1984–91). Bevor er OKRA gründete, arbeitete er als Designer für die Weltausstellung Floriade (1992). Martin Knuijt, 1966 geboren, und Christ-Jan van Rooij, 1963 geboren, sind die beiden anderen Mitbegründer von OKRA. Wim Voogt schloss sein Universitätsstudium der Garten- und Landschaftsgestaltung mit der Vertiefung Entwerfen an der Fachhochschule Van Hall Larenstein 1998 ab. Seitdem arbeitet er bei OKRA, anfangs als Designer und Illustrator, ab 2002 dann als Projektmanager. Seit 2010 ist er Direktor und weiterer Partner des Büros. Bart Dijk, 1980 geboren, arbeitete als Projektmanager bei dem Domplatz in Utrecht (2010, hier vorgestellt). Zu den Projekten von OKRA gehören New Wood Living, ein neues Konzept für den Hausbau in Waldgebieten (2001), die Nieuwveense Landen, ein Masterplan für eine 700 Hektar umfassende Stadtentwicklung in Meppel (2003–04), und der Afrikaanderplein (ebenfalls hier vorgestellt), eine Parkanlage in Rotterdam, die mit dem niederländischen Nationalpreis für den besten öffentlichen Freiraum (OAP) 2003–2005 ausgezeichnet wurde, alle in den Niederlanden.

Boudewijn Almekinders, né en 1966, a étudié l'architecture du paysage à l'Université agricole de Wageningen (1984–91) et a travaillé à la conception de l'exposition internationale Floriade en 1992 avant de cofonder l'agence **OKRA**. Martin Knuijt, né en 1966, et Christ-Jan van Rooij, né en 1963, sont les deux autres fondateurs d'OKRA. Wim Voogt, diplômé en 1998 de l'Université des sciences appliquées Van Hall Larenstein, où il a étudié l'aménagement de jardins et le paysagisme, a rejoint OKRA comme concepteur/illustrateur, puis directeur de projet à partir de 2002. Il est directeur et partenaire d'OKRA depuis 2010. Bart Dijk, né en 1980, a été directeur de projet de la Domplein à Utrecht (2010, publié ici). Parmi les réalisations d'OKRA aux Pays-Bas : un nouveau concept de logements en forêt (2001) ; Nieuwveense Landen, plan directeur pour une ville nouvelle de 700 hectares à Meppel (2003–04) et l'Afrikaanderplein (publié ici), un parc à Rotterdam qui a remporté le Prix national du meilleur espace public 2003–05 (OAP).

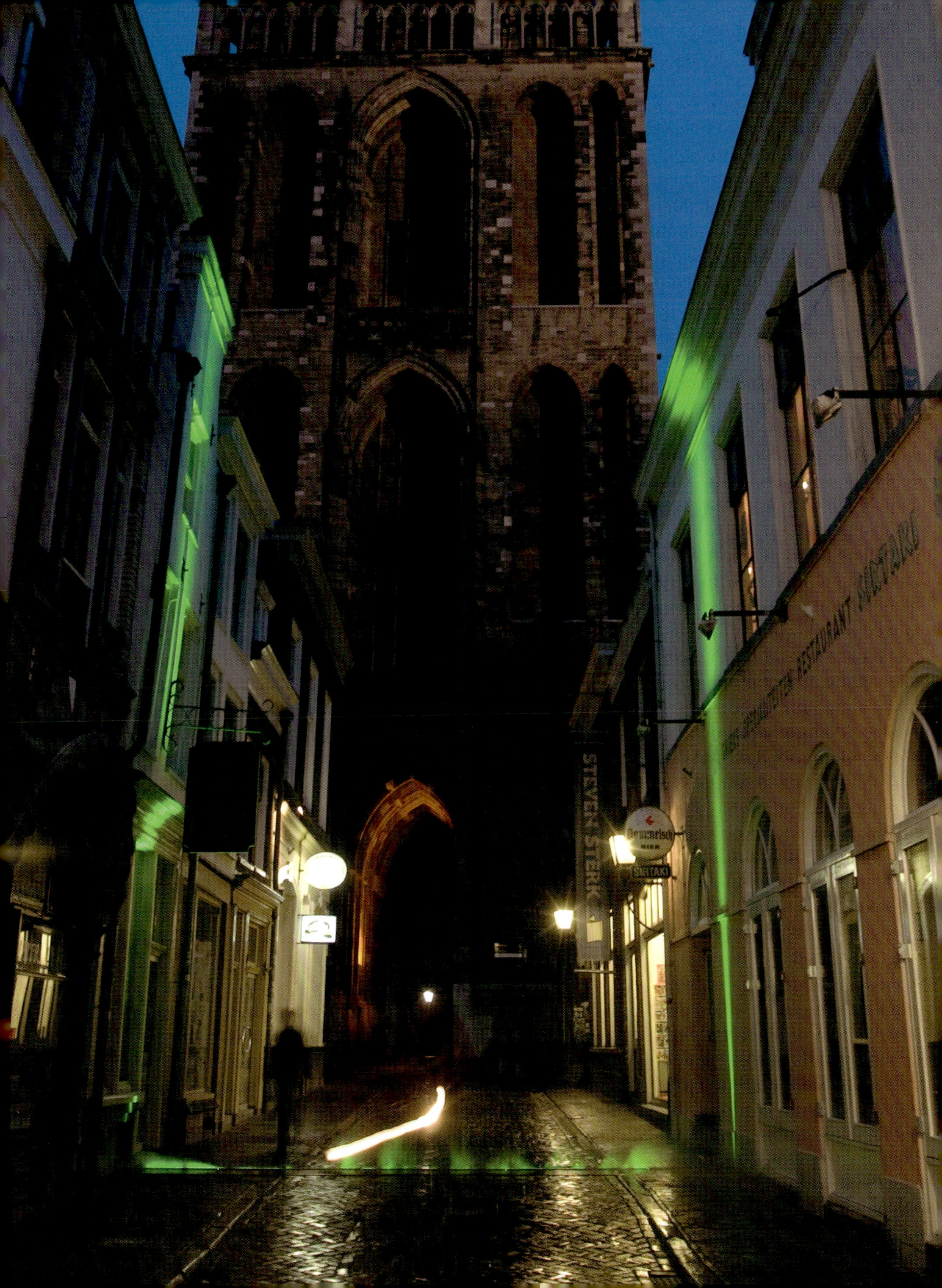
RESTAURANT SIRTAKI
STEVEN STERK
Dommelsch
BIER
SIRTAKI

DOMPLEIN
Utrecht, The Netherlands, 2010

Address: Domplein, 3512 JC Utrecht, The Netherlands
Area: 161 meters (wall length)
Client: Stichting Domplein 2013. Cost: €800 000

The **DOMPLEIN** is the cathedral square of Utrecht which is at the center of the history of the city. OKRA won the international competition for the renovation of the square. A line made of Corten steel marks the wall of the original Roman *castellum* that stood on the location. The designers explain: "Green light and wisps of mist emanate from the marking and the borders of the Roman empire are engraved on the steel. The marking is just as silent as the archaeological witness underground. The material and color are earthy and refer to times long past in the fertile river delta and wooded ruins dating back to the dark Middle Ages. Now and again a poetic message is signaled with the light, in a type of Roman Morse code. And on special days the light changes color, such as yellow on Catholic holidays and orange on the Queen's Birthday." LED lights controlled from a central installation with a computer are used in the design.

Der **DOMPLATZ** befindet sich an der Kathedrale von Utrecht mitten im historischen Stadtzentrum. OKRA gewann den internationalen Wettbewerb, der für die Neugestaltung des Platzes ausgeschrieben worden war. Eine Linie aus Cor-Ten-Stahl markiert die Mauer des römischen „castellum", das früher an dieser Stelle gestanden hatte. Die Planer erklären dazu: „Grünes Licht und Nebelschwaden strömen aus den Markierungen. In den Stahl sind die Grenzen des Römischen Reichs eingraviert. Die Markierung ist genauso geräuschlos wie die archäologischen Zeugnisse unter der Erde. Das Material und die Farbe sind erdnah und stellen einen Bezug her zu lang vergangenen Zeiten in dem fruchtbaren Flussdelta und zu Holzruinen aus dem Mittelalter. Dann und wann sendet das Licht eine poetische Botschaft – eine Art römisches Morsezeichen. An bestimmten Tagen wechselt das Licht seine Farbe, so wird es an katholischen Feiertagen gelb, und am Geburtstag der niederländischen Königin wechselt es zu Orange." Für diesen Entwurf wurden LED-Leuchten verwendet, die zentral von einem Computer gesteuert werden.

OKRA a remporté le concours international organisé pour la rénovation de la **DOMPLEIN**, la place qui s'étend devant la cathédrale d'Utrecht au cœur de la ville historique. Une ligne en acier Corten reprend le tracé des murs du *castellum* romain qui occupait jadis le même site. « Un halo vert et des filets de brume émanent de ce marquage, tandis que les limites de l'empire romain sont gravées dans l'acier. Le marquage est aussi silencieux que les vestiges archéologiques qui abondent dans le sol. Le matériau et la couleur sont de caractère organique et rappellent le lointain passé de ce delta fertile et les vestiges des constructions médiévales en bois. De temps en temps, un message poétique se glisse sous forme lumineuse, comme un code morse romain. Certains jours, l'éclairage change de couleur : jaune pour les fêtes catholiques et orange pour l'anniversaire de la reine. » L'éclairage, généré par des LED, est contrôlé par ordinateur à partir d'un poste central.

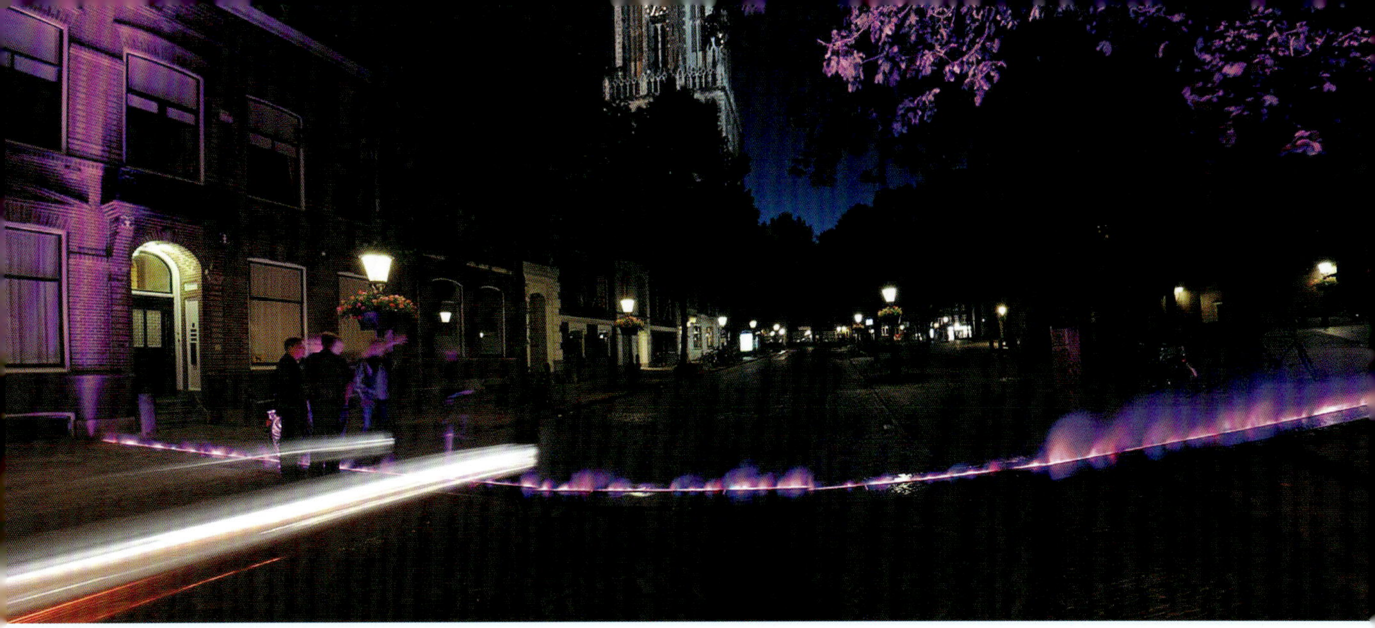

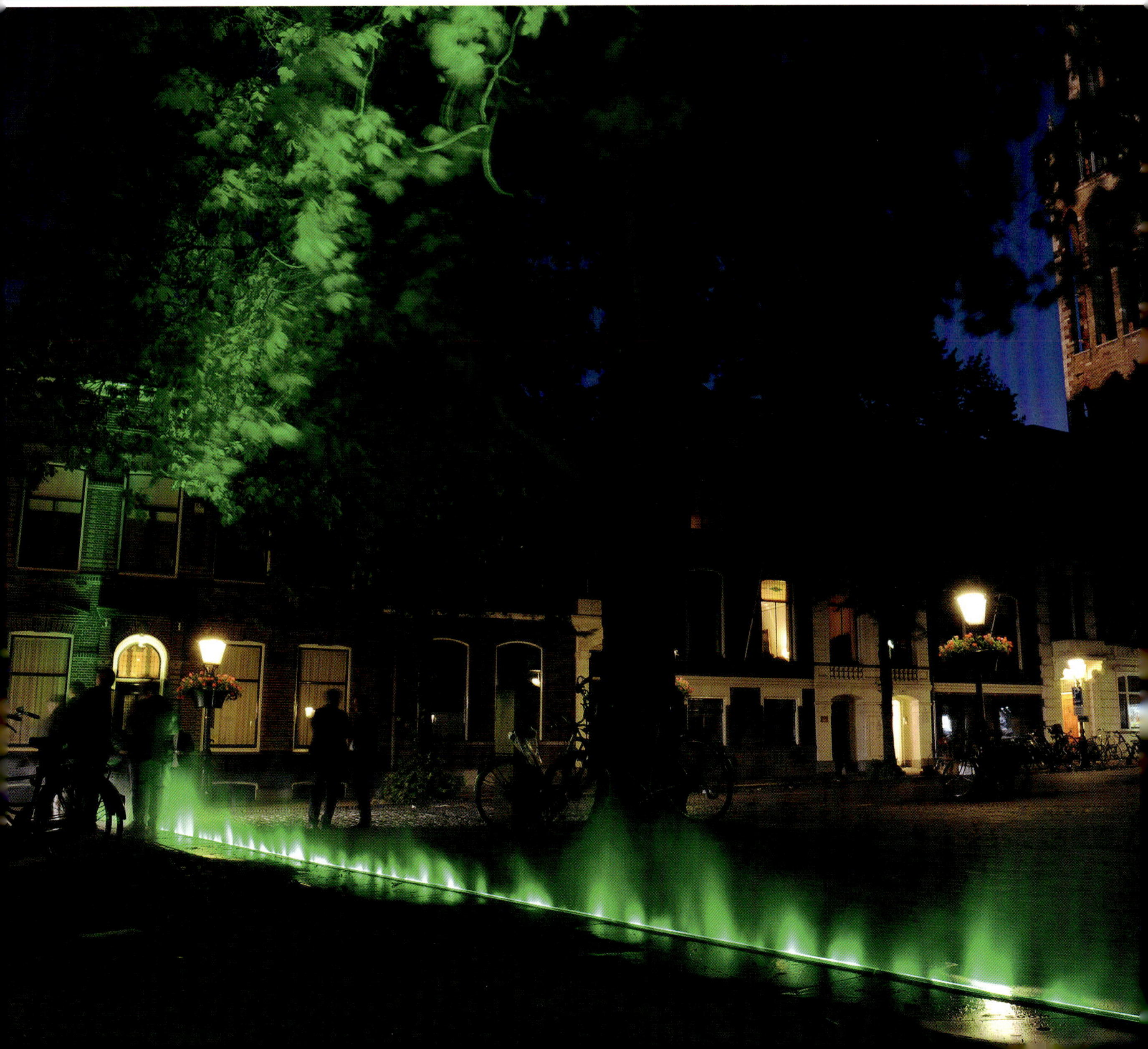

AFRIKAANDERPLEIN

Rotterdam, The Netherlands, 2003–05

Area: 5.6 hectares. Client: Rotterdam Council
Cost: €8.1 million

OKRA worked on the outdoor master plan for the Kop van Zuid area development in Rotterdam in the 1990s and was asked thereafter to work on the **AFRIKAANDERPLEIN**. The approach of the firm was to create "a free central area surrounded by a framework of specific functions that require a specific layout, such as the market, playground, and aviary. Three sides of this zone are intensely used. One side is quiet, the green oasis, where the botanical gardens and a mosque are located. A water feature with a single bridge ensures that this area stays quiet." The inner area of the park is surrounded by a fence in order to signify its function as the core of the project and to "ensure safety, sustainability, and quality of use." This specially designed fence can be opened during the day over a 30-meter stretch to improve accessibility. OKRA emphasizes that the populations using the space and living near it are of varied origin, a fact that contributed to the specifics of the design.

In den 1990er-Jahren bearbeitete OKRA den Masterplan für die Freiraumgestaltung des Stadtviertels Kop van Zuid in Rotterdam. Im Anschluss daran erhielt das Büro den Auftrag, die **AFRIKAANDERPLEIN** zu gestalten. Die Planer wollten „eine freie Mitte schaffen, die von vielen unterschiedlichen Funktionsbereichen mit jeweils spezifischer Gestaltung gerahmt wird, wie zum Beispiel der Marktplatz, Spielplätze und eine Voliere. Drei Seiten der Anlage werden intensiv genutzt. Eine Seite ist als ruhige, grüne Oase ausgewiesen. Hier befinden sich ein Botanischer Garten und eine Moschee. Ein vorgelagertes Wasserbecken, über das eine Brücke führt, garantiert, dass dieser Bereich ruhig bleibt." Das innere Areal dieses Parks wird von einem Zaun eingefasst, um zu signalisieren, dass es das Herzstück des Projektes ist, und um „Sicherheit, Nachhaltigkeit und Nutzungsqualität zu garantieren". Dieser eigens entworfene Zaun kann tagsüber auf einer Breite von 30 Metern geöffnet werden, damit der Bereich leicht zugänglich ist. OKRA betont, dass die Anwohner, die diesen Park nutzen, unterschiedlicher Herkunft sind, was die besonderen Details des Entwurfs mit beeinflusst hat.

OKRA, qui avait travaillé sur le plan directeur du quartier de Kop van Zuid à Rotterdam dans les années 1990, a ensuite été chargée d'intervenir sur le projet de l'**AFRIKAANDERPLEIN**. L'approche de l'agence a été de créer « une zone centrale libre entourée d'espaces présentant des fonctions spécifiques et exigeant des formes particulières, comme un marché, un terrain de jeux ou une volière. Trois côtés de la place sont soumis à un usage intensif, le dernier étant plus tranquille puisqu'il s'agit d'une oasis de verdure où sont regroupés une mosquée et des jardins botaniques. Un bassin que franchit une unique passerelle protège la tranquillité des lieux ». La partie intérieure du parc est entourée d'une clôture pour signifier clairement sa fonction de cœur du projet et « assurer la sécurité, la durabilité et la qualité des usages ». Cette barrière spécialement conçue s'ouvre en journée sur 30 m pour faciliter l'accès. Okra précise que les populations qui utilisent ce parc et vivent à proximité sont d'origines variées, ce qui a contribué à définir les particularités de ce projet.

The architects used a simple, angled design that cuts through the large open lawns of the square. The paths alternate with sharply defined low walls and a number of trees to enliven the otherwise very flat space.

Der schlichte Entwurf baut auf Wegen auf, die sich auf der offenen Rasenfläche des Platzes kreuzen. Im Wechselspiel dazu gibt es scharfkantige flache Mauern und einige Bäume, die den sonst sehr flachen Platz beleben.

Pour les allées, les architectes ont imaginé un plan simple qui découpe les vastes pelouses de la place. Les allées alternent avec des murets en angles obtus et quelques arbres qui animent cet espace très plat.

PALERM & TABARES DE NAVA

Palerm & Tabares de Nava Arquitectos
Calle 25 de Julio, n. 48
38004 Santa Cruz de Tenerife
Canary Islands
Spain

Tel: +34 922 24 75 70
E-mail: paltab@paltab.com
Web: www.paltab.com

JUAN MANUEL PALERM SALAZAR was born in Santa Cruz de Tenerife (Canary Islands, Spain) in 1957. **LEOPOLDO TABARES DE NAVA Y MARÍN** was born in La Laguna (Canary Islands, Spain) in 1958. They have been working together since 1986 and formed their current partnership in 1996. They place a particular emphasis on location and territory, with an interest in landscape and urban morphology. They have worked mainly in the Canary Islands, although they have also undertaken projects in Argentina, Costa Rica, Mexico, and the United States (Los Angeles), as well as in the cities of Gerona (Spain), and Trento and Venice (Italy). Their work includes the Biblioteca Pública del Estado (Las Palmas de Gran Canaria, Grand Canary, 1996–2002); Espacio Expositivo-Taller Fundación César Manrique (Teguise, Lanzarote, 1996–2004); the García Sanabria Park (Santa Cruz de Tenerife, Tenerife, 2004–06, published here); the House in La Laguna (San Cristóbal de La Laguna, Tenerife, 2004–07); a Theater, Service Center, and Sports Palace (Trento, Italy, 2007); and Barranco de Santos (Santa Cruz de Tenerife, Tenerife, 1997–2010), all in the Canary Islands, Spain, unless stated otherwise. Currently under construction is the Garañaña Urban Park (Arona, Tenerife, Canary Islands).

JUAN MANUEL PALERM SALAZAR wurde 1957 in Santa Cruz de Tenerife (Kanarische Inseln, Spanien) geboren. **LEOPOLDO TABARES DE NAVA Y MARÍN** wurde 1958 in La Laguna (Kanarische Inseln) geboren. Beide arbeiten seit 1986 zusammen und haben ihre derzeitige Büropartnerschaft 1996 gegründet. Ihr spezielles Interesse gilt dem Ort und dem Gelände, wobei auch Landschaft und urbane Morphologie wichtige Faktoren bei ihren Arbeiten sind. Sie sind vor allem auf den Kanarischen Inseln tätig, haben aber auch Projekte in Argentinien, Costa Rica, Mexiko und in den Vereinigten Staaten (Los Angeles) sowie in Gerona (Spanien), Trient und Venedig (Italien) ausgeführt. Zu ihren Projekten gehören die Biblioteca Pública del Estado (Las Palmas de Gran Canaria, 1996–2002), der Ausstellungsraum und das Atelier der Stiftung César Manrique (Teguise, Lanzarote, 1996–2004), die Neugestaltung des Parque García Sanabria (Santa Cruz de Tenerife, 2004–06, hier vorgestellt), das House in La Laguna (San Cristóbal de La Laguna, Teneriffa, 2004–07), ein Theater, Dienstleistungszentrum und Sportpalast in Trient (Italien, 2007) und der Barranco de Santos (Santa Cruz de Tenerife, 1997–2010), alle auf den Kanarischen Inseln, sofern nicht anders angegeben. Derzeit im Bau befindet sich das Projekt des Stadtparks von Garañaña (Arona, Teneriffa, Kanarische Inseln).

JUAN MANUEL PALERM SALAZAR est né à Santa Cruz de Tenerife (îles Canaries, Espagne) en 1957 et **LEOPOLDO TABARES DE NAVA Y MARÍN** à La Laguna (îles Canaries) en 1958. Ils travaillent ensemble depuis 1986 et ont fondé l'agence actuelle en 1996. Ils s'intéressent particulièrement à la situation et au territoire à travers la morphologie du paysage et de la ville. Ils ont essentiellement travaillé dans les îles Canaries, mais aussi en Argentine, au Costa Rica, au Mexique et aux États-Unis (Los Angeles), ainsi qu'à Gérone (Espagne), Trente et Venise (Italie). Parmi leurs réalisations (en Espagne, sauf mention contraire) : la bibliothèque municipale de Las Palmas (Grande Canarie, 1996–2002) l'espace d'exposition-atelier de la Fundación César Manrique (Teguise, Lanzarote, 1996–2004) ; le parc García Sanabria (Santa Cruz de Tenerife, Tenerife, 2004–06, publié ici) ; la Maison à La Laguna (San Cristóbal de La Laguna, Tenerife, 2004–07) ; un théâtre, un centre de services et un palais des sports (Trente, Italie, 2007) et le Barranco de Santos (Santa Cruz de Tenerife, Tenerife, 1997–2010). Ils ont actuellement en chantier le parc urbain de Garañaña (Arona, Tenerife, îles Canaries).

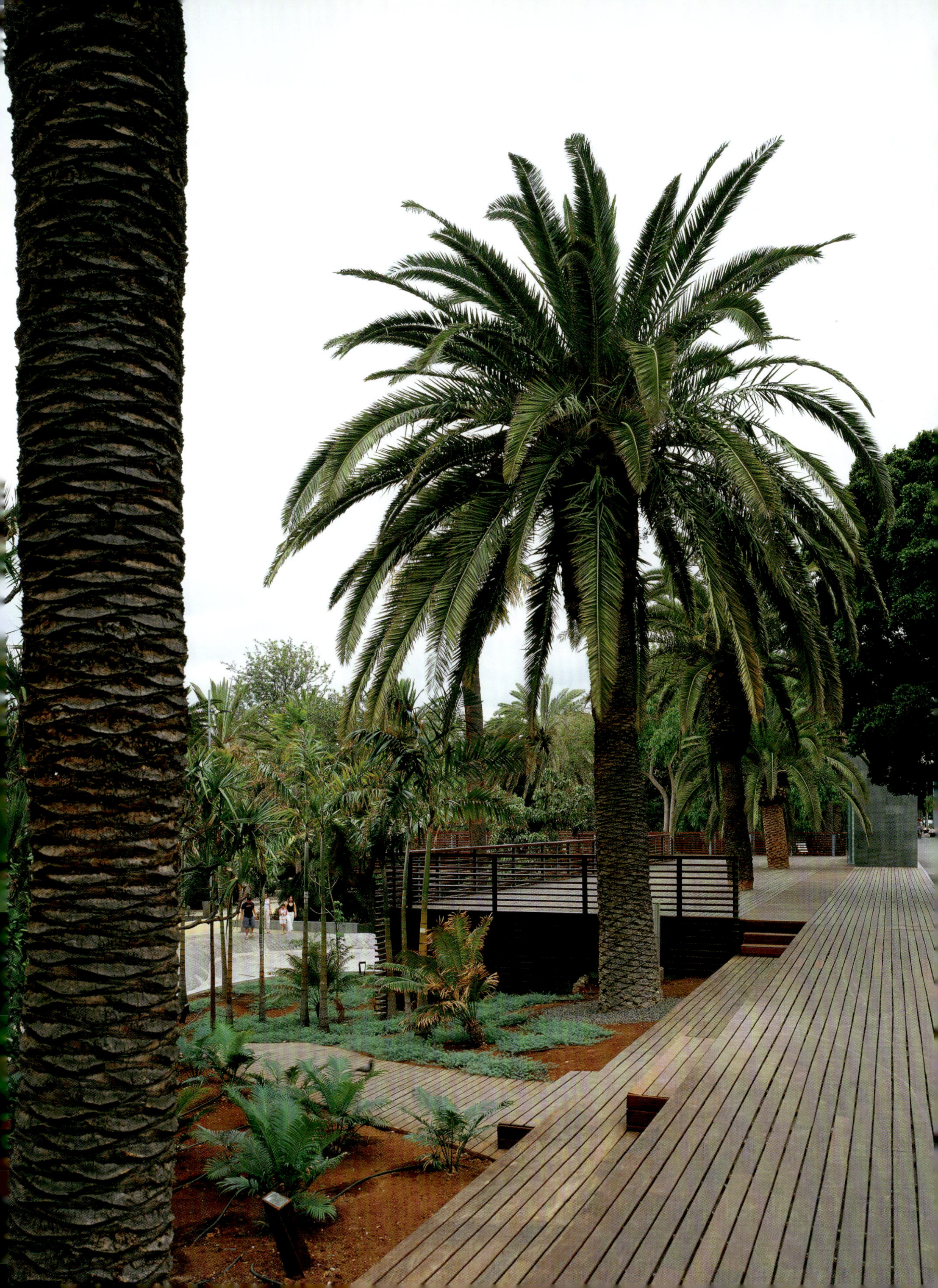

GARCÍA SANABRIA PARK

Santa Cruz de Tenerife, Canary Islands, Spain, 2004–06

Address: c/ Méndez Núñez, Numancia, Rambla y José Naveiras, Santa Cruz de Tenerife, Tenerife, Canary Islands, Spain. Area: 60 548 m². Client: City Council of Santa Cruz de Tenerife
Cost: not disclosed

The architects explain that their project "establishes a relationship between path and route systems to plantations and tree geometry." They attempt to "look for the naturalization of the components, accentuating the architectural and sculptural elements that will constitute artificial islands in a natural setting." A space with games for children was created on the natural slopes of the location of a former mini-golf course. A spiraling path with sculptural Corten steel lamps leads to the center of the park and its large fountain. A visitor information center and office were added near the main entrance to the **GARCÍA SANABRIA PARK**. Lightweight construction systems were used for a cafeteria–restaurant at the new upper entrance to the Park.

Die Architekten erklären, dass in ihrem Projekt „eine Beziehung zwischen der Form und Führung des Wegenetzes und der geometrischen Anordnung der Pflanzen und Bäume besteht". Sie versuchen, „die einzelnen Komponenten zu integrieren und die architektonischen und skulpturalen Elemente hervorzuheben, die dann wie künstliche Inseln in einem natürlichen Umfeld wirken". Auf den geneigten Flächen eines ehemaligen Minigolfplatzes ist ein Kinderspielplatz angelegt worden, ein sich spiralförmig windender Fußweg mit skulptural gestalteten Leuchten aus Cor-Ten-Stahl führt in das Zentrum des Parks mit seiner großen Springbrunnenanlage. In der Nähe des Haupteingangs zum **GARCÍA SANABRIA PARK** befindet sich nun ein Besucherinformationszentrum mit Büro. Die Cafeteria und das Restaurant nahe des neuen oberen Eingangs zum Park sind Leichtbaukonstruktionen.

Ce projet « établit une relation entre les systèmes de voies et d'allées, les plantations et la géométrie des arbres », expliquent les architectes, qui ont « cherché à naturaliser les composants du projet et à accentuer la présence d'éléments architecturaux et sculpturaux constituant des îles artificielles dans ce cadre naturel ». Une aire de jeux pour enfants a été aménagée sur les pentes d'un ancien terrain de minigolf. Un sentier en spirale, éclairé de lampadaires sculpturaux en acier Corten, conduit au centre du parc **GARCÍA SANABRIA**, marqué par une importante fontaine. Un centre d'information pour les visiteurs et des bureaux ont été édifiés près de l'entrée. Près de la nouvelle entrée supérieure, une cafétéria-restaurant a fait appel à des systèmes de construction légère.

As the plan to the right and the images on this double page show, the park features both long straight alleys and a curving walkway that allows visitors to circle the central fountain.

Wie der Plan rechts und die Fotos auf dieser Doppelseite zeigen, gibt es in diesem Park gerade Alleen und einen kurvig verlaufenden Weg, auf dem die Besucher den Springbrunnen in der Mitte der Anlage umrunden können.

Comme le montrent le plan de droite et les photos de cette double page, le parc s'organise selon de grands axes rectilignes et une allée circulaire qui entoure la fontaine centrale.

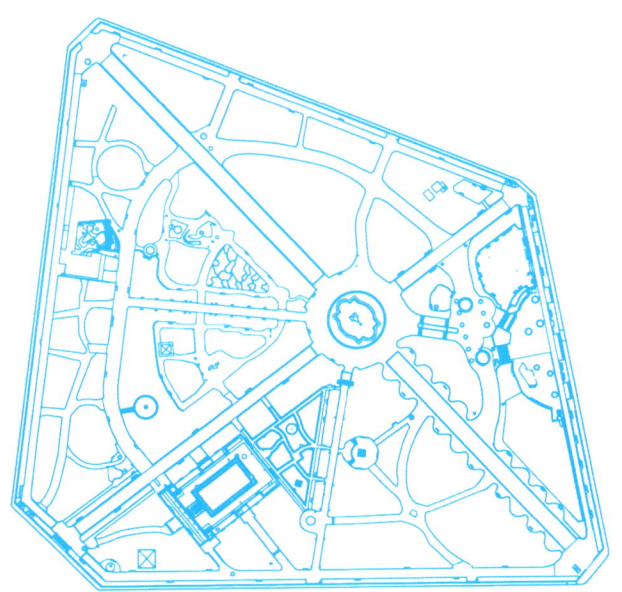

Above, an overall image of the park in its setting in Santa Cruz. Right, the architects employed lightweight construction for the new structures.

Oben, Blick auf den Park und sein Umfeld in Santa Cruz. Rechts: Die neuen Baukörper sind in Leichtbauweise erstellt worden.

Ci-dessus, image d'ensemble du parc dans le contexte de la ville de Santa Cruz. À droite : les architectes ont utilisé des constructions légères pour les nouvelles structures.

The Corten-steel lamps seen in the image to the left on the path were designed by the architects. Right (bottom) the central fountain in the park.

Die Leuchten aus Cor-Ten-Stahl am Weg auf dem Bild links sind ein Entwurf der Architekten. Rechts (unten) der Springbrunnen in der Mitte des Parks.

Les éclairages en acier Corten des allées (à gauche) ont été dessinés par les architectes. En bas à droite : la fontaine centrale.

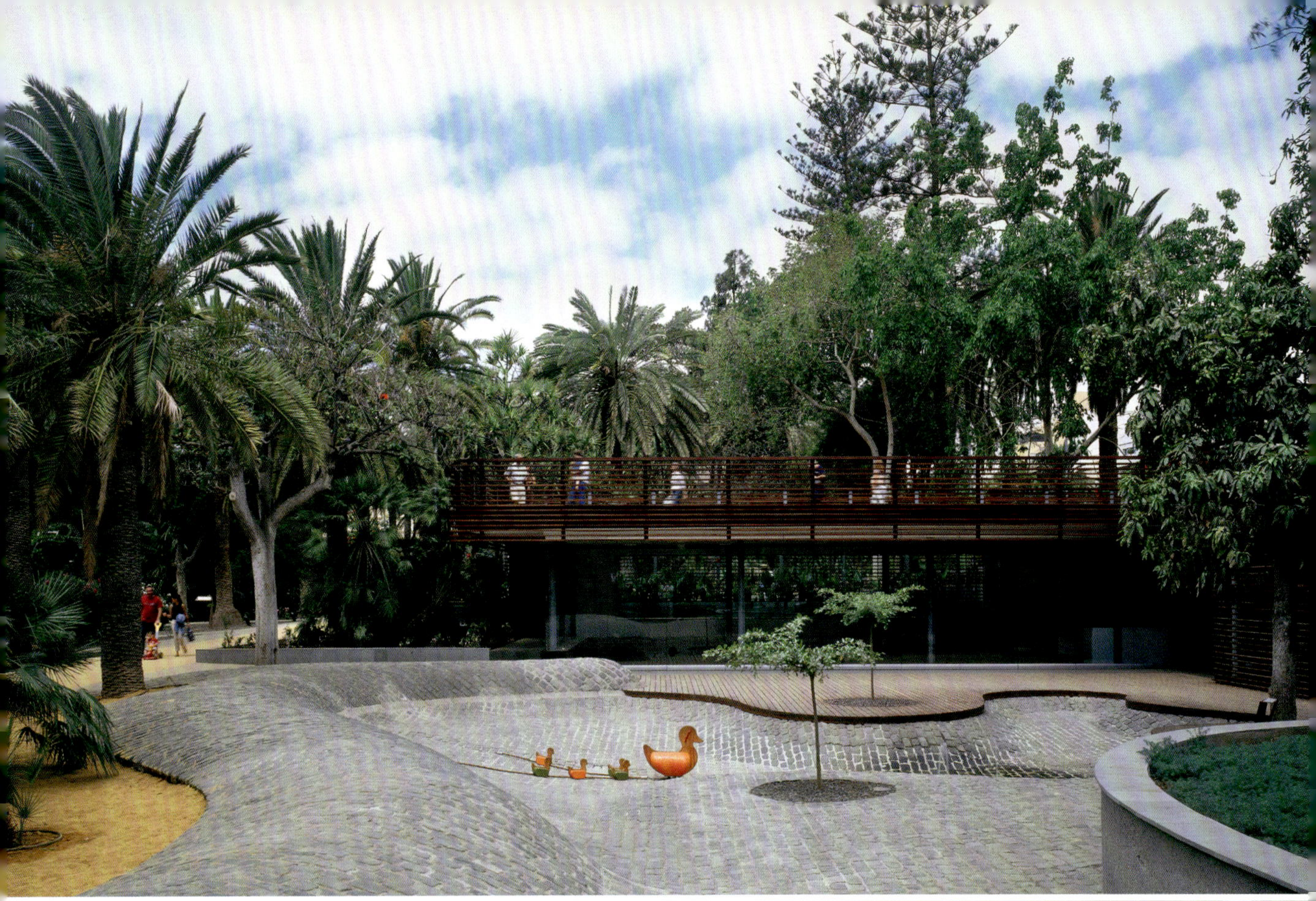
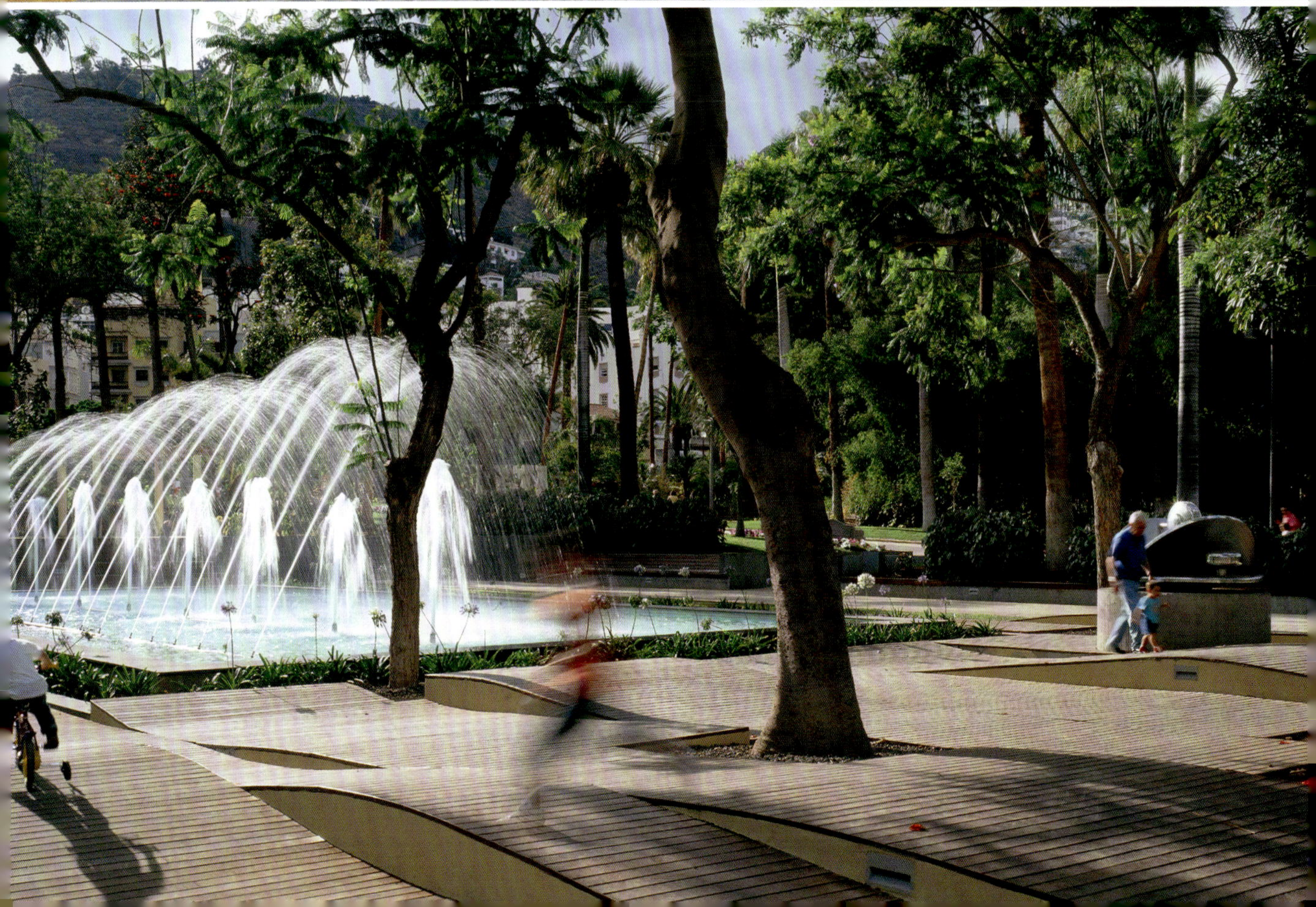

JOHN PAWSON

John Pawson
Unit B, 70–78 York Way
London N1 9AG
UK

Tel: +44 20 78 37 29 29 / Fax: +44 20 78 37 49 49
E-mail: email@johnpawson.co.uk
Web: www.johnpawson.com

Born in Halifax in central England in 1949, **JOHN PAWSON** attended Eton and worked in his family's textile mill before going to Japan for four years. On his return, he studied at the Architectural Association in London and set up his own firm in 1981. He has worked on numerous types of buildings including the flagship store for Calvin Klein in New York, airport lounges for Cathay Pacific airlines at the Chek Lap Kok Airport in Hong Kong, and an apartment for the author Bruce Chatwin. Pawson may be even better known to the general public because of his 1996 book *Minimum*. Some of his more recent work includes Lansdowne Lodge Apartments (London, UK, 2003); Hotel Puerta America in Madrid (Spain, 2005); the Tetsuka House (Tokyo, Japan, 2003–06); Calvin Klein Apartment (New York, USA, 2006); the Sackler Crossing in the Royal Botanic Gardens (Kew, London, UK, 2006, published here); work and renovation of a wing of the Monastery of Our Lady of Nový Dvůr (Bohemia, 2004; second phase 2009), ; the Martyrs Pavilion, Saint Edward's School (Oxford, UK, 2009); a church renovation concerning the Sacristy, Lateral, and Chapels (Monastery of Our Lady of Sept-Fons, Burgundy, France, 2009); and a number of apartments in New York (2009). In 2010 he realized the Stone House (Milan, Italy); and his work was the object of a solo exhibition, "John Pawson Plain Space" in London's Design Museum (UK). Current work includes the refurbishment of the Saint Moritz Church (Augsburg, Germany); renovation of the Former Commonwealth Institute (London, UK); as well as several houses in France, Greece, Portugal, Spain, the UK, and the USA.

JOHN PAWSON wurde 1949 in Halifax in Mittelengland geboren. Er ging in Eton zur Schule und arbeitete im Textilunternehmen seiner Familie, bevor er für vier Jahre nach Japan ging. Nach seiner Rückkehr studierte er an der Architectural Association in London und gründete 1981 sein eigenes Büro. Er hat zahlreiche Gebäude mit sehr unterschiedlichem Charakter realisiert, unter anderem den Flagshipstore von Calvin Klein in New York, die Airport Lounges für die Fluglinie Cathay Pacific im Flughafen Chek Lap Kok in Hongkong und ein Apartment für den Schriftsteller Bruce Chatwin. Pawson ist einem breiteren Publikum durch sein 1996 erschienenes Buch „Minimum" bekannt. Zu seinen neueren Projekten gehören die Lansdowne Lodge Apartments (London, 2003), das Hotel Puerta America in Madrid (2005), das Tetsuka House (Tokio, 2003–06), das Apartment für Calvin Klein (New York, 2006), die Brücke Sackler Crossing in den Royal Botanic Gardens (Kew, London, 2006, hier vorgestellt), Bau und Renovierung eines Gebäudeflügels des Klosters Unserer Lieben Frau in Nový Dvůr (Tschechien, 2004, zweiter Bauabschnitt 2009), der Martyrs Pavilion der Saint Edward's School (Oxford, 2009), die Kirchenrenovierung (Sakristei, Seitenschiff und Kapellen) des Klosters Notre-Dame de Sept-Fons in Burgund (Frankreich, 2009) sowie eine ganze Reihe von Apartments in New York (2009). 2010 baute er das Stone House in Mailand, und sein Gesamtwerk wurde in der Einzelausstellung „John Pawson Plain Space" im Design Museum in London gezeigt. Zu seinen aktuellen Projekten zählen die Renovierung der Moritzkirche in Augsburg und des ehemaligen Commonwealth Institute in London sowie mehrere Wohnhäuser in Frankreich, Griechenland, Portugal, Spanien, Großbritannien und den USA.

Né à Halifax (Royaume-Uni) en 1949, **JOHN PAWSON** a étudié à Eton et travaillé dans l'entreprise de textiles familiale avant de séjourner quatre ans au Japon. À son retour, il étudie à l'Architectural Association de Londres et crée son agence en 1981. Il est intervenu sur de nombreux types de projets dont le magasin amiral de Calvin Klein à New York, les salons de la compagnie Cathay Pacific à l'aéroport de Chek Lap Kok à Hong-Kong et un appartement pour l'écrivain Bruce Chatwin. Il est très connu du grand public pour son livre *Minimum* (1996). Parmi ses réalisations récentes : les Lansdowne Lodge Apartments (Londres, 2003) ; l'hôtel Puerta America à Madrid (2005) ; la maison Tetsuka (Tokyo, 2003–06) ; l'appartement de Calvin Klein (New York, 2006) ; la passerelle du Sackler Crossing dans les Royal Botanic Gardens (Kew, Londres, 2006, publiée ici) ; la construction et la rénovation d'une aile pour le monastère de Notre-Dame de Nový Dvůr (République tchèque, 2004 ; phase 2 en 2009) ; le Martyrs Pavilion, Saint Edward's School (Oxford, GB, 2009); la rénovation de la sacristie, des nefs latérales et chapelles de l'abbaye de Notre-Dame de Sept-Fons (Bourgogne, France, 2009) et plusieurs appartements à New York (2009). En 2010, il a construit la Stone House (Milan, Italie) et son œuvre a fait l'objet d'une exposition personnelle « John Pawson Plain Space » au Design Museum à Londres. Actuellement, il travaille à la restauration de l'église Sankt Moritz (Augsbourg, Allemagne) ; à la rénovation de l'ancien Commonwealth Institute (Londres), ainsi qu'à plusieurs projets de maisons en France, en Grèce, au Portugal, en Espagne, au Royaume-Uni et aux États-Unis.

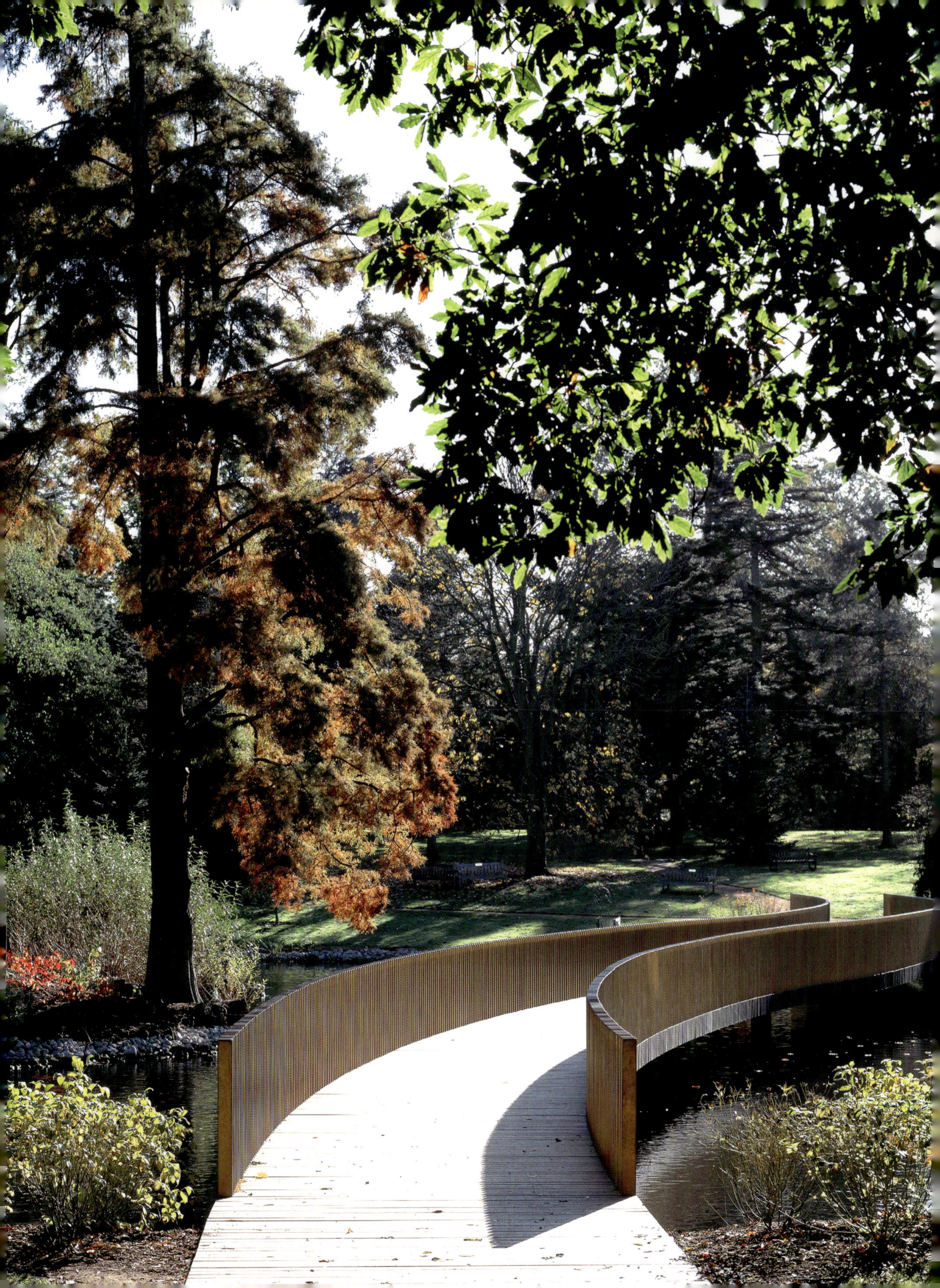

SACKLER CROSSING
Royal Botanic Gardens, Kew, London, UK, 2006

Address: Royal Botanic Gardens, Kew, Richmond, Surrey TW9 3AB, UK, +44 20 83 32 50 00, www.kew.org
Area: 70 meters (length), 3 meters (width). Client: Royal Botanic Gardens Kew
Cost: not disclosed. Collaboration: Ben Collins (Project Architect)

Located in southwest London, the Royal Botanic Gardens comprise an area of 121 hectares. Aside from its world-famous herbarium, or seed banks, the Gardens include a number of notable structures such as the Chinese pagoda built in 1761 by Sir William Chambers and the Palm House designed by Decimus Burton and built between 1844 and 1848. The Gardens receive approximately two million visitors a year and often commission contemporary architects to contribute to the institution, as was the case for the Davies Alpine House (Wilkinson Eyre, architects, 2006), or this more recent bridge by John Pawson. "The **SACKLER CROSSING** is a sculptural serpentine object," explains Pawson, "made from bronze and granite that is there to be looked at for its own beauty, but also to allow visitors to the gardens to experience the landscape from new and unexpected vantage points. Placed very low, the crossing seems to float across the water." John Pawson, in the context both of this historic place and its large public, employed unusually durable materials, a bronze alloy normally used for submarine propellers and large, flat planks of granite.

Die im Südwesten von London gelegenen Royal Botanic Gardens befinden sich auf einem Gelände von insgesamt 121 Hektar. Neben dem weltberühmten Herbarium und den Samenbanken befinden sich in diesem Botanischen Garten einige bemerkenswerte Bauwerke, wie die chinesische Pagode von Sir William Chambers aus dem Jahr 1761 und das von Decimus Burton entworfene Palmenhaus, das zwischen 1844 und 1848 errichtet wurde. Es kommen jährlich etwa zwei Millionen Besucher in die Botanischen Gärten. Die Verwaltung von Kew beauftragt häufig Architekten, einen Beitrag für ihre Institution beizusteuern. Zum Beispiel hat das Architekturbüro Wilkinson Eyre 2006 das Davies Alpine House gebaut und nun John Pawson eine Brücke. „Die Brücke **SACKLER CROSSING** ist ein skulpturales, schlangenlinienförmig gebogenes Objekt", sagt John Pawson. „Sie ist aus Bronze und Granit, und sie ist dazu bestimmt, wegen ihrer eigenen Schönheit betrachtet zu werden und dazu, dass Besucher des Botanischen Gartens die Landschaft aus neuen, unerwarteten Blickwinkeln erfahren können. Die Brücke liegt nur wenige Zentimeter über dem Wasserspiegel und scheint über dem Wasser zu schweben." John Pawson hat in Hinblick auf den historisch bedeutenden Ort und auf die zahlreichen Besucher ungewöhnlich feste Materialien gewählt – eine Bronzelegierung, die normalerweise für U-Boot-Schiffsschrauben verwendet wird, und große flache Granitplatten.

Situés au sud-ouest de Londres, les Royal Botanic Gardens de Kew s'étendent sur 121 ha. En dehors de leurs fameux herbier et banque de semences, ils comptent un certain nombre de constructions d'intérêt historique, dont la pagode chinoise édifiée en 1761 par William Chambers et la Palm House conçue par Decimus Bruton (1844–48). Les jardins reçoivent environ deux millions de visiteurs chaque année et passent régulièrement des commandes à des architectes contemporains, comme ce fut le cas pour la Davies Alpine House (Wilkinson Eyre Architects, 2006) ou pour ce récent petit pont signé John Pawson. « Le **SACKLER CROSSING** est un objet sculptural de forme serpentine fait de bronze et de granit, explique Pawson, et qui est là pour être admiré pour sa beauté intrinsèque, mais aussi pour permettre aux visiteurs des jardins de découvrir le paysage sous des angles nouveaux et inattendus. Positionné très bas, cet ouvrage semble flotter sur l'eau. » Dans le contexte de ces lieux historiques et vu le grand nombre de visiteurs, John Pawson a utilisé des matériaux d'une durabilité exceptionnelle : un alliage de bronze habituellement réservé à la fabrication des hélices de sous-marins et de grandes dalles de granit.

The Sackler Crossing is 70 meters long and three meters wide, yet in these images it appears to almost float above the water.

Die Brücke Sackler Crossing ist 70 Meter lang und drei Meter breit, auf diesen Bildern scheint sie über dem Wasser zu schweben.

Sur ces images, le Sackler Crossing qui mesure 70 mètres de long et trois de large, semble néanmoins flotter sur l'eau.

Above, a drawing shows the very low profile of the structure against the natural background of the Gardens. Despite being made with very durable materials (bronze and granite) the Sackler Crossing seems to be very light.

Die Zeichnung oben zeigt, dass die Brücke vor dem natürlichen Hintergrund des Botanischen Gartens kaum auffällt. Obwohl die Sackler Crossing aus sehr beständigen Materialien gebaut ist (Bronze und Granit) ist sie in ihrem Erscheinungsbild sehr leicht.

Ci-dessus, un dessin montre le profil surbaissé de l'ouvrage dans le contexte naturel des Kew Gardens. Bien que réalisé en matériaux très résistants (bronze et granit), il paraît très léger.

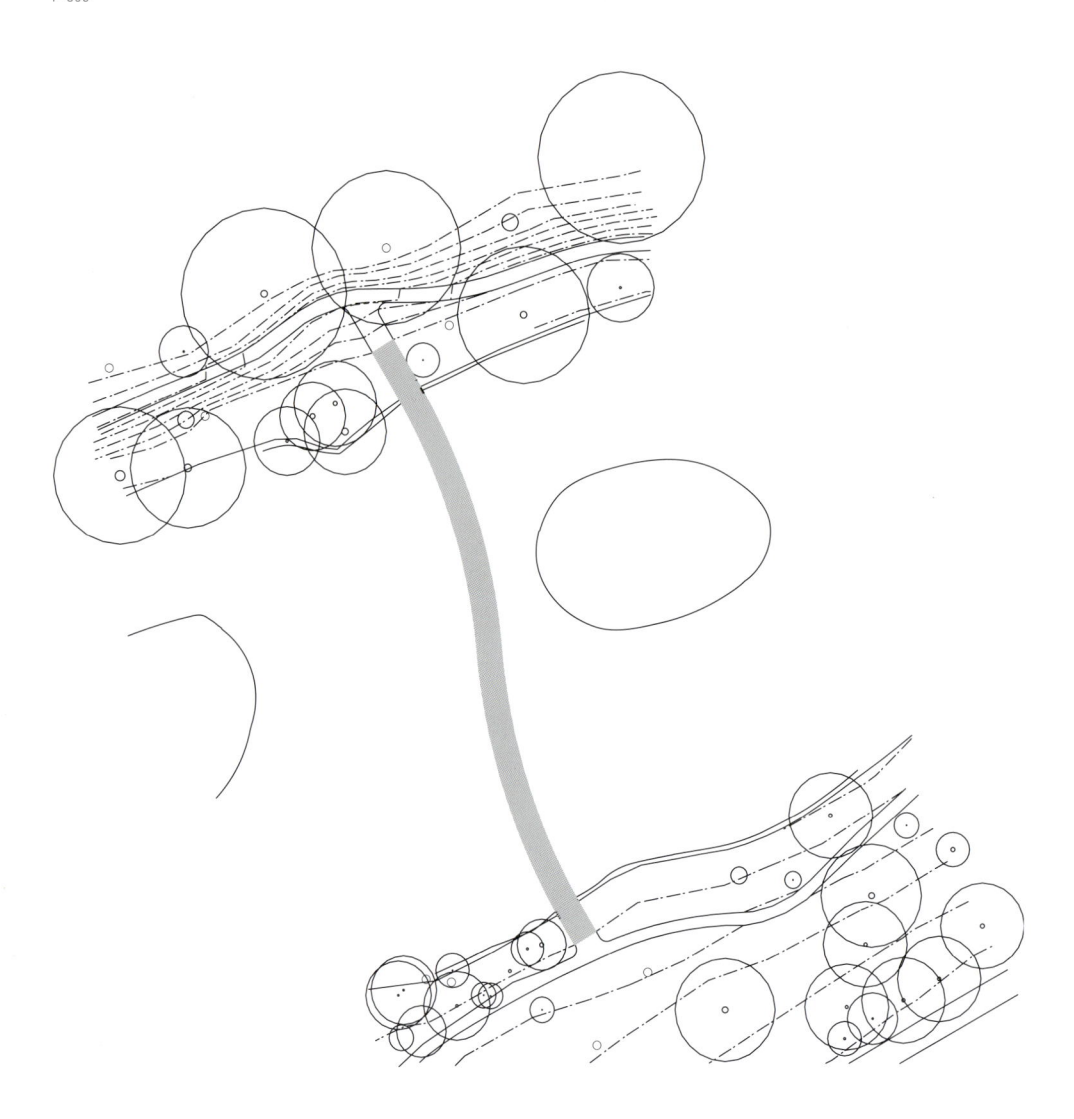

Above, a plan shows the bend in the bridge, while the image on the right page emphasizes its curvature. Right, the bronze fins that make up the actual barriers along the sides of the Sackler Crossing.

Der Grundriss oben zeigt die Krümmung der Brücke, die beim Bild auf der rechten Seite betont ist. Rechts, die Bronzelamellen, die das Geländer der Sackler Crossing bilden.

Ci-dessus, plan montrant la courbure de la passerelle, amplifiée par la photo de la page de droite. À droite, les ailettes de bronze qui constituent les rambardes latérales de l'ouvrage.

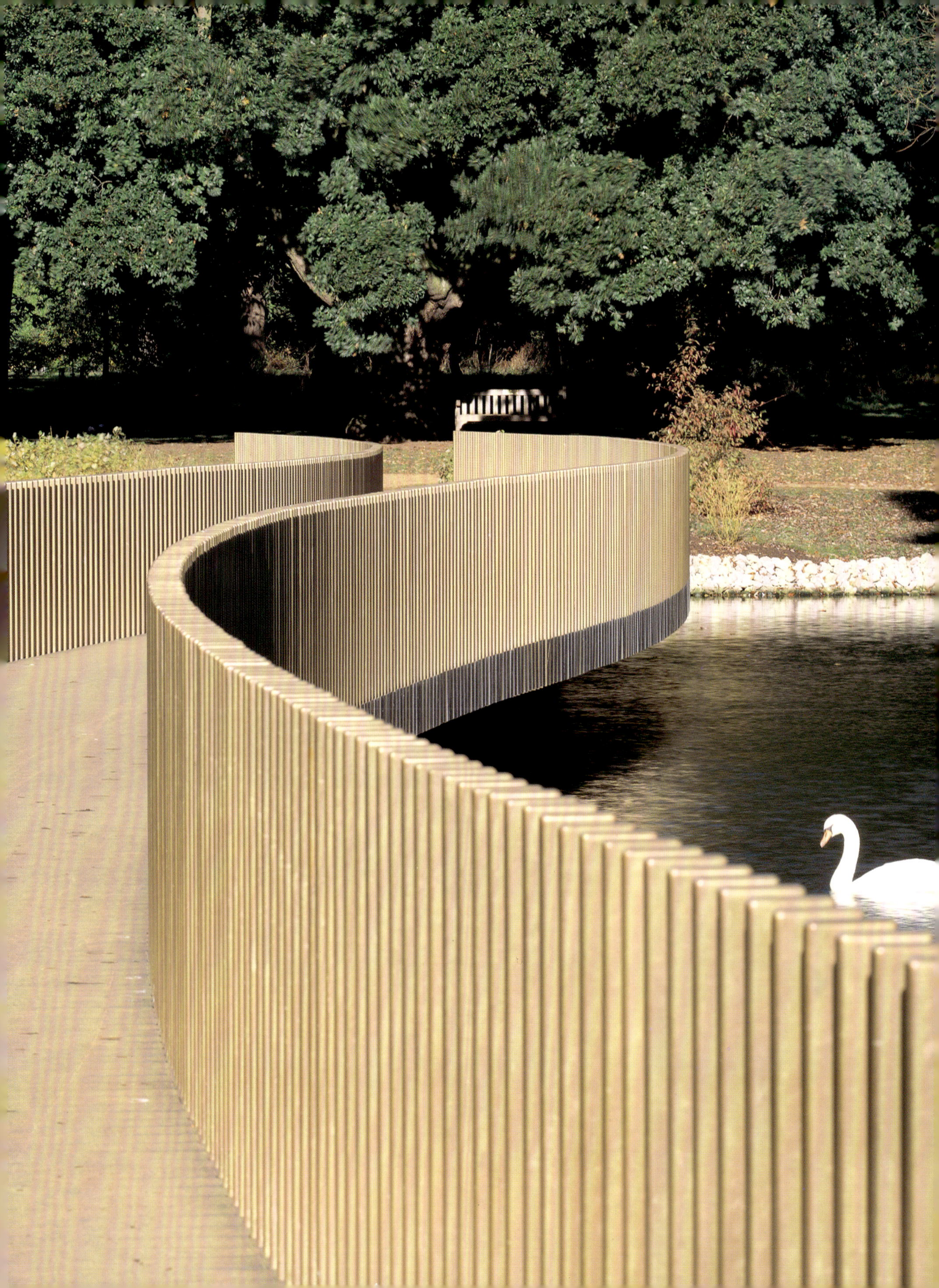

Renovation and Expansion of the
California Academy of Science

RENZO PIANO

Renzo Piano Building Workshop
Via P. Paolo Rubens 29, 16158 Genoa, Italy

Tel: +39 010 617 11 / Fax: +39 010 617 13 50
E-mail: italy@rpbw.com / Web: www.rpbw.com

RENZO PIANO was born in 1937 in Genoa, Italy. He studied at the University of Florence and at Milan's Polytechnic Institute (1964). He formed his own practice (Studio Piano) in 1965, associated with Richard Rogers (Piano & Rogers, 1971–78)—completing the Pompidou Center in Paris in 1977—and then worked with Peter Rice (Piano & Rice Associates, 1978–80), before creating the Renzo Piano Building Workshop in 1981 in Genoa and Paris. Piano received the RIBA Gold Medal in 1989. Built work after 2000 includes Maison Hermès (Tokyo, Japan, 1998–2001); Rome Auditorium (Italy, 1994–2002); conversion of the Lingotto Factory Complex (Turin, Italy, 1983–2003); the Padre Pio Pilgrimage Church (San Giovanni Rotondo, Foggia, Italy, 1991–2004); the Woodruff Arts Center Expansion (Atlanta, Georgia, USA, 1999–2005); the renovation and expansion of the Morgan Library (New York, New York, USA, 2000–06); and the New York Times Building (New York, New York, USA, 2005–07). Recently completed work includes the Broad Contemporary Art Museum (Phase 1 of the LACMA expansion, Los Angeles, California, USA, 2003–08); the California Academy of Sciences (San Francisco, California, USA, 2005–08, published here); the Modern Wing of the Art Institute of Chicago (Chicago, Illinois, USA, 2005–09); Saint Giles Court mixed-use development (London, UK, 2002–10); the Resnick Pavilion (Phase 2 of the LACMA expansion, Los Angeles, 2006–10); and the Poor Clare Monastery at Ronchamp (France, 2006–11). Ongoing work includes London Bridge Tower (London, UK, 2000–); the Stavros Niarchos Foundation Cultural Center (Athens, Greece, 2008–); Valletta City Gate (Valletta, Malta, 2008–); and the Botin Art Center (Santander, Spain, 2010–).

RENZO PIANO wurde 1937 in Genua, Italien, geboren. Er studierte bis 1964 an der Universität Florenz und am Polytechnikum in Mailand. 1965 gründete er sein eigenes Büro (Studio Piano), von 1971 bis 1978 leitete er mit Richard Rogers das Büro Piano & Rogers, das 1977 das Centre Pompidou in Paris fertigstellte. Von 1978 bis 1980 arbeitete Piano mit Peter Rice (Piano & Rice Associates), bevor er 1981 in Genua und Paris den Renzo Piano Building Workshop gründete. Renzo Piano erhielt 1989 die Goldmedaille des RIBA. Zu seinen nach 2000 entstandenen Werken gehören die Maison Hermès in Tokio (1998–2001), das Auditorium in Rom (1994–2002), der Umbau des Fabrikgebäudes Lingotto in Turin, Italien (1983–2003), die Pilgerkirche Padre Pio in San Giovanni Rotondo, Foggia, Italien (1991–2004), der Erweiterungsbau des Woodruff Arts Center in Atlanta, Georgia, USA (1999–2005), die Renovierung und Erweiterung der Morgan Library in New York (2000–06) und das New York Times Building in New York (2005–07). In letzter Zeit fertiggestellt wurden das Broad Contemporary Art Museum (1. Bauabschnitt der Erweiterung des LACMA in Los Angeles (2003–08), die California Academy of Sciences in San Francisco (Kalifornien, 2005–08, hier vorgestellt), der Modern Wing des Art Institute of Chicago (Illinois, USA, 2005–09), die Neubebauung mit Mischnutzung im Stadtteil Saint Giles Court (London, 2002–10), der Resnick Pavillon (2. Bauabschnitt der Erweiterung des LACMA in Los Angeles, 2006–10) und das Klarissenkloster in Ronchamp, Frankreich (2006–11). Momentan im Bau befinden sich das Hochhaus London Bridge Tower (seit 2000), das Kulturzentrum der Stavros Niarchos Foundation (Athen, Griechenland, seit 2008), das Valletta City Gate (Valletta, Malta, seit 2008) sowie das Kunstzentrum der Stiftung Botín in Santander, Spanien (seit 2010).

RENZO PIANO, né en 1937 à Gênes (Italie), a étudié à l'université de Florence et à l'Institut polytechnique de Milan (1964). Il crée son agence Studio Piano en 1965, puis s'associe à Richard Rogers (Piano & Rogers, 1971–78) et réalise le Centre Pompidou à Paris en 1977. Il collabore avec Peter Rice (Piano & Rice Associates, 1978–80) avant de fonder le Renzo Piano Building Workshop en 1981 à Gênes et à Paris. Il a reçu la médaille d'or du RIBA en 1989. Parmi ses réalisations après 2000 : la Maison Hermès (Tokyo, 1998–2001); l'auditorium Parco della Musica (Rome, 1994–2002); la reconversion du site industriel du Lingotto (Turin, Italie, 1983–2003); l'église de pèlerinage Padre Pio (San Giovanni Rotondo, Foggia, Italie, 1991–2004); l'extension du Centre d'art Woodruff (Atlanta, Géorgie, US, 1999–2005); la rénovation et l'agrandissement de la Morgan Library (New York, 2000–06) et le New York Times Building (New York, 2005–07). Parmi ses œuvres récentes : le Broad Contemporary Art Museum (Phase 1 de l'extension du LACMA, Los Angeles, 2003–08); la California Academy of Sciences (San Francisco, US, 2005–08, publiée ici); l'aile moderne de l'Art Institute of Chicago (Illinois, US, 2005–09); Saint Giles Court, immeubles mixtes (Londres, 2002–10); le Resnick Pavilion (Phase 2 de l'extension du LACMA, Los Angeles, 2006–10) et le monastère Sainte-Claire à Ronchamp (France, 2006–11). Il travaille actuellement au projet de la London Bridge Tower (Londres, 2000–); du Centre culturel de la Fondation Stavros Niarchos (Athènes, 2008–); de la porte de la ville de La Valette (Malte, 2008–) et le Centre d'art Botín (Santander, Espagne, 2010–).

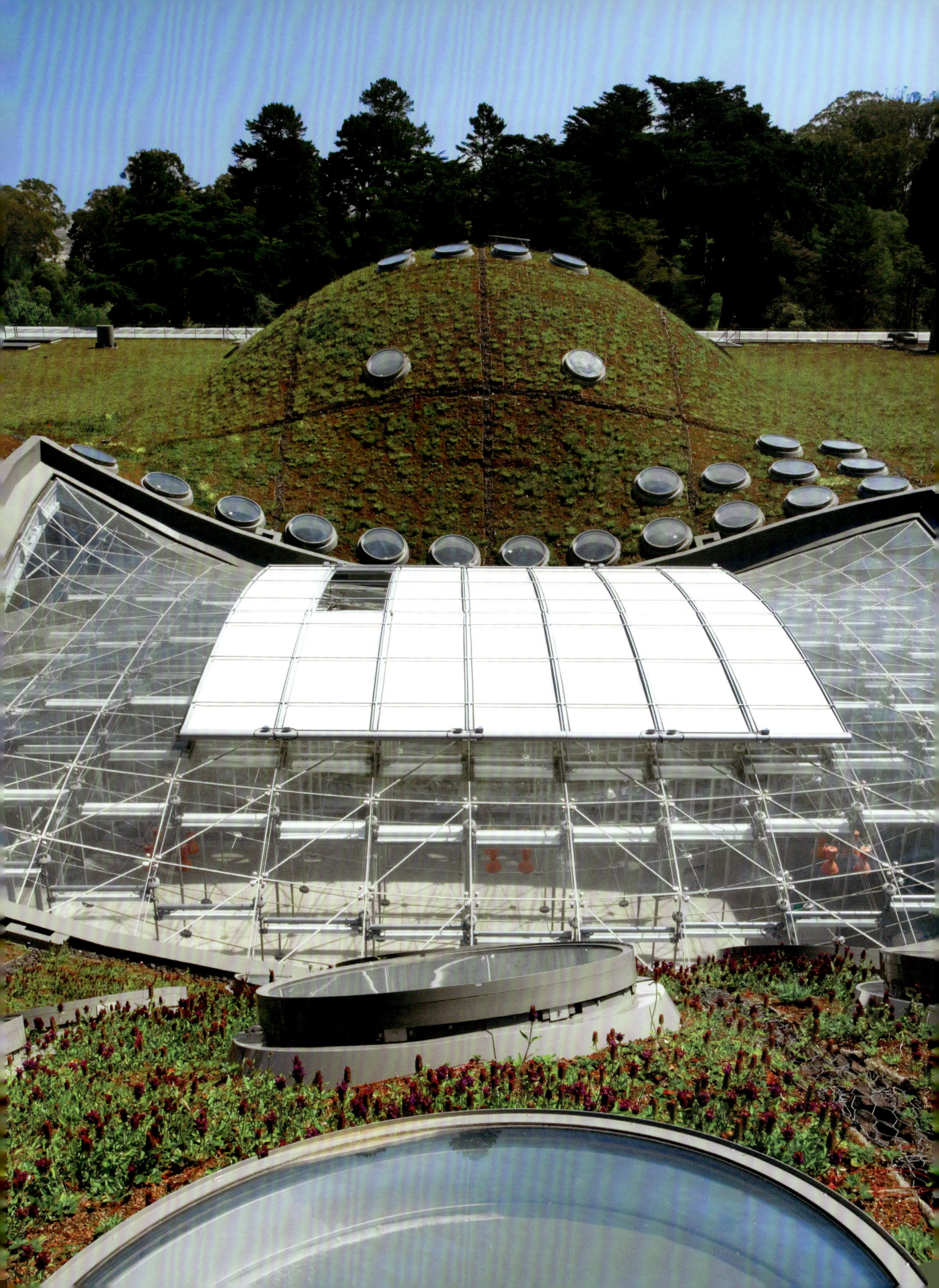

RENOVATION AND EXPANSION OF THE CALIFORNIA ACADEMY OF SCIENCES
San Francisco, California, USA, 2005–08

Address: 55 Music Concourse Drive, Golden Gate Park, San Francisco, CA 94118, USA, +1 415 379 8000, www.calacademy.org. Area: 74 322 m^2 (site), 34 374 m^2 (floor area). Client: California Academy of Sciences Cost: $370 million, including exhibition programand costs associated with the Academy's temporary housing Collaboration: Gordon H. Chung and Partners, San Francisco

The green roof of the California Academy of Sciences is one of the most outstanding features of the architecture, making up an artificial landscape of considerable proportions. Skylights admit overhead daylight into the building.

Das grüne Dach der California Academy of Sciences ist eines der auffälligsten Merkmale dieser Architektur. Es ist eine recht große künstliche Dachlandschaft. Über die Oberlichter gelangt Tageslicht in das Innere des Gebäudes.

La toiture végétalisée de la California Academy of Sciences est l'un des éléments les plus remarquables de ce projet. Elle constitue en soi un paysage artificiel de dimensions considérables. Des verrières assurent l'éclairage diurne de l'intérieur du bâtiment.

One of the 10 largest natural history museums in the world, the **CALIFORNIA ACADEMY OF SCIENCES** was founded in 1853. The institution declared: "The new CAS will be at the forefront of green building design, showcasing world-class architecture that fully integrates green building features to reflect its mission to protect the natural world." The completed structure has a LEED Platinum rating reflecting its strategies to conserve energy and to use environmentally friendly building materials. The undulating roof of the structure, with a surface of over one hectare, is covered with 1.8 million native California plants. Careful study of the plants themselves, but also of the seismic implications of a planted roof, was part of the preparation of this aspect of the design that is open to visitors. It is calculated that the design of the roof reduces temperatures inside the museum by about 6°C. A rainwater collection system is designed to store and reuse about 13 500 cubic meters of water each year, reused for irrigation and gray water. The roof's shape and, indeed, the entire design of the museum were conceived to form a continuum with the surrounding park environment. Intended for schoolchildren and the general public, the Academy focuses on education and research on conserving natural environments and habitats.

Die 1853 gegründete **CALIFORNIA ACADEMY OF SCIENCES** gehört zu den zehn größten Naturhistorischen Museen der Welt. Die Museumsleitung erkärt: „Die neue CAS wird an vorderster Stelle der nach ‚grünen' Maßstäben entworfenen Gebäude stehen, mit einer Weltklassearchitektur, die alle Merkmale des nachhaltigen Bauens in sich vereint und dadurch der Aufgabe des CAS entspricht, die Natur zu schützen." Das gesamte Gebäude ist wegen der Energiesparmaßnahmen und den umweltfreundlichen Baumaterialien nach der US-Klassifizierung für ökologisches Bauen, LEED, mit der höchsten Kategorie „Platinum" zertifiziert. Das gewellte Dach mit einer Fläche von etwa einem Hektar ist mit 1,8 Millionen in Kalifornien heimischen Pflanzen bedeckt und öffentlich zugänglich. Sorgfältige Pflanzenstudien, aber auch Untersuchungen bezüglich der Auswirkungen von Erdbeben auf das begrünte Dach wurden im Vorfeld durchgeführt. Laut den Berechnungen mindert die Dachgestaltung die Innenraumtemperaturen des Museums um etwa 6 °C. Dank eines Regenwassersammelsystems können etwa 13 500 Kubikmeter Grauwasser pro Jahr für Bewässerung wiederverwendet werden. Der Museumsentwurf einschließlich des Dachs bildet ein nahtloses Kontinuum mit der umliegenden Parklandschaft. Die Institution richtet sich sowohl an Schüler als auch das allgemeine Publikum und beschäftigt sich mit Bildung und Forschung über den Erhalt der natürlichen Umwelt und der Habitate.

L'un des dix plus grands musées d'histoire naturelle au monde, la **CALIFORNIA ACADEMY OF SCIENCES** date de 1853. Orientée vers le grand public et le public scolaire, elle se consacre à l'éducation et à la recherche sur la conservation des environnements et habitats naturels. Lors de la présentation du projet, l'institution a déclaré : « La nouvelle CAS sera à l'avant-garde de la construction écologique, la vitrine d'une architecture de niveau international, qui intègre pleinement les exigences de la construction durable dans sa mission de protéger le monde naturel. » Le bâtiment a obtenu la classification LEED Platine qui récompense les stratégies mises en œuvre pour conserver l'énergie et utiliser des matériaux de construction durables. Ouverte aux visiteurs, la toiture à ondulations de plus de un hectare a été plantée de plus de 1,8 million d'espèces californiennes. L'étude de ces végétaux, mais aussi des implications du poids élevé de la toiture en cas de tremblement de terre ont joué un rôle important. On a calculé que ce type de toit réduit la température à l'intérieur du musée de 6 °C environ. Un système de collecte des eaux de pluie permet de stocker et de réutiliser environ 13 500 m³ d'eau par an pour l'irrigation et les eaux sanitaires. La forme du toit et l'ensemble du musée forment un continuum avec le parc environnant.

RENZO PIANO

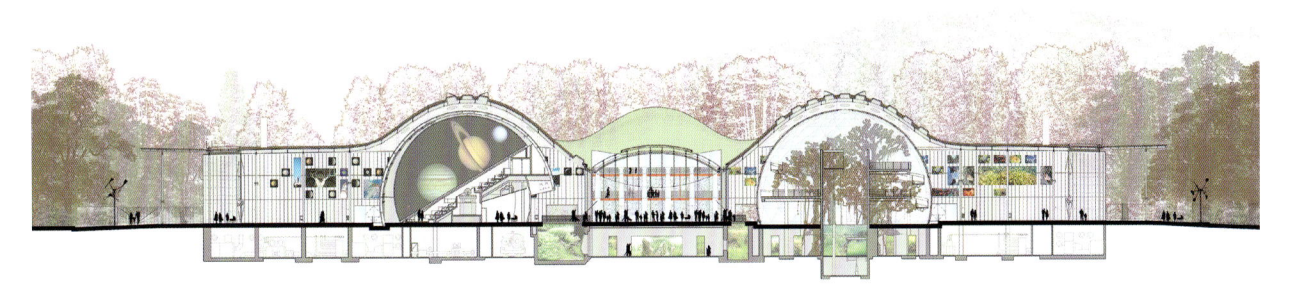

The building is itself located in a park and, as can be seen in the image and the elevation drawing above, greenery in various forms is present above and within the structure.

Das Gebäude liegt in einem Park, wie man auf dem Bild und der Fassadenabwicklung oben erkennen kann. Grün ist in verschiedenen Formen auf und im Gebäude vorhanden.

Le bâtiment est situé dans un parc, comme le montrent les images et le dessin de coupe ci-dessus. La verdure est présente sous toutes ses formes à l'intérieur comme alentour.

"Flowing Gardens," Xi'an World
Horticultural Fair 2011

PLASMA STUDIO

Plasma Studio
Unit 51 – Regents Studios
8 Andrews Road
London E8 4QN
UK

Tel: +44 20 78 12 98 75 / Fax: +44 87 04 86 55 63
E-mail: mail@plasmastudio.com
Web: www.plasmastudio.com

PLASMA STUDIO was founded by Eva Castro and Holger Kehne in London (UK) in 1999. Eva Castro studied architecture and urbanism at the Universidad Central de Venezuela and subsequently completed the Graduate Design program under Jeff Kipnis at the Architectural Association (AA) in London. She is Director of the AA Landscape Urbanism Program and Unit Master for its Diploma Unit 12. Holger Kehne studied architecture at the University of Applied Sciences in Münster, Germany, and at the University of East London. He is also a Unit Master for the Diploma Unit 12 at the AA. The office made its reputation through a number of small residential and refurbishment projects in London. The architects say: "The studio is best known for its architectural use of form and geometry. Shifts, folds, and bends create surface continuities that are never arbitrary but part of the spatial and structural organization." They won the Corus / Building Design "Young Architect of the Year Award" in 2002. They participated in the Hotel Puerta America project with architects like Jean Nouvel and Zaha Hadid (Madrid, Spain, 2005), and their recent work includes Esker House (San Candido, 2006); Tetris House, a multi-family residential compound (San Candido, 2007); Strata Hotel (Alto Adige, 2007); and Cube House (Sexten, 2005–08), all in Italy. They have recently completed "Flowing Gardens" (Xi'an World Horticultural Fair 2011, Xi'an, China, 2009–11, published here), while current projects include Loft Town (Chengdu, China); and Ordos 20+10 (Ordos, Inner Mongolia).

PLASMA STUDIO wurde 1999 von Eva Castro und Holger Kehne in London gegründet. Eva Castro studierte Architektur und Stadtplanung an der Universidad Central de Venezuela und graduierte im Fach Design bei Jeff Kipnis an der Architectural Association (AA) in London. Sie ist Leiterin des Studienfachs Landschaft und Stadtplanung an der AA und Unit Master für den Fachbereich Diploma Unit 12. Holger Kehne studierte Architektur an der Fachhochschule Münster und an der University of East London. Er ist ebenfalls Unit Master für den Fachbereich Diploma Unit 12 an der AA. Das Büro wurde durch eine Reihe kleinerer Wohnungsbau- und Sanierungsprojekte in London bekannt. Die Architekten sagen über ihr Büro: „Wir sind wegen der Art, wie unser Büro Form und Geometrie in Architektur umsetzt, bekannt. Schichten, Falten und Krümmungen schaffen kontinuierliche Oberflächen, die nie willkürlich, sondern immer ein Teil der räumlichen und konstruktiven Organisation sind." 2002 gewann das Büro den „Young Architect of the Year Award". Sie haben neben Architekten wie Jean Nouvel und Zaha Hadid an dem Projekt des Hotels Puerta America in Madrid (2005) teilgenommen. Zu ihren neueren Gebäuden in Italien gehören das Apartmenthaus Esker in San Candido (2006), das Mehrfamilienhaus Tetris in San Candido (2007), das Strata Hotel (Alto Adige, 2007) und das Haus Cube in Sexten (2005–08). Das neueste Projekt sind die „Flowing Gardens" (auf der Internationalen Gartenbauausstellung 2011 in Xi'an, China, hier vorgestellt). Zu den laufenden Projekten gehören die Loft Town in Chengdu (China) und Ordos 20+10 (Ordos, Innere Mongolei).

L'agence **PLASMA STUDIO** a été fondée par Eva Castro et Holger Kehne à Londres en 1999. Eva Castro a étudié l'architecture et l'urbanisme à l'Universidad Central du Venezuela, puis à l'Architectural Association (AA) à Londres (Graduate Design Program, sous la direction de Jeff Kipnis). Elle est directrice du Landscape Urbanism Program à l'AA et maître d'unité pour l'Unité 12 du diplôme. Holger Kehne a étudié l'architecture à l'Université des sciences appliquées de Münster (Allemagne) et à l'université de Londres-Est. Il est également maître d'unité pour l'Unité 12 du diplôme à l'AA. Plasma s'est fait remarquer par un certain nombre de petits projets résidentiels et de rénovation à Londres. « L'agence est particulièrement connue pour son utilisation architecturale des formes et de la géométrie. Glissements, plis et courbures créent des continuités de surface qui ne sont jamais arbitraires, mais font partie de l'organisation spatiale et structurelle », expliquent les architectes. Ils ont remporté le prix Corus/Building Design du jeune architecte de l'année en 2002 et ont participé au projet de l'hôtel Puerta America aux côtés d'architectes comme Jean Nouvel et Zaha Hadid (Madrid, 2005). Parmi leurs réalisations récentes, toutes en Italie : la maison Esker (San Candido, 2006) ; la maison Tetris, complexe résidentiel plurifamilial (San Candido, 2007) ; l'hôtel Strata (Haut-Adige, 2007) et la maison Cube (Sesto, 2005–08). Ils ont récemment achevé les « Flowing Gardens » (Exposition internationale d'horticulture de Xi'an 2011, Xi'an, Chine, 2009–11, publiés ici) et ils travaillent actuellement sur la Loft Town (Chengdu, Chine) et Ordos 20+10 (Ordos, Mongolie-Intérieure).

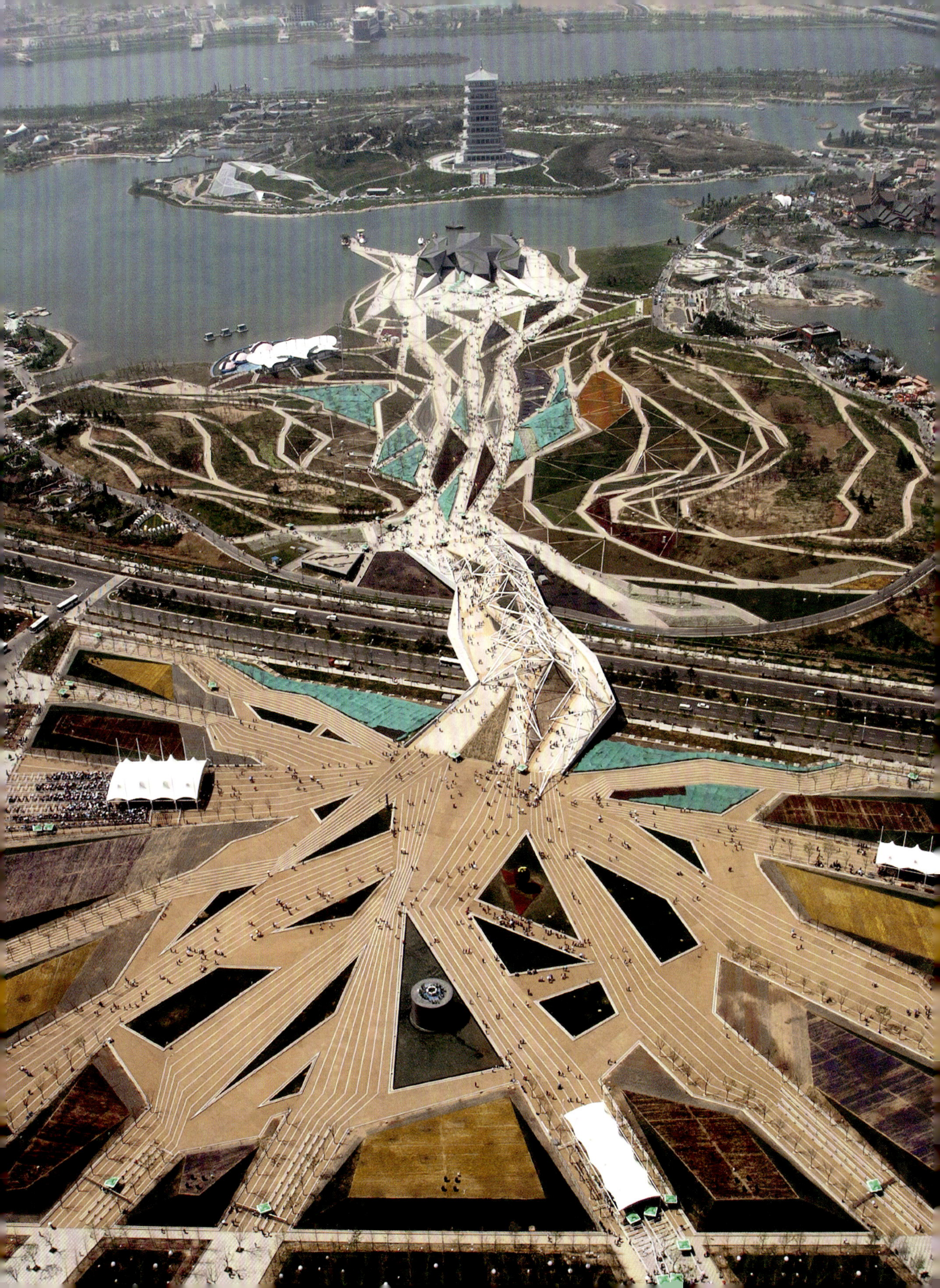

"FLOWING GARDENS," XI'AN WORLD HORTICULTURAL FAIR 2011

Xi'an, China, 2009-11

Address: Chan-Ba Ecological District, Xi'an 710024, China, +86 29 8359 6533, http://en.expo2011.cn/index.html. Area: 37 hectares (landscaped), 19 315 m² (buildings)
Client: Xi'an International Horticultural Exposition Investment (Group) Co., Ltd. Cost: not disclosed
Collaboration: Ulla Hell (Architect), Arup (Structural and Civil Engineering),
Groundlab (Landscape Design), Xiaowei Tong, Dongyun Liu

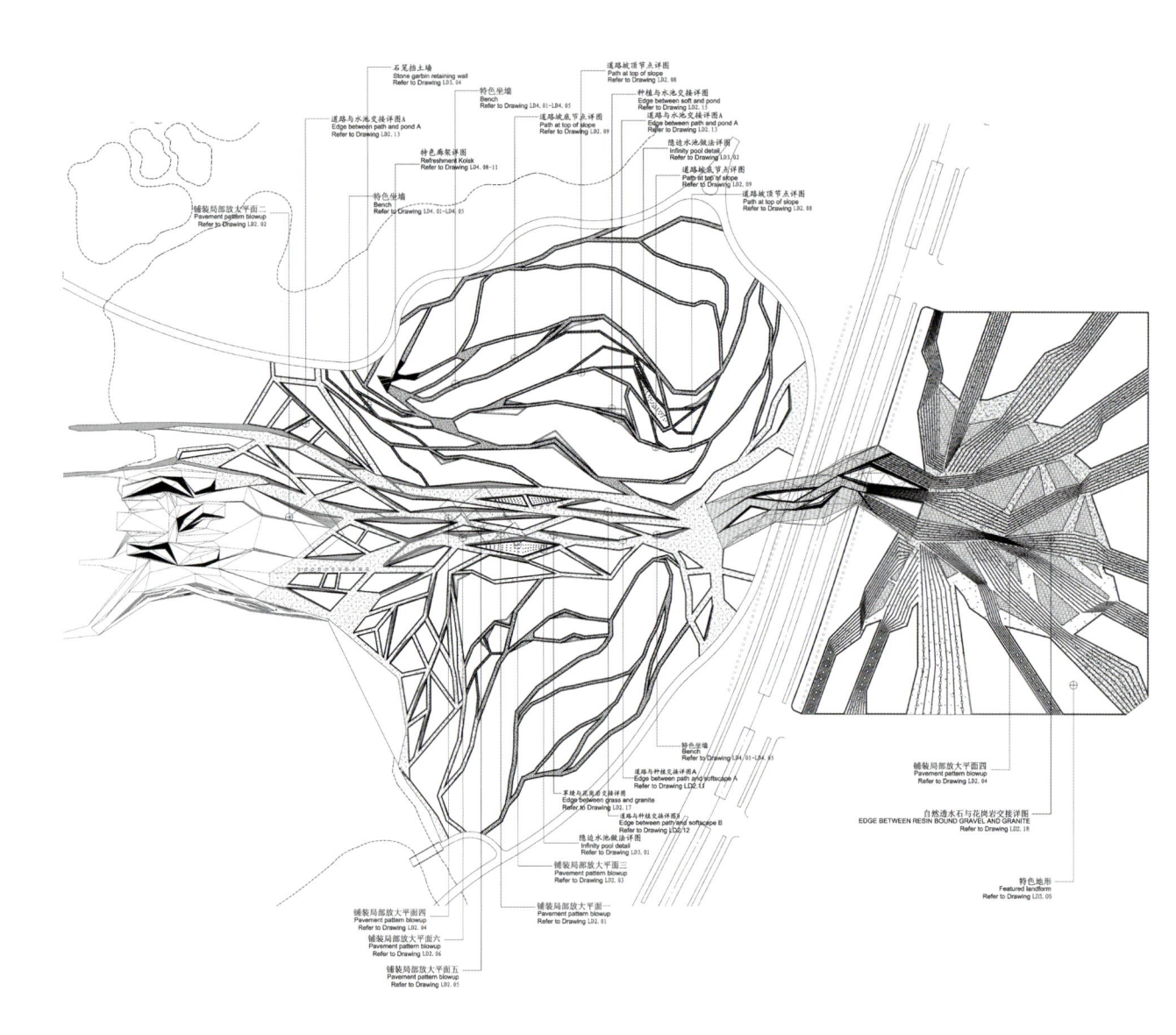

A plan (below) shows the intersecting paths that make up the organic design of the complex, leading to the main structure of the fairgrounds.

Der Plan unten zeigt die sich kreuzenden Wege, die die Grundlage des organischen Entwurfs sind und zum Hauptgebäude der Ausstellung führen

Le plan ci-dessous montre les intersections des allées qui donnent au projet sa forme et mènent vers le bâtiment principal de l'exposition.

PLASMA STUDIO

Plasma Studio has created not only an architectural environment, but also carried out the landscape design which is a continuation of the buildings in many senses.

Plasma Studio hat nicht nur ein architektonisches Umfeld geschaffen, sondern auch die Landschaftsplanung ausgeführt, die in vieler Hinsicht eine Fortsetzung der Gebäude darstellt.

Plasma Studio a non seulement créé un environnement architectural, mais a aussi réalisé un paysage, prolongement du bâtiment à plusieurs égards.

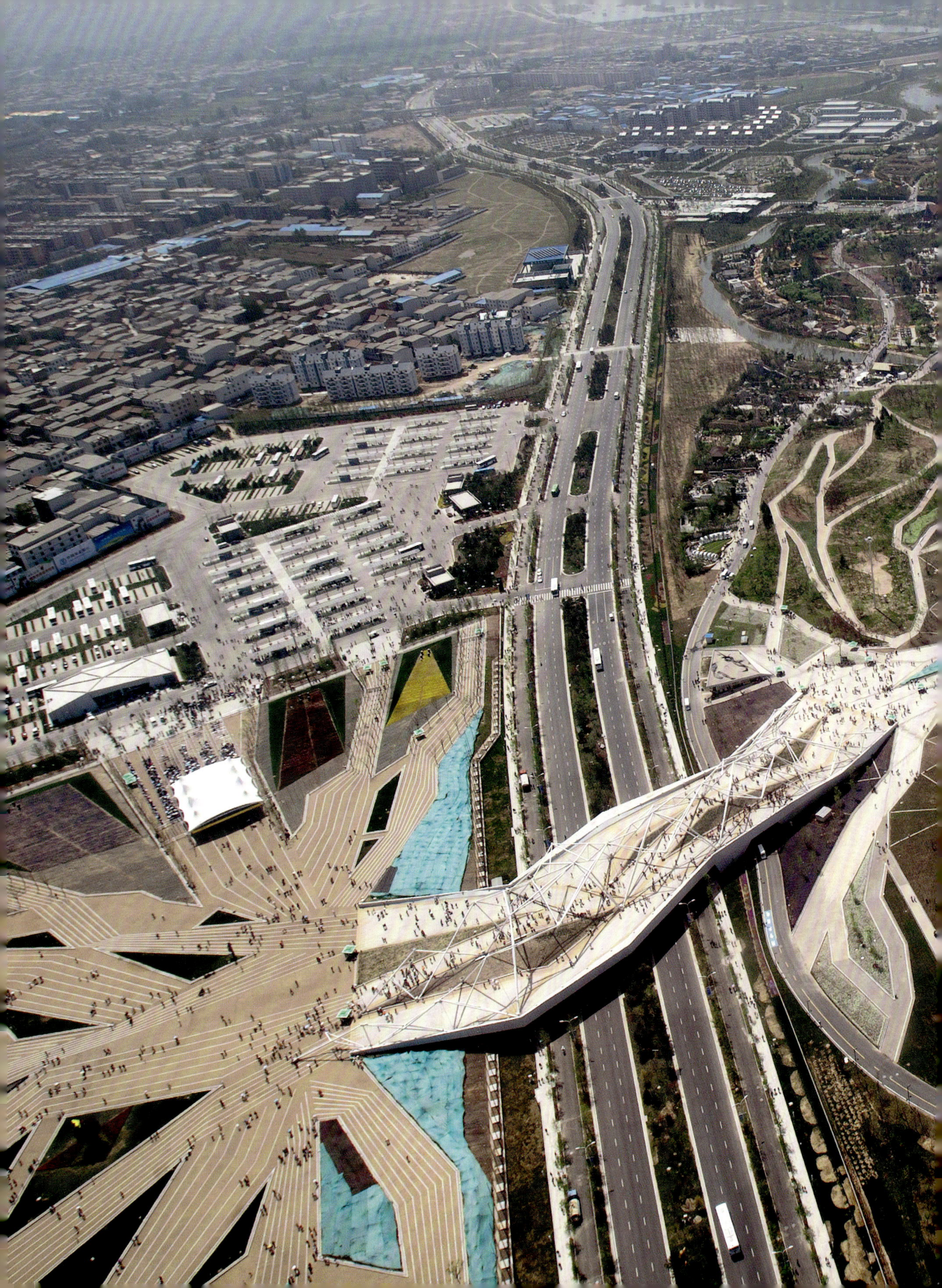

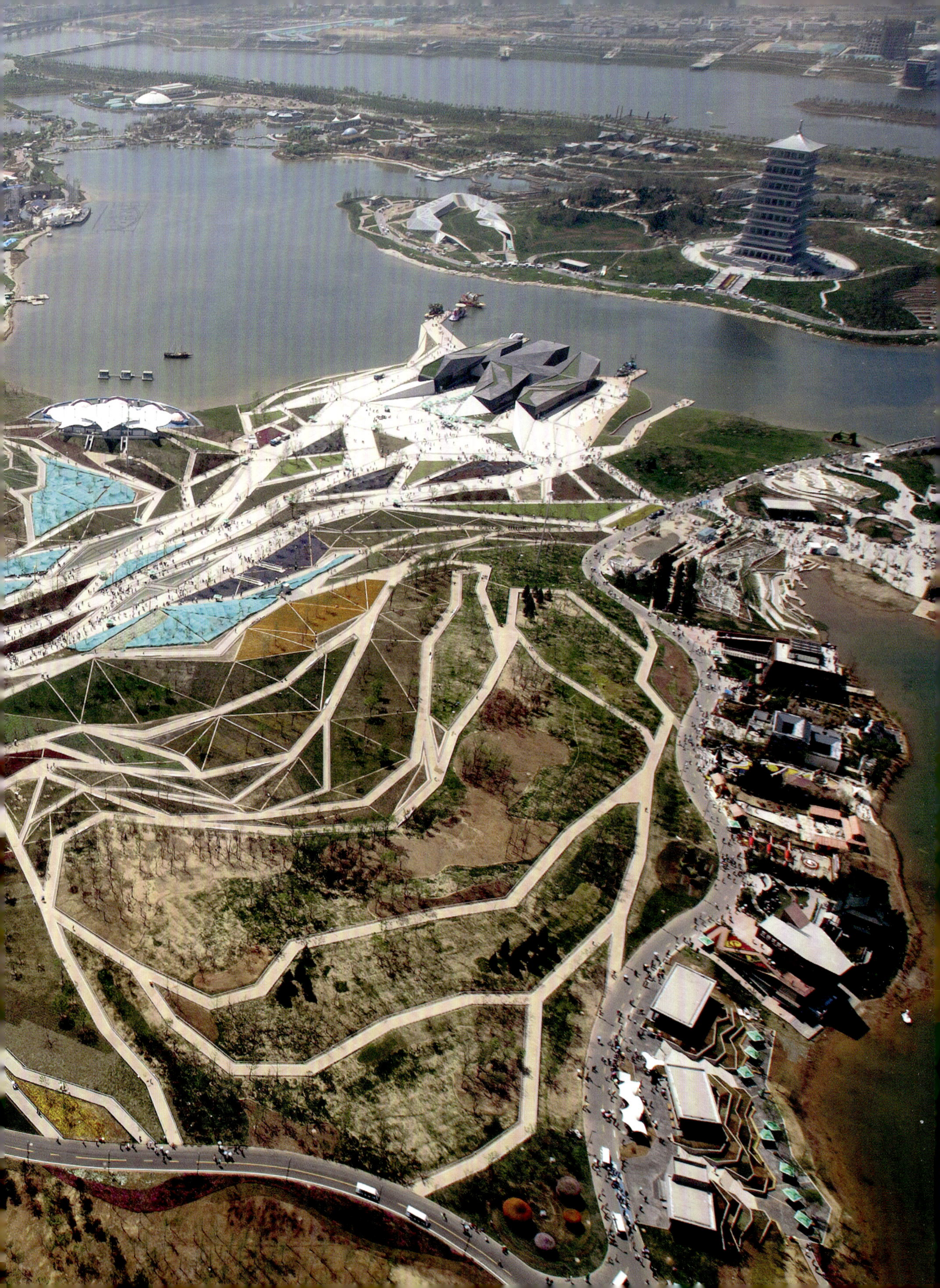

This project includes a 5000-square-meter Creativity Center, a 4000-square-meter Greenhouse, "formed like a precious crystal," the 3500-square-meter Guangyun Entrance Bridge which crosses the main road that bisects the site, and 37 hectares of landscaped areas. The architects state: "The project, titled **'FLOWING GARDENS'**, was generated as a synthesis of horticulture and technology where landscape and architecture converge at a sustainable and integral vision." A complex pattern of paths, inspired by natural forms, clearly links the landscape design and the architecture into a single entity. The Horticultural Fair, opened at the end of April 2011, was due to attract more than 20 million visitors in the first six months of operation. The overall plan of the complex is "strikingly similar to an estuary." The architects employ "flowing forms" to create an obvious continuity between the gardens and the buildings, between landscape and architecture. Located at the top of the South Hill, the Greenhouse allows visitors to see the "Flowing Gardens" and at the same time to view plants and flowers from four different climatic zones.

Zu diesem Projekt gehören ein 5000 Quadratmeter großes Kreativitätszentrum, ein 4000 Quadratmeter großes Gewächshaus, das „wie ein kostbarer Kristall geformt ist", die 3500 Quadratmeter große Guangyun-Eingangsbrücke, die über die Hauptstraße hinwegführt, die das Ausstellungsgelände zweiteilt, und 37 Hektar landschaftlich gestaltetes Gelände. Die Architekten kommentieren wie folgt: „Das Projekt ‚**FLOWING GARDENS**' ist als Synthese von Gartenkunst und Technologie entstanden, bei der Landschaft und Architektur als nachhaltige und ganzheitliche Vision ineinanderfließen." Ein komplexes Wegenetz, das von Naturformen inspiriert ist, verbindet den landschaftsplanerischen Entwurf und die Architektur zu einer Einheit. Auf der Internationalen Gartenbauausstellung, die Ende April 2011 eröffnet wurde, wurden mehr als 20 Millionen Besucher in den ersten sechs Monaten erwartet. Der Grundriss des Gesamtkomplexes erinnert sehr stark an eine Flussmündung. Die Architekten setzen „fließende Formen" ein, um eine sichtbare Kontinuität zwischen den Gartenanlagen und den Gebäuden herzustellen, zwischen der Landschaft und der Architektur. Vom Gewächshaus, das auf der Kuppe des Südhügels liegt, können die Besucher die „Flowing Gardens" von oben betrachten, und zugleich werden ihnen Pflanzen und Blumen aus vier unterschiedlichen Klimazonen vorgestellt.

Ce projet comprend un « centre de créativité » de 5000 m², une serre de 4000 m² « en forme de cristal précieux », le pont d'entrée Guangyun de 3500 m² qui franchit la route, coupant le site en deux parties, et 37 ha de terrain paysager. « Le projet, intitulé **"FLOWING GARDENS"**, est une synthèse d'horticulture et de technologie dans laquelle le paysage et l'architecture convergent vers une vision intégrante et durable », expliquent les architectes. Un réseau complexe d'allées, inspiré de formes naturelles, unifie les aménagements paysagers et les bâtiments en une seule entité. L'exposition d'horticulture, qui a ouvert ses portes fin avril 2011, devait attirer pas moins de 20 millions de visiteurs au cours des six premiers mois. Le plan d'ensemble du complexe est « étonnamment similaire à celui d'un estuaire ». Les architectes ont utilisé des « formes fluides » pour créer une continuité évidente entre les jardins et les bâtiments. Implantée au sommet de la colline sud, la serre permet aux visiteurs d'avoir une vue d'ensemble des jardins tout en découvrant des plantes et des fleurs de quatre zones climatiques différentes.

The cantilevered and folded forms of the architecture of the fairgrounds is very much in the spirit of today's "avant-garde" in architecture.

Die auskragenden, abgeknickten Bauten auf dem Ausstellungsgelände sind Beispiele für die heutige „Avantgarde" in der Architektur.

Les volumes pliés et en porte-à-faux du bâtiment du site de l'exposition sont bien dans l'esprit de l'architecture d'« avant garde » actuelle.

PLASMA STUDIO

The new architecture contrasts with the traditional forms of a neighboring pagoda (above). Below, elevations show the low, sloped, and angled shapes of the main pavilion.

Die neue Architektur steht in Kontrast zur traditionellen Form der Pagode nebenan (oben). Die Zeichnungen unten zeigen die niedrigen, leicht ansteigenden und abgeknickten Umrisse des Hauptpavillons.

L'architecture nouvelle contraste avec les formes traditionnelles d'une pagode voisine (ci-dessus). Ci-dessous, croquis du pavillon principal et de ses formes surbaissées, inclinées et plongeantes.

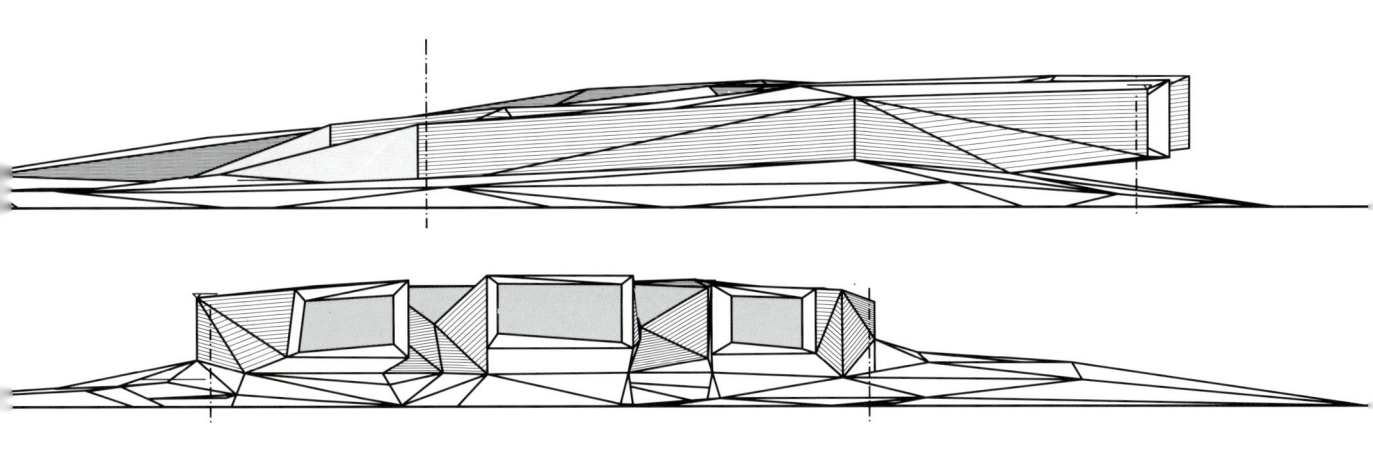

The greenhouses confirm the angled and flowing design of the rest of the grounds. Few examples of such sophisticated architecture have been realized in the context of projects destined to the general public.

Die Gewächshäuser verstärken den verwinkelten und fließenden Entwurf des übrigen Geländes. Es gibt nur wenige Beispiele, wo eine so anspruchsvolle Architektur im Zusammenhang mit Projekten für die breite Öffentlichkeit entstanden ist.

Les serres reprennent les formes fluctuantes et brisées du plan général. Peu d'exemples d'architecture aussi sophistiquée ont été réalisés pour des projets destinés au grand public.

The greenhouses (also seen left) are integrated into an overall plan that both flows and assumes sharp angles, a kind of cross between natural and geometric forms that is unexpected and quite successful here.

Die Gewächshäuser (siehe auch links) sind in einen Gesamtplan eingebunden, der sowohl fließende Linien als auch spitze Winkel in sich vereint. Der Entwurf ist eine Mischung aus natürlichen und geometrischen Formen, die überraschend und sehr gelungen umgesetzt sind.

Les serres (également page de gauche) sont intégrées dans le plan d'ensemble, lui aussi traité en un flux modifié par des ruptures anguleuses, en une sorte de mariage de formes géométriques et naturelles particulièrement réussi.

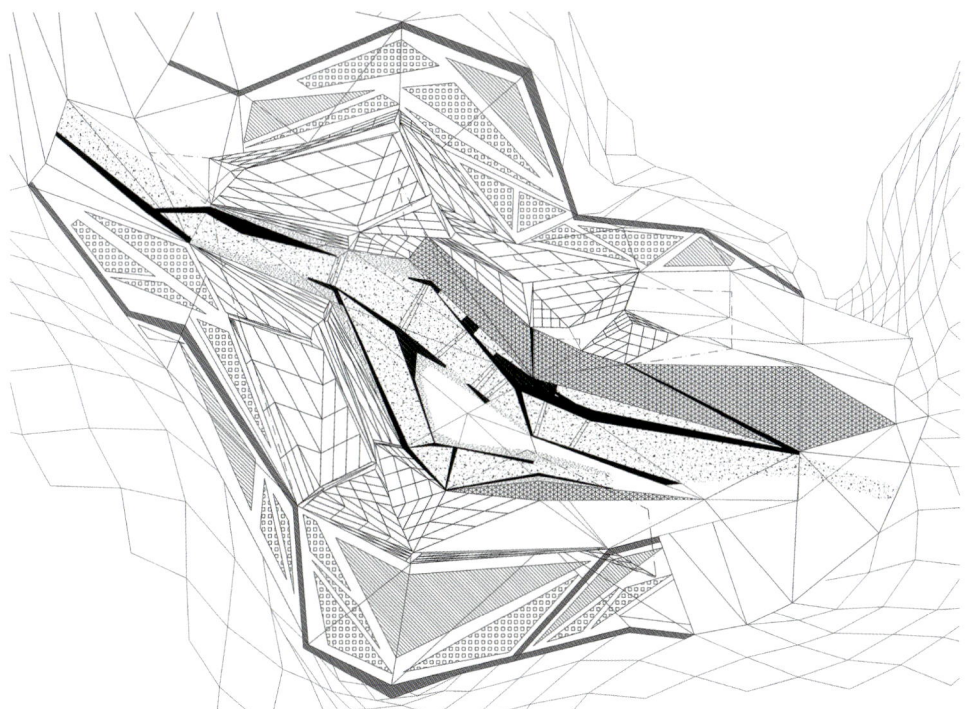

On this double page, images of the Guangyun Entrance to the complex show the innovative tensegrity structure designed with Arup "that appears as beams seemingly free-floating in space," according to the architects.

Auf dieser Doppelseite zeigen die Bilder vom Guangyun-Eingang die innovative Tensegrity-Konstruktion, die mit dem Büro Arup entworfen wurde und laut Aussage der Architekten „wie scheinbar im Raum frei schwebende Träger wirken".

Sur cette double page : images de l'entrée de Guangyun montrant la structure en tenségrité conçue par Arup « qui fait penser à des poutrelles flottant librement dans l'espace », commentent les architectes.

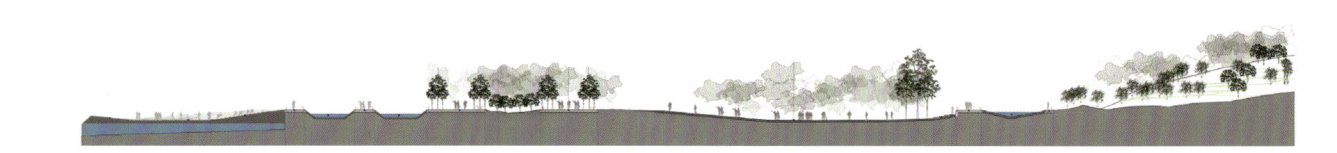

PLASMA STUDIO

Seen in a field of tulips, the Guangyun Entrance bridges over the main road that cuts through the site. With some points seven meters above ground level, the structure gives arriving visitors "an overview of the different zones of the Expo ahead."

Der Guangyun-Eingang, von einem Tulpenfeld gesehen, überquert die Hauptstraße, die das Gelände durchschneidet. Aussichtspunkte sieben Meter über dem Erdboden bieten „einen Überblick über die verschiedenen Bereiche des Expogeländes".

Vue d'un champ de tulipes, l'entrée de Guangyun forme un pont au-dessus de l'axe routier qui traverse le site. S'élevant jusqu'à 7 m, la structure donne aux visiteurs qui arrivent « une vue d'ensemble des différentes zones de l'exposition ».

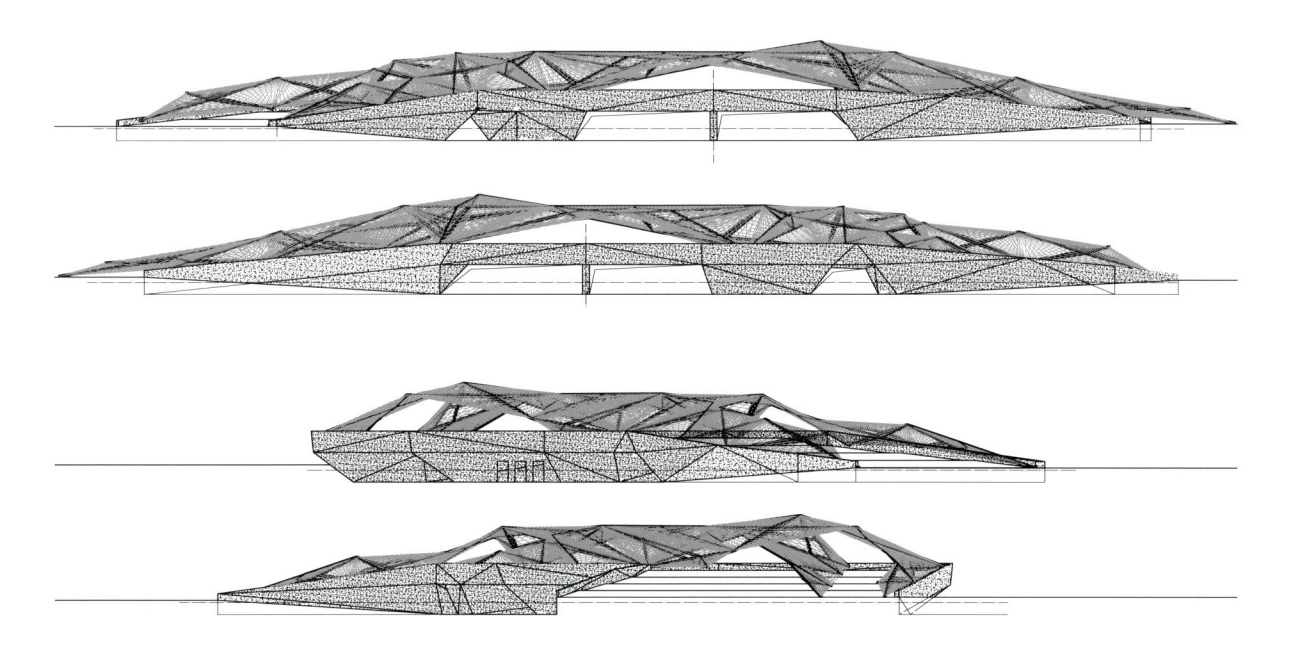

SHUNMYO MASUNO

Shunmyo Masuno
Japan Landscape Consultants Ltd.
Kenkohji 1–2–1 Baba
Tsurumi-ku
Yokohama 230–0076
Japan

E-mail: jlc@kenkohji.jp

SHUNMYO MASUNO was born in Yokohama, Japan, in 1953. He graduated from the Department of Agriculture of Tamagawa University in 1975. He became a Zen priest after studies at the Sohji-ji Temple in Kanagawa (1979). Shunmyo Masuno founded Japan Landscape Consultants Ltd in 1982, and became a professor of Tama Art University in 1998. He was appointed head priest of Kenkohji Temple in Yokohama in 2001. He has received the National Grand Prize from the Japanese Institute of Landscape Architecture. His work includes a Japanese Garden in Berlin (Germany, 2003); SanShin-Tei, One Kowloon Office Building (Hong Kong, China, 2006–07, published here); a private house (Germany, 2007); Gotanjyou-ji Temple (Hukui, Japan, 2009); and the Samukawa Shrine (Kanagawa, Japan, 2009).

SHUNMYO MASUNO wurde 1953 in Yokohama, Japan, geboren. Er machte 1975 seinen Universitätsabschluss an der Landwirtschaftlichen Fakultät der Universität Tamagawa. Nachdem er am Tempel Soji-ji in Kanagawa studiert hatte, wurde er 1979 Zen-Priester. Shunmyo Masuno gründete 1982 das Büro Landscape Consultants Ltd, 1998 wurde er Professor an der Kunstuniversität Tama. 2001 wurde er zum Oberpriester des Tempels Kenkoh-ji in Yokohama ernannt. Das Japanese Institute of Landscape Architecture, der Verband der japanischen Landschaftsarchitekten, hat ihn mit dem Großen Nationalpreis ausgezeichnet. Zu seinen Arbeiten gehört der Japanische Garten in Berlin (2003), der Garten und die Innenarchitektur des Bürogebäudes SanShin-Tei in One Kowloon (Hongkong, 2006–07, hier vorgestellt), ein privates Wohnhaus in Deutschland (2007), der Tempel Gotanjyou-ji (Hukui, Japan, 2009) und der Samukawa-Schrein (Kanagawa, Japan, 2009).

SHUNMYO MASUNO, né à Yokohama (Japon) en 1953, est diplômé du département d'agriculture de l'université Tamagawa (1975). Devenu prêtre zen après des études au temple Sohji-ji à Kanagawa (1979), il fonde l'agence Japan Landscape Consultants Ltd en 1982 et devient professeur à l'Université des beaux-arts Tama en 1998. Il a été nommé supérieur du temple Kenko-ji de Yokohama en 2001. Il a reçu le grand prix national de l'Institut japonais de l'architecture du paysage. Parmi ses réalisations : un jardin japonais à Berlin (2003) ; le jardin SanShin-Tei pour l'immeuble de bureaux One Kowloon (Hong-Kong, 2006–07, publié ici) ; une résidence privée (Allemagne, 2007) ; le temple de Gotanjyou-ji (Hukui, Japon, 2009) et le sanctuaire de Samukawa (Kanagawa, Japon, 2009).

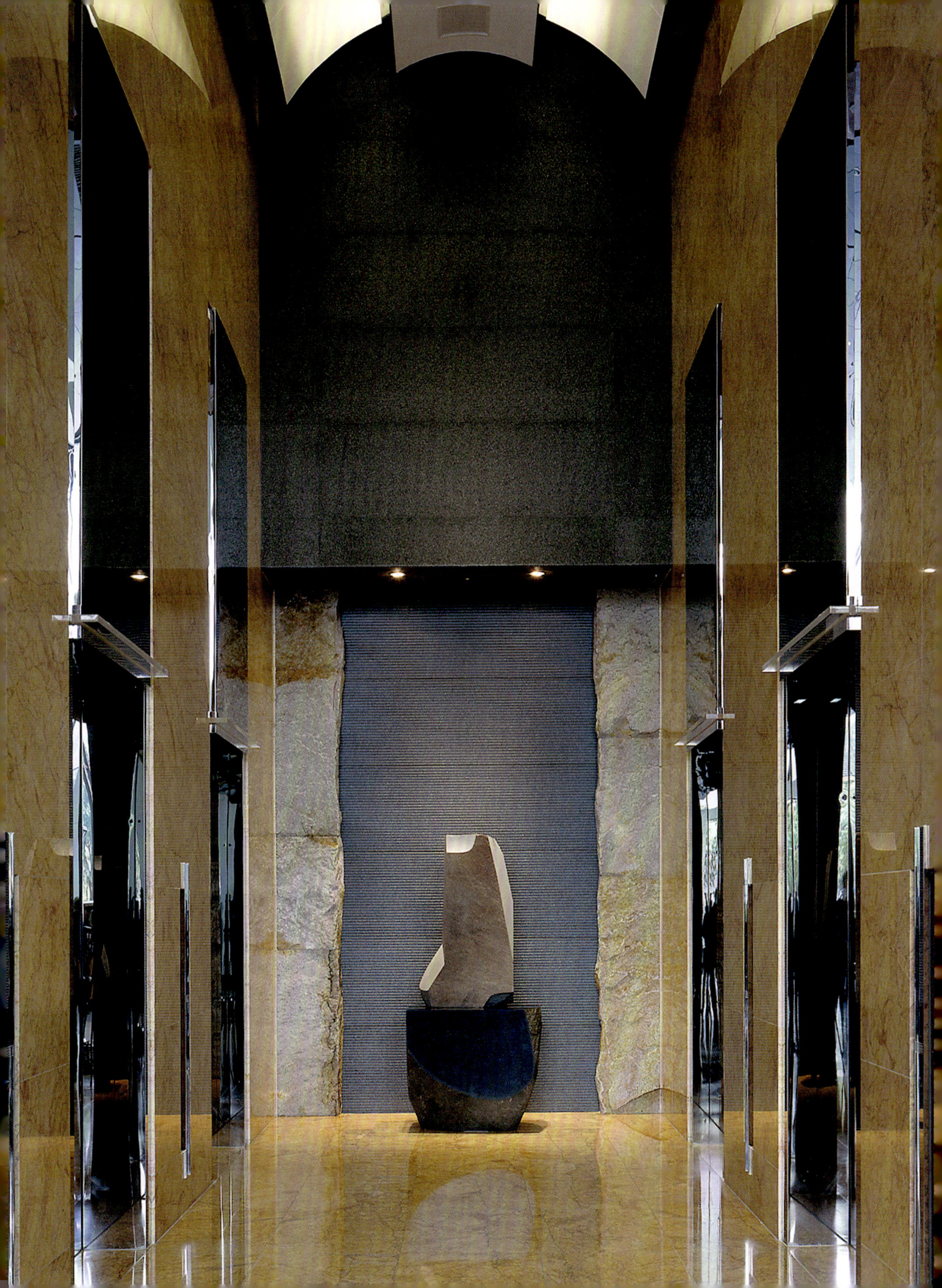

SANSHIN-TEI

One Kowloon Office Building, Hong Kong, China, 2006–07

Address: Wang Yuen Street, Kowloon Bay, Hong Kong
Area: 1545 m² (entrance), 554 m² (lobby). Client: Glorious Sun Properties Ltd.
Cost: not disclosed

Shunmyo Masuno explains the name **SANSHIN-TEI**: "The theme of SanShin-Tei is derived from the Zen word *sanshin* (*san* means three, *shin* means mind), and the 'three mind' means 'joy mind,' 'mature mind,' and 'great mind.' Originally, *sanshin* shows an attitude, which we should treasure for living in this world, and it may not refer only to individuals, but could also refer to the company's existence itself in modern society." The idea of the installation is that the *sanshin* spirit should flow from the upper floors to the lower floors and into the lobby lounge, radiating out from there. The lobby lounge is set in a 13-meter-high atrium, and includes a 9.5-meter-high waterfall. For the outside garden, a large block of granite marks a transition from the "artificial" inside the building to the "natural" outside. The Zen priest concludes: "We designed SanShin-Tei with a wish that all the people who work in this building could always have an attitude to live as an individual and also be aware of the social meaning of existence as a company."

Shunmyo Masuno erklärt den Namen **SANSHIN-TEI**: „Das Thema ist von dem Zen-Wort ‚sanshin' abgeleitet (san bedeutet drei und shin Geisteshaltung), und die ‚drei Geisteshaltungen' bedeuten ‚freudiger Geist', ‚reifer Geist' und ‚großer Geist'. Ursprünglich verdeutlicht ‚sanshin' eine Geisteshaltung, die wir für das Leben in dieser Welt hoch schätzen und bewahren sollten. Sie bezieht sich nicht nur auf Individuen, sondern kann auch auf die Existenz des Unternehmens in der modernen Gesellschaft angewendet werden." Der Grundgedanke ist, dass der Geist des „sanshin" von den oberen in die unteren Geschosse und die Eingangslobby fließt und von dort ausstrahlen kann. Die Eingangslobby befindet sich in einem 13 Meter hohen Atrium mit einem 9,5 Meter hohen Wasserfall. Im äußeren Garten markiert ein großer Granitblock den Übergang vom „Künstlichen" des Innenraums zum „Natürlichen" des Außenraums. Der Zen-Priester schließt mit den Worten: „Wir haben SanShin-Tei mit dem Anliegen entworfen, dass alle Menschen, die in diesem Gebäude arbeiten, immer eine Haltung haben mögen, die es ihnen erlaubt, als Individuum zu leben und sich gleichzeitig bewusst zu sein, welche gesellschaftliche Bedeutung die Existenz des Unternehmens hat."

Shunmyo Masuno explique le nom **SANSHIN-TEI** : « Le thème de SanShin-Tei vient du terme zen *sanshin* (*san* signifiant "trois" et *shin*, "esprit"), dans lequel "trois esprits" signifie "joie de l'esprit", "esprit mûr" et "esprit élevé". À l'origine, *sanshin* exprime une attitude que nous aurions intérêt à préserver si nous voulons vivre dans ce monde et qui ne se réfère pas seulement aux individus, mais aussi à l'existence de l'entreprise au sein de la société moderne. » L'idée est que l'esprit *sanshin* s'écoule des niveaux supérieurs vers les niveaux inférieurs jusqu'au hall d'entrée, d'où il irradie. Le hall d'entrée occupe un atrium de 13 m de haut, animé par une cascade de 9,5 m de haut. Dans la partie extérieure du jardin, un important bloc de granit marque la transition entre l'installation « artificielle » de l'intérieur de l'immeuble et les arrangements « naturels » à l'extérieur. Le prêtre zen conclut en précisant : « Nous avons conçu SanShin-Tei en souhaitant que tous ceux qui travaillent dans cet immeuble puissent toujours aborder la vie en tant qu'individus et être conscients de la signification sociale de l'existence en tant que membre d'une entreprise. »

Within the context of a modern office building, Shunmyo Masuno has created a landscape that calls on Zen tradition without fundamentally contradicting the architectural environment.

In einem modernen Bürogebäude hat Shunmyo Masuno eine Landschaftsgestaltung ausgeführt, die sich auf die Tradition des Zen beruft, ohne einen fundamentalen Widerspruch zum architektonischen Umfeld aufzubauen.

Dans le contexte d'un immeuble de bureaux contemporain, Shunmyo Masuno a créé un paysage qui s'appuie sur la tradition zen sans se heurter pour autant à son environnement architectural.

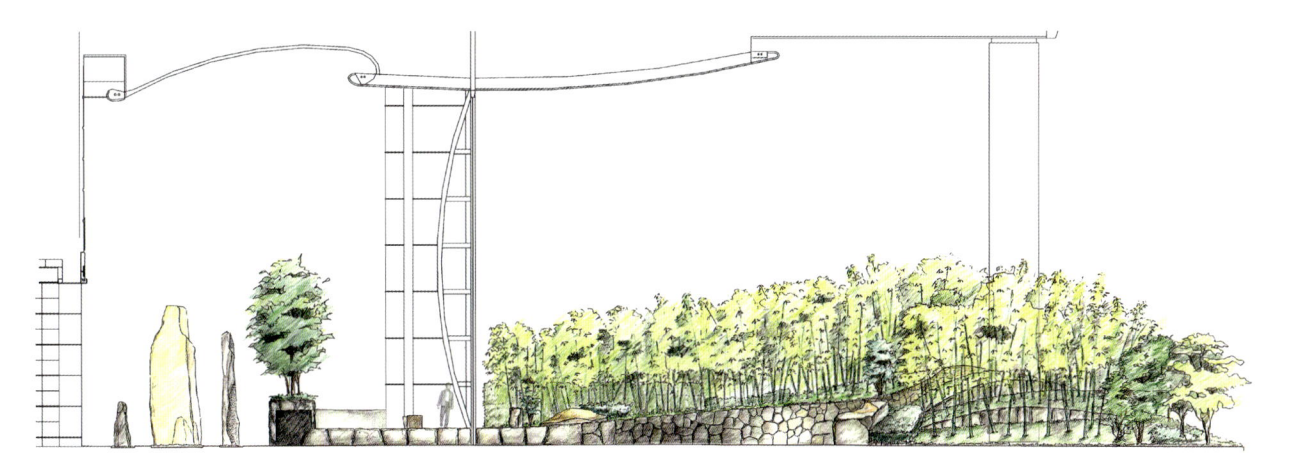

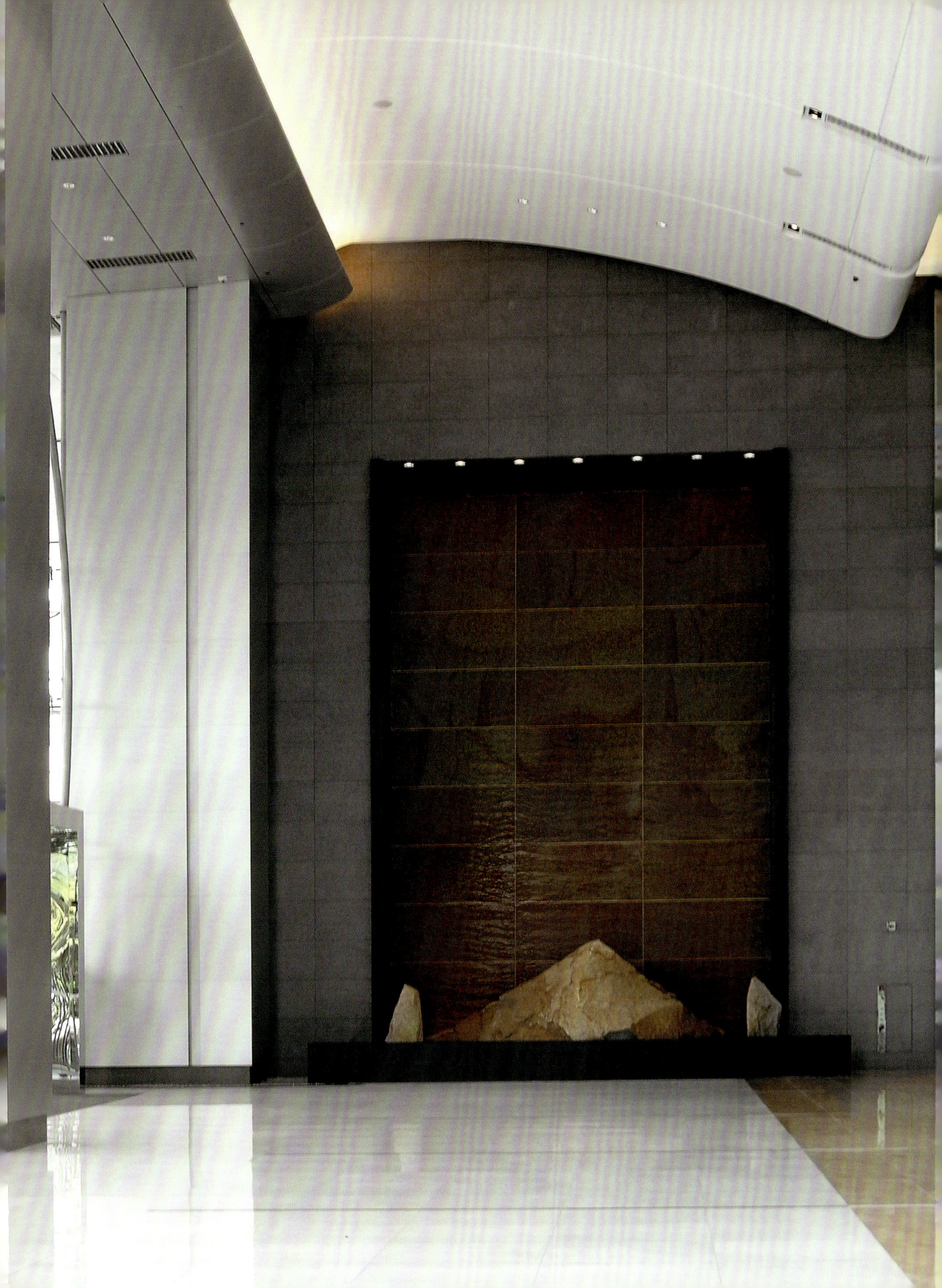

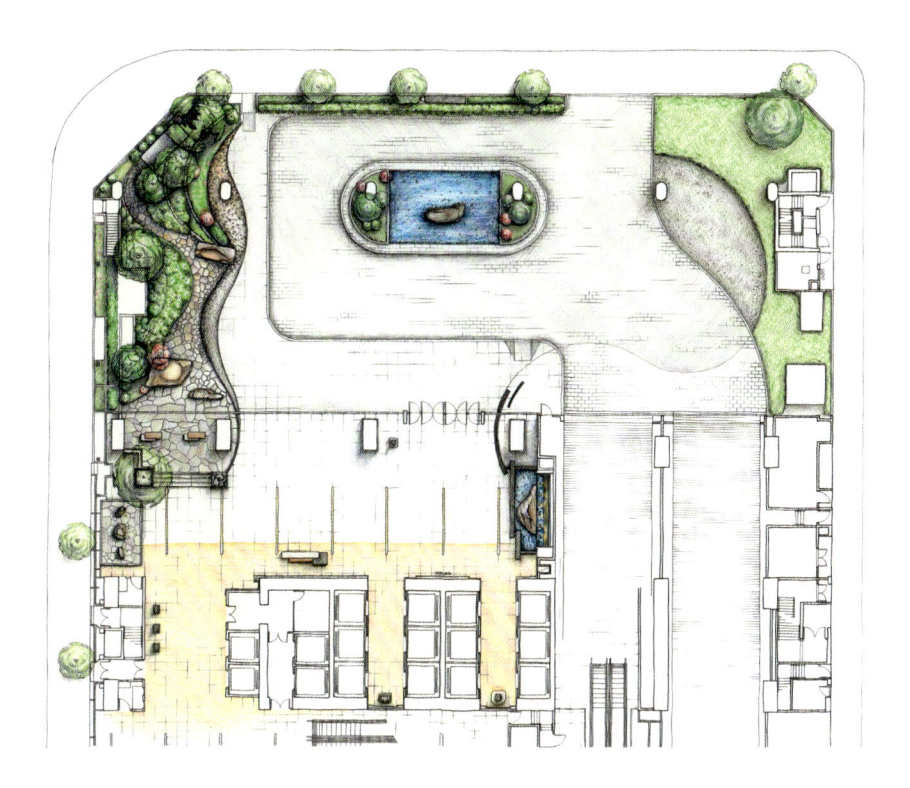

The plan above shows that the intervention of Shunmyo Masuno concerns both the environment of the building and some interior spaces seen here.

Der Grundriss oben zeigt, dass der Entwurf von Shunmyo Masuno sowohl das Umfeld des Gebäudes betrifft als auch einige der hier abgebildeten Innenräume.

Le plan ci-dessus montre que l'intervention de Shunmyo Masuno porte à la fois sur l'environnement de l'immeuble et sur certains espaces intérieurs.

KEN SMITH

WORKSHOP: Ken Smith Landscape Architect
450 West 31st Street, Fifth Floor
New York, NY 10001, USA

Tel: +1 212 791 3595 / Fax: +1 212 732 1793
E-mail: info@kensmithworkshop.com

KEN SMITH was born in 1953, and graduated from Iowa State University with a B.S. in Landscape Architecture (Ames, Iowa, 1975). He attended the Harvard GSD (Master of Landscape Architecture program, 1986). He worked as a Landscape Architect for the State Conservation Commission in Iowa (1979–84), as a consultant for the Department of Environmental Management (Massachusetts, 1984–86), in the office of Peter Walker and Martha Schwartz (New York, San Francisco, 1986–89) and with Martha Schwartz Ken Smith David Meyer Inc. (San Francisco, 1990–92), before creating his present firm, Ken Smith Landscape Architect, in New York. His current and recent landscape work includes Lever House Landscape Restoration (New York, New York, 2000); the MoMA Decorative Rooftop (Museum of Modern Art, New York, New York, 2004–05); 7 World Trade Center, Triangle Park (New York, New York, 2002–06); 40 Central Park South, Courtyard Garden (New York, New York, 2005–06, published here); H-12 Office Complex (Hyderabad, India, 2007); Santa Fe Railyard Park and Plaza (Santa Fe, New Mexico, 2006–08, also published here); 17 State Street Plaza (New York, New York, 2008); and the East River Waterfront (New York, New York, 2006–). Other current work includes Brooklyn Academy of Music Cultural District Public Space and Streetscape (Brooklyn, New York, 2006–); the Croton Water Treatment Plant (Bronx, New York, 2006–); and Orange County Great Park (Irvine, California, 2007–, part of which is published here), all in the USA unless stated otherwise.

KEN SMITH wurde 1953 geboren und schloss 1975 sein Studium an der Iowa State University mit einem Bachelor in Landschaftsarchitektur ab (Ames, Iowa). 1986 machte er seinen Master in Landschaftsarchitektur an der Harvard GSD. Er arbeitete von 1979 bis 1984 als Landschaftsarchitekt bei der State Conservation Commission in Iowa, von 1984 bis 1986 als Berater beim Department of Environmental Management (Massachusetts), von 1986 bis 1989 im Büro von Peter Walker und Martha Schwartz (New York und San Francisco) und von 1990 bis 1992 im Büro Martha Schwartz Ken Smith David Meyer Inc. (San Francisco). Danach gründete er sein eigenes Büro Ken Smith Landscape Architect in New York. Zu seinen aktuellen landschaftsarchitektonischen Projekten gehören die Restaurierung der Gartenanlagen des Lever House (New York, 2000), der Dachgarten des MoMA (Museum of Modern Art, New York, 2004–05), 7 World Trade Center, Triangle Park (New York, 2002–06), 40 Central Park South, Innenhofgestaltung (New York, 2005–06, hier vorgestellt), der Bürokomplex H-12 (Hyderabad, Indien, 2007), die Park- und Platzanlage auf den ehemaligen Gleisanlagen in Santa Fe (New Mexico, 2006–08, ebenfalls hier vorgestellt), die 17 State Street Plaza (New York, 2008) und die Ufergestaltung des East River (New York, seit 2006). Als weitere laufende Projekte sind zu nennen der öffentliche Freiraum und die Straßengestaltung im Kulturviertel der Musikakademie von Brooklyn (Brooklyn, New York, seit 2006), die Wasseraufbereitungsanlage Croton Water Treatment Plant (Bronx, New York, seit 2006) und der Great Park von Orange County (Irvine, Kalifornien, seit 2007, von dem ein Teil hier ebenfalls vorgestellt wird), alle in den USA, sofern nicht anders angegeben.

Né en 1953, **KEN SMITH** est titulaire d'un B.S. de paysagisme de l'université d'Iowa (Ames, Iowa, 1975). Il a également étudié à la Harvard GSD (Master of Landscape Architecture Program, 1986). Il a été architecte paysagiste pour la Commission de préservation de l'État d'Iowa (1979–84) et consultant pour le Département de gestion environnementale (Massachusetts, 1984–86). Il a travaillé pour l'agence de Peter Walker et Martha Schwartz (New York, San Francisco, 1986–89) et pour Martha Schwartz Ken Smith David Meyer Inc. (San Francisco, 1990–92), avant de créer sa propre agence, Ken Smith Landscape Architect, à New York. Parmi ses interventions récentes (toutes aux États-Unis, sauf mention contraire) : la restauration de l'aménagement paysager de Lever House (New York, 2000) ; la toiture paysagée du MoMA (Museum of Modern Art, New York, 2004–05) ; le parc triangulaire du 7 World Trade Center (New York, 2002–06) ; le jardin de la cour du 40 Central Park South (New York, 2005–06, publié ici) ; le complexe de bureaux H-12 (Hyderabad, Inde, 2007) ; le Railyard Park et Plaza de Santa Fe (Nouveau-Mexique, 2006–08, publié ici) ; la 17 State Street Plaza (New York, 2008) ; l'aménagement des berges de l'East River (New York, 2006–) ; les espaces publics et l'aménagement des rues du Brooklyn Academy of Music Cultural District (Brooklyn, New York, 2006–) ; les installations de traitement des eaux de Croton (Bronx, New York, 2006–) et le Orange County Great Park (Irvine, Californie, 2007–, partiellement publié ici).

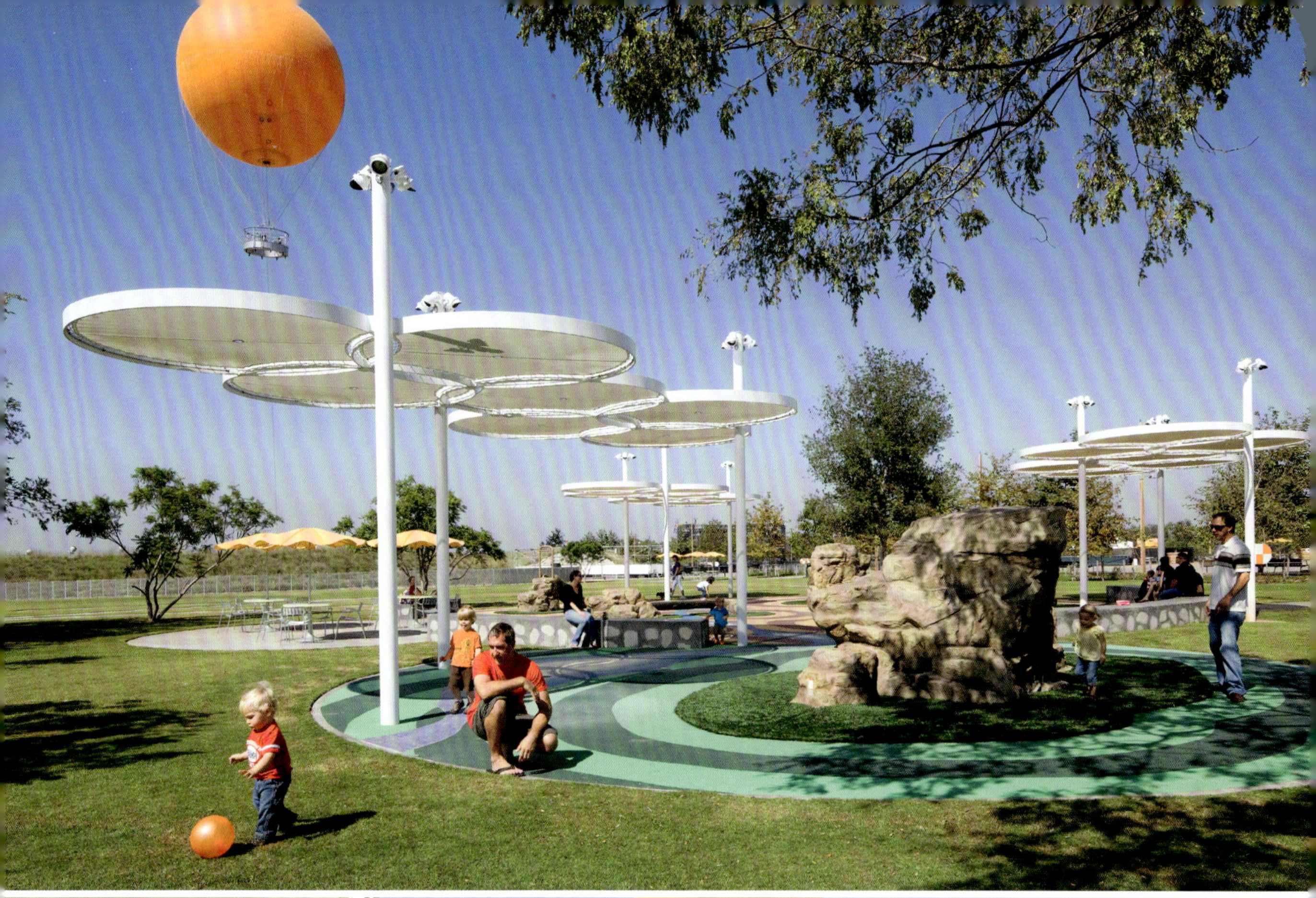
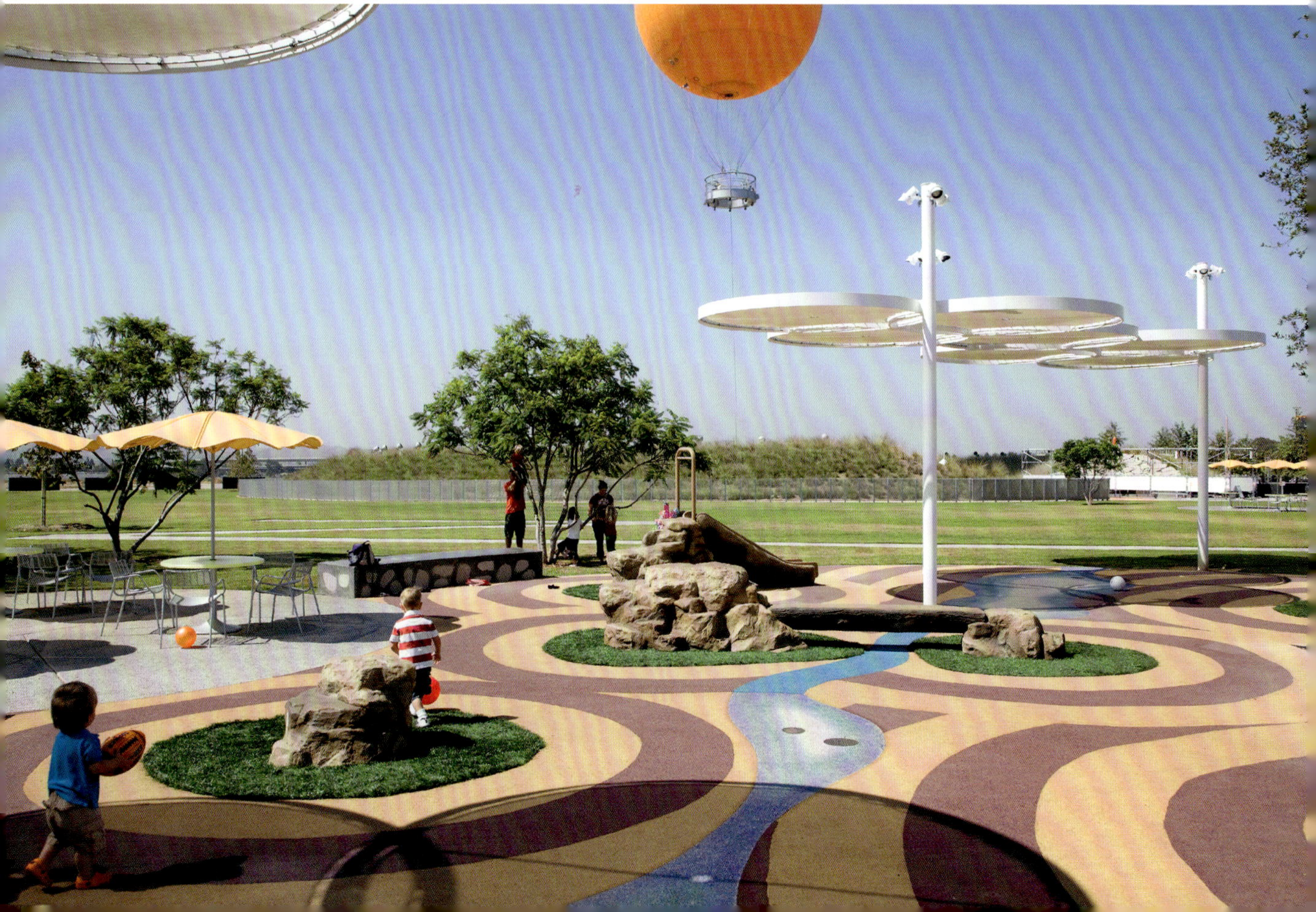

KIDS ROCK

Children's Play Environment, Orange County Great Park, Irvine, California, USA, 2010

Address: 14430 'C' Street, Irvine, CA, USA, +1 866 829 3829 (Visitors Center), www.ocgp.org
Area: 511 m². Client: Orange County Great Park Corporation. Cost: $1.1 million

KIDS ROCK is a small component of the much larger (545 hectare) Orange County Great Park development that Ken Smith has undertaken. As he explains: "It is envisioned as the first of many small environmental play areas that will eventually be located throughout the Great Park. The basic metaphor for Kids Rock is that of the rain cycle and the clouds coming over the mountains and dropping rain into canyons and streams flowing to the ocean. The play area is intended to draw on the larger environmental themes of the park well as iconic landform features of the area." Structures inspired by clouds provide shade and use misting devices to cool the air. Caves, a tunnel, slides, and other spaces for children were fabricated with rock. Insets provide information on local wildlife. The designers further explain that "the ground plane is composed of concentric rings of safety surfacing that represent the pattern of water waves around rockeries that occur along the Pacific surf. Decorative glass paving forms a stream that cuts across the play area and under the feature bridge. Recycled concrete from demolition of the site's former airfield runways were incorporated into the seating elements at the perimeter of the play area."

KINDS ROCK ist ein kleines Element in dem 545 Hektar großen Orange County Great Park, der von Ken Smith angelegt worden ist. Er erklärt: „Dies ist der erste von vielen kleinen Umweltspielplätzen, die überall im Great Park verteilt sein werden. Die Grundmetapher von Kids Rock ist der Regenkreislauf – die Wolken kommen über die Berge und lassen den Regen in die Schluchten und Flüsse fallen, die ins Meer fließen. Der Spielbereich soll auf die großen Umweltthemen im Park Bezug nehmen und zugleich mit der ikonografischen Landschaftsgestaltung eine eigene Identität bekommen." Strukturen, die an Wolken erinnern, spenden Schatten und kühlen mit Sprühnebel die Luft. Höhlen, ein Tunnel, Rutschen und anderes Spielgerät sind hier aus Fels gearbeitet. Kleine Tafeln bieten Informationen über die örtliche Tierwelt. Die Planer erklären: „Der Grundriss besteht aus konzentrischen Ringen aus unfallsicheren Bodenbelägen, die wie Wellen an den Felsen der Pazifikküste gestaltet sind. Dekoratives Glaspflaster bildet einen Fluss, der durch den Spielbereich und unter der ‚Brücke' hindurchfließt. Am Rand des Spielbereichs sind Sitzelemente aus recyceltem Beton platziert, der von dem Abriss der Start- und Landebahn stammt, die sich hier früher befunden hat."

KIDS ROCK n'est qu'une partie du vaste Orange County Great Park (545 ha), aménagé par Smith. « C'est la première d'une série de petites aires de jeux d'esprit environnemental qui trouveront peu à peu leur place dans le parc. La métaphore simple qui a présidé à la conception de Kids Rock est celle du cycle de la pluie et des nuages qui, bloqués par les montagnes, alimentent les rivières et les fleuves s'écoulant ensuite jusqu'à la mer. L'aire de jeux s'appuie donc sur les thèmes environnementaux de ce parc ainsi que sur les formes de paysage caractéristiques des lieux », précise Ken Smith. Les parasols métalliques inspirés des nuages sont dotés de brumisateurs pour rafraîchir l'air ambiant. Des grottes, un tunnel, des toboggans et autres aménagements ont été réalisés à l'aide de rochers. Des panneaux fournissent des informations sur la vie sauvage de la région. « Le sol est marqué d'anneaux concentriques qui représentent les ondes se formant autour des récifs de la côte du Pacifique. Un ruisseau en pavés de verre décoratifs s'écoule à travers l'aire de jeux et passe sous un pont. Des morceaux de béton recyclé, issus de la démolition des anciennes installations de l'aérodrome dont le parc a pris la place, ont été incorporés dans les banquettes implantées tout autour de cette zone ludique », expliquent les designers.

The 500-square-meter park calls on themes related to the cycles of water, from rain and the mountains to the ocean.

Dieses 500 Quadratmeter große Projekt befasst sich mit Themen, die mit dem Wasserkreislauf in Verbindung stehen – vom Regen und den Bergen bis zum Ozean.

Le projet de ce parc de 500 m² s'appuie sur des thèmes liés au cycle de l'eau de pluie, des montagnes à la mer.

40 CENTRAL PARK SOUTH
New York, New York, USA, 2006

Area: 581 m². Client: not disclosed
Cost: not disclosed

Within a relatively limited interstitial space, Ken Smith mixes areas covered in blue glass, crushed white marble, and even recycled black rubber shards with more traditional planting.

In einem relativ beengten Zwischenraum mischt Ken Smith Flächen aus blauem Glas, gebrochenem weißem Marmor und recycelten schwarzen Gummistücken mit einer eher traditionellen Bepflanzung.

Dans un espace interstitiel relativement limité, Ken Smith associe des parterres de verre bleu, de marbre concassé et même de déchets de pneus recyclés à des plantations plus traditionnelles.

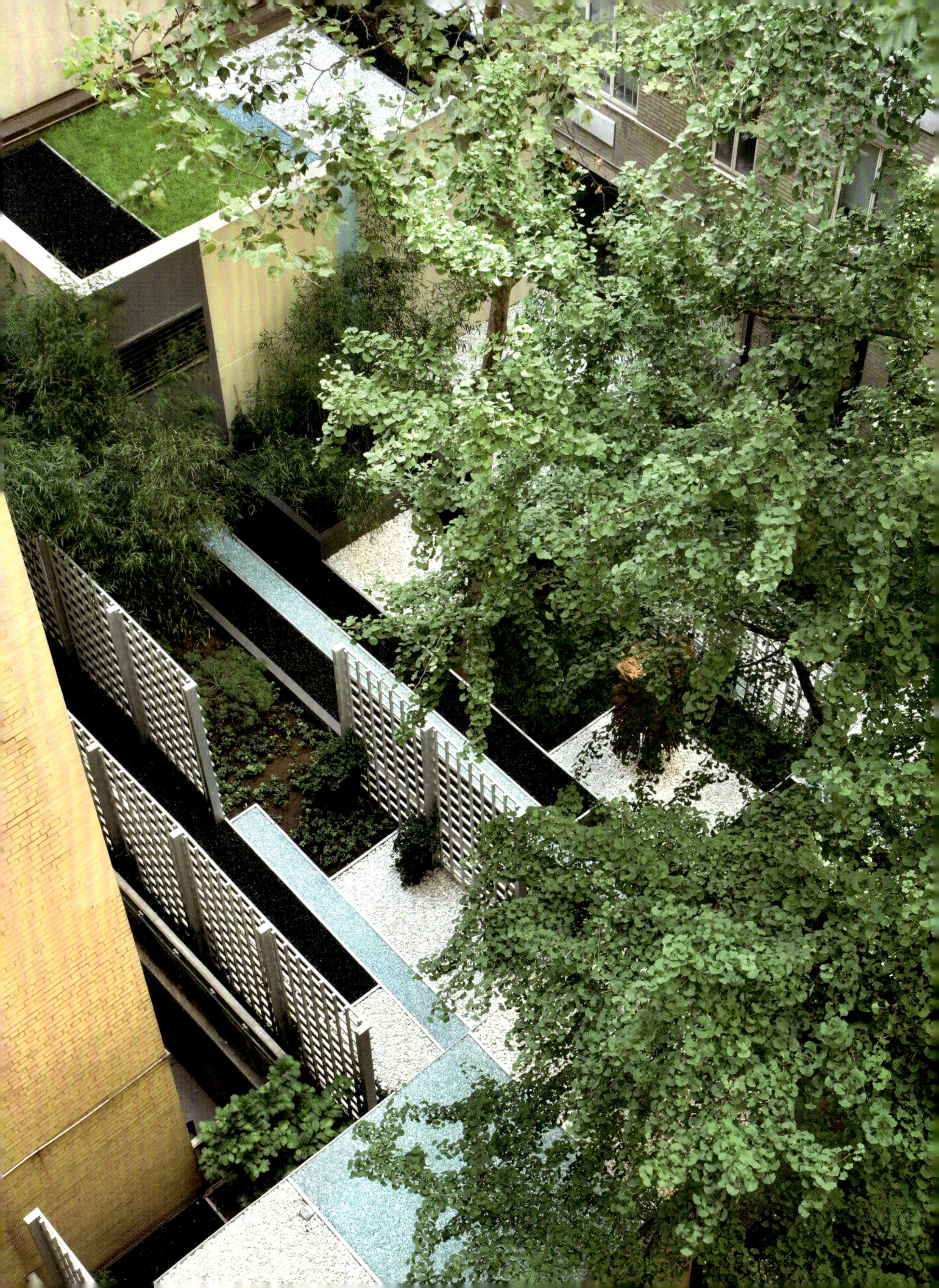

Ken Smith designed the rooftop gardens of New York's Museum of Modern Art after the renovations and construction of new structures by Yoshio Taniguchi (2003–05) using plastic rocks and artificial trees. He explains: "Shortly after completing the rooftop garden at the Museum of Modern Art, I was asked to design a small courtyard garden just south of Central Park. Set between two postwar apartment buildings facing 58th and 59th Streets, the garden shares a material palette similar to MoMA's rooftop but without its irony and double entendre." Intended to be seen from the neighboring apartment buildings, the garden expresses a "strong linearity" with its crushed white marble, recycled black rubber shards, and underlit blue glass marble bands. Ken Smith continues: "My strategy was to simplify the ground plane to reveal the organic structure of the existing sycamore and locust trees. I think of them as sculptures that stand out against the grid of the buildings and their windows. We sited ornamental plantings of Japanese maple, bamboo, and magnolia for background color and texture. The older trees shelter three metal sculptures that include an Isamu Noguchi torso from the late 1940s, a mid-century bronze by Chaim Gross, and a recent large bronze, *Guardian II*, by Michele Oka Doner."

Ken Smith entwarf den Dachgarten des New Yorker Museum of Modern Art, nachdem die Bau- und Sanierungsarbeiten von Yoshio Taniguchi abgeschlossen waren (2003–05). Für die Gestaltung des Dachgartens setzte er Kunststofffelsen und künstliche Bäume ein. Er erklärt: „Kurz nachdem ich den Dachgarten des Museum of Modern Art fertiggestellt hatte, beauftragte man mich mit der Gestaltung eines Innenhofs südlich des Central Park. Der Garten liegt zwischen zwei Gebäuden aus der Nachkriegszeit, die an der 58. und 59. Straße liegen. Ich habe in dem Garten ähnliche Materialien verwendet wie beim Dachgarten des MoMA, allerdings habe ich auf das ironische Augenzwinkern und die Doppeldeutigkeit verzichtet." Der Garten, der von den umliegenden Apartmenthäusern eingesehen werden soll, zeichnet sich durch seine strenge Linearität aus, die durch ein Streifenmuster aus weißem Marmorsplitt, Stücken aus recyceltem schwarzem Gummi und aus von unten beleuchteten blauen Glasmurmeln entsteht. Ken Smith sagt weiter: „Meine Strategie war, den Grundriss so einfach wie möglich zu gestalten, damit die organische Struktur des vorhandenen Bergahorns und der Robinien zur Geltung kommt. Diese Bäume sind für mich Skulpturen, die einen Kontrapunkt zu dem Raster der Gebäude und der Fenster bilden. Wir haben als dekorative Pflanzungen japanischen Ahorn, Bambus und Magnolien eingesetzt, um Hintergrundfarben und Texturen zu bekomen. Im Schatten der älteren Bäume sind drei Metallskulpturen platziert, ein Torso von Isamu Noguchi aus den späten 1940er-Jahren, eine Bronzeskulptur von Chaim Gross aus der Mitte des vergangenen Jahrhunderts und eine neue große Bronzeskulptur, ‚Guardian II', von Michele Oka Doner."

Ken Smith avait réalisé l'aménagement des jardins en toiture du MoMA de New York suite aux opérations de rénovation et d'agrandissement de Yoshio Taniguchi (2003–05) avec des rochers en plastique et des arbres artificiels. « Peu après avoir achevé le jardin de la toiture du MoMA, on m'a demandé une petite cour-jardin pour un immeuble au sud de Central Park. Pris entre deux immeubles d'appartements d'après-guerre, face à la 58ᵉ et à la 59ᵉ Rue, le jardin a été réalisé avec des matériaux similaires à ceux de la terrasse du MoMA, sans ironie ni double sens », explique Ken Smith. Conçu pour être vu des immeubles voisins, il exprime une « forte linéarité » à travers des bandes de marbre blanc broyé, de débris de caoutchouc noir et de verre poli bleu rétroéclairés. « Ma stratégie a été de simplifier le plan au sol pour mettre en valeur la structure organique des sycomores et des caroubiers. Je les ai pensés comme des sculptures se détachant de la trame des immeubles et de leurs fenêtres. Nous avons planté des érables du Japon, des bambous et des magnolias pour apporter une couleur et une texture de fond. Les arbres anciens protègent trois sculptures en métal : un torse d'Isamu Noguchi de la fin des années 1940, un bronze de Chaim Gross des années 1950 et un autre récent, *Guardian II*, de Michele Oka Doner. »

The design of the garden is strictly geometric, with long rectangular bands dominating the formal composition, where plants occupy only part of the space.

Dieser Garten ist nach einem streng geometrischen Entwurf angelegt, mit langen rechteckigen Streifen als dominierende Kompositionselemente. Pflanzen sind nur in einem kleinen Bereich dieses Freiraums vorhanden.

Le plan du jardin est strictement géométrique. Sa composition formaliste est dominée par de longs parterres rectangulaires et les plantes n'occupent qu'une partie de l'espace.

SANTA FE RAILYARD PARK AND PLAZA
Santa Fe, New Mexico, USA, 2006–08

Address: corner of Guadalupe Street and Paseo De Peralta Santa Fe, NM 87501, USA, www.railyardpark.org
Area: 4.85 hectares. Client: The Trust for Public Land. Cost: $13 million
Collaboration: Frederic Schwartz, Mary Miss (Artist)

An "iconic water tank" (below) serves as a point of collection for water harvested from rain that is used to irrigate the park area. The overall plan of the park and plaza is seen on the right page.

In einem „typischen Wassertank" (unten) wird das Regenwasser für die Bewässerung des Parks gesammelt. Der Gesamtlageplan von Park und Plaza ist auf der rechten Seite zu sehen.

Un « réservoir d'eau typique » (ci-dessous) récupère les eaux de pluies qui servent à l'irrigation du parc. Un plan d'ensemble du parc et de la place est visible sur la page droite.

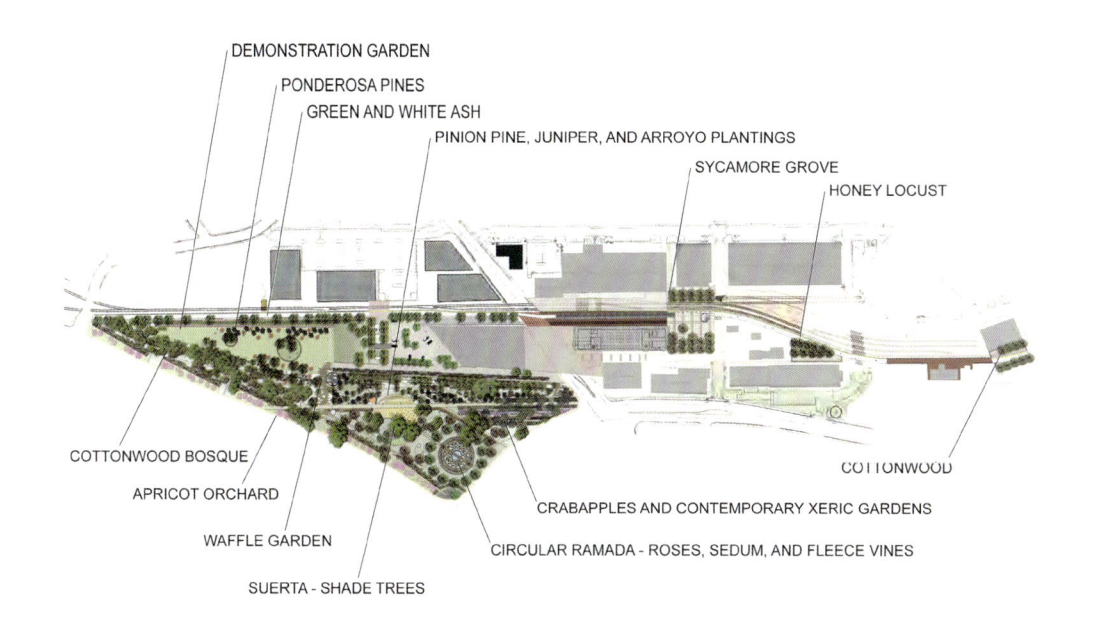

KEN SMITH

- DEMONSTRATION GARDEN
- PONDEROSA PINES
- GREEN AND WHITE ASH
- PINION PINE, JUNIPER, AND ARROYO PLANTINGS
- SYCAMORE GROVE
- HONEY LOCUST
- COTTONWOOD BOSQUE
- APRICOT ORCHARD
- WAFFLE GARDEN
- SUERTA - SHADE TREES
- CRABAPPLES AND CONTEMPORARY XERIC GARDENS
- CIRCULAR RAMADA - ROSES, SEDUM, AND FLEECE VINES
- COTTONWOOD

The history of Santa Fe as a railway point of passage goes back to the late 19th century. Ken Smith was called on, in collaboration with the artist Mary Miss and the architect Frederic Schwartz, to redevelop railyards located near downtown Santa Fe. The resulting park includes a plaza, a tree-shaded promenade, and "sophisticated water conservation features." According to Ken Smith, the "design draws on traditional northern New Mexican traditions but the expression is contemporary. Along the paths, we designed benches of simple large wood rectangles, local bricks, and other local material… As the centerpiece of a newly revitalized mixed-use district, the park makes strong connections with the neighborhood and cultural institutions including SITE Santa Fe, the Santa Fe Farmers Market, El Museo Cultural de Santa Fe, Warehouse 21, and other local institutions." A 400-year-old irrigation ditch runs through the site, a fact that Ken Smith seized on to emphasize the continuity of the stewardship of the landscape by local residents. Smith states: "Throughout the park, water is captured from neighboring roof areas, stored, and used as a visible element in the park design. An iconic water tank in the new plaza is the central water storage component for harvested water. Beneath it, a drip fountain recalls the watershed of the Santa Fe River and this harvested water supports xeric plantings, native grasses, and garden environments."

Die Geschichte von Santa Fe als bedeutendem Eisenbahndrehkreuz geht bis in das späte 19. Jahrhundert zurück. Ken Smith wurde gemeinsam mit der Künstlerin Mary Miss und dem Architekten Frederic Schwartz eingeladen, das Bahngelände nahe der Innenstadt von Santa Fe neu zu gestalten. Zum neuen Park gehören eine Plaza, eine schattenspendende Allee und „raffinierte Maßnahmen zum Gewässerschutz". Smith erläutert, dass der Entwurf „auf die Traditionen von New Mexico Bezug nimmt, aber in seiner Umsetzung zeitgenössisch ist. Für die Wege haben wir einfache, große, rechteckige Bänke aus Holz, lokalem Klinker und anderen ortstypischem Material entworfen … Der Park bildet das Zentrum eines sanierten Stadtviertels mit Mischnutzung und stellt enge Verbindungen mit der Nachbarschaft und Kulturinstitutionen wie der SITE Santa Fe, dem Bauernmarkt von Santa Fe, dem Museo Cultural de Santa Fe, dem Warehouse 21 und anderen örtlichen Einrichtungen her." Durch das Gelände verläuft ein 400 Jahre alter Bewässerungsgraben, für Ken Smith ein Symbol dafür, dass die Menschen hier schon immer die umgebende Landschaft gestaltet haben. Er sagt: „Im gesamten Park wird das Wasser von den Gebäudedächern gesammelt und als sichtbares Gestaltungselement eingesetzt. Ein typischer Wassertank auf dem neuen Platz dient als Reservoir für den Park. Darunter greift ein Brunnen das Motiv des Santa Fe River und seines Einzugsgebiets auf. Das hier gespeicherte Wasser versorgt die Wüstenpflanzen, die heimischen Gräser und die Gartenanlagen."

L'histoire de Santa Fe en tant que nœud ferroviaire remonte à la fin du XIXe siècle. Ken Smith a été sollicité, avec l'artiste Mary Miss et l'architecte Frederic Schwartz, pour reconvertir des friches ferroviaires à proximité du centre-ville. Le nouveau parc comprend une place, une promenade ombragée par des arbres et « des systèmes de stockage de l'eau sophistiqués ». Ken Smith explique que « le projet s'appuie sur les traditions du nord du Nouveau-Mexique, mais dans une expression contemporaine. Le long des allées, nous avons dessiné des banquettes qui sont de simples grands rectangles en bois, brique rouge et autres matériaux locaux… Élément central d'un quartier d'usage mixte récemment revitalisé, le parc établi de solides connexions avec son voisinage et des institutions culturelles comme SITE Santa Fe, le Santa Fe Farmers Market, El Museo Cultural de Santa Fe ou Warehouse 21 ». Smith s'est emparé d'un fossé d'irrigation vieux de quatre siècles qui traversait le site pour souligner la continuité de l'appropriation du paysage par les habitants. « Dans l'ensemble du parc, l'eau est récupérée à partir des toitures avoisinantes, stockée et réutilisée sous forme visible. Un réservoir de forme typique, sur la nouvelle place, constitue l'élément central du stockage des eaux de récupération. En dessous, une fontaine rappelle la Santa Fe River et l'eau récupérée sert à arroser des plantes xérophiles, des graminées de variétés locales et des environnements traités en jardins. »

A 400-year-old irrigation ditch is integrated into the park site. As is often the case, Ken Smith employs a palette of different materials, as seen in these images, emphasizing not only different colors but also different tactile or textural effects.

Ein 400 Jahre alter Bewässerungsgraben wurde in den Park integriert. Wie so oft, verwendet Smith viele verschiedene Materialien, wie auf den Fotos zu erkennen, um unterschiedliche Farben und die Wirkung verschiedener Texturen hervorzuheben

Un fossé d'irrigation datant de quatre siècles a été intégré au site. Comme souvent, Ken Smith s'est servi d'une palette de matériaux très différents, qui met en valeur non seulement des couleurs mais aussi des effets de textures différents.

The overall design is a subtle combination of areas that might have occurred naturally and others that are most clearly part of a willful design.

Der Gesamtplan ist eine subtile Kombination von ursprünglich wirkenden Bereichen mit anderen, denen ein bewusster Entwurf zugrunde liegt.

Le plan d'ensemble combine subtilement des zones qui semblent naturelles et d'autres relevant davantage de la volonté du paysagiste.

Mixing stone, gravel, plants, and wood, the designers create a stimulating and unexpected environment, a park full of surprises.

Mit dieser Mischung von Naturstein, Schotter, Pflanzen und Holz schaffen die Planer ein stimulierendes und unerwartetes Ambiente, einen Park voller Überraschungen.

En mélangeant la pierre, le gravier, les plantes et le bois, les paysagistes ont créé un environnement à la fois stimulant, inattendu et plein de surprises.

TAYLOR CULLITY LETHLEAN

Taylor Cullity Lethlean
385 Drummond Street
Carlton, VIC 3053
Australia

Tel: +61 39380 4344
Fax: +61 39348 1232
E-mail: agata.k@tcl.net.au
Web: tcl.net.au

Taylor Cullity Lethlean is directed by three principals, Kevin Taylor, Kate Cullity, and Perry Lethlean. **KEVIN TAYLOR** was born in 1953 in Clarence Park, Australia. He graduated from RMIT in Landscape Architecture and from the University of South Australia in Architecture. **KATE CULLITY** was born in 1956 in Perth and received degrees in Botany and Education from the University of Western Australia. **PERRY LETHLEAN** graduated from RMIT in Landscape Architecture (Bachelor of Applied Science) in 1985. He received a Master of Urban Design from the same university in 1992, and his Ph.D. in 2010. They founded Taylor Cullity Lethlean in 1991. Their work includes the Geelong Waterfront Development not far from Melbourne (Geelong, 1999–2001); Birrarung Marr (Melbourne, 2002); the Kangaroo Island National Park, located on Kangaroo Island 112 kilometers southwest of Adelaide (1999–2003); noise barriers for the Craigieburn Bypass Freeway (Melbourne, 2003); Tidbinbilla Nature Reserve, located between the Tidbinbilla and Gibraltar Ranges to the south of Canberra (2005); and Manly Corso (Sydney, 2005). The firm is currently working on the Canberra International Arboretum (with TGZ Architects, 2005–); Lonsdale Street (Dandenong, Victoria, 2007–); the Auckland Waterfront (New Zealand, 2009–); and the Royal Botanic Gardens published here (Cranbourne, Melbourne, 1995; Stage 1 complete; Stage 2 ongoing).

Das Büro Taylor Cullity Lethlean wird von drei Partnern geführt: Kevin Taylor, Kate Cullity, und Perry Lethlean. **KEVIN TAYLOR** wurde 1953 in Clarence Park, Australien, geboren. Er machte seinen Universitätsabschluss an der RMIT in Landschaftsarchitektur sowie einen Abschluss als Architekt an der University of South Australia. **KATE CULLITY** wurde 1956 in Perth geboren und schloss ihr Studium in Botanik und Erziehungswissenschaften an der University of Western Australia ab. **PERRY LETHLEAN** schloss 1985 sein Studum der Landschaftsarchitektur an der RMIT (Bachelor of Applied Science) ab. Dort machte er 1992 den Master in Stadtplanung und promovierte 2010. Das Büro Taylor Cullity Lethlean wurde 1991 gegründet. Zu ihren Arbeiten gehören die Neugestaltung und Bebauung der Meeresuferzone von Geelong, das nicht weit von Melbourne entfernt ist (1999–2001), Birrarung Marr (Melbourne, 2002), der Kangaroo Island National Park auf der Insel Kangaroo Island, 112 Kilometer südwestlich von Adelaide (1999–2003), Lärmschutzwände an der Ringstraße um Craigieburn (Melbourne, 2003), das Tidbinbilla-Naturreservat, das zwischen dem Tidbinbilla und den Gibraltar Ranges südlich von Canberra (2005) liegt, und die Gestaltung der Einkaufsstraße Manly Corso in Sydney (2005). Derzeit realisiert das Büro das Internationale Arboretum von Canberra (mit dem Architekturbüro TGZ Architects, seit 2005), die Gestaltung der Lonsdale Street (Dandenong, Victoria, seit 2007), die Ufergestaltung von Auckland (Neuseeland, seit 2009) und den hier vorgestellten Botanischen Garten (Cranbourne, Melbourne, Bauabschnitt 1 wurde 1995 fertiggestellt, Bauabschnitt 2 ist gerade im Bau).

L'agence Taylor Cullity Lethlean est dirigée par trois associés : Kevin Taylor, Kate Cullity et Perry Lethlean. **KEVIN TAYLOR**, né en 1953 à Clarence Park (Australie), est titulaire d'un diplôme d'architecture du paysage du RMIT (Melbourne) et d'un diplôme d'architecture de l'université d'Australie du Sud. **KATE CULLITY**, née en 1956 à Perth, est diplômée en botanique et en éducation de l'université d'Australie occidentale. **PERRY LETHLEAN** est diplômé en architecture du paysage du RMIT (Bachelor of Applied Science, 1985) et titulaire d'un M.A. (1992) et d'un Ph.D. (2010) d'urbanisme de la même université. Ils ont fondé Taylor Cullity Lethlean en 1991. Parmi leurs réalisations en Australie, sauf mention contraire : l'aménagement du front de mer de Geelong, près de Melbourne (Geelong, 1999–2001) ; le parc de Birrarung Marr (Melbourne, 2002) ; le parc national de Kangaroo Island, 112 km au sud-ouest d'Adélaïde (1999–2003) ; des barrières antibruit pour la rocade de Craigieburn (Melbourne, 2003) ; la réserve naturelle de Tidbinbilla, entre les chaînes de montagnes de Tidbinbilla et de Gibraltar, au sud de Canberra (2005) et le Manly Corso (Sydney, 2005). L'agence travaille actuellement sur le projet du Canberra International Arboretum (avec TGZ Architects, 2005–) ; la Lonsdale Street (Dandenong, Victoria, 2007–) ; le front de mer d'Auckland (Nouvelle-Zélande, 2009–) et les Royal Botanic Gardens (Cranbourne, Melbourne, 1995 ; phase 1 achevée, phase 2 en cours, publiés ici).

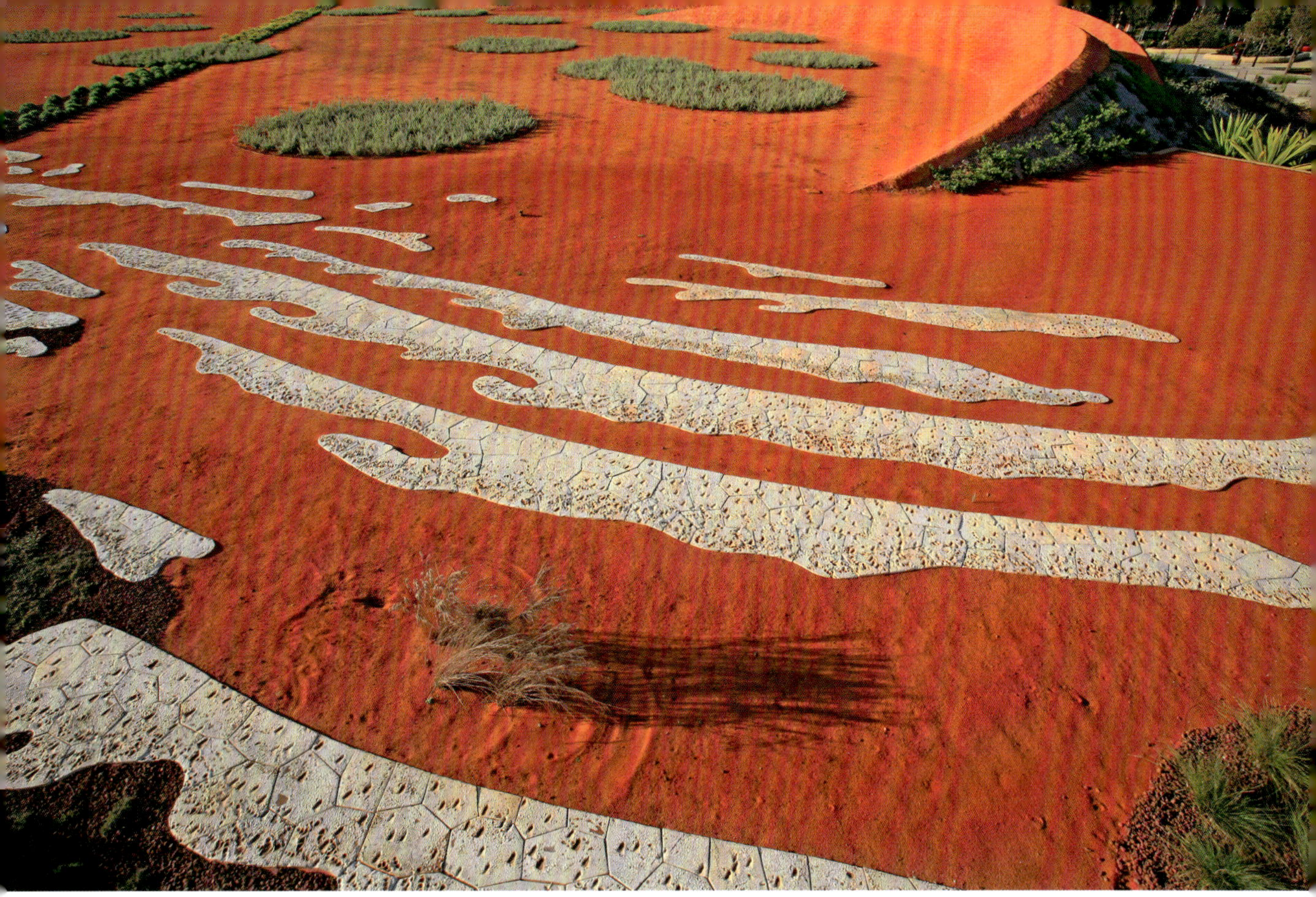

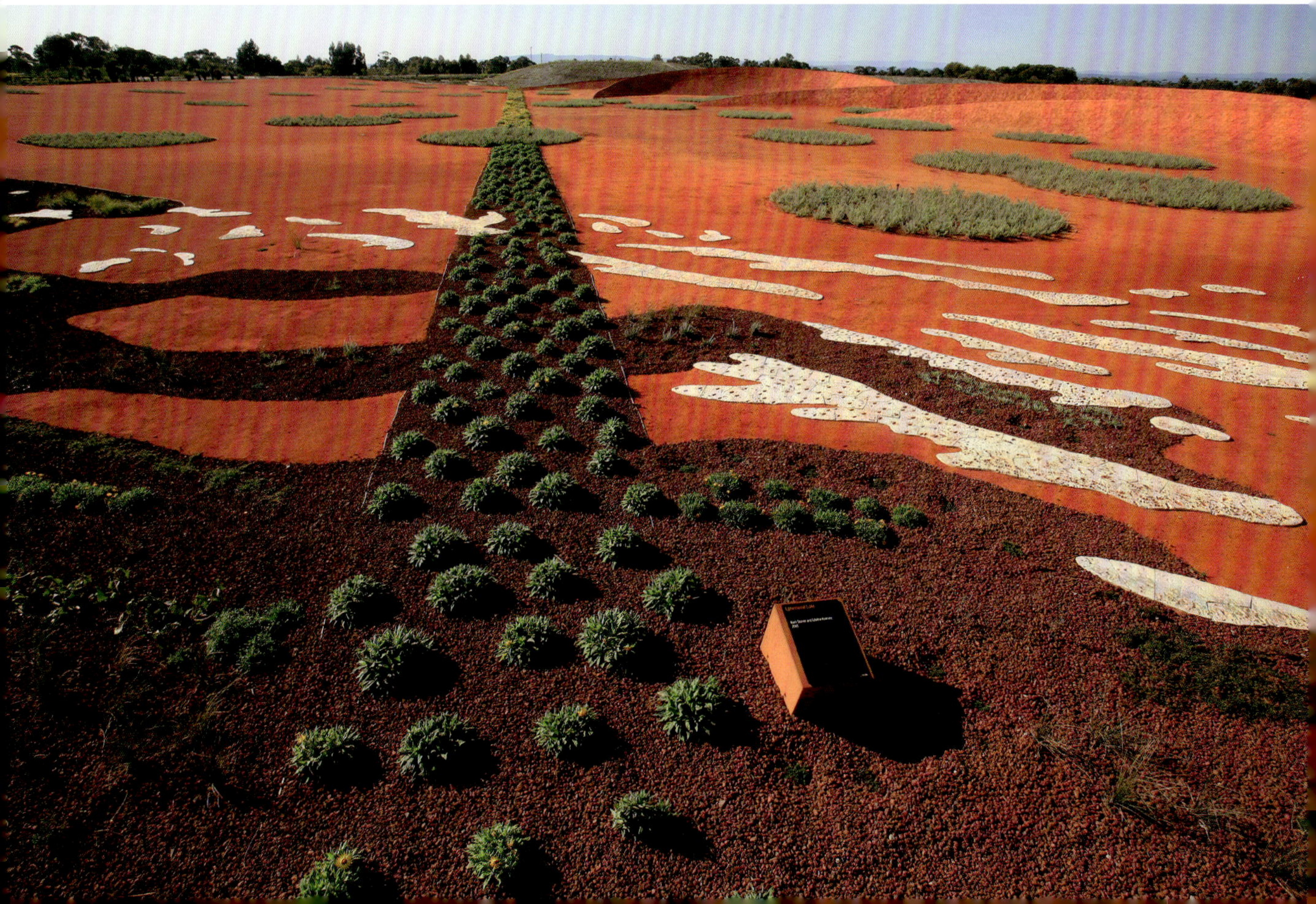

ROYAL BOTANIC GARDENS

Cranbourne, Melbourne, Australia, 1995 (Stage 1 complete; Stage 2 ongoing)

Address: 1000 Ballarto Road, Cranbourne, VIC 3977, Australia, +61 3 5990 2200,
www.rbg.vic.gov.au. Area: 1.25 hectares. Client: Royal Botanic Gardens. Cost: $18 million
Collaboration: Paul Thompson, Mark Stoner, Edwina Kearney, Greg Clark (Artists)

This project, called the Australian Garden, is located 30 kilometers south of Melbourne at the **ROYAL BOTANIC GARDENS** in Cranbourne. The designers point out that usually botanic gardens are based on European predecessors. "The Australian Garden by contrast," they say, "uses the Australian landscape as its inspiration to create a sequence of powerful sculptural and artistic landscape experiences that recognize its diversity, breadth of scale, and wonderful contrasts. The project seeks to stimulate and educate visitors of the potential use and diversity of Australian flora." They have sought to express the tension between Australian reverence for nature and the continual drive to modify it. Thus the western side of the garden is more free flowing and inspired directly by nature, while the eastern side contains "highly designed exhibition gardens" by various landscape architects. Logically, the center of the project is the Sand Garden, evoking the "dry red center of the Australian continent." Water features with variable rates of flow are used as the "mediating element" between these contrasting views of Australian nature. Biodiversity, sustainability, and other ecological concerns are integrated into the park design. Artworks by Greg Clark and Mark Stoner are included in the park as are structures by Kerstin Thompson and Greg Burgess. Based on the success of the first phase of this project, the client asked Taylor Cullity Lethlean and Paul Thompson to work on a second phase expansion.

Dieses Projekt, der Australische Garten, liegt 30 Kilometer südlich von Melbourne auf dem Gelände der **ROYAL BOTANIC GARDENS** in Cranbourne. Die Planer weisen darauf hin, dass sich Botanische Gärten normalerweise an europäischen Vorbildern orientieren. „Der Australische Garten dagegen nutzt die Landschaft Australiens als Inspiration, um eine Folge von kraftvollen skulpturalen und künstlerischen Landschaftserfahrungen mit ihrer gesamten Vielfalt, ihren unterschiedlichen Maßstäben und den wunderbaren Kontrasten zu schaffen. Das Projekt versucht, die Besucher anzuregen und ihnen die Potenziale und Vielfalt der australischen Flora zu zeigen." Die Planer wollen die Spannung aufzeigen zwischen der Verehrung der Natur und dem ständigen Drang, eben diese zu verändern. So ist der westliche Gartenbereich eher in freien, unmittelbar von der Natur beeinflussten Formen angelegt, wohingegen im östlichen Bereich „durchgestylte Ausstellungsgärten" von mehreren Landschaftsarchitekten präsentiert werden. Den Mittelpunkt dieses Projekts bildet logischerweise der „Sand Garden", der an die „trockene rote Mitte des australischen Kontinents" erinnert. Als „vermittelnde Elemente" zwischen diesen gegensätzlichen Betrachtungsweisen der australischen Natur dienen unterschiedliche Wasserelemente. Biodiversität, Nachhaltigkeit und andere ökologische Aspekte sind Bestandteile des Parkentwurfs. Im Park befinden sich Werke von Greg Clark und Mark Stoner sowie Bauten von Kerstin Thompson und Greg Burgess. Da der erste Bauabschnitt des Projektes ein großer Erfolg war, beauftragte der Bauherr das Büro Taylor Cullity Lethlean und Paul Thompson mit der Ausarbeitung einer zweiten Erweiterungsphase.

Ce projet, l'Australian Garden, a été réalisé dans les **ROYAL BOTANIC GARDENS** de Cranbourne, à 30 km au sud de Melbourne. Ses créateurs notent que les jardins botaniques s'appuient en général sur le modèle établi par leurs prédécesseurs européens. « Par contraste, expliquent-ils, l'Australian Garden s'inspire du paysage australien pour créer une séquence d'explorations éloquentes du paysage, sculpturales et artistiques, qui prennent en compte sa diversité, l'ampleur de son échelle et ses merveilleux contrastes. Le projet veut stimuler le visiteur et lui montrer les usages potentiels et la diversité de la flore australienne. » L'agence a cherché à exprimer la tension entre le respect des Australiens pour la nature et leur tendance permanente à la modifier. Le côté ouest du jardin est traité en flux plus libres et directement inspirés de la nature, tandis que le côté est regroupe « des jardins d'exposition très dessinés » par divers architectes paysagistes. Dans une logique géographique, le centre du projet est le Sand Garden qui évoque « la région centrale sèche et rouge du continent australien ». Divers bassins et éléments aquatiques servent d'« éléments de médiation » entre ces visions contrastées de la nature australienne. La biodiversité, la durabilité et d'autres préoccupations écologiques ont été intégrées dans le projet, de même que des œuvres d'art de Greg Clark et Mark Stoner et des constructions de Kerstin Thompson et de Greg Burgess. Face au succès de la première phase de ce projet, les Royal Botanic Gardens ont chargé Taylor Cullity Lethlean et Paul Thompson de travailler sur une seconde phase.

An overall plan of the project and an aerial view (seen from a different angle) show the attention of the designers to creating a variety of different environments, some more "painterly" as seen here, and others more "natural."

Der Gesamtplan des Projekts und die Luftaufnahme, die aus einem anderen Blickwinkel gemacht wurde, zeigen, mit welcher Sorgfalt die Planer vorgegangen sind, um eine Vielfalt unterschiedlicher Naturumfelder darzustellen. Manche sind eher „malerisch" wie hier, andere eher „naturnah" umgesetzt.

Un plan d'ensemble du projet et une vue aérienne mettent en évidence l'attention portée par les paysagistes à la création d'environnements différents et variés, certains plus « picturaux » comme ici, et d'autres plus « naturels ».

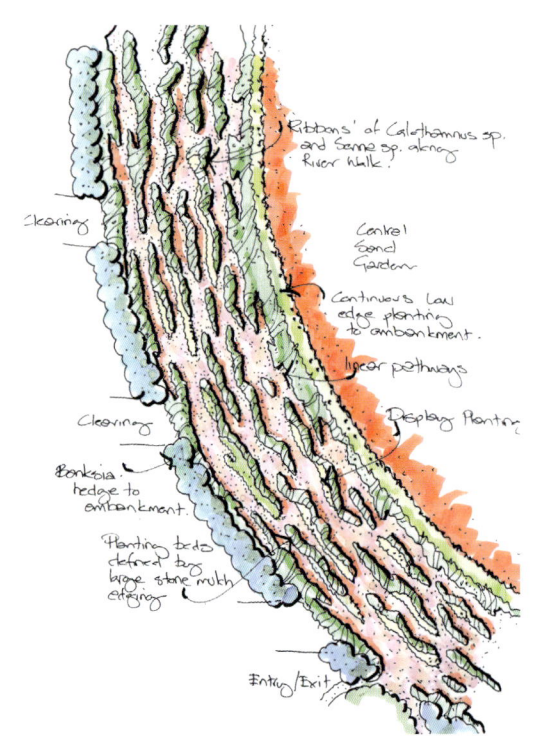

The brief for the project requires that it should "explore and illustrate the role of native flora in shaping the nature of Australia; display native flora in creative ways; and celebrate the role of Australian plants in Australian life and culture."

In der Auftragsbeschreibung für dieses Projekt wurde verlangt, es solle „die Rolle der heimischen Flora untersuchen und darstellen, wie sie der Natur in Australien ihre Gestalt gibt; die heimische Flora in kreativer Art und Weise vorstellen und die Rolle der australischen Pflanzen im Leben und in der Kultur Australiens herausarbeiten".

Le cahier des charges du projet spécifiait qu'il devait « explorer et illustrer le rôle de la flore locale dans la nature australienne ; présenter cette flore de manières créatives et célébrer le rôle des plantes australiennes dans la vie et la culture du pays ».

The designers state: "Utilizing 100 000 species of flora, some never before seen in cultivation, the garden illustrates the enormous potential of our flora in creating distinctive, bold, and memorable garden experiences."

Die Planer stellen fest: „Mit 100 000 Pflanzenarten, von denen einige nie zuvor kultiviert wurden, ist dieser Garten mit seinen unverwechselbaren, kühnen und unvergesslichen Erlebnissen Beweis für das enorme Potenzial unserer Pflanzenwelt."

« Mettant en scène 100 000 espèces de plantes, dont certaines encore jamais cultivées, le jardin illustre l'énorme potentiel de notre flore de créer des opportunités de découverte originales, audacieuses et inoubliables », précisent les paysagistes.

LUIS VALLEJO

LVEP
Paseo de los Servales, 4 Urb. Ciudalcampo
San Sebastian de los Reyes
28707 Madrid
Spain

Tel: + 34 91 657 09 54 / Fax: + 34 91 657 03 52
E-mail: paisajismo@luisvallejo.com
Web: www.luisvallejo.com

LUIS VALLEJO GARCÍA-MAURIÑO was born in Madrid, Spain, in 1954. He created his own office, Luis Vallejo Estudio de Paisajismo (LVEP), in Madrid in 1986. He states that his projects, ranging from private gardens to large campuses such as that of the Ciudad Grupo Santander published here, have in common "natural elements—vegetal and mineral—always forming an axis that branches out, develops, and converts the project into a reality; from a private garden, to a public plaza or a hospital." Luis Vallejo is also the Curator and Director of the Municipal Bonsai Museum in Alcobendas, the Curator of the Bonsai collections at the Royal Botanical Gardens in Madrid, and at the Parla Botanical Gardens–Bonsai Museum also in Madrid. He also manages and owns the company Arceval/Jardineria S. L., which is in charge of executing his garden projects. In 2008, Luis Vallejo was awarded "The Order of the Rising Sun" by HM the Emperor of Japan for his role in promoting Japanese culture in other countries. Aside from the campus of the Santander Group City (Madrid, Spain, 2002–05, published here), Vallejo has completed the Río Hortega University Hospital (Valladolid, Spain, 2007–09); the landscaping of the Royal Mansour Hotel (Marrakech, Morocco, 2005–10); and a large number of private gardens in Spain. Current work includes the Port Aventura Theme Park (Salou, Tarragona, Spain, 2005–under construction); and the 400-hectare Oman Botanic Garden (Muscat, Oman, 2007–under construction).

LUIS VALLEJO GARCÍA-MAURIÑO wurde 1954 in Madrid geboren. Er gründete 1986 dort sein eigenes Büro Luis Vallejo Estudio de Paisajismo (LVEP). Er sagt, dass alle seine Projekte, vom Privatgarten bis zu großen Anlagen wie der hier vorgestellte Campus der Santander-Gruppe, immer „natürliche – pflanzliche und mineralische – Elemente aufweisen, die eine Achse bilden, die sich verzweigt, entwickelt und einen Entwurf in die Realität eines Privatgartens, eines öffentlichen Platzes oder das Gelände eines Krankenhauses verwandelt". Luis Vallejo ist außerdem Kurator und Direktor des Museo Municipal de Bonsái in Alcobendas, Kurator der Colección de Bonsái del Real Jardín Botánico in Madrid und Kurator der Bonsai-Sammlung des Jardín Botánico von Parla, Provinz Madrid. Er ist außerdem Eigentümer des Garten- und Landschaftsbaubetriebes Arceval/Jardineria S. L., der seine Gartenprojekte ausführt. 2008 wurde Luis Vallejo vom Kaiser von Japan mit dem Orden der aufgehenden Sonne für seine Verdienste um die Förderung der japanischen Kultur in anderen Ländern ausgezeichnet. Neben dem Campus für die Santander-Gruppe in Madrid (2002–05) hat Vallejo die Außenanlagen für die Universitätsklinik Río Hortega in Valladolid, Spanien (2007–09), die Außenanlagen des Hotels Royal Mansour in Marrakesch, Marokko (2005–10), sowie viele Privatgärten in Spanien ausgeführt. Zu seinen derzeitigen Projekten gehören der Themenpark Port Aventura (Salou, Tarragona, Spanien, seit 2005 in Ausführung befindlich) sowie der 400 Hektar große Botanische Garten von Oman in Maskat, Oman (seit 2007 in Ausführung befindlich).

LUIS VALLEJO GARCÍA-MAURIÑO, né à Madrid en 1954, crée son agence Luis Vallejo Estudio de Paisajismo (LVEP) à Madrid en 1986. Il explique que ses projets, allant de jardins privés à de vastes campus comme celui de la Ciudad Grupo Santander publié ici, ont en commun « des éléments naturels – végétaux et minéraux – pour constituer à chaque fois un axe qui se diversifie et transforme le projet en réalité, que ce soit un jardin privé, une place publique ou l'environnement d'un hôpital ». Luis Vallejo est conservateur et directeur du Museo Municipal de Bonsái d'Alcobendas, conservateur de la collection de bonsaïs du Real Jardín Botánico de Madrid et du Jardín Botánico Museo Bonsái de Parla, également à Madrid. Il possède et dirige par ailleurs la société Arceval/Jardineria S. L., chargée de l'exécution de ses projets de jardins. En 2008, il a reçu l'ordre du Soleil levant de l'empereur du Japon pour son rôle dans la promotion de la culture japonaise à l'étranger. En dehors du campus de la Ciudad Grupo Santander (Madrid, 2002–05, publié ici), Vallejo a réalisé l'environnement paysager de l'hôpital universitaire Río Hortega (Valladolid, Espagne, 2007–09) ; celui de l'hôtel Royal Mansour (Marrakech, Maroc, 2005–10) et un grand nombre de jardins privés en Espagne. Il travaille actuellement aux projets du parc d'attractions Port Aventura (Salou, Tarragone, Espagne, 2005–en cours) et du Oman Botanic Garden de 400 ha (Mascate, Oman, 2007–en cours).

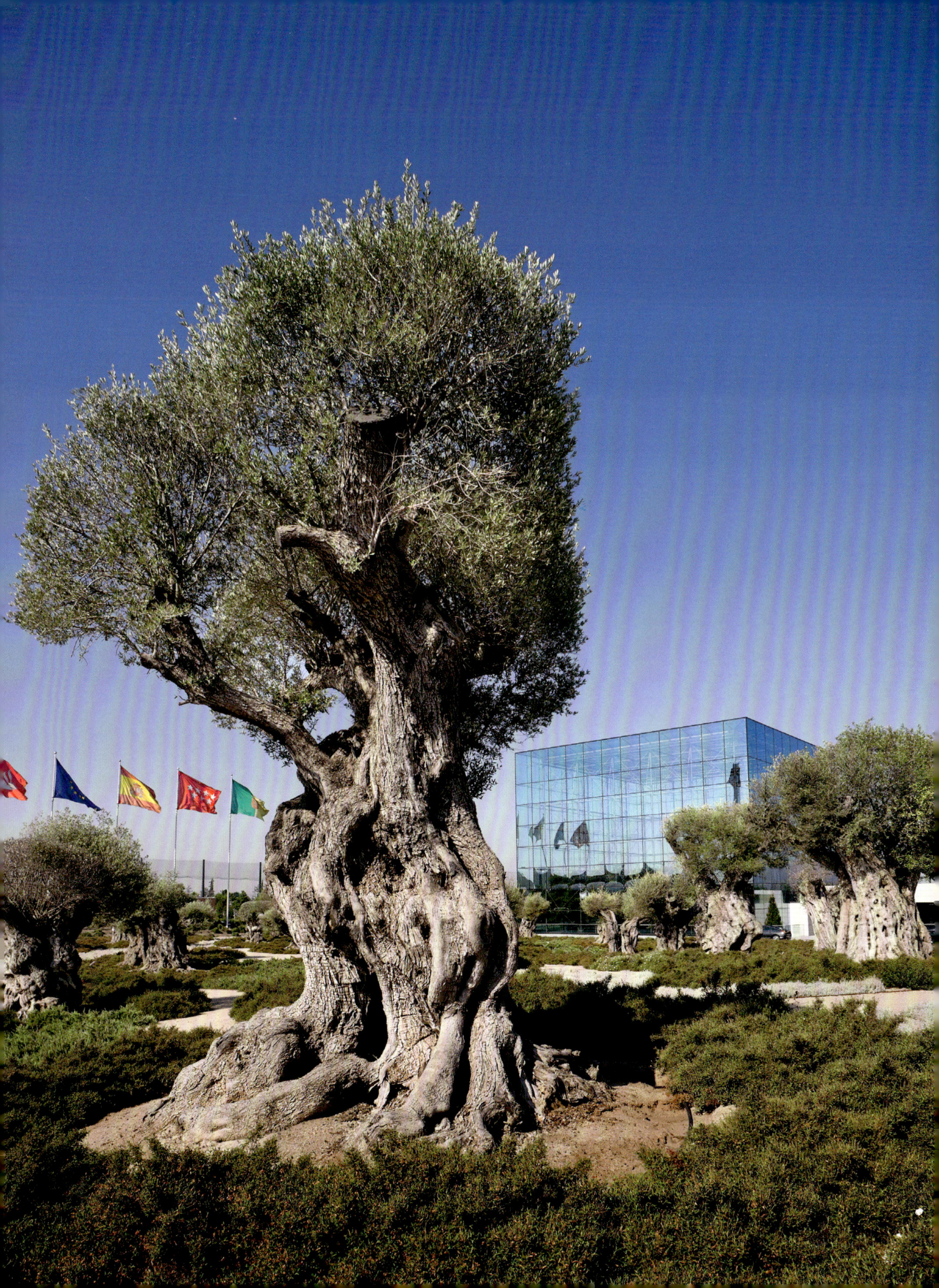

SANTANDER GROUP CITY CAMPUS
Boadilla del Monte, Madrid, Spain, 2002–05

Address: Boadilla del Monte, Madrid, Spain, www.santander.com / www.fundacionbancosantander.com
Area: 250 hectares, 170 hectares of gardens and green areas. Client: Banco Santander. Cost: not disclosed
Collaboration: Roche Dinkeloo Architects, designers of the Ciudad Banco Santander, completed in 2004

Vast areas of landscaped space surround the buildings on the campus—with the glass cube of the entrance buildings seen in the background, an unusual variety of olive tree is planted in the midst of this area.

Weitläufige, gestaltete Landschaftsräume umgeben die Gebäude auf dem Campus. Im Hintergrund erkennt man den Glaskubus des Eingangsgebäudes, in der Mitte eine ungewöhnliche Form von Olivenbäumen.

De vastes zones paysagées entourent les bâtiments du campus (dont on aperçoit le cube de verre de l'entrée dans le fond), sur lesquelles sont plantées une grande variété d'espèces d'oliviers.

The Ciudad Grupo Santander, whose buildings were designed essentially by Kevin Roche, was built in relatively arid, rolling hills 15 kilometers northwest of Madrid. Luis Vallejo explains that the landscaping of the complex "commences with the main building that is anchored on a wide esplanade." A grid of parterres, conceived on the basis of plans by Roche, creates a rhythm of green and flowering plants that flows throughout the complex. Vallejo refers to the idea of an "instant garden, a landscape completed as soon as it is planted," making use, amongst other plants, of a series of eight-meter-tall cypresses interspersed with specifically selected ancient Tuscan olive trees. The sculptural presence of these remarkable trees certainly contrasts with Roche's modernist architectural design, but also highlights the buildings. Meticulous planning went into the garden design and its implementation as soon as the building construction progress allowed. Vallejo's objectives were to "create harmony, proportion, balance, and uniformity, without falling prey to monotony." The roof decks of the peripheral buildings are an important part of the overall project, as is the distant landscape. The result is "an essentially textured garden that is also visual and pictorial." Work on a 16-hectare public park across the road from the Ciudad, also designed by Luis Vallejo, is currently being completed.

Die Ciudad Grupo Santander, deren Gebäude im Wesentlichen von Kevin Roche entworfen sind, liegt etwa 15 Kilometer nordwestlich von Madrid auf einem recht trockenen, hügeligen Gelände. Luis Vallejo erklärt, dass die landschaftliche Gestaltung vom Hauptgebäude ausgeht, dem eine großzügige Promenade vorgelagert ist. Das Raster von Parterregärten, die auf der Grundlage der Grundrisse von Roche entstanden sind, schafft einen Rhythmus von grünen und blühenden Pflanzen, der sich durch den gesamten Komplex zieht. Vallejo wollte hier einen „Garten schaffen, der mit Abschluss der Pflanzungen bereits fertig ist". Dazu pflanzte er unter anderem eine Reihe acht Meter hoher Zypressen, zwischen die er alterhrwürdige toskanische Olivenbäume setzte. Die skulpturale Präsenz dieser ungewöhnlichen Bäume bildet einen Kontrast zu der modernen Architektur von Kevin Roche und betont die Gebäude zugleich. Die Gartenanlagen wurden sorgfältig geplant und ausgeführt, sobald der Baufortschritt es erlaubte. Vallejo wollte „Harmonie, Ausgewogenheit und Einheitlichkeit schaffen, ohne dabei in Monotonie zu verfallen". Die Dachterrassen der am Rand liegenden Gebäude sind ebenso ein wesentliches Merkmal des Gesamtprojektes wie die weiter entfernte Landschaft. Das Ergebnis ist ein „durchstrukturierter Garten, der auch visuelle und malerische Qualitäten hat". Derzeit arbeitet Luis Vallejo an der Fertigstellung eines ebenfalls von ihm entworfenen, 16 Hektar großen öffentlichen Parks auf der anderen Straßenseite der Ciudad Grupo Santander.

La Ciudad Grupo Santander, dont les bâtiments ont été conçus par Kevin Roche, a été édifiée dans un paysage de collines relativement arides, à 15 km au nord-ouest de Madrid. Luis Vallejo explique que l'aménagement paysager du complexe « part du bâtiment principal, dressé sur une vaste esplanade ». Une trame de parterres, prévue par les plans de Roche, crée une alternance de pelouses et de plantations fleuries qui couvre la totalité du complexe. Vallejo évoque « un jardin instantané, un paysage achevé dès qu'il est planté », dans lequel il a notamment mis en place une ligne de cyprès de 8 m de haut intercalés avec d'anciens oliviers de Toscane spécialement sélectionnés. La présence sculpturale de ces arbres remarquables contraste avec l'architecture contemporaine moderniste de Roche, qu'elle contribue également à mettre en valeur. Une programmation méticuleuse a présidé à ce projet et à sa mise en place, au fur et à mesure que le chantier de construction le permettait. L'objectif de Vallejo était de créer une harmonie, des proportions, un équilibre et une uniformité sans tomber dans la monotonie. Les toits végétalisés des bâtiments périphériques jouent un rôle fondamental dans ce projet, tout comme le paysage au loin. Au final, il s'agit d'« un jardin essentiellement texturé, visuel et graphique ». Luis Vallejo intervient actuellement sur un parc public de 16 ha de l'autre côté de la route qui dessert la Ciudad Santander.

A plan of the Ciudad on the left page can be compared to the aerial view above, with exceptionally large gardens surrounded by the more typical arid zones that exist around Madrid visible in the distance.

Man kann den Grundriss des Campus auf der linken Seite mit der Luftaufnahme oben vergleichen und erkennt, dass die außergewöhnlich große Anlage von den für die Umgebung von Madrid typischen ariden Zonen umgeben ist, die man in der Ferne sieht.

Le plan de la Ciudad (page de gauche) peut être rapproché de la vue aérienne ci-dessus. Les jardins de taille exceptionnelle sont entourés d'une zone aride typique de la région de Madrid que l'on aperçoit au loin.

The central, round building that houses the offices of the bank's directors is, like the rest of the complex, surrounded by carefully designed planted areas.

In dem runden Hauptgebäude befinden sich die Büros der Direktoren der Bank; es ist, wie alle anderen Bauten, von sorgfältig geplanten Grünanlagen umgeben.

Le bâtiment circulaire central, qui est celui de la direction, est lui aussi entouré de jardins minutieusement dessinés.

Within the immediate area of the office buildings, planting is often more precise and geometric than in the periphery, where more exuberant strategies are used for the landscaping.

Die Bepflanzung in unmittelbarer Nähe der Bürogebäude ist häufig präziser und geometrischer als am Rand. Dort ist das Grün eher üppig und weniger streng.

Les plantations sont plus dessinées et plus régulières à proximité des bâtiments administratifs qu'en périphérie où des stratégies plus libres ont été mises en œuvre.

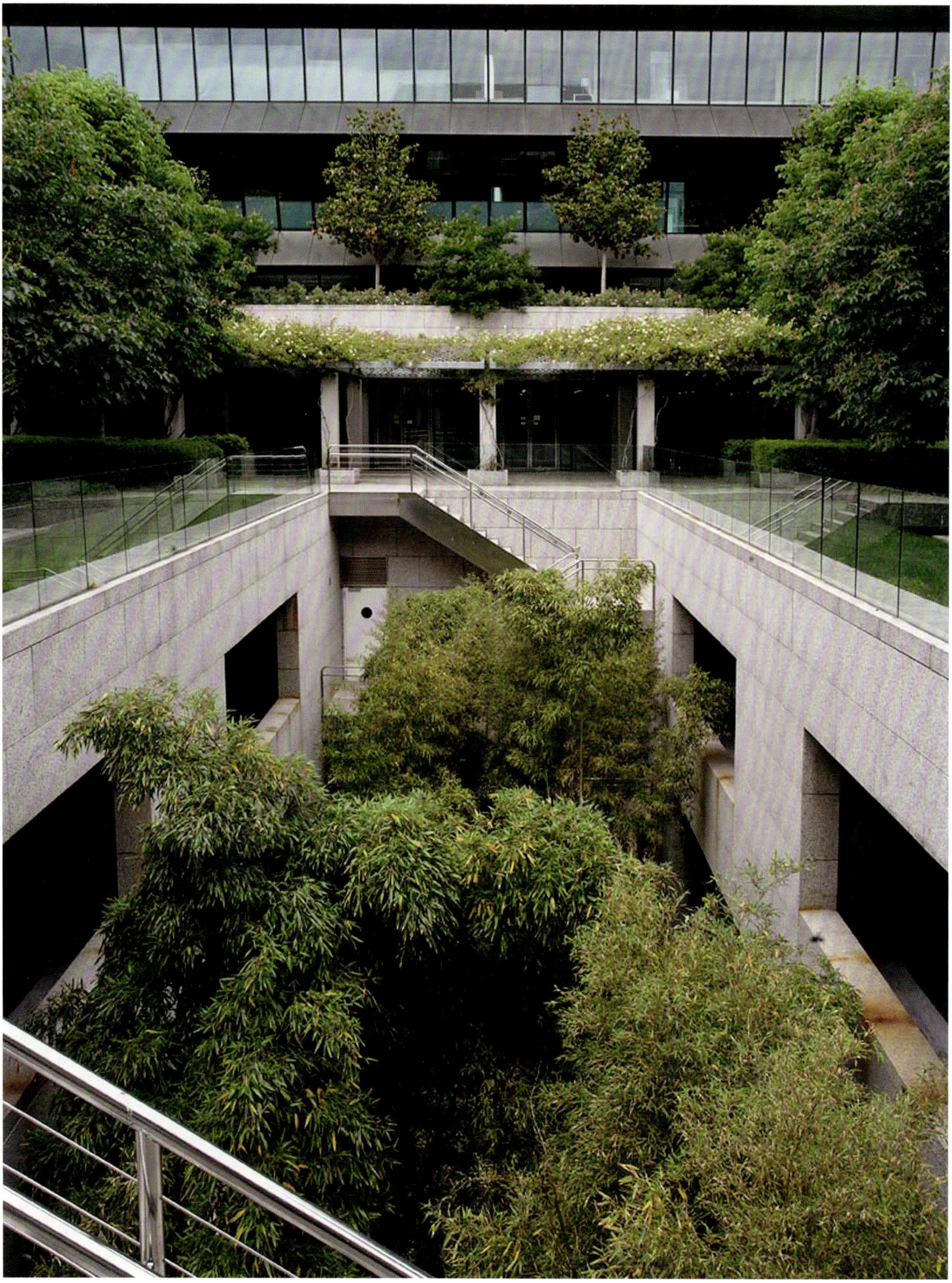

A careful consideration of the varieties of plants that do well in the hot climate of Madrid together with a large gardening staff assure that the Santander landscape remains luxuriant and relaxing for the bank staff.

Eine sorgfältige Auswahl der Pflanzen, die in dem extremen Madrider Klima gedeihen können, und ein großer Stab von Gärtnern gewährleisten, dass die Landschaft des Santander Group City Campus ihre Üppigkeit behält und den Bankangestellten einen Erholungsraum bietet.

Le choix de variétés adaptées au climat très chaud en été du plateau madrilène doublé d'une multitude de jardiniers font que le paysage de la Ciudad reste toujours aussi luxuriant et reposant.

MICHAEL VAN VALKENBURGH

Michael Van Valkenburgh Associates, Inc.,
Landscape Architects, PC
16 Court Street, 11th Floor
Brooklyn, NY 11241, USA

Tel: +1 718 243 2044 / Fax: +1 718 243 1293
E-mail: mvva_ny@mvvainc.com / Web: www.mvvainc.com

MICHAEL VAN VALKENBURGH received a B.S. degree from the Cornell University College of Agriculture (Ithaca, New York, 1973) and a Master of Fine Arts in Landscape Architecture from the University of Illinois (Champaign/Urbana, 1977). He oversees both the New York and Cambridge (Massachusetts) offices of the firm he founded in 1982— Michael Van Valkenburgh Associates, Inc. (MVVA)—and is involved in some way in every project. Other firm Principals are **MATTHEW URBANSKI**, who is a lead designer for many of the firm's public projects, and **LAURA SOLANO**, who is a specialist in landscape technology. Matthew Urbanski joined MVVA in 1989 after receiving a Master of Landscape Architecture degree from Harvard the same year. Laura Solano has worked at MVVA since 1991. She received a Bachelor of Landscape Architecture from Ohio State University in 1983. Their work includes Tahari Courtyards (Millburn, New Jersey, 2002–03); Alumnae Valley Landscape Restoration, Wellesley College (Wellesley, Massachusetts, 2001–05, published here); the Connecticut Water Treatment Facility (with Steven Holl; New Haven, Connecticut, 2001–05, also published here); Teardrop Park (New York, New York, 1999–2006, also published here); the ASLA Green Roof (Washington, D.C., 2005–06); and Harvard Yard Restoration (Cambridge, Massachusetts, 1993–2009). Ongoing work of the firm includes Brooklyn Bridge Park (Brooklyn, New York, 2003–); Princeton University (master plan and various projects, Princeton, New Jersey, 2006–); North Grant Park (Chicago, Illinois, 2009–); and CityArchRiver 2015 (Saint Louis, Missouri, 2010–), all in the USA.

MICHAEL VAN VALKENBURGH machte 1973 seinen Bachelor am College of Agriculture der Cornell University (Ithaca, New York) und 1977 seinen Master in Landschaftsarchitektur an der University of Illinois (Champaign/Urbana). Er leitet beide von ihm 1982 gegründeten Büros in New York und in Cambridge (Massachusetts) – Michael Van Valkenburgh Associates, Inc. (MVVA) – und ist in jedes Projekt eingebunden. Zu seinen Büropartnern gehören **MATTHEW URBANSKI**, Chefdesigner für viele öffentliche Projekte des Büros, und **LAURA SOLANO**, Spezialistin für Landschaftstechnologie. Matthew Urbanski kam nach Abschluss seines Masters in Landschaftsarchitektur in Harvard 1989 zu MVVA. Laura Solano arbeitet dort seit 1991. Sie machte 1983 ihren Bachelor in Landschaftsarchitektur an der Ohio State University. Zu den von MVVA ausgeführten Projekten gehören die Tahari Courtyards (Millburn, New Jersey, 2002–03), die Sanierung des Alumnae Valley am Wellesley College (Wellesley, Massachusetts, 2001–05, hier vorgestellt), die ebenfalls hier vorgestellte Wasseraufbereitungsanlage Connecticut Water Treatment Facility (mit Steven Holl, New Haven, Connecticut, 2001–05), der Teardrop Park (New York, 1999–2006, auch hier vorgestellt), das Gründach der American Society of Landscape Architects (ASLA) in Washington, D. C. (2005–06), und die Restaurierung des Parks Harvard Yard (Cambridge, Massachusetts, 1993–2009). Zu den derzeit laufenden Projekten des Büros zählen der Brooklyn Bridge Park (Brooklyn, New York, seit 2003), der Masterplan und mehrere Projekte für die Princeton University (Princeton, New Jersey, seit 2006), der North Grant Park (Chicago, Illinois, seit 2009) und die Parkanlage CityArchRiver 2015 (Saint Louis, Missouri, seit 2010), alle in den USA.

MICHAEL VAN VALKENBURGH est titulaire d'un B.S. du Cornell University College of Agriculture (Ithaca, New York, 1973) et d'un M.F.A. d'architecture du paysage de l'université de l'Illinois (Champaign/Urbana, 1977). Il dirige les bureaux de New York et de Cambridge (Massachusetts) de son agence – Michael Van Valkenburgh Associates, Inc. (MVVA) – et s'implique dans chaque projet. Les autres dirigeants sont **MATTHEW URBANSKI**, concepteur de projets pour le secteur public, et **LAURA SOLANO**, spécialiste des technologies du paysage. Matthew Urbanski a rejoint MVVA en 1989 après avoir obtenu son Master of Landscape Architecture à Harvard. Laura Solano, à MVVA depuis 1991, est titulaire d'un B.A. en architecture du paysage de l'université de l'Ohio (1983). Parmi leurs réalisations aux États-Unis : les Tahari Courtyards (Millburn, New Jersey, 2002–03) ; la restauration paysagère de l'Alumnae Valley au Wellesley College (Wellesley, Massachusetts, 2001–05, publié ici) ; la Connecticut Water Treatment Facility (avec Steven Holl, New Haven, Connecticut, 2001–05, publiée ici) ; le Teardrop Park (New York, 1999–2006, publié ici) ; le toit vert de l'ASLA (Washington, 2005–06) et la restauration du Harvard Yard (Cambridge, Massachusetts, 1993–2009). L'agence travaille actuellement sur le Brooklyn Bridge Park (Brooklyn, New York, 2003–) ; l'université de Princeton (plan directeur et divers projets, Princeton, New Jersey, 2006–) ; le North Grant Park (Chicago, Illinois, 2009–) et le CityArchRiver 2015 (Saint Louis, Missouri, 2010–).

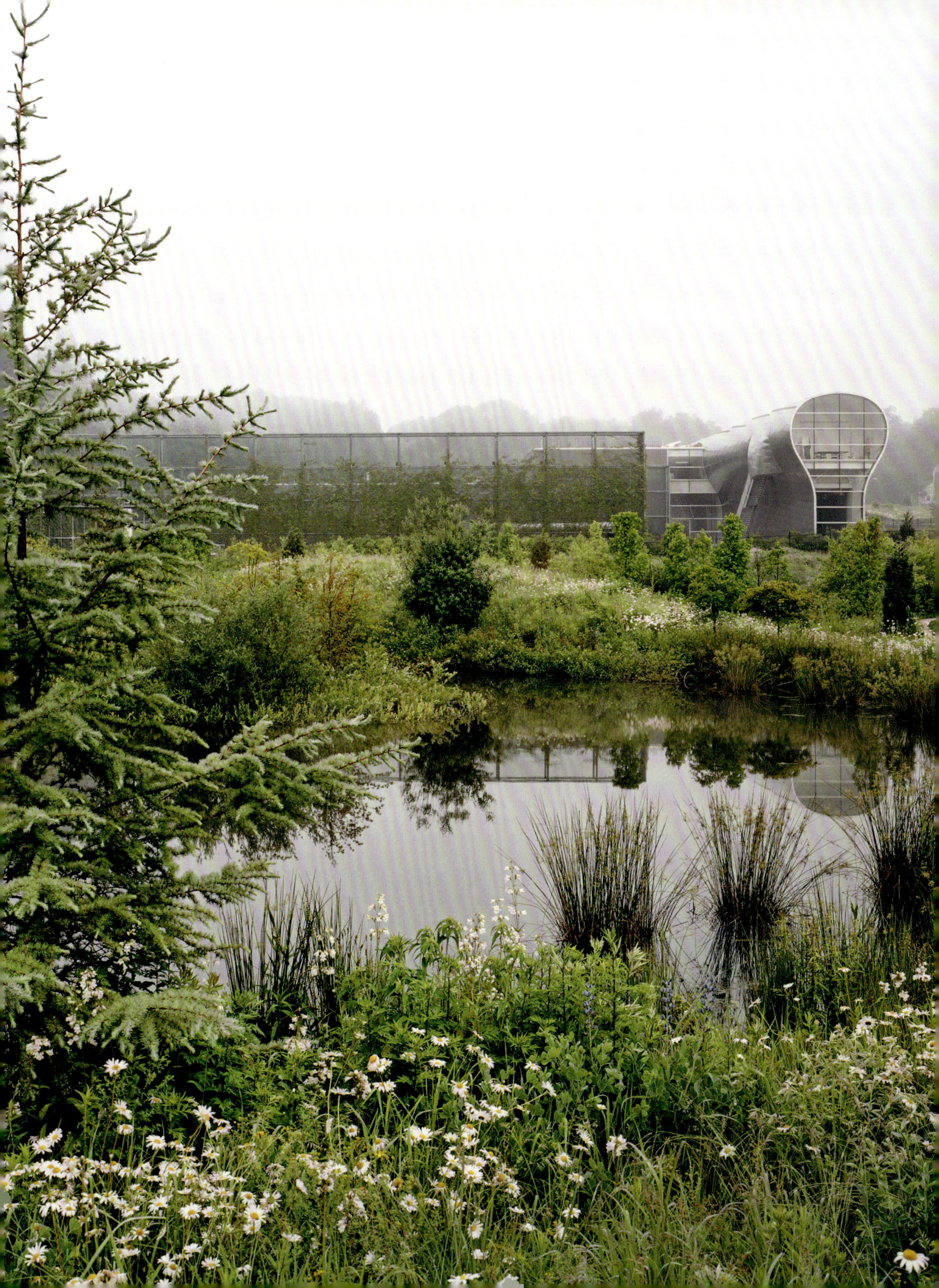

CONNECTICUT WATER TREATMENT FACILITY

New Haven, Connecticut, USA, 2001–05

Area: 5.67 hectares. Client: South Central Connecticut Regional Water Authority
Cost: $2.88 million. Collaboration: A. Paul Seck (Senior Associate), Robert Rock (Associate)

This project is located on the outskirts of New Haven. The facility concerned is a reserve water source for the South Central Connecticut Regional Water authority, drawing water from Lake Whitney. The architect of the facility was Steven Holl. MVVA was given a limited budget of approximately $5 per square foot for the landscaping. The landscape architects state: "The use of the most elemental of landscape architectural tools—soil, water, and plants—offsets the sleek form of the facility building. The design creates topographical variety and interest through sustainable reuse of excavated soil. Swales replace a traditional engineered drainage system. The planting program, inspired by restoration ecology, is at once primal and sophisticated in its extent and complexity." Stormwater and runoff from the facility roof are run through the landscape and filtered. MVVA used native species that require no fertilizers or pesticides to thrive. They also sought to create "seasonal variation in color and texture" with the plant selection.

Dieses Projekt befindet sich außerhalb von New Haven. Die Anlage dient als Wasserreservoir für die Regionale Wasserbehörde South Central Connecticut, die das Wasser aus dem Whitney-See bezieht. Architekt des Gebäudes war Steven Holl. MVVA wurde nur ein beschränktes Budget von etwa 5 US-Dollar pro Quadratmeter zugestanden. Die Landschaftsarchitekten bemerken dazu: „Die Verwendung der grundlegenden Gestaltungselemente der Landschaftsarchitekten – Erde, Wasser und Pflanzen – dient als Ausgleich für die schlanke Form des Gebäudes. Der Entwurf schafft eine topografische Vielfalt und ist wegen der nachhaltigen Wiederverwendung des Aushubmaterials interessant. Bodensenken ersetzen das herkömmliche technische Drainagesystem. Das Bepflanzungsprogramm beruft sich auf die Ökologie der Sanierung und ist in seinem Umfang und seiner Komplexität sowohl einfach als auch sorgfältig durchdacht." Das Regen- und Abflusswasser vom Gebäudedach wird in die Naturlandschaft geleitet und gefiltert. MVVA verwendete heimische Pflanzen, die weder Dünger noch Pestizide benötigen, um zu gedeihen. Man versuchte auch durch die Auswahl der Pflanzen je nach Jahreszeit einen Wechsel von Farben und Texturen zu erreichen.

„Le projet est situé dans la banlieue de New Haven. Il portait sur une réserve d'eau naturelle, pompée dans le lac Whitney par le Département régional des eaux du centre-sud du Connecticut. Les bâtiments des installations techniques ont été réalisés par Steven Holl. MVVA disposait d'un budget limité d'environ 5 $ par mètre carré pour l'aménagement paysager. « Le recours aux outils les plus basiques de l'architecture paysagère – la terre, l'eau et les plantes – vient contrebalancer la forme lisse du bâtiment. Le projet apporte une diversité topographique et suscite un intérêt visuel grâce à la réutilisation de la terre extraite lors du creusement du sol. Des baissières remplacent les systèmes traditionnels de drainage. Le programme des plantations, inspiré de principes écologiques de restauration, est à la fois élémentaire et sophistiqué dans sa complexité et son étendue », explique l'architecte. MVVA a utilisé des espèces végétales locales qui ne demandent ni engrais ni pesticides et ont permis de mettre en scène des « variations saisonnières de couleurs et de textures ». Les eaux de pluie et celles récupérées par l'intermédiaire du toit du bâtiment sont canalisées et filtrées à travers le paysage.

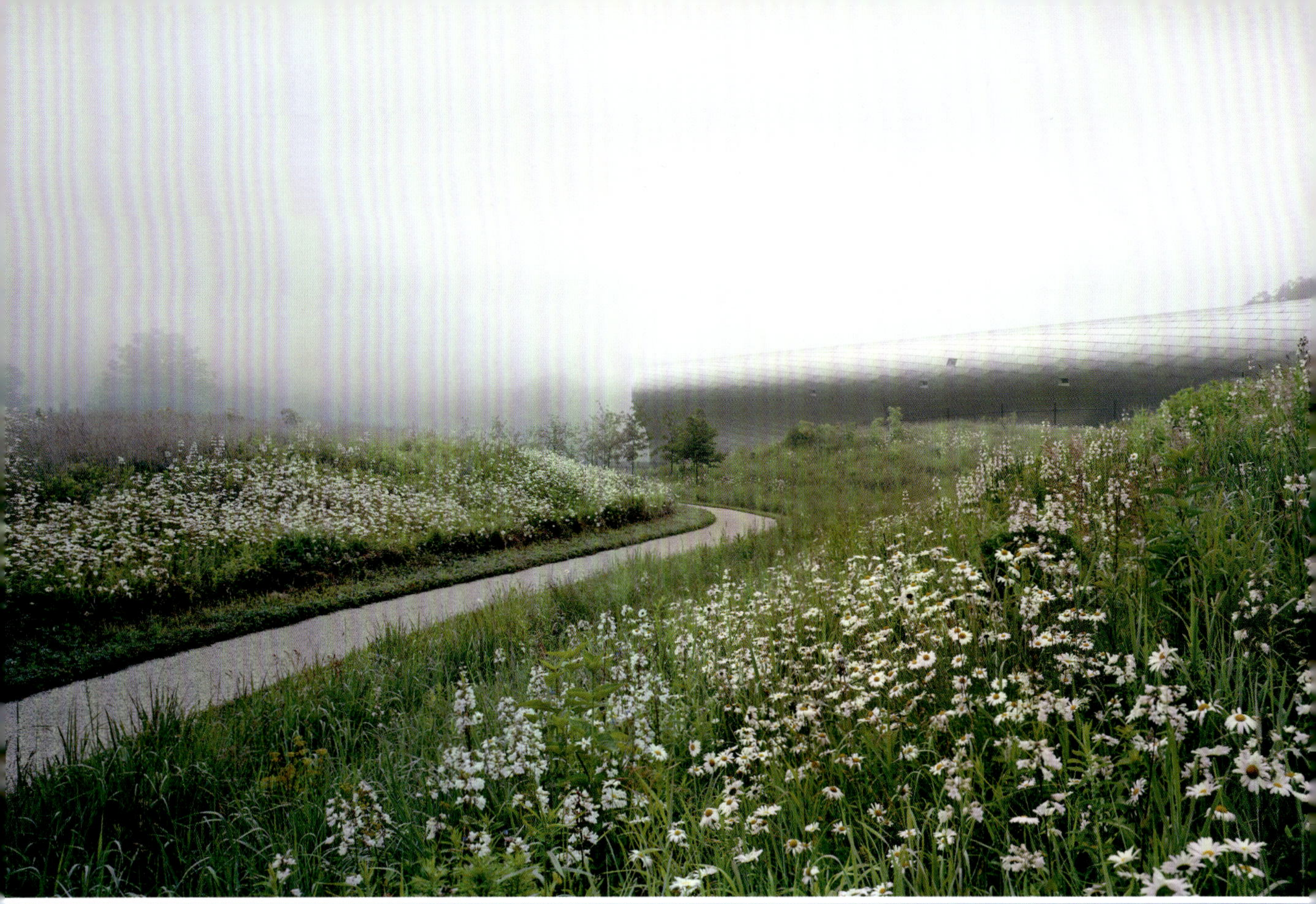

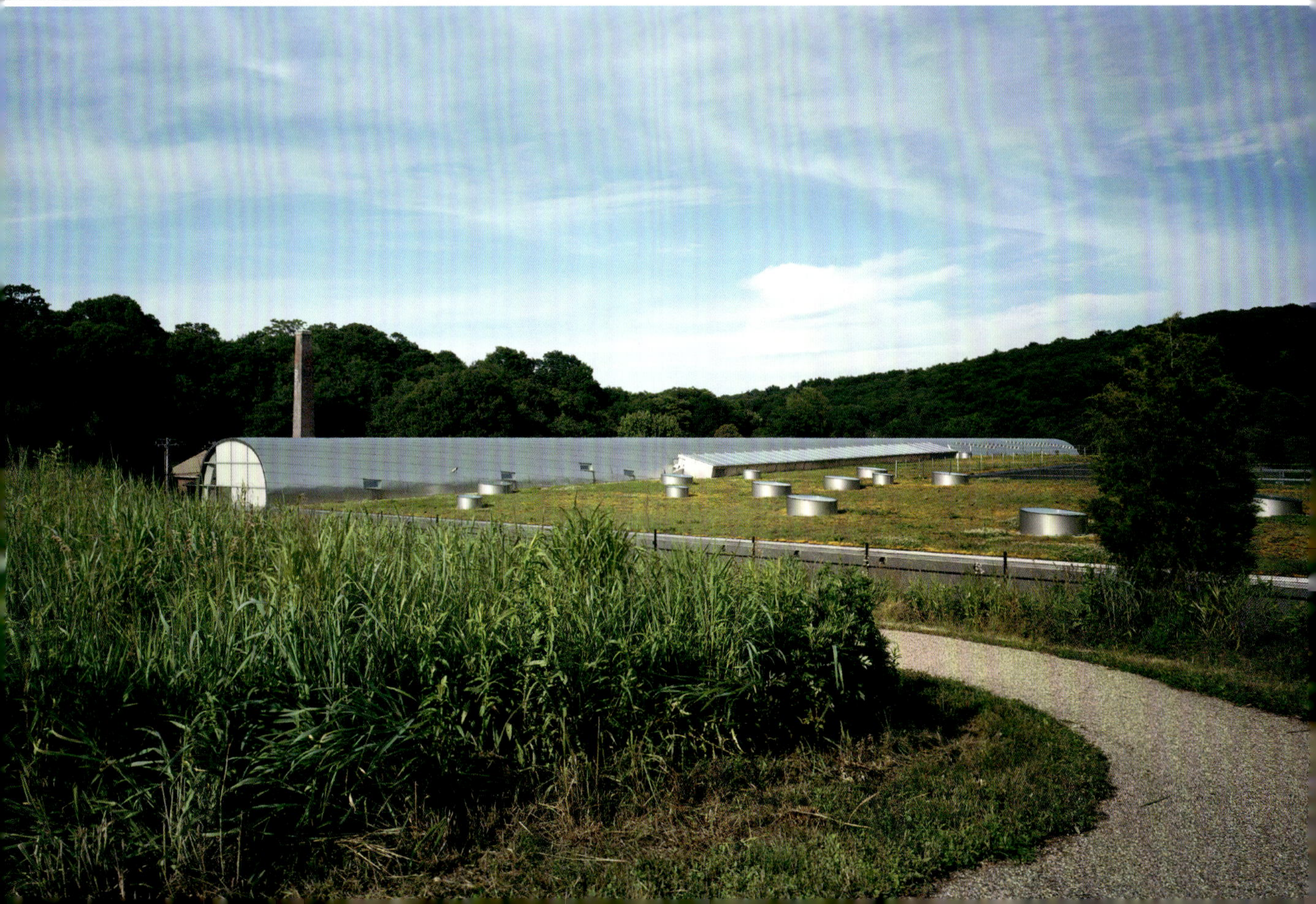

ALUMNAE VALLEY LANDSCAPE RESTORATION

Wellesley College, Wellesley, Massachusetts, USA, 2001–05

Address: Wellesley College, 106 Central Street, Wellesley, MA 02481, USA, +1 781 283 1000, www.wellesley.edu
Area: 5.46 hectares. Client: Wellesley College. Cost: not disclosed

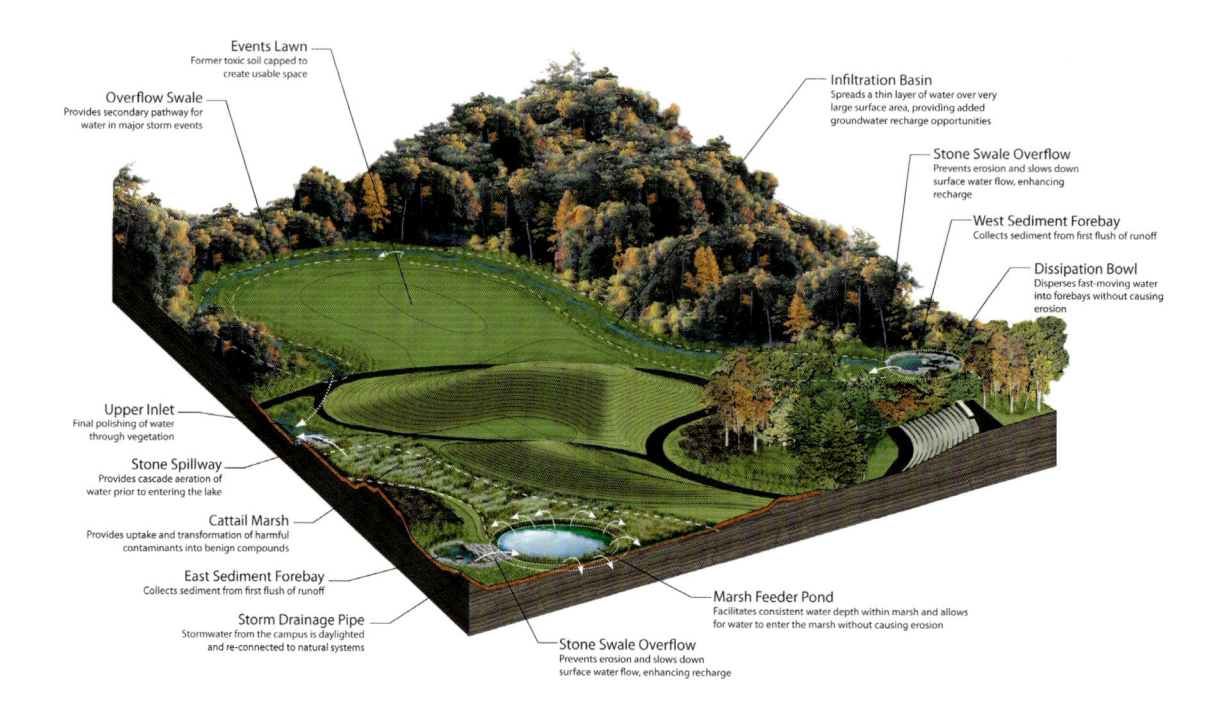

Events Lawn — Former toxic soil capped to create usable space
Overflow Swale — Provides secondary pathway for water in major storm events
Infiltration Basin — Spreads a thin layer of water over very large surface area, providing added groundwater recharge opportunities
Stone Swale Overflow — Prevents erosion and slows down surface water flow, enhancing recharge
West Sediment Forebay — Collects sediment from first flush of runoff
Dissipation Bowl — Disperses fast-moving water into forebays without causing erosion
Upper Inlet — Final polishing of water through vegetation
Stone Spillway — Provides cascade aeration of water prior to entering the lake
Cattail Marsh — Provides uptake and transformation of harmful contaminants into benign compounds
East Sediment Forebay — Collects sediment from first flush of runoff
Storm Drainage Pipe — Stormwater from the campus is daylighted and re-connected to natural systems
Marsh Feeder Pond — Facilitates consistent water depth within marsh and allows for water to enter the marsh without causing erosion
Stone Swale Overflow — Prevents erosion and slows down surface water flow, enhancing recharge

    In 1997, MVVA was engaged by Wellesley College to create a master plan for the campus. At that time, the firm singled out one area, a parking facility for 175 cars, as a potential development zone. The subsequent construction of a car parking facility allowed MVVA to turn this space into a restored landscape, freed of pollutants in the soil and reconnected to the natural water flows of the beautiful campus area. MVVA states: "The restored **ALUMNAE VALLEY** again becomes part of the natural valley hydrological system that structures the form of the Wellesley campus. Not merely a restoration, the reconceptualization of the site included an understanding of its historical function: from glacial valley to industrial dumping ground to parking lot to a valley restored yet informed by its previous incarnations. Its use of topography as both a means of design solution and experiential enhancement underscores a landscape that is at once willfully artificial and unabashedly picturesque."

    1997 wurde das Büro MVVA vom Wellesley College beauftragt, einen Masterplan für den Universitätscampus auszuarbeiten. Damals wählte das Büro einen Bereich aus, auf dem sich ein Parkplatz mit 175 Stellplätzen befand. Durch den Bau eines Parkhauses konnte MVVA diesen Freiraum renaturieren, von Altlasten im Boden befreien und ihn wieder mit den natürlichen Wasserläufen verbinden, die den schönen Campus durchfließen. MVVA sagt dazu: „Das sanierte **ALUMNAE VALLEY** ist wieder zu einem Bestandteil des natürlichen hydrologischen Systems im Tal geworden, das den Campus von Wellesley strukturiert. Es handelt sich nicht einfach um eine Sanierung, sondern es ging auch um eine neue Konzeption für das Gelände und um das Verständnis seiner historischen Funktion: vom eiszeitlichen Tal zur Müllhalde für Industriemüll und zum Parkplatz und dann wieder zu einem renaturierten Tal, dessen frühere Nutzungen ablesbar bleiben. Die Topografie des Tals wurde einerseits als Gestaltungsmittel des Entwurfs genutzt, aber auch als empirisches Mittel der Steigerung, sie hebt eine Landschaft hervor, die mit voller Absicht beides ist: künstlich und ausgesprochen malerisch."

    En 1997, MVVA avait été appelé par le Wellesley College pour établir le plan directeur de son campus. L'agence avait alors noté qu'il existait une aire de parking pour 175 voitures qui possédait un potentiel de développement intéressant. La construction ultérieure d'un autre parking a permis à MVVA de transformer les lieux en un paysage au sol dépollué et reconnecté aux écoulements naturels des eaux du superbe campus. Pour MVVA : « L'**ALUMNAE VALLEY** fait à nouveau partie du système hydrologique naturel qui structure le campus de Wellesley. Plus qu'une restauration, cette reconceptualisation du site s'est appuyée sur une compréhension nouvelle de sa fonction historique : il est passé d'une vallée glaciaire à une friche industrielle, puis à un parking avant de récupérer son statut de vallée, restaurée mais enrichie de ses incarnations antérieures. L'utilisation de la topographie à la fois comme solution de conception et comme outil d'amélioration empirique du lieu a permis de souligner un paysage volontairement artificiel, tout étant résolument pittoresque. »

MICHAEL VAN VALKENBURGH

As the image above shows, the intervention of the designers allows these spaces on the Wellesley campus to assume a natural appearance that they may well have had before a parking facility stood here.

Wie das Bild oben zeigt, haben die Landschaftsplaner den Campus von Wellesley in ein naturnahes Areal zurückverwandelt, so wie es ausgesehen haben mag, bevor hier der Parkplatz gebaut wurde.

Comme le montre la photographie ci-dessus, l'intervention des paysagistes permet à ces espaces du campus de Wellesley de revêtir l'aspect naturel qu'ils auraient pu avoir avant qu'un parking n'y soit installé.

Already in the midst of a beautiful natural environment, Wellesley's landscaped spaces gained a natural, almost "untended" appearance thanks to the intelligent intervention of MVVA.

Der Campus von Wellesley liegt in einer wunderschönen Naturumgebung. Dank der neuen, klugen Landschaftsgestaltung durch das Büro MVVA erhält er ein natürliches, fast „wildes" Aussehen.

Au cœur d'un environnement naturel splendide, ces nouveaux aménagements paysagers ont pris un aspect naturel, presque « sauvage », grâce aux interventions intelligentes de MVVA.

Paths and an "Events Lawn" assure that the landscaped area remains usable for outdoor activities.

Wege und eine Rasenfläche für Veranstaltungen tragen dazu bei, dass der landschaftlich gestaltete Bereich auch für Outdoor-Aktivitäten nutzbar bleibt.

Grâce à des allées et aux « pelouses pour évènements », le paysage reste adapté aux activités en plein air.

TEARDROP PARK

Battery Park City, New York, New York, USA, 1999–2006

Address: between Warren Street and Murray Street, east of River Terrace, south Manhattan,
Battery Park City, New York, NY, USA, www.bpcparks.org. Area: 0.71 hectares
Client: The Hugh L. Carey Battery Park City Authority, New York, New York;
Battery Park City Parks Conservancy, New York, New York. Cost: not disclosed

The plan (right) shows the space visible on the right page (rotated 90°). Here, paths and grass surfaces alternate with more densely planted borders.

Der Grundriss (rechts) zeigt den auf dem Foto rechts abgebildeten Bereich (um 90 Grad gedreht). Wege und Rasenflächen wechseln sich mit dicht bepflanzten Beeten ab.

Le plan (à droite) montre la partie aménagée de la page de gauche (pivotée à 90°). Des allées et des pelouses alternent avec des parterres plus densément plantés.

Intended primarily as a space for children, this public park area was formerly a featureless landfill area on the edge of the Hudson River. Wind and shade conditions resulting from nearby buildings provided challenges for the design. Sustainable materials and installation methods were part of the project, with the active encouragement of the client. The designers explain: "Throughout the park, including the 8.2-meter-high, 51-meter-long stacked blue stone Ice-Water Wall, stone selection was limited to those that could be quarried within 800 kilometers of the park site." MVVA further explains that they designed **TEARDROP PARK** to counter the trend that has led plants to be banished from children's parks in favor of equipment. "Specific features, like the Ice-Water Wall," they state, "the Marsh with its access path scaled to children, the steeply sloped planted areas, groves of trees, and the Water Play rocks, as well as the stone Reading Circle placed where there is also an outward view to the Hudson River, celebrate the expressive potential of the natural materials of landscape construction while reinventing the idea of nature play in the city."

Dieser öffentliche Park, in dem ursprünglich ein Freiraum für Kinder entstehen sollte, war früher eine gesichtslose Deponie in der Nähe des Hudson River. Der Wind und der Schattenwurf der umstehenden Gebäude waren eine große Herausforderung für die Planer. Das Projekt sollte mit nachhaltigen Materialien und Bautechniken ausgeführt werden, dies wurde vom Bauherren aktiv gefördert. Die Planer erklären ihren Entwurf wie folgt: „Im gesamten Park, und dazu gehört auch die 8,2 Meter hohe und 51 Meter lange, aus geschichtetem Blaustein errichtete Mauer, die ‚Eiswassermauer', wurde darauf geachtet, dass alle im Park verbauten Steine aus Steinbrüchen stammen, die nicht weiter als 800 Kilometer vom Park entfernt sind." MVVA erklärt weiter, dass man den **TEARDROP PARK** entgegen dem Trend entworfen hat, wonach Pflanzen von den Spielplätzen verbannt werden und stattdessen nur in Spielgeräte investiert wird. „Besondere Elemente wie die Eiswassermauer, das Sumpfareal mit seinem kindgerechten Zugangsweg, die steilen bepflanzten Böschungen, die Wäldchen, die Wasserspielfelsen und der Lesekreis aus Stein, von dem man sogar Aussicht auf den Hudson hat, stellen unter Beweis, dass natürliche Materialien in der Landschaftsgestaltung über eine große Ausdruckskraft verfügen, wenn man sich von Neuem darauf einlässt, mitten in der Stadt die Natur spielen zu lassen."

Conçu au départ comme un lieu réservé aux enfants, ce parc public était auparavant une zone de décharge en bordure de l'Hudson. Le projet devait répondre au défi posé par l'ombre et l'écoulement des vents provoqués par la présence de grands immeubles. Des méthodes et des matériaux durables ont été mis en œuvre, avec l'encouragement actif du client. L'agence explique que « pour l'ensemble du parc, y compris le Ice-Water Wall de 51 m de long et 8,2 m de haut en blocs de pierre empilés, le choix des pierres s'est limité à ce que l'on pouvait trouver à moins de 800 km ». Le **TEARDROP PARK** a été conçu à l'encontre de la tendance actuelle, qui limite la présence des plantations dans les jardins destinés aux enfants au profit de multiples équipements. L'agence précise : « Le Ice-Water Wall, le Marsh, dont le chemin d'accès est à l'échelle des enfants, les zones de plantations en escalier, les bosquets d'arbres et les rochers du Water Play, ou encore le Reading Circle en pierre, placé là pour bénéficier d'une vue sur l'Hudson : autant d'installations spécifiques qui célèbrent le potentiel expressif des matériaux naturels utilisés dans la construction de ce paysage, tout en réinventant l'idée de jouer dans la nature, mais en pleine ville. »

MICHAEL VAN VALKENBURGH

The Water Play Rocks and the Ice-Water Wall seen on this double page are part of the features that give this park a variety of appearances and possible activities that appeal both to children and adults.

Die auf dieser Doppelseite abgebildeten Wasserspielfelsen und die Eiswassermauer sind wesentliche Elemente des Parks, die ihm Vielfalt verleihen und sowohl Kinder als auch Erwachsene zu Aktivitäten anregen.

Les rochers du Water Play et le Ice-Water Wall reproduits sur cette double page font partie des équipements qui confèrent à ce parc urbain une variété d'aspect et d'activités appréciée autant des enfants que des adultes.

MICHAEL VAN VALKENBURGH

WEISS/MANFREDI

Weiss/Manfredi Architecture/Landscape/Urbanism
200 Hudson Street, 10th floor
New York, NY 10013, USA

Tel: +1 212 760 9002 / Fax: +1 212 760 9003
Web: www.weissmanfredi.com

MARION WEISS received her B.Arch from the University of Virginia, and her M.Arch degree from Yale University. Prior to establishing her present firm, she was a designer for Cesar Pelli & Associates, Smith-Miller-Hawkinson, and Mitchell/Giurgola Architects. She is a Design Partner of Weiss/Manfredi and a Professor of Architecture at the University of Pennsylvania's Penn School of Design. She has also taught Design studios at Yale University, Cornell University, and the Harvard GSD. **MICHAEL MANFREDI** was born in Trieste, Italy. He received his B.Arch degree from the University of Notre Dame and his M.Arch at Cornell University, where he studied under Colin Rowe. He then worked in the offices of Richard Meier and Mitchell/Giurgola Architects. He is a Design Partner of Weiss/Manfredi and has taught at Princeton University, the University of Pennsylvania, Yale University, the Harvard GSD, and Cornell University, where he is currently a Professor of Architecture. As well as the Olympic Sculpture Park (Seattle, Washington, 2001–07, published here), recent projects include the Barnard College Diana Center (New York, New York, 2003–10), a multiuse arts center, and winner of an invited competition and a 2008 Progressive Architecture Award. Current projects include the Brooklyn Botanic Garden Visitor Center (New York, New York, 2004–12), which will achieve LEED Gold certification; the Khrishna P. Singh Center for Nanotechnology at the University of Pennsylvania (Philadelphia, 2008–13); and Hunters Point South, a waterfront park on the East River in Queens, New York (New York, 2009–13), all in the USA.

MARION WEISS machte ihren Bachelor in Architektur an der University of Virginia und ihren Master an der Yale University. Bevor sie ihr jetziges Büro gründete, war sie Entwurfsarchitektin in den Büros Cesar Pelli & Associates, Smith-Miller-Hawkinson und Mitchell/Giurgola Architects. Im Büro Weiss/Manfredi ist sie als Entwurfsarchitektin tätig. Marion Weiss ist Professorin für Architektur an der Penn School of Design der Pennsylvania University. Sie hat zudem Design an den Universitäten Yale, Cornell und der Harvard GSD unterrichtet. **MICHAEL MANFREDI** wurde in Triest, Italien, geboren. Er machte seinen Bachelor in Architektur an der University of Notre Dame (South Bend, Indiana) und seinen Master an der Cornell University, wo er bei Colin Rowe studierte. Danach arbeitete er bei Richard Meier und Mitchell/Giurgola Architects. Heute ist er als Entwurfsarchitekt Partner im Büro Weiss/Manfredi. Er lehrte an der Princeton University, der University of Pennsylvania, Yale University, der Harvard GSD und der Cornell University, wo er derzeit Professor für Architektur ist. Zu den neueren Projekten von Weiss/Manfredi gehören der Olympic Sculpture Park (Seattle, Washington, 2001–07, hier vorgestellt) und das Barnard College Diana Center (New York, 2003–10), ein Kunstzentrum mit einer Mehrfachnutzung. Für den Entwurf erhielten Weiss/Manfredi den ersten Preis bei einem eingeladenen Wettbewerb und 2008 die Auszeichnung für fortschrittliche Architektur. Derzeit in Ausführung befindliche Projekte sind das Besucherzentrum des Brooklyn Botanic Garden (New York, 2004–12), das die Auszeichnung für ökologisches Bauen, LEED, in Gold erhalten wird, das Khrishna P. Singh Center für Nanotechnologie an der University of Pennsylvania (Philadelphia, 2008–13) und Hunters Point South, ein Uferpark am East River im Stadtteil Queens, New York (2009–13), alle in den USA.

MARION WEISS est titulaire d'un B.Arch. de l'université de Virginie et d'un M. Arch. de l'université Yale. Avant de créer son agence actuelle, elle a travaillé pour Cesar Pelli & Associates, Smith-Miller-Hawkinson et Mitchell/Giurgola Architects. Elle est partenaire chargée de projets chez Weiss/Manfredi et professeur d'architecture à l'École de design de l'université de Pennsylvanie. Elle a également dirigé des ateliers de conception à l'université Yale, à l'université Cornell et à la Harvard GSD. **MICHAEL MANFREDI**, né à Trieste en Italie, est titulaire d'un B.Arch. de l'université de Notre Dame (South Bend, Indiana) et d'un M.Arch. de l'université Cornell, où il a étudié sous la direction de Colin Rowe. Il a ensuite travaillé dans les agences de Richard Meier et de Mitchell/Giurgola Architects. Il a enseigné à l'université de Princeton, à l'université de Pennsylvanie, à l'université de Yale, à la Harvard GSD et à l'université Cornell, où il est actuellement professeur d'architecture. En dehors de l'Olympic Sculpture Park (Seattle, Washington, 2001–07, publié ici), ses projets récents comptent notamment le Diana Center du Barnard College (New York, 2003–10), un centre d'art polyvalent qui a remporté un concours sur invitation et le Prix de l'architecture progressiste 2008. Ils travaillent actuellement sur le Visitor Center du Brooklyn Botanic Garden (New York, 2004–12), qui obtiendra la certification LEED Or ; le Khrishna P. Singh Center for Nanotechnology de l'université de Pennsylvanie (Philadelphie, US, 2008–13) et le Hunters Point South, un parc en bordure de l'East River dans le Queens (New York, 2009–13).

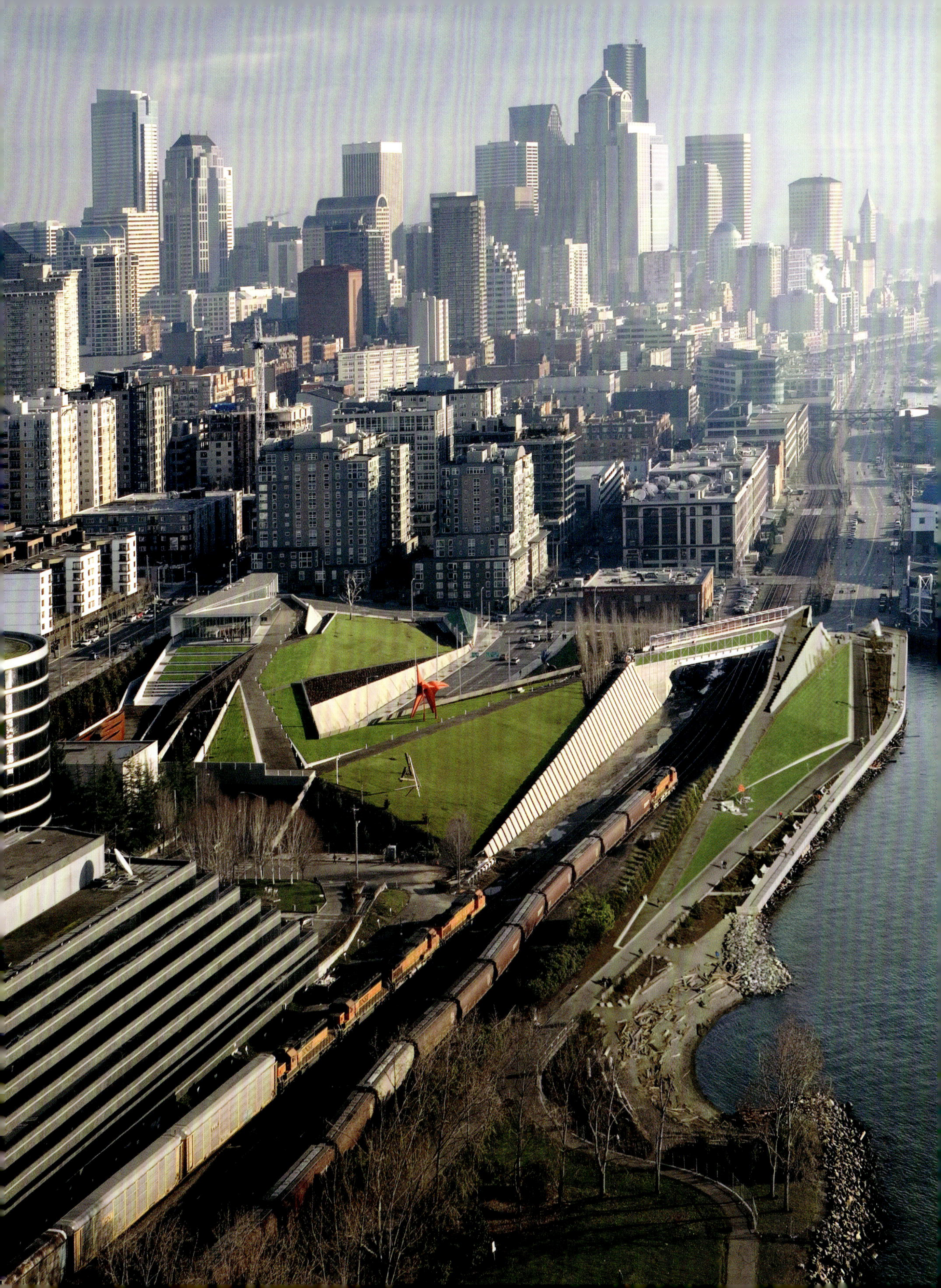

OLYMPIC SCULPTURE PARK

Seattle, Washington, USA, 2001–07

Address: Seattle Art Museum, 2901 Western Avenue, Seattle, WA 98121, USA, +1 206 654 3100 / +1 206 332 1377, www.seattleartmuseum.org. Area: 3.6 hectares (park), 3159 m² (exhibition pavilion and garage)
Client: Seattle Art Museum. Cost: $32 million. Collaboration: Christopher Ballentine
(Project Manager), Todd Hoehn, Yehre Suh (Project Architects)

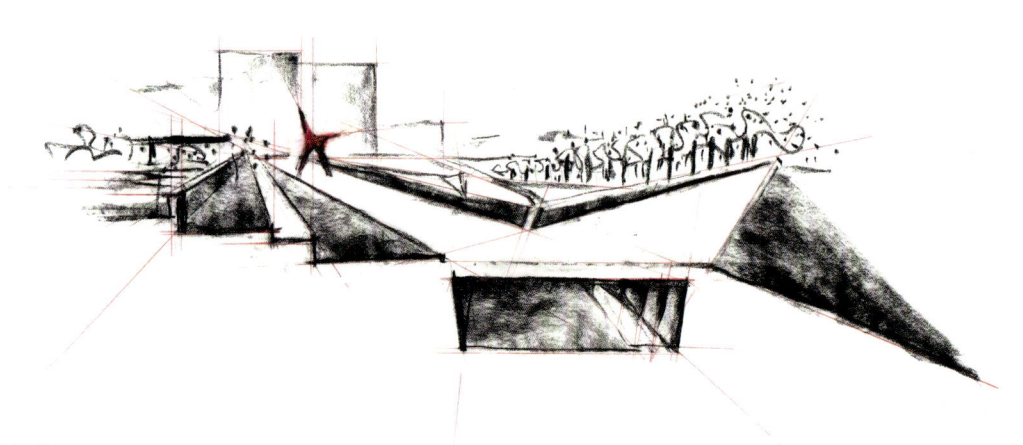

The citation for this project's 2008 AIA Honor Award explains the context: "This project is located on Seattle's last undeveloped waterfront property, sliced by train tracks and an arterial road. The design connects three separate sites with an uninterrupted Z-shaped 'green' platform, descending over 12 meters from the city to the water, capitalizing on views of the skyline and Elliot Bay and rising over existing infrastructure to reconnect the urban core to the revitalized waterfront." Formerly used by Union Oil of California as an oil transfer facility, the site required the removal of 120 000 tons of contaminated soil, capped with 51 000 cubic meters of clean fill, recuperated in part from the Seattle Art Museum's expansion site. It is used for the exhibition of sculptures. The park includes an exhibition pavilion and its paths offer views of both the distant Olympic Mountains and of the city and beaches. The architects write: "As a 'landscape for art,' the **OLYMPIC SCULPTURE PARK** defines a new experience for modern and contemporary art outside the museum walls. The topographically varied park provides diverse settings for sculpture of multiple scales. Deliberately open-ended, the design invites new interpretations of art and environmental engagement, reconnecting the fractured relationships of art, landscape, and urban life."

In der Laudatio zu dem Ehrenpreis der AIA, mit dem dieses Projekt 2008 ausgezeichnet wurde, wird der Kontext erklärt: „Das Projekt befindet sich auf der letzten unbebauten Uferzone von Seattle. Sie wird von Eisenbahngleisen und einer Hauptverkehrsstraße durchschnitten. Der Entwurf verbindet drei voneinander getrennte Gelände mit einer durchgehenden, ‚grünen', Z-förmigen Plattform, die ein zwölf Meter hohes Gefälle zwischen der Innenstadt und dem Meer überwindet. Von hier wird der Anblick der Skyline von Seattle und der Blick auf die Elliot Bay besonders betont. Der Steg überquert die vorhandene Infrastruktur und verbindet die Stadtmitte mit einer neubelebten Uferzone." Das Gelände wurde früher von der Firma Union Oil of California als Erdölumschlagplatz genutzt. 120 000 Tonnen kontaminierten Bodens mussten abtransportiert werden, und das Areal wurde mit 51 000 Kubikmetern sauberen Materials vom Erweiterungsbau des Seattle Art Museum abgedeckt. Zu dem Park, in dem nun Skulpturen ausgestellt werden, gehört ein Ausstellungspavillon, und von den Wegen hat man Aussicht auf die Olympic Mountains in der Ferne, auf die Stadt und auf die Strände. Die Architekten schreiben dazu: „Als ‚Kunstlandschaft' eröffnet der **OLYMPIC SCULPTURE PARK** eine neue Erfahrung von moderner und zeitgenössischer Kunst außerhalb der Museumsmauern. Durch die Lebendigkeit der Topografie entstehen vielfältige Standorte für Skulpturen in sehr unterschiedlichem Maßstab. Der Entwurf ist bewusst offen gehalten und lädt ein, Kunst und Engagement für den Umweltschutz neu zu interpretieren, indem er die ansonsten getrennten Bereiche Kunst, Landschaftsgestaltung und urbanes Leben miteinander verbindet."

« Ce projet s'étend sur les derniers espaces non encore construits du front de mer de Seattle, sillonné de voies ferrées et de grandes artères. Il relie trois parcelles séparées au moyen d'une plate-forme "verte" continue en forme de Z, sur un dénivelé de 12 m. Elle offre des perspectives sur le panorama urbain et Elliot Bay, et s'élève au-dessus des infrastructures existantes pour reconnecter le cœur urbain au front de mer rénové » (citation du jury de l'AIA, qui a accordé un prix d'honneur au projet en 2008). Naguère utilisé par l'Union Oil of California pour le transfert du pétrole, le terrain a dû être nettoyé de 120 000 tonnes de déblais contaminés, puis il a été recouvert de 51 000 m³ de terre propre, récupérée en partie sur le chantier de l'extension du Seattle Art Museum. Ce parc de sculptures comprend également un pavillon d'exposition et des allées qui offrent des vues sur les montagnes Olympiques au loin, et sur la ville et les plages. « En tant que "paysage conçu pour l'art", expliquent les architectes, l'**OLYMPIC SCULPTURE PARK** établit une nouvelle forme de découverte de l'art moderne et contemporain hors les murs. Ce parc de topographie variée propose de multiples lieux où présenter des sculptures d'échelles diverses. Délibérément non finalisé, le plan invite à de nouvelles interprétations de l'art et de sa présence dans l'environnement, en renouant des relations fracturées entre les œuvres d'art, le paysage et la vie urbaine. »

Along the waterfront, the Olympic Sculpture Park offers open spaces and viewing for visitors of such works as the large Alexander Calder stabile seen in these images. Above, a sketch shows the road on which the park sits.

Der am Ufer gelegene Olympic Sculpture Park bietet dem Besucher offene Freiräume und Blicke auf Großskulpturen, wie dieses Stabile von Alexander Calder (rechts). Die Zeichnung oben zeigt, wie der Park über der vierspurigen Straße angelegt ist.

Situé en front de mer, l'Olympic Sculpture Park offre des espaces ouverts et présente des œuvres comme ce grand stabile d'Alexander Calder. Ci-dessus, un croquis montre la voie au-dessus de laquelle le parc a été aménagé.

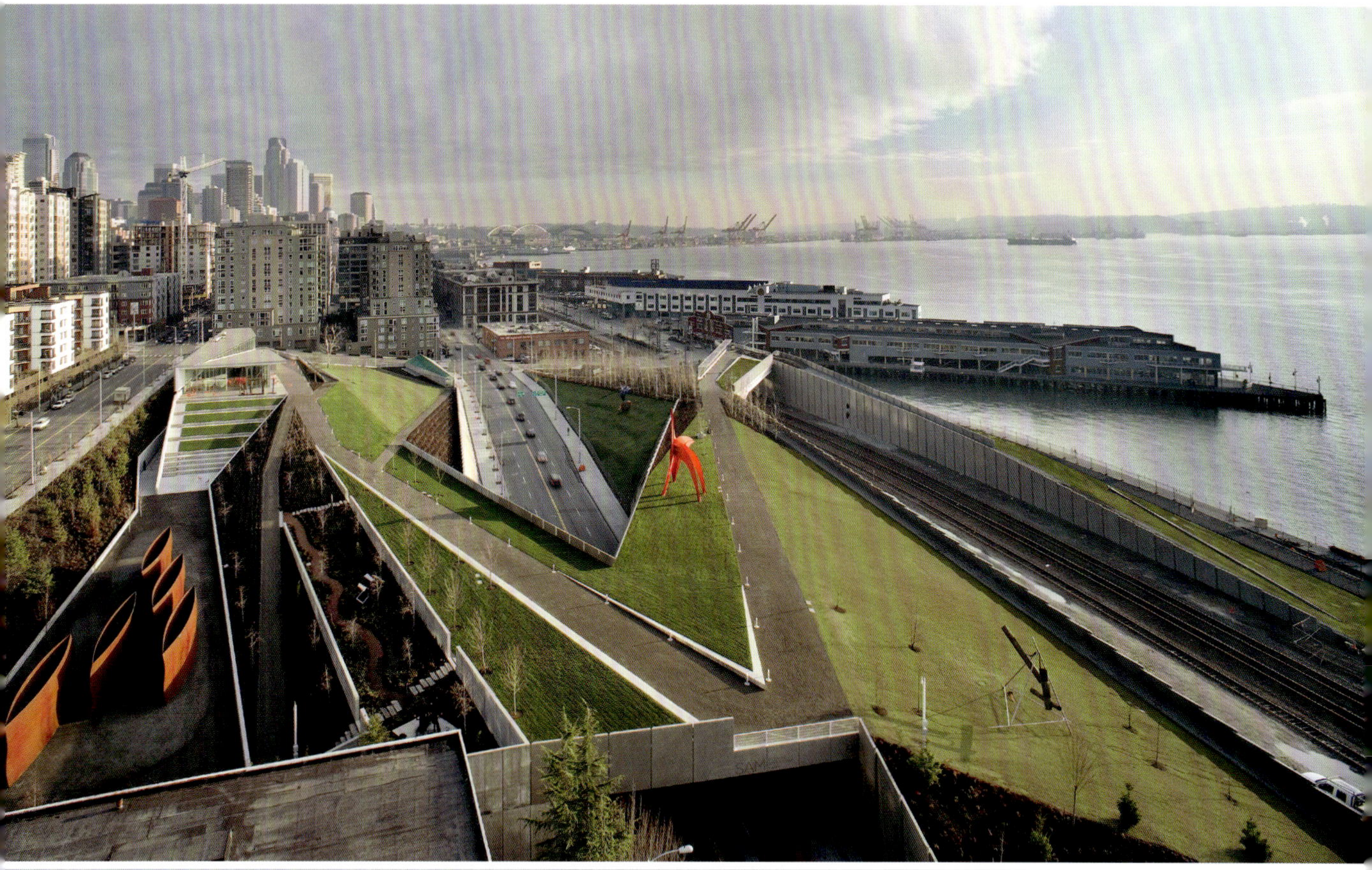

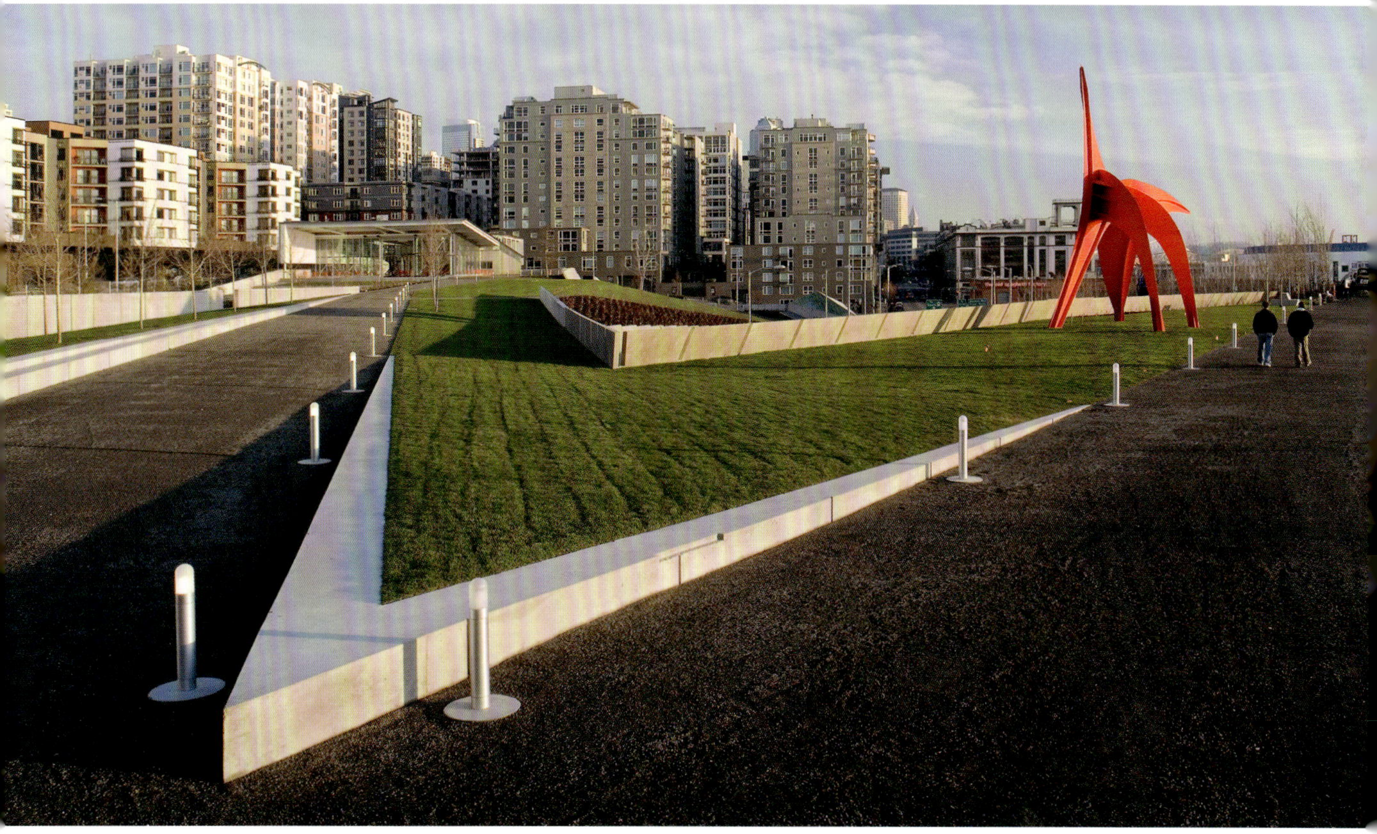

Seen in the drawings below and in the images on this double page, the park permits visitors to view art and the striking mountain landscape along the water.

Die Zeichnungen unten und die Bilder auf dieser Doppelseite verdeutlichen, wie die Besucher die Kunst und die fantastische Berglandschaft am Wasser genießen können.

Comme le montrent les plans ci-dessous et les photos de cette double page, le parc offre aux visiteurs des perspectives sur le superbe paysage de montagnes et sur des œuvres d'art en bordure de l'eau.

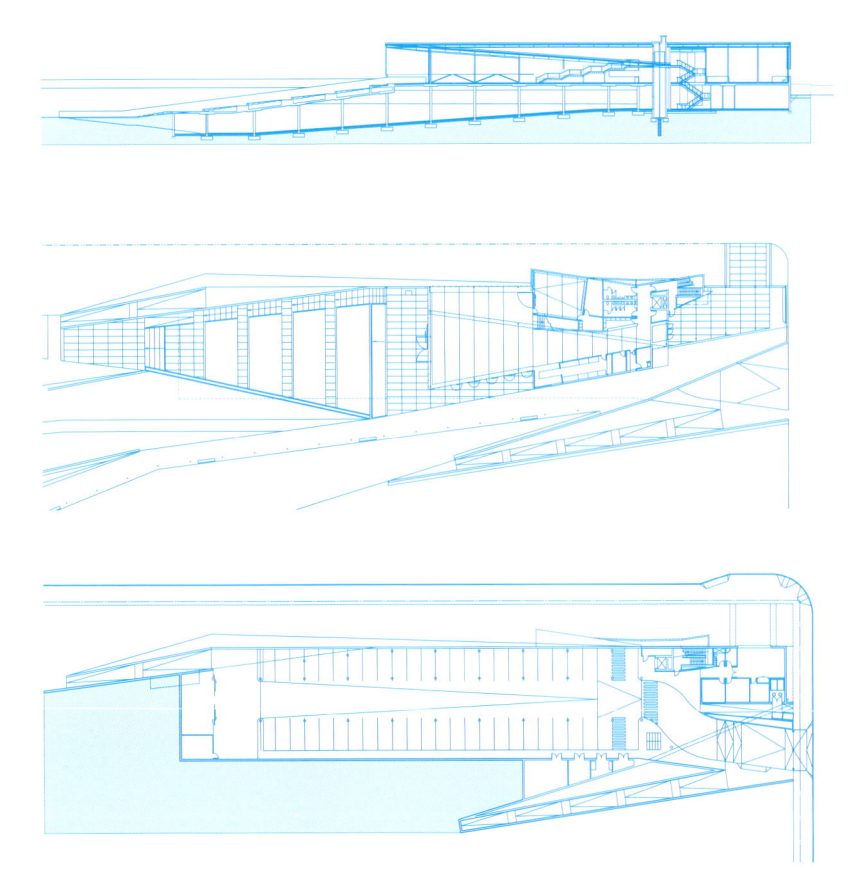

The advantage of the park is that it brings contemporary art to a broader public that might not venture into many museums. It, of course, also recovers an otherwise difficult-to-use waterfront site.

Bemerkenswert an diesem Park ist, dass er zeitgenössische Kunst einem breiteren Publikum nahebringt, das normalerweise nicht zu den Museumsgängern gehört. Außerdem wertet er eine problematisch zu nutzende Uferzone auf.

Un des intérêts de ce parc est de rapprocher de l'art contemporain un grand public qui ne se rend pas souvent dans les musées. Et de réhabiliter un terrain en front de mer particulièrement difficile à exploiter.

WEST 8

West 8 Urban Design & Landscape Architecture
Schiehaven 13M, 3024 EC Rotterdam, The Netherlands

Tel: +31 10 485 58 01 / Fax: +31 10 485 63 23
E-mail: west8@west8.com / Web: www.west8.com

"The knowledge that contemporary landscape is for the major part artificial, made up of different components—designed and undersigned—allows West 8 the freedom to respond by positioning its own narrative spaces. The basic ingredients are ecology, infrastructure, weather conditions, building programs, and people. The goal is to incorporate awareness of these various aspects in a playful, optimistic manner stimulating the desire to conquer and take possession of a space." It is in these terms that the Dutch firm describes its activity, centered in urban design and landscape architecture. **ADRIAAN GEUZE**, born in 1960 (Dordrecht), received his Master's in Landscape Architecture in 1987 (Agricultural University of Wageningen). He was a cofounder of West 8 the same year. **EDZO BINDELS** is a Principal and Partner of West 8 Urban Design & Landscape Architecture. Born in Monnickendam, the Netherlands, in 1969, he attended the University of Technology, Delft (Urban Design, 1987–2002). **MARTIN BIEWENGA**, born in Smallingerland, the Netherlands, in 1969, received his Master's in Urban Design from the Technical University of Delft in 1996. Since 1996 he has worked as a project director on large-scale projects for West 8, and since 2003 he has been a Partner of the firm. **JAMIE MASLYN LARSON**, born in New York in 1969, is co-director of the New York office of West 8. She received her Master's in Landscape Architecture from Utah State University (Logan, UT, 1997). Their recent work includes the master plan for Playa de Palma (Mallorca, Spain, 2008–10); Madrid RIO (Madrid, Spain, 2007–11, published here); Lincoln Park (Miami Beach, Florida, USA, 2009–11); Waterfront Toronto (Toronto, Canada, 2006–, also published here); and Governors Island Park and Public Space Master Plan New York (New York, USA, 2012, construction Phase 1, also published here).

„Das Wissen, dass heutige Landschaft zum größten Teil etwas Künstliches und aus unterschiedlichen – geplanten und ungeplanten – Komponenten zusammengestellt ist, gibt West 8 die Freiheit, eigene erzählerische Räume zu positionieren. Die Grundbestandteile sind Ökologie, Infrastruktur, Wetterbedingungen, Bebauungsprogramme und die Menschen. Es ist das Ziel, das Bewusstsein für diese unterschiedlichen Aspekte in einer spielerischen, optimistischen Weise zu verinnerlichen und den Wunsch zu erwecken, einen Raum zu erobern und ihn in Besitz zu nehmen." So beschreibt das niederländische Büro seine Tätigkeit in der Stadtplanung und Landschaftsarchitektur. **ADRIAAN GEUZE** wurde 1960 in Dordrecht geboren und machte 1987 seinen Master in Landschaftsarchitektur an der Landwirtschaftlichen Hochschule Wageningen. Im gleichen Jahr gehörte er zu den Mitbegründern von West 8. **EDZO BINDELS** ist ebenfalls Partner und Direktor von West 8. Er wurde 1969 in Monnickendam, Niederlande, geboren und studierte von 1987 bis 2002 Stadtplanung an der Technischen Universität Delft. **MARTIN BIEWENGA**, 1969 in Smallingerland, Niederlande, geboren, machte 1996 seinen Master in Stadtplanung an der Technischen Universität Delft. Seit 1996 arbeitet er als Leiter für Großprojekte bei West 8 und ist seit 2003 Büropartner. **JAMIE MASLYN LARSON** wurde 1969 in New York geboren und ist Kodirektorin des New Yorker Büros von West 8. Sie machte 1997 ihren Master in Landschaftsarchitektur an der Utah State University in Logan. Zu den neueren Projekten gehören der Masterplan für die Playa de Palma (Mallorca, Spanien, 2008–10), Madrid RIO (Madrid, 2007–11, hier vorgestellt), der Lincoln Park (Miami Beach, Florida, USA, 2009–11), die Ufergestaltung in Toronto (Kanada, seit 2006, ebenfalls hier vorgestellt) sowie der Governors Island Park und der Masterplan für öffentliche Freiräume in New York (2012, Ausführungsphase 1, ebenfalls hier vorgestellt).

« Être consciente de la grande part d'artificiel dans le paysage contemporain, composé de différents éléments – conçus ou non –, donne à l'agence West 8 la liberté de répondre en situant ses propres espaces narratifs. Les ingrédients de base en sont l'écologie, l'infrastructure, les conditions atmosphériques, les programmes de construction et les usagers. L'objectif est d'intégrer la conscience de ces divers aspects de façon optimiste et ludique, en stimulant le désir de conquérir et de prendre possession de l'espace. » C'est ainsi que l'agence néerlandaise décrit son activité, centrée sur l'urbanisme et l'architecture du paysage. **ADRIAAN GEUZE**, né en 1960 à Dordrecht, titulaire d'un M.A. en architecture du paysage (Université agricole de Wageningen, 1987), a participé à la fondation de West 8 la même année. **EDZO BINDELS** est l'un des dirigeants et partenaires de l'agence. Né à Monnickendam (Pays-Bas) en 1969, il a étudié l'urbanisme à l'Université de technologie de Delft (1987–2002). **MARTIN BIEWENGA**, né à Smallingerland (Pays-Bas) en 1969, a obtenu son Master en urbanisme à l'Université de technologie de Delft en 1996. Depuis, il est responsable de projet sur d'importantes interventions de West 8, dont il partenaire depuis 2003. **JAMIE MASLYN LARSON**, née à New York en 1969, est codirectrice du bureau de l'agence à New York. Elle a obtenu son Master en architecture du paysage à l'université de l'Utah (Logan, 1997). Parmi leurs réalisations récentes : le plan directeur de Playa de Palma (Majorque, Espagne, 2008–10) ; le Madrid RIO (2007–11, publié ici) ; le Lincoln Park (Miami Beach, Floride, US, 2009–11) ; le Waterfront Toronto (Toronto, Canada, 2006–, publié ici) et le plan directeur du parc et des espaces publics de Governors Island (New York, 2012, phase 1 de construction, publié ici).

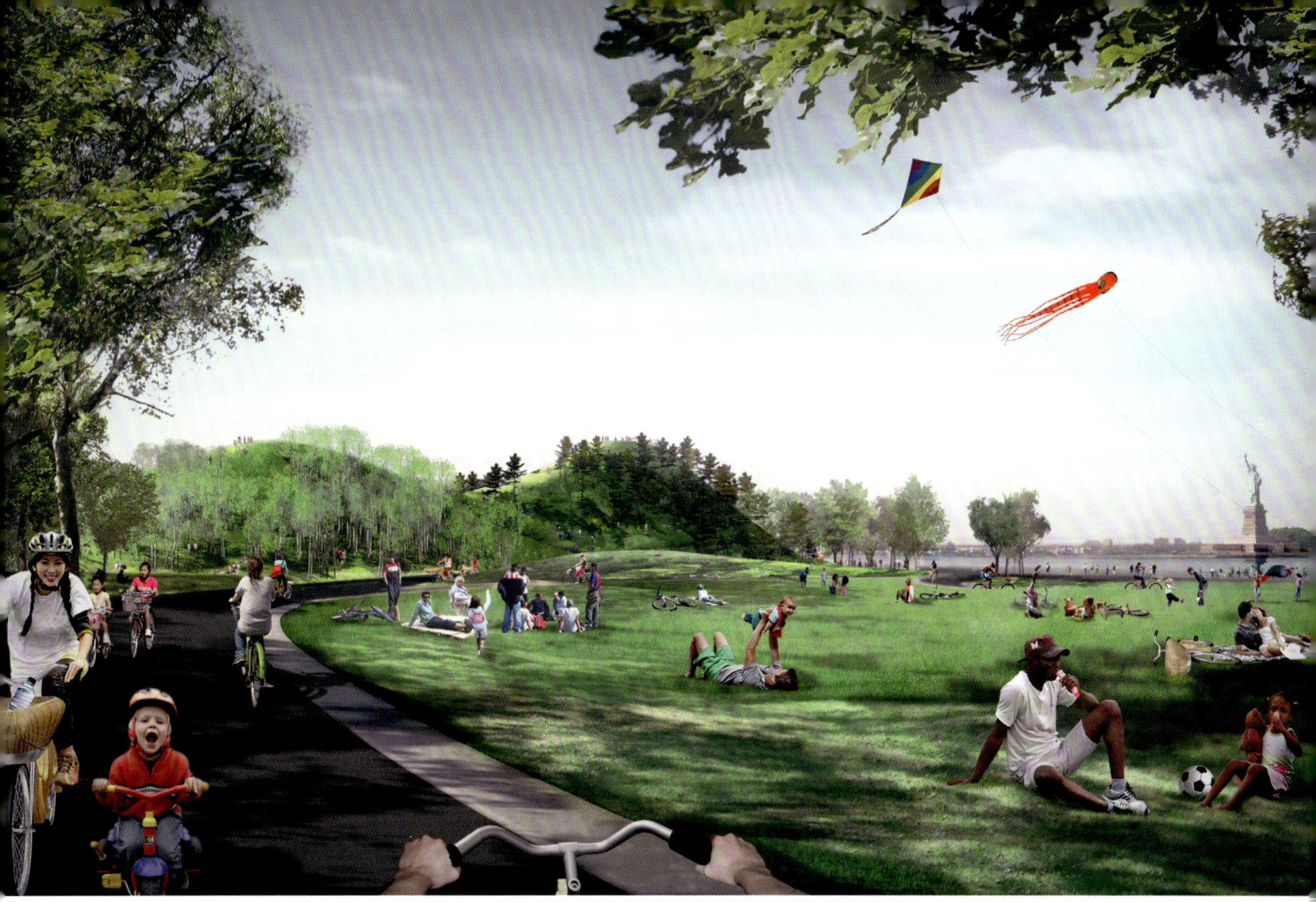
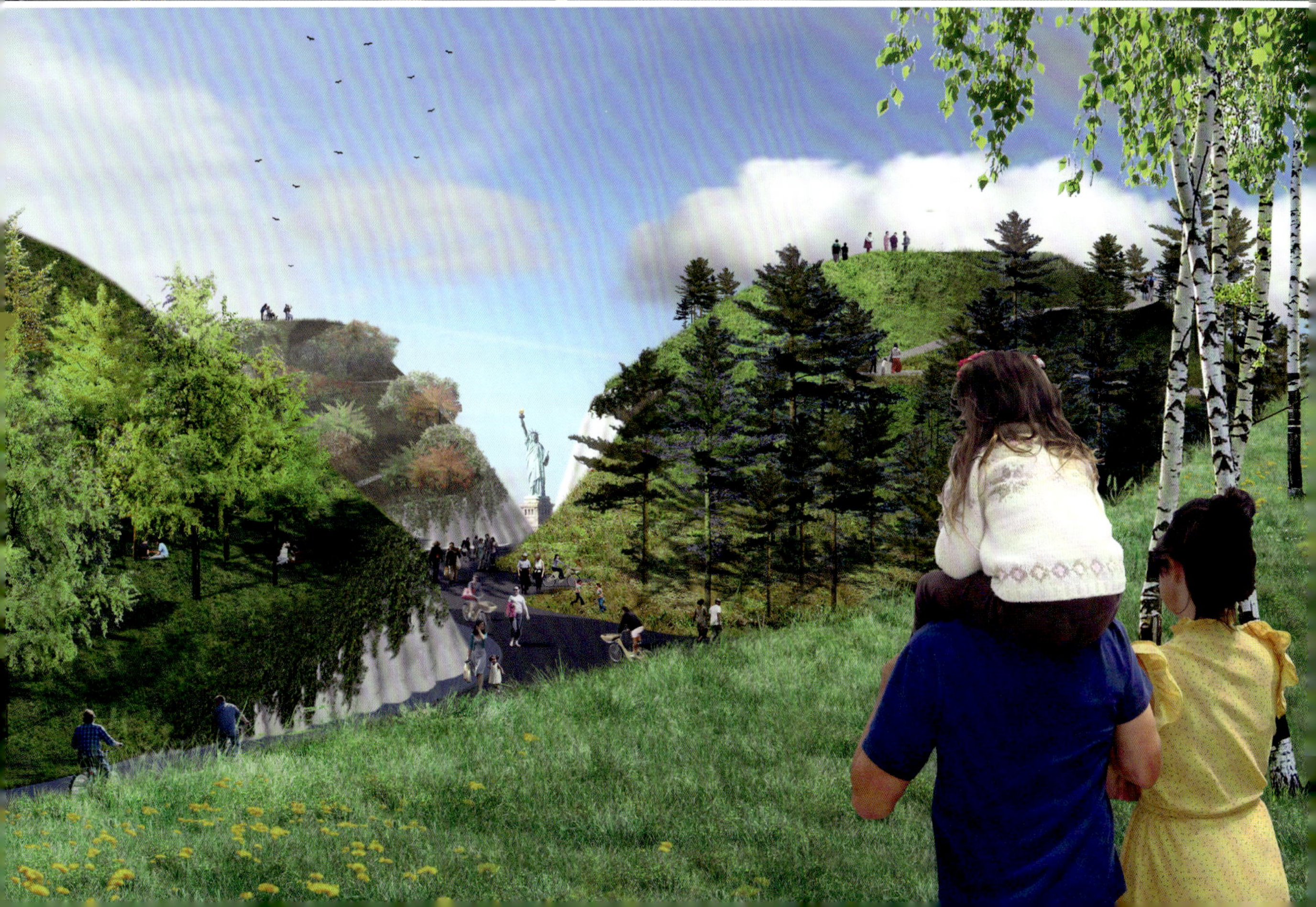

GOVERNORS ISLAND PARK AND PUBLIC SPACE MASTER PLAN

New York, New York, USA, 2012 (Phase 1)

Address: Governors Island, New York, NY 10004, USA, www.govisland.com. Area: 35.2 hectares
Client: The Trust for Governors Island. Cost: not disclosed. Collaboration: in association with
Rogers Marvel Architects, Diller Scofidio + Renfro, Mathews Nielsen, Urban Design +

These perspective drawings show the site of the island vis-à-vis Manhattan and the Statue of Liberty (below). Part of the scheme involves "rejuvenating" existing landscapes.

Diese Perspektivzeichnungen zeigen die Lage der Insel vor Manhattan und der Freiheitsstatue (unten). Teil der Aufgabe war es, vorhandene Landschaftsgestaltungen zu „verjüngen".

Ces perspectives dessinées montrent le site de l'île face à Manhattan et à la statue de la Liberté (en bas). Une partie du projet portait sur la « régénération » de paysages existants.

For almost two centuries, Governors Island was a military base for the US Army and Coast Guard. In 2003 the Federal government sold most of the Island to the State of New York for one dollar. Today, the Governors Island Preservation and Education Corporation (GIPEC) oversees 61 hectares of the Island, while the National Park Service manages the balance: the 9-hectare Governors Island National Monument which includes two 1812-era forts. In 2006, GIPEC launched an international design competition for the design of the Island's park and public spaces. The goal of the park and public space design is to create a comprehensive design for 35.2 hectares of open green space, including rejuvenating existing landscapes in the National Historic District, transforming 16 hectares in the southern half of the island, and creating a 3.5-kilometer Great Promenade along the waterfront, offering spectacular views of New York Harbor. As the designers explain: "West 8's winning completion entry was founded on a richly layered conceptual approach, responsive to the promise of the Island's opportunities and the diverse interests of New Yorkers."

Governors Island diente fast 200 Jahre der US-Army als Militärbasis und der Küstenwache als Stützpunkt. 2003 verkaufte die Bundesregierung den größten Teil der Insel für einen Dollar an den Staat New York. Heute betreut die Governors Island Preservation and Education Corporation (GIPEC) 61 Hektar und der National Park Service die neun Hektar des denkmalgeschützten Governors Island National Monument, zu dem zwei 1812 errichtete Festungen gehören. 2006 schrieb die GIPEC einen internationalen Wettbewerb für die Gestaltung des Inselparks und der öffentlichen Freiräume aus. Ziel ist es, für 35,2 Hektar ein einheitliches Gesamtkonzept mit offenen Grünräumen zu entwickeln. Dazu gehören auch Verjüngungsmaßnahmen im National Historic District, die Neugestaltung der 16 Hektar großen Fläche im Süden der Insel und die Anlage der 3,5 Kilometer langen Uferpromenade mit atemberaubenden Blicken auf den Hafen von New York. Die Planer erklären: „Unserem Entwurf, der mit dem ersten Preis ausgezeichnet wurde, liegt ein vielschichtiges Konzept zugrunde, das die vielversprechenden Möglichkeiten der Insel ausschöpft und den unterschiedlichen Interessen der New Yorker entgegenkommt."

Pendant près de deux siècles, Governors Island a été une base de l'armée et des garde-côtes américains. En 2003, le gouvernement fédéral en a vendu la plus grande partie à l'État de New York pour un dollar symbolique. Aujourd'hui, la Governors Island Preservation and Education Corporation (GIPEC) contrôle 61 ha de la surface de l'île et le National Park Service les 9 ha du monument national de Governors Island, qui comprend deux forts datant de la guerre de 1812. En 2006, la GIPEC a lancé un concours international pour un parc et des espaces publics sur 35,2 ha d'espaces verts libres. Le projet comprenait la rénovation des aménagements paysagers existants du quartier historique national de 16 ha au sud de l'île et la création d'une grande promenade de 3,5 km le long du front de mer, qui offre des vues spectaculaires sur le port de New York. Pour l'équipe qui a remporté le concours : « La proposition gagnante de West 8 s'appuyait sur une approche conceptuelle aux multiples strates, qui exploitait les opportunités offertes par l'île et répondait aux intérêts très diversifiés des New-Yorkais. »

A plan of the island shows the areas concerned in green with an existing fort visible at the broader end of the island. Right, computer perspectives of the finished park areas.

Auf dem Grundriss der Insel sind die betroffenen Bereiche grün markiert, man sieht die Festung in dem breiteren Teil der Insel. Rechts: Computerperspektiven der fertigen Parkareale.

Ci-dessus, un plan de l'île montre les zones concernées en vert et l'ancien fort existant en haut à droite. À droite : image de synthèse du projet achevé.

WEST 8

MADRID RIO
Madrid, Spain, 2007–11

Area: 80 hectares. Client: Madrid City Government
Cost: €280 million

The City of Madrid decided to put the part of the M 30 motorway closest to the old city center underground. West 8, winners of a 2005 invited international competition, created the master plan for a six-kilometer-long section of the new urban area thus generated along the banks of the Manzanares River. A total of 47 subprojects have since been developed within the master plan area. A number of these projects have been carried out already, including the Salon de Pinos, a linear green space planted with 8000 pines on top of the motorway tunnel; the Avenida de Portugal (2007), a major road into the center of Madrid was placed in a tunnel and planted with different types of cherry trees; the Huerta de la Partida (2007–09), "a modern interpretation of the orchard"; the Parque de Arganzuela (2011) based on the "canalized" waters of the Manzanares; and the Puente Cascara "designed as a massive concrete dome with a rough texture," with 100 cables and a thin steel deck.

Die Stadt Madrid entschloss sich, einen Teil der Stadtautobahn M 30 in dem Bereich, der der Altstadt am nächsten ist, in den Untergrund zu verlegen. West 8 gewann 2005 einen internationalen eingeladenen Wettbewerb. Das Büro arbeitete einen Masterplan für einen sechs Kilometer langen Abschnitt aus, in dem ein neues städtisches Areal am Ufer des Manzanares entstehen sollte. Inzwischen sind im Bereich des Masterplans 47 Unterprojekte geplant, von denen einige realisiert worden sind, wie der Salón de Pinos, ein linear verlaufender grüner Freiraum, in dem 8000 Pinien oben auf dem Stadtautobahntunnel gepflanzt wurden. Die Avenida de Portugal, eine der Hauptzufahrtstraßen in das Stadtzentrum von Madrid, wurde ebenfalls in einen Tunnel gelegt (2007) und die entstandene Fläche mit verschiedenen Zierkirschensorten bepflanzt. Es entstanden die Huerta de la Partida (2007–09), „eine moderne Interpretation eines Obstbaumgartens", der Parque de Arganzuela (2011), dessen Hauptmotiv der „kanalisierte" Wasserlauf des Manzanares ist, und die Puentes Cascara, Brücken, „die als massive Betonkuppel mit einer groben Oberflächenstruktur entworfen sind", mit 100 Stahlseilen und einer Brückentafel aus dünnem Stahl.

Suite à la décision de la ville de Madrid de faire passer en souterrain la partie de l'autoroute M 30 la plus proche du centre-ville, West 8 a remporté le concours international organisé en 2005 et a conçu le plan directeur d'une section de 6 km de cette nouvelle zone urbaine le long de la Manzanares. Au total, 47 projets ont été proposés dans le cadre de ce plan directeur. Un certain nombre ont été déjà réalisés, dont le Salón de Pinos, espace vert linéaire planté de 8000 pins au-dessus du tunnel de l'autoroute ; l'Avenida de Portugal (2007), grand axe du centre de la capitale également mis sous tunnel et dont la surface est désormais plantée de différentes variétés de cerisiers ; la Huerta de la Partida (2007–09), une « interprétation moderne du verger » ; le Parque de Arganzuela (2011), qui utilise les eaux « canalisées » de la Manzanares et le Puente Cascara, avec son « dôme de béton massif de texture brute » et son tablier en acier soutenu par 100 câbles.

A plan of part of the area concerned (above) and, on the right page, two pictures of the Avenida de Portugal.

Plan eines Teils des betroffenen Areals (oben) und auf der rechten Seite zwei Fotos der Avenida de Portugal.

Plan d'une partie de la zone concernée (ci-dessus) et, page de droite, deux photos de l'Avenida de Portugal.

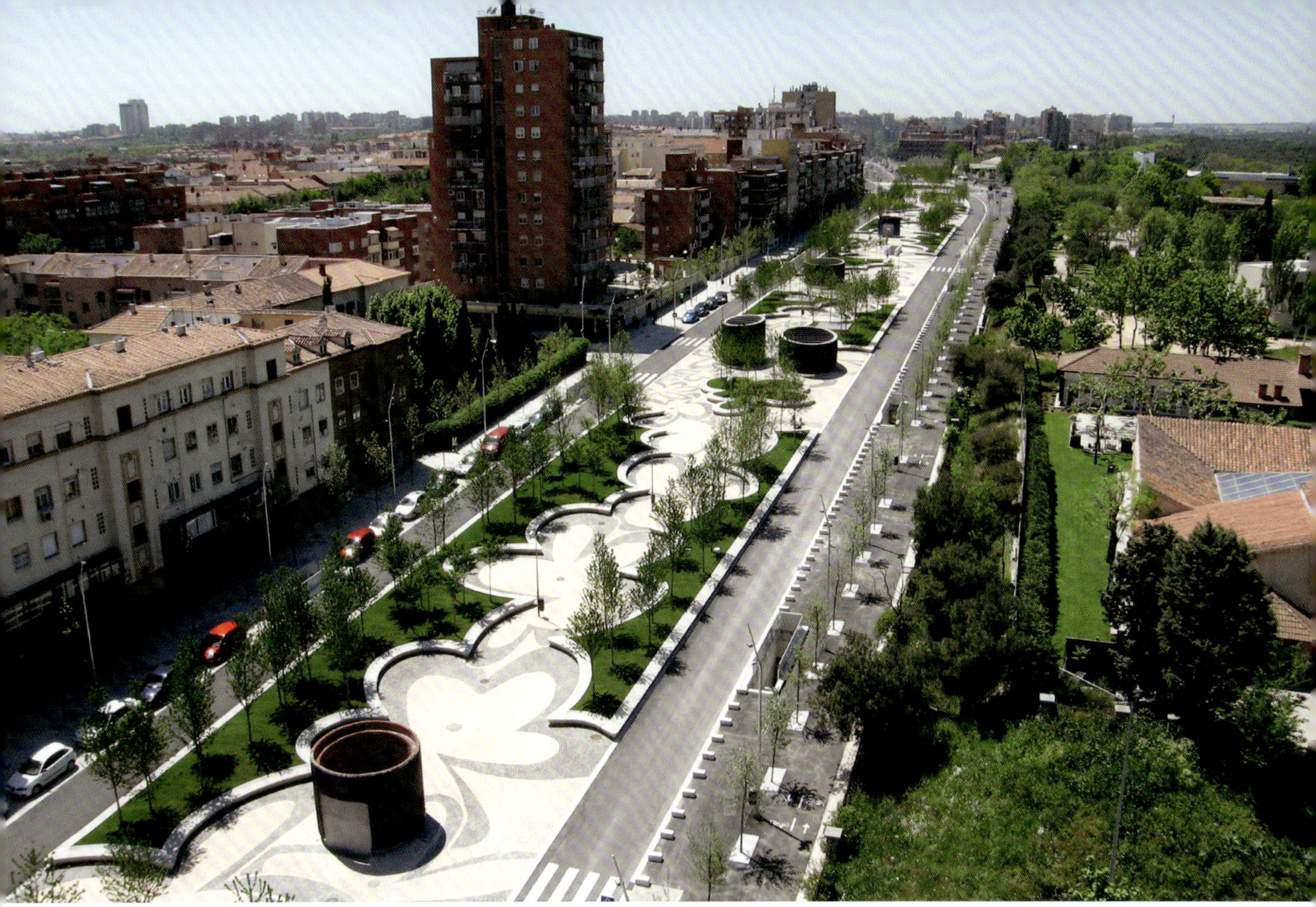
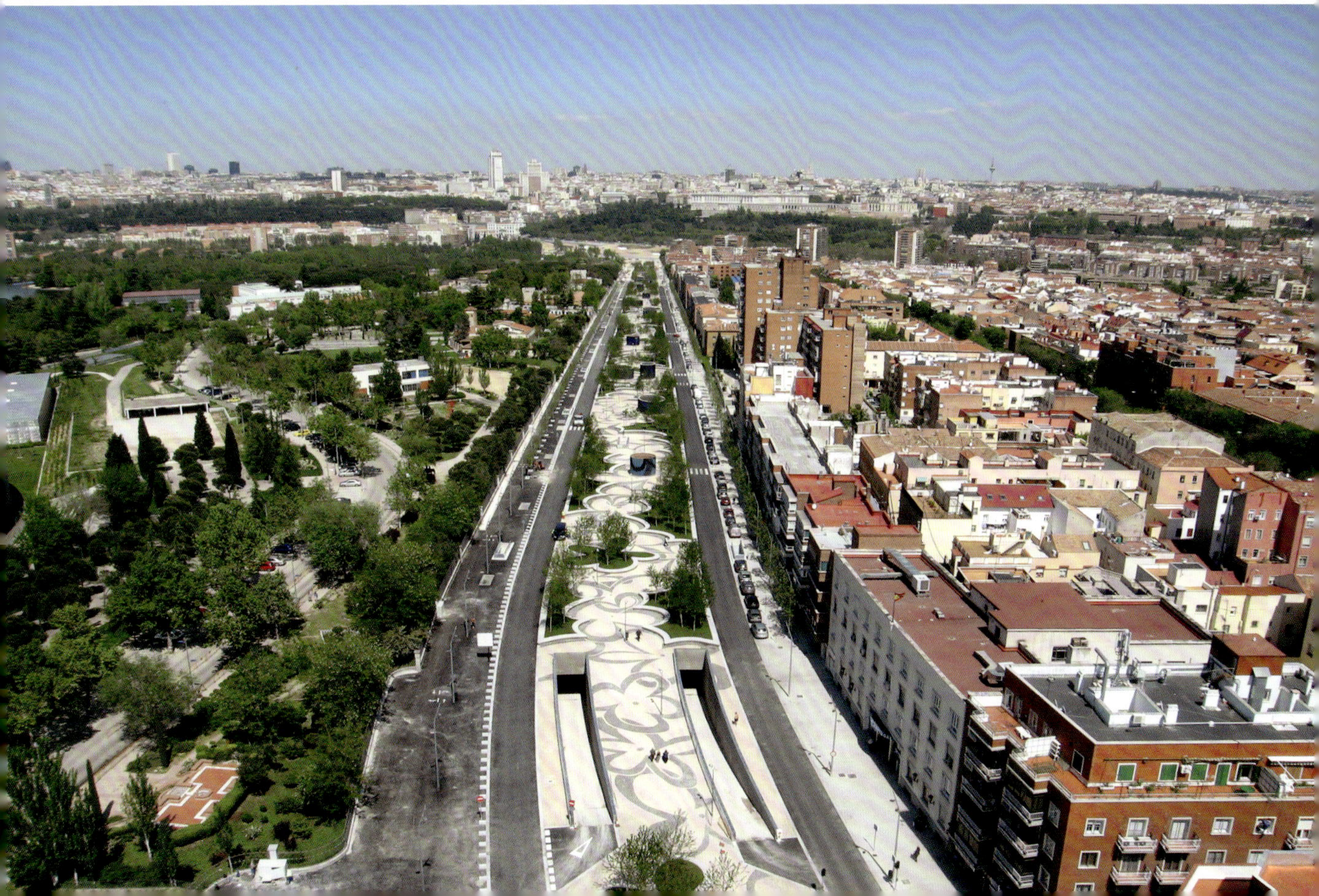

A map shows the Jardines del Puente de Toledo, Salón de Pinos, and Jardines del Puente de Segóvia, together with images of spaces designed by West 8 in the midst of Madrid.

Ein Auszug aus dem Stadtplan zeigt die Jardines del Puente de Toledo, den Salón de Pinos und die Jardines del Puente de Segóvia, daneben stehen Fotos der von West 8 mitten in Madrid entworfenen Parkanlagen.

La carte montre les Jardines del Puente de Toledo, le Salón de Pinos et les Jardines del Puente de Segóvia. Les photographies sont des exemples de réalisations de West 8 au centre de Madrid.

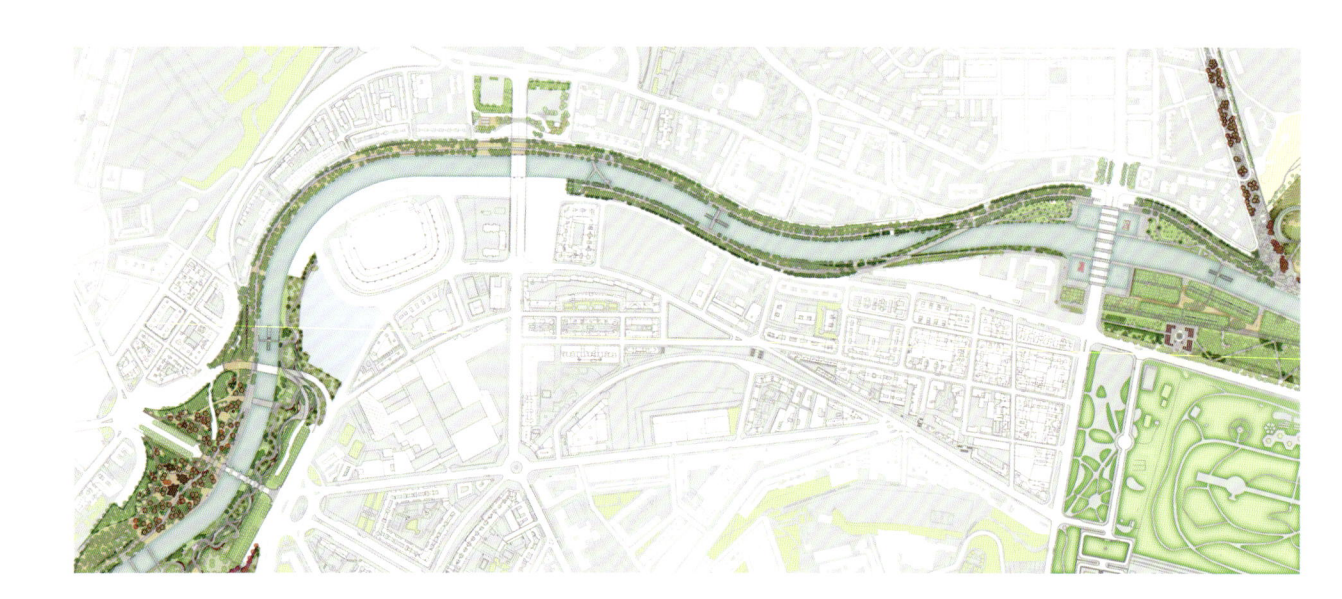

WEST 8

Above, a map of the Parque de la Arganzuela, Matadero, Jardines del Puente de Toledo, and Salón de Pinos.

Oben, Plan mit dem Parque de la Arganzuela, dem Kulturzentrum Matadero, den Jardines del Puente de Toledo und dem Salón de Pinos.

Ci-dessus, une carte sur laquelle figurent le Parque de la Arganzuela, le Matadero, les Jardines del Puente de Toledo et le Salón de Pinos.

Above, and left, the Salón de Pinos section of the project with its walkways and trees.

Oben und links, der Bereich des Salón de Pinos mit seinen Fußgängerwegen und den Pinien.

Ci-dessus et à gauche, la section du Salón de Pinos aux allées bordées d'arbres.

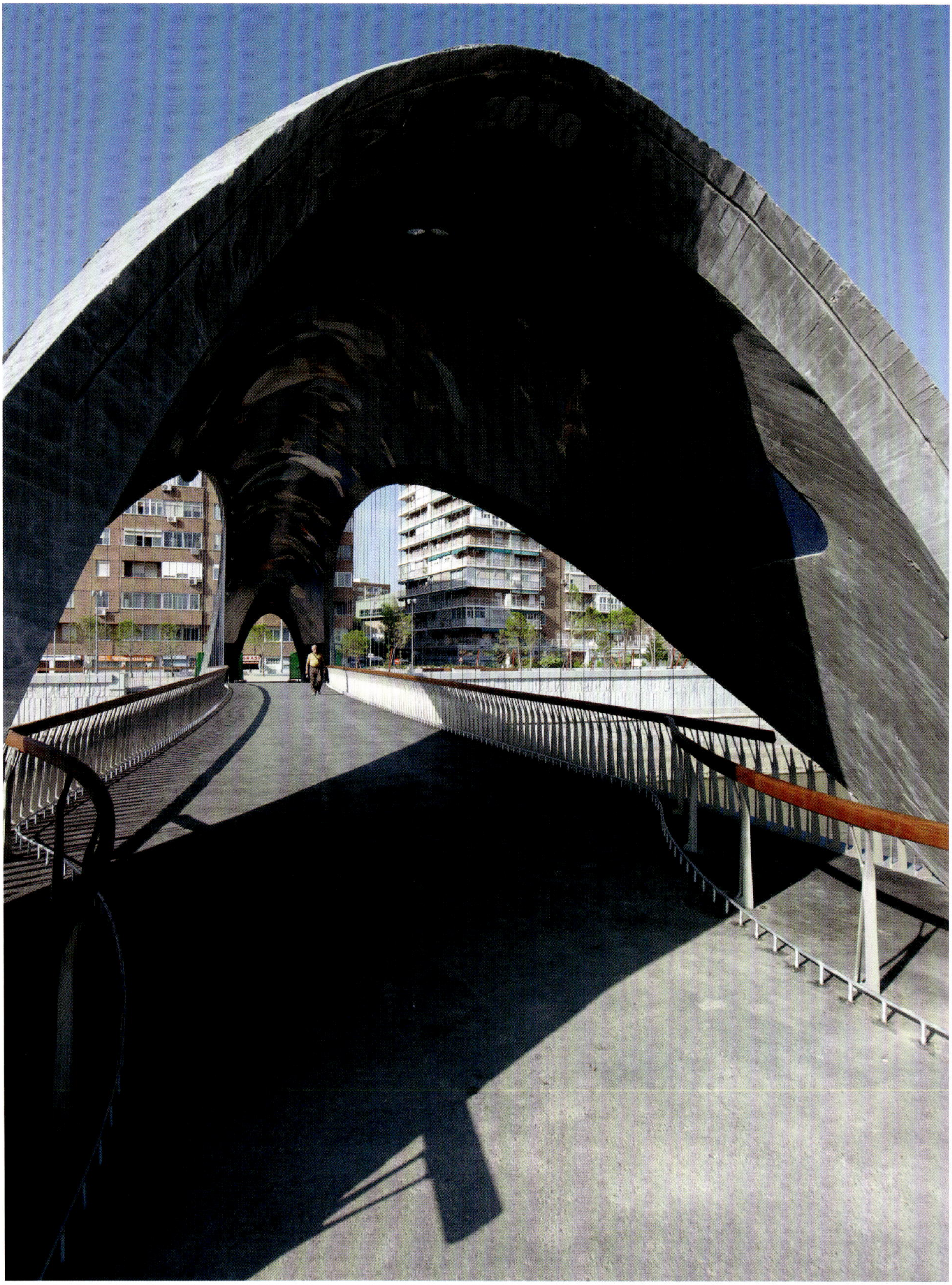

On this double page, the Puente Cáscara pedestrian bridges are rough-textured concrete dome structures. Their ceilings have mosaic designs by the Spanish artist Daniel Canogar.

Auf dieser Doppelseite: Die Fußgängerbrücken Puentes Cáscara sind als Betonkuppelkonstruktionen mit einer groben Oberflächenstruktur ausgeführt. Die Decken sind mit Mosaiken nach einem Entwurf des spanischen Künstlers Daniel Canogar gestaltet.

Sur cette double page, les passerelles piétonnières du Puente Cáscara sont des coupoles de béton brut. Leur plafond est décoré de mosaïques de l'artiste espagnol Daniel Canogar.

TORONTO CENTRAL WATERFRONT

Toronto, Canada, 2006–

Address: www.waterfrontoronto.ca
Area: 3.5 kilometers (length). Client: Waterfront Toronto
Cost: $190 million

The Wavedecks designed by West 8 are "a series of timber structures that explore variations of a simple articulation in the change in level between Queens Quay Boulevard and Lake Ontario along the Toronto Central Waterfront."

Die von West 8 entworfenen Wavedecks sind „Holzkonstruktionen an der Toronto Central Waterfront zwischen dem Queens Quay Boulevard und dem Ontariosee, mit denen man die Abweichungen erfahren kann, die durch eine geringfügige Änderung des Winkels entstehen".

Les « plates-formes en vague » de West 8 sont « une série de structures en bois qui explorent les variantes d'une articulation simple dans le changement de niveau entre le Queens Quay Boulevard et le lac Ontario le long de la rive du centre de Toronto. »

The waterfront area runs along a narrow strip between the water and city roads, as seen in the image to the right.

Die Uferzone führt als schmaler Streifen zwischen dem Wasser und den Innenstadtstraßen entlang, wie man auf dem Foto rechts erkennen kann.

La promenade court le long d'une étroite bande entre le lac et la voirie urbaine (image de droite).

The **CENTRAL WATERFRONT** is a 3.5-kilometer stretch of the Lake Ontario shoreline immediately adjacent to the downtown business district of the city. West 8 responded to a design competition launched by Waterfront Toronto with a scheme to unify design language along the shorefront and to more firmly connect the city to the lake. The designers proposed to create several different areas or approaches: "The Primary Waterfront—a continuous water's edge promenade with a series of pedestrian bridges; the Secondary Waterfront—a recalibrated Queens Quay Boulevard with a new urban promenade and public spaces at the heads-of-slips; the Floating Waterfront—a series of floating elements that offer new boat moorings and public spaces in relation to the lake; and the Cultures of the City—connections from Toronto's diverse neighborhoods toward the waterfront." West 8 presented a temporary installation showing their ideas for Queens Quay Boulevard during the summer of 2006. A series of timber "wavedecks" explore the new connections between the city and its lake.

Die **CENTRAL WATERFRONT** ist ein 3,5 Kilometer langer Uferbereich am Ontariosee, der unmittelbar an das innerstädtische Geschäftsviertel von Toronto angrenzt. West 8 gewann den von Waterfront Toronto ausgeschriebenen Wettbewerb mit einem Entwurf, der die gesamte Uferzone einheitlich gestaltete und die Verbindung zwischen der Stadt und dem See sehr viel enger knüpfte. Die Planer schlugen vor, mehrere unterschiedliche Bereiche zu schaffen: „Die Erste Uferzone (Primary Waterfront), eine durchgehende Uferpromenade mit einer Reihe von Fußgängerbrücken, die Zweite Uferzone (Secondary Waterfront), der umgestaltete Queens Quay Boulevard mit einer neuen urbanen Promenade und öffentlichen Freiräumen an den Kopfenden, das Schwimmende Ufer (Floating Waterfront), eine Reihe schwimmender Elemente, mit zusätzlichen Bootsliegeplätzen und Freiräumen, die zum See orientiert sind, sowie die Cultures of the City – die Verbindungen mehrerer Stadtviertel von Toronto mit dem Seeufer." West 8 präsentierte eine temporäre Installation, anhand derer das Büro seine Vorstellungen für die Gestaltung des Queens Quay Boulevard im Sommer 2006 darstellte. Auf wellenförmigen Holzdecks entdeckt man die neuen Verbindungen zwischen der Stadt und dem See.

Le **CENTRAL WATERFRONT** est une section de 3,5 km de la côte du lac Ontario, adjacente aux quartiers d'affaires du centre-ville. Pour le concours lancé par Waterfront Toronto, West 8 a proposé un plan qui unifie le langage conceptuel des aménagements et relie plus nettement la ville au lac. L'agence a proposé de créer plusieurs zones ou approches différentes : « Le Primary Waterfront, une promenade continue le long de l'eau, ponctuée d'une succession de passerelles piétonnières ; le Secondary Waterfront, un Queens Quay Boulevard recalibré assorti d'une nouvelle promenade urbaine et d'espaces publics ; le Floating Waterfront, une série d'éléments flottants qui offrent de nouveaux postes d'amarrage pour les bateaux et des espaces publics davantage liés au lac ; et les Cultures of the City, des liaisons entre divers quartiers de la ville et le lac. » West 8 a mis en place une installation temporaire pour présenter ses idées pour le Queens Quay Boulevard pendant l'été 2006. Une série de « plates-formes en forme de vague » en bois propose de nouveaux rapports entre la ville et son lac.

Above, the Rees Wavedeck to some extent recalls the tradition of "boardwalks" often seen in the eastern part of North America.

Das Rees Wavedeck (oben) erinnert ein wenig an die Tradition der „boardwalks", der hölzernen Uferpromenaden, die man oft in Neuengland sieht.

Ci-dessus, le Rees Wavedeck rappelle la tradition des « promenades en planches » que l'on trouve souvent sur la côte est de l'Amérique du Nord.

Above, the Simcoe Wavedeck in a nighttime view. Right, "aquatic ecological design is integrated in the foreshore design."

Oben, das Simcoe Wavedeck bei Nacht. Rechts, „zu dem Entwurf der Uferzone gehört auch eine Planung für das Ökosystem des Wassers".

Ci-dessus le Simcoe Wavedeck vu la nuit. À droite « un projet écologique aquatique intégré dans l'aménagement de la rive ».

Ernsting's Family Headquarters

WIRTZ INTERNATIONAL

Wirtz International N. V.
Botermelkdijk 464
2900 Schoten
Belgium

Tel: +32 3 680 13 22
E-mail: info@wirtznv.be
Web: www.wirtznv.be

Born in 1924 in Shoten, Belgium, Jacques Wirtz created his own garden design firm in 1948 in Antwerp. His sons Martin and Peter joined the firm in 1986 and 1990. Perhaps the best known of the interventions of **WIRTZ INTERNATIONAL** is his Carrousel garden at the Louvre just opposite I. M. Pei's Pyramid (1990–93). Wirtz created the garden for Arquitectonica's Banque de Luxembourg building (Luxembourg, 1989–94). Other work by the firm includes the gardens for the Belgian Pavilion at the Osaka World Exposition in 1970; for the European pavilions at the Technology Expo at Tsukuba in 1985; and redesigning the gardens of the Elysée Palace (Paris, 1992). They have also completed Jubilee Park (Canary Wharf, London, 2000); Ernsting's Family Headquarters (Coesfeld-Lette, Germany, 2000–02, published here); NATO Headquarters (Brussels, Belgium, 2006); and gardens for the Novartis Campus (Basel, Switzerland, 2009). Recently Wirtz International won the competition for the master plan design for the European headquarters in Brussels (with Atelier Christian de Portzamparc; Belgium, 2009); and the competition for the extension of the Kunsthaus in Zurich (with David Chipperfield; Switzerland, 2009).

Jacques Wirtz wurde 1924 in Schoten, Belgien, geboren und gründete 1948 sein Büro für Gartengestaltung in Antwerpen. Seine Söhne Martin und Peter traten 1986 bzw. 1990 in das Büro ein. Das wahrscheinlich bekannteste Projekt des Büros **WIRTZ INTERNATIONAL** ist der an den Innenhof des Louvre anschließende Jardin du Carrousel gegenüber der Pyramide von Ieoh Ming Pei (1990–93). Das Büro Wirtz gestaltete außerdem den Garten für die von Arquitectonica entworfene Banque de Luxembourg (Luxemburg, 1989–94). Weitere Projekte von Wirtz sind die Gärten des Belgischen Pavillons auf der Weltausstellung in Osaka 1970, die der europäischen Pavillons auf der Technologie-Expo in Tsukuba 1985 und die Neugestaltung der Gartenanlagen des Elysée-Palastes (Paris, 1992). Wirtz hat den Jubilee Park im Londoner Stadtteil Canary Wharf (2000) realisiert und die Gartenanlagen um die Hauptverwaltung der Firma Ernsting's Family in Coesfeld-Lette (2000–02, hier vorgestellt), die Gartenanlagen des NATO-Hauptquartiers in Brüssel (2006) und die Gärten des Novartis Campus in Basel (2009). Das Büro Wirtz International gewann 2009 den Wettbewerb für den Entwurf des Masterplans für die Hauptverwaltung der Europäischen Union in Brüssel (mit dem Atelier Christian de Portzamparc) und den Wettbewerb für den Erweiterungsbau für das Kunsthaus Zürich (mit David Chipperfield).

Né en 1924 à Shoten (Belgique), Jacques Wirtz a créé son entreprise d'aménagement de jardins à Anvers en 1948. Ses fils Martin et Peter ont rejoint l'agence en 1986 et 1990. L'intervention de **WIRTZ INTERNATIONAL** la plus connue est sans doute le jardin du Carrousel, face à la pyramide du Louvre de I. M. Pei (1990–93). Parmi les autres réalisations de l'agence : le jardin de la Banque de Luxembourg, Arquitectonica, (Luxembourg, 1989–94) ; les jardins du pavillon de la Belgique à l'Exposition universelle d'Osaka en 1970 ; les jardins des pavillons européens à l'Exposition de technologie de Tsukuba en 1985 ; la rénovation des jardins du palais de l'Élysée (Paris, 1992). Les Wirtz ont également réalisé le Jubilee Park (Canary Wharf, Londres, 2000) ; les jardins du siège de la famille Ernsting (Coesfeld-Lette, Allemagne, 2000–02, publiés ici) ; les jardins du siège de l'OTAN (Bruxelles, 2006) et les jardins du campus Novartis (Bâle, Suisse, 2009). Récemment, Wirtz International a remporté le concours pour le plan directeur du siège de l'Union européenne à Bruxelles (avec l'Atelier Christian de Portzamparc, 2009) et le concours pour l'extension de la Kunsthaus de Zurich (avec David Chipperfield, Suisse, 2009).

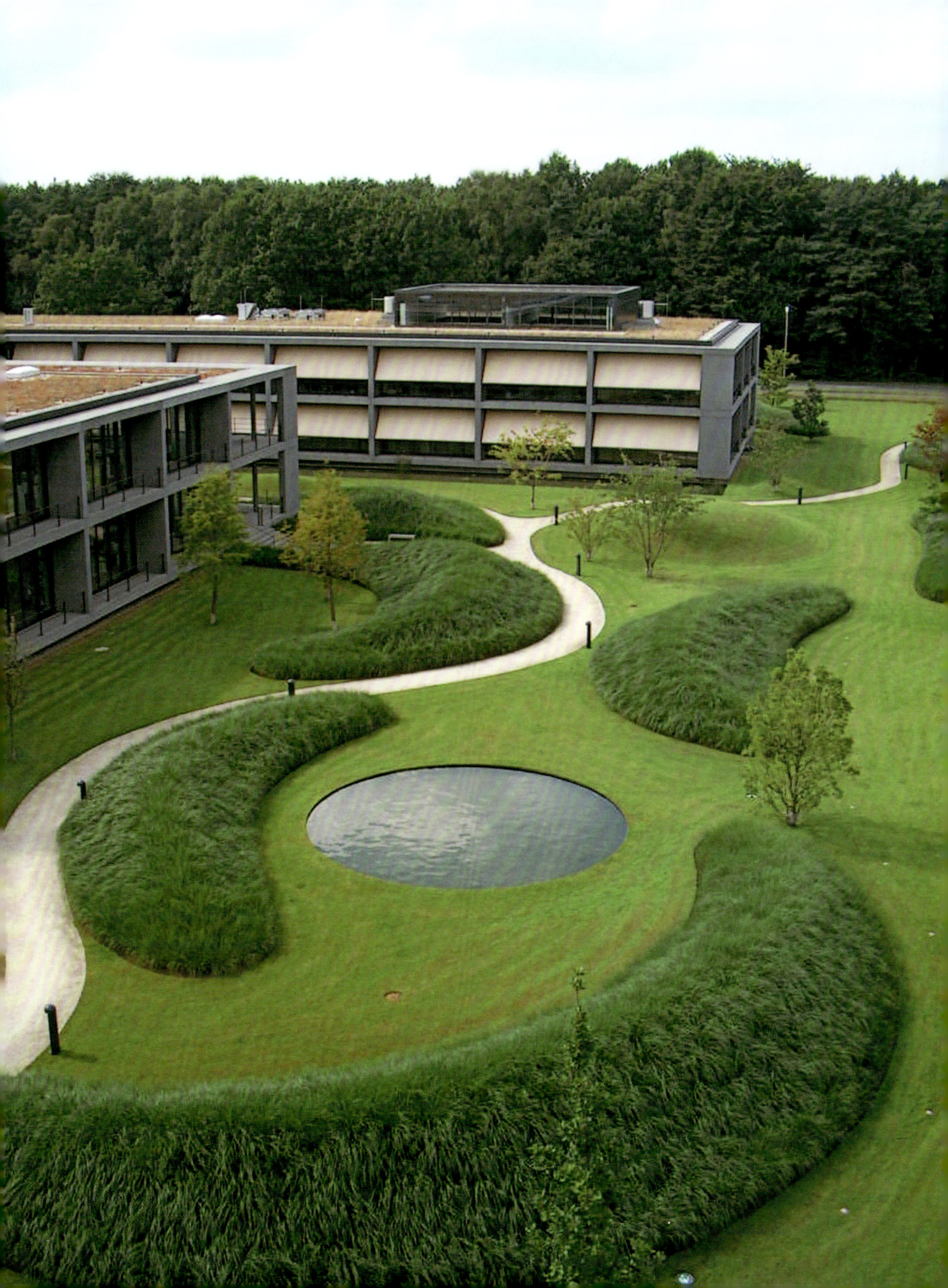

ERNSTING'S FAMILY HEADQUARTERS
Coesfeld-Lette, Germany, 2000–02

Address: Ernsting's family GmbH & Co. KG, Industriestraße 1, 48653 Coesfeld-Lette, Germany, www.ernstings-family.de. Area: 2.5 hectares. Client: Ernsting's. Cost: €1.5 million

The firm was given the task of landscaping the Ernsting campus in Coesfeld-Lette, west of Münster in the Nordrhein Westfalen region of Germany, where David Chipperfield, Santiago Calatrava, and Johannes Schilling have built structures for the clothing retailer. As the designers explain: "The challenge here was to bind all these buildings together in one campus atmosphere of beauty and well-being. The basic structure consists of curvilinear paths winding through a landscape of grass mounds, great cushions of ornamental grasses, water pieces, and a tree collection. The quiet and soft character of the landscape allows the geometric buildings to speak and live together in a park." Careful attention was paid to the selection of plants, including such unexpected species as zelkova trees (*Zelkova serrata*). Chipperfield's Ernsting Service Center (1998–2001) completes the Ernsting compound, now brought together by the Wirtz garden. Though Wirtz began work in Coesfeld just as Chipperfield was finishing, there is a comfortable rapport between architecture and the garden.

Dem Büro Wirtz wurde der Auftrag für die landschaftliche Gestaltung des Firmensitzes von Ernsting in Coesfeld-Lette, westlich von Münster in Nordrhein-Westfalen, übertragen. David Chipperfield, Santiago Calatrava und Johannes Schilling haben hier Gebäude für den Bekleidungseinzelhändler errichtet. Die Planer erklären: „Die Herausforderung bei diesem Auftrag lag darin, all diese unterschiedlichen Gebäude miteinander zu verbinden und eine Campus-Atmosphäre von Schönheit und Ausgeglichenheit herzustellen, in der man sich wohlfühlen kann. Die Grundstruktur bilden geschwungene Wege, die sich durch eine Landschaft aus Grashügeln schlängeln, große Kissen aus Ziergräsern, Wasserbecken und eine Baumsammlung. Der ruhige und sanfte Charakter der Landschaft lässt die geometrischen Gebäude miteinander in einen Dialog treten und verbindet sie in diesem Park." Besondere Sorgfalt wurde der Pflanzenauswahl gewidmet, wie zum Beispiel solch ungewöhnlichen Arten wie den japanischen Zelkoven (Zelkova serrata). Das Ernsting Service Center von Chipperfield (1998–2001) vervollständigt den Ernsting-Komplex, der durch den Garten von Wirtz zu einer Einheit gefunden hat. Obwohl das Büro Wirtz erst mit der Gartengestaltung begann, als Chipperfield gerade sein Gebäude fertiggestellt hatte, herrscht eine wunderbare Beziehung zwischen der Architektur und dem Garten.

L'agence Wirtz International a été chargée des aménagements paysagers du site Ernsting à Coesfeld-Lette, à l'est de Münster, en Rhénanie-du-Nord-Westphalie, où David Chipperfield, Santiago Calatrava et Johannes Schilling avaient déjà construit divers bâtiments pour l'entreprise de confection. « Le défi était de réunir tous ces bâtiments en créant une ambiance unique de beauté et de bien-être sur le site. La structure de base se compose de sentiers serpentant dans un paysage fait de parterres gazonnés, de coussins de graminées, de plans d'eau et d'arbres. Le caractère arrondi et doux du paysage met bien en évidence la géométrie des bâtiments, qui s'expriment ensemble dans un cadre comparable à un parc », explique l'agence. Une attention particulière a été portée à la sélection des plantes, qui comprend des espèces inattendues comme les zelkovas (*Zelkova serrata*). Le Ernsting Service Center construit par Chipperfield (1998–2001) complète cet ensemble, désormais unifié par le jardin des Wirtz. Même si ceux-ci sont intervenus après la fin du chantier de Chipperfield, on apprécie un rapport agréable entre l'architecture et le jardin.

The landscaping relies on trees and ornamental grasses on densely planted, oblong mounds together with water features to unite this group of buildings designed by famous architects.

Zentrale Elemente dieser Landschaftsgestaltung sind Bäume und Ziergräser, die sehr dicht auf leicht gekrümmten, länglichen Hügeln gepflanzt sind. Zusammen mit den Wasserflächen vereinen sie die Gebäude, die von berühmten Architekten entworfen wurden.

L'aménagement paysager s'appuie sur la plantation d'arbres et de graminées ornementales sur des bermes oblongues, ainsi que d'éléments aquatiques pour unifier cet ensemble de bâtiments conçu par des architectes célèbres.

The Central Office Building on this corporate campus was designed by David Chipperfield, with warehouses by Santiago Calatrava and Schilling Architekten.

Das Gebäude der Hauptverwaltung ist von David Chipperfield entworfen, die Lagerhäuser wurden von Santiago Calatrava und Schilling Architekten geplant.

L'immeuble de bureaux central de ce site d'affaires a été conçu par David Chipperfield et certains entrepôts par Santiago Calatrava et Schilling Architekten.

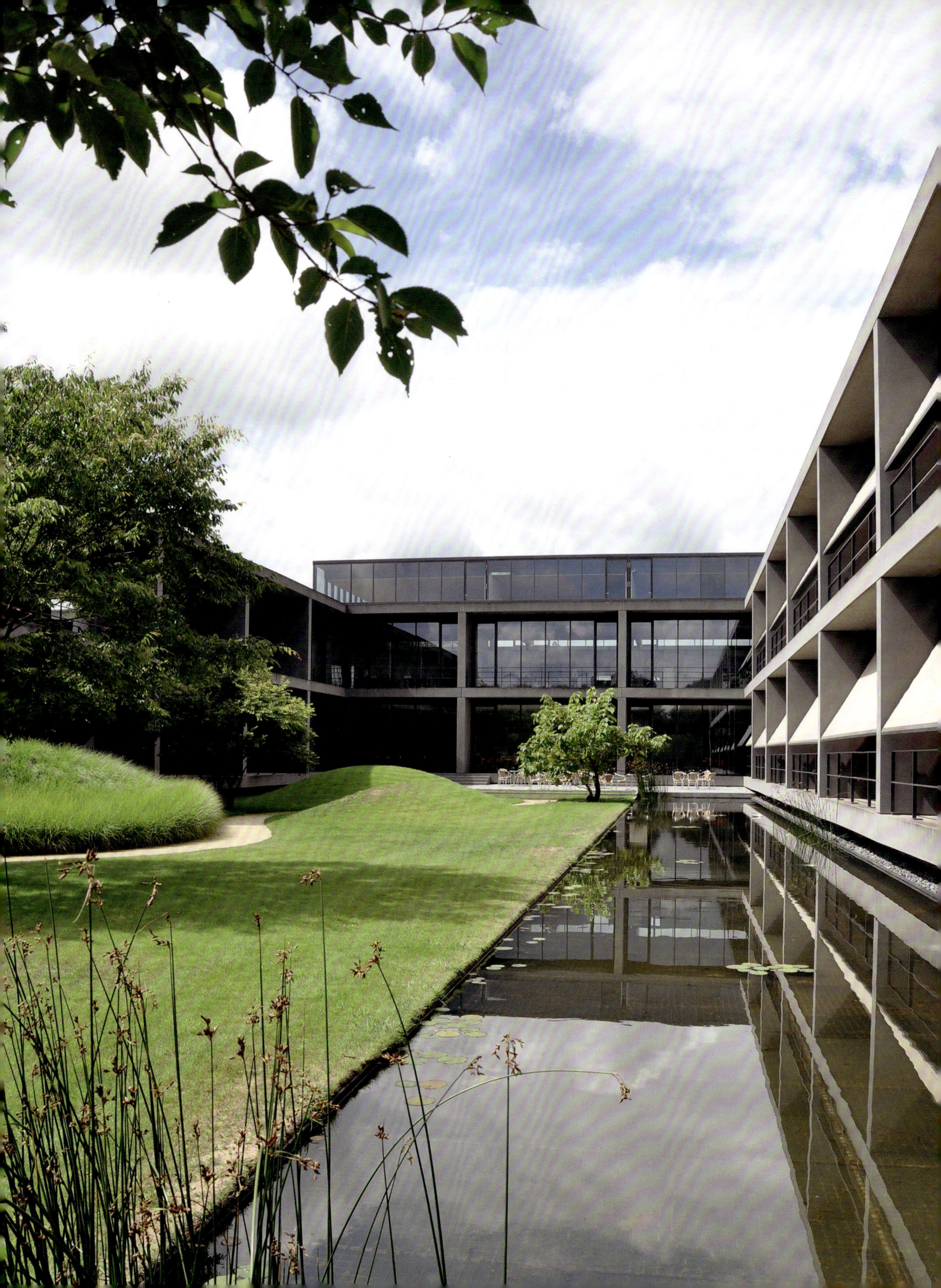

WORK ARCHITECTURE COMPANY

WORK Architecture Company
156 Ludlow Street, 3rd Floor
New York, NY 10002
USA

Tel: +1 212 228 1333
Fax: +1 212 228 1674
E-mail: office@work.ac
Web: www.work.ac

Born in Beirut, Lebanon, **AMALE ANDRAOS** received her B. Arch at McGill University (1996) and her Master's degree from Harvard University (1999). She worked with Rem Koolhaas / OMA (Rotterdam, 1999–2003), before founding WORK AC in New York in 2003. **DAN WOOD**, born in Rhode Island, received his Bachelor's degree (in film theory) from the University of Pennsylvania (1989) and his M.Arch from Columbia University (1992). He lived in Paris and the Netherlands, before moving to New York in 2002. He worked with Rem Koolhaas / OMA (Rotterdam, 1994–2000) and was President and Founder of AMO, Inc. in New York (2000–03) and a Partner with Rem Koolhaas / OMA in New York (2000–03), before cofounding WORK AC with Amale Andraos. Their work includes the Diane von Furstenberg Studio Headquarters (New York, 2007); Public Farm 1 (Long Island City, New York, 2008, published here); Wild West Side (New York, New York, 2008); designs for the Shenzhen Metro Tower (Shenzhen, China, 2010); Wuhan University Library (Wuhan, China, 2010): Kew Gardens Hills Library (Queens, New York, 2011); Children's Museum of the Arts (New York, 2011); and Edible Schoolyard, PS 216 (Brooklyn, New York, 2011), all in the USA unless stated otherwise.

AMALE ANDRAOS wurde in Beirut geboren und machte 1996 ihren Bachelor in Architektur an der McGill University und 1999 ihren Master an der Harvard Universität. Bevor sie 2003 ihr eigenes Büro WORK AC in New York gründete, arbeitete sie von 1999 bis 2003 im Büro Rem Koolhaas/OMA in Rotterdam. **DAN WOOD**, der in Rhode Island geboren wurde, machte 1989 seinen Bacholor in Filmtheorie an der University of Pennsylvania und 1992 seinen Master in Architektur an der Columbia University. Bevor er nach New York ging, lebte er in Paris und in den Niederlanden. Von 1994 bis 2000 war er Mitarbeiter im Büro von Rem Koolhaas in Rotterdam und von 2000 bis 2003 sowohl Geschäftsführer und Gründer der AMO, Inc., in New York als auch Büropartner von Rem Koolhaas/OMA in dessen New Yorker Büro, bis er mit Amale Andraos WORK AC gründete. Zu ihren Arbeiten gehören die Diane von Furstenberg Studio Headquarters in New York (2007), die Public Farm 1 in Long Island City, New York, 2008 (hier vorgestellt), die Wild West Side in New York (2008), Entwürfe für das Hochhaus Shenzhen Metro Tower in Shenzhen, China (2010), die Universitätsbibliothek der Universität in Wuhan, China (2010), die Kew Gardens Hills Library in Queens, New York (2011), das Kinderkunstmuseum in New York (2011) und der Essbare Schulhof, PS 216 (Brooklyn, New York, 2011). Alle Projekte befinden sich in den USA, sofern nicht anders angegeben.

Née à Beyrouth, **AMALE ANDRAOS** est titulaire d'un B.Arch. de l'université McGill (1996) et d'un M.Arch de l'université Harvard (1999). Elle a travaillé pour Rem Koolhaas/OMA (Rotterdam, Pays-Bas, 1999–2003), avant de fonder l'agence WORK AC à New York en 2003. **DAN WOOD**, né dans le Rhode Island, est titulaire d'un B.A. en théorie du cinéma de l'université de Pennsylvanie (1989) et d'un M.Arch de l'université Columbia (1992). Il a vécu à Paris et aux Pays-Bas avant de s'installer à New York en 2002. Il a travaillé pour Rem Koolhaas/OMA (Rotterdam, Pays-Bas, 1994–2000) et a été président-fondateur d'AMO, Inc. à New York (2000–03) et partenaire de Rem Koolhaas/OMA à New York (2000–03), avant de fonder WORK AC avec Amale Andraos. Parmi leurs réalisations : le siège de Diane von Furstenberg Studio (New York, 2007) ; Public Farm 1 (Long Island City, New York, 2008, publiée ici) ; Wild West Side (New York, 2008) ; des installations pour la tour Metro de Shenzhen (Shenzhen, Chine, 2010) ; la bibliothèque de l'université de Wuhan (Wuhan, Chine, 2010) : la bibliothèque de Kew Gardens Hills (Queens, New York, 2011) ; le Children's Museum of the Arts (New York, 2011) et le projet Edible Schoolyard de la PS 216 (Brooklyn, New York, 2011).

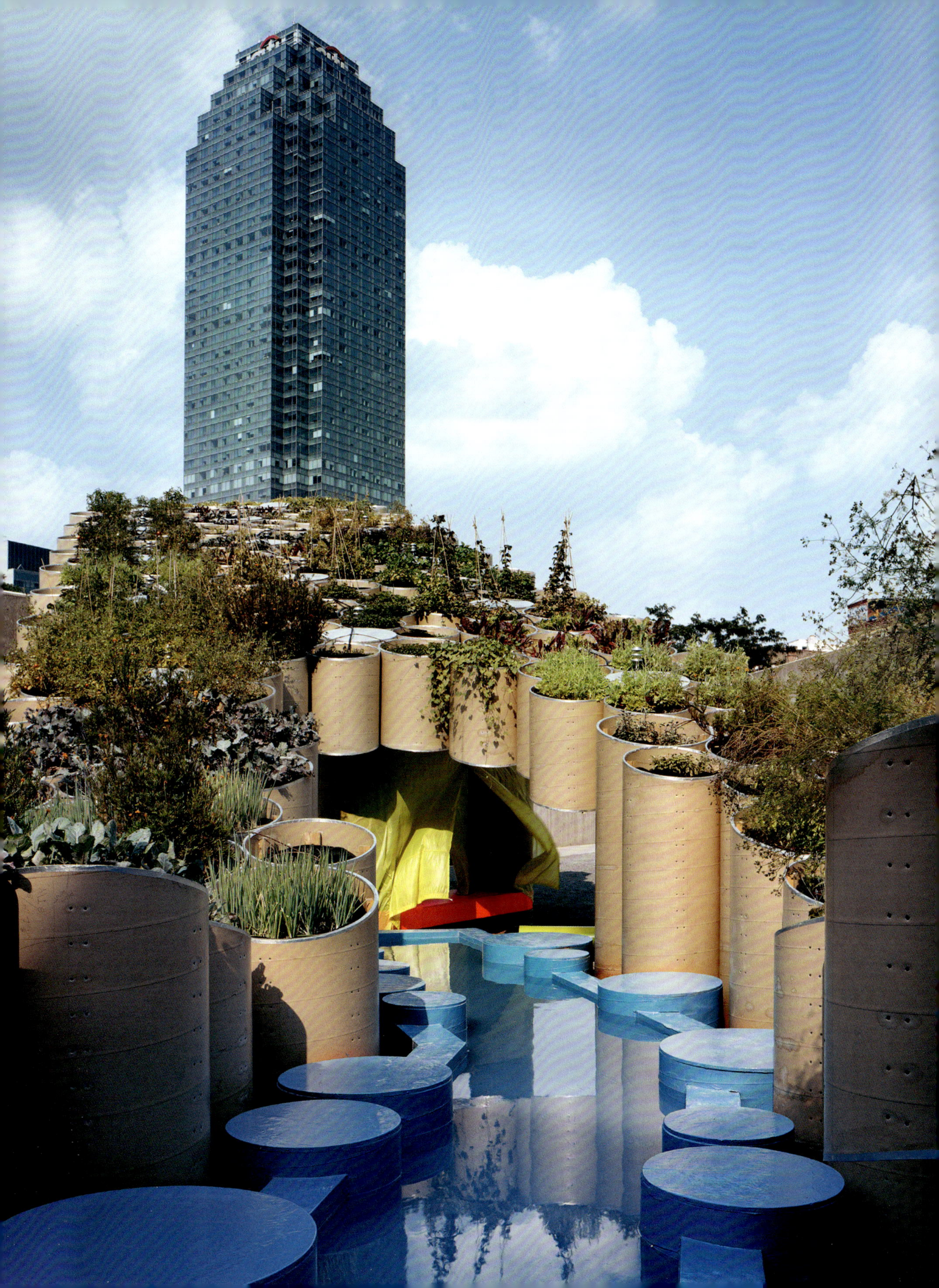

PUBLIC FARM 1

Long Island City, New York, USA, 2008

Address: 22–25 Jackson Avenue at the intersection of 46th Avenue,
Long Island City, New York, NY 11101, USA, www.ps1.org. Area: 1011 m²
Client: Museum of Modern Art (MoMA). Cost: $180 000

This project was the winning entry for the 2008 MoMA/PS1 Young Architect Program. Built in the P.S.1 Contemporary Art Center's courtyards, the temporary installation introduced a quarter acre fully functioning urban farm in the form of a folded plane made of structural cardboard tubes. The project was built entirely with recyclable materials, powered by solar energy, and irrigated by a rooftop rainwater collection system. The farm actually produced over 50 varieties of organic fruit, vegetables, and herbs that were used by the museum's café, served at special events, and harvested directly by visitors. Each of these crops was planted in six tubes of varying diameter arrayed around a central tube that was part of the structural support. The columns were also used to house videos about farms, solar pow-ered fans, and a mobile phone charger. The architects state: "As a live urban farm, PF1 was a testament to the possibilities of rural engagement in urban environments and proposed that cities be reinvented to become a more complete and integrated system capable of producing their own food, producing their own power, and reusing their own water while creating new shared spaces for social interaction and public pleasure."

Dieser Entwurf gewann 2008 den Wettbewerb MoMA/PS1 Young Architect Program. Er wurde als temporäre Installation in den Innenhöfen des P.S.1 Contemporary Art Center angelegt. Auf gut 1000 Quadratmetern wurde ein voll funktionierender urbaner Bauernhof in Form einer abgeknickten Ebene angelegt, deren Konstruktion aus Papphören bestand. Das gesamte Projekt wurde aus recycelfähigen Materialien errichtet, mit Solarenergie gespeist und durch ein Regenwassersammelsystem bewässert. Die Farm produzierte mehr als 50 Sorten Obst, Gemüse und Kräuter, die im Museumscafé bei besonderen Ereignissen serviert und von den Besuchern geerntet wurden. Alle Pflanzen wurden jeweils in sechs unterschiedlich großen Röhren angebaut, die um eine zentrale Röhre angeordnet waren, die Teil der Konstruktion war. In den Papphören waren auch Bildschirme integriert, auf denen Videos über Bauernhöfe abgespielt wurden oder mit Solarenergie angetriebene Ventilatoren und eine Ladestation für Mobiltelefone. Die Architekten meinen: „PF1 stellte die Möglichkeit unter Beweis, Landwirtschaft in einem urbanen Umfeld zu betreiben, und es war eine Anregung, dass sich die Städte zu einem System entwickeln können, das in der Lage ist, sich mit Essen und Energie selbst zu versorgen und das vor Ort anfallende Wasser wiederzuverwenden. Zugleich entstehen neue gemeinschaftlich genutzte Freiräume, die zum sozialen Dialog einladen und als öffentliche Erholungsräume dienen."

Ce projet a remporté le concours du programme des jeunes architectes du MoMA/PS1 en 2008. Réalisée dans les cours du P.S.1 Contemporary Art Center, cette installation temporaire a recréé une exploitation agricole pleinement opérationnelle d'environ 1000 m² sur un plan incliné plié en tubes de carton structurel. Le projet, réalisé en matériaux recyclables, était alimenté par l'énergie solaire et irrigué par les eaux récupérées des toitures. La ferme a produit plus de 50 variétés de fruits, de légumes et d'herbes, servis dans le café du musée lors de manifestations ou directement récoltés par les visiteurs. Chaque variété était plantée dans six tubes de divers diamètres, disposés autour d'un tube central solidaire du système structurel. Les colonnes étaient également utilisées pour proposer des vidéos sur les fermes, des ventilateurs solaires et un chargeur de téléphones mobiles. « En tant que ferme urbaine vivante, PF1 illustrait la possibilité d'une présence de la ruralité dans des environnements urbains et proposait que les villes se réinventent pour devenir des systèmes plus complets et plus intégrés, capables de produire leur propre nourriture, leur énergie et réutiliser leurs eaux, tout en créant de nouveaux espaces de partage visant à l'interaction sociale et au plaisir des visiteurs », ont expliqué les architectes.

WORK ARCHITECTURE COMPANY

The array of cardboard tubes that forms the Public Farm 1 project are essentially placed on an angled plane that leads up from and down to the ground level.

Les tubes de carton qui constituent la Public Farm 1 sont pour l'essentiel disposés sur un plan doublement incliné, descendant vers le sol et en remontant.

Die Pappröhren des Projekts Public Farm 1 sind größtenteils auf einer doppelt geneigten Ebene angeordnet.

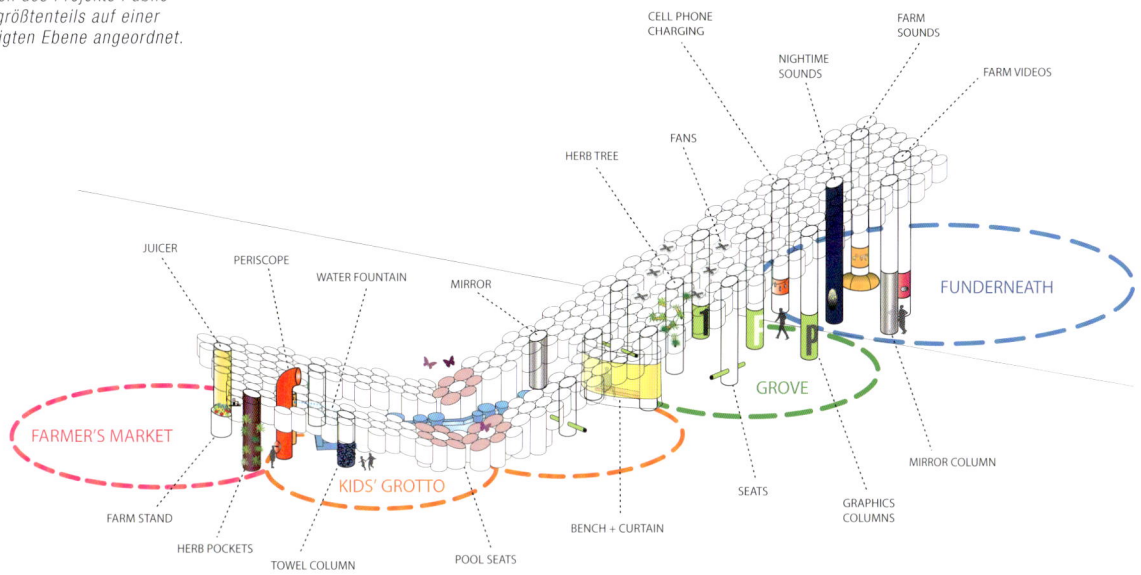

FARMER'S MARKET — JUICER, PERISCOPE, WATER FOUNTAIN, FARM STAND, HERB POCKETS

KIDS' GROTTO — TOWEL COLUMN, POOL SEATS, MIRROR, BENCH + CURTAIN

GROVE — HERB TREE, FANS, SEATS, GRAPHICS COLUMNS, MIRROR COLUMN

FUNDERNEATH — CELL PHONE CHARGING, NIGHTTIME SOUNDS, FARM SOUNDS, FARM VIDEOS

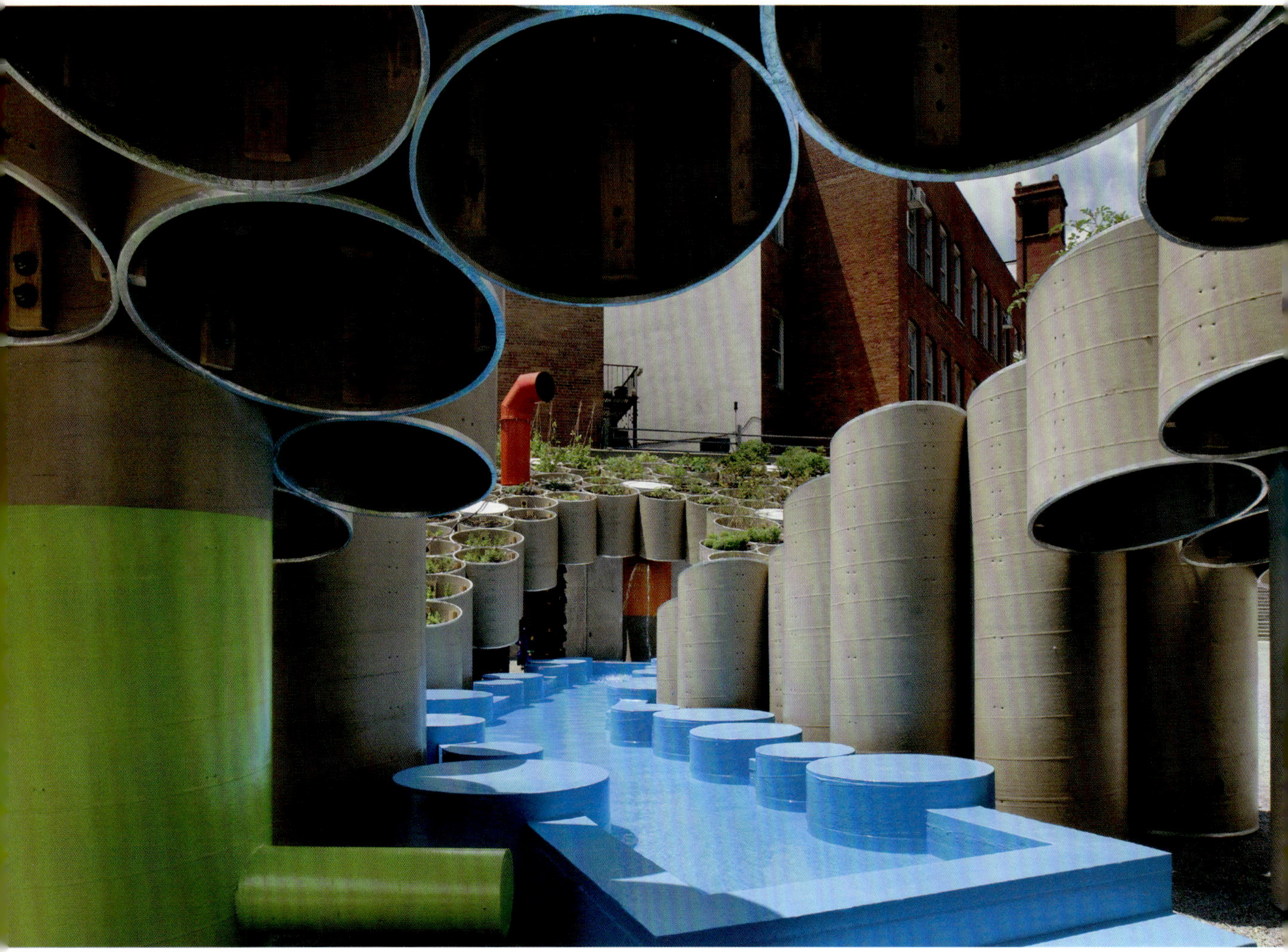

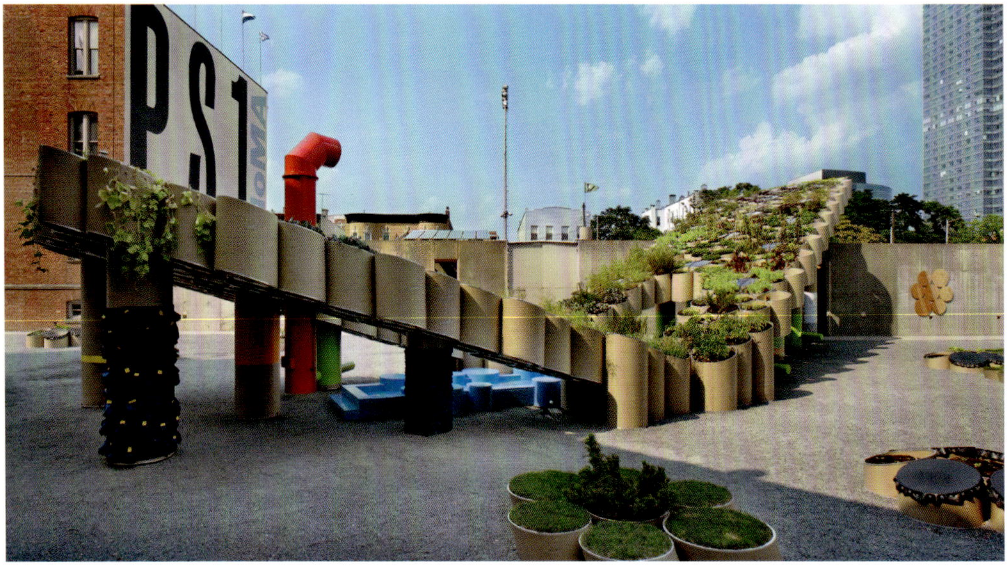
PS1

WORK ARCHITECTURE COMPANY

The installation made use of a rooftop rainwater collection system and featured 50 different vegetables, fruits, and herbs.

Für die Installation wurde ein Regenwassersammelsystem genutzt, mit dem das Dachflächenwasser aufgefangen wurde, und es gab 50 verschiedene Gemüse- und Obstsorten sowie Küchenkräuter.

L'installation fonctionnait à l'aide d'un système de récupération des eaux de pluie et permettait de cultiver une cinquantaine de variétés de légumes, de fruits et d'herbes.

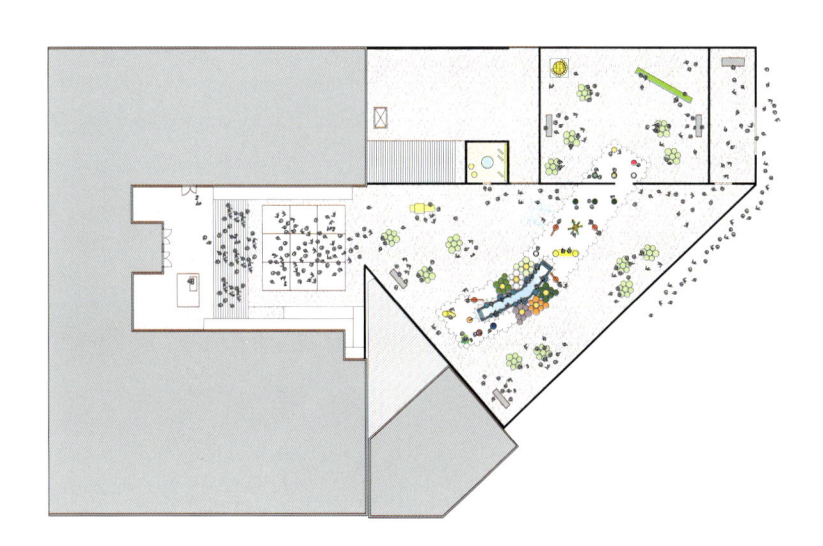

P 410

A diagram on the right shows the careful selection of plants as outlined by the designers. This project might be considered to be a cross between architecture, landscape design, and ecological consciousness-raising.

Das Diagramm rechts zeigt die sorgfältige Pflanzenauswahl, auf die die Planer hinweisen. Dieses Projekt kann man als eine Mischung aus Architektur, Landschaftsgestaltung und Vermittlung ökologischen Bewusstseins bezeichnen.

Le plan de droite montre la sélection de plantes voulue par les aménageurs. Ce projet est au croisement de l'architecture, de l'aménagement paysager et de la prise de conscience des enjeux écologiques.

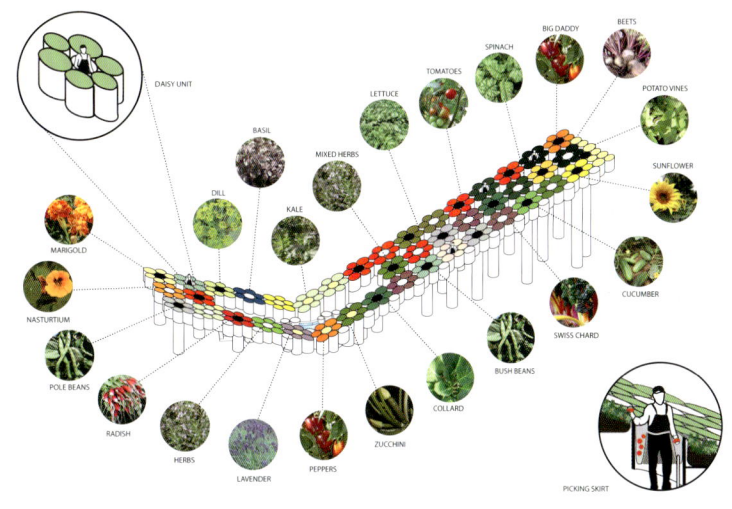

INDEX OF BUILDINGS, NAMES AND PLACES

0-9
1111 Lincoln Road, Miami Beach, Florida, USA — 190
40 Central Park South, New York, New York, USA — 10, 25, 40, 336

A
Adriana Varejão Gallery, Inhotim Contemporary Art Center,
  Brumadinho, Brazil — 98
AECOM — 11, 26, 41, 42, 50
Afrikaanderplein, Rotterdam, The Netherlands — 12, 27, 42, 294
Alumnae Valley Landscape Restoration, Wellesley College,
  Wellesley, Massachusetts, USA — 366
Ando, Tadao — 18, 34, 49, 60
Anglona Paleobotanic Park, North Sardinia, Italy — 216
Archeological Museum of Praça Nova do Castelo de São Jorge,
  Costa do Castelo, Portugal — 88
Australia
  Melbourne, Cranbourne, Royal Botanic Gardens — 346

B
Balmori Associates — 7, 8, 22, 38, 70
Blanc, Patrick — 8, 21, 23, 32, 38, 39, 76
Brazil
  Brumadinho, Inhotim Contemporary Art Center, Adriana Varejão Gallery — 98

C
Cabecera Park, Valencia, Spain — 244
Caffarena, M. / Cobos, V. / Alcaraz, G. / Delgado, G. — 82
Canada
  Quebec, L'autre rive (The Other Bank) — 17, 18, 33, 48, 254
  Toronto, Toronto Central Waterfront — 394
Carrilho da Graça, João Luís — 15, 16, 30, 32, 45, 46, 88
Centenary Park, Punta San García, Algeciras, Spain — 82
Cerviño Lopez, Rodrigo — 98
Chile
  Casablanca Valley, Izaro Estate — 172
  Casablanca Valley, Morandé Winery Productive Services — 178
China
  Hong Kong, One Kowloon Office Building, SanShin-Tei — 15, 31, 45, 46, 326
  Shenzhen, Vanke Center / Horizontal Skyscraper — 160
  Xi'an, "Flowing Gardens," Xi'an World Horticultural Fair 2011 — 314

City of Culture of Galicia, Santiago de Compostela, Spain — 13, 28, 43, 118
Civic Space Park, Phoenix, Arizona, USA — 11, 26, 41, 56
Coconut Grove, Florida Garden, Coconut Grove, Florida, USA — 196
Community Gardens, Chiasso, Switzerland — 282
Connecticut Water Treatment Facility,
  New Haven, Connecticut, USA — 14, 15, 29, 30, 44, 45, 362
Cuchillitos de Tristán Park, Santa Cruz de Tenerife, Canary Islands, Spain — 234
Cyprus
  Nicosia, Eleftheria Square Redesign — 150

D
Denmark
  Herning, HEART: Herning Museum of Contemporary Art — 168
Diana, Princess of Wales Memorial Fountain, Hyde Park, London, UK — 144
Diller Scofidio + Renfro — 10, 12, 13, 24, 25, 27, 28, 40, 42, 43, 104, 182
Djurovic, Vladimir — 9, 23, 38, 39, 108
Domplein, Utrecht, The Netherlands — 290

E
Eisenman, Peter — 13, 28, 43, 118
Eleftheria Square Redesign, Nicosia, Cyprus — 150
Eleven Minute Line, Wanås, Sweden — 17, 18, 33, 34, 48, 212
Ernsting's Family Headquarters, Coesfeld-Lette, Germany — 398

F
"Flowing Gardens," Xi'an World Horticultural Fair 2011, Xi'an, China — 314
France
  Aix-en-Provence, Max Juvénal Bridge — 8, 22, 38, 80

G
García Sanabria Park, Santa Cruz de Tenerife, Canary Islands, Spain — 296
Georg-Büchner-Plaza, Darmstadt, Germany — 202
Germany
  Coesfeld-Lette, Ernsting's Family Headquarters — 398
  Darmstadt, Georg-Büchner-Plaza — 202
  Hamburg, HafenCity Public Spaces — 10, 25, 40, 41, 250
Glavovic Studio — 124
Governors Island Park and Public Space Master Plan,
  New York, New York, USA — 380

INDEX

Gustafson Guthrie Nichol 16, 17, 32, 33, 47, 130
Gustafson Porter 140

H
Hadid, Zaha 150
HafenCity Public Spaces, Hamburg, Germany 10, 25, 40, 41, 250
Hariri Memorial Garden, Beirut, Lebanon 9, 23, 39, 108
HEART: Herning Museum of Contemporary Art, Herning, Denmark 168
Herzog & de Meuron 48, 154
Holl, Steven 14, 15, 18, 29, 30, 34, 44, 45, 49, 160
Hurtado, Martin 172
Hypar Pavilion Lawn,
  New York, New York, USA 10, 13, 24, 25, 28, 40, 41, 43, 104

I
Italy
  North Sardinia, Anglona Paleobotanic Park 216
Izaro Estate, Casablanca Valley, Chile 172

J
James Corner Field Operations 12, 27, 42, 182
Japan
  Kagawa, Naoshima, Lee Ufan Museum 66
  Osaka, Higashiosaka, Shiba Ryotaro Memorial Museum 60
  Shiga, Shigaraki, Miho Museum Gardens 14, 15, 30, 45, 258
Jungles, Raymond 190

K
Kids Rock, Children's Play Environment, Orange County Great Park,
  Irvine, California 332

L
L'autre rive (The Other Bank), Quebec, Canada 16, 18, 33, 48, 254
Lebanon
  Beirut, Hariri Memorial Garden 9, 23, 39, 108
  Faqra, Salame Residence 38, 114
Lederer + Ragnarsdóttir + Oei / Helmut Hornstein 202
Lee Ufan Museum, Naoshima, Kagawa, Japan 66
Lin, Maya 17, 18, 32, 33, 34, 35, 47, 48, 49, 208
Lurie Garden, Chicago, Illinois, USA 16, 32, 47, 130

M
Maciocco, Giovanni 216
Madrid RIO, Madrid, Spain 12, 27, 42, 386
Malaysia
  Kuala Lumpur, YTL Residence Garden 286
Maltzan, Michael 222
Max Juvénal Bridge, Aix-en-Provence, France 8, 22, 38, 80
Mecanoo 228
Menis, Fernando 234
Miguel, Eduardo de 242
Miho Museum Gardens, Shigaraki, Shiga, Japan 14, 15, 30, 45, 258
Miralles Tagliabue EMBT 10, 25, 40, 41, 248
Morandé Winery Productive Services, Casablanca Valley, Chile 178
Mosbach, Catherine 17, 18, 33, 34, 48, 254

N
Nakamura, Yoshiaki 14, 15, 30, 45, 258
NBGM Landscape Architects 13, 28, 29, 43, 264
The Netherlands
  Delft, Mekel Park, TU Campus 228
  Rotterdam, Afrikaanderplein 12, 27, 42, 294
  Utrecht, Domplein 290
Neves, Victor 46, 270
Nieto Sobejano 11, 25, 26, 41, 276

O
Officina del Paesaggio 282
OKRA 12, 27, 42, 290
Old Market Square, Nottingham, UK 140
Olympic Sculpture Park, Seattle, Washington, USA 374
Orchid Waltz, National Theater Concert Hall, Taipei, Taiwan 8, 21, 23, 38, 76

P
Palerm & Tabares de Nava 296
Pawson, John 16, 31, 32, 46, 302
Pedestrian Bridge, Carpinteira River,
  Covilhã, Portugal 15, 16, 30, 31, 45, 46, 94
Piano, Renzo 13, 16, 29, 32, 44, 47, 308
Pier Head and Canal Link, Liverpool, UK 11, 26, 41, 50
Plasma Studio 314

INDEX OF BUILDINGS, NAMES AND PLACES

Playa Vista Park and Bandshell, Los Angeles, USA	222
Plaza de España, Santa Cruz de Tenerife, Canary Islands, Spain	48, 154
Plaza de Santa Bárbara, Madrid, Spain	11, 25, 26, 41, 276
Portugal	
Costa do Castelo, Archeological Museum of Praça Nova do Castelo de São Jorge	88
Covilhã, Carpinteira River, Pedestrian Bridge	94
Esposende, Reorganization of the Riverside of Esposende	46, 270
Public Farm 1, Long Island City, New York, USA	9, 24, 38, 39, 404

R

Renovation and Expansion of the California Academy of Sciences, San Francisco, California, USA	13, 29, 44, 308
Reorganization of the Riverside of Esposende, Esposende, Portugal	46, 270
Robert and Arlene Kogod Courtyard, Washington, D.C., USA	16, 32, 33, 47, 136
Royal Botanic Gardens, Cranbourne, Melbourne, Australia	346

S

Sackler Crossing, Royal Botanic Gardens, Kew, London, UK	16, 31, 32, 46, 302
Salame Residence, Faqra, Lebanon	38, 114
SanShin-Tei, One Kowloon Office Building, Hong Kong, China	15, 31, 45, 46, 326
Santa Fe Railyard Park and Plaza, Santa Fe, New Mexico, USA	340
Santander Group City Campus, Boadilla del Monte, Madrid, Spain	352
Shiba Ryotaro Memorial Museum, Higashiosaka, Osaka, Japan	60
Shunmyo Masuno	15, 16, 31, 45, 46, 326
Smith, Ken	10, 25, 40, 332
South Africa	
Tshwane, Pretoria, Salvokop, The Freedom Park	13, 28, 29, 43, 264
Spain	
Algeciras, Punta San García, Centenary Park	82
Bilbao, The Garden That Climbs the Stairs	7, 8, 22, 38, 70
Canary Islands, Santa Cruz de Tenerife, Cuchillitos de Tristán Park	234
Canary Islands, Santa Cruz de Tenerife, García Sanabria Park	296
Canary Islands, Santa Cruz de Tenerife, Plaza de España	48, 154
Madrid, Boadilla del Monte, Santander Group City Campus	352
Madrid, Madrid RIO	12, 27, 42, 386
Madrid, Plaza de Santa Bárbara	11, 25, 26, 41, 276
Santiago de Compostela, City of Culture of Galicia	13, 28, 43, 118
Valencia, Cabecera Park	244
Storm King Wavefield, Mountainville, New York, USA	17, 33, 48, 208
Sweden	
Wanås, Eleven Minute Line	17, 18, 33, 34, 48, 212
Switzerland	
Chiasso, Community Gardens	282

T

Taiwan	
Taipei, National Theater Concert Hall, Orchid Waltz	8, 21, 23, 38, 76
Taylor Cullity Lethlean	346
Teardrop Park, Battery Park City, New York, New York, USA	9, 24, 40, 370
The Freedom Park, Salvokop, Tshwane, Pretoria, South Africa	13, 28, 29, 43, 264
The Garden That Climbs the Stairs, Bilbao, Spain	7, 8, 22, 38, 70
The High Line, New York, New York, USA	12, 13, 27, 28, 42, 43, 182
Toronto Central Waterfront, Toronto, Canada	394
TU Campus, Mekel Park, Delft, The Netherlands	228

U

UK	
Hyde Park, London, Diana, Princess of Wales Memorial Fountain	144
Kew, London, Royal Botanic Gardens, Sackler Crossing	302
Liverpool, Pier Head and Canal Link	11, 26, 41, 50
Nottingham, Old Market Square	140
USA	
Arizona, Phoenix, Civic Space Park	10, 26, 41, 56
California, Irvine, Orange County Great Park, Kids Rock, Children's Play Environment	332
California, Los Angeles, Playa Vista Park and Bandshell	222
California, San Francisco, Renovation and Expansion of the California Academy of Sciences	13, 29, 44, 308
Connecticut, New Haven, Connecticut Water Treatment Facility	14, 15, 29, 30, 44, 45, 362
Florida, Coconut Grove, Florida Garden, Coconut Grove	196
Florida, Hollywood, Young Circle ArtsPark	124
Florida, Miami Beach, 1111 Lincoln Road	190
Illinois, Chicago, Lurie Garden	16, 32, 47, 130
Massachusetts, Wellesley, Wellesley College, Alumnae Valley Landscape Restoration	366
New Mexico, Santa Fe, Santa Fe Railyard Park and Plaza	340

New York, Long Island City, Public Farm 1	9, 24, 38, 39, 404
New York, Mountainville, Storm King Wavefield	17, 33, 48, 208
New York, New York, 40 Central Park South	10, 25, 40, 336
New York, New York, Battery Park City, Teardrop Park	9, 24, 40, 370
New York, New York, Governors Island Park and Public Space Master Plan	380
New York, New York, Hypar Pavilion Lawn	10, 24, 25, 40, 104
New York, New York, The High Line	12, 13, 27, 28, 42, 43, 182
Washington, D.C., Robert and Arlene Kogod Courtyard	16, 32, 33, 47, 136
Washington, Seattle, Olympic Sculpture Park	374

V

Vallejo, Luis	352
Van Valkenburgh, Michael	9, 10, 14, 24, 29, 30, 40, 44, 45, 362
Vanke Center / Horizontal Skyscraper, Shenzhen, China	14, 29, 44, 160

W

Weiss/Manfredi	374
West 8	12, 27, 42, 380
Wirtz International	398
WORK Architecture Company	9, 24, 38, 39, 404

Y

Young Circle ArtsPark, Hollywood, Florida, USA	124
YTL Residence Garden, Kuala Lumpur, Malaysia	286

CREDITS

PHOTO CREDITS — **2, 7** © Iwan Baan / **8** © Patrick Blanc / **11** © AECOM 2011 Photography by David Lloyd / **12** © Jeroen Musch / **13** © Courtesy of Eisenman Architects / **14** © Elizabeth Felicella / **15** © Carrilho da Graça arquitectos / **16** © Nic Lehoux / **17** © Courtesy of Jerry Thompson and Storm King Art Center / **18** © Mosbach Catherine / **21** © Patrick Blanc / **23** © Matteo Piazza / **25** © Iwan Baan / **26** © Roland Halbe / **29** © Tristan McLaren/VIEW / **30** © Kohseki Co., Ltd. / **32** © Edmund Sumner/VIEW / **33** © Nigel Young/Foster + Partners / **37** © Matteo Piazza / **39** © WORK Architecture Company / **40** © Roland Halbe / **42** © Ben ter Mull / **44** © Steven Holl Arquitects / **45** © Japan Landscape Consultants Ltd. / **46** © Victor Neves / **48** © Roland Halbe / **50** © AECOM / **51-59** © AECOM 2011 Photography by David Lloyd / **60-65** © Tadao Ando Architect & Associates / **66-69** © Shigeo Ogawa / **70, 74** © Balmori Associates Inc. / **71-73, 75** © Iwan Baan / **76-81** © Patrick Blanc / **82** © M. Caffarena, V. Cobos, G. Alcaraz, G. Delgado / **83-87** © Jesus Granada/Bisimages / **88, 94-97** © Carrilho da Graça arquitectos / **89-93** © Leonardo Finotti / **98** © Tacoa Arquitetos Associados / **99-103** © Leonardo Finotti / **104** © Courtesy of Diller Scofidio + Renfro / **105-107** © Iwan Baan / **108** © Vladimir Djurovic Landscape Architecture / **109-117** © Matteo Piazza / **118, 120** © Courtesy of Eisenman Architects / **119, 121-123** © Roland Halbe / **124** © Glavovic Studio Inc. / **125-129** © Robin Hill / **131** © Nic Lehoux / **130, 132, 134 bottom right-135** © Gustafson Guthrie Nichol / **133** © Linda Oyama Bryan / **134 top** © Piet Oudolf / **134 bottom left** © Juan Rois / **136-139** © Nigel Young/Foster + Partners / **140** © Gustafson Porter / **141-142** © Martine Hamilton Knight/Builtvision / **143, 146-148** © Gustafson Porter / **144-145, 149** © Hélène Binet / **150-153** © Courtesy of Zaha Hadid Architects / **154** © Adriano A. Biondo / **155-159** © Roland Halbe / **160-161, 164-165, 166 bottom-167, 171** © Steven Holl Architects / **162-163, 166 top, 168-170** © Iwan Baan / **172** © Martin Hurtado Arquitectos / **173-181** © Leonardo Finotti / **182** © James Corner Field Operations / **183-189** © Iwan Baan / **190** © Raymond Jungles, Inc. / **191-195** © Steven Brooke Studios / **196-201** © Annie Schlechter / **202** © Zooey Braun / **203-207** © Roland Halbe / **208** © Walter Smith / **209-211** © Courtesy of Jerry Thompson and Storm King Art Center / **212-215** © Anders Norrsell / **216** © Giovanni Maciocco / **217-221** © Gianni Calaresu / **222** © Monica Nowens / **223-227** © Iwan Baan / **228** © Mecanoo architecten / **229-233** © Christian Richters / **235, 240-241** © Roland Halbe / **234, 236-339** © Menis Arquitectos / **242** © Eduardo de Miguel / **243-247** © Duccio Malagamba / **248** © Miralles Tagliabue EMBT / **249-253** © Roland Halbe / **254-257** © Mosbach Catherine / **258-263** © Kohseki Co., Ltd. / **264** © NBGM Landscape Architects / **265-269** © Tristan McLaren/VIEW / **270–275** © Victor Neves / **276** © Nieto Sobejano Arquitectos SLP / **277-281** © Roland Halbe / **282-285** © Officina del Paesaggio / **286-289** © Roland Halbe / **290-291, 293 top, 294, 295 bottom right** © OKRA landscape architects / **292, 293 bottom, 295 top-bottom left** © Ben ter Mull / **296** © Palerm & Tabares de Nava / **297-301** © Roland Halbe / **302** © John Pawson Ltd / **303-307** © Edmund Sumner/VIEW / **308** © Renzo Piano Building Workshop / **309-311** © Nic Lehoux / **312-313** © Renzo Piano Building Workshop / **314, 316-317, 321, 323-325** © Plasma Studio / **315, 318-319** © ChinaFotoPress/Getty Images / **320, 322** © Cristóbal Palma / **326-331** © Japan Landscape Consultants Ltd. / **332-335** © Ken Smith Landscape Architect / **336-345** © Peter Mauss/Esto / **346** © Taylor Cullity Lethlean / **347, 349** © Peter Hyatt/fabpics / **348, 350-351** © Taylor Cullity Lethlean / **352** © LVEP / **353-361** © Miquel Tres / **362** © Michael Van Valkenburgh Associates, Inc. / **363-373** © Elizabeth Felicella / **374-379** © Weiss/Manfredi / **380-387** © West 8 urban design and landscape architecture / **388-393** © Jeroen Musch / **394-397** © Courtesy of Waterfront Toronto / **398-403** © Wirtz International N. V. / **404** © Andy French / **405-411** © WORK Architecture Company

CREDITS FOR PLANS / DRAWINGS / CAD DOCUMENTS — **54, 59** © AECOM / **62-63, 69** © Tadao Ando Architect & Associates / **75** © Balmori Associates Inc. / **78, 81** © Patrick Blanc / **85-87** © M. Caffarena, V. Cobos, G. Alcaraz, G. Delgado / **90, 92, 95, 97** © Carrilho da Graça arquitectos / **101-102** © Tacoa Arquitetos Associados / **110, 112, 115-116** © Vladimir Djurovic Landscape Architecture / **121-123** © Courtesy of Eisenman Architects / **126** © Glavovic Studio Inc. / **133, 137** © Gustafson Guthrie Nichol / **143, 149** © Gustafson Porter / **151-153** © Courtesy of Zaha Hadid Architects / **158** © Herzog & de Meuron / **163, 165** © Steven Holl Architects / **174, 177-178, 181** © Martin Hurtado Arquitectos / **192, 196, 198** © Raymond Jungles, Inc. / **207** © Lederer+Ragnarsdóttir+Oei / **218, 220** © Giovanni Maciocco / **224, 227** © Michael Maltzan Architecture, Inc. / **230** © Mecanoo architecten / **239-240** © Menis Arquitectos / **244, 246** © Eduardo de Miguel / **253** © Miralles Tagliabue EMBT / **257** © Mosbach Catherine / **275** © Victor Neves / **279-280** © Nieto Sobejano Arquitectos SLP / **285, 287** © Officina del Paesaggio / **299** © Palerm & Tabares de Nava / **305-306** © John Pawson Ltd / **313** © Renzo Piano Building Workshop / **316, 323-325** © Plasma Studio / **329, 331** © Japan Landscape Consultants Ltd. / **341** © Ken Smith Landscape Architect / **349-350** © Taylor Cullity Lethlean / **356** © LVEP / **364, 366, 370** © Michael Van Valkenburgh Associates, Inc. / **376, 378** © Weiss/Manfredi / **381-386, 388, 390** © West 8 urban design and landscape architecture / **401** © Wirtz International N. V. / **407, 409, 411** © WORK Architecture Company